Advances in Geodesy

Edited by
Erik W. Grafarend
Richard H. Rapp

Selected Papers from Reviews of Geophysics and Space Physics

American Geophysical Union

Selected papers from *Reviews of Geophysics and Space Physics*

Library of Congress Cataloging in Publication Data
Main entry under title:

Advances in geodesy.

1. Geodesy. I. Grafarend, Erik W. II. Rapp, Richard H. III. Reviews of geophysics and space physics.
QB281.A28 1984 526'.1 84-11138
ISBN 0-87590-235-9

Printed in the United States of America

CONTENTS

Foreword

In the past several years, significant advances have been made in solving geodetic-related problems from both a theoretical and an application point of view. Many of these advances have been documented in papers that have appeared in *Reviews of Geophysics and Space Physics.* These papers have appeared in many different issues of this journal. For many students and researchers the availability of these papers is not convenient. We have therefore collected for this volume those papers of geodetic interest in order to give them an audience wider than when the first publication took place. The selection of these papers is not meant to imply that geodetic developments have only been reported in the *Reviews of Geophysics and Space Physics.* To the contrary, many important papers have been published in the *Journal of Geophysical Research, Bulletin Géodésique, Manuscripta Geodetica,* and other journals. However, we restricted this selection to articles that appeared in *RGSP.* Most of the articles were developed while one of us (E.W.G.) was the geodesy Associate Editor of *Reviews of Geophysics and Space Physics.*

The papers have been organized into four parts. The first part is devoted to geodetic theory. The central problem, where the earth's surface becomes singular, is treated by F. Bocchio. The new and exciting solution of the geodetic boundary value problem in gravity space is reviewed by F. Sanso. Part two is centered around the geodetic estimation procedures. H. Moritz summarizes least squares collocation. A review on the state of the art of optimal design of geodetic networks is presented by G. Schmitt. F. P. Bretherton and J. C. McWilliams treat estimation from irregular arrays. A. Tarantola and B. Valette treat the nonlinear inverse problem by using a least squares criterion. Next, part three deals with the gravity field. G. Anger presents a systematic approach of the inverse gravimetric source problem through extremal measures. A comparison study of various methods for the detailed representation of the earth's gravity field has been carried out by C. Tscherning. E. Groten discusses the advantages of various local and global gravity field representations, while H. F. Schmidt investigates the advantages of sampling functions and the finite element method. An extensive review of planar geodetic covariance functions is presented by S. Meier. The monument of the gravity field of terrestrial planets, the long-wavelength anomalies and tectonics, has been built up by R. J. Philipps and K. Lambeck. Last, but not least, applications are reviewed in part four. J. Kouba summarizes geodetic and geodynamic satellite Doppler positioning. P. Vanicek, R. O. Castle, and E. I. Balazs discuss space-time geodetic leveling and its applications. A. Dermanis and E. Livieratos apply deformation analysis in geodesy and geodynamics. The new field of inertial surveying is summarized by K. P. Schwarz. C. Wunsch and E. M. Gaposchkin review the use of satellite altimetry to determine the general circulation of the oceans with a view to geoid improvement. A progress report on the application of satellite altimetry for the observation of the mesoscale variability and general circulation of the oceans is given by L. L. Fu. Finally, E. W. Schwiderski highlights the charting of global ocean tides.

The authors of the various review papers have to be thanked for all their expert work on various current geodetic problems. May a wide distribution of their results be a proper reward.

Erik W. Grafarend
Richard H. Rapp

Foreword

REVIEWS OF GEOPHYSICS AND SPACE PHYSICS, VOL. 20, NO. 3, PAGES 399–409, AUGUST 1982

Geodetic Singularities

F. Bocchio

Istituto di Geodesia e Geofisica, Università di Trieste, Trieste, Italy

Both general and local coordinate systems for geodesy are usually defined in terms of the gravity vector and on the basis of the assumption of the 'convexity' of the equipotential surfaces of the gravity field, which ensures the uniqueness of the coordinates of geodetic type. Two formulations of the singularity condition which have been deduced starting from two different approaches to the geodetic singularity problem, that is, the local-operational and the global-geometrical approaches, are discussed. A geodetic singularity problem and the related inverse problem are also discussed for a homogeneous and a nonhomogeneous planet; in particular, the existence of geodetic singularities is shown in connection with a perturbing field with a very simple geometry, and the density distribution is deduced which gives rise to geodetic singularities in the external gravity field of the planet. The structure of the gravity field in the neighborhood of the singularities is investigated in detail together with the behavior of the gradients of the disturbances in the geodetic coordinates, and the scale of the perturbed zone is deduced in connection with a given perturbation.

General and Local Coordinates for Gravity Fields

In the external gravity field we may consider two different kinds of coordinate systems: systems of general coordinates and systems of local coordinates. As far as the former are concerned, a set of coordinates could be (Φ, Λ, W), that is, astronomical latitude and longitude and geopotential. The system of local coordinates is usually a Cartesian system based on the north, east, and zenith directions at any point in the gravity field. This system, also known as the 'local astronomical' or 'natural' system, cannot form the basis of a general system of coordinates unless the gravity field is rotationally symmetric [*Marussi,* 1951], which in general is not the case. In the early 1970's a closer insight into the geometrical structure of the gravity field and of the related coordinate systems showed that some geodetic problems could be better understood within the frame of a generalized Riemannian geometry and, in particular, of the concept of 'torsion spaces' [*Bocchio,* 1970, 1974; *Grafarend,* 1971, 1975; *Grossman,* 1974]. In particular, it was shown that the differentials (dx, dy, dz) in the local astronomical system are 'anholonomic differentials,' that is, $\oint dx, \oint dy, \oint dz \neq 0$, and the geodetic relevance of the matrix of integrating factors transforming anholonomic differentials into holonomic differentials of the type $(d\Phi, d\Lambda, dW)$ was revealed by Grafarend. It will be shown that the matrix of integrating factors also plays a central role in connection with the geodetic singularity problem.

In what follows, a system of general coordinates of the type (Φ, Λ, W) and a local Cartesian system (x, y, z) of astronomical type will be considered, and the pertaining geometrical and mechanical properties of the gravity field which are required to properly introduce the geodetic singularity problem will be evaluated.

If $\mathbf{i}_r^*(r = 1, 2, 3)$ denotes the local Cartesian astronomical triads and $\mathbf{X}_r(r = 1, 2, 3)$ the fixed Cartesian triad (see the notation list at the end of the paper), the astronomical latitude Φ and longitude Λ are defined in terms of the direction cosines of the unit normal $\mathbf{i}_3^*$ to a W surface by means of the relations

$$\begin{aligned} \cos\Phi\cos\Lambda &= \mathbf{i}_3^*\cdot\mathbf{X}_1 \\ \cos\Phi\sin\Lambda &= \mathbf{i}_3^*\cdot\mathbf{X}_2 \\ \sin\Phi &= \mathbf{i}_3^*\cdot\mathbf{X}_3 \end{aligned} \tag{1}$$

We can now express the local astronomical system in terms of the fixed Cartesian system as follows:

$$\begin{aligned} \mathbf{i}_1^* &= -\mathbf{X}_1\sin\Phi\cos\Lambda - \mathbf{X}_2\sin\Phi\sin\Lambda + \mathbf{X}_3\cos\Phi \\ \mathbf{i}_2^* &= -\mathbf{X}_1\sin\Lambda + \mathbf{X}_2\cos\Lambda \\ \mathbf{i}_3^* &= \mathbf{X}_1\cos\Phi\cos\Lambda + \mathbf{X}_2\cos\Phi\sin\Lambda + \mathbf{X}_3\sin\Phi \end{aligned} \tag{2}$$

If G denotes the transformation matrix in (2), that is, $\mathbf{i}_r^* = G\mathbf{X}_r$, we have

$$\mathbf{X}_r = G^T\mathbf{i}_r^* \tag{3}$$

G being orthogonal. If we consider the covariant derivative of the equations (2) and take into account that the vectors $\mathbf{X}_r$ are constant under covariant differentiation, it can be shown [*Bocchio,* 1981*a*] that the gradients of the coordinates are given by

$$\begin{aligned} \Phi_s &= \nu_{rs}\lambda^r \\ \Lambda_s &= \nu_{rs}\mu^r\sec\Phi \\ W_s &= -g\nu_s \end{aligned} \tag{4}$$

where $(\lambda^r, \mu^r, \nu^r)$ and $(\lambda_r, \mu_r, \nu_r)$ are the sets of the contravariant and covariant components, respectively, of the triads in the (Φ, Λ, W) system. In other words if $\mathbf{v}_r^*$ and $\mathbf{v}^{r*}$ denote the base vectors and the reciprocal base vectors of the (Φ, Λ, W) system, the following relations hold:

$$\begin{aligned} \lambda^r &= \mathbf{v}^{r*}\cdot\mathbf{i}_1^* \\ \mu^r &= \mathbf{v}^{r*}\cdot\mathbf{i}_2^* \\ \nu^r &= \mathbf{v}^{r*}\cdot\mathbf{i}_3^* \end{aligned} \tag{5}$$

$$\begin{aligned} \lambda_r &= \mathbf{v}_r^*\cdot\mathbf{i}_1^* \\ \mu_r &= \mathbf{v}_r^*\cdot\mathbf{i}_2^* \\ \nu_r &= \mathbf{v}_r^*\cdot\mathbf{i}_3^* \end{aligned} \tag{6}$$

$$\mathbf{i}_1{}^* = \lambda^r \mathbf{v}_r{}^* = \lambda_r \mathbf{v}^{r*}$$
$$\mathbf{i}_2{}^* = \mu^r \mathbf{v}_r{}^* = \mu_r \mathbf{v}^{r*} \qquad (7)$$
$$\mathbf{i}_3{}^* = \nu^r \mathbf{v}_r{}^* = \nu_r \mathbf{v}^{r*}$$

We have also

$$\lambda^r = \left(\frac{\partial \Phi}{\partial S_\Phi}, \frac{\partial \Lambda}{\partial S_\Phi}, \frac{\partial W}{\partial S_\Phi}\right)$$
$$\mu^r = \left(\frac{\partial \Phi}{\partial S_\Lambda}, \frac{\partial \Lambda}{\partial S_\Lambda}, \frac{\partial W}{\partial S_\Lambda}\right) \qquad (8)$$
$$\nu^r = \left(\frac{\partial \Phi}{\partial S_W}, \frac{\partial \Lambda}{\partial S_W}, \frac{\partial W}{\partial S_W}\right)$$

where S_Φ, S_Λ, S_W denote elements of length in the direction of the (Φ, Λ, W) lines, respectively. We can write

$$\partial\Phi/\partial S_\Phi = \Phi_s \lambda^s \quad \partial\Lambda/\partial S_\Phi = \Lambda_s \lambda^s \quad \partial W/\partial S_\Phi = W_s \lambda^s$$
$$\partial\Phi/\partial S_\Lambda = \Phi_s \mu^s \quad \partial\Lambda/\partial S_\Lambda = \Lambda_s \mu^s \quad \partial W/\partial S_\Lambda = W_s \mu^s \qquad (9)$$
$$\partial\Phi/\partial S_W = \Phi_s \nu^s \quad \partial\Lambda/\partial S_W = \Lambda_s \nu^s \quad \partial W/\partial S_W = W_s \nu^s$$

It can be shown that taking into account the orthogonality relations in the $\mathbf{i}_r{}^*$ system, that is, $\lambda_r\mu^r = \lambda_r\nu^r = \nu_r\mu^r = 0$, we have

$$\lambda^r = (k_1, t_1 \sec \Phi, 0)$$
$$\mu^r = (t_1, k_2 \sec \Phi, 0) \qquad (10)$$
$$\nu^r = (\gamma_1, \gamma_2 \sec \Phi, -g)$$

where we have put

$$\nu_{rs}\lambda^r\lambda^s = k_1$$
$$\nu_{rs}\mu^r\mu^s = k_2$$
$$\nu_{rs}\lambda^r\mu^s = t_1 \qquad (11)$$
$$\nu_{rs}\lambda^r\nu^s = \gamma_1$$
$$\nu_{rs}\mu^r\nu^s = \gamma_2$$

k_1 and k_2 being the normal curvature of a W surface in the meridian and parallel directions respectively, and t_1 the geodetic torsion of the W surface in the meridian direction. It can be shown that γ_1 and γ_2 are the arc rate of change of ln g in the meridian and parallel directions, respectively, that is,

$$\gamma_1 = (\ln g)_s \lambda^s$$
$$\gamma_2 = (\ln g)_s \mu^s \qquad (12)$$

If we consider the third of equations (4), that is, $W_r = -g\nu_r$, we have, by covariant differentiation,

$$W_{rs} = -g_s\nu_r - g\nu_{rs} \qquad (13)$$

and therefore, from (11) and (15), taking into account that $\mathbf{i}_r{}^* \cdot \mathbf{i}_s{}^* = \delta_{rs}$,

$$g^{-1}W_{rs}\lambda^r\lambda^s = -k_1$$
$$g^{-1}W_{rs}\mu^r\mu^s = -k_2$$
$$g^{-1}W_{rs}\lambda^r\mu^s = -t_1 \qquad (14)$$
$$g^{-1}W_{rs}\lambda^r\nu^s = -\gamma_1$$
$$g^{-1}W_{rs}\mu^r\nu^s = -\gamma_2$$

where W_{rs} is the tensor of gravity gradients. Equations (14) illustrate the strict connection between dynamical aspects (left-hand side) and geometrical aspects (right-hand side) and show that all the parameters defining the geometry of the gravity field can be deduced from the tensor of gravity gradients.

The covariant components of the local astronomical triads in the (Φ, Λ, W) system can be deduced by taking into account the relation

$$\lambda^r\lambda_s + \mu^r\mu_s + \nu^r\nu_s = \delta_s{}^r \qquad (15)$$

We have

$$\lambda_r = \left(\frac{k_2}{K}, -\frac{t_1 \cos \Phi}{K}, \frac{\gamma_1 k_2 - t_1\gamma_2}{gK}\right)$$
$$\mu_r = \left(-\frac{t_1}{K}, \frac{k_1 \cos \Phi}{K}, \frac{k_1\gamma_2 - t_1\gamma_1}{gK}\right) \qquad (16)$$
$$\nu_r = (0, 0, -g^{-1})$$

where $K = k_1k_2 - t_1{}^2$ is the Gaussian curvature of the W surfaces. If we write

$$\mathbf{i}_r{}^* = A\mathbf{v}_r{}^* \qquad \mathbf{v}_r{}^* = A^{-1}\mathbf{i}_r{}^* = B^T\mathbf{i}_r{}^* \qquad (17)$$
$$\mathbf{i}_r{}^* = B\mathbf{v}^{r*} \qquad \mathbf{v}^{r*} = B^{-1}\mathbf{i}_r{}^* = A^T\mathbf{i}_r{}^* \qquad (18)$$

where $AB^T = I$, we have thus, considering (5), (10), and (16), that the transformation matrices relating the system of local Cartesian coordinates and the system of general coordinates, or, in other words, the matrices relating the anholonomic differentials (dx, dy, dz) to the holonomic differentials ($d\Phi$, $d\Lambda$, dW), are given by

$$A = \begin{Vmatrix} k_1 & t_1 \sec \Phi & 0 \\ t_1 & k_2 \sec \Phi & 0 \\ \gamma_1 & \gamma_2 \sec \Phi & -g \end{Vmatrix} \qquad (19)$$

$$B = \begin{Vmatrix} \frac{k_2}{K} & -\frac{t_1 \cos \Phi}{K} & \frac{\gamma_1 k_2 - t_1\gamma_2}{gK} \\ -\frac{t_1}{K} & \frac{k_1 \cos \Phi}{K} & \frac{k_1\gamma_2 - t_1\gamma_1}{gK} \\ 0 & 0 & -g^{-1} \end{Vmatrix} \qquad (20)$$

Notice that the transformations (17) and (18) are nonsingular only if

$$\det A \neq 0 \qquad (21)$$

From (19) it follows that

$$\det A = -gK \sec \Phi \qquad (22)$$

a result which will prove to be very important in connection with the geodetic singularity problem, since (22) is the Jacobian J of the transformation between local Cartesian coordinates and (Φ, Λ, W) coordinates, that is,

$$J = \text{grad}\ \Phi \times \text{grad}\ \Lambda \cdot \text{grad}\ W$$
$$= -gK \sec \Phi \qquad (23)$$

The metric tensor $g_{rs}{}^*$ of the (Φ, Λ, W) system is given, in terms of the base vectors, by the relations

$$g_{rs}{}^* = \mathbf{v}_r{}^* \cdot \mathbf{v}_s{}^* \qquad (24)$$

It can be shown, taking into account the previous results, that we obtain

$$g_{11} = \frac{t_1^2 + k_2^2}{K^2} \tag{25a}$$

$$g_{12} = -\frac{t_1(k_1 + k_2)\cos\Phi}{K^2} \tag{25b}$$

$$g_{13} = \frac{k_2(\gamma_1 k_2 - t_1\gamma_2) - t_1(k_1\gamma_2 - t_1\gamma_1)}{gK^2} \tag{25c}$$

$$g_{22} = \frac{(t_1^2 + k_1^2)\cos^2\Phi}{K^2} \tag{25d}$$

$$g_{23} = \frac{[k_1(k_1\gamma_2 - t_1\gamma_1) - t_1(\gamma_1 k_2 - t_1\gamma_2)]\cos\Phi}{gK^2} \tag{25e}$$

$$g_{33} = \frac{1}{g^2}\left[1 + \frac{(\gamma_1 k_2 - t_1\gamma_2)^2 + (k_1\gamma_2 - t_1\gamma_1)^2}{K^2}\right] \tag{25f}$$

The Christoffel's symbols

$$\begin{Bmatrix} k \\ ij \end{Bmatrix}$$

which relate the base vectors of the general system at different points of the field, are given by

$$\begin{Bmatrix} k \\ ij \end{Bmatrix} = \frac{\partial \mathbf{v}_i^*}{\partial y_j}\cdot \mathbf{v}^{k*} \qquad y^1 = \Phi \quad y^2 = \Lambda \quad y^3 = W \tag{26}$$

Since it can be shown that

$$\frac{\partial \mathbf{v}_1^*}{\partial \Phi} = \frac{\partial}{\partial \Phi}\left(\frac{k_2}{K}\right)\mathbf{i}_1^* - \frac{\partial}{\partial \Phi}\left(\frac{t_1}{K}\right)\mathbf{i}_2^* - \frac{k_2}{K}\mathbf{i}_3^* \tag{27a}$$

$$\frac{\partial \mathbf{v}_1^*}{\partial \Lambda} = \left[\frac{\partial}{\partial \Lambda}\left(\frac{k_2}{K}\right) + \frac{t_1 \sin\Phi}{K}\right]\mathbf{i}_1^* - \left[\frac{\partial}{\partial \Lambda}\left(\frac{t_1}{K}\right) + \frac{k_2\sin\Phi}{K}\right]\mathbf{i}_2^* + \frac{t_1\cos\Phi}{K}\mathbf{i}_3^* \tag{27b}$$

$$\frac{\partial \mathbf{v}_1^*}{\partial W} = \frac{\partial}{\partial W}\left(\frac{k_2}{K}\right)\mathbf{i}_1^* - \frac{\partial}{\partial W}\left(\frac{t_1}{K}\right)\mathbf{i}_2^* \tag{27c}$$

$$\frac{\partial \mathbf{v}_2^*}{\partial \Phi} = -\frac{\partial}{\partial \Phi}\left(\frac{t_1\cos\Phi}{K}\right)\mathbf{i}_1^* + \left[\cos\Phi\frac{\partial}{\partial \Phi}\left(\frac{k_1}{K}\right) - \frac{k_1\sin\Phi}{K}\right]\mathbf{i}_2^* + \frac{t_1\cos\Phi}{K}\mathbf{i}_3^* \tag{28a}$$

$$\frac{\partial \mathbf{v}_2^*}{\partial \Lambda} = \cos\Phi\left\{\left[\frac{k_1\sin\Phi}{K} - \frac{\partial}{\partial \Lambda}\left(\frac{t_1}{K}\right)\right]\mathbf{i}_1^* + \left[\frac{\partial}{\partial \Lambda}\left(\frac{k_1}{K}\right) + \frac{t_1\sin\Phi}{K}\right]\mathbf{i}_2^* - \frac{k_1\cos\Phi}{K}\mathbf{i}_3^*\right\} \tag{28b}$$

$$\frac{\partial \mathbf{v}_2^*}{\partial W} = \cos\Phi\left[-\frac{\partial}{\partial W}\left(\frac{t_1}{K}\right)\mathbf{i}_1^* + \frac{\partial}{\partial W}\left(\frac{k_1}{K}\right)\mathbf{i}_2^*\right] \tag{28c}$$

$$\frac{\partial \mathbf{v}_3^*}{\partial \Phi} = -\left[\frac{\partial}{\partial \Phi}\left(\frac{t_1\gamma_2 - \gamma_1 k_2}{gK}\right) + \frac{1}{g}\right]\mathbf{i}_1^* - \frac{\partial}{\partial \Phi}\left(\frac{t_1\gamma_1 - k_1\gamma_2}{gK}\right)\mathbf{i}_2^* \tag{29a}$$

$$\frac{\partial \mathbf{v}_3^*}{\partial \Lambda} = -\left[\frac{\partial}{\partial \Lambda}\left(\frac{t_1\gamma_2 - \gamma_1 k_2}{gK}\right) + \frac{(t_1\gamma_1 - k_1\gamma_2)}{gK}\sin\Phi\right]\mathbf{i}_1^* - \left[\frac{\partial}{\partial \Lambda}\left(\frac{t_1\gamma_1 - k_1\gamma_2}{gK}\right) - \frac{(t_1\gamma_2 - \gamma_1 k_2)}{gK}\sin\Phi + \frac{\cos\Phi}{g}\right]\mathbf{i}_2^* \tag{29b}$$

$$\frac{\partial \mathbf{v}_3^*}{\partial W} = -\frac{\partial}{\partial W}\left(\frac{t_1\gamma_2 - \gamma_1 k_2}{gK}\right)\mathbf{i}_1^* - \frac{\partial}{\partial W}\left(\frac{t_1\gamma_1 - k_1\gamma_2}{gK}\right)\mathbf{i}_2^* - \frac{\partial}{\partial W}\left(\frac{1}{g}\right)\mathbf{i}_3^* \tag{29c}$$

we obtain

$$\begin{Bmatrix} 1 \\ 11 \end{Bmatrix} = k_1\frac{\partial}{\partial \Phi}\left(\frac{k_2}{K}\right) - t_1\frac{\partial}{\partial \Phi}\left(\frac{t_1}{K}\right) - \frac{\gamma_1 k_2}{K} \tag{30a}$$

$$\begin{Bmatrix} 1 \\ 12 \end{Bmatrix} = k_1\left[\frac{\partial}{\partial \Lambda}\left(\frac{k_2}{K}\right) + \frac{t_1\sin\Phi}{K}\right] - t_1\left[\frac{\partial}{\partial \Lambda}\left(\frac{t_1}{K}\right) + \frac{k_2\sin\Phi}{K}\right] + \frac{\gamma_1 t_1\cos\Phi}{K} \tag{30b}$$

$$\begin{Bmatrix} 1 \\ 13 \end{Bmatrix} = k_1\frac{\partial}{\partial W}\left(\frac{k_2}{K}\right) - t_1\frac{\partial}{\partial W}\left(\frac{t_1}{K}\right) \tag{30c}$$

$$\begin{Bmatrix} 1 \\ 22 \end{Bmatrix} = \cos\Phi\left\{k_1\left[\frac{k_1}{K}\sin\Phi - \frac{\partial}{\partial \Lambda}\left(\frac{t_1}{K}\right)\right] + t_1\left[\frac{\partial}{\partial \Lambda}\left(\frac{k_1}{K}\right) + \frac{t_1\sin\Phi}{K}\right] - \frac{k_1\gamma_1\cos\Phi}{K}\right\} \tag{30d}$$

$$\begin{Bmatrix} 1 \\ 23 \end{Bmatrix} = \cos\Phi\left[-k_1\frac{\partial}{\partial W}\left(\frac{t_1}{K}\right) + t_1\frac{\partial}{\partial W}\left(\frac{k_1}{K}\right)\right] \tag{30e}$$

$$\begin{Bmatrix} 1 \\ 33 \end{Bmatrix} = k_1\frac{\partial}{\partial W}\left(\frac{\gamma_1 k_2 - t_1\gamma_2}{gK}\right) - t_1\frac{\partial}{\partial W}\left(\frac{t_1\gamma_1 - k_1\gamma_2}{gK}\right) - \gamma_1\frac{\partial}{\partial W}\left(\frac{1}{g}\right) \tag{30f}$$

$$\begin{Bmatrix} 2 \\ 11 \end{Bmatrix} = \sec\Phi\left[t_1\frac{\partial}{\partial \Phi}\left(\frac{k_2}{K}\right) - k_2\frac{\partial}{\partial \Phi}\left(\frac{t_1}{K}\right) - \frac{\gamma_2 k_2}{K}\right] \tag{30g}$$

$$\begin{Bmatrix} 2 \\ 12 \end{Bmatrix} = \sec \Phi \left\{ t_1 \left[\frac{\partial}{\partial \Lambda} \left(\frac{k_2}{K} \right) + \frac{t_1 \sin \Phi}{K} \right] - k_2 \left[\frac{\partial}{\partial \Lambda} \left(\frac{t_1}{K} \right) + \frac{k_2 \sin \Phi}{K} \right] + \frac{\gamma_2 t_1 \cos \Phi}{K} \right\} \tag{30h}$$

$$\begin{Bmatrix} 2 \\ 22 \end{Bmatrix} = t_1 \left[\frac{k_1 \sin \Phi}{K} - \frac{\partial}{\partial \Lambda} \left(\frac{t_1}{K} \right) \right] + k_2 \left[\frac{\partial}{\partial \Lambda} \left(\frac{k_1}{K} \right) + \frac{t_1 \sin \Phi}{K} \right] - \frac{\gamma_2 k_1 \cos \Phi}{K} \tag{30i}$$

$$\begin{Bmatrix} 2 \\ 13 \end{Bmatrix} = \sec \Phi \left[t_1 \frac{\partial}{\partial W} \left(\frac{k_2}{K} \right) - k_2 \frac{\partial}{\partial W} \left(\frac{t_1}{K} \right) \right] \tag{30j}$$

$$\begin{Bmatrix} 2 \\ 23 \end{Bmatrix} = - t_1 \frac{\partial}{\partial W} \left(\frac{t_1}{K} \right) + k_2 \frac{\partial}{\partial W} \left(\frac{k_1}{K} \right) \tag{30k}$$

$$\begin{Bmatrix} 2 \\ 33 \end{Bmatrix} = \sec \Phi \left[-t_1 \frac{\partial}{\partial W} \left(\frac{t_1 \gamma_2 - \gamma_1 k_2}{gK} \right) - k_2 \frac{\partial}{\partial W} \left(\frac{t_1 \gamma_1 - k_1 \gamma_2}{gK} \right) - \gamma_2 \frac{\partial}{\partial W} \left(\frac{1}{g} \right) \right] \tag{30l}$$

$$\begin{Bmatrix} 3 \\ 11 \end{Bmatrix} = \frac{g k_2}{K} \qquad \begin{Bmatrix} 3 \\ 12 \end{Bmatrix} = - \frac{g t_1 \cos \Phi}{K} \tag{30m}$$

$$\begin{Bmatrix} 3 \\ 22 \end{Bmatrix} = \frac{g k_1 \cos^2 \Phi}{K} \qquad \begin{Bmatrix} 3 \\ 33 \end{Bmatrix} = - \frac{\partial}{\partial W} \ln g \tag{30n}$$

We observe that it can be shown that [*Marussi*, 1951]

$$W_{rs} = - \begin{Bmatrix} 3 \\ rs \end{Bmatrix}$$

The integrability conditions for the existence of the coordinate surfaces Φ, Λ, W, known also as 'Mainardi-Codazzi equations', are

$$\frac{\partial \mathbf{v}_1^*}{\partial \Lambda} = \frac{\partial \mathbf{v}_2^*}{\partial \Phi} \qquad \frac{\partial \mathbf{v}_2^*}{\partial W} = \frac{\partial \mathbf{v}_3^*}{\partial \Lambda} \qquad \frac{\partial \mathbf{v}_1^*}{\partial W} = \frac{\partial \mathbf{v}_3^*}{\partial \Phi} \tag{31}$$

which, considering (27), (28), and (29), can be written explicitly as follows:

$$\cos \Phi \frac{\partial}{\partial \Phi} \left(\frac{t_1}{K} \right) + \frac{\partial}{\partial \Lambda} \left(\frac{k_2}{K} \right) = 0 \tag{32a}$$

$$\cos \Phi \frac{\partial}{\partial \Phi} \left(\frac{k_1}{K} \right) + \frac{\partial}{\partial \Lambda} \left(\frac{t_1}{K} \right) = \frac{(k_1 - k_2) \sin \Phi}{K} \tag{32b}$$

$$\frac{\partial}{\partial \Phi} \left(\frac{t_1 \gamma_2 - \gamma_1 k_2}{gK} \right) + \frac{\partial}{\partial W} \left(\frac{k_2}{K} \right) = - \frac{1}{g} \tag{32c}$$

$$\frac{\partial}{\partial \Phi} \left(\frac{t_1 \gamma_1 - k_1 \gamma_2}{gK} \right) - \frac{\partial}{\partial W} \left(\frac{t_1}{K} \right) = 0 \tag{32d}$$

$$\frac{\partial}{\partial \Lambda} \left(\frac{t_1 \gamma_2 - \gamma_1 k_2}{gK} \right) - \cos \Phi \frac{\partial}{\partial W} \left(\frac{t_1}{K} \right) = - \frac{(t_1 \gamma_1 - k_1 \gamma_2) \sin \Phi}{gK} \tag{32e}$$

$$\frac{\partial}{\partial \Lambda} \left(\frac{t_1 \gamma_1 - k_1 \gamma_2}{gK} \right) + \cos \Phi \frac{\partial}{\partial W} \left(\frac{k_1}{K} \right) = \frac{(t_1 \gamma_2 - \gamma_1 k_2) \sin \Phi}{gK} - \frac{\cos \Phi}{g} \tag{32f}$$

If we consider now the 'normal' or ellipsoidal field referred to geodetic latitude and longitude ϕ and λ and normal potential U, the transformation matrices (19) and (20) take the form

$$C = \begin{Vmatrix} \dfrac{1}{\rho} & 0 & 0 \\ 0 & \dfrac{1}{N \cos \phi} & 0 \\ \dfrac{1}{\rho} \dfrac{\partial \ln \gamma}{\partial \phi} & 0 & -\gamma \end{Vmatrix} \tag{33}$$

$$D = \begin{Vmatrix} \rho & 0 & \dfrac{1}{\gamma} \dfrac{\partial \ln \gamma}{\partial \phi} \\ 0 & N \cos \phi & 0 \\ 0 & 0 & -\dfrac{1}{\gamma} \end{Vmatrix} \tag{34}$$

with $C^T D = I$. We have thus

$$\mathbf{v}^r = C^T \mathbf{i}_r \qquad \mathbf{v}_r = D^T \mathbf{i}_r \tag{35}$$

where $\mathbf{v}_r$, $\mathbf{i}_r$ denote the base vectors of the (ϕ, λ, U) system and of the associated local Cartesian system, respectively. Notice that in deducing (33) and (34) the fact that for an ellipsoidal field

$$t_1 = \gamma_2 = 0 \tag{36}$$

has been taken into account. Furthermore, in this case we have

$$\gamma_1 = \frac{1}{\rho} \frac{\partial \ln \gamma}{\partial \phi} \tag{37}$$

The metric tensor of the normal field is thus

$$g_{rs} = \begin{Vmatrix} \rho^2 & 0 & \dfrac{\rho}{\gamma} \dfrac{\partial \ln \gamma}{\partial \phi} \\ 0 & N^2 \cos^2 \phi & 0 \\ \dfrac{\rho}{\gamma} \dfrac{\partial \ln \gamma}{\partial \phi} & 0 & \dfrac{1}{\gamma^2} \left[1 + \left(\dfrac{\partial \ln \gamma}{\partial \phi} \right)^2 \right] \end{Vmatrix} \tag{38}$$

The only nonidentically satisfied Mainardi-Codazzi equations can be deduced from (32*b*)–(32*d*). We have

$$\frac{\partial N}{\partial U} = -\frac{1}{\gamma} + tg\phi \frac{\partial}{\partial \phi}\left(\frac{1}{\gamma}\right) \tag{39a}$$

$$\frac{\partial \rho}{\partial U} = -\frac{1}{\gamma} - \frac{\partial^2}{\partial \phi^2}\left(\frac{1}{\gamma}\right) \tag{39b}$$

$$\frac{\partial N}{\partial \phi} = (N - \rho)tg\phi \tag{39c}$$

Notice that the integrability conditions (39*a*) and (39*b*), together with Brun's formula

$$\frac{\partial \gamma}{\partial U} = \frac{2\omega^2}{\gamma} + \frac{1}{N} + \frac{1}{\rho}\left[1 + \left(\frac{\partial \ln \gamma}{\partial \phi}\right)^2\right] \tag{40}$$

give a means for the upward expansion of ρ, N, γ and the other relevant quantities of the three-dimensional normal ellipsoidal field [*Bocchio,* 1977, 1978].

If we denote by ξ, ε, T the disturbances in latitude, longitude, and potential, respectively, we can write

$$\mathbf{v}^{r*} = \mathbf{v}^r + \mathbf{\Delta}^r \tag{41}$$

where

$$\mathbf{\Delta}^r \equiv \text{grad}\, \Delta \qquad \Delta = [\xi, \varepsilon, T]^T \tag{42}$$

If we now denote by E the matrix relating the local astronomical triads $\mathbf{i}_r{}^*$ and the local geodetic triads $\mathbf{i}_r$, that is,

$$\mathbf{i}_r{}^* = E\mathbf{i}_r \qquad E^{-1} = E^T \tag{43}$$

we have, since $\mathbf{v}^r = C^T\mathbf{i}_r$,

$$\mathbf{v}^r = C^TE^T\mathbf{i}_r{}^* \tag{44}$$

We can write

$$\mathbf{\Delta}^r = F\mathbf{v}^r \tag{45}$$

where

$$F = \begin{Vmatrix} \dfrac{\partial \xi}{\partial \phi} & \dfrac{\partial \xi}{\partial \lambda} & \dfrac{\partial \xi}{\partial U} \\[2ex] \dfrac{\partial \varepsilon}{\partial \phi} & \dfrac{\partial \varepsilon}{\partial \lambda} & \dfrac{\partial \varepsilon}{\partial U} \\[2ex] \dfrac{\partial T}{\partial \phi} & \dfrac{\partial T}{\partial \lambda} & \dfrac{\partial T}{\partial U} \end{Vmatrix} \tag{46}$$

Since we have also

$$\mathbf{\Delta}^r = FC^T\mathbf{i}_r = FC^TE^T\mathbf{i}_r{}^* \tag{47}$$

it follows from (41), (44), and (47) that

$$\begin{aligned} \mathbf{v}_r{}^* &= C^TE^T\mathbf{i}_r{}^* + FC^TE^T\mathbf{i}_r{}^* \\ &= (I + F)C^TE^T\mathbf{i}_r{}^* \end{aligned} \tag{48}$$

But we can also write, from (18), $\mathbf{v}^{r*} = B^{-1}\mathbf{i}_r{}^* = A^T\mathbf{i}_r{}^*$, and thus

$$A^T = (I + F)C^TE^T = \begin{Vmatrix} k_1 & t_1 & \gamma_1 \\ t_1 \sec\phi & k_2 \sec\phi & \gamma_2 \sec\phi \\ 0 & 0 & -g \end{Vmatrix} \tag{49}$$

Since $\det E^T = 1$, we have

$$\det A = \det A^T = \det (I + F) \det C^T \tag{50}$$

and the condition (23) can thus be written, since $\det C^T \neq 0$, as

$$\det (I + F) \neq 0 \tag{51}$$

SOME FORMULATIONS OF THE SINGULARITY CONDITION

When the matrix of integrating factors (19) has no full rank, the condition (21) is not satisfied, and we have thus

$$\det A = -gK \sec \Phi = 0 \tag{52}$$

which is thus the singularity condition for the coordinates (Φ, Λ, W). The problem arising from (52) is referred to as the 'geodetic singularity problem' [*Grafarend,* 1975, 1978*a, b, c*].

In a local-operational approach to the geodetic singularity problem, *Livieratos* [1976, 1977*a, b,* 1978, 1979*a, b*] expresses the singularity condition in terms of the gradients of the disturbances in the geodetic coordinates. His arguments can be summarized as follows.

If we consider (50), we see that the singularity condition $\det A \neq 0$ is equivalent to the condition (51), which can be written explicitly if one takes (46) into account. Therefore the singularity condition $gK \sec \Phi = 0$ is equivalent to the condition

$$\det \begin{Vmatrix} 1 + \dfrac{\partial \xi}{\partial \phi} & \dfrac{\partial \xi}{\partial \lambda} & \dfrac{\partial \xi}{\partial U} \\[2ex] \dfrac{\partial \varepsilon}{\partial \phi} & 1 + \dfrac{\partial \varepsilon}{\partial \lambda} & \dfrac{\partial \varepsilon}{\partial U} \\[2ex] \dfrac{\partial T}{\partial \phi} & \dfrac{\partial T}{\partial \lambda} & 1 + \dfrac{\partial T}{\partial U} \end{Vmatrix} = 0 \tag{53}$$

or, equivalently,

$$\begin{aligned} &\frac{\partial \xi}{\partial \phi} + \frac{\partial \varepsilon}{\partial \lambda} + \frac{\partial T}{\partial U} + \frac{\partial \xi}{\partial \phi}\frac{\partial \varepsilon}{\partial y} + \frac{\partial \xi}{\partial \phi}\frac{\partial T}{\partial U} + \frac{\partial \varepsilon}{\partial \lambda}\frac{\partial T}{\partial U} - \frac{\partial \xi}{\partial \lambda}\frac{\partial \varepsilon}{\partial \phi} \\ &- \frac{\partial \xi}{\partial U}\frac{\partial T}{\partial \phi} - \frac{\partial \varepsilon}{\partial U}\frac{\partial T}{\partial \lambda} + \frac{\partial \xi}{\partial \phi}\frac{\partial \varepsilon}{\partial \lambda}\frac{\partial T}{\partial U} + \frac{\partial \xi}{\partial \lambda}\frac{\partial \varepsilon}{\partial U}\frac{\partial T}{\partial \phi} \\ &+ \frac{\partial \xi}{\partial U}\frac{\partial \varepsilon}{\partial \phi}\frac{\partial T}{\partial \lambda} - \frac{\partial \xi}{\partial \phi}\frac{\partial \varepsilon}{\partial U}\frac{\partial T}{\partial \lambda} - \frac{\partial \xi}{\partial \lambda}\frac{\partial \varepsilon}{\partial \phi}\frac{\partial T}{\partial U} - \frac{\partial \xi}{\partial U}\frac{\partial \varepsilon}{\partial \lambda}\frac{\partial T}{\partial \phi} = -1 \end{aligned} \tag{54}$$

Livieratos [1976] gives a first-order approximation of the singularity condition (54) which nevertheless cannot be used in singularity regions, since, as will be shown in the next section, the gradients of the disturbances ξ, ε, T are in this case of the order of unity and some of the products in (54) thus cannot be neglected.

The approach by *Grossman* [1976, 1978] to the geodetic singularity problem has a global-geometrical character. He introduces the concept of the 'trapped' geoid G_0, defined as a geoid contained within a surface G_t, $t > 0$, whose area is strictly smaller than the area of G_0, the subscript t indicating differences in potential energy. He shows that a sufficient condition for G_0 to be trapped is

$$\iint_{G_0} \frac{H}{g}\, dA < 0 \tag{55}$$

H being the mean curvature of the geoid G_0 (defined so that it is positive when calculated with respect to the outward normal on a round sphere), dA the element of area, and g the scalar gravity. Starting from a theorem by H. Hopf, according to which a point exists on a closed surface where both principal curvatures are positive, Grossman argues that if a surface has $\iint H\, dA < 0$, there must be a point at which $H < 0$, whence at least one principal curvature is negative there. If one then considers the surface curve joining these two points, at least one point exists on the curve where a principal curvature vanishes. It thus follows that infinitely many parabolic singularities can be found on such a surface.

Grossman also gives the example of a 'fantasy planet' made up of two massive parallel rods of finite length and at a finite distance apart surrounded by a fluid of negligible mass compared to that of the rods and shows that the geoid of this planet is a 'trapped equipotential' of the planet. As will be shown in the next section, it is possible to find singularities in connection with mass distributions which are closer to the domain of geodesy or, more generally, of planetary geodesy. Nevertheless, even though the validity of the convexity assumption with regard to the equipotentials of the actual gravity field was questioned more than a century ago by *Bruns* [1878], the question concerning the occurrence of geodetic singularities in the external gravity field of the earth is still unresolved despite recent research [*Bocchio*, 1981*a*, *b*, 1982] pointing out that interest should be devoted to the 'geodetic relevance' of geodetic singularities, when they occur, considering the 'scale' of the singularity zones in connection with the generating perturbation in the mass distribution.

It is worth mentioning that *Grossman* [1976] quotes a discussion with E. Grafarend, which is also reported by *Grafarend* [1978*c*], in which the absolute value of gravity was proposed as a coordinate in addition to the geopotential W in order to find a transformation which should have less singularities when compared with the classical one.

A DIRECT AND AN INVERSE GEODETIC SINGULARITY PROBLEM

In this section it will be shown that geodetic singularities can be generated in the external gravity field of a spherical planet by a perturbing cavity with a spherical shape and independent of the cavity radius. A homogeneous and a nonhomogeneous density distribution will be considered, and the related inverse geodetic singularity problem will be solved. We will confine ourselves to the case of a nonrotating planet, since geodetic interest is mainly related to singularitires of gravitational rather than of kinematical origin.

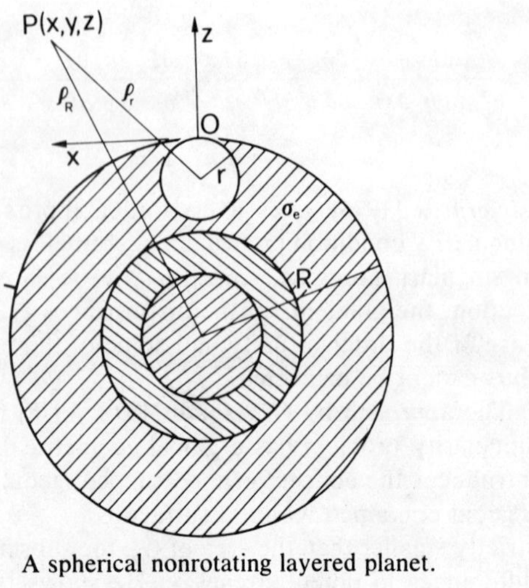

Fig. 1. A spherical nonrotating layered planet.

Consider a spherical nonrotating layered planet of radius R (Figure 1) with a spherically symmetrical density distribution, which we assume to be uniform within the single layers, and a spherical cavity of radius r buried in the external layer. Let σ_m be the mean density of the unperturbed planet, that is, of the planet without the cavity, and σ_e the density of the external spherical shell in which the cavity is buried. Let the unperturbed planet be referred to 'normal' or 'geodetic' coordinates (ϕ, λ, U), and let $W = U + T = U - V_r$ be the gravity potential of the perturbed planet, where U is the normal potential and $V_r = -T$ is the potential of a homogeneous sphere of radius r and density σ_e. It is easy to show that with reference to a local Cartesian system with the x axis northward, the y axis eastward, and the z axis upward we have (see notation list)

$$W = \tfrac{4}{3}\pi k(R^3\sigma_m\rho_R^{-1} - r^3\sigma_e\rho_r^{-1}) \tag{56}$$

$$W_1 \equiv \partial W/\partial x = -(\gamma\rho_R^{-1} + \delta\gamma\rho_r^{-1})x \tag{57a}$$

$$W_2 \equiv \partial W/\partial y = -(\gamma\rho_R^{-1} + \delta\gamma\rho_r^{-1})y \tag{57b}$$

$$W_3 \equiv \partial W/\partial z = -\gamma\rho_R^{-1}(R + z) - \delta\gamma\rho_r^{-1}(r + z) \tag{57c}$$

$$W_{11} \equiv \partial^2 W/\partial x^2 = -\gamma\rho_R^{-1}(1 - 3x^2\rho_R^{-2}) - \delta\gamma\rho_r^{-1}(1 - 3x^2\rho_r^{-2}) \tag{58a}$$

$$W_{22} \equiv \partial^2 W/\partial y^2 = -\gamma\rho_R^{-1}(1 - 3y^2\rho_R^{-2}) - \delta\gamma\rho_r^{-1}(1 - 3y^2\rho_r^{-2}) \tag{58b}$$

$$W_{33} \equiv \partial^2 W/\partial z^2 = -\gamma\rho_R^{-1}[1 - 3(R + z)^2\rho_R^{-2}] - \delta\gamma\rho_r^{-1}[1 - 3(r + z)^2\rho_r^{-2}] \tag{58c}$$

$$W_{12} \equiv \partial^2 W/\partial x\, \partial y = 3(\gamma\rho_R^{-3} + \delta\gamma\rho_r^{-3})xy \tag{58d}$$

$$W_{13} \equiv \partial^2 W/\partial x\, \partial z = 3[\gamma\rho_R^{-3}(R + z) + \delta\gamma\rho_r^{-3}(r + z)]x \tag{58e}$$

$$W_{23} \equiv \partial^2 W/\partial y\, \partial z = 3[\gamma\rho_R^{-3}(\mathrm{R} + z) + \delta\gamma\rho_r^{-3}(r + z)]y \tag{58f}$$

where, in the present case,

$$\gamma = \tfrac{4}{3}\pi kR^3\sigma_m\rho_R^{-2} \qquad \delta_\gamma = -\tfrac{4}{3}\pi kr^3\sigma_e\rho_r^{-2} \tag{59}$$

are the expressions of the unperturbed or 'normal' gravity and the gravity disturbance due to the cavity, respectively. Let us assume (x, y) as surface coordinates on the equipotential surfaces $W(x, y, z) =$ const. Since tensor notation will be used in this section, the notation x^r $(r = 1, 2, 3)$ will be adopted for the coordinates (x, y, z).

The first ground form of the W surfaces referred to (x^1, x^2) coordinates is given by

$$ds^2 = a_{\alpha\beta}\, dx^\alpha\, dx^\beta \qquad \alpha, \beta = 1, 2 \tag{60}$$

where $a_{\alpha\beta}$ is the metric tensor of the W surfaces referred to (x^1, x^2). Notice that Greek and Latin letters are here used for two- and three-dimensional quantities, respectively. Since we have

$$a_{\alpha\beta} = \gamma_{rs}x_\alpha{}^r x_\beta{}^s \qquad r, s = 1, 2, 3 \qquad \alpha, \beta = 1, 2 \tag{61}$$

where γ_{rs} is the metric tensor of the Cartesian system x^r, and thus $\gamma_{rs} = \delta_{rs}$, and

$$x_\alpha{}^r \equiv \partial x^r/\partial x^\alpha \quad (62)$$

It is easy to show that the tensor $x_\alpha{}^r$ is given by

$$x_\alpha{}^r = \begin{Vmatrix} 1 & 0 \\ 0 & 1 \\ -W_1/W_3 & -W_2/W_3 \end{Vmatrix} \quad (63)$$

and from (61) we thus obtain

$$a_{\alpha\beta} = \begin{Vmatrix} \dfrac{W_1^2 + W_3^2}{W_3^2} & \dfrac{W_1 W_2}{W_3^2} \\ \dfrac{W_1 W_2}{W_3^2} & \dfrac{W_2^2 + W_3^2}{W_3^2} \end{Vmatrix} \quad (64)$$

It can be shown that the contravariant form of the metric tensor (64) is given by

$$a^{\alpha\beta} = \begin{Vmatrix} \dfrac{W_2^2 + W_3^2}{W_1^2 + W_2^2 + W_3^2} & -\dfrac{W_1 W_2}{W_1^2 + W_2^2 + W_3^2} \\ -\dfrac{W_1 W_2}{W_1^2 + W_2^2 + W_3^2} & \dfrac{W_1^2 + W_3^2}{W_1^2 + W_2^2 + W_3^2} \end{Vmatrix} \quad (65)$$

The x^r coordinates being Cartesian in the present case, it can be shown that the tensor of the second ground form of the W surfaces can be deduced by the expression

$$b_{\alpha\beta} = \tfrac{1}{2}\varepsilon^{\sigma\tau}\varepsilon_{mnp}\frac{\partial^2 x^m}{\partial x^\alpha \partial x^\beta} x_\sigma{}^n x_\tau{}^p \quad (66)$$

where $\varepsilon^{\sigma\tau}$, ε_{mnp} are ε system [*McConnell,* 1947] on the W surfaces and in the Cartesian system, respectively. We obtain

$$b_{\alpha\beta} = \begin{Vmatrix} \dfrac{W_1 W_{13} - W_{11} W_3}{W_3(W_1^2 + W_2^2 + W_3^2)^{1/2}} & \dfrac{W_1 W_{23} - W_{12} W_3}{W_3(W_1^2 + W_2^2 + W_3^2)^{1/2}} \\ \dfrac{W_1 W_{23} - W_{12} W_3}{W_3(W_1^2 + W_2^2 + W_3^2)^{1/2}} & \dfrac{W_2 W_{23} - W_{22} W_3}{W_3(W_1^2 + W_2^2 + W_3^2)^{1/2}} \end{Vmatrix} \quad (67)$$

The total curvature of the W surfaces can be evaluated by means of the relation

$$K = \frac{b_{11}b_{22} - b_{12}^2}{a_{11}a_{22} - a_{12}^2} \quad (68)$$

and thus, taking into account (64) and (67),

$$K = \frac{(W_1 W_{13} - W_{11} W_3)(W_2 W_{23} - W_{22} W_3) - (W_1 W_{23} - W_{12} W_3)^2}{(W_1^2 + W_2^2 + W_3^2)^2} \quad (69)$$

In terms of the tensors of the first and second ground forms of a surface the mean curvature can be defined [*Finzi and Pastori,* 1949] by

$$H = -b_{\alpha\beta}a^{\alpha\beta} \quad (70)$$

Notice that according to the definition (70), H is positive when calculated with respect to the outward normal on a round sphere. The mean curvature of the W surfaces is thus given, considering (65) and (67), by

$$H = [(W_2^2 + W_3^2)(W_{11}W_3 - W_1 W_{13}) + 2W_1 W_2(W_1 W_{23} - W_{12}W_3) + (W_1^2 + W_3^2)(W_{22}W_3 - W_2 W_{23})] \cdot [W_3(W_1^2 + W_2^2 + W_3^2)^{3/2}]^{-1} \quad (71)$$

Notice that at the origin 0 of the Cartesian system we have, from (57) and (58),

$$W_1 = W_2 = W_{12} = W_{13} = W_{23} = 0 \quad (72)$$

and thus (69) and (71) reduce to

$$K_0 \equiv K(0, 0, 0) = \frac{W_{11}W_{22}}{W_3^2}$$

$$H_0 \equiv H(0, 0, 0) = \frac{W_{11} + W_{22}}{W_3} \quad (73)$$

in which, since $W_{11} = W_{22}$ on the z axis, $W_{11}/W_3 = W_{22}/W_3$ are the principal curvatures of the W surfaces at the origin.

Since the gravity field is rotationally symmetrical with respect to the z axis, we can investigate its structure by simply confining ourselves, for example, to the (x, z) plane where $W_2 = W_{12} = W_{23} = 0$; (69) and (71) thus reduce to

$$K(x, 0, z) = \frac{(W_{11}W_3 - W_1 W_{13})W_{22}W_3}{(W_1^2 + W_3^2)^2} \quad (74)$$

$$H(x, 0, z) = \frac{(W_{11}W_3 - W_1 W_{13})W_3 + (W_1^2 + W_3^2)W_{22}}{(W_1^2 + W_3^2)^{3/2}} \quad (75)$$

Since the singularity condition for coordinates of the type astronomical latitude Φ, astronomical longitude Λ, actual potential W can be written as

$$\text{grad } \Phi \times \text{grad } \Lambda \cdot \text{grad } W = 0 \quad (76)$$

and since it can be shown that

$$\text{grad } \Phi \times \text{grad } \Lambda \cdot \text{grad } W = -\frac{gK}{\cos \Phi} \quad (77)$$

where $g = \gamma + \delta\gamma$ is the scalar gravity, the singularity condition is thus expressed by the vanishing of the total curvature of the W surfaces. In the (x, z) plane the singularity condition is, from (74)

$$\frac{(W_{11}W_3 - W_1W_{13})W_{22}W_3}{(W_1^2 + W_3^2)^2} = 0 \tag{78}$$

in which the explicit functional dependence of the condition (78) on the Cartesian coordinates is shown in (57) and (58). We investigate now whether geodetic singularities exist in the external field of a nonhomogeneous planet, in connection with the perturbation previously considered, and under which conditions for the density distribution they might occur.

A sufficient condition for the singularity condition (78) to be satisfied is that $W_{22} = 0$; from (58b) we see that on the z axis this condition reads

$$W_{22}(0, 0, z) = -\tfrac{4}{3}\pi k[R^3(R + z)^{-3}\sigma_m - r^3(r + z)^{-3}\sigma_e] = 0 \tag{79}$$

which has the solution

$$z = \frac{r[(\sigma_e\sigma_m^{-1})^{1/3} - 1]}{1 - rR^{-1}(\sigma_e\sigma_m^{-1})^{1/3}} \tag{80}$$

In order to confine ourselves only to meaningful density distributions the hypothesis $\sigma_e\sigma_m^{-1} < Rr^{-1}$ will be assumed, and thus $W_{22}(0, 0, z) = 0$ for $z > 0$, that is, in the external field, if

$$\sigma_e\sigma_m^{-1} > 1 \tag{81}$$

We can thus say that geodetic singularities can occur in the external field of a nonhomogeneous planet due to a spherical cavity buried in its external layer if the density σ_e of the latter is greater than the mean density σ_m of the planet.

The singular point on the z axis with the coordinate (80) will be denoted by T (Figures 2 and 3); it can be shown that it is the 'uppermost' geodetic singularity.

Notice that in the homogeneous case, $\sigma_m = \sigma_e = \sigma$ and thus W_{22} vanishes at the origin 0 of the coordinate system (x, y, z); that is, T is on the surface of the planet.

A qualitative explanation of the fact that in the homogeneous case, 0 is a parabolic singularity (a property which seems to have scarce appeal to an intuitive approach, particularly in connection with the fact that it is independent of the dimensions of the cavity) can be given taking into

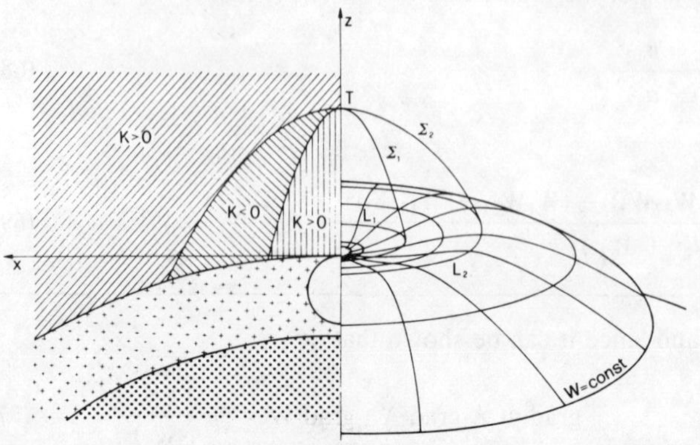

Fig. 2. Total curvature of the equipotential surfaces and qualitative description of the singularity zone in the case $\sigma_e > \sigma_m$.

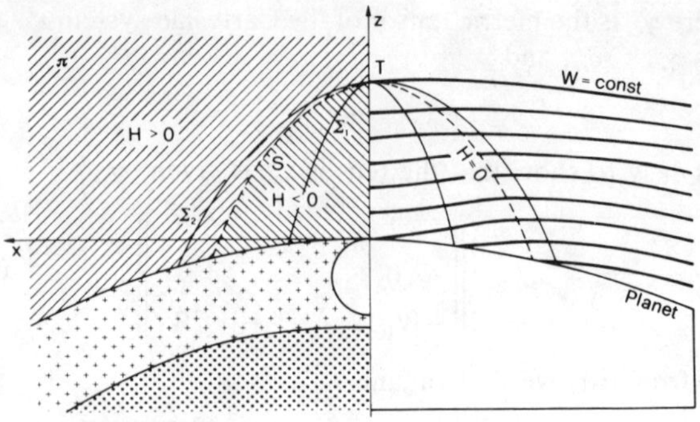

Fig. 3. Mean curvature of the equipotential surfaces in the singularity zone in the case $\sigma_e > \sigma_m$.

account an elementary property of the gravity gradients due to a spherical homogeneous body, that is, that the maximum value of the gravity gradients is in this case independent of the dimensions of the body, depends only on its density, and is found on the body surface. Since the point 0 is at the same time on the planet surface and on the cavity surface and since the components W_{ij} of the tensor of gravity gradients are given by the difference between the gravity gradients due to the unperturbed planet and the gravity gradients due to a homogeneous sphere with the same density as the planet and the same radius as the cavity, it follows that at 0, W_{ij} vanishes.

In the homogeneous case the singularity in 0 can also be interpreted from another point of view. It can in fact be shown in potential theory [*Kellogg*, 1953; *Edwards*, 1964] that within a spherical cavity in a spherical homogenous self-gravitating body the field of forces is uniform with an intensity given by $\frac{4}{3}\pi k\sigma$ times the distance between the centers of the spheres and with the direction of the line joining their centers (Appendix A). If we consider the bounding surface of the cavity tangent to the body surface, as is the case for our planet, and consider that gravity is continuous everywhere under the conditions considered here and, in particular, across the bounding surfaces, we can thus understand why the external gravity field of the planet is locally uniform at 0. If the effect of the cavity is to give rise to a singular point on the surface of the planet, it must be pointed out that the cavity itself is a singular region as far as the gravity field inside it is concerned. From a pictorial point of view the point 0 could be looked on in the homogeneous case as the gate of a 'gravitationally flat land,' since within the cavity the gravity field is the same everywhere and the equipotential surfaces reduce to parallel planes.

Since the existence of geodetic singularities has been proved in the nonhomogeneous case under the condition (81), we now investigate the structure and geometry of the gravity field in the neighborhood of the perturbing source. In this connection, consider the expression of the total and mean curvature (74) and (75) of the W surfaces in the (x, z) plane. Since here $W_3(W_1^2 + W_3^2)^2$ does not change sign, the zeroes and sign of the total curvature (74) are thus determined by the product

$$(W_{11}W_3 - W_1W_{13})W_{22} \tag{82}$$

We can see that at the origin 0 of the Cartesian coordinate system we have

$$(W_{11}W_3 - W_1W_{13})_0 = (\tfrac{4}{3}\pi k)^2(R\sigma_m - r\sigma_e)(\sigma_m - \sigma_e) < 0 \tag{83}$$

On the other side it can be shown that for $x \gg R$ we have

$$W_{11}W_3 - W_1W_{13} \approx R\gamma^2\rho_R^{-2}(1 + \gamma^{-1}\delta\gamma) + r\delta\gamma^2\rho_r^{-2} > 0 \tag{84}$$

and it is thus proved, because of the continuity of the functions in (82) in the external field of the planet, tht $W_{11}W_3 - W_1W_{13}$ vanishes on a curve in the (x, z) plane, that is, K vanishes on a surface Σ_1 in the external gravity field of the planet.

With similar reasoning we can see that at the origin 0,

$$(W_{22})_0 = \tfrac{4}{3}\pi k(\sigma_e - \sigma_m) > 0 \tag{85}$$

and we can easily see that for $x = D \gg R$ we have

$$W_{22} \approx -\tfrac{4}{3}\pi k D^{-3}(R^3\sigma_m - r^3\sigma_e) < 0 \tag{86}$$

W_{22} thus vanishes on a curve in the (x, z) plane, and K also vanishes on a surface Σ_2 in the planet's external space.

Detailed investigations worked out with the aid of numerical techniques show that the qualitative structure of the field is as shown synthetically in Figures 2 and 3. In this connection we can summarize the results obtained as follows. As far as the total curvature of the W surfaces is concerned, two singular surfaces Σ_1 and Σ_2 are found to be made up of parabolic points of the field, that is, points of vanishing total curvature of the W surfaces. In particular, Σ_2 is the singular surface where the gravity vector has everywhere the same direction, namely, the direction of the z axis, whence on the Σ_1 singular surface we have, with

$$W_{11}W_3 - W_1W_{13} = W_3^2 \frac{\partial}{\partial x}\left(\frac{W_1}{W_3}\right) \tag{87}$$

that the following relation holds on the curve of the (x, z) plane where the left side of (87) vanishes:

$$\frac{\partial}{\partial x}\left(\frac{W_1}{W_3}\right) = 0 \tag{88}$$

that is, the gravity field has a locally constant direction on the surface Σ_1, or, in other words, Σ_1 is the 'locus' of the parabolic 'parallels' of the W surfaces considered as surfaces of revolution around the z axis. As far as an interpretation of the local and global aspects of the singularities (88) and of the surface Σ_1 is concerned, we can say that given a point P' on one side as a W surface with respect to its parabolic parallel, we can find a point P'' on the other side of the surface such that the gravity vectors in P' and P'' are parallel.

We point out that the region between the singular 'domes' Σ_1 and Σ_2 is iperbolic in character, that is, the total curvature of the W surfaces is there negative, whence the field 'outside' Σ_1 and 'inside' Σ_2 is elliptical in character.

If we now consider (71) or (81), it could be shown that (which is also confirmed by numerical computations) the mean curvature of the W surfaces vanishes on a surface S between the surfaces Σ_2 and Σ_1 (Figure 3).

It is interesting to observe, considering (73) and (85), that at the origin 0 of the coordinate system (x, y, z) we have

$$H_0 = \frac{2(\sigma_m - \sigma_e)}{(R\sigma_m - r\sigma_e)} \tag{89}$$

and thus, in the present case,

$$H_0 < 0 \tag{90}$$

It is worth mentioning that in the region inside S the vertical gravity gradient is reversed, that is, gravity increases in magnitude with height. This can easily be seen by considering (58c). We have, with $g = |W_3|$ on the z axis,

$$\partial g/\partial z = -\tfrac{8}{3}\pi k[R^3(R + z)^{-3}\sigma_m - r^3(r + z)^{-3}\sigma_e] \tag{91}$$

which at $z = 0$ gives

$$(\partial g/\partial z)_0 = \tfrac{8}{3}\pi k(\sigma_e - \sigma_m) > 0 \tag{92}$$

a result which could also be obtained by means of Bruns' formula $\partial g/\partial z = -gH$ if we take into account only (89) and the fact that at the origin we have

$$g_0 \equiv g(0, 0, 0) = (|W_3|)_0 = \tfrac{4}{3}\pi k(R\sigma_m - r\sigma_e) \tag{93}$$

As far as the scale of the singularity zone is concerned, an example with $\sigma_m = 5\ \text{g cm}^{-3}$, $\sigma_e = 6\ \text{g cm}^{-3}$, and $r = R/10$ is illustrated in Figure 4.

If (ϕ, λ, U) are geodetic coordinates defined with reference to the gravity field of the unperturbed planet, it can be shown (Appendix B) that at the origin of the (x, y, z) coordinate system we have, in the homogeneous case

$$\frac{\partial\xi}{\partial\phi} = \frac{\partial\varepsilon}{\partial\lambda} = -1 \qquad \frac{\partial T}{\partial U} = -\frac{r}{R} \tag{94}$$

$$\frac{\partial\xi}{\partial\lambda} = \frac{\partial\xi}{\partial U} = \frac{\partial\varepsilon}{\partial\phi} = \frac{\partial\varepsilon}{\partial U} = \frac{\partial T}{\partial\phi} = \frac{\partial T}{\partial\lambda} = 0 \tag{95}$$

and thus in the homogeneous case the condition (54) is satisfied at the origin 0.

It is interesting to evaluate the gradients of the disturbances in the geodetic coordinates along the z axis in the nonhomogeneous case. We have

$$\frac{\partial\xi}{\partial\phi} = -\frac{(R + z)}{\gamma}\frac{\partial^2 T}{\partial x^2} = \frac{(R + z)\delta\gamma}{\gamma\rho_r} \tag{96}$$

$$\frac{\partial\varepsilon}{\partial\lambda} = -\frac{(R + z)\cos\phi}{\gamma}\frac{\partial^2 T}{\partial x\,\partial y} = \frac{(R + z)\delta\gamma}{\gamma\rho_r} \tag{97}$$

$$\frac{\partial T}{\partial U} = -\frac{1}{\gamma}\frac{\partial T}{\partial z} = \frac{(r + z)\delta\gamma}{\gamma\rho_r} \tag{98}$$

If we observe that the relations (95) are valid not only at the origin 0 but also along the z axis, that on the z axis we have

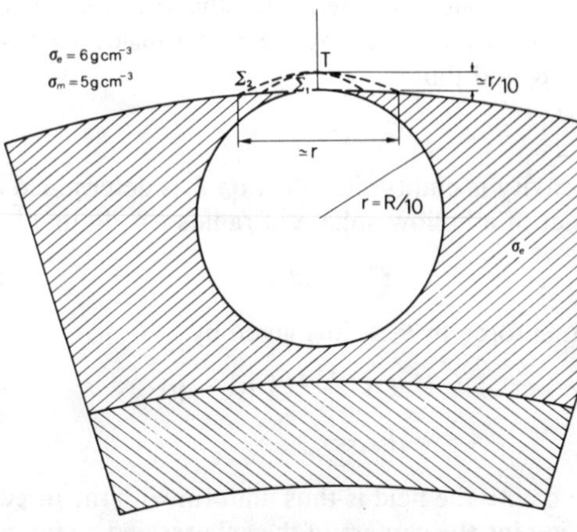

Fig. 4. Scale of the singularity zone.

$$\frac{\delta\gamma}{\gamma\rho_r} = -\frac{r^3\sigma_e}{R^3\sigma_m}\frac{(R+z)^2}{(r+z)^3} \quad (99)$$

and that at the point T of the z axis, $W_{22} = W_{11} - W_1W_{13} = 0$, we have, from (79),

$$\left(\frac{r+z}{R+z}\right)^3 = \frac{r^3\sigma_e}{R^3\sigma_m} \quad (100)$$

and the gradients (96), (97), and (98) evaluated on the point T therefore give

$$\partial\xi/\partial\phi = \partial\varepsilon/\partial\lambda = -1 \quad (101)$$

$$\partial T/\partial U = -(r+z)/(R+z) \quad (102)$$

If we now consider (101) and (102), we can easily check that the condition (54) is satisfied on the 'top' T of the singular domes Σ_2 and Σ_1.

A Comment on Singularities in Gravity Coordinates

The singularity condition for coordinates of the type (g_x, g_y, g_z), where g_x, g_y, and g_z are the Cartesian components of the gravity vector, is the vanishing of the determinant of the matrix of gravity gradients:

$$\det \|W_{rs}\| = 0 \quad (103)$$

The condition (103) is generally more strict than the singularity condition for (Φ, Λ, W) coordinates, $K = 0$, since it is associated with the fact that at different points in space the gravity vectors coincide, whence the condition for parabolic singularities is associated to parallelism between gravity vectors at different points. This can easily be seen if we consider the singularity condition for gravity coordinates (103) with reference to the case considered in the previous section. In the (x, z) plane we have

$$\det \|W_{rs}\| = W_{22}(W_{11}W_{33} - W_{13}^2) = 0 \quad (104)$$

and therefore it follows that whereas the Σ_2 surface, which is singular for (Φ, Λ, W) coordinates, is singular also for gravity coordinates, which is due to the fact that on L_2 type curves on the W surfaces the gravity vectors coincide, this is not the case for the Σ_1 singular surface, since it is associated with parallel but not necessarily equal gravity vectors on opposite sides of L_1 type curves on the W surface.

Appendix A

It is well known that if the cavity were filled with matter with the same density σ as the planet, the force $\mathbf{f}$ in P (Figure 5) would be determined only by the matter inside a sphere of radius $\rho = |\boldsymbol{\rho}|$ and thus

$$\mathbf{f}' = -\tfrac{4}{3}\pi k\sigma\boldsymbol{\rho} \quad (A1)$$

Since the cavity is empty, its effect on P is determined only by the effect of a hollow sphere of radius $l = |\mathbf{l}|$, that is,

$$\mathbf{f}'' = \tfrac{4}{3}\pi k\sigma\mathbf{l} \quad (A2)$$

The resulting force in P is thus given by

$$\mathbf{f} = \mathbf{f}' + \mathbf{f}'' = \tfrac{4}{3}\pi k\sigma(\mathbf{l} - \boldsymbol{\rho}) = -\tfrac{4}{3}\pi k\sigma\mathbf{s} \quad (A3)$$

Within the cavity the field is thus uniform with the direction of the line joining the centers of the spheres and an intensity

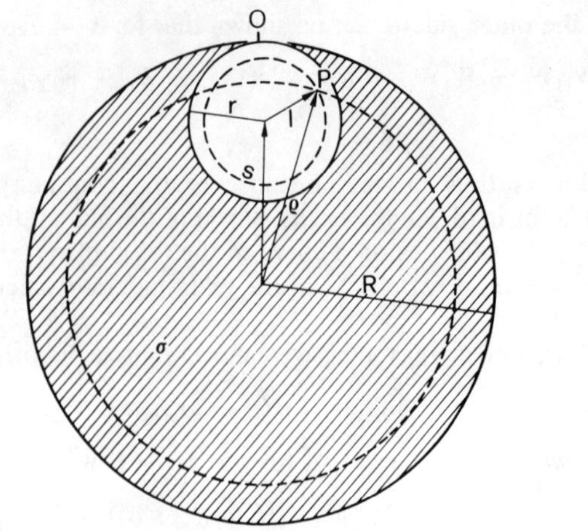

Fig. 5. Gravity field inside a spherical cavity in a homogeneous planet.

$$f = |\mathbf{f}| = \tfrac{4}{3}\pi k\sigma s = \tfrac{4}{3}\pi k\sigma(R - r) \quad (A4)$$

where $s = |\mathbf{s}|$.

Appendix B

Gradients of the disturbances ξ, ε, T in the neighborhood of the origin in the homogeneous case are

$$\frac{\partial\xi}{\partial\phi} = -\frac{R}{\gamma}\frac{\partial^2 T}{\partial x^2} = \frac{R\delta\gamma}{\gamma\rho_r}\left(1 - \frac{3x^2}{\rho_r^2}\right) \quad (B1)$$

$$\frac{\partial\xi}{\partial\lambda} = -\frac{R\cos\phi}{\gamma}\frac{\partial^2 T}{\partial x\,\partial y} = -\frac{3R\delta\gamma\cos\phi}{\gamma\rho_r^3}xy \quad (B2)$$

$$\frac{\partial\xi}{\partial U} = \frac{1}{\gamma^2}\frac{\partial^2 T}{\partial x\,\partial z} = \frac{3\delta\gamma}{\gamma^2\rho_r^3}x(r+z) \quad (B3)$$

$$\frac{\partial\varepsilon}{\partial\phi} = -\frac{R}{\gamma\cos\phi}\frac{\partial^2 T}{\partial x\,\partial y} = -\frac{3R\delta\gamma}{\gamma\cos\phi\rho_r^3}xy \quad (B4)$$

$$\frac{\partial\varepsilon}{\partial\lambda} = -\frac{R}{\gamma}\frac{\partial^2 T}{\partial y^2} = \frac{R}{\gamma}\frac{\delta\gamma}{\rho_r}\left(1 - \frac{3y^2}{\rho_r^2}\right) \quad (B5)$$

$$\frac{\partial\varepsilon}{\partial U} = -\frac{1}{\gamma^2\cos\phi}\frac{\partial^2 T}{\partial y\,\partial z} = -\frac{3\delta\gamma}{\gamma^2\cos\phi\rho_r^3}y(r+z) \quad (B6)$$

$$\frac{\partial T}{\partial\phi} = R\frac{\partial T}{\partial x} = -\frac{R\delta\gamma}{\rho_r}x \quad (B7)$$

$$\frac{\partial T}{\partial\lambda} = R\cos\phi\frac{\partial T}{\partial y} = -\frac{R\delta\gamma\cos\phi}{\rho_r}y \quad (B8)$$

$$\frac{\partial T}{\partial U} = -\frac{1}{\gamma}\frac{\partial T}{\partial z} = \frac{\partial\gamma}{\gamma\rho_r}(r+z) \quad (B9)$$

Notation

W_{rs} tensor of gravity gradients.
γ 'normal' gravity.
$\delta\gamma$ gravity disturbance.
g actual gravity ($g = \gamma + \delta\gamma$).

(Φ, Λ, W)	astronomical latitude and longitude and geopotential.
(ϕ, λ, U)	geodetic latitude and longitude and normal potential.
ξ, ε	disturbances in latitude and longitude.
K, H	Gaussian and mean curvature of the equipotential surfaces of the actual field.
σ_m	mean density of the unperturbed planet.
σ_e	density of the external layer of the planet.
ρ_R	distance of a point $P(x, y, z)$ external to the planet from the center of the planet.
ρ_r	distance of $P(x, y, z)$ from the center of the cavity.
δ_{rs}, δ_r^s	Kronecker's symbol.
k	gravitational constant.
$\mathbf{v}_r^*, \mathbf{v}^{r*}$	base vectors and reciprocal base vectors of the (Φ, Λ, W) system.
$\mathbf{i}_r^*$	local Cartesian astronomical triads.
g_{rs}^*	metric tensor of the (Φ, Λ, W) system.
$\mathbf{v}_r, \mathbf{v}^r$	base vectors and reciprocal base vectors of the (ϕ, λ, U) system.
$\mathbf{i}_r$	local Cartesian geodetic triads.
g_{rs}	metric tensor of the (ϕ, λ, U) system.
$\mathbf{X}_r$	fixed Cartesian triad with $\mathbf{X}_3$ parallel to the earth's rotation axis.

REFERENCES

Bocchio, F., Su alcune applicazioni di interesse geodetico delle connessioni non simmetriche, *Atti Accad. Naz. Lincei Cl. Sci. Fis. Mat. Nat. Rend., 48,* 343–351, 1970.

Bocchio, F., From differential geodesy to differential geophysics, *Geophys. J. R. Astron. Soc., 39,* 1–10, 1974.

Bocchio, F., First order numerical computation of the three-dimensional normal ellipsoidal field, *Geophys. J. R. Astron. Soc., 49,* 633–644, 1977.

Bocchio, F., Numerical computation of the three dimensional normal ellipsoidal field with a 10^{-2} m and 10^{-2} mgal accuracy, *Geophys. J. R. Astron. Soc., 55,* 503–506, 1978.

Bocchio, F., Representations of the gravity field of the earth, *Quat. Geod., 1,* 1–40, 1981*a*.

Bocchio, F., An inverse geodetic singularity problem, *Geophys. J. R. Astron. Soc., 67,* 181–188, 1981*b*.

Bocchio, F., Geodetic singularities in the gravity field of a nonhomogeneous planet, *Geophys. J. R. Astron. Soc., 68,* 643–652, 1982.

Bruns, H., *Die Figur der Erde,* P. Stankiewicz, Berlin, 1878.

Edwards, H. W., *Analytic and Vector Mechanics,* p. 248, Dover, New York, 1964.

Finzi, B., and M. Pastori, *Calcolo Tensoriale e Applicazioni,* p. 233, Zanichelli, Bologna, Italy, 1949.

Grafarend, E., The object of anholonomity and a generalized Riemannian geometry for geodesy, *Boll. Geofis. Teor. Appl., 13,* 241–253, 1971.

Grafarend, E., Three dimensional geodesy, I, The holonomity problem, *Z. Vermess.,* no. 100, 270–281, 1975.

Grafarend, E., The definition of the telluroid, *Bull. Geod., 52*(1), 25–37, 1978*a*.

Grafarend, E., Dreidimensionale geodätische Abbildungs gleichungen und die Näherungsfigur der Erde, *Z. Vermess.,* no. 3, 132–140, 1978*b*.

Grafarend, E., Marussian geodesy, *Boll. Geod. Sci. Affini.,* no. 2-3, 201–247, 1978*c*.

Grossman, N., Holonomic measurables in geodesy, *J. Geophys. Res., 79,* 689–694, 1974.

Grossman, N., The nature of the space near the earth, *Boll. Geod. Sci. Affini,* no. 4, 413–424, 1976.

Grossman, N., Is the geoid a trapped surface?, *Boll. Geod. Sci. Affini,* no. 2-3, 173–183, 1978.

Kellogg, O. D., *Foundations of Potential Theory,* p. 20, Dover, New York, 1953.

Livieratos, E., On the geodetic singularity problem, *Manuscr. Geod., 1,* 269–292, 1976.

Livieratos, E., A decomposition theorem for the transformation of local physical frames with geodetic interest, *Geophys. J. R. Astron. Soc., 50,* 739–742, 1977*a*.

Livieratos, E., Connection in the actual gravity field of the earth in terms of the disturbing potential, *Atti Accad. Naz. Lincei Cl. Sci. Fis. Mat. Nat. Rend., 42,* 523–528, 1977*b*.

Livieratos, E., On the introduction of local physical quantities of the gravity field of the earth in space positioning, in *The Use of Artificial Satellites for Geodesy and Geophysics,* vol. II, edited by G. Veis and E. Livieratos, pp. 588–599, National Technical University, Athens, 1978.

Livieratos, E., Some topics in operational geometric differential geodesy, *Boll. Geod. Sci. Affini,* no. 2, 260–285, 1979*a*.

Livieratos, E., On some structural characteristics of the mechanical and geometrical properties of the gravity field of the earth in terms of the disturbing potential, *Boll. Geod. Sci. Affini,* no. 4, 541–552, 1979*b*.

Marussi, A., *Fondamenti di Geodesia Intrinseca, Comm. Geod. Ital. Mem. 7,* Tamburini, Milan, 1951.

McConnell, A. J., *Applications of the Absolute Differential Calculus,* Blackie, London, 1947.

(Received February 2, 1982;
accepted February 22, 1982.)

REVIEWS OF GEOPHYSICS AND SPACE PHYSICS, VOL. 19, NO. 3, PAGES 437–449, AUGUST 1981

Recent Advances in the Theory of the Geodetic Boundary Value Problem

FERNANDO SANSO'

Politecnico di Milano, Istituto di Topografia, Fotogrammetria e Geofisica, Milan, Italy

After a short discussion on the general meaning of boundary value problems in physical geodesy the most classical Molodensky's problem is considered. Recent advancements in the study of this problem, in both the classical and the gravity space approach, are discussed. In particular, some new theorems are presented, one of which also refers to the dependence of the solution on variations of the choice of the telluroid. Geodetic boundary value problems of different types are mentioned: one of these, the altimetry-gravimetry problem, is also of valuable practical importance.

1. The Role of Boundary Value Problems in Physical Geodesy

In physical geodesy we usually deal with geometrical quantities (like distances and angles) and with physical quantities (like potential and modulus of the gravity) referred to the space outside the masses. The original point of physical geodesy is that the two kinds of quantities are strictly related one to the other, so that achievements in the knowledge of the geometry of the space improve our information on the gravity field and vice versa. How to combine measurements of both kinds in order to get the best representation of the gravity field is a matter of discussion among geodesists, and in this direction a lot of good original work has been done.

However, there are very important cases in which only the strictly necessary data are considered as input for the geodetic problem: for instance, if we give the shape of the boundary (geometry) and the values of the potential (physics) on it, we have a Dirichlet problem; if we give the values of the potential as well as those of the gravity vector (only physics) all over the (unknown) boundary, we have the Molodensky problem, and so on. These boundary value problems are of great importance in geodesy exactly because of their limiting role: in fact, they tell us what minimum information is necessary in order to perform a complete reconstruction of the gravity field, up to the boundary. Moreover, by studying the mathematical properties of their solutions we can say whether these problems are well posed or not, which is of maximum importance when we go to numerical solutions. As a counterexample it is enough to mention that if we know the potential even in a small sphere in its field of regularity, then it is known everywhere, although this analytic continuation is extremely unstable.

Summarizing, we can say that the role of the boundary value problem in geodesy is that of representing a sure theoretical frame of reference in dealing with practical (maybe even more general) problems.

2. The Classical Modern Geodetic Boundary Value Problem

The classical modern geodetic boundary value problem (g.b.v.p.), which is named after the famous geodesist M. S. Molodensky, may be formulated as follows: 'given at all points of the physical earth's surface S, the gravity potential w and the gravity vector g, to determine the surface S' [see *Moritz,* 1978].

This problem is classical in that its input data are obtainable by classical geodetic measurements like spirit leveling, gravimetry, and determination of astronomical coordinates; it is modern, that is, rigorous, in that it does not need the introduction of any extra data on internal mass density like those required by reduction techniques (think, for example, of Stokes' problem). Moreover, we call Molodensky's problem 'classical' to distinguish it from another approach to g.b.v.p. which will be presented in section 4.

Indeed, in order to confront this problem we need to make a number of simplifying hypotheses. First, we want to deal with the gravity field produced only by the earth's inner masses, and to this end we must correct w and g for the attraction of the atmosphere and for the direct gravitational effects of other bodies, the moon and sun mainly. Second, we shall consider the earth as an undeformable body, uniformly rotating around a fixed axis passing through its center of mass. This further hypothesis excludes accounting for the indirect effects of celestial bodies, like tidal effects: in short, we consider the steady field of a steady state earth. Although simplified in this way, the g.b.v.p. is still very complicated: in fact, it is a free boundary value problem (that is, the boundary itself is unknown) and also an oblique derivative problem, since we know the gradient of w along plumb lines which are not orthogonal to the earth's surface. This problem is still far from being solved in general.

The first natural step in attacking such a problem was to linearize it; this has been accomplished in a general and satisfactory way by T. Krarup in his letters on g.b.v.p. (There is only one point in Krarup's derivation which is not perfectly rigorous; this is the implicit extension of the field of harmonicity of the anomalous potential T down to the telluroid under the pretext that T is a small quantity of the first order. This small drawback is overcome when the linearization is performed by means of a derivative with respect to a continuous parameter, as in *Hörmander*'s [1975] paper.) Krarup's reasoning starts from a reference potential ('normal' potential) and a reference surface S_0 (telluroid) which were supposed to be near the actual potential w and the actual surface of the earth S. (In this, Krarup has been helped by the pioneering work of Marussi, who first studied extensively the mappings of the actual gravity field on the normal one.) If g_0 is the reference gravity of S_0, ξ is any field (of small modulus) representing the separation between S_0 and S (see Figure 1), and T as usual indicates the anomalous potential $T = w - w_0$, we must have, to first order in ξ or T,

$$\begin{aligned} g(x) &= g(x_0 + \xi) = g(x_0) + (\partial g/\partial x)_0 \cdot \xi \\ &= g_0(x_0) + \nabla_x T(x_0) + (\partial g_0/\partial x)_0 \cdot \xi \end{aligned} \tag{1}$$

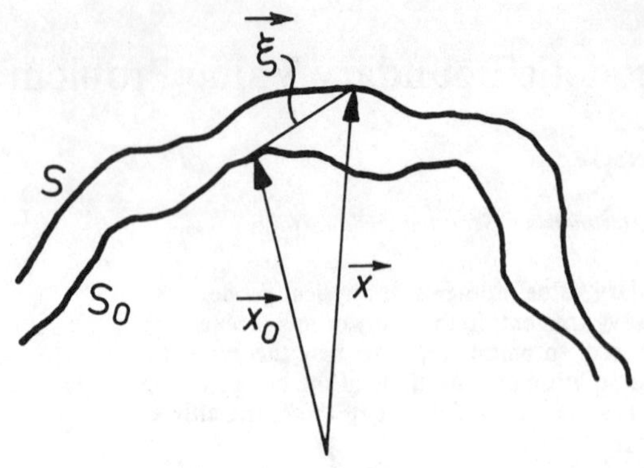

Fig. 1. Correspondence between telluroid and earth surface.

$$\begin{aligned} w(x) &= w(x_0 + \xi) = w(x_0) + g \cdot \xi \\ &= w_0(x_0) + T(x_0) + g_0 \cdot \xi \end{aligned} \tag{2}$$

By $(\partial g_0/\partial x)$ we mean the Jacobian of g_0 with respect to x, that is, the matrix

$$\left[\frac{\partial g_{0i}}{\partial x_k}\right] = \left[\frac{\partial^2 w_0}{\partial x_i \partial x_k}\right]$$

which is supposed to be regular in a region near S_0.

Solving (2) for ξ and substituting into (2), we get

$$\xi = (\partial g_0/\partial x)_0^{-1}[\delta g - \nabla_x T(x_0)] \tag{3}$$

$$\delta w - g_0 \cdot (\partial g_0/\partial x)_0^{-1}\delta g = T(x_0) - g_0 \cdot (\partial g_0/\partial x)_0^{-1}\nabla_x T(x_0) \tag{4}$$

where we have introduced the potential anomaly δw and the gravity anomaly δg in the traditional way for geodesy. Equation (4) is the famous boundary part of the linearized g.b.v.p.; the complete formulation of the b.v.p. also takes into account the harmonic character of the anomalous potential, and hence it reads (putting $\delta w - g_0 \cdot (\partial g_0/\partial x)_0^{-1}\delta g = f$)

$$\Delta T(x) = 0 \qquad \text{outside} \quad S_0 \tag{5}$$

$$T(x_0) - g_0 \cdot (\partial g_0/\partial x)_0^{-1}\nabla T(x_0) = f \qquad \text{on} \quad S_0 \tag{6}$$

It should be pointed out that by assuming T to be harmonic we implicitly admit that the reference pontential w_0 contains the centrifugal term $\frac{1}{2}\omega^2(x_1^2 + x_2^2)$; in this sense, $w_0 = M/r$ is not acceptable, while $w_0 = M/r + \frac{1}{2}\omega^2(x_1^2 + x_2^2)$ is permitted.

Since, as it will be clarified in a moment, the vector $(\partial g_0/\partial x)^{-1}g_0$ is not pointing in general in the same direction as the normal to S_0, we are in front of an oblique derivative problem; however, within the linearization the free boundary parameters ξ have been eliminated, and the problem is now of fixed boundary type. More specifically, let us examine the term (Einstein's convention is assumed to hold)

$$g_0 \cdot (\partial g_0/\partial x)^{-1}\nabla T = g_{0i}(\partial g_0/\partial x)_{ik}^{-1}T_{,k} \tag{7}$$

Since

$$\left(\frac{\partial g_0}{\partial x}\right)^{-1}_{ik} = \left(\frac{\partial x}{\partial g_0}\right)_{ik} = \frac{\partial x_i}{\partial g_{0k}} = \left(\frac{\partial g_0}{\partial x}\right)^{-1}_{ki} = \frac{\partial x_k}{\partial g_{0i}}$$

we get from (7)

$$\begin{aligned} g_0 \cdot (\partial g_0/\partial x)\nabla T &= g_{0i}(\partial x_k/\partial g_{0i})(\partial T/\partial x_k) \\ &= g_{0i}(\partial T/\partial g_{0i}) \end{aligned} \tag{8}$$

If we perform the further change of variables

$$\begin{aligned} g_{01} &= g_0 \cos\phi \sin\lambda \\ g_{02} &= g_0 \cos\phi \cos\lambda \\ g_{03} &= g_0 \sin\phi \end{aligned} \tag{9}$$

where ϕ and λ are geographical latitude and longitude and g_0 is the reference gravity modulus, we obtain straightforwardly

$$g_0 \cdot (\partial g_0/\partial x)^{-1}\nabla T = g_0(\partial T/\partial g_0) \tag{10}$$

Formula (10) yields two interesting conclusions. The first is that the derivative in the boundary operator (6) is taken along the coordinate line 'g_0,' that is, along the line ϕ = const, λ = const, which, following Marussi, is called the isozenithal line of the reference field. Such a line is, in general, not orthogonal to the telluroid S_0, whence the oblique derivative character of the b.v.p. The second conclusion is that the boundary operator is mostly simplified if we express it in 'gravity coordinates,' so that we are logically compelled to investigate a possible representation of the g.b.v.p. in such coordinates.

The line of reasoning just presented was first proposed by *Moritz* [1978] and successively, but independently by *Sanso'* [1980].

Before concluding this section we must account for two important modifications of the formulation of the linearized g.b.v.p., (5) and (6). In fact, after the first analysis of the most simple cases, like Stokes' problem, geodesists soon realized that there cannot be a unique solution to this problem, nor does there always exist a solution whatever the value of the known term f. As a matter of fact, let us put

$$\tilde{T} = c \cdot g_0 = c_j g_{0j} = c_i w_{0,j} \tag{11}$$

where c is any vector with three constant components c_i. If we neglect the rotational effects ($\omega^2 = 0$), w_0 is regular at infinity, and it is evident from (11) that

$$\Delta\tilde{T} = 0$$

Moreover, let us compute the term

$$\begin{aligned} \left(\frac{\partial g_0}{\partial x}\right)^{-1}\nabla(c_j g_{0j})|_k &= \left(\frac{\partial g_0}{\partial x}\right)^{-1}_{ki}\frac{\partial}{\partial x_i}(c_j g_{0j}) \\ &= \frac{\partial x_i}{\partial g_{0k}}\frac{\partial}{\partial x_i}(c_j g_{0j}) = \frac{\partial}{\partial g_{0k}}(c_j g_{0j}) = c_k \end{aligned} \tag{12}$$

We have then

$$\tilde{T} - g_0 \cdot \left(\frac{\partial g_0}{\partial x}\right)^{-1}\nabla\tilde{T} = c_j g_{0j} - g_{0k}c_k = 0 \tag{13}$$

In other words,

$$\tilde{T} = c \cdot g_0$$

is an eigenfunction of the homogeneous problem, whatever the value of the vector c: this entails the existence of at least three independent eigenfunctions of the homogeneous problem, namely, $w_{0,j}$ (j = 1, 2, 3). Whether the number of independent solutions of the homogeneous problem is three or more is a matter of a deep analysis which has not yet been concluded for the general case.

However, between geodesists it is commonly guessed that there cannot be more than three such eigenfunctions; this is also because (11) has a precise physical interpretation in terms of invariance of Molodensky's problem under translations and we are not able to see any other invariance of the problem.

It is evident, in fact, that if $S \equiv x(\alpha)$ is the surface resulting from the solution of Molodensky's problem and $w(x)$ is the potential harmonic outside S, such that

$$w|_S = w(\alpha)$$

$$\nabla w|_S = g(\alpha)$$

then also

$$S' \equiv \{c + x(\alpha)\}$$

$$w' = w(x - c)$$

is a solution of the same problem in the sense that

$$w'|_{S'} = w(\alpha)$$

$$\nabla w'|_{S'} = g(\alpha)$$

This is confirmed in our linearized version, since if T is a solution of (5) and (6) and if ξ, computed by (3), gives the corresponding surface S, then also

$$T' = T - c \cdot g_0$$

is a solution of (5) and (6), and the corresponding surface S' will be given by

$$\begin{aligned}\xi' &= (\partial g_0/\partial x)^{-1}[\delta g - \nabla T + \nabla(c \cdot g_0)] \\ &= \xi + (\partial g_0/\partial x)^{-1}\nabla(c \cdot g_0) \\ &= \xi + c\end{aligned}$$

that is, S' will be equal to S translated by c.

In order to get free of this indetermination of the solution it is a tradition in physical geodesy to impose supplementary conditions. The following asymptotic relation usually serves as such:

$$T = (\delta M/r) + O(r^{-3})$$

which says that the origin of the coordinates has been placed at the barycenter of the masses generating T; this condition is obviously not invariant under translation and is therefore suitable to determine the free value of $c = (c_1, c_2, c_3)$.

It must be noticed that for the sake of simplicity here and throughout this review the gravitational constant G of Newton's law is assumed to be equal to 1. Therefore the term δM has the meaning of the difference of the total mass M of the earth minus the total mass M_0 of the normal gravity field. Since M is not a datum of the problem, in principle, we cannot assume $\delta M = 0$. From the general theory of partial differential equations we know that the oblique derivative problems, like (5) and (6), enjoy a mathematical property which is called Fredholm's alternative: following this theory, if the homogeneous problem ($f = 0$) has three independent eigensolutions, then the general problem (5) and (6) has a solution if and only if the known term f satisfies three supplementary conditions. These are not easily written in explicit form unless we known the eigensolutions of the adjoint homogeneous problem, which we do not. In order to surmount the obstacle we can introduce three extra unknowns in the boundary operator, in order to equalize the balance of unknowns and conditions. This has been done by Hörmander, who modified (6) in the form

$$T - g_0 \cdot (\partial g_0/\partial x)^{-1}\nabla T = f + \sum_1^3 a_j A_j$$

where the a_j are three unknown constants and the A_j have been taken as the first-order harmonics computed on S_0,

$$A_j = \left.\frac{Y_{1j}}{r^2}\right|_{S_0}$$

The author has also considered the choice

$$A_j = g_{0j}|_{S_0}$$

If one wants to grasp the reason for introducing this extra term, one can think of Stokes' problem, where the elimination of first-order harmonics from the gravity anomaly δg is necessary in order to get a solution. Summarizing, we can claim the linearized g.b.v.p. to be as follows: to find the anomalous potential T and the constant vector $a = (a_1, a_2, a_3)$ such that

$$\Delta T = 0 \text{ outside } S_0 \tag{14}$$

$$T - g_0 \cdot (\partial g_0/\partial x)^{-1}\nabla T = f + \sum a_j A_j \text{ on } S_0 \tag{15}$$

$$T = (\delta M/r) + O(r^{-3}) \quad \text{at } \infty \tag{16}$$

The problem is not solved in general, but there are considerable hints that there should exist a unique solution for every f in a fair suitable space and for not too rough choices of S_0.

3. Recent Advancements in the Solution of the Classical G.B.V.P.

As we have emphasized many times in the preceding sections, we are not yet able to answer the three fundamental questions regarding the g.b.v.p.:

1. Does the solutin exist for every datum in a fair functional class?
2. Is the solution unique?
3. Does the solution depend continuously on the data?

However, in recent years there have been new theoretical achievements: these are especially due to Hörmander, who after a careful analysis of the linearized g.b.v.p. was the first to derive a result for the nonlinear problem. The new results arise from perturbation techniques, so that they cannot yield general statements. We shall present here in short a result for the linearized g.b.v.p. which in some respects is better than Hörmander's. This will be achieved in two stages.

The simple g.b.v.p. This is the linearized g.b.v.p. when the following steps are taken: (1) we assume as reference potential w_0 the function

$$w_0 = (M/r) + \tfrac{1}{2}\omega^2(x_1^2 + x_2^2) \tag{17}$$

so that

$$w_{0,i} = g_{0i} = -\frac{Mx_i}{r^3} + \omega^2(x_1\delta_{1i} + x_2\delta_{2i}) \tag{18}$$

$$w_{0,ik} = -\frac{M}{r^3}\left[\delta_{ik} - 3\frac{x_i x_k}{r^2}\right] + \omega^2(\delta_{1i}\delta_{1k} + \delta_{2i}\delta_{2k}) \tag{19}$$

and (2) we consider the centrifugal part (the one depending on ω^2) in the field

$$m_{0i} = w_{0,ik}{}^{-1} g_{0k} \tag{20}$$

as negligible, if computed on the telluroid S_0, in comparison with the gravitational part. By means of these two hypotheses (the so-called spherical approximation) we have

$$m_0 = (\partial g_0/\partial x)^{-1} g_0 = -\tfrac{1}{2}x \tag{21}$$

so that the boundary operator (15) becomes

$$T + \tfrac{1}{2}x \cdot \nabla T = f + \sum a_j A_j \tag{22}$$

or

$$T + \tfrac{1}{2}r(\partial T/\partial r) = f + \sum a_j A_j \tag{23}$$

The relevant point in this model is that since T has to be harmonic, $T + \frac{1}{2}x \cdot \nabla T$ has to be harmonic too. Therefore we can first solve the Dirichlet problem

$$\Delta u = 0$$

$$u|_{S_0} = f + \Sigma a_j A_j \qquad A_j = Y_{1j}/r^2$$

remove the three first-order harmonics in u by using the three arbitrary constants (a_1, a_2, a_3), and then solve (23) as a simple differential equation, thus obtaining the usual upward continuation formulae [cf. *Moritz,* 1978]. Notice that for any given continuous f the three constants $\{a_j\}$ are henceforth fixed bounded values, or, in more strict terminology, they are three continuous functionals of f. Let us now take into consideration the regularity of the solution. In this respect we recall that a Banach space $C_{n+\epsilon}(B)$ means the space of all functions defined in the set B and with finite norm

$$\|f\|_{C_{n+\epsilon}(B)} = \|f\|_{C_n(B)}$$

$$+ \sup_{x,y\in B} \sum_{\alpha_1 + \cdots \alpha_n = n} \left| \frac{\partial^n f(x)}{\partial x_1^{\alpha_1} \cdots \partial x_n^{\alpha_n}} - \frac{\partial^n f(y)}{\partial x_1^{\alpha_1} \cdots \partial x_n^{\alpha_n}} \right| |x-y|^{-\epsilon}$$

$$\|f\|_{C_n(B)} = \sup_{x\in B} |f(x)| + \sup_{x\in B} \sum_{\alpha_1 + \cdots + \alpha_n = n} \left| \frac{\partial^n f(x)}{\partial x_1^{\alpha_1} \cdots \partial x_n^{\alpha_n}} \right|$$

By analogy, a surface S_0 is said to be of class $A_{n+\epsilon}$ if it is described by an atlas each chart of which is a function of class $C_{n+\epsilon}$. If we assume f to be in $C_{1+\epsilon}(S_0)$, it is known that u is also of class $C_{1+\epsilon}$ outside and on S_0 [cf. *Miranda,* 1970, p. 167]. Since on the other hand, assuming for the sake of simplicity $\delta M = 0$ (cf. (16)) so that $u = O(r^{-3})$, we have

$$-\tfrac{1}{2}r^2 T = \int_r^{-\infty} su(s)\, ds$$

it is apparent that T must belong to $C_{1+\epsilon}(S_0)$ too. But in this case, from the boundary relation

$$r(\partial T/\partial r) = f + \sum a_j A_j - T$$

we have

$$(\partial T/\partial r)|_{S_0} = h \in C_{1+\epsilon}(S_0)$$

Hence we can claim T to be a solution of the regular oblique derivative problem:

$$\Delta T = 0$$

$$(\partial T/\partial r)|_{S_0} = h \qquad h \in C_{1+\epsilon}(S_0) \tag{24}$$

If the angle θ between r and the normal n to S_0 satisfies the inequality

$$|\theta| \le (\pi/2) - \delta \qquad \delta > 0 \quad \text{fixed} \tag{25}$$

the corresponding homogeneous problem

$$\Delta T = 0$$

$$(\partial T/\partial r)|_{S_0} = 0$$

is known to have the solution [cf. *Miranda,* 1970, p. 11]

$$T = \text{const}$$

However, if we add the condition $T \to 0$ for $x \to \infty$, we see that the unique solution is $T = 0$. Consequently, the problem (24) admits a unique solution, namely, the T which we are studying, and this solution is of class $C_{2+\epsilon}$ on and outside S_0, as it is known from the theory of regular oblique derivative problems [cf. *Fiorenza,* 1966; *Miranda,* 1970, p. 179]. Summarizing, we have proved the following theorem.

Theorem 1: Let the surface S_0 be of class $A_{2+\epsilon}$ and satisfy condition (25), and let the known term f in

$$T + \tfrac{1}{2}x \cdot \nabla T = f + \sum a_j A_j \tag{26}$$

be of class $C_{1+\epsilon}(S_0)$; then there exist three bounded values (a_1, a_2, a_3) and a (unique) function T harmonic outside S_0 and of class $C_{2+\epsilon}$, there satisfying (26) and the further condition

$$T = (\delta M/r) + 0(r^{-3})$$

From the theorem we can claim the existence of three bounded linear functional F_j such that

$$a_j = \langle F_j, f \rangle \tag{27}$$

and of a linear operator G such that

$$T = Gf \tag{28}$$

The pairing $\langle\ ,\ \rangle$ is here the natural pairing for the space $C_{1+\epsilon}(S_0)$ to which f belongs. If we restrict the relation (28) to the boundary S_0, we can say in view of theorem 1 that G is a continuous transformation of $C_{1+\epsilon}(S_0)$ into $C_{2+\epsilon}(S_0)$, that is, that

$$\|T\|_{C_{2+\epsilon}(S_0)} \leqq \|G\| \cdot \|f\|_{C_{1+\epsilon}(S_0)} \tag{29}$$

This was the result we needed in order to take the second step.

The g.b.v.p. in almost spherical approximation. By almost spherical approximation we mean that the reference potential w_0 used in the linearization of the g.b.v.p. is 'near' the spherical potential $\tilde{w}_0 = M/r$ in such a way that the reference (isozenithal) field

$$m_0 = (\partial g_0/\partial x)^{-1} g_0$$

is near, in a suitable space, to the corresponding spherical field

$$\tilde{m}_0 = (\partial \tilde{g}_0/\partial x)^{-1} \tilde{g}_0 = -\tfrac{1}{2}x$$

if computed on the surface S_0.

The space in which we measure the distance between m_0 and $\tilde{m}_0$ is $C_{1+\epsilon}(S_0)$ in this paper: indeed, this is not the only possible choice. It is worth emphasizing that in our definition it is not required that S_0 be near a sphere: we only ask that m_0 be near $\tilde{m}_0$ on S_0. In this sense the potential of a rotating sphere

$$w_0 = (M/r) + \tfrac{1}{2}\omega^2 (x_1{}^2 + x_2{}^2)$$

is almost spherical on a surface of the kind of S_0 in Figure 2 if its mean radius R is of the same order as the radius of the earth. Also, the normal potential (i.e., one which is constant on the reference standard ellipsoid) is almost spherical on telluroids (surfaces near the physical surface of the earth), although to determine 'how near' m_0 and $\tilde{m}_0$ are is in this case a matter of a specific analysis to be performed later.

What we want to show is that if m_0 is near enough $\tilde{m}_0$, then

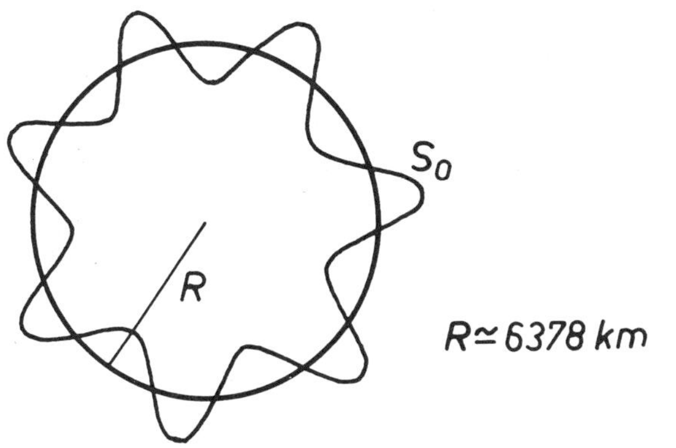

Fig. 2. Telluroid and mean earth's sphere.

the b.v.p.

$$\Delta T = 0$$
$$T - m_0 \cdot \nabla T = f + \sum a_j A_j \quad \text{on} \quad S_0 \tag{30}$$
$$T = \delta M/r + O(r^{-3})$$

has a unique solution at the same conditions as those of theorem 1. In fact, (30) is equivalent to

$$\Delta T = 0$$
$$T - \tilde{m}_0 \cdot \nabla T = (m_0 - \tilde{m}_0) \cdot \nabla T + f + \sum a_j A_j \quad \text{on} \quad S_0 \tag{31}$$
$$T = \delta M/r + O(r^{-3})$$

Furthermore, for any solution of class $C_{2+\epsilon}$ of (31) we have also

$$\nabla T|_{S_0} \in C_{1+\epsilon}(S_0) \qquad (m_0 - \tilde{m}_0) \cdot \nabla T|_{S_0} \in C_{1+\epsilon}(S_0)$$

In fact, $C_{n+\epsilon}$ is a Banach algebra, that is,

$$||f \cdot g||_{C_{n+\epsilon}} \leq ||f||_{C_{n+\epsilon}} \cdot ||g||_{C_{n+\epsilon}} \qquad f, g \in C_{n+\epsilon}$$

Therefore by applying theorem 1 we get the equivalent formulation, with T harmonic outside S_0,

$$a_j = \langle F_j, (m_0 - \tilde{m}_0) \cdot \nabla T + f \rangle \tag{32}$$
$$T = G(m_0 - \tilde{m}_0) \cdot \nabla T + Gf \tag{33}$$

To be more precise, we note that in (32) and (33) by $(m_0 - \tilde{m}_0) \cdot \nabla T$ we mean the trace of this function on S_0.

Now we want to transform (33) into an equation in the space $C_{2+\epsilon}(S_0)$: to this end we should note that ∇T is the spatial gradient which is not defined only by the values $T|_{S_0}$ but requires the further consideration of the harmonic property of T. If we want to express ∇T in terms of only the boundary values T_{S_0}, we can proceed as follows:

1. By solving a Dirichlet problem we can extend T_{S_0} outside S_0; let

$$T = DT_{S_0}$$

We know that the norm of T in $C_{2+\epsilon}$, as a spatial function, can be majorized by the norm of T_{S_0} in $C_{2+\epsilon}$ on S_0:

$$||T||_{C_{2+\epsilon}} \leq K_1 ||T_{S_0}||_{C_{2+\epsilon}(S_0)}$$

2. We can now compute $\nabla T = \nabla DT_{S_0}$ and then restrict this vector field to the surface S_0 (take the trace, in mathematical terminology): we shall then find

$$||\nabla DT_{S_0}||_{C_{1+\epsilon}(S_0)} \leq K_2 ||DT_{S_0}||_{C_{2+\epsilon}} \leq K_3 ||T_{S_0}||_{C_{2+\epsilon}(S_0)} \tag{34}$$

or, in other words, we see ∇D as a continuous transformation of $C_{2+\epsilon}(S_0)$ into $C_{1+\epsilon}(S_0)$. It is now clear that (33) can be written as

$$T_{S_0} = G(m_0 - \tilde{m}_0) \cdot \nabla DT_{S_0} + Gf \tag{35}$$

and this can be viewed as a transformation of $C_{2+\epsilon}(S_0)$ into itself. In fact, for any $f \in C_{1+\epsilon}(S_0)$ we have $Gf \in C_{2+\epsilon}(S_0)$; moreover, by dint of (29) and (34),

$$\begin{aligned} &||G(m_0 - \tilde{m}_0) \cdot \nabla DT_{S_0}||_{C_{2+\epsilon}(S_0)} \\ &\leq ||G|| \cdot ||(m_0 - \tilde{m}_0) \cdot \nabla DT_{S_0}||_{C_{1+\epsilon}(S_0)} \\ &\leq ||G|| \cdot ||m_0 - \tilde{m}_0|| \cdot K_3 \cdot ||T_{S_0}||_{C_{2+\epsilon}(S_0)} \end{aligned} \tag{36}$$

This relation tells that the operator

$$A = G(m_0 - \tilde{m}_0) \cdot \nabla D$$

continuously transforms $C_{2+\epsilon}(S_0)$ into itself and that its norm satisfies the inequality

$$||A|| \leq K_4 \cdot ||m_0 - \tilde{m}_0||_{C_{1+\epsilon}(S_0)} \tag{37}$$

Therefore if m_0 is near enough $\tilde{m}_0$ in $C_{1+\epsilon}(S_0)$, we have certainly

$$||A|| < 1$$

so that (35) becomes the equation of a contraction of which it is known that the solution exists and is unique. Since the correspondence between harmonic functions of class $C_{2+\epsilon}$ outside S_0 and their traces of class $C_{2+\epsilon}$ on S_0 is one to one, we have clearly proved the existence and uniqueness of the solution of the original problem (30). Furthermore, since by the contraction theorem, T_{S_0} depends continuously on $f \in C_{1+\epsilon}(S_0)$, we also have that T_{S_0}, and then T, depends continuously on $f \in C_{1+\epsilon}(S_0)$. Finally, by using (37) it is straightforward to deduce that T_{S_0}, and hence T, depends continuously on $m_0 \in C_{1+\epsilon}(S_0)$.

Summarizing, we have proved the following theorem.

Theorem 2: Under the conditions of theorem 1 for S_0 and f and if a vector field m_0 is given on S_0 in such a way that $m_0 \in C_{1+\epsilon}(S_0)$ and $||m_0 - \tilde{m}_0||_{C_{1+\epsilon}(S_0)}$ is suitably small, there exist a unique harmonic function T and a unique vector $a = (a_1, a_2, a_3)$ such that the boundary relation

$$T - m_0 \cdot \nabla T = f + \sum a_j A_j$$

and the asymptotic condition

$$T = (\delta M/r) + O(r^{-3})$$

are satisfied. The solution (T, a) depends continuously on the data (f, m_0) in their respective spaces.

This theorem is essentially an application of an analogous technique used by *Sanso* [1979] in the so-called gravity space approach. At this point some remarks are necessary.

1. This result is better than Hörmander's, especially in that we did not need to impose many conditions on S_0, apart from (9). On the other hand, we are not able to give an explicit quantitative meaning to the relation

$$K_4 \parallel m_0 - \tilde{m}_0 \parallel_{C_{1+\epsilon}(S_0)} < 1$$

whereas Hörmander was able to arrive at a numerical estimate for his constants.

2. By means of the generalized Bruns' relation (equation

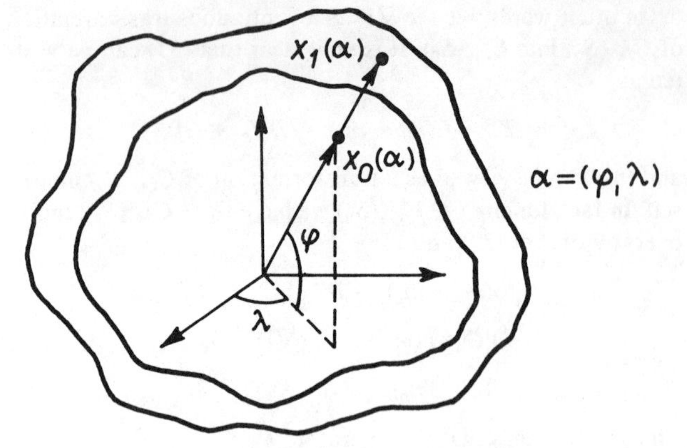

Fig. 3. Family of star-shaped surfaces.

3)) one can construct the surface S_1, which was our original unknown, by adding to the vector x_0 (spanning S_0) the vector ξ. This last, however, depends linearly on $\nabla T|_{S_0}$, which is a $C_{1+\epsilon}(S_0)$ function; consequently, ξ itself is in general of class $C_{1+\epsilon}(S_0)$, and the new surface S_1 will be less regular than S_0 and not of class $A_{2+\epsilon}$. This is a bothersome inconvenience if one tries to iterate the reasoning starting from S_1 as a better approximation of the true physical surface of the earth, S. This is also the reason why Hörmander had to use a smoothing technique combined with the Nash method. For this reason, too, Hörmander's treatment is rather complicated and will not be reproduced here. On the other hand, this drawback can be surmounted by the gravity-space approach.

3. The linearization process presented in section 2 starts by assigning an approximate (normal) potential w_0 and an approximate surface S_0 (telluroid): since both w_0 and S_0 are somewhat arbitrary, it is important to ascertain the stability of the solution under variations of these data. It is, in fact, evident that we could not accept a solution which would undergo wild changes if we gave mild perturbations to w_0 and S_0. The dependence of T and subsequently of ξ on w_0 passes through the dependence on m_0, which has already been analyzed.

Turning to the dependence on the boundary S_0, one can easily realize that this problem is much more complicated. Taking advantage of some simplifications, it can be stated as follows: given two star-shaped surfaces (see Figure 3) $S_0 \equiv \{x_0(\alpha)\}$ and $S_1 \equiv \{x_1(\alpha)\}$ and the family $S_t \equiv \{(1-t)x_0(\alpha) + tx_2(\alpha); 0 \leq t \leq 1\}$, to determine the continuous dependence on t of the field

$$\xi_t = (\partial g_0/\partial x)^{-1}[\delta g - \nabla T_t]|_{S_t}$$

where T_t solves the g.b.v.p. with telluroid S_t. Naturally, it is enough to state that $\xi_t \rightarrow \xi_1$ when $t \rightarrow 1$, in a suitable sense. If $(\partial g_0/\partial x)$ is a smooth function of x, ξ_t has the same properties as ∇T, so that we can concentrate our attention on this last factor. Furthermore, since the continuous dependence of T_t on f has already been studied, we reduce the problem to a study of the behavior of $\nabla T_t|_{S_t}$ as a function of $\alpha = (\phi, \lambda)$, T_t being the solution of

$$\nabla T_t = 0 \qquad \text{outside} \quad S_t$$

$$T_t - m_0 \cdot \nabla T_t = f + \sum a_j A_j \qquad \text{on} \quad S_t \tag{38}$$

$$T_t = (\delta M_t/r) + O(r^{-3})$$

Note that in (38), f is a fixed function of α. If we assume the simple case to hold

$$m_0 = \tilde{m}_0 = -\tfrac{1}{2}x$$

then an interesting theorem can be stated, the proof of which is given in the appendix.

Theorem 3: If $x_0(\alpha)$ and $x_1(\alpha)$ are of class $H^{1,\infty}$ and f is of class $H^{1/2}(\sigma)$, then we have

$$\lim_{t\rightarrow 1} \int_\sigma |\nabla T_t|_{S_t} - \nabla T_1|_{S_1}|^2 \, d\alpha = 0 \tag{39}$$

the integral taking place on the unit sphere.

By $x(\alpha) \in H^{1,\infty}$ we mean that the first derivatives of $x(\alpha)$ are essentially bounded (for example, a triangulated surface is in this class). By $f(\alpha) \in H^{1/2}$ we mean that there exists a function $u \in H^{1,2}$ (i.e., such that $\int_\Omega (u^2 + |\nabla u|^2) \, d\Omega < +\infty$, $\partial\Omega = S$) which assumes the values $f(\alpha)$ on S. Equation (39) will hold in particular if $x_0(\alpha)$ and $x_1(\alpha)$ are of class C_1 and f is of class C_λ ($\lambda > \frac{1}{2}$). It is clear that (39) is a kind of mean continuity and a geodesist might consider it too weak; however, it is worth noting that a sequence of polygonal approximations of S can fit uniformly the conditions of theorem 3, but not more restrictive ones. In any case, from (39) it is easy to derive the following theorem.

Theorem 4: If $x_0(\alpha)$ and $x_1(\alpha)$ are of class $C_{2+\epsilon}$ and f is of class $C_{1+\epsilon}$, then we have

$$\lim_{t\rightarrow 1} \sup_{\alpha\in\sigma} |\nabla T_t|_{S_t} - \nabla T_1|_{S_1}| = 0 \tag{40}$$

In fact, if these conditions are satisfied, $T_t|_{S_t}$ is in $C_{2+\epsilon}$, and $\nabla T_t|_{S_t}$ is in $C_{1+\epsilon}$: more precisely,

$$\sup_{\alpha\in\sigma} |\nabla T_t|_{S_t}| \leq C \qquad t \quad 0 \leq t \leq 1$$

But then, choosing any $t_n \rightarrow 1$, we have that $\nabla T_n|_{S_{tn}}$ is a compact sequence in C_1. Since, on the other hand, $\nabla T_{t_n}|_{S_{tn}}$ must tend to $\nabla T_1|_{S_1}$ in L_2, according to (39), the C_1 limit of $\nabla T_n|_{S_n}$ is necessarily $\nabla T_1|_{S_1}$: hence the conclusion (40).

Concluding this section, we can say that in our opinion the time is ripe for more general (i.e., nonperturbative) achievements on the topic of linearized g.b.v.p., but we cannot discuss this subject further in this review.

4. The Gravity Space Approach

Recently, a new approach to the g.b.v.p. has been proposed. This was also suggested in section 2 when we discovered the particularly simple form of the boundary operator in gravity coordinates (see (10)). The new approach, in fact, lies in describing the problem in gravity coordinates, that is, considering the gravity vector's components $g = (g_{x_1}, g_{x_2}, g_{x_3})$ as coordinates of the point at which they are computed. In order to avoid unnecessary complications the coordinate transformation

$$g = g(x) \tag{41}$$

has been considered as a transformation of $R^3(x)$ into $R^3(g)$, the first being the ordinary space, the second the gravity space. This mapping, to be acceptable as coordinate transformation, should be one to one on the whole space external to the earth: this is the most serious limitation of the method, since it results necessarily in suppressing the centrifugal acceleration. In fact, it is enough to think of a homogeneous rotating sphere to understand that on the equatorial plane we are able to find a circle on which $g = 0$, the one to one mapping thus being broken. For this reason we shall consider a nonrotating model of the earth; that is, we shall suppose that the following data are supplied as input for the g.b.v.p.: the gravitational potential $u(\alpha)$ on the (unknown) surface S and the corresponding gravitational vector $g(\alpha)$. In other words we suppose that from the usual data the centrifugal part has been filtered; this can also be accomplished in practice by using an approximate surface to compute $\frac{1}{2}\omega^2(x_1{}^2 + x_2{}^2)$ and $\omega^2\,(x_1\mathbf{e}_1 + x_2\mathbf{e}_2)$, which are small quantities anyway because of the presence of the factor ω^2. Once this problem is settled, we proceed to a closer analysis of the transformation (41) and of its inverse: as we shall see, in this way we shall discover surprising elements of symmetry. In fact, what can we say about (41) at first sight?

1. It is of gradient type, that is, there exists a function u such that

$$g_j(x) = \partial u/\partial x_j \tag{42}$$

2. It is one to one by hypothesis, so that the Marussi condition

$$\det U = \det[\partial^2 u/\partial x_i \partial x_j] \neq 0 \tag{43}$$

must be satisfied, apart from points at infinity; a sufficient condition for $g(x)$ being one to one is that the field $g(x)$ behave monotonically along isozenithals.

3. It is a solenoidal field (indeed, we are considering only the outer space); in fact, since by assumption, $\omega^2 = 0$, we have

$$(\partial/\partial x_j) g_j(x) = \Delta u = 0 \tag{44}$$

We should notice that (44) can also be written as

$$\mathrm{Tr}\; U = 0 \tag{45}$$

Now let us go to the inverse transformation

$$x = x(g) \tag{46}$$

The main point to be stated is that (46) is of gradient type. This can be done by observing that the Jacobian of the transformation (46) is nothing but

$$(\partial x/\partial g) = (\partial g/\partial x)^{-1} = U^{-1} \tag{47}$$

Since V is a symmetric matrix, so is U^{-1}, and consequently,

$$(\partial x_i/\partial g_k) = (\partial x_k/\partial g_i)$$

In other words, x is an irrotational field, and by Poincaré lemma it is of gradient type. The function $\psi(g)$ such that

$$x_j = \partial\psi/\partial g_j \tag{48}$$

has been called the adjoint potential and is the protagonist of gravity space theory. This can also be seen in an elegant way by using the symbolism of external differential calculus. In fact, from $du = g_j\, dx_j$ we get $d(du) = dg_j \wedge dx_j = 0$: but this can also be written as $d(x_j\, dg_j) = 0$, from which we deduce $x_j\, dg_j = d\psi$.

A crucial point is to find the relation between ψ and u. Let us, for instance, derive u with respect to g_j: we have

$$\frac{\partial u}{\partial g_j} = \frac{\partial u}{\partial x_k} \cdot \frac{\partial x_k}{\partial g_j} = g_k \frac{\partial}{\partial g_j}\frac{\partial \psi}{\partial g_k}$$

$$= \frac{\partial}{\partial g_j}\left(g_k \frac{\partial\psi}{\partial g_k}\right) - \frac{\partial\psi}{\partial g_j}$$

Hence we obtain

$$u = g_k(\partial\psi/\partial g_k) - \psi + c \tag{49}$$

The arbitrary constant c is eliminated by specifying the value of ψ at a point. The usual choice is to put

$$\lim_{g\to 0} \psi = \lim_{x\to\infty} \psi = 0 \tag{50}$$

To this end we first invert (49), noting that by dint of (48) we can write

$$\psi = x \cdot g - u + c = x_i(\partial u/\partial x_i) - u + c \tag{51}$$

Hence we observe that when $g \to 0$, correspondingly, $x \to \infty$, $u = O(r^{-1})$, and $g = O(r^{-2})$, so that from (50) and (51) we derive $c = 0$; that is, (49) becomes

$$u = g_k \frac{\partial\psi}{\partial g_k} - \psi = g\frac{\partial\psi}{\partial g} - \psi = g\psi' - \psi \tag{52}$$

where we have put $\psi' = (\partial\psi/\partial g)$. Concerning the Jacobian of (46), we clearly have

$$\left[\frac{\partial x_i}{\partial g_k}\right] = \left[\frac{\partial^2\psi}{\partial g_i \partial g_k}\right] = \Psi = U^{-1} \tag{53}$$

so that the regularity of det U clearly yields the regularity of $\det \Psi = (\det U)^{-1}$. Moreover, the relation (53) is most important to account for the harmonic character of u; in fact, from (43) and (53) we get straightforwardly the basic field equation in gravity space

$$\mathrm{Tr}\;\Psi^{-1} = \mathrm{Tr}\; U = 0 \tag{54}$$

It is a simple matter of algebra to prove that

$$\mathrm{Tr}\;\Psi^{-1} = (\det\Psi)^{-1}[(\mathrm{Tr}\;\Psi)^2 - \mathrm{Tr}\;\Psi^2]$$

so that (54) can also be written as

$$\mathrm{Tr}\;\Psi^2 - (\mathrm{Tr}\;\Psi)^2 = 0 \tag{55}$$

The basic equation (55) has to be fulfilled outside the surface S or, in terms of gravity space, inside the surface $G \equiv \{g(\alpha)\}$ (see Figure 4).

It is worth noting that while S is unknown, G is known by hypothesis. In other words, we do not know S in x coordinates, although S in g coordinates is part of the data of the g.b.v.p. To (55), which is a nonlinear partial differential equation in a fixed domain D, we must add a boundary relation: this is derived by imposing the condition that (52) hold on the (known!) boundary G

$$g\psi' - \psi|_G = u(\alpha)$$

Summarizing, we have stated that in the known domain D in gravity space the following nonlinear oblique derivative problem must hold for ψ:

$$\begin{aligned} &\mathrm{Tr}\;\Psi^2 - (\mathrm{Tr}\;\Psi)^2 = 0 \\ &g\psi' - \psi|_G = u(\alpha) \end{aligned} \tag{56}$$

The classical indeterminacy of the g.b.v.p. is nicely illustrated by (56); in fact, it is evident that if $\bar\psi$ is a solution of (56), then so is

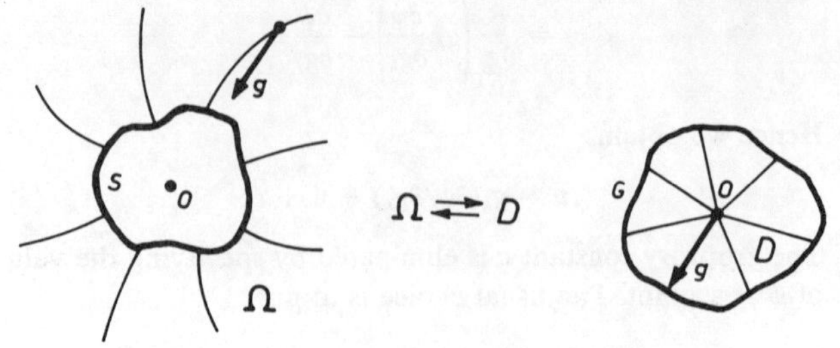

Fig. 4. The geometry-gravity space mapping.

$$\psi = \bar{\psi} + a \cdot g$$

where a is a constant vector. On the other hand, if we compute the surface S starting from ψ, we get

$$x(\alpha) = \nabla_g \psi|_G = (\nabla_g \bar{\psi} + a)|_G = \bar{x}(\alpha) + a$$

that is, the same surface as derived from $\bar{\psi}$, translated by a. In order to fix a particular solution we can use exactly the same condition that we used in geometry space, namely, to put the origin of the x coordinates at the barycenter of the masses. As we have seen, this results in the asymptotic relation

$$u = (M/r) + O(r^{-3})$$

to which we add

$$g = (M/r^2) + O(r^{-4}) \tag{57}$$

Consequently, making use of (51) with $c = 0$, we get

$$\psi = -2(M/r) + O(r^{-3}) \tag{58}$$

Inverting (57), we find

$$r = M^{1/2} g^{-1/2} + O(g^{1/2}) \tag{59}$$

and inserting (59) into (58), we derive the gravity space form of the uniqueness condition, that is

$$\psi = -2M^{1/2} g^{1/2} + O(g^{3/2}) \tag{60}$$

It is evident that by applying (60) we can eliminate the arbitrary constant vector $a = (a_1, a_2, a_3)$ in the class of solutions of (56), $\psi = \bar{\psi} + a \cdot g$. Therefore the asymptotic relation (60) must be considered as part of the definition of the g.b.v.p. in gravity space: on the other hand, also in analogy with the treatment in section 2, it is convenient to alter the boundary relation, introducing a vector of three arbitrary constants to balance the number of conditions implicit in (60).

We arrive in this way at the general formulation: find ψ and $c = (c_1, c_2, c_3)$ such that

$$\begin{aligned} &\mathrm{Tr}\,\Psi^2 - (\mathrm{Tr}\,\Psi)^2 = 0 \qquad \text{in } D \\ &g\psi' - \psi = u + c \cdot g \qquad \text{on } G \\ &\psi = -2M^{1/2} g^{1/2} + O(g^{3/2}) \end{aligned} \tag{61}$$

Equation (61) has been extensively studied by the author and also by K. J. Witsch. The most simple results are obtained in spherical approximation, from which interesting general conclusions may also be drawn. The spherical approximation is defined as the linear problem derived from (61) by assuming the adjoint potential ψ to be near the spherical adjoint potential

$$\psi_0 = -2M^{1/2} g^{1/2} \tag{62}$$

Equation (62) shows the adjoint potential corresponding to the spherical potential

$$u = M/r \qquad |g| = M/r^2$$

as is easy to verify from (57) and (58). The matrix Ψ_0 of second derivatives of (62) is

$$\Psi_0 = -M^{1/2} g^{-3/2}(I - \tfrac{3}{2}P) \qquad P = [g_i g_k / g^2]$$

and obviously it fulfills (61). Now if we put

$$\psi = \psi_0 + \delta\psi \qquad \Psi = \Psi_0 + \delta\Psi$$

after some algebra, from (61) we get the equivalent problem

$$\begin{aligned} &\mathrm{Tr}\,(I + 3P)\delta\Psi = M^{-1/2} g^{3/2} [\mathrm{Tr}\,\delta\Psi^2 - (\mathrm{Tr}\,\delta\Psi)^2] \\ &g\delta\psi' - \delta\psi = \delta u + \delta c \cdot g \\ &\delta\psi = -2Kg^{1/2} + O(g^{3/2}) \end{aligned} \tag{63}$$

The constant $K = \delta(M^{1/2})$ must necessarily be taken into account because by hypothesis, M is not included between the data and hence we can not suppose that the constant $M^{1/2}$ in $\psi_0 = -2M^{1/2} g^{1/2}$ is such as to eliminate the component $O(g^{1/2})$ in $\delta\psi = \psi - \psi_0$. Before linearizing (63), which is a very easy task, it is preferable to go one step further and take advantage of a transformation which yields a version of (63) regularized at the origin: namely, we shall put

$$\gamma = |g|^{-1/2} g \qquad \delta\phi = |g|^{-1/2} \delta\psi \tag{64}$$

After some lengthy computations one arrives at a problem of the type

$$\begin{aligned} &\Delta\delta\phi = Q(\delta\phi) \\ &\gamma(\partial\delta\phi/\partial\gamma) - \delta\phi = 2(\delta u/\gamma) + 2\delta c \cdot \gamma \\ &\delta\phi = -2K + O(\gamma^2) \end{aligned} \tag{65}$$

where $Q(\delta\phi)$ is a quadratic operator of which it is here enough to know that

$$\| Q(\delta\phi) \|_{C_\epsilon} \leqq C \| \delta\phi \|_{C_{2+\epsilon}}^2$$

The norms are indeed referred to functions of the variables $\gamma = (\gamma_1, \gamma_2, \gamma_3)$ considered in the domain D. Now in order to linearize (65) we must simply retain the linear part of the first relation, the second and the third being already linear. Then we get

$$\begin{aligned} &\Delta\delta\phi = 0 \\ &\gamma(\partial\delta\phi/\partial\gamma) - \delta\phi = 2(\delta u/\gamma) + 2\delta c \cdot \gamma \\ &\nabla\delta\phi(0) = 0 \end{aligned} \tag{66}$$

The last condition is evidently equivalent to

$$\delta\phi = -2K + O(\gamma^2)$$

It is really striking that we have arrived at a problem which is very much of the same nature as the 'simple g.b.v.p.': as a matter of fact, (66) can be treated in the same way and reduced to a Dirichlet problem, the only difference being that the latter refers to a compact domain D, while the former refers to the complement of a compact domain, Ω. The existence and the uniqueness of the solution can be derived from the following theorem.

Theorem 5: Given a domain D of class $A_{2+\epsilon}$ and for every f in $C_\epsilon(D)$ and δu in $C_{1+\epsilon}(G)$, the problem

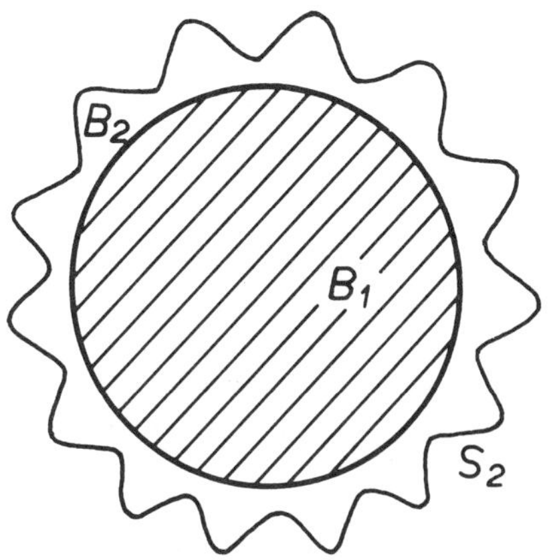

Fig. 5. Planel with massive core and light, rough outer layer.

$$\Delta\delta\phi = f$$
$$\gamma(\partial\delta\phi/\partial\gamma) - \delta\phi = 2(\delta u/\gamma) + 2\delta c \cdot \gamma$$
$$\nabla\delta\phi(0) = 0$$

admits a unique solution $\delta\phi$ in $C_{2+\epsilon}(D)$ and δc in R^3.

As we can see, as far as the linearized problem is involved, we have not gained very much by the gravity space approach. However, turning to the nonlinear problem, we must note that once a first-order $\delta\phi_1$ is obtained by solving (66), we can use it in (65) to iterate the process

$$\Delta\delta\phi_2 = Q(\delta\phi_1)$$
$$\gamma(\partial\delta\phi_2/\partial\gamma) - \delta\phi_2 = 2(\delta u/\gamma) + 2\delta c_2 \cdot \gamma \tag{67}$$
$$\nabla\delta\phi_2(0) = 0$$

As matter of fact, since

$$\| Q(\delta\phi_1) \|_{C_\epsilon} \leqq C \| \delta\phi_1 \|_{C_{2+\epsilon}}{}^2$$

the known term in (67) is in C_ϵ, and we can again apply theorem 5 to get a new solution $\delta\phi_2$, and so on. As we promised in section 3, the difficulty of losing regularity by iterating is overcome by the gravity space approach. Concerning the convergence of the sequence $\{\delta\phi_n\}$, that is, the existence of the solution of the nonlinear problem (65), we can mention the following theorem.

Theorem 6: Let the domain D of class $A_{2+\epsilon}$ and δu of class $C_{1+\epsilon}(G)$; then the nonlinear problem

$$\Delta\delta\phi = Q(\delta\phi)$$
$$\gamma(\partial\delta\phi/\partial\gamma) - \delta\phi = 2(\delta u/\gamma) + 2\delta c \cdot \gamma$$
$$\nabla\delta\phi(0) = 0$$

admits a unique solution $(\delta\phi, \delta c) \in C_{2+\epsilon}(D) \times R^3$ on condition that

$$\left\| \frac{\delta u}{\gamma} \right\|_{C_{1+\epsilon}(G)} \tag{68}$$

is small enough.

Two remarks on the theorem are required.

1. The result obtained in theorem 6 refers to the nonlinear problem, and it can be proved starting from the linear problem by an almost elementary application of Newton's method. On the contrary, in geometry space the transition from the linear to the nonlinear problem is a very hard task.

2. The crucial condition (68) refers to $C_{1+\epsilon}$ norm of

$$\frac{\delta u(\alpha)}{\gamma(\alpha)} = g^{-1/2}(\alpha)\{u(\alpha) - u_0(\alpha)\}$$
$$= g^{-1/2}(\alpha)\{u(\alpha) - M^{1/2}g^{1/2}(\alpha)\}$$

We have gained one derivative with respect to Hörmander's theory; that is, we can be sure of the existence and uniqueness of the solution even under less restrictive conditions on the datum δu. Moreover, we do not need to impose any a priori bound directly related to the inclination of the terrain. To understand the implication of this statement, one can imagine a physical body comprising two parts: an inner homogeneous heavy sphere B_1 of total mass M_1 and an external crust B_2, composed of very light material and with a smooth but rapidly oscillating boundary (see Figure 5). The gravitational potential produced by this body is

$$u(x) = \frac{M_1}{r_x} + \int_{B_2} \frac{\rho_2(y)}{|x-y|}\, dB_2(y)$$

and it is not difficult to prove that by suitably reducing the density ρ_2 and the mass M_2 with respect to M_1 we have a difference

$$u(x)g^{-1/2} - M_1{}^{1/2}$$

as small as we like at the boundary, since to first order this difference is, with obvious symbolism,

$$ug^{-1/2} - M_1{}^{1/2} = M_1{}^{1/2}\left(-\frac{1}{2}\frac{g_{2r}}{g_1} + \frac{u_2}{u_1}\right)$$

This means that for a suitable ratio M_2/M_1 we can apply the gravity space theory to derive S_2 from $u(\alpha)$, $g(\alpha)$, whereas Hörmander's theory is not applicable because of the strong inclinations of S_2. Since this scheme is not too far from reality, we think that the achievements of theorem 6 are significant.

5. ALTIMETRY-GRAVIMETRY AND OTHER B.V. PROBLEMS

As we noted in section 1, we can consider a boundary value problem of geodetic relevance as the limiting case, describing a continuous coverage of the boundary by homogeneous geodetic measurements. By this we mean that contiguous measurements are of the same kind: that is, we do not take the limit of a situation where we alternately measure different quantities. However, we shall admit that in different regions the measurements may be of different kinds, as in the altimetry-gravimetry problem. The classical g.b.v.p. that we have examined in the previous sections is, so to say, the 'simplest' from the geodetic point of view: in fact, it requires as input data only quantities obtainable by relatively simple geodetic measurements, like gravimetry or spirit leveling.

As a counterexample it is enough to think that if one wants to arrive at a Dirichlet problem, which is certainly a simpler problem from the mathematical point of view, one should supply suitable measurements in order to determine the shape of the boundary S: this is certainly not a simple task if we do not want to assume an a priori model of the global gravity field. On the other hand, in practice, we use many schemes of geodetic measurements, and it would be desirable to arrive at a definition of the corresponding boundary value problem: however, most of them are so complicated that it is hardly

TABLE 1. Main g.b.v.p.

Problem	Data	Unknown	Linear Form
Dirichlet	w, S	g_n	the same
Neumann	g_n, S	w	the same
Molodensky	w, g	S	oblique derivative
Altimetry-gravimetry	w_1, g_1 w_2, S_2	S_1 g_{2n}	oblique derivative–Dirichlet

conceivable that they will ever be used. However, one of them, namely, the altimetry-gravimetry problem, is simple enough and also of valuable practical importance: it will be presented next. Then we shall mention two other cases more as examples than as practical problems.

The Altimetry-Gravimetry Problem

As we know, gravity data at present are available mainly for the continents; although they are not uniformly distributed, we can assume that we have or we shall have in the future a sufficient coverage of this area with gravimetric data because of the relative ease in performing such measurements on solid ground. Things are different on the seas, where we cannot forecast, for the immediate future, an analogous coverage. Fortunately, modern geodetic science has allowed, through space techniques, the direct measurement of the height of a satellite flying over the sea. This also results in the knowledge of the corresponding geoidal undulation N, that is, of the height of the sea surface over the reference ellipsoid. Consequently, we can schematize the situation by assuming that we know the shape of the surface of the sea, which we shall call S_2, and, if we identify S_2 with a part of the geoid, also the potential on S_2, $w_2 = \text{const}$. On the surface of the continents, which we shall call S_1, assumed to be unknown, we propose to find both the potential $w_1(\alpha)$ and the gravity vector $g_1(\alpha)$.

This is the altimetry-gravimetry problem, and it appears to be a boundary problem of mixed type: in part a Dirichlet problem on a known surface (S_2) and in part an oblique derivative problem on a free boundary (S_1), for the Laplace equation. Indeed, in its general form it is rather difficult to study.

As a first step we can proceed by linearizing it. First, we can choose an approximating surface S_0 which will be composed of two parts, one (S_{01}) approximating the continents, the other (S_{02}) coinciding with the sea:

$$S_{02} = S_2$$

Then we must fix a reference (normal) potential w_0, for instance, $w_0 = M/r$, whence on S_{01} we can repeat the same analysis reported in section 2, arriving at the boundary relation

$$T - m_0 \cdot \nabla T|_{S_{01}} = f \qquad (69)$$

where

$$m_0 = (\partial g_0/\partial x)^{-1} g_0 \qquad f = \delta w - m_0 \cdot \delta g$$

and where δw and δg are the potential and the gravity anomaly, respectively. In the spherical approximation, that is, when $w_0 = M/r$, instead of (69) we get

$$T + \tfrac{1}{2} r \left. \frac{\partial T}{\partial r} \right|_{S_{01}} = f \qquad (70)$$

In any case, the whole linearized problem is formulated as follows: find T regular at infinity such that

$$\Delta T = 0 \quad \text{in} \quad \Omega_0 \qquad \partial\Omega_\cup \quad S_0 = S_{01} \cup S_{02}$$

$$T - m_0 \cdot \nabla T = f \quad \text{on} \quad S_{01} \qquad (71)$$

$$T = h \quad \text{on} \quad S_{02}$$

As we see, it is a mixed problem of Dirichlet–oblique derivative type. The problem is by no means simpler even in the case of the spherical approximation (70): in this case, however, it is relatively easy to prove that the alternative holds. But we cannot go into details here. It is better to observe that no arbitrary constants are introduced in (71): this should be connected to the fact that now the problem is not invariant under translations, since a part of the boundary is fixed. Conversely, we cannot impose any asymptotic relation on T, as we did for Molodensky's problem.

We conclude this short discussion of the altimetry-gravimetry problem by observing that this completes the range of any possible b.v.p. involving boundary values of T and its first derivatives only. For this purpose we can perform the problems in Table 1, which is due to H. Moritz.

Gradiometry and Isozenithal Gradiometry

To conclude this section, we mention two more examples. The first corresponds to the case where we give for the unknown boundary both $g(\alpha)$ and the gradiometric derivative $\partial g/\partial h$. For this we can elaborate an explicit form in terms of w: putting $W = [\partial^2 w/\partial x_i \partial x_k]$,

$$2g^2(\partial g/\partial h) = g(\partial g^2/\partial h) = |g| g_k (\partial/\partial h) g_k = g_k w_{,ki} g_i = g \cdot Wg$$
$$= f(\alpha) \qquad \text{known on } S$$

Then the problem can be linearized: we first introduce a surface S_0 (gravimetric telluroid)

$$g_0(x_0) = g(x)$$

so that

$$\xi = -(\partial g_0/\partial x)^{-1} \nabla T \qquad \xi_i = -M_{ij} T_{,j}$$

Then we find

$$\delta f = f(\alpha) - f_0(\alpha) = g \cdot \{W(x) - W_0(x_0)\} g$$
$$= g_l g_k \{T_{,lk} - (M_{ij} T_{,j}) w_{0,ilk}\} \qquad (72)$$

If (72) is assumed to be the boundary relation for a harmonic function T, we arrive at a new b.v.p.: this is rather unusual, since the boundary operator involves second derivatives of the unknown function and this is not natural for solutions of an equation of the second order. By this we mean that most of the mathematical literature has treated only first-order b.v.p. for equations of the second order. This is because for boundary operators of the first order we can apply Green's formulae, which are extremely useful in studying the properties of the solutions, whereas for boundary operators of higher order no such formulae are available.

One may even suspect that there is no general solution to a problem of this kind; however, a rapid inspection of (72) in the spherical approximation will convince us of the contrary. In spherical approximation, in fact, (72) becomes

$$\delta f = (M^2/r^6)\{r^2(\partial^2 T/\partial r^2) + 3r(\partial T/\partial r)\} \qquad (73)$$

and by applying the b.v.p. (73) to a spherical case one gets the conclusion. Taking advantage of the fact that

$$r^2(\partial^2 T/\partial r^2) + 2r(\partial T/\partial r) + \Delta_\alpha T = 0$$

one could eliminate the second-order radial derivative, but one pays the price of introducing second-order lateral derivatives.

It is curious to observe that the sought solution, in the case of a spherical boundary, exists only if the known term is devoid of first-order harmonics, so that we are led back to the classical indetermination

$$T = \bar{T} + \sum a_j(Y_{1j}(\alpha)/r^2)$$

This is indeed related to the invariance of the problem under translations.

The last example we shall take into account considers as data the gravity vector $g(\alpha)$ and the derivative of the gravity modulus in the direction of the isozenithal (Φ, Λ) = const, with respect to the natural coordinate w (this quantity has been taken into consideration by A. Marussi in his famous equations of intrinsic geodesy). In principle, this can be measured by taking two points P, P', one near the other along the isozenithal, and measuring g, g' at the extremes: then we have $dw = \frac{1}{2}(g + g') \cdot \overline{PP'}$. Since, however, the isozenithal is very near the vertical, we can take, in first approximatin, $dw = g\, ds$, so that

$$\frac{\partial g}{\partial w} = \lim \frac{g' - g}{w' - w} \simeq \frac{dg}{g\, ds} = \frac{(\partial g/\partial h)}{g}$$

Since we have, indeed, along the isozenithal,

$$\partial w/\partial g = (\partial g/\partial w)^{-1}$$

$\partial w/\partial g = w'$, over the unknown surface S, can be considered as a datum as well. Consequently, the problem has a very simple translation into the gravity space language: in fact, recalling that

$$w = g\psi' - \psi$$

we get immediately, on a known surface $G \equiv \{g(\alpha)\}$,

$$w' = g\psi'' = f(\alpha) \tag{74}$$

Equation (74) stands as a second-order oblique derivative problem for the solution of the nonlinear equation

$$\text{Tr } \Psi^2 - (\text{Tr } \Psi)^2 = 0 \tag{75}$$

in the known domain $D(\partial D = G)$ in the gravity space. It is nice to notice that once more the translational invariance of the problem leads to the classical indetermination

$$\psi = \bar{\psi} + a \cdot g$$

since

$$\partial^2(a \cdot g)/\partial g^2 = 0$$

6. CONCLUSIONS

Molodensky's problem has been confirmed once more as the simplest and thus most fundamental g.b.v.p. The mathematical analysis of the properties of this problem is making progress; theorems like those mentioned in sections 3 and 4 are already considering realistic conditions.

In particular, the study of the dependence of the solution on variations of the boundary is considered of great importance.

Generalization of such theorems to more general Hilbert spaces is at hand, although a decisive step toward solution of the nonlinear problem in general (no limitations on the norms of the data!) is not yet in sight. In attacking this problem the gravity space approach seems to be more promising because of its greater formal simplicity (no loss of regularity in iterating).

Beyond the classical Molodensky problem, other b.v.p.'s are interesting for geodetic purposes: one of them, namely, the altimetry-gravimetry problem, is also of considerable practical value and should be thoroughly studied from both points of view: theoretical (existence, uniqueness of the solution) and applied (approximation of the solution).

APPENDIX

The aim of this appendix is to prove the following theorem: Let us consider the star-shaped domains

$$\Omega_n = \{r \geq r_n(\alpha)\} \qquad \Omega = \{r \geq r(\alpha)\}$$

and suppose that their boundaries

$$S_n = \partial\Omega_n \qquad S = \partial\Omega$$

are of class $H^{1,\infty}$, that is,

$$\sup |\nabla_\alpha r_n(\alpha)|, \sup |\nabla_\alpha r(\alpha)| \leq \text{const}$$

(The sup here should be understood as the essential superior in L^∞.)

Let us further suppose that $r_n(\alpha)$ tends monotonically to $r(\alpha)$ in $H^{1,\infty}$: this hypothesis, although not necessary, yields a useful simplification of the proof of the theorem.

If a function f is given on the unitary sphere σ so that $f \in H^{1/2}(\sigma)$, it is known that it is possible to define the functions

$$\begin{aligned} &\Delta u^{(n)} = 0 \quad \text{in } \Omega_n && \Delta u = 0 \quad \text{in } \Omega \\ &u^{(n)} = f + \sum a_j^{(n)} A_j \quad \text{on } S_n && u = f + \sum a_j A_j \quad \text{on } S \end{aligned}$$

where

$$A_j = Y_{1j}(\alpha) r^{-2},$$

$$u^{(n)} = (a_n/r) + O(r^{-3}) \qquad u = (a/r) + O(r^{-3})$$

and we have

$$u^{(n)} \in H^{1,2}(\Omega_n) \qquad u \in H^{1,2}(\Omega)$$

Let us now define the potentials $T^{(n)}$ and T through the relations

$$T^{(n)} + \tfrac{1}{2} r \frac{\partial T^{(n)}}{\partial r} = u^{(n)} \qquad T + \tfrac{1}{2} r \frac{\partial T}{\partial r} = u \tag{A1}$$

If we write $T_{,l}^{(n)}$ and $T_{,l}$ for the lth derivatives of $T^{(n)}$ and T, we have

$$\int_\sigma |T_{,l}^{(n)}|_{S_n} - T_{,l}|_S|^2 \, d\alpha \xrightarrow{n\to\infty} 0 \tag{A2}$$

In order to prove the theorem we need two lemmata.

Lemma 1: Let Ω_n, Ω, $u^{(n)}$, and u be as in the theorem, and let us put $Ru^{(n)}$ for the restriction of $u^{(n)}$ to Ω: then we have

$$\lim_{n\to\infty} \| Ru^{(n)} - u \|_{H^{1,2}(\Omega)} = \lim_{n\to\infty} \int_\Omega \sum_l |u_{,l}^{(n)} - u_{,l}|^2 \, d\Omega = 0 \tag{A3}$$

Let us first observe that we are in a space of harmonic functions regular at infinity ($u = O(r^{-1})$), so that this is a true norm; that is, $\| u \|_{H^{1,2}} = 0 \rightarrow u = 0$. In fact, we have

$$\int_\Omega |\nabla u|^2 \, d\Omega = 0 \rightarrow u = \text{const} \rightarrow u = 0$$

This lemma is essentially a theorem of continuous dependence of the solution of Dirichlet's problem with obvious modifications to account for the side conditions: for its proof we refer to the literature [e.g., *Necas,* 1967].

Lemma 2: Let h be a harmonic function in a domain $\Omega(H^{1,\infty})$ with boundary $S = \{r = r(\alpha)\}$: if $u = O(r^{-4})$, we can define the function $g(\alpha)$ (function on the unit sphere σ) as

$$g(\alpha) = \int_{r(\alpha)}^{+\infty} ds \, s^2 h(s, \alpha) \tag{A4}$$

The following majorization holds:

$$\|g(\alpha)\|_{L_\sigma}{}^2 \leq C\|h\|_{L^2(\Omega)} \tag{A5}$$

The proof of (A5) can be reached in two steps. First, we suppose $r(\alpha) = \bar{R} = \text{const}$. But then we have

$$h(s, \alpha) = \sum_{n\geq 3,m} a_{nm} \left(\frac{\bar{R}}{s}\right)^{n+1} Y_{n,m}(\alpha)$$

From (A4) we get

$$g(\alpha) = \int_{\bar{R}}^{+\infty} ds \, s^2 h(s, \alpha) = \bar{R}^3 \sum_{n\geq 3,m} \frac{a_{nm}}{n-2} Y_{n,m}(\alpha)$$

so that if the $Y_{nm}(a)$ are fully normalized spherical harmonics,

$$\| g(\alpha) \|_{L_\sigma}{}^2 = \int g^2(\alpha) \, d\alpha = \bar{R}^6 \sum_{n\geq 3,m} \frac{a_{nm}{}^2}{(n-2)^2} \tag{A6}$$

On the other hand,

$$\| h \|_{L^2(\Omega)}{}^2 = \int_{\bar{R}}^{+\infty} ds \, s^2 \int a\alpha h^2(s, \alpha)$$

$$= \bar{R}^3 \sum_{n\geq 3,m} \frac{a_{nm}{}^2}{2n+1} \tag{A7}$$

A direct comparison of (A7) with (A6) gives the result in the case $r(\alpha) = \bar{R}$. It is interesting to observe that from (A6) and (A7) we could easily derive

$$\| g \|_{H_\sigma{}^{1/2}} \leq C \| h \|_{L^2(\Omega)}$$

This inequality, if generalized to any regular boundary S, yields a new approach to the linearized g.b.v.p.

Let us now consider the general case. By taking any $\bar{R} > r(\alpha)$ we have

$$|g(\alpha)|^2 \leqq 2\left|\left(\int_{r(\alpha)}^{\bar{R}} ds \, s^2 h(s, \alpha)\right)^2 + \left(\int_{\bar{R}}^{+\infty} ds \, s^2 h(s, \alpha)\right)^2\right|$$

$$\leqq 2\left|\int_{r(\alpha)}^{\bar{R}} ds \, s^2 \cdot \int_{r(\alpha)}^{\bar{R}} ds \, s^2 h^2(s, \alpha) + \left(\int_{\bar{R}}^{+\infty} ds \, s^2 h(s, \alpha)\right)^2\right|$$

$$\leqq \tfrac{2}{3} \bar{R}^3 \int_{r(\alpha)}^{\bar{R}} ds \, s^2 h^2(s, \alpha) + 2\left(\int_{\bar{R}}^{+\infty} ds \, s^2 h(s, \alpha)\right)^2$$

Integrating this inequality over the unit sphere and taking into account that for $r \geq \bar{R}$ the theorem is already proved, we get

$$\int |g(\alpha)|^2 \, d\alpha \leqq \frac{2\bar{R}^3}{3} \int d\alpha \int_{r(\alpha)}^{\bar{R}} ds \, s^2 h^2(s, \alpha)$$

$$+ 2C \int d\alpha \int_{\bar{R}}^{+\infty} ds \, s^2 h^2(s, \alpha)$$

$$\leqq C' \int d\alpha \int_{r(\alpha)}^{+\infty} ds \, s^2 h^2(s, \alpha)$$

QED.

Now let us consider the theorem and note that if we put

$$v^{(n)} = u^{(n)} - \frac{a_n}{r} \qquad v = u - \frac{a}{r} \qquad v_n, v = O(r^{-3})$$

from (A1) we get

$$T + \tfrac{1}{2} r(\partial T/\partial r) = v + (a/r)$$

and analogously for $T^{(n)}$. But then we can put

$$[T - (2a/r)] + \tfrac{1}{2} r(\partial/\partial r)[T - (2a/r)] = v$$

It is easy to realize that $[T - (2a/r)]$ is $O(r^{-3})$ like v. Therefore we can write

$$T^{(n)} = (2a_n/r) + Z^{(n)} \qquad T = (2a/r) + Z$$

where

$$Z^{(n)} + \tfrac{1}{2} r(\partial/\partial r)Z^{(n)} = v^{(n)} \qquad Z + \tfrac{1}{2} r(\partial/\partial r)Z = v \tag{A8}$$

$$Z^{(n)}, Z, v^{(n)}, v = O(r^{-3}) \tag{A9}$$

In this way we can conveniently split the analysis of $T_{,l}{}^{(n)} - T_{,l}$ into two parts:

$$T_{,l}{}^{(n)}|_{S_n} - T_{,l}|_S = \left(-\left.\frac{2a_n x_l}{r^3}\right|_{S_n}\right) - \left(-\left.\frac{2ax_l}{r^3}\right|_S\right) + Z_{,l}{}^{(n)}|_{S_n} - Z_{,l}|_S \tag{A10}$$

As for the first part, since it is clear that

$$\lim_{n\to\infty} \left.\frac{x_l}{r^3}\right|_{S_n} = \left.\frac{x_l}{r^3}\right|_S$$

uniformly on the unit sphere, it suffices to consider the difference $a_n - a$. By taking any radius $\bar{R}$ such that

$$\bar{R} > r(\alpha) \geq r_n(\alpha)$$

we can say that

$$a_n = \frac{\bar{R}}{4\pi} \int u^{(n)} (\bar{R}, \alpha) \, d\alpha \qquad 2 = \frac{\bar{R}}{4\pi} \int u(\bar{R}, \alpha) \, d\alpha$$

Consequently, we can write

$$|a_n - a|^2 \leqq \frac{\bar{R}^2}{16\pi^2} \int d\alpha \cdot \int d\alpha \, |u^{(n)}(\bar{R}, \alpha) - u(\bar{R}, \alpha)|^2$$

$$= \frac{\bar{R}^2}{4\pi} \int d\alpha \left|\int_{\bar{R}}^{+\infty} ds \, \frac{\partial}{\partial s}[u^{(n)}(s, \alpha) - u(s, \alpha)]\right|^2$$

$$\leqq \frac{\bar{R}^2}{4\pi} \int d\alpha \left\{\int_{\bar{R}}^{+\infty} \frac{ds}{s^2}\right\} \cdot \left\{\int_{\bar{R}}^{+\infty} ds \, s^2 \left(\frac{\partial u^{(n)}}{\partial s} - \frac{\partial u}{\partial s}\right)^2\right\}$$

$$\leqq \frac{\bar{R}}{4\pi} \int d\alpha \int_{\bar{R}}^{+\infty} ds \, s^2 |\nabla u^{(n)} - \nabla u|^2$$

$$\leqq \frac{\bar{R}}{4\pi} \| Ru^{(n)} - u \|_{H^{1,2}(\Omega)} \tag{A11}$$

By dint of lemma 1 the last term tends to zero when $n \to \infty$, and the same is true for $a_n - a$.

Let us now examine $Z_{,l}^{(n)} - Z_{,l}$. Starting from (A8) and differentiating with respect to x_l, we obtain

$$3Z_{,l}^{(n)} + r\frac{\partial}{\partial r}Z_{,l}^{(n)} = 2v_{,l}^{(n)} \qquad 3Z_{,l} + r\frac{\partial}{\partial r}Z_{,l} = 2v_{,l} \tag{A12}$$

The integrating factor of these equations is r^2; from (A9) we get

$$Z_{,l}^{(n)}, v_{,l}^{(n)}, Z_{,l}, v_{,l} = O(r^{-4})$$

so that (A12) can be directly integrated and we arrive at the relations

$$Z_{,l}^{(n)}|_{S_n} = -\frac{2}{r_n^3(\alpha)}\int_{r_n(\alpha)}^{\infty} ds\, s^2\, v_{,l}^{(n)} \tag{A13}$$

$$Z_{,l}|_S = -\frac{2}{r^3(\alpha)}\int_{r(\alpha)}^{\infty} ds\, s^2\, v_{,l} \tag{A14}$$

Taking the difference (A13) minus (A14) and conveniently splitting it, we can write

$$\begin{aligned}|Z_{,l}^{(n)}|_{S_n} - Z_{,l}|_S| &\leq \frac{2}{r_n^3(\alpha)}\left|\int_{r_n(\alpha)}^{r(\alpha)} ds\, s^2\, v_{,l}^{(n)}\right| \\ &+ \left|\frac{2}{r_n^3(\alpha)} - \frac{2}{r^3(\alpha)}\right|\left|\int_{r(\alpha)}^{+\infty} ds\, s^2\, v_{,l}^{(n)}\right| \\ &+ \frac{2}{r^3(\alpha)}\left|\int_{r(\alpha)}^{+\infty} ds\, s^2\,(v_{,l}^{(n)} - v_{,l})\right| \\ &= I_1(\alpha) + I_2(\alpha) + I_3(\alpha)\end{aligned}$$

Taking into account that

$$r(\alpha) \geq r_n(\alpha) > K(\text{const})$$

for the first term we have

$$\begin{aligned}\int I_1^2(\alpha)\, d\alpha &\leq \frac{4}{K^6}\int d\alpha\left\{\int_{r_n(\alpha)}^{r(\alpha)} ds\, s^2\right\}\cdot\left\{\int_{r_n(\alpha)}^{r(\alpha)} ds\, s^2 v_{,l}^{(n)2}\right\} \\ &= O\left[\sup_\alpha |r(\alpha) - r_n(\alpha)| \cdot \|v^{(n)}\|_{H^{1,2}(\Omega_n)}^2\right]\end{aligned}$$

Since

$$\|v^{(n)}\|_{H^{1,2}(\Omega_n)}^2 \leq C\,\|f\|_{H^{1/2}(S_n)}^2 \leq C'\|f\|_{H^{1/2}(\sigma)}^2$$

It is clear that

$$\int I_1^2(\alpha)\, d\alpha = O\left[\sup_\alpha |r(\alpha) - r_n(\alpha)|\right] \xrightarrow{n\to\infty} 0$$

As for the second term, since by dint of lemma 2,

$$\int d\alpha\left|\int_{r(\alpha)}^{+\infty} ds\, s^2 v_{,l}^{(n)}\right|^2 \leq \int_\Omega v_{,l}^{(n)2}\, d\Omega \leq \|v\|_{H^{1,2}(\Omega_n)}^2 \leq \text{const}$$

we see immediately that

$$\int I_2^2(\alpha)\, d\alpha = O\left[\sup_\alpha |r(\alpha) - r_n(\alpha)|\right] \xrightarrow{n\to\infty} 0$$

Finally, for the third term, by using lemma 2 we have

$$\begin{aligned}\int I_3^2(\alpha)\, d\alpha &\leq \frac{4}{K^6}\int d\alpha\left|\int_{r(\alpha)}^{+\infty} ds\, s^2(v_{,l}^{(n)} - v_{,l})\right|^2 \\ &\leq C''\int_\Omega d\Omega |v_{,l}^{(n)} - v_{,l}|^2 \\ &\leq C''\|Rv^{(n)} - v\|_{H^{1,2}(\Omega)}^2\end{aligned}$$

but then, recalling lemma 1, we obtain

$$\int I_3^2(\alpha)\, d\alpha \xrightarrow{n\to\infty} 0$$

Summarizing, we can claim that $Z_{,l}^{(n)}|_{S_n}$ tends to $Z_{,l|S}$ in $L^2(\sigma)$, so that, also recalling (A11), the same is true for $T_{,l}^{(n)}|_{S_n}$ with respect to $T_{,l|S}$ and the theorem is proved.

BIBLIOGRAPHY

Cryer, C. W., *A Bibliography of Free Boundary Problems,* University of Wisconsin Press, Madison, 1977.

Fiorenza, R., Sui problemi di derivata obliqua per le equazioni ellittiche quasi lineari, *Ric. Mat. Napoli, 15,* 74–108, 1966.

Gerontopoulos, P., Molodensky's problem in the plane, Ph.D. thesis, Inst. of Phys. Geod., Tech. Univ. of Graz, Graz, Austria, 1978.

Grafarend, E., and W. Niemeier, The free non-linear boundary value problem of physical geodesy, *Bull. Geod., 45,* 243–260, 1971.

Heiskanen, W. W., and H. Moritz, *Physical Geodesy,* W. H. Freeman, San Francisco, Calif., 1967.

Holota, P., Determination of the free boundary in potential theory, the boundary problem of physical geodesy, paper presented at 3rd International Symposium on Geodesy and Physics of the Earth, Zentralinst. für Phys. der. Erde, Weimar, East Germany, Oct. 1976.

Holota, P., The boundary problem of physical geodesy after the appearance of Lars Hörmander's paper: Present state, report, Res. Inst. of Geod., Topogr. and Cartogr. of Prague, Prague, 1978.

Hörmander, L., The boundary value problem of physical geodesy, *Rep. 9,* Mittag Leffler Inst., Stockholm, Sweden, 1975.

Miranda, C., *Partial Differential Equations of Elliptic Type,* Springer, New York, 1970.

Moritz, H., Recent developments in the geodetic boundary value problem, report, Dep. of Geod. Sci., Ohio State Univ., Columbus, 1978.

Nakiboglu, S. M., Variational formulation of the geodetic boundary value problem, *Boll. Geod., 2,* 93–105, 1978.

Necas, J., *Les Méthodes Directes Dans la Théorie des Équations Elliptiques,* Editions de l'Académie Tchécoslovaque des Sciences, Prague, 1967.

Petrovskaya, M. S., Solution of the geodetic boundary value problem, *Bull. Geod., 53*(1), 37–51, 1979.

Sanso', F., The geodetic boundary value problem in gravity space, *Atti R. Accad. Naz. Lincei Mem. Cl. Sci. Fis. Mat. Nat., 16,* 1–52, 1977.

Sanso', F., Molodensky's problem in gravity space: A review of first results, *Bull. Geod., 52*(1), 59–70, 1978.

Sanso', F., On the local solvability of Molodensky's problem in gravity space, *Manuscr. Geod., 3,* 157–227, 1979.

Sanso', F., Dual relations in geometry and gravity space, *Z. Vermess., 6,* 279–290, 1980.

Witsch, K. J., Ein schiefes Randwertproblem zu einer elliptischen Differentialgleichung zweiter Ordnung mit einer expliziten L_z-Abschätzung für die zweiten Ableitungen, report, Inst. für Angew. Math. der Univ. Bonn, Bonn, 1978*a*.

Witsch, K. J., The geodetic boundary value problem: A uniqueness result, report, Inst. für Angew. Math. der Univ. Bonn, Bonn, 1978*b*.

(Received August 6, 1980;
revised January 5, 1981;
accepted January 19, 1981.)

Least-Squares Collocation

HELMUT MORITZ

Technical University, Graz, Austria
Ohio State University, Columbus, Ohio 43210

Least-squares collocation is a mathematical technique for determining the earth's figure and gravitational field by a combination of heterogeneous data of different kinds. The same formulas may be interpreted in very different ways: as the solution of a geophysical inverse problem, as a statistical estimation method combining least-squares adjustment and least-squares prediction, and as an analytical approximation to the earth's potential by means of harmonic functions. The present review article attempts a unified presentation of the subject with emphasis on theory.

INTRODUCTION

Since the review articles by *Kaula* [1963, 1967] the subject of estimating the earth's gravitational field from measurements of different types, quality, and distribution has developed considerably. Least-squares collocation has started from the subject of interpolation of gravity anomalies by least-squares prediction and has been generalized to arbitrary data. The statistical origin of collocation has sometimes overshadowed the fact that this method can also be considered as a purely analytical approximation method.

The present article attempts a review of the subject from a theoretical point of view. The subject is first approached from approximation theory, before embarking on statistical interpretation. The list of references is intended to give a representative sample of all principal trends. Its object is to guide the readers, rather than provide a complete documentation. It is essentially restricted to literature published within the last 10 years; earlier literature is given by *Kaula* [1963, 1967] and *Heiskanen and Moritz* [1967].

The fundamental publication on collocation is a paper by *Krarup* [1969]; a comprehensive elementary presentation is given by *Moritz* [1973]; a compact review paper from the stochastic process point of view with a very complete bibliography is by *Grafarend* [1976]; and a good picture of the present status is provided by a collection of lectures presented at a summer school in Ramsau, Austria, in August 1977 [*Moritz and Sünkel*, 1978].

1. APPROXIMATE REPRESENTATION OF THE EARTH'S GRAVITATIONAL FIELD

The earth's external gravitational potential V is conventionally split up as follows:

$$V = V_R + T \tag{1}$$

where V_R, the reference potential or normal potential, is usually taken as the external potential of a reference ellipsoid, which is given by a simple closed analytical expression [cf. *Heiskanen and Moritz*, 1967, p. 66]; thus V_R can be considered known. The quantity

$$T = V - V_R \tag{2}$$

called the disturbing potential, or the anomalous potential, is small but quite irregular. Therefore its analytical representation poses some problems, with which we shall be concerned in the following.

Clearly, such an analytical representation can only be approximate. We shall try to approximate T by a linear combination f of suitable base functions $\phi_1, \phi_2, \phi_3, \cdots, \phi_q$:

$$T(P) \doteq f(P) = \sum_{k=1}^{q} b_k\phi_k(P) \tag{3}$$

P denoting the space point at which these functions are being considered and b_k designating suitable coefficients.

Since T is harmonic outside the earth's surface, it is appropriate to select as base functions ϕ_k also harmonic functions, satisfying Laplace's equation

$$\Delta\phi_k = 0 \tag{4}$$

Otherwise, many different choices are possible. The ϕ_k may, for instance, be spherical harmonics or the potential of point masses suitably distributed below the earth's surface. The coefficients b_k are chosen so as to satisfy certain conditions.

Interpolation. Assume, for instance, that we are given errorless values of T at q space points P_i. Then we may postulate that the approximation f to T, as given by (3), exactly reproduces T at the q given points. On putting

$$f(P_i) = T(P_i) = f_i \qquad i = 1, 2, \cdots, q \tag{5}$$

we thus have the conditions

$$\sum_{k=1}^{q} A_{ik}b_k = f_i \qquad A_{ik} = \phi_k(P_i) \tag{6}$$

They form q linear equations for the q unknowns b_k, which can be solved uniquely provided they are linearly independent.

Collocation. A generalization of the interpolation problem is the case in which q values of linear functionals L_1T, L_2T, $\cdots$, L_qT of T are to be reproduced. Such functionals are, for instance, values of deflection of the vertical components

$$\xi = -\frac{1}{G}\frac{\partial T}{\partial x} \qquad \eta = -\frac{1}{G}\frac{\partial T}{\partial y} \tag{7}$$

and gravity anomalies

$$\Delta g = -\frac{\partial T}{\partial z} - \frac{2}{R}T \tag{8}$$

at certain data points; here xyz represents a local coordinate system, the z axis being vertical and directed upward and the x axis pointing north, and R = 6370 km is the mean radius of the earth.

On putting

$$L_if = L_iT = l_i \tag{9}$$

we thus get from (3)

$$\sum_{k=1}^{q} B_{ik}b_k = l_i \qquad B_{ik} = L_i\phi_k \tag{10}$$

which is a $q \times q$ linear system for b_k quite similar to (6). This method for fitting an analytical approximation to q given functionals is called collocation and is frequently used in numerical mathematics [cf. *Collatz,* 1966, p. 29].

It is clear that interpolation is a special case of collocation if the functionals L_if are the 'evaluation functionals' $f(P_i)$.

We thus see that in both interpolation and collocation, for any assumed system of base functions ϕ_k the determination of the coefficients b_k always involves the solution of a $q \times q$ system of linear equations (which will, in general, not be symmetric).

2. LEAST-SQUARES INTERPOLATION AND COLLOCATION

Least-squares interpolation. The base functions $\phi_i(P)$ for interpolation may be selected in such a way that the rms interpolation error m_P is minimized; here m_P is defined by

$$m_P^2 = M\{\epsilon_P^2\} \tag{11}$$

where

$$\epsilon_P = T(P) - f(P) \tag{12}$$

is the local interpolation error (it is zero at the q data points P_i). The symbol M denotes a suitable mean or average (see below).

The condition

$$m_P = \min \tag{13}$$

leads to

$$\phi_k(P) = K(P, P_k) \tag{14}$$

where the covariance function $K(P, Q)$ is defined by

$$K(P, Q) = M\{T(P)T(Q)\} \tag{15}$$

M being the averaging operator already used above. Thus in (6),

$$A_{ik} = K(P_i, P_k) = C_{ik} \tag{16}$$

and we can solve for b_k and substitute into (3). On putting

$$\phi_k(P) = K(P, P_k) = C_{Pk} \tag{17}$$

the result is

$$f(P) = [C_{P1}\ C_{P2}\ \cdots\ C_{Pq}] \begin{bmatrix} C_{11} & C_{12} & \cdots & C_{1q} \\ C_{21} & C_{22} & \cdots & C_{2q} \\ \vdots & \vdots & & \vdots \\ C_{q1} & C_{q2} & \cdots & C_{qq} \end{bmatrix}^{-1} \begin{bmatrix} f_1 \\ f_2 \\ \vdots \\ f_q \end{bmatrix} \tag{18}$$

This least-squares interpolation formula is frequently used for gravity interpolation, on taking $f = \Delta g$ instead of T; compare with the derivation of *Heiskanen and Moritz* [1967, p. 268]. It is, of course, nothing but the Wiener-Kolmogorov prediction formula, well known from the theory of stochastic processes [cf. *Grafarend,* 1975*a*; *Liebelt,* 1967, p. 138].

Least-squares collocation. Here the minimum principle (13), the least rms interpolation error, leads to

$$\phi_k(P) = L_k{}^QK(P, Q) \tag{19}$$

where $L_k{}^Q$ means that the linear functional L_k is applied to the variable Q. Thus in (10),

$$B_{ik} = L_i{}^PL_k{}^QK(P, Q) = C_{ik} \tag{20}$$

On putting

$$\phi_k(P) = L_k{}^QK(P, Q) = C_{Pk} \tag{21}$$

we get formally the same formula (18) as before but with $f_1, f_2, \cdots, f_q$ replaced by $l_1, l_2, \cdots, l_q$:

$$f(P) = [C_{P1}\ C_{P2}\ \cdots\ C_{Pq}] \begin{bmatrix} C_{11} & C_{12} & \cdots & C_{1q} \\ C_{21} & C_{22} & \cdots & C_{2q} \\ \vdots & \vdots & & \vdots \\ C_{q1} & C_{q2} & \cdots & C_{qq} \end{bmatrix}^{-1} \begin{bmatrix} l_1 \\ l_2 \\ \vdots \\ l_q \end{bmatrix} \tag{22}$$

The application of the simple least-squares prediction formula (18) to problems involving the gravitational field is rather limited, essentially to the interpolation of gravity anomalies and similar tasks. On the other hand, the least-squares collocation formula has a considerably wider scope of application because it involves as data arbitrary linear functionals of T. In fact, geodetic data are of many different types: horizontal and vertical angles, distances, gravity measurements, astronomical observations of latitude and longitude, satellite observations of many kinds, second-order gravity gradients, etc. All these geodetic data obviously share the property that they depend on the earth's gravitational field; therefore on linearization all of them may be expressed as linear functionals of the anomalous potential T, for example, (7) and (8), so as to serve as data for an optimal determination of T by (22).

Least-squares interpolation and collocation share with all representation methods of type (3) the disadvantage that a (usually large) $q \times q$ matrix has to be inverted; it has, however, the advantage that the matrix to be inverted here, being a covariance matrix, is symmetric, in contrast to the general case of a nonsymmetric matrix in (6) and (10).

So far, we have supposed that the data are errorless. It is easy, however, to generalize least-squares collocation to the case in which there are random measuring errors and also systematic parameters to be determined. This will be done in section 4; in section 3 we shall consider the problem of the covariance function, which obviously plays a fundamental role here.

3. COVARIANCE FUNCTIONS

As we have seen, the base functions (17) and (21) are directly related to covariance functions. If a simple analytical form for $K(P, Q)$ is chosen, then the base functions will be analytically simple functions, which is desirable in many applications. In fact, the method described in the preceding section works with any function $K(P, Q)$ that is symmetric in P and Q, harmonic as a function of both P and Q, and positive definite (see below). Thus this method can be considered purely as an analytical approximation without statistical interpretation, the function $K(P, Q)$ being a kernel function in a Hilbert space with reproducing kernel [*Meschkowski,* 1962, p. 114; *Krarup,* 1969; *Tscherning,* 1975*b*, 1977].

Choosing the kernel function $K(P, Q)$ to be the covariance function of T becomes essential only if we wish to achieve minimum rms interpolation error. On the other hand, the exact covariance function of the disturbing potential T, even if it could be determined exactly, would probably not have a simple analytical form. By fitting a suitable expression to the empirical covariance data, one may hope to reconcile the two desiderata, simplicity and optimality, in a practically acceptable way.

In (15) we have introduced the covariance function as

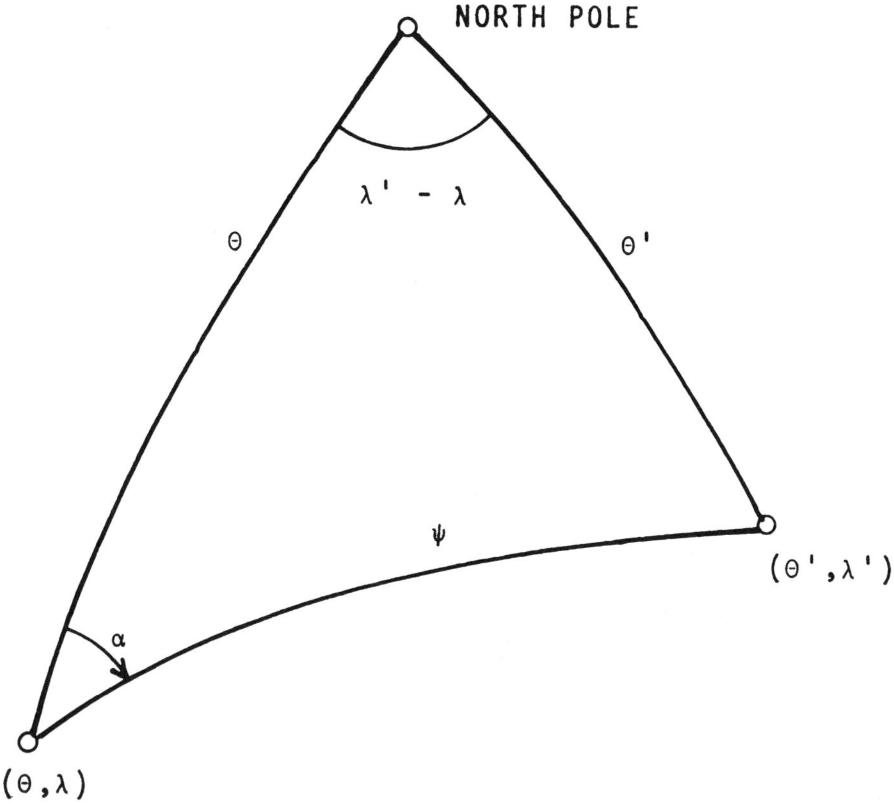

Fig. 1. The basic spherical triangle.

$$K(P, Q) = M\{T(P)T(Q)\} \tag{23}$$

If we should consider a statistical ensemble of various 'sample earths' with similar statistical properties, then we could regard M as an ensemble average or a statistical expectation. This would correspond to the interpretation of the anomalous gravitational field as a stochastic process [*Grafarend,* 1975*b*, 1976].

It may appear preferable not to introduce other sample earths and to work only with the anomalous gravitational field of the actual earth. Then the operator M must be some kind of global (or, possibly, regional) average.

We may start with the case in which both points P and Q are on the surface of a sphere $r = R$ representing a mean terrestrial sphere. If we prescribe M to be a homogeneous and isotropic operator, then the function $K(P, Q)$ will be a function only of the spherical distance ψ between P and Q:

$$K(P, Q) = K(\psi) = \frac{1}{8\pi^2} \int_{\lambda=0}^{2\pi} \int_{\theta=0}^{\pi} \int_{\alpha=0}^{2\pi} T(\theta, \lambda) \cdot T(\theta', \lambda') \sin\theta \, d\theta \, d\lambda \, d\alpha \tag{24}$$

where (θ, λ) and (θ', λ') denote points on the sphere that have a spherical distance ψ, such that (Figure 1)

$$\cos\psi = \cos\theta \cos\theta' + \sin\theta \sin\theta' \cos(\lambda' - \lambda) \tag{25}$$

Both points (θ, λ) and (θ', λ') are variable but subject to the condition that their distance ψ remains constant. The azimuth α may also be expressed in terms of (θ, λ) and (θ', λ') by well-known formulas of spherical trigonometry.

The integration over (θ, λ) in (24) expresses homogeneity, and the integration over the azimuth α denotes isotropy. The two concepts, homogeneity and isotropy, are obviously closely interrelated in the case of the sphere. The three parameters θ, λ, and α may be interpreted as three Eulerian angles of a rotation, so that the operator M defined by (24) is an average over rotation group space.

A spherical harmonic expansion of the covariance function gives

$$K(\psi) = \sum_{n=2}^{\infty} k_n P_n(\cos\psi) \tag{26}$$

where the $P_n(\cos\psi)$ are Legendre's polynomials. If the anomalous potential T on the sphere $r = R$ is likewise expanded into spherical harmonics:

$$T(\theta, \lambda) = \sum_{n=2}^{\infty} \sum_{m=0}^{n} (\bar{a}_{nm} \cos m\lambda + \bar{b}_{nm} \sin m\lambda) \bar{P}_{nm}(\cos\theta) \tag{27}$$

$\bar{P}_{nm}(\cos\theta)$ denoting the fully normalized Legendre's functions, then the coefficients k_n of K are related to the coefficients $\bar{a}_{nm}$ and $\bar{b}_{nm}$ of T by [*Kaula,* 1959]

$$k_n = \sum_{m=0}^{n} (\bar{a}_{nm}^2 + \bar{b}_{nm}^2) \tag{28}$$

[cf. *Heiskanen and Moritz,* 1967, section 7-3]; there we have considered the covariance function $C(\psi)$ of Δg instead of the covariance function $K(\psi)$ of T.

The extension of the function (26) into space outside the sphere $r = R$ follows uniquely if we note that the function $K(P, Q)$ in space must be harmonic both with respect to P and with respect to Q, i.e., it must satisfy Laplace's equation whether it is considered as a function of point P or as a function of point

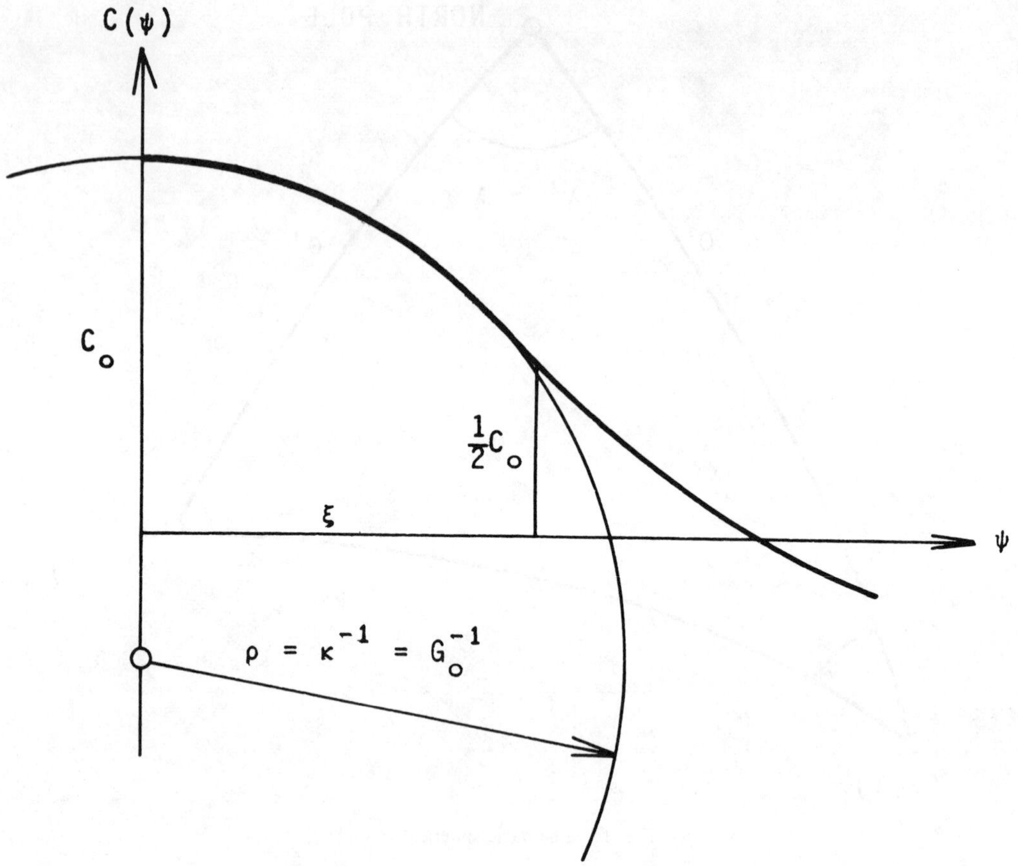

Fig. 2. Local parameters of a covariance function.

Q (this is a simple consequence of the harmonicity of T). We thus get, in space,

$$K(P, Q) = \sum_{n=2}^{\infty} k_n \left(\frac{R^2}{r_P r_Q}\right)^{n+1} P_n(\cos\psi) \qquad (29)$$

r_P and r_Q being the radius vectors of points P and Q, respectively.

The corresponding covariance function of the gravity anomaly Δg is given by

$$C(P, Q) = \sum_{n=2}^{\infty} c_n \left(\frac{R^2}{r_P r_Q}\right)^{n+2} P_n(\cos\psi) \qquad (30)$$

where

$$c_n = (n-1)^2 R^{-2} k_n \qquad (31)$$

This follows immediately from the well-known spherical harmonic relation between T and Δg [*Heiskanen and Moritz*, 1967, p. 97] and from the fact that $r\Delta g$ is harmonic in space [*Heiskanen and Moritz*, 1967, p. 90]. In this way, from one of the functions $C(P, Q)$ and $K(P, Q)$ it is possible to compute the other.

All covariance functions, as well as all covariance matrices, must be positive definite. Positive definiteness of a function is equivalent to the nonnegativity of the spectrum [cf. *Papoulis*, 1965, p. 349]. In the case of functions defined on a sphere, the spectral representation is the expansion into spherical harmonics. Therefore positive definiteness of the functions K and C is equivalent to the nonnegativity of all coefficients k_n and c_n, which is guaranteed by (28) and (31).

Empirical determination. If the gravity anomaly Δg were known over the whole earth, then $C(P, Q)$ would be obtained by integrating the product $\Delta g(\theta, \lambda)\Delta g(\theta', \lambda')$ according to (24), with T replaced by Δg. In view of the gaps in the global gravity coverage, one resorts to suitable sampling techniques. The lower degree k_n (up to $n = 20$, say) can be obtained, by (28), from coefficients $\bar{a}_{nm}$ and $\bar{b}_{nm}$ determined by a combination of satellite and gravimetric data. Estimates from empirical gravity anomaly fields give certain 'local' parameters of the gravity anomaly covariance function.

It has turned out that the gravity anomaly covariance function $C(\psi)$, the function (30) restricted to the sphere $r = R$, can be characterized, to an accuracy that is sufficient for many applications, by three 'essential' parameters (Figure 2):

Variance

$$C_0 = C(0) = \text{var}(\Delta g) \qquad (32)$$

Correlation length ξ defined by

$$C(\xi) = \tfrac{1}{2}C_0 \qquad (33)$$

Gradient variance

$$G_0 = \text{var}(\partial\Delta g/\partial s) \qquad (34)$$

$\partial\Delta g/\partial s$ denoting the gradient of Δg along a horizontal direction; also,

$$G_0 = \tfrac{1}{2}\,\text{var}(\partial\Delta g/\partial h) \qquad (35)$$

where $\partial\Delta g/\partial h$ denotes the vertical gradient of Δg. The parameter G_0 equals the curvature κ of the curve $C(\psi)$ at $\psi = 0$:

$$\kappa = 1/\rho = G_0 \tag{36}$$

[cf. *Moritz*, 1976*b*, section 3].

A good estimate for the global variance C_0 of the gravity anomaly is [*Tscherning and Rapp*, 1974, p. 20]

$$C_0 = 1795 \text{ mGal}^2 \tag{37}$$

The other two local parameters are far less reliable; one might take

Correlation length

$$\xi = 70 \text{ km} \tag{38a}$$

Gradient variance

$$G_0 = 2 \text{ mGal}^2 \text{ km}^{-2} \tag{38b}$$

Tscherning and Rapp [1974] have modeled the general trend of the lower degree variances c_3 to c_{20}, as well as C_0, by means of the expression

$$C(P, Q) = \alpha \sum_{n=3}^{\infty} \frac{n-1}{(n-2)(n+B)} \sigma^{n+2} \cdot \left(\frac{R^2}{r_P r_Q}\right)^{n+2} P_n(\cos \psi) \tag{39}$$

which contains the three free parameters α, B, and σ. If in addition to the lower degree variances and to C_0, one wishes to model also the local characteristics G_0 and ξ, one may use a model

$$C(P, Q) = \alpha_1 \sum_{3}^{\infty} \frac{n-1}{n+A} \sigma_1^{n+2} \left(\frac{R^2}{r_P r_Q}\right)^{n+2} P_n(\cos \psi) + \alpha_2 \sum_{3}^{\infty} \frac{n-1}{(n-2)(n+B)} \sigma_2^{n+2} \left(\frac{R^2}{r_P r_Q}\right)^{n+2} P_n(\cos \psi) \tag{40}$$

which contains six parameters, α_1, α_2, A, B, σ_1, and σ_2 [*Moritz*, 1977].

The only global empirical point anomaly covariance function available so far seems still to be *Kaula*'s [1959] determination; the model by *Tscherning and Rapp* [1974] has been derived from 1° × 1° mean anomalies together with a point variance C_0. A new reliable determination of a worldwide covariance function that represents both global and local features is highly desirable.

Plane covariance models. For certain limited purposes, such as interpolation, one may approximate the terrestrial sphere locally by a plane. Let (x, y) denote rectangular coordinates in the plane, and let the points $P(x_P, y_P)$ and $Q(x_Q, y_Q)$ be restricted to the plane. Then homogeneity and isotropy imply that the covariance function (say, of Δg) can depend only on the plane distance $s = PQ$, given by

$$s^2 = (x_P - x_Q)^2 + (y_P - y_Q)^2 \tag{41}$$

Possible analytical expressions (they must be positive definite) are, for instance,

$$C(s) = C_0 e^{-A^2 s^2} \tag{42}$$

$$C(s) = C_0/(1 + B^2 s^2)^m \tag{43}$$

with the suitable constants A, B, C_0, and m. An example is Hirvonen's regional covariance function for Ohio, which is (43) with $C_0 = 337 \text{ mGal}^2$, $B^{-1} = 40$ km, and $m = 1$ [*Heiskanen and Moritz*, 1967, p. 255].

If the points P and Q are no longer restricted to the xy plane but may lie in the upper half space (the space above the xy plane), then the spatial covariance function must be harmonic as a function of both (x_P, y_P, z_P) and (x_Q, y_Q, z_Q), the z axis being, of course, normal to the xy plane and positive upward. The model (42) cannot easily be extended into the upper half space so as to obtain a harmonic function. For the model (43) this is possible for $m = \frac{1}{2}$ and $m = \frac{3}{2}$:

$$C(P, Q) = \frac{C_0 b}{[s^2 + (z_P + z_Q + b)^2]^{1/2}} \tag{44}$$

and

$$C(P, Q) = \frac{C_0 b^2 (z_A + z_B + b)}{[s^2 + (z_P + z_Q + b)^2]^{3/2}} \tag{45}$$

respectively. Here, b is a constant, and s is again given by (41). For $z_A = z_B = 0$ the functions (44) and (45) reduce to (43), with $B = b^{-1}$ [cf. *Moritz*, 1976*b*, pp. 41–42].

Other possible plane covariance functions are derived from the Markov model:

$$C(s) = C_0 e^{-s/D} \tag{46}$$

They are called higher-order Markov models:

$$C(s) = C_0[1 + (s/D)]e^{-s/D} \tag{47}$$

$$C(s) = C_0[1 + (s/D) + (s^2/3D^2)]e^{-s/D} \tag{48}$$

For discussions and uses of them, see *Jordan* [1972], *Kasper* [1971], and *Shaw et al.* [1969].

Empirically, values of the covariance functions can be determined by suitable sampling methods. A simple procedure is as follows. Cover the plane region under consideration by a square grid of width a, and assume the region itself to be a rectangle of sides Ma and Na, M and N being integers. Then the values of Δg at the grid points form a rectangular matrix $\Delta g_{i,j}$, where $i = 1, 2, \cdots, M$ and $j = 1, 2, \cdots, N$. We get 'north-south' covariances

$$C'(ka) = \frac{1}{(M-k)N} \sum_{j=1}^{N} \left(\sum_{i=k+1}^{M} \Delta g_{i-k,j} \Delta g_{i,j} \right) \tag{49a}$$

and 'east-west' covariances

$$C''(ka) = \frac{1}{M(N-k)} \sum_{i=1}^{M} \left(\sum_{j=k+1}^{N} \Delta g_{i,j-k} \Delta g_{i,j} \right) \tag{49b}$$

the weighted mean of which forms the desired estimates:

$$C(ka) = \frac{(M-k)NC'(ka) + M(N-k)C''(ka)}{(M-k)N + M(N-k)} \tag{50}$$

Thus we obtain estimates for the covariance function $C(s)$ for $s = 0, a, 2a, 3a, \cdots$, and we can try to fit an analytical expression to these discrete values. A Fortran program for covariance computation is given by *Rapp* [1966].

Nonisotropy. So far, we have assumed the covariance functions for T and Δg to be isotropic, that is, independent of azimuth. Such functions have been used almost exclusively so far. Nonisotropic covariance functions have been studied by several authors [cf. *Grafarend*, 1976; *Morrison*, 1977; *Rummel and Schwarz*, 1977].

Various other aspects of covariance functions are discussed by *Bellaire* [1977], *Groten* [1970], *Kubackova* [1974, 1976], *Meissl* [1971], *Tscherning* [1973, 1976], and *Vyskocil* [1970].

4. CONSIDERATION OF RANDOM ERRORS AND SYSTEMATIC PARAMETERS

Collocation with random errors. The collocation formula (22) is valid if the observed 'functionals' l_i are errorless. Frequently, however, random measuring errors n_i (the 'noise') must be taken into account, so we now have

$$l_i = s_i + n_i \tag{51}$$

where s_i (the 'signal') denotes the errorless part of l_i and represents the purely gravitational effect (e.g., errorless gravity anomalies or deflections of the vertical).

Let $[C_{ij}]$ denote, as before, the signal covariance matrix:

$$C_{ij} = M\{s_i s_j\} \tag{52}$$

to be derived from the covariance function (section 3) by 'covariance propagation' according to (20). Let, similarly, $[D_{ij}]$ denote the error covariance matrix, which is the variance-covariance matrix of the measuring errors n_i (only this matrix is considered in usual least-squares adjustment). Then the matrix

$$[\bar{C}_{ij}] = [C_{ij} + D_{ij}] \tag{53}$$

represents the total covariance matrix of the vector of observed functionals l_i.

It is now quite remarkable that the only modification of (22) to take measuring errors into account is the replacement of the matrix $[C_{ij}]$ by the total covariance matrix $[\bar{C}_{ij}]$ as given by (53); the covariances C_{Pk} remain unchanged.

We can also easily free ourselves from the requirement that the function $f(P)$ to be estimated is the anomalous potential. In fact, we can use (22), or its modification for random errors just mentioned, for the estimation of any quantity of the anomalous gravity field such as gravity anomalies, geoidal height, deflections of the vertical, spherical harmonic coefficients, etc. All such quantities s are linear functionals of T; symbolically,

$$s = L^P T(P) \tag{54}$$

Applying the linear functional L^P to both sides of (22), we see that $f(P)$ is replaced by the signal $\hat{s}$ to be estimated and C_{Pk} must be replaced by

$$C_{sk} = L^P C_{Pk} = L^P L_k^Q K(P, Q) \tag{55}$$

everything else remaining unchanged.

Together with the introduction of $\bar{C}_{ij}$ instead of C_{ij} this gives finally

$$\hat{s} = [C_{s1}\, C_{s2} \cdots C_{sq}] \begin{bmatrix} \bar{C}_{11} & \bar{C}_{12} & \cdots & \bar{C}_{1q} \\ \bar{C}_{21} & \bar{C}_{22} & \cdots & \bar{C}_{2q} \\ \vdots & \vdots & & \vdots \\ \bar{C}_{q1} & \bar{C}_{q2} & \cdots & \bar{C}_{qq} \end{bmatrix}^{-1} \begin{bmatrix} l_1 \\ l_2 \\ \vdots \\ l_q \end{bmatrix} \tag{56}$$

In abbreviated matrix form this may be written

$$\hat{s} = \mathbf{C}_{sl}\bar{\mathbf{C}}_{ll}^{-1}\mathbf{l} \tag{57}$$

which again has the form of the Wiener-Kolmogorov prediction formula but applied to heterogeneous quantities. In this form the simple prediction formula is really quite universal.

Let us mention some of its properties. The formula (57) again minimizes the estimation error of s; it provides a best linear estimate in the statistical sense: an unbiased linear estimate of minimum variance. Clearly, (56) represents a linear combination of signal covariances

$$\hat{s} = \sum_{k=1}^{q} b_k C_{sk} \tag{58}$$

It is immediately seen that measuring errors do not affect the C_{sk}, which are purely analytical; the statistics of measuring errors influences only the 'best' estimation of the coefficients b_k in the sense just mentioned.

There is thus a 'clean' mathematical model at the base of the collocation formula (56), represented by the analytical character of the covariance function $K(P, Q)$ and of the signal covariances C_{ij}, C_{Pk}, and C_{sk} derived from it by the 'covariance propagation' formulas (20), (21), and (55); the latter are now seen to carry the precise functional structure of the terrestrial gravitational field.

Let us briefly compare the present model and 'errorless' or 'exact' collocation as described in section 2. Both models are formally identical; they differ, however, in the underlying mechanism. In exact collocation the kernel function $K(P, Q)$ may be taken as a covariance function to obtain a statistically best estimate, or the kernel function may be selected freely to possess certain desirable analytical characteristics such as simplicity; in either case the q coefficients b_k are completely determined by the condition that the interpolation function exactly fits q given functionals. In collocation with noise the kernel function $K(P, Q)$ may again be taken as a covariance function of the anomalous gravity field; the coefficients b_k are now determined in such a way that the effect of random measuring errors in the given functionals is filtered out as much as possible: the functionals are no longer exactly reproduced.

Since the basic work by *Krarup* [1969], such collocation models have been widely applied for estimating the geoid by a combination of various data and for other purposes [cf. *Grafarend,* 1971; *Grafarend and Offermanns,* 1975; *Heitz and Tscherning,* 1972; *Lachapelle,* 1975; *Moritz,* 1970; *Rapp,* 1973, 1974, 1975; *Tscherning,* 1975*b*; *Zieliński,* 1975].

The general least-squares collocation model. So far, we have assumed that we deal only with quantities that have zero average over the sphere, such as the elements of the anomalous gravitational field. This restriction must be removed if collocation is to be applicable to more general geodetic problems.

We shall therefore consider observations x_i that include a 'systematic' part besides the 'random' part (51):

$$x_i = \sum_{r=1}^{m} A_{ir}X_r + s_i + n_i \qquad i = 1, 2, \cdots, q \tag{59}$$

or, in matrix notation,

$$\mathbf{x} = \mathbf{AX} + \mathbf{s} + \mathbf{n} \tag{60}$$

The vector **x** comprises the q measured quantities, and **s** and **n** are the signal and noise parts as in (51). The new component is **AX**, where the vector **X** comprises m systematic nonrandom parameters and **A** is a known $q \times m$ matrix, $m < q$. In general, the linear relation **AX** will be a linearization of an originally nonlinear function.

This model is generally enough to encompass all conceivable geodetic measurements. In fact, any geodetic measurement can be split into three parts, corresponding to (60): (1) a systematic part **AX** involving, on the one hand, the ellipsoidal reference system and, on the other hand, other parameters and systematic errors, (2) a 'random' part **s** expressing the effect of the anomalous gravity field, and (3) random measuring errors **n**.

As an example, consider a measurement of gravity g. Here, **AX** represents normal gravity γ, depending on the parameters of the reference ellipsoid, as well as systematic errors such as gravimeter drift; **s** is the gravity anomaly Δg; and **n** stands for the measuring error [cf. *Moritz,* 1969].

In this case the best linear estimate is

$$\hat{\mathbf{X}} = (\mathbf{A}^T\mathbf{C}_{xx}^{-1}\mathbf{A})^{-1}\mathbf{A}^T\mathbf{C}_{xx}^{-1}\mathbf{x} \tag{61}$$

$$\hat{\mathbf{s}} = \mathbf{C}_{sx}\mathbf{C}_{xx}^{-1}(\mathbf{x} - \mathbf{A}\hat{\mathbf{X}}) \tag{62}$$

where $\mathbf{C}_{xx}$ is the matrix (53):

$$\mathbf{C}_{xx} = \mathbf{C}_{ll} = \bar{\mathbf{C}} = [C_{ij} + D_{ij}] \tag{63}$$

Equation (61) is analogous to classical least-squares adjustment by parameters [cf. *Bjerhammar,* 1973, p. 123] except that now the covariance matrix $\mathbf{C}_{xx}$ includes the covariances of the signal as well as those of the measuring errors. Equation (62) is an obvious generalization of (57) to the case in which the average of **x** is no longer zero but **AX**. The vector $\hat{\mathbf{s}}$ comprises one or several estimated signals $\hat{s}_\alpha$ ($\alpha = 1, 2, \cdots, p$); consequently, $\mathbf{C}_{sx}$ is a $p \times q$ matrix consisting of p rows of the form $[C_{s1}, C_{s2}, \cdots, C_{sq}]$ as entering in (56). The estimated $\hat{s}_\alpha$ can be completely different from the s_i in (59), or they can also be estimates for them.

These formulas are an extension of the corresponding problem for time series [*Grenander and Rosenblatt,* 1957, p. 87]. Similar models have already been used by *Kaula* [1963].

The estimates (61) and (62), like all least-squares estimates, satisfy two different minimum principles: (1) least variances of the estimated quantities $\hat{\mathbf{X}}$ and $\hat{\mathbf{s}}$, which thus represent statistically best estimates, and (2) a minimum condition

$$\mathbf{v}^T\mathbf{K}^{-1}\mathbf{v} = \min \tag{64}$$

where the $p + q$ vector **v** consists of the p estimated signals $\hat{s}_\alpha$ and the q measuring errors n_i (or rather their estimates $\hat{n}_i$) and **K** is the covariance matrix of **v**.

The present model may be regarded as a combination of least-squares adjustment and least-squares prediction into a unified scheme. It provides a comprehensive method for an optimal combination of all geodetic data—classical angle and distance measurements, astronomical observations, gravity measurements, satellite data of different kinds, etc.—to obtain the geometric position of points on the earth's surface as well as the gravitational field. In practice, the amount of data that can be combined is limited by the size of the matrix that can be inverted on the computer. This requires suitable representative selection of the data and some working 'from the large to the small' in several steps.

Accuracy. The variance-covariance matrices $\mathbf{E}_{XX}$ and $\mathbf{E}_{ss}$ describing the accuracy of the estimates $\hat{\mathbf{X}}$ and $\hat{\mathbf{s}}$ are given by

$$\mathbf{E}_{XX} = (\mathbf{A}^T\mathbf{C}_{xx}^{-1}\mathbf{A})^{-1} \tag{65}$$

$$\mathbf{E}_{ss} = \mathbf{C}_{ss} - \mathbf{C}_{sx}\,\mathbf{C}_{xx}^{-1}\,\mathbf{C}_{sx}^T + \mathbf{C}_{sx}\,\mathbf{C}_{xx}^{-1}\mathbf{A}\mathbf{E}_{XX}\,\mathbf{A}^T\mathbf{C}_{xx}^{-1}\mathbf{C}_{sx}^T \tag{66}$$

The matrix $\mathbf{C}_{ss}$ represents the signal covariance matrix for the vector $\hat{\mathbf{s}} = [\hat{s}_\alpha]$. If the true values s_α of the estimates $\hat{s}_\alpha$ are given by

$$s_\alpha = L_\alpha^P T(P) \tag{67}$$

then the elements $C_{\alpha\beta}$ of the matrix $\mathbf{C}_{ss}$ are given by

$$C_{\alpha\beta} = L_\alpha^P L_\beta^Q K(P, Q) \tag{68}$$

in analogy to (20).

A derivation of these formulas is given by *Moritz* [1973, section 3]. In addition to their use in estimating the accuracy of computed quantities, they may also be applied for feasibility studies such as the planning of surveys. They do not require actual data, but they presuppose good estimates of the covariances involved. Applications have been made, for instance, to the planning of gravity surveys [*Tscherning,* 1975*a*] and to feasibility studies involving aerial gradiometry [*Schwarz,* 1976*b*, 1977] and satellite gradiometry [*Kryński et al.*, 1977]. Literature on the relation between least-squares adjustment and collocation includes papers by *Blaha* [1976], *Koch* [1977], *Rummel* [1976], *Schwarz* [1976*a*], and *Wolf* [1974]. Collocation methods have been applied in photogrammetry [*Kraus and Mikhail,* 1972], in meteorology [*Baussus von Luetzow,* 1973], and in other fields. Related methods have been discussed, among others, by *Bjerhammar* [1975*a*, *b*, 1976], *Groten* [1973, 1974], *Monget and Albuisson* [1971], and *Thomas and Heller* [1976]. Relations between collocation and geodetic boundary value problems have been investigated by *Krarup* [1969], *Bjerhammar* [1975*a*, 1976], *Moritz* [1976*a*], and *Neyman* [1974, 1975, 1977]. A very general Fortran program for applications of collocation has been given by *Tscherning* [1974].

5. RELATION TO GEOPHYSICAL INVERSE PROBLEMS

A complete mathematical description of the irregular gravitational field inevitably involves infinitely many parameters, for instance, the set of all coefficients in the spherical harmonic expansion of the disturbing potential T; let us arrange these coefficients into an infinite sequence

$$s_1, s_2, s_3, \cdots \tag{69}$$

Since they refer to the anomalous gravity field, the 'signal field,' they themselves have the character of signals.

Every quantity l_i of the anomalous gravity field can be expanded into a series of spherical harmonics provided this series converges:

$$l_i = \sum_{r=1}^{\infty} b_{ir}s_r \tag{70}$$

Let us again have q observations, so that $i = 1, 2, \cdots, q$. Each observation l_i may be expanded in the form (70). We thus have a system of q linear equations for infinitely many unknowns. This is clearly an underdetermined problem: There are infinitely many possible sets of solutions.

A unique best (in the statistical sense outlined above) linear solution is provided by least-squares collocation. On admitting measuring errors n_i the system (70) is replaced by

$$\sum_{r=1}^{\infty} b_{ir}s_r + n_i = l_i \qquad i = 1, 2, \cdots, q \tag{71}$$

or, in matrix symbolism,

$$\mathbf{Bs} + \mathbf{n} = \mathbf{l} \tag{72}$$

B being a matrix with q rows and infinitely many columns. Denote the signal covariance matrix of the infinite vector **s** by **K**; this matrix is found to be an infinite diagonal matrix whose diagonal elements are proportional to the k_n in (29). Let the covariance matrix of the measuring errors be denoted by **D**, and let **s** and **n** be uncorrelated, as usual. Then the application of covariance propagation to (72) gives readily

$$\mathbf{C}_{ll} = \mathbf{BKB}^T + \mathbf{D} \tag{73}$$

$$\mathbf{C}_{sl} = \mathbf{KB}^T \tag{74}$$

for the autocovariance matrix $\mathbf{C}_{ll}$ of $\mathbf{l}$ and the cross-covariance matrix $\mathbf{C}_{sl}$ between $\mathbf{s}$ and $\mathbf{l}$. Then (57) gives for the estimate $\hat{\mathbf{s}}$ of the spherical harmonic coefficient vector $\mathbf{s}$

$$\hat{\mathbf{s}} = \mathbf{KB}^T(\mathbf{BKB}^T + \mathbf{D})^{-1}\mathbf{l} \tag{75}$$

For $\mathbf{D} = \mathbf{0}$ we get the usual solution of the errorless equation (70) with $\mathbf{n} = \mathbf{0}$ by means of generalized inverses, $\mathbf{KB}^T(\mathbf{BKB}^T)^{-1}$ being a generalized inverse of $\mathbf{B}$ and different $\mathbf{K}$ corresponding to different inverses.

The mathematical model (70) or (72) and the solution (75) have also been suggested for geophysical inverse problems [cf. *Burkhard and Jackson*, 1976; *Wiggins*, 1972, pp. 260–261]. In fact, the determination of the earth's external gravity field from observations also has the character of an underdetermined 'improperly posed' problem: the finite number of observations can never fully determine the infinite number of parameters of the gravity field. Here we mention also the Hilbert space approach of *Backus* [1970].

The solution (75) has favorable numerical properties, especially stability. In the case of the terrestrial gravity field the covariance matrices $\mathbf{K}$ and $\mathbf{D}$ are not just auxiliary mathematical quantities introduced to obtain a convenient solution; they admit of a physical definition in terms of the statistics of the anomalous gravitational field and are, in principle, determinable by observation.

The accuracy of the estimated spherical harmonic coefficients can be estimated by (66) with $\mathbf{A} = \mathbf{0}$ and $\mathbf{C}_{ss} = \mathbf{K}$. If $\mathbf{E}_{ss}$ and $\mathbf{C}_{ss}$ are simultaneously transformed into diagonal form, then the comparison of the respective diagonal terms provides an objective criterion on whether a certain coefficient s_r can be meaningfully determined from the data or not.

The present model has been applied to the determination of zonal and, in part, also tesseral harmonics by *Balmino et al.* [1976], *Moritz and Schwarz* [1973], *Reigber and Ilk* [1976], and *Schwarz* [1974, 1975*a*, *b*].

6. Analytical and Statistical Aspects

Least-squares collocation has its roots in many fields: (1) least-squares estimation, (2) stochastic processes and spectral analysis, (3) approximation theory, (4) functional analysis, especially the theory of Hilbert spaces with kernel functions, (5) potential theory, and (6) inverse and improperly posed problems. These fields are frequently overlapping but sufficiently characteristic to be mentioned. All of these 'many facets of collocation' must be taken into consideration if a one-sided treatment is to be avoided.

Most obvious is the relation to classical least-squares estimation. In fact, collocation models bear resemblance to conventional adjustment models. The geometrical interpretation in terms of simple geometrical operations, projection on subspaces, is similar. Similar too is the invariance with respect to linear transformations, which is fundamental, since it ensures consistency: whichever quantity we compute by collocation, it always refers to the same gravity field. The characteristic difference is the infinite number of spectral parameters necessary to characterize the gravitational field fully. Hence one must operate in an infinite-dimensional Hilbert space rather than in finite linear spaces; this is a subtle but essential difference. It is here that the theory of stochastic processes and their spectral analysis by a spherical harmonic representation enters.

Least-squares estimation and stochastic processes provide a very convenient mathematical formalism and terminology and also a statistical interpretation of the results. However, this statistical interpretation may sometimes have been too much in the foreground and may have hidden the equally significant analytical aspects of collocation. Therefore we have attempted here to approach collocation from the point of view of approximation theory, working with a kernel function $K(P, Q)$ which can, in principle, be arbitrary; we take $K(P, Q)$ to be a statistically defined covariance function only to get a best estimate. However, as we have said, it is fully legitimate to take for $K(P, Q)$ any positive definite harmonic function; then collocation may be regarded as a purely analytic interpolation in a Hilbert space with a kernel function, even without stochastic interpretation. (There is also a relation between stochastic processes and kernel functions which dates back to *Parzen* [1961].) In fact, how seriously the statistical interpretation is taken is a matter of controversy and also of personal taste. Formulas such as (61) and (62) remain the same; only the interpretation of the estimated variances and covariances (65) and (66) and the best character of the estimates depend on statistics.

It is tempting to consider the anomalous gravity field as a stochastic process on a sphere. The main argument in favor of such an interpretation, as we have mentioned, is the obvious interpretation of covariances, prediction formulas, etc., in terms of stochastic process theory. There are, however, two counterarguments. First, there is only one earth; a probability space of possible earths may appear physically unnatural, although it is logically unobjectionable. Second, *Lauritzen* [1973] has proved that it is impossible to find a stochastic process harmonic outside the sphere that is both Gaussian and ergodic. This does not preclude non-Gaussian ergodic processes but may enhance the argument of unnaturality. Therefore it has been proposed that the stochastic process interpretation be dropped in favor of an interpretation in terms of Wiener's 'covariance analysis of individual functions' [cf. *Doob*, 1949, section 1], interpreting the averaging operator M not as a stochastic expectation (average over a probability space of many functions) but as the average of one function over rotation group space (24), that is, over the whole earth and over all directions. This allows the statistical interpretation of the variances of estimated quantities as average accuracies over an area. Observation errors, which do have a probability distribution and a stochastic expectation E, are readily incorporated by regarding as the total averaging operator the commutative product $EM = ME$ [*Moritz*, 1973, sections 8 and 9]. More about analytical and stochastic aspects of collocation is given by *Dermanis* [1976], *Eeg and Krarup* [1975], and *Tscherning* [1977].

It should be pointed out that the non-Gaussian character of the anomalous gravitational field implies that the 'best linear' estimate is not the absolutely best estimate: nonlinear estimates may still reduce the variance of the result. Therefore nonlinear prediction has been suggested, among others, by *Grafarend* [1972] and *Kaula* [1966].

Potential theory enters most obviously, but relatively superficially, through the harmonic character of the spatial covariance function. Subtler but perhaps even more important is the fact that the physical structure of the anomalous gravitational field with its precise mathematical relations between its quantities—T, Δg, ξ, η, and N—is reflected in linear operations on the analytical base functions $\phi_k(P)$. The statistical interpretation of these operations as 'covariance propagation' should not veil the fact that they are essentially analytical.

References

Backus, G., Inference from inadequate and inaccurate data, I, II, *Proc. Nat. Acad. Sci. U.S.*, *65*, 1–7, 281–287, 1970.

Balmino, G., C. Reigber, and B. Moynot, The Grim 2 earth gravity field model, *Publ. Deut. Geod. Komm. A, 86,* 34 pp., 1976.

Baussus von Luetzow, H., A review of collocation methods in physical geodesy and meteorology (abstract), *Eos Trans. AGU, 54,* 1067–1068, 1973.

Bellaire, R. G., Correlation functions on the upper half space, *Bull. Geod., 51,* 149–161, 1977.

Bjerhammar, A., *Theory of Errors and Generalized Matrix Inverses,* 420 pp., Elsevier, New York, 1973.

Bjerhammar, A., Discrete solutions of the boundary value problem in physical geodesy, *Tellus, 27,* 97–106, 1975*a*.

Bjerhammar, A., A general model for optimal prediction and filtering in the linear space, *Method. Verfahren Math. Phys., 14,* 29–46, 1975*b*.

Bjerhammar, A., Discrete approaches to the solution of the boundary value problem in physical geodesy, *Boll. Geod. Sci. Affini, 34,* 185–240, 1976.

Blaha, G., The least-squares collocation from the adjustment point of view and related topics, *Rep. AFGL-TR-76-0073,* 201 pp., Air Force Geophys. Lab., Hanscomb Air Force Base, Mass., 1976.

Burkhard, N., and D. D. Jackson, Application of stabilized linear inverse theory to gravity data, *J. Geophys. Res., 81,* 1513–1518, 1976.

Collatz, L., *The Numerical Treatment of Differential Equations,* 568 pp., Springer, New York, 1966.

Dermanis, A., Probabilistic and deterministic aspects of linear estimation in geodesy, *Rep. 244,* 167 pp., Dep. of Geod. Sci., Ohio State Univ., Columbus, 1976.

Doob, J. L., Time series and harmonic analysis, in *Proceedings of the Berkeley Symposium on Mathematical Statistics and Probability,* edited by J. Neyman, p. 303, University of California Press, Berkeley, 1949.

Eeg, J., and T. Krarup, Integrated geodesy, *Method. Verfahren Math. Phys., 13,* 77–123, 1975.

Grafarend, E., A combined gravimetric-astrogeodetic method for telluroid and vertical deflection analysis, *Publ. Deut. Geod. Komm. B, 188,* 23–36, 1971.

Grafarend, E., Nichtlineare Prädiktion, *Z. Vermess., 97,* 245–255, 1972.

Grafarend, E., Geodetic prediction concepts, *Method. Verfahren Math. Phys., 13,* 161–200, 1975*a*.

Grafarend, E., Geodetic stochastic processes, *Method. Verfahren Math. Phys., 14,* 1–27, 1975*b*.

Grafarend, E., Geodetic applications of stochastic processes, *Phys. Earth Planet. Interiors, 12,* 151–179, 1976.

Grafarend, E., and G. Offermanns, Eine Lotabweichungskarte Westdeutschlands nach einem geodätisch konsistenten Kolmogorov-Wiener-Modell, *Publ. Deut. Geod. Komm. A, 82,* 60 pp., 1975.

Grenander, V., and M. Rosenblatt, *Statistical Analysis of Stationary Time Series,* 300 pp., John Wiley, New York, 1957.

Groten, E., On the covariance function of the terrain correction, *Stud. Geophys. Geod., 14,* 264–269, 1970.

Groten, E., Some numerical aspects of least-squares autoregression prediction, *Boll. Geod. Sci. Affini, 4,* 331–357, 1973.

Groten, E., Combination solutions in geoid computations, in *Proceedings of the International Symposium on Applications of Marine Geodesy,* pp. 357–369, Marine Technology Society, Washington, D. C., 1974.

Heiskanen, W. A., and H. Moritz, *Physical Geodesy,* 364 pp., W. H. Freeman, San Francisco, Calif., 1967.

Heitz, S., and C. Tscherning, Comparison of two methods of astrogeodetic geoid determination based on least-squares prediction and collocation, *Tellus,* 24, 271–276, 1972.

Jordan, S. K., Self-consistent statistical models for the gravity anomaly, vertical deflections, and undulation of the geoid, *J. Geophys. Res., 77,* 3660–3670, 1972.

Kasper, J. F., A second-order Markov gravity anomaly model, *J. Geophys. Res., 76,* 7844–7849, 1971.

Kaula, W. M., Statistical and harmonic analysis of gravity, *J. Geophys. Res., 64,* 2401–2421, 1959.

Kaula, W. M., Determination of the earth's gravitational field, *Rev. Geophys. Space Phys., 1,* 507–551, 1963.

Kaula, W. M., Global harmonic and statistical analysis of gravimetry, in *Gravity Anomalies: Unsurveyed Areas, Geophys. Monogr. Ser.,* vol. 9, pp. 58–67, AGU, Washington, D. C., 1966.

Kaula, W. M., Theory of statistical analysis of data distributed over a sphere, *Rev. Geophys. Space Phys., 5,* 83–107, 1967.

Koch, K. R., Least squares adjustment and collocation, *Bull. Geod., 51,* 127–135, 1977.

Krarup, T., A contribution to the mathematical foundation of physical geodesy, *Publ. 44,* 80 pp., Dan. Geod. Inst., Copenhagen, 1969.

Kraus, K., and E. Mikhail, Linear least-squares interpolation, *Photogramm. Eng., 38,* 1016–1029, 1972.

Kryński, J., H. Noe, K. P. Schwarz, and H. Sünkel, Numerical studies and programs for interpolation and collocation, *Publ. Geod. Inst. Tech. Univ. Graz, 26,* 67 pp., 1977.

Kubackova, L., Autocovariance function and the spectral density of the anomalous gravitational field transformed by an integral operator of convolution type, *Stud. Geophys. Geod., 18,* 329–338, 1974.

Kubackova, L., The fundamental statistical properties of the Vening Meinesz and Stokes integral transformation, *Stud. Geophys. Geod., 20,* 209–212, 1976.

Lachapelle, G., Determination of the geoid using heterogeneous data, *Publ. Geod. Inst. Tech. Univ. Graz, 19,* 122 pp., 1975.

Lauritzen, S., The probabilistic background of some statistical methods in physical geodesy, *Publ. 48,* 96 pp., Dan. Geod. Inst., Copenhagen, 1973.

Liebelt, P. B., *An Introduction to Optimal Estimation,* 273 pp., Addison-Wesley, Reading, Mass., 1967.

Meissl, P., A study of covariance functions related to the earth's disturbing potential, *Rep. 151,* 88 pp., Dep. of Geod. Sci., Ohio State Univ., Columbus, 1971.

Meschkowski, H., *Hilbertsche Räume mit Kernfunktion,* 256 pp., Springer, New York, 1962.

Monget, J. M., and M. Albuisson, A new statistical treatment of gravity data, *Bull. Geod., 102,* 451–466, 1971.

Moritz, H., A general theory of gravity processing, *Rep. 122,* 58 pp., Dep. of Geod. Sci., Ohio State Univ., Columbus, 1969.

Moritz, H., Least-squares estimation in physical geodesy, *Publ. Deut. Geod. Komm. A, 69,* 34 pp., 1970.

Moritz, H., Least-squares collocation, *Publ. Deut. Geod. Komm. A, 75,* 91 pp., 1973.

Moritz, H., Integral formulas and collocation, *Manuscripta Geod., 1,* 1–40, 1976*a*.

Moritz, H., Covariance functions in least-squares collocation, *Rep. 240,* 79 pp., Dep. of Geod. Sci., Ohio State Univ., Columbus, 1976*b*.

Moritz, H., On the computation of a global covariance model, *Rep. 255,* 36 pp., Dep. of Geod. Sci., Ohio State Univ., Columbus, 1977.

Moritz, H., and K. P. Schwarz, On the computation of spherical harmonics from satellite observations, *Boll. Geod. Sci. Affini, 32,* 185–200, 1973.

Moritz, H., and H. Sünkel (Eds.), *Approximation Methods in Geodesy,* Wichmann, Karlsruhe, West Germany, in press, 1978.

Morrison, F., Azimuth-dependent statistics for interpolating geodetic data, *Bull. Geod., 51,* 105–118, 1977.

Neyman, Yu. M., Probabilistic modification of Stokes' formula for the computation of height anomalies (in Russian), *Geod. Aerofotosyemka,* 21–24, 1974.

Neyman, Yu. M., On the regularization of Molodensky's boundary value problem (in Russian), *Geod. Aerofotosyemka,* 57–64, 1975.

Neyman, Yu. M., A variational method of solving discrete problems of physical geodesy (in Russian), *Geod. Aerofotosyemka,* 21–27, 1977.

Papoulis, A., *Probability, Random Variables, and Stochastic Processes,* 583 pp., McGraw-Hill, New York, 1965.

Parzen, E., An approach to time series analysis, *Ann. Math. Statist., 32,* 951–989, 1961.

Rapp, R. H., A Fortran program for the computation of localized gravity anomaly statistics, *Rep. 74,* 38 pp., Dep. of Geod. Sci., Ohio State Univ., Columbus, 1966.

Rapp, R. H., Numerical results from the combination of gravimetric and satellite data using the principles of least-squares collocation, *Rep. 200,* 58 pp., Dep. of Geod. Sci., Ohio State Univ., Columbus, 1973.

Rapp, R. H., Gravity anomaly recovery from satellite altimetry data using least-squares collocation techniques, *Rep. 220,* 25 pp., Dep. of Geod. Sci., Ohio State Univ., Columbus, 1974.

Rapp, R. H., Comparison of least-squares and collocation estimated potential coefficients, *Method. Verfahren Math. Phys., 14,* 133–148, 1975.

Reigber, C., and K. H. Ilk, Vergleich von Resonanzparameterbestimmungen mittels Ausgleichung und Kollokation, *Z. Vermess., 101,* 59–67, 1976.

Rummel, R., A model comparison in least-squares collocation, *Bull. Geod., 50,* 181–192, 1976.

Rummel, R., and K. P. Schwarz, On the nonhomogeneity of the global covariance function, *Bull. Geod.*, *51*, 93–103, 1977.

Schwarz, K. P., Tesseral harmonic coefficients and station coordinates from satellite observations by collocation, *Rep. 217*, 42 pp., Dep. of Geod. Sci., Ohio State Univ., Columbus, 1974.

Schwarz, K. P., Application of collocation: Spherical harmonics from satellite observations, *Method. Verfahren Math. Phys.*, *14*, 111–131, 1975*a*.

Schwarz, K. P., Zonale Kugelfunktionskoeffizienten aus Satellitendaten durch Kollokation, *Publ. Deut. Geod. Komm. C*, *209*, 58 pp., 1975*b*.

Schwarz, K. P., Least-squares collocation for large systems, *Boll. Geod. Sci. Affini*, *35*, 309–324, 1976*a*.

Schwarz, K. P., Geodetic accuracies obtainable from measurements of first and second order gravitational gradients, *Rep. 242*, 56 pp., Dep. of Geod. Sci., Ohio State Univ., Columbus, 1976*b*.

Schwarz, K. P., Capabilities of airborne gradiometry for gravity estimation, *Boll. Geod. Sci. Affini*, *36*, 195–214, 1977.

Shaw, L., I. Paul, and P. Henrikson, Statistical models for the vertical deflection from gravity-anomaly models, *J. Geophys. Res.*, *74*, 4259–4265, 1969.

Thomas, S. W., and W. G. Heller, Efficient estimation techniques for integrated gravity data processing, *Rep. AFGL-TR-76-0232*, 75 pp., Air Force Geophys. Lab., Hanscom Air Force Base, Mass., 1976.

Tscherning, C. C., On the relation between the variation of degree variances and the variation of the anomalous potential, *Boll. Geod. Sci. Affini*, *32*, 149–158, 1973.

Tscherning, C. C., A Fortran IV program for the determination of the anomalous potential using stepwise least-squares collocation, *Rep. 212*, 62 pp., Dep. of Geod. Sci., Ohio State Univ., Columbus, 1974.

Tscherning, C. C., Application of collocation for the planning of gravity surveys, *Bull. Geod.*, *116*, 183–198, 1975*a*.

Tscherning, C. C., Application of collocation: Determination of a local approximation to the anomalous potential of the earth using 'exact' astro-gravimetric collocation, *Method. Verfahren Math. Phys.*, *14*, 83–110, 1975*b*.

Tscherning, C. C., Covariance expressions for second and lower order derivatives of the anomalous potential, *Rep. 225*, 62 pp., Dep. of Geod. Sci., Ohio State Univ., Columbus, 1976.

Tscherning, C. C., A note on the choice of norm when using collocation for the computation of approximations to the anomalous potential, *Bull. Geod.*, *51*, 137–147, 1977.

Tscherning, C. C., and R. H. Rapp, Closed covariance expressions for gravity anomalies, geoid undulations, and deflections of the vertical implied by anomaly degree variance models, *Rep. 208*, 89 pp., Dep. of Geod. Sci., Ohio State Univ., Columbus, 1974.

Vyskocil, V., On the covariance and structure functions of the anomalous gravity field, *Stud. Geophys. Geod.*, *14*, 174–179, 1970.

Wiggins, R. A., The general linear inverse problem: Implication of surface waves and free oscillations for earth structure, *Rev. Geophys. Space Phys.*, *10*, 251–285, 1972.

Wolf, H., Über verallgemeinerte Kollokation, *Z. Vermess.*, *99*, 475–478, 1974.

Zieliński, J. B., Solution of the downward continuation problem by collocation, *Bull. Geod.*, *117*, 267–277, 1975.

(Received December 8, 1977;
accepted March 16, 1978.)

REVIEWS OF GEOPHYSICS AND SPACE PHYSICS, VOL. 20, NO. 4, PAGES 877-884, NOVEMBER 1982

OPTIMIZATION OF GEODETIC NETWORKS

Günter Schmitt

Geodetic Institute, University of Karlsruhe
D-7500 Karlsruhe, Federal Republic of Germany

Abstract. It is the aim of this paper to give a concise review of the current state of research, the trends, and the outstanding problems in the optimization of geodetic networks. The datum problem, which has to be solved in the computational stage, is referred to as the zero-order design problem. In the real design stage, three classes of problems have arisen: the configuration problem as the first-order design problem, the weight problem as the second-order design problem, and the optimal improvement of an existing network as the third-order design problem. Furthermore, all subproblems must be analyzed within the three general objectives of geodetic networks: precision, reliability, and cost.

General Remarks

In any review of the optimization of geodetic networks, it is imperative that due recognition and heed be paid to the pioneering work of Schreiber. Starting with his studies on the optimal distribution of angle measurements in the base extension figure of Göttingen, Schreiber [1882] created the classical standard example of network optimization. Of highest importance was the recognition that it is possible to estimate the precision of a network before any observations are made, provided the configuration of the network is known.

When one is working with a geodetic network, there are two occasions when consideration of certain optimization topics may advantageously take place. These are, first, the design stage, in which one has to decide on the configuration (point situation and types of observations) and the distribution of observational work in the network, and, second, the adjustment stage, where one has to decide on an optimal datum using all available information of a relative and an absolute nature. Both are valid, not only for new planned networks but also when existing networks are extended. A classification of the different optimization problems has been proposed by Grafarend [1974]. This classification has proved useful during the last decade despite some weaknesses.

The datum problem is a search for an optimal coordinate system and is called the zero-order design problem. The first-order design problem is to be understood as the configuration problem, where the positioning of the points and the observation plan have to be optimized, provided that the precision of the observations is known a priori. The weight problem, which is the optimal distribution of observational work in a fixed configuration, is called the second-order design problem. A further class is the third-order design problem, which is defined as the optimal improvement of an existing network by insertion of additional points and/or additional observations. This classification is obvious if one regards the free elements in the formula of a least squares adjustment as variation of coordinates [Schmitt, 1981a]. Denoting the configuration or design matrix as A, and the weight matrix P as the inverse of the cofactor matrix Q_l of the observations, then the cofactor matrix Q_x of the unknown parameters (mainly coordinates) can be derived as a generalized inverse of the normal equations:

$$(A^T PA)^- = Q_x$$

With the notation $(\ )^-$ both free $(\ )^+$ and constrained $(\ )^{-1}$ networks are allowed. The above descriptive classification can now be characterized by fixed and free parameters:

Design	Fixed Parameters	Free Parameters
Zero order	A, P	x, Q_x
First order	P, Q_x	A
Second order	A, Q_x	P
Third order	Q_x	A, P (partly free)

The characterization of Q_x as a fixed parameter implies that the object of the optimization problem is formulated in the elements of or as a function of Q_x.

Most of the real optimization problems include subproblems of different orders. So one has to proceed according to the principle that when a first-, second-, or third-order design problem is to be solved, a concept for the solution of the zero-order design problem must be given. This can be expressed most succinctly by stating that the result of a weight optimization is dependent on the datum of the network.

Optimization means minimizing or maximizing an objective function, which represents the criteria adopted to define the 'quality of the network'. Three general criteria are used to evaluate this quality: precision, reliability, and economy.

A network can be designed in such a way that (1) the postulated precision of the network elements, and of the derived quantities, can be realized, (2) it is as sensitive as possible, that is, it allows the detection of outliers in the measurements on the one hand and the detection of movements in deformation networks on the other, and (3) the marking of the points and the performance of the measurements are satisfying some cost criteria.

These requirements act in such a way that they have a contrary influence on the optimal design of a network [Wolf, 1970]. One can say, generally, that the quality criteria may have different weights in the different orders of design. The realization of reliability is mainly the task of the first-order design, whereas the second-order design particularly requires the optimization of the precision; the minimization of the costs, compared with the other objectives, can be interpreted as a secondary transformation problem [Schmitt, 1979b].

Concerning the three quality criteria, precision is most suited to be formulated as an analytical risk function. In older works, only scalar functions as functions of the eigenvalues of the variance-covariance matrix of the unknowns or of derived quantities have been used, trace and determinant being preferred. An extensive catalogue of risk functions is given by Grafarend and Harland [1973]. It is based on detailed studies of accuracy measures in multidimensional point fields and their statistical properties [Grafarend, 1972].

Quite another type of risk functions are criterion matrices, which have had great success in the last decade. Criterion matrices have been introduced in geodesy as quality measures for networks and as artificial variance-covariance matrices for network densification by Baarda [1971, 1973]. In the optimization of networks, criterion matrices are to be understood as ideal variance-covariance matrices that are to be approximated by the design as best as possible [Grafarend and Schaffrin, 1979].

An analytical formulation of reliability as a risk function seems to be hardly realizable. Until now it has been used as a decision factor in simulation processes [Augath, 1976, 1979] and as a secondary restriction in the weight optimization [van Mierlo, 1981]. Wimmer [1981a] tries to optimize the external reliability (influence of undetected errors in the observations on the unknown parameters) by standardizing the spectrum of eigenvalues of the criterion matrix.

A lot of effort has been expended to bring cost criteria into a mathematical form and to mix them with other criteria, for example, with precision functions [Pelzer, 1972; Wolf, 1970]. Köhr [1966, 1967b], especially, has expressed cost factors in terms of efficiency numbers for the instruments used. After splitting off constant terms for driving, station occupying, and beaconing, the objectives of some optimization strategies are optimal repetition numbers for the measurements [Milbert, 1979; Pelzer, 1967, 1969].

The numerical methods of solution in the differend orders of design are dependent upon the mathematical formulation of the design problem and upon the shape of the risk function. The oldest solution strategy is the network simulation method. A large number of design variants (the number is only restricted by the available computer time) are generated by variation of the free parameters, and the designs are compared to the objectives. The variation of the parameters can be arranged by a random process (Monte-Carlo design [Schmitt, 1977]) or by a selection process, which is most effective if one can use interactive computer graphics [Conzett et al., 1981; Nickerson et al., 1976].

The advantage of simulation is its suitability for first-, second-, and third-order design problems and the possibility of using arbitrary decision criteria for the choice of an optimal design, without the need for a mathematical formulation of the criteria as risk functions. The disadvantage of these digital simulation processes is the large amount of computer time required. A very interesting approach is therefore the on-line solution in a physical analog network, for example, an electric one, called 'hybrid design' [Grafarend et al., 1979], provided that an analogous verification of the objective function is possible. In the second-order design problem with criterion matrices it is possible to write the problem as a system of linear and/or nonlinear equations. Here, solutions can be derived by the use of generalized inverses, standard algorithms of linear or nonlinear programming, or mixed methods.

Zero-Order Design (ZOD)

The solution of the datum problem, especially in 'free networks', is possible by different approaches. A free network is defined as a network in which only its internal shape is given by measurements of a relative nature, for instance, gravity and height differences in one dimension, bearings, angles and distances or distance relations in two dimensions, and solid angles and distances in three-dimensional networks. The typical degrees of freedom in such networks (displacements, rotations, and scales) can be eliminated by absolute measurements. To fix the datum, one can introduce absolute gravity measurements into a one-dimensional network and Doppler observations in two- and three-dimensional networks. The orientation can be fixed in a two-dimensional network by astronomical azimuths and in a three-dimensional one by azimuths and zenith distances. In both cases the scale is fixed by distance measurements. Detailed studies on the optimal number and distribution of such additional observations have been made by Ashkenazi and Cross [1972], Cross and Ashkenazi [1979], and Sarközy [1981].

If no absolute information is available, the datum defects are eliminated by providing additional constraints, which lead, in the classical concept, to special solutions with coordinates and corresponding variance-covariance matrices, which refer to a datum base or a S system [Baarda, 1971]. In the modern concept of free net adjustment the final datum information is drawn from the approximate coordinates of all net points, computed to an arbitrary datum. This concept of the so-called 'inner solution' gives unique results, which can be characterized by the following three remarks. The inner solution is the minimum-norm least squares solution of the singular equations $Ax - l = v$, $v^TPv \rightarrow \min$, $x^Tx \rightarrow \min$; its cofactor matrix Q_x, which represents the 'inner accuracy' and has minimal trace, is the pseudoinverse $(A^TPA)^+ = N^+$ of the normal matrix; the restriction $x^Tx \rightarrow \min$ can be interpreted geometrically, and between adjusted and approximate coordinates a Helmert-like transformation relation is valid. For detailed proofs and interpretations, see Meissl [1964, 1965, 1969], Mittermayer [1971, 1972a,b], Pelzer [1971, 1974a,b], Pope [1973], and Schmid and Scherrer [1980]; for the theory of generalized inverses, see Ben-Israel and Greville [1974], Bjerhammar [1973], and Rao and Mitra [1971]. A general solution for the computation of the pseudoinverse, based on a singular value decomposition, is shown, for example, by Golub and Reinsch [1970].

Pope [1973] discusses the three strategies that are mainly applied in geodesy. The first method is the direct inversion of the normal equation matrix, which has been regularized by additive modification, followed by the subtraction of the additive modification. The second and most applied method is the pivoted inversion of the normals, which are regularized by bordering with the condition equations for $x^Tx \rightarrow \min$, where N^+ is the resulting submatrix (this strategy has the advantage that sparse techniques can be applied!). The third possibility is the transformation from a special solution and its cofactor matrix to the inner one [Alpâr, 1978; Meissl, 1969; Pelzer, 1971; Wolf, 1973]. The last case can be seen as a special case of the theory of S transformations of solutions and corresponding cofactor matrices between arbitrary datum systems. This theory was founded by Baarda [1971, 1973] and extended by Molenaar [1981]; see also van Mierlo [1980] and Grafarend et al. [1982].

Modifications of the minimum-norm concept in free nets are of increasing importance in deformation networks, where the results of different measuring epochs are to be referred to a common datum. Here the condition $x^Tx \rightarrow \min$ is restricted to a subset of points, in which no movements are expected [Caspary, 1978; Koch, 1980; Pope, 1973].

A further method for fixing the datum of a network, neglecting the trivial case of a constrained adjustment, is the introduction of the coordinates of the connection points together with their variance-covariance matrix. Typical applications are combinations between terrestrial networks and Doppler points and network densification [Cooper and Leahy, 1978; Matthias, 1981; van Mierlo, 1982]. If no real information about the connection points is available from a previous adjustment, Baarda [1973] proposes to introduce an artificial variance-covariance matrix.

In general, two different aspects of ZOD must be clearly distinguished. One problem is the geometrical interpretation of the results; the other problem is the statistical valuation of the datum solutions, namely, estimability and testability (see, for example, Grafarend and Schaffrin[1974], Koch [1980], and Pelzer [1974a,b]).

First-Order Design (FOD)

The justification of FOD has long been questioned, especially for high-order triangulations, where the topographical realities restrict any significant variation in the location of points and in the observation plan. Classical FOD examples are studies on the optimal configuration of geodetic standard problems like intersection and resection. The objectives have been the trace and determinant of the cofactor matrix of the new point as well as isotropy (circle-shaped error ellipse) [Baetslê, 1970; Bähr, 1974; Reissmann, 1957; Schmitt, 1975]. These studies have been deepened by Sarközy [1979] and Sarközy and Markus [1979]; Grafarend and Harland [1973] extended them to simultaneous inclusion of two and more points using mixed objective functions to maximize the point accuracies. Kelm [1976, 1979] proposed the use of sequential optimization procedures, which contain elements of both analytical and trial methods. He uses a graph method for the decomposition of the variance-covariance matrix of a network into its local parts. Koch [1981] solves the configuration problem by a least squares approximation of a criterion matrix. The regions in which the points are allowed to be shifted with regard to the topography are introduced as inequality constraints. This linear complementarity problem is solved by an algorithm, such as described by Schaffrin [1981b]. Further studies to develop closed analytical methods for the FOD seem to have stagnated because of the complication of this nonlinear problem, whereas interactive software systems working in a simulation mode have been established successfully [Conzett et al., 1981; Nickerson et al., 1976].

Nevertheless, FOD has made progress for special geodetic tasks. Concerning the deformation analysis, Schmitt [1981b] points to the necessity of an optimal definition of the object or building by the point field of the deformation network, paying particular attention to the static aspects and the expected directions of movements. For the detection of vertical crustal movements in leveling nets, Niemeier [1981] shows examples in which he varies the shape of the network and the density of points and discusses the optimality properties of the designs with respect to characteristic tectonic deformation models.

Dermanis and Mueller [1979] have examined various configurations of the existing and future VLBI-stations in different datum systems and discussed the optimum quality of the geometry of the network and of the derived earth rotation parameters. Drewes [1981] has carried out detailed studies of an optimal datum and an optimal configuration of a global network designed for the determination of plate dynamics. Both studies are examples of a mixed ZOD/FOD problem.

Second-Order Design (SOD)

A large number of important developments have been made in the SOD in the last decades. The classical example for an optimal weight distribution is the base extension net, as shown in the introduction. It is an example of the derivation of one function from a network with maximal accuracy for minimal expense. Similar general considerations have been shown by Köhr [1966, 1967a] and for the collective estimation of several functions, for example, by Gleinsvik [1964] and Köhr [1971]. They use a successive trace minimization of the cofactor matrix of the functions.

Many publications deal with simulation algorithms for accuracy maximization and cost minimization for homogeneously shaped trilateration networks [Gleinsvik, 1965a], for polygonal traverses [Gleinsvik, 1965b], and for arbitrary networks [Allman and Hoar, 1973; Dracup and Fronczek, 1979; Korobochkin and Zaytseva, 1968; Kubacek and Pazman, 1973; Neymann, 1966; Schmitt, 1977, 1978].

Earlier analytical solutions use standard algorithms of linear or nonlinear programing. Milbert [1979] minimizes costs or repetition numbers of the measurements by a piecewise linear approximation method with accuracy restrictions in the form of pointwise coordinate variance sums. Pelzer [1967, 1969] calculates the optimal repetition numbers for distance measurements in artificial and real trilateration networks by an integer programing method. Herzog [1972] looks for an optimal distribution of a constant sum of observational work by a modified Rosen algorithm. The risk function is the trace of the cofactor matrix of the coordinates or of derived quantities. The restrictions are the sum of repetition numbers and a minimal repetition number for each observation. Correlations between the observations of a station can be included in his concept. Heister [1978] applies a discrete dynamical optimization algorithm to the two standard problems: trace minimization (maximal accuracy) with a constant sum of observational costs and cost minimization with accuracy restrictions. Augath [1976, 1979] applies a nonlinear integer programing method with restrictions for reliability parameters to the design of large networks.

Sprinsky [1974] creates a criterion matrix from user requirements (accuracy of point position and estimable quantities) and reduces a maximal configuration where he takes the values of the correlation coefficients between the observations as a criterion.

The first analytical approximation of a criterion matrix as an ideal variance-covariance matrix of coordinates has been carried out by Bossler et al. [1973]. In a regular model network, $(A^TPA)^{-1} = Q_x$ is inverted to $A^TPA = Q_x^{-1} =: P_x$, written as a system of linear equations $(A^T \otimes A^T)p=q$, $p = \text{vec } P$, $Q = \text{vec } P_x$, and solved by the pseudoinverse $p = (A^T \otimes A^T)^+q$ ($\otimes$ is the Kronecker product; operation vec means a columnwise matrix to vector mapping). The result is a fully occupied weight matrix of the observations, which is unsuited for realization by practical measurements. For that reason, all following studies have been restricted to uncorrelated observations (diagonal design). The consequence is, of course, that most optimization problems, especially in free networks, not have a consistent solution. Grafarend [1975] formulates the diagonal SOD problem in terms of inequalities $(A^T \odot A^T)p \leq q$ ($\odot$ is the Khatri-Rao product [Rao and Mitra, 1971]) with weight restrictions $p \geq 0$. His solution by means of the simplex algorithm ('simplex design') is to be understood as a 'better-than' approximation of the criterion matrix. The fact that the simplex design trends to a strong reduction of the observation plan, frequently to configurations which are no more redundant, was the reason for proposing a least-squares approximation of $(A^T \odot A^T)p = q+v$ by $p = (A^T \odot A^T)^+q$, where v is the inconsistency parameter [Schmitt, 1978, 1979b; Schmitt et al., 1978]. On the basis of defect analyses of $(A^T \odot A^T)$ it is possible to write effective and fast algorithms, at least for special types of networks [Schmitt, 1979à, 1980].

An alternative concept, developed by Schaffrin et al. [1977], is the 'canonical design'. He uses the singular value decomposition $P_x = UDU^T$, where D is the diagonal matrix of the eigenvalues and U the matrix of the corresponding eigenvectors of P_x, written columnwise. Multiplication of $A^TPA = UDU^T$ from the left side by U^T and from the right side by U results in $(AU)^TP(AU) = Z^TPZ = D$ (U is orthogonal). This can be transformed into $(Z^T \odot Z^T)p = d$, $P = \text{diag } p$, $d = \text{vec } D$. This approach is suited to proving the existence of a consistent design [Schaffrin, 1979]. Schaffrin [1981a,b] and Schaffrin et al. [1981] follow this concept and propose better-than approximations by aid of a complementarity algorithm, which allows additional restrictions for positive weights and cost and reliability restrictions.

All the approaches reviewed so far have been approximations not of the criterion matrix Q_x itself, but of its inverse P_x. Wimmer [1981a,b] proposes the direct approximation of the criterion matrix by $(QA^TP)P^{-1}(QA^TP)^T = HP^{-1}H^T = Q_x$, where $Q = (A^TPA)^-$, in which case $p = (H \odot H)^+q$, $P = \text{diag } p$, $q = \text{vec } Q_x$. The solution is derived by an iterative algorithm, because P is part of Q.

In the first SOD publications, distance networks were used as examples, because two additional problems arise with the optimization of direction observations. On the one hand, an individual weight for each direction is important for the estimation of the quality and the reliability of a network, but on the other hand, it is necessary with practical measurements to introduce a common or group weight for all directions of one set. The second problem is that heterogeneous parameters in the adjustment model, coordinate and orientation unknowns, complicate the setting up of appropriate criterion matrices. Schmitt [1981a] proposes therefore a design in three steps with the elimination of the orientation inknowns in the approach of group weights and with an 'extracted', related only to coordinates Khatri-Rao product in the approach of individual weights.

The SOD has also been applied to one-dimensional networks, leveling nets, and gravity nets. Cross and Thapa [1979] and Cross and Whiting [1980] use the simplex design to optimize the observation plan and weights in leveling nets. In a similar way the first-order gravity network of Germany has been optimized by Boedecker [1979]. Wenzel [1977, 1979] has applied a sequential optimization to the same network. His method can be interpreted within the concept of the third-order design.

Criterion Matrices

One central problem in the SOD is the design of appropriate criterion matrices. In general, a two-dimensional geodetic network is said to be optimal if it has the properties of both homogeneity and isotropy, in which case all point error ellipses degenerate into circles with equal radius. The most simple matrix, which leads to these optimal properties, is the unit matrix; but is has the disadvantage of setting the covariances between all coordinates to zero. In recent years, Taylor-Karman structured matrices whose coefficients are calculated from correlation functions for longitudinal and transverse errors [Grafarend, 1972] have succeeded in many applications. The basic assumption is the interpretation of a network as a realization of a stochastic process with expectation values equal to zero and correlation functions of the absolute coordinates. The theory is described in detail by Grafarend and Schaffrin [1979], and there is an extensive review of planar geodetic covariance functions by Meier [1981].

Grafarend and Schaffrin [1979] and Schaffrin and Grafarend [1982] have studied in detail the conversion between correlation functions and criterion matrices for absolute coordinates and derived quantities as coordinate differences, distances, angles, etc.

It remains to be mentioned that if the longitudinal and lateral correlation functions are equal, the error situation is called completely isotropic or 'chaotic', in which case all relative error ellipses are circles also.

In the refinement of SOD methods, special attention is paid to the avoidance of certain model errors in the different approaches, for example, the postulation of rank identity between the configuration matrix A and the criterion matrix Q_x and the derivation of criterion matrices for estimable quantities. Taylor-Karman structured criterion matrices, which are calculated from positive definite correlation functions, have full rank. Their use in the SOD of free networks is correct only after an S transformation to a projected criterion matrix with a right defect [Schaffrin, 1981b; Schaffrin et al., 1981]. If a criterion matrix Q_f of estimable quantities is given [Schaffrin and Grafarend, 1982], a transformation to an 'allocated' criterion matrix Q_x as input for the usual SOD algorithms is neccessary, for example, by $Q_x := F^+Q_x(F^T)^+$, $f = F_x$ [Schaffrin, 1981b; Schmitt, 1980].

Not for all problems are the properties of homogeneity, and especially isotropy, to be valued as optimal. In the design of tunnel networks, or of deformation nets with preferred directions of movements, general approaches on the level of correlation functions are possible, for example, under the postulates of axis or point symmetry [Meier, 1981]. The direct formulation of criterion matrices from special user requirements (elements of error ellipses, accuracy of derived quantities) has been studied by Banov [1981a,b], Cross and Thapa [1979], Penev [1981], and Sprinsky [1974].

Least squares approximations in the SOD need appropriate or realizable criterion matrices. Investigations of real correlations in model networks with homogeneous shape harmonize with theoretical structures of the Taylor-Karman type [Wimmer, 1981a] and can be used to determine the characteristic distances of correlation functions.

Third-Order Design (THOD)

Very few practical applications are known which solve the THOD problem in a fundamental way, although Bähr [1973] discusses adjustment procedures for the extension of existing networks, which avoid a new computation of the complete network, and Grafarend [1979] tries to solve the THOD problem with criterion matrices. But if one is restricting the THOD on the variation of the observation plan, one has to speak about sequential optimization, which has been considered by Wolf [1970] for two special cases. The 'network construction' starts with a minimal configuration of observations which fixes the geometry without redundancy and extends the observation plan by those measurements that bring the largest contribution to the objective functions. The 'network reduction' operates in the opposite way: starting with a maximal configuration with all possible observations, it follows a sequential elimination of those measurements that have the smallest influence on the objective function. The network reduction is consequently used by Wenzel [1977, 1979] to optimize gravity networks, where the trace of the cofactor matrix of the gravity values as an objective function has been generated by the Neumann series.

Review Papers

A review paper on the optimization of geodetic networks cannot close withoug referring to some general publications which are of mostly tutorial character, such as those by Alberda [1980], Ayan [1981], Cross and Whiting [1981], and Grafarend [1974]. Special reference should be made to the first textbook on this topic [Grafarend et al., 1979].

References

Alberda, J. E., A review of analysis techniques for engineering survey control schemes, paper presented at the Industrial and Engineering Survey Conference, London, 1980.

Allman, J. S., and G. J. Hoar, Optimization of geodetic networks, Surv. Rev., 22, 11-22, 1973.

Alpár, G., A straight presentation of the adjustment of so-called free geodetic networks, Acta Geod. Geophys. Montan. Acad. Sci. Hung., 13(1-2), 195-199, 1978.

Ashkenazi, V., and P. Cross, Strength analysis of block VI of the European triangulation, Bull. Geod., 103, 5-24, 1972.

Augath, W., Untersuchungen zum Aufbau geodätischer Lagenetze, Rep. 72, Wiss. Arb. der Lehrstühle für Geod., Photogramm. und Kartogr. an der Tech. Univ. Hannover, Hannover, Federal Republic of Germany, 1976.

Augath, W., Über den Aufbau geodätischer Lagenetze nach Zuverlässigkeitskriterien, in Optimization of Design and Computation of Control Networks, Proceedings of the IAG Symposium in Sopron 1977, edited by F. Halmos and J. Somogyi, 403-417, Akademiai Kiado, Budapest, Hungary, 1979.

Ayan, T., Jeodezik aglarin optimizasyonu, habilitation thesis, Tech. Univ., Istanbul, Turkey, 1981.

Baarda, W., S-transformations and criterion matrices in the adjustment of geodetic networks, report, Comput. Centre of the Delft Geod. Inst., Delft, Netherlands, 1971.

Baarda, W., S-transformations and criterion matrices, Neth. Geod. Comm. Publ. Geod., New Ser., 5(1), 1973.

Baetslé, P. L., Quelques problèmes élêmentaires d'optimilisation dans le levés planimétriques, Bull. Trimest. Soc. Belg. Photogramm., 102, 3, 1970.

Bähr, H. G., Zum Ausgleichungsverfahren bei der Ergänzung bestehender geodätischer Netze, Rep. C 187, Dtsch. Geod. Komm., Munich, 1973.

Bähr, H. G., Optimale Punktanordnung für vorgegebene Beobachtungen, Allg. Vermess. Nachr., 81, 300-304, 1974.

Banov, B., Optimization of the weights of observations of geodetic networks, in Proceedings of F.I.G. Symposium Automated Processing of Surveying Data, November 1980, Varna, Bulgaria, Int. Fed. of Surv., 41-54, 1981a.

Banov, B., Optimization of networks, in Proceedings of the F.I.G. XVI International Congress, Commission 5, Montreux, Switzerland, Int. Fed. of Surv., 507.4/1-507.4/9, 1981b.

Ben-Israel, A., and T. N. Greville, Generalized Inverses, John Whiley & Sons, New York, 1974.

Bjerhammar, A., Theory of Errors and Generalized

Matrix Inverses, Elsevier Scientific Publishing Company, New York, 1973.

Boedecker, G., The development of an observation scheme for the variance-covariance matrix of the unknowns, in Optimization of Design and Computation of Control Networks, Proceedings of the IAG Symposium in Sopron 1977, edited by F. Halmos and J. Somogyi, 635-640, Akademiai Kiado, Budapest, Hungary, 1979.

Bossler, J. D., E. Grafarend, and R. Kelm, Optimal design of geodetic nets, 2, J. Geophys. Res., 78, 5887-5897, 1973.

Caspary, W., Zur Lösung singulärer Ausgleichungsmodelle durch Bedingungsgleichungen, Allg. Vermess. Nachr., 85, 81-87, 1978.

Conzett, F., A. Frank, and C. Misslin, Interaktive Triangulation, in Ingenieurvermessung 80, Beiträge zum VIII. Internationalen Kurs für Ingenieurvermessung, Zürich 1980, edited by R. Conzett, H.J. Matthias and H. Schmid, Dümmler Verlag, A7/1-A7/12, Bonn, 1981.

Cooper, M. A. R., and F. J. Leahy, An adjustment of a second-order network, Surv. Rev., 24, 224-233, 1978.

Cross, P. A., and V. Ashkenazi, Internal consistency of national geodetic control networks, in Optimization of Design and Computation of Control Networks, Proceedings of the IAG-Symposium in Sopron 1977, edited by F. Halmos and J. Somogyi, 473-485, Akademiai Kiado, Budapest, Hungary, 1979.

Cross, P. A., and K. Thapa, The optimal design of levelling networks, Surv. Rev., 25, 68-79, 1979.

Cross, P. A., and B. M. Whiting, The design of national vertical control networks, in Proceedings of the Second International Symposium on Problems Related to the Redefinition of North American Vertical Geodetic Networks, Canadian Institute of Surveying, 419-432, Ottawa, Canada, 1980.

Cross, P. A., and B. M. Whiting, The design of engineering networks, paper presented at Seminar on Engineering Surveying, University of Calgary, Canada, 1981.

Dermanis, A., and I. I. Mueller, Earth rotation and network geometry optimization for very long baseline interferometers, in Optimization of Design and Computation of Control Networks, Proceedings of the IAG Symposium in Sopron 1977, edited by F. Halmos and J. Somogyi, 67-94, Akademiai Kiado, Budapest, Hungary, 1979.

Dracup, J. F., and C. J. Fronczek, An immediate solution to the optimization problem, in Optimization of Design and Computation of Control networks, Proceedings of the IAG Symposium in Sopron 1977, edited by F. Halmos and J. Somogyi, 117-131, Akademiai Kiado, Budapest, Hungary, 1979.

Drewes, H., Design of a global geodetic network for geodynamics, Proceedings of the IAG-Symposium on Geodetic Networks and computations in München 1980, edited by R. Sigl, Rep. C 258/II, 51-59, Dtsch. Geod. Komm., Munich, 1981.

Gleinsvik, P., Die günstigste Gewichtsverteilung, Bull. Geod., 74, 295-306, 1964.

Gleinsvik, P., Die Gewichtsverteilung bei der Trilateration, Allg. Vermess. Nachr., 72, 309-320, 1965a.

Gleinsvik, P., Die Gewichtsverteilung in polygonalen Zügen, Z. Vermessungswesen, 90, 37-46, 1965b.

Golub, G. H., and C. Reinsch, Singular value decomposition and least squares solutions, Numer. Math., 14, 403-420, 1970.

Grafarend, E., Genauigkeitsmasse geodätischer Netze, Rep. A 73, Dtsch. Geod. Komm., Munich, 1972.

Grafarend, E., Optimization of geodetic networks, Boll. Geod. Sci. Affini, 33, 351-406, 1974.

Grafarend, E., Second order design of geodetic nets, Z. Vermessungswesen, 100, 158-168, 1975.

Grafarend, E., Third-order design of geodetic networks, in Optimization of Design and Computation of Control Networks, Proceedings of the IAG-Symposium in Sopron 1977, edited by F. Halmos and J. Somogyi, 133-149, Akademiai Kiado, Budapest, Hungary, 1979.

Grafarend, E., and P. Harland, Optimales Design geodätischer Netze I, Rep. A 74, Dtsch. Geod. Komm., Munich, 1973.

Grafarend, E., and B. Schaffrin, Unbiased free net adjustment, Surv. Rev., 22, 200-218, 1974.

Grafarend, E., and B. Schaffrin, Kriterion Matrizen, I, Zweidimensionale homogene und isotrope geodätische Netze, Z. Vermessungswesen, 104, 133-149, 1979.

Grafarend, E., et al., Optimierung geodätischer Messoperationen, Herbert Wichmann Verlag, Karlsruhe, Federal Rebpulic of Germany, 1979.

Grafarend, E., E. H. Knickmeyer, and B. Schaffrin, Geodätische Datumtransformationen, Z. Vermessungswesen, 107, 15-25, 1982.

Heister, H., Die diskrete dynamische Optimierung und ihre Anwendung beim geodätischen Netzentwurf, Allg. Vermess. Nachr., 85, 64-81, 1978.

Herzog, H., Zur Optimierung geodätischer Beobachtungen, Rep. C 180, Dtsch. Geod. Komm., Munich, 1972.

Kelm, R., Zur analytischen Optimierung geodätischer Netze, Rep. C 220, Dtsch. Geod. Komm., Munich, 1976.

Kelm, R., Graph methods within the optimal design of geodetic nets, in Optimization of Design and Computation of Control Networks, Proceedings of the IAG Symposium in Sopron 1977, edited by F. Halmos and J. Somogyi, 151-162, Akademiai Kiado, Budapest, Hungary, 1979.

Koch, K.-R., Parameterschätzung und Hypothesentest in linearen Modellen, Dümmler Verlag, Bonn, 1980.

Koch, K.-R., Optimization of the configuration of geodetic networks, Proceedings of the IAG Symposium on Geodetic Networks and Computations in München 1980, edited by R. Sigl, Rep. C 258/III, 82-89, Dtsch. Geod. Komm., Munich, 1981.

Köhr, J., Über die optimale Programmierung von Messungen unter Berücksichtigung von Wirtschaftlichkeitsprinzipien, Z. Vermessungswesen, 91, 105-116, 1966.

Köhr, J., Die Optimierung von Messungen auf Kostengrundlage, Z. Vermessungswesen, 92, 92-97, 1967a.

Köhr, J., Der allgemeine Fall der optimalen Programmierung von Messungen, Z. Vermessungswesen, 92, 381-393, 1967b.

Köhr, J., Über den günstigsten Schnittwinkel der Bestimmungskreise beim einfachen Rückwärtsschnitt, Z. Vermessungswesen, 96, 14-23, 1971.

Korobochkin, M., and N. Zaytseva, Methods for planning geodetic systems, Geod. Aerophotogr., Engl. Transl., 4, 216-218, 1968.

Kubacek, L., and A. Pazman, Optimum networks,

paper presented at the IAG Symposium on Computational Methods in Geometrical Geodesy, Int. Assoc. Geod., Oxford, England, 1973.

Matthias, H. J., Allgemeine vermittelnde Netzausgleichung, Rep. 30, Inst. für Geod. und Photogramm. der Eidgenoess. Tech. Hochsch. Zürich, Zürich, Switzerland, 1981.

Meier, S., Planar geodetic covariance functions, Rev. Geophys. Space Phys., 19, 673-686, 1981.

Meissl, P., Über die Verformungsfehler eines Systems von endlich vielen Punkten, Oesterr. Z. Vermessungswesen, 52, 105-109, 1964.

Meissl, P., Über die innere Genauigkeit dreidimensionaler Punkthaufen, Z. Vermessungswesen, 90, 108-118, 1965.

Meissl, P., Zusammenfassung und Ausbau der inneren Fehlertheorie eines Punkthaufens, Beiträge zur Theorie der geodätischen Netze im Raum, Rep. A 61, Dtsch. Geod. Komm., Munich, 1969.

Milbert, D. G., Optimization of horizontal Control networks by nonlinear programing, NOAA Tech. Rep. NOS 79 NGS 12, Natl. Oceanic and Atmos. Admin., Rockville, Md., 1979.

Mittermayer, E., Eine Verallgemeinerung der Methode der kleinsten Quadrate zur Ausgleichung freier Netze, Z. Vermessungswesen, 96, 401-410, 1971.

Mittermayer, E., Zur Ausgleichung freier Netze, Z. Vermessungswesen, 97, 481-489, 1972a.

Mittermayer, E., A generalisation of the least-squares method for the adjustment of free networks, Bull. Geod., 104, 139-157, 1972b.

Molenaar, M., A further inquiry into the theory of S-transformations and criterion matrices, dissertation, Tech. Univ.,Delft, Netherlands, 1981.

Neymann, Y., An algorithm for planning a geodetic net on an electronic computer, Geod. Aerophotogr., Engl. Transl., 6, 325-329, 1966.

Nickerson, B. G., et al., Design of survey networks using interactive computer graphics, in Proceedings ACSM Annual Meeting, Washington D. C., 1976.

Niemeier, W., and G. Rohde , On the optimization of levelling networks with respect to the determination of crustal movements, Proceedings of the IAG-Symposium on Geodetic Networks and Computations in München 1980, edited by R. Sigl, Rep. C 258V, 148-160, Munich, 1981.

Pelzer, H., Verteilung von Streckenmessungen in trigonometrischen Netzen, Allg. Vermess. Nachr., 74, 273-280, 1967.

Pelzer, H., Die Genauigkeit elektromagnetisch gemessener Streckennetze, insbesondere im Flachland und über küstennahen Gewässern, Rep. C 136, Dtsch. Geod. Komm., Munich, 1969.

Pelzer, H., Zur Analyse geodätischer Deformationsmessungen, Rep. C 164, Dtsch. Geod. Komm., Munich, 1971.

Pelzer, H., Zur Behandlung singulärer Ausgleichungsaufgaben I, Z. Vermessungswesen, 99, 181-194, 1974a.

Pelzer, H., Zur Behandlung singulärer Ausgleichungsaufgaben II, Z. Vermessungswesen, 99, 479-488, 1974b.

Penev, P., Network optimization in order to determine deformations of dams, in Proceedings F.I.G. Symposium on Automated Processing of Surveying Data, November 1980, Varna, Bulgaria, Int. Fed. of Surv., 209-220, 1981.

Pope, A., The use of the solution space in the analysis of geodetic network adjustments, paper presented at the IAG Symposium on Computational Methods in Geometrical Geodesy, Int. Assoc. Geod., Oxford, England, 1973.

Rao, C. R., and S. K. Mitra, Generalized Inverse of Matrices and Its Applications, John Wiley & Sons, New York, 1971.

Reissmann, G., Die Genauigkeitsmasse bei der Punktbestimmung, habilitation thesis, Tech. Univ. Dresden, German Democratic Republic, 1957.

Sarközy, F., Komplexe optimale Planung von Bergmessnetzen, in Proceedings of the IV Internationales Symposium für Markscheidewesen, Eurogress Aachen, Aachen, Federal Republic of Germany, 1979.

Sarközy, F., Optimierungsaufgaben in geodätischen Netzen von grossem Umfang, in Proceedings of F.I.G. XVI International Congress, Commission 5, Int. Fed. of Surv., 507.2/1-507.2/11, Montreux, Switzerland, 1981.

Sarközy, F., and B. Markus, Optimization of horizontal control points, Optimization of Design and Computation of Control Networks, Proceedings of the IAG Symposium in Sopron 1977, edited by F. Halmos and J. Somogyi, 163-173, Akademiai Kiado, Budapest, Hungary, 1979.

Schaffrin, B., A study of the second order design problem in geodetic networks, Optimization of Design and Computation of Control Networks, Proceedings of the IAG Symposium in Sopron 1977, edited by F. Halmos and J. Somogyi, 175-177, Akademiai Kiado, Budapest, Hungary, 1979.

Schaffrin, B., Ausgleichung mit Bedingunsg-Ungleichungen, Allg. Vermess. Nachr., 88, 227-238, 1981a.

Schaffrin, B., Some proposals concerning the diagonal second order design of geodetic networks, Manuscr. Geod., 6, 303-326, 1981b.

Schaffrin, B., and E. Grafarend, Kriterion-Matrizen, II, Zweidimensionale homogene und isotrope geodätische Netze, Z. Vermessungswesen, 107, 183-193, 1982.

Schaffrin, B., E. Grafarend and G. Schmitt, Kanonisches Design geodätischer Netze I, Manuscr. Geod., 2, 263-306, 1977.

Schaffrin, B., F. Krumm and D. Fritsch, Positiv-diagonale Genauigkeitsoptimierung von Realnetzen über den Komplementaritäts-Algorithmus, in Ingenieurvermessung 80, Beiträge zum VIII. Internationalen Kurs für Ingenieurvermessung, Zürich, 1980, edited by R. Conzett, H.J. Matthias and H. Schmid, A15/1-A15/19, Dümmler Verlag, Bonn, 1981.

Schmid, H. H., and R. Scherrer, Vom freien zum gelagerten Netz, Rep. 29, Inst. für Geod. und Photogramm. der Eidgenoess. Tech. Hochsch. Zürich, Zürich, Switzerland, 1980.

Schmitt, G., Optimaler Schnittwinkel der Bestimmungsstrecken beim einfachen Bogenschnitt, Allg. Vermess. Nachr., 82, 226-230, 1975.

Schmitt, G., Monte-Carlo-Design geodätischer Netze, Allg. Vermess. Nachr., 84, 87-94, 1977.

Schmitt, G., Gewichtsoptimierung bei Mehrpunkteinschaltung mit Streckenmessung, Allg. Vermess. Nachr., 85, 1-15, 1978.

Schmitt, G., Zur Numerik des Designs zweiter Ordnung, Rep. C 256, Dtsch. Geod. Komm., Munich, 1979a.

Schmitt, G., Experiences with the second order problem in theoretical and practical geodetic networks, in Optimization of Design and Computation of Control Networks, Proceedings of the IAG Symposium in Sopron 1977, edited by F. Halmos and J. Somogyi, 179-206, Akademiai Kiado, Budapest, Hungary, 1979b.

Schmitt, G., Second order design of free distance networks, considering different types of criterion matrices, Bull. Geod., 54, 531-543, 1980.

Schmitt, G., Zur Gewichtsoptimierung in Richtungsnetzen, in Ingenieurvermessung 80, Beiträge zum VIII. Internationalen Kurs für Ingenieurvermessung, Zürich, 1980, edited by R. Conzett, H.J. Matthias and H. Schmit, A9/1-A9/11, Dümmler Verlag, Bonn, 1981a.

Schmitt, G., Das Design, 2, Ordnung geodätischer Netze, Proceedings F.I.G. Symposium Automated Processing of Surveying Data, November 1980, Varna, Bulgaria, Int. Fed. of Surv., 23-40, 1981b.

Schmitt, G., E. Grafarend and B. Schaffrin, Kanonisches Design geodätischer Netze II, Manuscr. Geod., 3, 1-22, 1978.

Schreiber, O., Anordnung der Winkelbeobachtungen im Göttinger Basisnetz, Z. Vermessungswesen, 7, 129-161, 1882.

Sprinsky, W. H., The design of special purpose horizontal geodetic control networks, Ph.D. thesis, The Ohio State Univ., Columbus, 1974.

van Mierlo, J., Free network adjustment and S-transformations, Rep. B 252, Dtsch. Geod. Komm., Munich, 1980.

van Mierlo, J., Second order design: Precision and reliability aspects, Allg. Vermess. Nachr., 88, 95-101, 1981.

van Mierlo, J., Design and adjustment of densification networks, Proc. Conf. Afr. Surv. 7th, 1982.

Wenzel, H.-G., Zur Optimierung von Schwerenetzen, Z. Vermessungswesen, 102, 452-457, 1977.

Wenzel, H.-G., Zur Optimierung von Schwerenetzen, in Optimization of Design and Computation of Control Networks, Proceedings of the IAG Symposium in Sopron 1977, edited by F. Halmos and J. Somogyi, 721-728, Akademiai Kiado, Budapest, Hungary, 1979.

Wimmer, H., Ein Beitrag zur Gewichtsoptimierung geodätischer Netze, Rep. C 269, Dtsch. Geod. Komm., Munich, 1981a.

Wimmer, H., Second order design of geodetic networks by an iterative approximation of a given criterion matrix, Proceedings of the IAG-Symposium on Geodetic Networks and Computations in München 1980, edited by R. Sigl, Rep. C258/III, 112-127, Dtsch. Geod. Komm., Munich 1981b.

Wolf, H., Polarität und Optimierung bei freien und eingeschalteten geodätischen Netzen, Allg. Vermess. Nachr., 77, 291-298, 1970.

Wolf, H., Die Helmert-Inverse bei freien geodätischen Netzen, Z. Vermessungswesen, 98, 396-398, 1973.

(Received June 1, 1982;
accepted July 7, 1982.)

REVIEWS OF GEOPHYSICS AND SPACE PHYSICS, VOL. 18, NO. 4, PAGES 789–812, NOVEMBER 1980

Estimations From Irregular Arrays

FRANCIS P. BRETHERTON AND JAMES C. MCWILLIAMS

National Center for Atmospheric Research, Boulder, Colorado 80307

A general approach is presented for estimating statistics of a random process from a finite data set from an irregularly spaced array of measurement locations. Particular attention is focused upon estimation of the spectrum under circumstances where it is smooth (i.e., broadband). A procedure is proposed based upon the technique of optimal linear estimation or generalized least squares fits in order to estimate coefficients of spectral features of interest to the investigator. Selection of which features are appropriate for estimation from a given array and set of measurements is made within the formalism of Bayesian statistical inference. The converse problem is also addressed: given a set of spectral features and a fixed number of possible measurements, which configuration of measurement locations will yield the most information about the amplitudes of those features? Finally, an assessment is made of the efficiency (i.e., the bias and accuracy) of various previously proposed methods of spectral estimation from irregular arrays.

1. INTRODUCTION

One of the fundamental problems of measurement in the geophysical sciences is to estimate the statistics of a random process from a severely limited sample of data. When a field $\phi(\mathbf{x})$ is measured at a set of irregularly spaced observation points $\{\mathbf{x}_i, i = 1, \cdots, N\}$ in one or more dimensions, the following questions often arise:

1. How well can an individual realization of $\phi(\mathbf{x})$ be interpolated between the observation points?
2. What are the mean

$$A_\mathbf{x} \equiv \langle \phi(\mathbf{x}) \rangle$$

and the spatial covariance function

$$C_{\mathbf{xy}} \equiv \langle [\phi(\mathbf{x}) - \langle \phi(\mathbf{x}) \rangle][\phi(\mathbf{y}) - \langle \phi(\mathbf{y}) \rangle] \rangle$$

averaged over many independent realizations of $\phi(\mathbf{x})$?

3. If $\phi(\mathbf{x})$ also varies continuously in time and time sequences are measured at the points $\mathbf{x}_i$, what is the frequency–wave number spectrum for ϕ?
4. Given a limited number of instruments N, what is the optimum array configuration for answering each of the questions 1–3?

This paper presents a general approach to answering such questions, with particular attention to questions 2 and 4, illustrating how they may be treated by statistical estimation theory. Question 3 is easily reduced to question 2, and given $A_\mathbf{x}$ and $C_{\mathbf{xy}}$, the answer to question 1 is readily calculable according to well-known theory.

The heart of the analysis is a quantitative estimate of the uncertainties inherent in any inference of $C_{\mathbf{xy}}$ from observations of a limited sample of realizations of ϕ. For each pair of points $\mathbf{x}_i$ and $\mathbf{x}_j$ the product of observed values, $(\phi_i - \langle \phi_i \rangle)(\phi_j - \langle \phi_j \rangle)$, averaged over the available realizations, provides a raw estimate of the covariance between these points, but it is limited in scope to these points only and is subject to sampling errors. The approach here is to fit these raw estimates in an optimal manner to one of a family of model covariance functions containing only a limited, though initially excessively large, number of undetermined parameters. Such a family consists of a first guess $\tilde{C}_{\mathbf{xy}}$ chosen by the investigator and a set of features of specified shape but unknown magnitude which define ways in which the estimated $C_{\mathbf{xy}}$ may differ from the first guess. Attention is concentrated on circumstances in which a significant amount about the statistics of $\phi(\mathbf{x})$ is known a priori, so that a reasonable first guess can be adduced, and the amount of available data is enough to make a substantial improvement a posteriori, refining the first guess and greatly reducing the uncertainties in it. Under such circumstances a linear model of many possible refinements to $\tilde{C}$ is a reasonable starting point. Such a fitting procedure organizes the information contained in the data and, within the context of the model family, provides both a best estimate of $C_{\mathbf{xy}}$ and, most important, an explicit measure of the uncertainty in that estimate. The latter can be used by the investigator to limit more effectively the range of plausible covariance features under consideration in the model.

A key step in the argument is to relate fourth-order moments of the variable ϕ to second-order ones. We shall do this by supposing that the random process is Gaussian or, more specifically, that the fourth-order cumulants of $\phi(\mathbf{x})$ vanish. In the absence of more information this is a common assumption; however, we know from experience that it is often not correct for geophysical fields. In that event the fitting procedure described below is still valid, though it may not be completely optimal and the uncertainty estimates may be too low. Nevertheless, the latter are probably still qualitatively correct and a guide to reasonable analysis procedures. Given sufficient data, the discrepancies can be analyzed and crudely corrected.

Of course, in any inferential procedure the framework of prior knowledge and assumptions brought to the problem by the investigator, here codified in the model family, is at least as important as the data themselves. For example, the assertion that the statistics of $\phi(\mathbf{x})$ are stationary can never be deduced from observations. It can merely be postulated as a working hypothesis, with tests conducted whether in specific respects it is significantly inconsistent with the observations. The objective here is to expose this framework and the judgments it implies as explicitly as possible and then to evaluate systematically the finite amount of new information which is provided by the data set. It is important to understand how the final estimate of $C_{\mathbf{xy}}$ is sensitive on the one hand to the observations and on the other hand to the prior assumptions and to devise adequate tests for the consistency of those assumptions.

Such an analysis provides a standard against which other

approaches to estimating the covariance function (or, equivalently for a stationary process, the spectrum) can be judged. Provided the prior knowledge and available data set are sufficient to allow reasonably secure inferences, the bias and sampling sensitivity of other methods can be compared systematically, using linear perturbation theory. It will be shown that after correction for bias some extant techniques are essentially equivalent in performance to the technique proposed here but that others are intrinsically less efficient.

One advantage of such an explicit analysis is that it is possible to calculate in advance of actually taking observations the probable ability of a given array to provide the information desired by the investigator. Indeed, given a specific set of model features to be considered, the locations $\{\mathbf{x}_i\}$ of a given number of observation points may be varied in such a way as to maximize the information likely to be obtained about those features; that is, the array design can be optimized. Of course, many other considerations—such as including redundancy to protect against measurement losses, accommodating logistical constraints, and being prepared for surprises in the phenomena—should also enter into the final decision on the design of a major field experiment. Nevertheless, once properly formulated, even quite extensive computer simulations tend to be inexpensive in relation to geophysical field measurements, and prior statistical assessment against specific assumptions and objectives is a potentially important element in this decision-making process.

After a brief summary in section 2 of the available answers to questions 1 and 3, a discussion of other approaches to spectral estimation in sections 3 and 4, and a discussion of generalized least squares fits in section 5, the central concepts in our approach are introduced through an example in section 6 and then elaborated upon somewhat more formally in section 7. Question 4 is addressed in section 8, and several summary remarks are made in section 9. For further details the reader may consult the earlier report of this work [*Bretherton and McWilliams,* 1979].

2. MAPPING AND FREQUENCY-WAVE NUMBER ESTIMATION

A treatment of question 1 on the interpolation of $\phi(\mathbf{x})$ dates back to the work of Gauss in the nineteenth century, though it has been rediscovered more recently in several geophysical disciplines [e.g., *Gandin,* 1965; *Bretherton et al.,* 1976]. We suppose that an observation at point $\mathbf{x}_i$ can be considered as the sum

$$\psi_i = \phi(\mathbf{x}_i) + \epsilon_i$$

of the true value there and a random noise or error ϵ_i, of zero mean and uncorrelated with ϕ. We seek the value of ϕ at a position $\mathbf{x}$ between data points estimated as a linear combination of the observations $\{\psi_i\}$,

$$\hat{\phi}_{\mathbf{x}} = \sum_{i=1}^{N} \alpha_{\mathbf{x}i}\psi_i + \beta_{\mathbf{x}} \tag{1}$$

The choice of coefficients, $\alpha_{\mathbf{x}i}$ and $\beta_{\mathbf{x}}$, which minimizes the mean square error $\langle \{\phi(\mathbf{x}) - \hat{\phi}_{\mathbf{x}}\}^2 \rangle$ is

$$\alpha_{\mathbf{x}i} = \sum_j C_{\mathbf{x}j}\,[(C+U)^{-1}]_{ji} \qquad \beta_{\mathbf{x}} = A_{\mathbf{x}} - \sum_j \alpha_{\mathbf{x}j}A_j \tag{2}$$

where C_{ij} is the positive definite $N \times N$ matrix of covariances between the values of ϕ at points $\mathbf{x}_i$ and $\mathbf{x}_j$, $C_{\mathbf{x}j}$ is the covariance between ϕ at the interpolation point and the observation at $\mathbf{x}_j$, U_{ij} is the error covariance matrix $\langle \epsilon_i\epsilon_j \rangle$ associated with point measurements, and $(C + U)^{-1}$ is the inverse of the matrix $C + U$. This result is a straightforward consequence of completing the square in α and β in the expression

$$\left\langle \left[\phi(\mathbf{x}) - \sum \alpha_{\mathbf{x}i}\psi_i - \beta_{\mathbf{x}} \right]^2 \right\rangle$$

It not only provides an objective interpolation procedure which among linear estimators cannot be bettered on average, but it also yields an estimate of the residual uncertainties in the interpolated values in the error covariance function

$$\langle [\phi(\mathbf{x}) - \hat{\phi}_{\mathbf{x}}][\phi(\mathbf{y}) - \hat{\phi}_{\mathbf{y}}] \rangle = C_{\mathbf{x}\mathbf{y}} - \sum_{i,j} C_{\mathbf{x}i}\,[(C+U)^{-1}]_{ij}C_{\mathbf{y}j} \tag{3}$$

Thus if the mean $A_{\mathbf{x}}$, covariance function $C_{\mathbf{x}\mathbf{y}}$, and error covariance function $U_{\mathbf{x}\mathbf{y}}$ are known, the implications of the partial sampling of $\psi(\mathbf{x})$ at only N discrete points can be summarized without reference to any particular data set. A convenient form is a plot of the residual variance against $\mathbf{x}$. This may be computed by setting $\mathbf{y} = \mathbf{x}$ in (3), though it should be noted that for some purposes it is the correlation between interpolation errors at different points which is most significant. Thus if the primary purpose of an observational array is to map individual realizations of $\phi(\mathbf{x})$, a prior analysis of the performance of different array designs $\{\mathbf{x}_i\}$ is straightforward, provided the statistics of the field to be mapped have already been determined. This procedure can also be generalized for optimal mapping of functionals of $\phi(\mathbf{x})$, based upon observations $\{\psi_i\}$.

In practice, the measurement noise ϵ_i included in an observation of a physical field $\psi(\mathbf{x})$ often arises from fluctuations on scales too small to be resolved by the array; $\phi(\mathbf{x})$ is then a spatially smoothed version of ψ, with the degree of smoothing set by the investigator. One approach is to prepare estimates $\hat{\psi}(\mathbf{x})$ for the total field by using the total covariance and no noise, which are then smoothed to get $\hat{\phi}(\mathbf{x})$. Alternatively but equivalently, the total covariance is separated into a smoothed coherent part $C_{\mathbf{x}\mathbf{y}}$ and a fine scale residual $U_{\mathbf{x}\mathbf{y}}$, which is treated as independent noise a priori. In each case the final estimate and the uncertainties are given by (2) and (3).

In another variant of this analysis the modes of variation permitted to $\phi(\mathbf{x})$ are a linear subspace

$$\phi(\mathbf{x}) = \sum_{p=1}^{P} \lambda_p f_p(\mathbf{x}) \tag{4}$$

of the total variations permitted to $\psi(\mathbf{x})$, defined by the basis set of functions $f_p(\mathbf{x})$ and the coefficients λ_p. If the contributions

$$C_{ij} = \sum_{p,q} f_p(x_i)\,\langle \lambda_p\lambda_q \rangle\, f_q(x_j)$$

from this variation to the total covariance of the observations are much larger than the contribution of U_{ij} due to the noise, it can be shown after some manipulation that (2) and (3) become approximately

$$\alpha_{\mathbf{x}i} = \sum_{p,q} f_p(\mathbf{x})(R^{-1})_{pq} \left[\sum_j f_q(x_j)\,(U^{-1})_{ji} \right] \tag{5}$$

and

$$\langle [\phi(\mathbf{x}) - \hat{\phi}_{\mathbf{x}}][\phi(\mathbf{y}) - \hat{\phi}_{\mathbf{y}}] \rangle = \sum_{p,q} f_p(\mathbf{x})\,(R^{-1})_{pq}\, f_q(\mathbf{y}) \tag{6}$$

where R is the $P \times P$ matrix with elements

$$R_{pq} = \sum_{i,j} f_p(x_i)(U^{-1})_{ij} f_q(x_j) \tag{7}$$

These expressions are identical to those derived by optimal generalized least squares fit (see section 5). They are a special case of optimal interpolation when the uncertainty U_{ij} due to noise is much smaller than that C_{ij} due to the variability of the parameters $\{\lambda_p\}$.

Question 3 frequently arises when the observational array is intended to detect propagating waves, as, for example, in the Large Aperture Seismic Array [*Green et al.*, 1965], designed to pinpoint earthquakes and underground nuclear explosions. Records from each of the individual sensors, lasting a time T many wave periods long, permit decomposition by frequency according to standard techniques of discrete Fourier transforms; however, inference of the spatial structure in each frequency band requires a different approach. As was pointed out by *Capon* [1969], this problem is equivalent to question 2 with $\phi(\mathbf{x})$ replaced by the complex amplitude $\phi_\omega(\mathbf{x})$ of the Fourier component with frequency ω and $C_{\mathbf{xy}}$ by the Hermitian form $\langle \phi_\omega(\mathbf{x})\phi_\omega{}^*(\mathbf{y})\rangle$, where the asterisk denotes complex conjugation. Amplitudes for frequencies differing by $\Delta\omega = 2\pi/T$ are independent; hence averaging over M neighboring frequencies is equivalent to sampling over M realizations of $\phi(\mathbf{x})$, provided that the mean structure does not vary significantly with frequency within this band.

3. Other Approaches to Spectral Estimation

The simplest case for question 2 arises when the statistics are believed to be spatially stationary (i.e., independent of $\mathbf{x}$). We will concentrate on this case, although the governing principles can be applied more generally. Here the mean A can be estimated with relatively high precision and subtracted at the outset [*Bretherton et al.*, 1976, section 2.3], and the covariance function $C_{\mathbf{xy}}$ is a function of $\mathbf{x} - \mathbf{y}$ alone:

$$C_{\mathbf{xy}} = \int d\boldsymbol{\kappa} \cos [\boldsymbol{\kappa} \cdot (\mathbf{x} - \mathbf{y})]E(\boldsymbol{\kappa}) \tag{8}$$

where the spectral density $E(\boldsymbol{\kappa})$ is nonnegative. The problem, then, is to estimate $E(\boldsymbol{\kappa})$ from M independent realizations of data values $\{\phi_i\}$ at the N observation points $\{\mathbf{x}_i\}$. M may be quite large or as small as unity, but it is finite. Specification of $E(\boldsymbol{\kappa})$ is equivalent to specification of $C_{\mathbf{xy}}$ from (8) and, for a stationary Gaussian process, of all the statistical properties of $\phi(\mathbf{x})$ (when A is also known). Of course, the stationarity of the process should not be taken for granted, and to the extent that the amount of data reasonably allows it, an important part of the analysis procedure should be to test for inconsistencies with this assumption.

Previous investigators have used a wide variety of techniques to estimate spectra from irregularly spaced observations. Ultimately, they all are based upon raw estimates of the pairwise covariances

$$(C^M)_{ij} \equiv \langle(\phi_i - \langle\phi_i\rangle)(\phi_j - \langle\phi_j\rangle)\rangle^M \tag{9}$$

averaged over the M realizations, with an estimate for the mean $\langle\phi_i\rangle$ subtracted. Given a very large ensemble of samples, all drawn from the same statistical population, the ensemble average of $(C^M)_{ij}$ should equal the true covariance $C_{\mathbf{x}_i\mathbf{x}_j}$ appropriate to the separation $|\mathbf{x}_i - \mathbf{x}_j|$ between the pair (i, j). However, in any one data sample, which is all the investigator has available, the raw estimates include a random sampling error, or noise, which severely limits the inferences which can be drawn from them.

According to *Davis and Regier* [1977] there are three basic types of methods for estimating spectra from a given array: a priori, data adaptive, and model fitting. A thesis of this paper is that although philosophically these methods may look very different, in the circumstances of most interest there is much in common between them, and they can, and should, be subjected to quantitative intercomparison.

a. A Priori Method

In the a priori approach, the spectral estimate $\mathscr{E}(\boldsymbol{\kappa})$ is a linear combination of the raw covariances

$$\mathscr{E}(\boldsymbol{\kappa}) = \sum_{i,j} \alpha_{ij}(\boldsymbol{\kappa})(C^M)_{ij} \tag{10}$$

where the α_{ij} are chosen independently of the observations or any assumptions about the nature of the process being measured. If the number of realizations M is large or if many widely separated pairs (i, j) are averaged together where the associated lag distances $|\mathbf{x}_i - \mathbf{x}_j|$ are similar, then $\mathscr{E}(\boldsymbol{\kappa})$ approximates with fluctuations an ensemble mean value $\bar{\mathscr{E}}(\boldsymbol{\kappa})$, related to the true spectrum $E(\boldsymbol{\kappa})$ by

$$\bar{\mathscr{E}}(\boldsymbol{\kappa}) = \int d\boldsymbol{\kappa}' \; W(\boldsymbol{\kappa}, \boldsymbol{\kappa}')E(\boldsymbol{\kappa}') \tag{11}$$

where

$$W(\boldsymbol{\kappa}, \boldsymbol{\kappa}') = \sum_{i,j} \alpha_{ij}(\boldsymbol{\kappa}) \cos [\boldsymbol{\kappa}' \cdot (\mathbf{x}_i - \mathbf{x}_j)] \tag{12}$$

is a 'window function' that determines the inherent resolution, aliasing, and bias of the spectral estimate. There are various ways of selecting the coefficients $\alpha_{ij}(\boldsymbol{\kappa})$, leading to different window functions and different sensitivities of $\mathscr{E} - \bar{\mathscr{E}}$ to sampling fluctuations in $(C^M)_{ij}$. For example, they may be derived from the output of a linear filter acting on the process $\phi(\mathbf{x})$, so that they have the form

$$\alpha_{ij} = \{\beta_i \exp [(-1)^{1/2} \boldsymbol{\kappa} \cdot \mathbf{x}_i]\} \{\beta_j \exp [(-1)^{1/2} \boldsymbol{\kappa} \cdot \mathbf{x}_j]\}^* \tag{13}$$

[*Thompson*, 1971], or they may be chosen to make $W(\boldsymbol{\kappa}, \boldsymbol{\kappa}')$ resemble as closely as possible for each $\boldsymbol{\kappa}$ a delta function with a narrow peak around $\boldsymbol{\kappa}' = \boldsymbol{\kappa}$ and minimal side lobes [*Davis and Regier*, 1977]. Only superficially different is the interpolation method, upon which are based most estimates of the kinetic energy spectrum in the atmosphere [e.g., *Saltzman and Fletcher*, 1962]. In this technique, $\phi(\mathbf{x})$ is interpolated between data points, the resulting continuous function is Fourier-transformed, and the power in the Fourier coefficients is averaged over neighboring wave number bands. If, as is nearly always the case, the interpolated values are linearly dependent upon the ϕ_i (though not necessarily optimally in the sense of section 2) and linear procedures are used for smoothing the spectrum, the resulting estimate has the form of (10) for some $\alpha_{ij}(\boldsymbol{\kappa})$. Another widely used approach is the 'bin method' [e.g., *Julian and Cline*, 1974], in which raw covariance estimates for pairs with similar spacing $|\mathbf{x}_i - \mathbf{x}_j|$ are first composited to reduce scatter, then a smooth curve $\hat{C}(\xi)$ is fitted, and finally $\hat{C}(\xi)$ is Fourier-transformed to yield $\mathscr{E}(\boldsymbol{\kappa})$. In all of these examples, a definite prescription is given or can be inferred for $\alpha_{ij}(\boldsymbol{\kappa})$, and the outcome of the analysis is $\mathscr{E}(\boldsymbol{\kappa})$, which is a noisy linear functional of the true spectrum $E(\boldsymbol{\kappa})$. Inversion of the equation

$$\mathscr{E}(\boldsymbol{\kappa}) = \int W(\boldsymbol{\kappa}, \boldsymbol{\kappa}')\hat{E}(\boldsymbol{\kappa}') \, d\boldsymbol{\kappa}' \tag{14}$$

to obtain an unbiased (though still noisy) estimate $\hat{E}(\boldsymbol{\kappa})$ is usually treated qualitatively or left tacit.

b. Data Adaptive Method

An example of a data adaptive scheme is the maximum likelihood method of *Capon* [1969], in which the estimate is

$$\mathscr{E}(\boldsymbol{\kappa}) = \left\{ \sum_{i,j} [\exp(-1)^{1/2}\boldsymbol{\kappa}\cdot\mathbf{x}_i] [(C^M)^{-1}]_{ij} [\exp(-1)^{1/2}\boldsymbol{\kappa}\cdot\mathbf{x}_j]^* \right\}^{-1} \quad (15)$$

involving the inverse $(C^M)^{-1}$ of the matrix of raw covariance estimates. An estimate of this type is highly nonlinear in the $(C^M)_{ij}$, but it can still be expressed in the form of (10), where the $\alpha_{ij}(\boldsymbol{\kappa})$ are themselves functions of the data being analyzed. For the rationale behind (15) the reader is referred to the original source, but for present purposes we suppose that the number of realizations M is large, so that

$$(C^M)_{ij} = C_{ij} + (C')_{ij} \quad (16)$$

where $(C')_{ij}$ is fairly small. This is usually reasonable in the frequency–wave number application where this method was devised and is probably prerequisite for successful applications. Expanding in powers of C',

$$\bar{\mathscr{E}}(\boldsymbol{\kappa}) = \left\{ \sum_{i,j} [\exp(-1)^{1/2}\boldsymbol{\kappa}\cdot\mathbf{x}_i](C^{-1})_{ij}[\exp(-1)^{1/2}\boldsymbol{\kappa}\cdot\mathbf{x}_j]^* \right\}^{-1} \quad (17)$$

$$\mathscr{E}(\boldsymbol{\kappa}) = \bar{\mathscr{E}}(\boldsymbol{\kappa}) + \bar{\mathscr{E}}(\boldsymbol{\kappa})^2 \left\{ \sum_{i,j} \sum_{i',j'} [\exp(-1)^{1/2}\boldsymbol{\kappa}\cdot\mathbf{x}_i](C^{-1})_{ii'} \cdot (C')_{i'j'}(C^{-1})_{j'j} [\exp(-1)^{1/2}\boldsymbol{\kappa}\cdot\mathbf{x}_j] \right\} + O(C'^2) \quad (18)$$

Thus to this approximation there is a definite expression $\bar{\mathscr{E}}(\boldsymbol{\kappa})$ for the ensemble mean of the estimates $\mathscr{E}(\boldsymbol{\kappa})$. Sampling fluctuations $\mathscr{E} - \bar{\mathscr{E}}$ can be written as a linear combination of those in C^M and can be analyzed analogously with the a priori methods. The same approach can be applied to other data adaptive schemes, such as that of *Davis and Regier* [1977], provided the number of data is large enough for sampling variations to be fairly small. Since $\bar{\mathscr{E}}(\boldsymbol{\kappa})$ is not in general equal to the true spectral density $E(\boldsymbol{\kappa})$ from which C_{ij} is derived, the maximum likelihood estimator is biased, but it will be seen later that after this bias has been corrected, the sampling errors are mimimal for large M (see section 9).

c. Model-Fitting Method

In the model-fitting approach, inference of $E(\boldsymbol{\kappa})$ is dealt with directly by assuming an analytic form for the spectrum and then choosing the undetermined coefficients so that the implied covariances best fit the data. A good example is the work of *Jones* [1977], in which the matrix of model covariances is written as a definite function $C_{ij}(\boldsymbol{\lambda})$ of a finite dimensional vector parameter $\boldsymbol{\lambda} = \{\lambda_1, \cdots, \lambda_P\}$. If $\phi(\mathbf{x})$ is Gaussian with zero mean, every finite subset $\{\phi_i\}$ is joint normally distributed, with a probability density for obtaining a given set of observations $\{\phi_i\}$ having the following proportionality:

$$\rho(\phi|\boldsymbol{\lambda}) = \alpha\{\det C(\boldsymbol{\lambda})\}^{-1/2} \exp\left\{-\frac{1}{2}\sum_{i,j}\phi_i[C(\boldsymbol{\lambda})^{-1}]_{ij}\phi_j\right\} \quad (19)$$

where det C is the determinant of the matrix C and α is a normalizing constant. For a sample of M independent realizations the product $\phi_i\phi_j$ in the exponent should be replaced by the raw covariance $(C^M)_{ij}$. Assuming that a priori all values of the parameter $\boldsymbol{\lambda}$ are equally likely, one uses Bayes' theorem [e.g., *Jeffreys,* 1961] to infer that given a set of observations $\{\phi\}$, the a posteriori most likely value of $\boldsymbol{\lambda}$ is that which makes $\rho(\phi|\boldsymbol{\lambda})$ a maximum. This value is then found numerically by minimizing

$$\frac{1}{2}\left\{\sum_{i,j}[C(\boldsymbol{\lambda})^{-1}]_{ij}(C^M)_{ij} + \log\det C(\boldsymbol{\lambda})\right\} \quad (20)$$

This is, of course, a nonlinear procedure; however, when M is large enough or there are sufficient observation points $\{\mathbf{x}_i\}$ in relation to the dimensionality P of $\boldsymbol{\lambda}$, it too can be analyzed in terms of small sampling fluctuations $\boldsymbol{\lambda}'$ about a mean $\langle\boldsymbol{\lambda}\rangle$ determined by replacing $(C^M)_{ij}$ by C_{ij} in (20). It can be shown that, ignoring terms $O(C'^2)$, the pth component of $\boldsymbol{\lambda}'$ satisfies

$$\lambda_p' = \sum_q (R^{-1})_{pq} \sum_{i,j}\sum_{i',j'} \frac{1}{2}\frac{\partial C_{ij}}{\partial\lambda_q}(C^{-1})_{ii'}(C^{-1})_{jj'}(C')_{i'j'} \quad (21)$$

where R^{-1} is the inverse of the $P \times P$ matrix

$$R_{pq} \equiv \sum_{i,j}\sum_{i',j'}\frac{1}{2}\frac{\partial C_{ij}}{\partial\lambda_p}(C^{-1})_{ii'}(C^{-1})_{jj'}\frac{\partial C_{i'j'}}{\partial\lambda_q} \quad (22)$$

The derivatives $\partial C/\partial\boldsymbol{\lambda}$ and the inverse C^{-1} are here evaluated at $\boldsymbol{\lambda} = \langle\boldsymbol{\lambda}\rangle$. This expression may be complicated, as well as computationally cumbersome for large numbers of data, but the important fact is that there exists a linear relation between fluctuations in the raw covariances and fluctuations in the estimates, with coefficients which can in principle be evaluated explicitly.

The model-fitting approach has been criticized on the grounds that the outcome is largely determined by the initial assumptions about the form of the spectrum and that the true spectrum is almost certainly not of this form. We believe that such criticisms are not necessarily valid, provided that a proper balance is maintained between the possibilities permitted within the model family and the information available from the data. Some sort of model fitting, implicit or explicit, is inherent in every approach to spectral estimation, being required for the inversion of (14) or its equivalent to obtain an unbiased estimate. An explicit discussion of this inversion and its interpretation is highly desirable. We will return to this issue in section 9.

4. SPECTRAL ESTIMATION FOR UNIFORMLY SPACED ARRAYS

Before developing further the model-fitting approach, consider the well-known situation of uniformly spaced arrays. For simplicity, assume that space is one dimensional. The observation points are at

$$x_j = (j-1)\,\Delta x \qquad j = 1, \cdots, N \quad (23)$$

An accepted technique [e.g., *Cooley and Tukey,* 1965] is to treat each realization separately, first subtracting from $\{\phi_j\}$ an estimated mean and linear trend, leaving ϕ_j for j near 1 roughly continuous with ϕ_j for j near N. Next perform a discrete Fourier transform

$$a_k = \frac{1}{N}\sum_{j=1}^{N}\left\{\exp\left[-(-1)^{1/2}\frac{2\pi}{N}jk\right]\right\}\phi_j \quad (24)$$

to obtain Fourier amplitudes a_k associated with each harmonic, $k = 0, \pm1, \pm2, \cdots, \pm N/2$, of the array length $N\Delta x$ (note

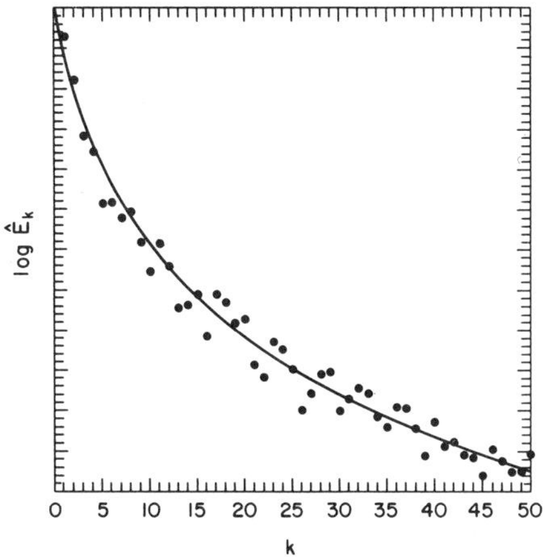

Fig. 1. Schematic plot of individual estimates of log $\hat{E}_k$ for $k = 0, 1, \cdots, 50$ from a 100-point, equally spaced array. No adjacent band averaging has been applied (i.e., $m = 1$). A smooth curve has been drawn through the scattered estimates as discussed in the text.

that the corresponding radian wave numbers are $\kappa = 0, \pm 2\pi/N\Delta x, \cdots, \pm\pi/\Delta x$). For each harmonic a raw estimate of the spectral density is provided by

$$e_k = \frac{N\Delta x}{\pi} |a_k|^2 \tag{25}$$

except for $k = 0$ or $N/2$, where e_k is half this value. To reduce scatter, e_k is usually averaged both over M realizations and in neighboring harmonics to yield

$$\hat{E}_k = \frac{1}{m} \sum_{n=-(m-1)/2}^{(m-1)/2} \langle \{|e_{k+n}|^2 + |e_{-k+n}|^2\} \rangle^M \tag{26}$$

for k between $(m - 1)/2$ and $(N/2) - [(m - 1)/2]$. These estimates are then plotted as in Figure 1, with log $\hat{E}_k$ versus some measure of k, and a smooth curve is fitted to the estimates out to the Nyquist wave number, $k = N/2$ or $\kappa = \pi/\Delta x$.

Plotting the logarithm of $\hat{E}$, rather than $\hat{E}$ itself or some other function, is not incidental. Because the data $\{\phi_i\}$ differ from realization to realization, the estimates $\hat{E}_k$ are subject to sampling fluctuations. However, for a Gaussian process it is well known that the fluctuations in nonoverlapping wave number bands are independent, with an ensemble mean equal to the true value

$$\langle \hat{E}_k \rangle = E_k \tag{27}$$

and a variance

$$\langle \{\hat{E}_k'\}^2 \rangle = \frac{1}{Mm} E_k^2 \tag{28}$$

Thus when the product Mm is reasonably large, so that the square of the relative fluctuations $(\hat{E}'/E)$ may be neglected in the expression for log $\hat{E}$, we obtain

$$\langle \log \hat{E}_k \rangle \sim \log E_k + O[(\hat{E}'/E)^2] \tag{29}$$

and

$$\langle \{(\log \hat{E}_k)'\}^2 \rangle \sim \frac{1}{Mm} \{1 + O[(\hat{E}'/E)^2]\} \tag{30}$$

The logarithmic plot thus ensures that the sampling errors have the same statistical magnitude uniformly for all wave numbers, while at each k the mean of an ensemble of estimated values is approximately equal to the true value. A smooth curve fitted to such data by simple least squares is then the most reliable final estimate that can be obtained; it is unbiased and has minimal uncertainty. Since an approximation to a least squares fit is readily performed by eye, this second, formal step of assuming a functional form for log $E(\kappa)$ and explicitly minimizing the sum of squared deviations is often omitted, though quantitative applications of the inferred spectrum are almost always in terms of an analytic model.

Even this standard procedure is not devoid of traps for the unwary investigator. The ensemble mean estimate of $\hat{E}(\kappa)$ is only equal to the true spectral density at that wave number if the array spacing Δx is sufficiently small. Otherwise power at other wave numbers related to κ by the formula $l\pi/\Delta x \pm \kappa$, for any integer l, is also included in $\hat{E}(\kappa)$; this is an irrecoverable aliasing or confounding of the spectral density at widely different wave numbers which cannot be disentangled from observations at the given array points by any analysis procedure, however ingenious. If the step of subtracting a linear trend is omitted, the fact that the array is only of finite length can lead to serious errors, with power on very large scales being redistributed through all wave numbers with a spectrum proportional to κ^{-2}. However, the most important issue is trade-off between precision and resolution, reflected in (30), where the uncertainty in each wave number band varies as $1/\mathrm{m}^{1/2}$, whereas the width of independent bands is proportional to m. Fitting a smooth curve effectively performs additional averaging, reducing the uncertainty but increasing the wave number width of features which can be distinguished. This trade-off is a fundamental consequence of the finite data sample and must always be borne in mind when interpreting the fitted spectrum and its discrepancies with theories of the true spectrum $E(\kappa)$.

5. GENERALIZED LEAST SQUARES FITS

Central to the easy interpretation of the plot in Figure 1 are the lack of bias and the independence of the estimates for different wave numbers. With irregularly spaced observations it is in general impossible to process the data so as to preserve the latter aspect; nevertheless, given an unbiased estimator, the covariances between the sampling uncertainties can still be calculated and allowed for in the second-stage fitting of a smooth curve to the plotted points, using generalized rather than simple least squares. Since the properties of generalized least squares fits are crucial in the sequel, we will elaborate somewhat, although the following equations are in essence applications of the approach to optimal estimation discussed in section 2 and discussed more fundamentally by *Liebelt* [1967].

Suppose data values y_i at points x_i have the form

$$y_i = \sum_{p=1} \lambda_p f_p(x_i) + \epsilon_i \tag{31}$$

where the functions $f_p(x)$ are known but the parameters $\boldsymbol{\lambda} = \{\lambda_p\}$, $p = 1, \cdots, P$, are to be determined. Here ϵ_i is a random noise which is independent of the λ_p, with mean zero and the covariance

$$\langle \epsilon_i \epsilon_j \rangle \equiv U_{ij} \tag{32}$$

A generalized least squares fit for the vector $\boldsymbol{\lambda}$ is that which minimizes the quadratic form

$$W(\boldsymbol{\lambda}) \equiv \frac{1}{2} \sum_{i,j} \left[y_i - \sum_p \lambda_p f_p(x_i) \right] W_{ij} \left[y_j - \sum_q \lambda_q f_q(x_j) \right] \tag{33}$$

where the W_{ij} are a set of weights forming a positive definite matrix. This minimum occurs when λ satisfies the P linear equations

$$\sum_q R_{pq}\hat{\lambda}_q = \sum_{i,j} f_p(x_i)W_{ij}y_j \quad (34)$$

where

$$R_{pq} = \sum_{i,j} f_p(x_i)W_{ij}f_q(x_j) \quad (35)$$

The matrix R is positive definite, since W is, and hence can be inverted. This yields an estimate $\hat{\lambda}$, which is a definite linear combination of the data $\{y_i\}$, and a corresponding fitted curve

$$\hat{y}(x) = \sum_{p=1}^{P} \hat{\lambda}_p f_p(x) \quad (36)$$

Because of the noise $\{\epsilon_i\}$ in the data this is only an estimate of the true curve $y(x)$, but $\hat{y}(x)$ is a random variable with calculable statistical properties.

The most important of these is that the estimator is intrinsically unbiased as a consequence of its linearity and the zero noise mean: if the true curve $y(x)$ does indeed have the form (36), then the average of $\hat{y}(x)$, over many samples of data which differ only in the noise $\{\epsilon_i\}$, is equal to the true value

$$\langle \hat{y}(x) \rangle = y(x) \quad (37)$$

A second important property is its efficiency; that is, the random uncertainty in the outcome should be as small as possible. This requires a particular choice of weights W_{ij}, specifically

$$W_{ij} = (U^{-1})_{ij} \quad (38)$$

where U^{-1} is the inverse of the noise covariance matrix (32). In this case the $P \times P$ fitting matrix (35) becomes

$$R_{pq} = \sum_{i,j} f_p(x_i)(U^{-1})_{ij} f_q(x_j) \quad (39)$$

and the residual uncertainties in the fitting parameters are

$$\langle(\hat{\lambda}_p - \langle\hat{\lambda}_p\rangle)(\hat{\lambda}_q - \langle\hat{\lambda}_q\rangle)\rangle = (R^{-1})_{pq} \quad (40)$$

which implies that the mean square fitting error at point x is

$$\langle[\hat{y}(x) - y(x)]^2\rangle = \sum_{p,q} f_p(x)(R^{-1})_{pq}f_q(x) \quad (41)$$

This is not only the least quadratic fitting error which can be achieved in the generalized least squares formalism (i.e., minimizing $W(\lambda)$ from (33)), but it is also the least error for any unbiased fit of the form (36) where the $\hat{\lambda}$ depend linearly upon the data $\{y_i\}$. The smaller the noise, the larger is the magnitude of the matrix elements in R, and the smaller is the residual uncertainty in the estimate $\hat{\lambda}$. We will call such a choice of weights optimal. Note that (34), (39), and (41) are equivalent to the previously derived equations (5)–(7). It should also be noted that in the case where the noise at different data points is uncorrelated with spatially uniform variance,

$$U_{ij} = U\delta_{ij} \quad (42)$$

the preceding equations reduce to the expressions for simple least squares.

The last result that we will need concerns the effect of applying two fitting procedures in tandem, using the output estimate from the first fitting as input data to the second, in which a smaller number of variables are to be estimated. If the functional forms $g_q(x)$ used in the second stage are linear combinations of the $f_p(x)$ used in the first and if optimum weights are used throughout, the final outcome (i.e., the estimates for the coefficients of the $g_q(x)$) is identical to what it would have been had the $g_q(x)$ been fitted to the original data in one operation. Thus there is no loss of information in breaking optimal fitting into two or more stages, which greatly increases its utility as a way of organizing data and selecting the final model functions $g_q(x)$ prior to drawing final conclusions.

6. AN EXAMPLE: THE FIRST-STAGE FIT

The approach to estimating covariances will now be illustrated by an example. The analysis has several stages. The first is to organize the observations to provide an assessment of their information content and a first rough estimate of the spectrum. The second stage is to refine the domain and resolution of this estimate, compromising between the desire for precision and the limited information available. Interspersed are a number of cross checks as to whether the data are significantly inconsistent with key assumptions made in the analysis.

a. *Spectral Representation*

We assume that $\phi(x)$ represents a real random process in one dimension which is believed to be stationary with zero mean. On the basis of previous experience an investigator feels that a reasonable first guess for the spectrum would be $\tilde{E}(\kappa)$ proportional to κ^{-2}, except perhaps as $\kappa \to 0$, so that $d\phi/dx$ is white noise. Substantial interest lies in whether the spectrum is in fact a power law and, if so, whether -2 is the correct power. The array available for measurements is shown in Figure 2. It has $N = 20$ observation points, a total length of $\Delta x_{\max} = 28.8$ units, and a minimum spacing of $\Delta x_{\min} = 0.16$. About one third of the points are clustered in a quasi-uniform array at about this minimum spacing; the remainder are much more widely spread. An explanation of why an array with these properties was chosen is deferred to section 8; here we simply remark that for the questions of interest this is an appropriate array. The data set consists of values from $M = 25$ independent realizations, one set of which is plotted in Figure 2 with linear interpolation between data values to improve the clarity of the display. Linear interpolation is an optimal interpolation (in the sense of section 2) for processes with spectra proportional to κ^{-2}. In fact, these data were generated in the computer from a known spectrum using a pseudo random number generator, but this information is only used to compare with the final estimate.

To analyze this situation, we begin by choosing a family of model spectra which has greater flexibility and more undetermined parameters than are likely to be warranted by the number of data available, leaving to a second stage the fitting of a smooth curve with the appropriate number of degrees of freedom which will be the final estimate. The model family has the form

$$E(\kappa; \nu) = \tilde{E}(\kappa)\left\{1 + \sum_{s=1}^{S} \nu_s F^s(\kappa)\right\} \quad (43)$$

The features $F^s(\kappa)$ are of fixed shape and amplitude, but the S coefficients ν_s are to be determined. As the vector $\nu = \{\nu_s\}$ varies, it generates a linear space of possible relative deviations

$[E(\kappa;\, \nu) - \tilde{E}(\kappa)]/\tilde{E}(\kappa)$ of the model spectrum from the first guess. In principle, if S is large enough, $E(\kappa;\, \nu)$ can be quite arbitrary, but as a practical matter we need a reasonable, finite basis representation of the possibilities under review. These should include not only those spectral features in which we are interested but also those which must be guarded against (because, if present in unexpected amplitude, they would give rise to misleading conclusions if ignored). We adopt a representation which is local in wave number: each basis element $F^s(\kappa)$ is nonzero only in a narrow band of wave numbers centered on some κ_s but is everywhere continuously differentiable in κ. Together the $F^s(\kappa)$ span the whole range of wave numbers for which $\tilde{E}(\kappa)$ is nonzero. They are normalized so that if each ν_s were equal to unity, $E(\kappa)$ would everywhere equal $2\tilde{E}(\kappa)$; that is,

$$\sum_{s=1}^{S} F^s(\kappa) = 1 \tag{44}$$

for all κ. An important decision for the investigator is how the resolution available from S basis elements should be distributed in wave number (i.e., how the κ_s should be distributed). Because of the underlying interest in spectra which are powers of κ, as well as the wide range of spatial intervals in the array, we opt over a central range of wave numbers for uniform resolution in the logarithm of κ rather than κ itself, with reduced resolution outside the central range.

Note that the total number of observations is only $N \cdot M = 500$. If ideally distributed, they might support estimates in, say, five wave number bands with 100 degrees of freedom (note that were these 500 observations from a uniformly spaced array, with adjacent band averaging as in section 4 to produce five estimates, then from (28) the precision in each would be ±14%). Accordingly, we set the initial resolution (i.e., number of quantities to be estimated) intentionally too high by the choice $S = 31$. We also note that the array is incapable of resolving wave numbers much larger than $\pi/\Delta x_{\min} = 19$ or much smaller than $\pi/\Delta x_{\max} = 0.11$, and we set the central range for logarithmic resolution as $(\kappa_{\min}, \kappa_{\max}) = (0.05, 45)$. These parameters are deliberately extended beyond the array capabilities to allow insights into possible aliasing and into trade-offs between precision and resolution. To avoid computational problems, $\tilde{E}(\kappa)$ is made finite at the origin and band limited (with a large band limit κ_*),

$$\begin{aligned} \tilde{E}(\kappa) &= \frac{2\kappa_0}{\pi(\kappa_0{}^2 + \kappa^2)} \qquad \kappa \le \kappa_* \\ \tilde{E}(\kappa) &= 0 \qquad \kappa > \kappa_* \end{aligned} \tag{45}$$

where $\kappa_0 = 0.04$, $\kappa_* = 1200$, and the normalization in (45) is such that

$$\int_0^\infty \tilde{E}(\kappa)\, d\kappa = 1 \tag{46}$$

(i.e., in the first-guess spectrum, $\phi(x)$ is assumed to have unit variance).

The features $F^s(\kappa)$ are defined in terms of an auxiliary function $\sigma(\kappa)$, the attention function, which increases monotonically from 0 to S while κ varies from 0 to κ_* (or infinity), the derivative $d\sigma/d\kappa$ at each κ being proportional to the desired resolution.

In this case,

$$\begin{aligned} \sigma(\kappa) &= S w_0 \cdot [\kappa/\kappa_{\min}] \qquad 0 \le \kappa \le \kappa_{\min} \\ \sigma(\kappa) &= S w_0 \cdot [1 + \log(\kappa/\kappa_{\min})] \qquad \kappa_{\min} \le \kappa \le \kappa_{\max} \\ \sigma(\kappa) &= S w_0 \cdot [2 + \log(\kappa_{\max}/\kappa_{\min}) - \kappa_{\max}/\kappa] \qquad \kappa_{\max} \le \kappa \le \kappa_* \\ \sigma(\kappa) &= S \qquad \kappa \ge \kappa_* \end{aligned} \tag{47}$$

where

$$w_0 = 1/[2 + \log(\kappa_{\max}/\kappa_{\min}) - \kappa_{\max}/\kappa_*] \tag{48}$$

which has the numerical value of 0.11. Figure 3 is a plot of $\sigma(\kappa)$ and its derivative as functions of $\log \kappa$. For each integer $s = 1, \cdots, S$ the feature $F^s(\kappa)$ is a B spline of order 2, centered at $\kappa = \kappa_s$, where $\sigma(\kappa_s) = s - \frac{1}{2}$. Regarded as a function of $\Sigma = \sigma(\kappa) - s + \frac{1}{2}$, the B spline is defined by

$$\begin{aligned} B(\Sigma) &= \tfrac{3}{4} - \Sigma^2 \qquad |\Sigma| \le \tfrac{1}{2} \qquad \sigma \epsilon [0, S] \\ B(\Sigma) &= \tfrac{1}{2}\Sigma^2 - \tfrac{3}{2}|\Sigma| + \tfrac{9}{8} \qquad \tfrac{1}{2} \le |\Sigma| \le \tfrac{3}{2} \qquad \sigma \epsilon [0, 2] \\ B(\Sigma) &= 0 \qquad \text{otherwise} \end{aligned} \tag{49}$$

for $s = 2, \cdots, S - 1$. In order to preserve the property (44), alternative forms for $B(\Sigma)$ must be used for $s = 1$ and S: For $s = 1$,

$$\begin{aligned} B(\Sigma) &= -\tfrac{1}{2}\Sigma^2 - \tfrac{1}{2}\Sigma + \tfrac{7}{8} \qquad -\tfrac{1}{2} \le \Sigma \le \tfrac{1}{2} \\ B(\Sigma) &= \tfrac{1}{2}\Sigma^2 - \tfrac{3}{2}\Sigma + \tfrac{9}{8} \qquad \tfrac{1}{2} \le \Sigma \le \tfrac{3}{2} \\ B(\Sigma) &= 0 \qquad \Sigma > \tfrac{3}{2} \end{aligned}$$

and for $s = S$,

$$\begin{aligned} B(\Sigma) &= 0 \qquad \Sigma < -\tfrac{3}{2} \\ B(\Sigma) &= \tfrac{1}{2}\Sigma^2 + \tfrac{3}{2}\Sigma + \tfrac{9}{8} \qquad -\tfrac{3}{2} \le \Sigma \le -\tfrac{1}{2} \\ B(\Sigma) &= -\tfrac{1}{2}\Sigma^2 + \tfrac{1}{2}\Sigma + \tfrac{7}{8} \qquad -\tfrac{1}{2} \le \Sigma \le \tfrac{1}{2} \end{aligned} \tag{50}$$

Figure 3*b* shows the $F^s(\kappa)$ as functions of $\sigma(\kappa)$. Note that these features are centered at half-integer values of σ. A nonuniform κ axis is also drawn as an abscissa in Figure 3*b*. Table 1 lists the values of the κ_s about which the features are centered.

The precise functional forms of the attention function and basis elements are technical details, and for most purposes the only concern need be with the coefficients ν_s which measure the relative deviation of $E(\kappa;\, \nu)$ from $\tilde{E}(\kappa)$ in the vicinity of wave number κ_s. Thus in future we will refer to the S-dimensional vector $\boldsymbol{\nu}$ as the relative spectral deviation and work only within this finite set of possibilities.

b. Fitting Procedure

According to (8) the model spectrum (43) is equivalent to a model of the covariance functions C_{xy} and, in particular, to

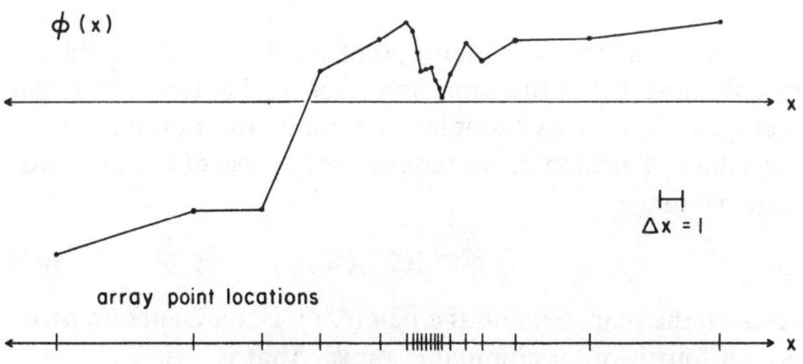

Fig. 2. A 20-point irregular array $\{x_i\}$ and a random realization of $\{\phi(x_i)\}$ with data values connected by straight lines.

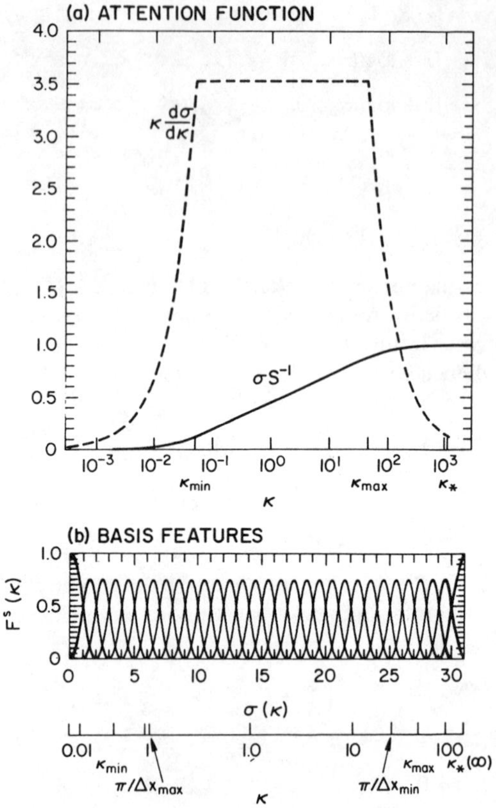

Fig. 3. (*a*) Attention function $\sigma(\kappa)$ from (47) and its derivative (note that $d\sigma/d\log\kappa = \kappa\, d\sigma/d\kappa$) versus $\log\kappa$. (*b*) Basis features $F^s(\kappa)$ versus $\sigma(\kappa)$ for all s values as defined by the B spline formulas (49) and (50). Also shown is an abscissa of the corresponding κ values.

the individual covariances between each pair of observation points. Thus for each $i, j = 1, \cdots, N$,

$$C_{ij}(\nu) = \tilde{C}_{ij} + \sum_{s=1}^{S} \nu_s C_{ij}^{s} \tag{51}$$

which is a linear function of ν. The first-guess covariance $\tilde{C}$ and a few of the basis covariances C^s are plotted as a function of $\xi = |x - y|$ in Figure 4. Note that the C^s are nearly all similar in shape except for a contraction of scale and a reduction of amplitude as s increases. This is a consequence of the power law form for $\tilde{E}(\kappa)$ and the logarithmic attention function $\sigma(\kappa)$.

Suppose now that the $C_{ij}(\nu)$ are to be fitted by generalized least squares to the raw covariances $(C^M)_{ij}$ obtained by averaging the products $\phi_i\phi_j$ over the M realizations in the data sample. The data points are the N^2 pairs (i, j), for each of which

$$(C^M)_{ij} = C_{ij} + (C')_{ij} \tag{52}$$

where C_{ij} is the true covariance, supposedly a function of $|x_i - x_j|$ only, and C_{ij}' is the sampling error, which has zero mean averaged over many samples. To apply the optimal fitting procedure of section 5, we require the inverse of the error covariance matrix

$$U_{ij;i'j'} \equiv \langle (C')_{ij}(C')_{i'j'} \rangle \tag{53}$$

between the pair (i, j) and the pair (i', j'). For a Gaussian process, all fourth-order cumulants vanish; that is,

$$\langle \phi_i\phi_j\phi_{i'}\phi_{j'} \rangle = \langle \phi_i\phi_j \rangle\langle \phi_{i'}\phi_{j'} \rangle + \langle \phi_i\phi_{i'} \rangle\langle \phi_j\phi_{j'} \rangle + \langle \phi_i\phi_{j'} \rangle\langle \phi_{i'}\phi_j \rangle$$

whence it is easily shown that

$$U_{ij;i'j'} = \frac{1}{M}\{C_{ii'}C_{jj'} + C_{ij'}C_{i'j}\} \tag{54}$$

This is an $N^2 \times N^2$ matrix. If we adopt the convention that U will be multiplied only by expressions which are symmetric within the index pairs (i, j) and (i', j'), then U may be inverted to give

$$(U^{-1})_{ij;i'j'} = \frac{M}{4}\{(C^{-1})_{ii'}(C^{-1})_{jj'} + (C^{-1})_{ij'}(C^{-1})_{i'j}\}$$

$$= \frac{M}{2}\{(C^{-1})_{ii'}(C^{-1})_{jj'}\} \tag{55}$$

Substituting now into (34) and (35), with weights as in (38), we obtain the optimal fit $\hat{\nu}$ as the solution of the S linear equations

$$\sum_t R_{st}\hat{\nu}_t = \frac{M}{2}\sum_{i,j}\sum_{i',j'} C_{ij}^{s}(C^{-1})_{ii'}(C^{-1})_{jj'}[(C^M)_{i'j'} - \tilde{C}_{i'j'}] \tag{56}$$

where

$$R_{st} = \frac{M}{2}\sum_{i,j}\sum_{i',j'} C_{ij}^{s}(C^{-1})_{ii'}(C^{-1})_{jj'}C_{i'j'}^{t} \tag{57}$$

The residual uncertainty, as in (40), is here

$$\langle(\hat{\nu}_s - \langle\hat{\nu}\rangle_s)(\hat{\nu}_t - \langle\hat{\nu}\rangle_t)\rangle = (R^{-1})_{st} \tag{58}$$

The expressions (56) and (57) involve extensive manipulations of $N \times N$ matrices, where $N = 20$ for the present example. Though somewhat lengthy, their computation is quite

TABLE 1. Central Wave Numbers for the $F^s(\kappa)$

s	σ_s	κ_s
1	0.5	0.007
2	1.5	0.0212
3	2.5	0.0353
4	3.5	0.0495
5	4.5	0.0657
6	5.5	0.0871
7	6.5	0.116
8	7.5	0.153
9	8.5	0.203
10	9.5	0.270
11	10.5	0.358
12	11.5	0.475
13	12.5	0.630
14	13.5	0.836
15	14.5	1.11
16	15.5	1.47
17	16.5	1.95
18	17.5	2.59
19	18.5	3.44
20	19.5	4.56
21	20.5	6.05
22	21.5	8.03
23	22.5	10.65
24	23.5	14.13
25	24.5	18.75
26	25.5	24.88
27	26.5	33.01
28	27.5	43.80
29	28.5	60.46
30	29.5	97.49
31	30.5	251.6

For each $s = 1, \cdots, S(= 31)$ the central value of $\sigma(\kappa)$ is $\sigma_s = s - \frac{1}{2}$, and the corresponding κ_s is determined from (47).

straightforward, with one exception. The difficulty is that the estimator $\hat{\nu}$ from (56) requires prior knowledge of C, which is unavailable (it is, after all, the quantity being estimated). As an approximation we use the first guess $\tilde{C}$, replacing R by

$$\tilde{R}_{st} = \frac{M}{2} \sum_{i,j} \sum_{i',j'} C_{ij}{}^s (\tilde{C}^{-1})_{ii'} (\tilde{C}^{-1})_{jj'} C_{i'j'}{}^t \quad (59)$$

with a corresponding change in the right-hand side of (56). This means that the weights W are no longer quite optimal, though the estimator $\hat{\nu}$ is still unbiased, and the residual uncertainty is not precisely given by $\tilde{R}^{-1}$. However, in the spirit of the true spectrum being a refinement on the first guess, the error involved in this approximation should be small. A better approximation may be had by iterating, using the estimate $C(\hat{\nu})$ from the first iteration to calculate the weights for the second, and so on. For a strictly Gaussian process an analysis of this formal asymptotic expansion in powers of $1/M$ is given by *Bretherton and McWilliams* [1979, section 2.5], who show that for moderately large M it almost always converges rapidly. In view of doubts about the extent to which $\phi(x)$ is likely in practice to be strictly Gaussian, we will not here pursue this refinement, preferring a strictly linear estimator which is unbiased even if somewhat less than optimal. We will accept $\tilde{R}^{-1}$ as a good enough basis for estimates of uncertainty.

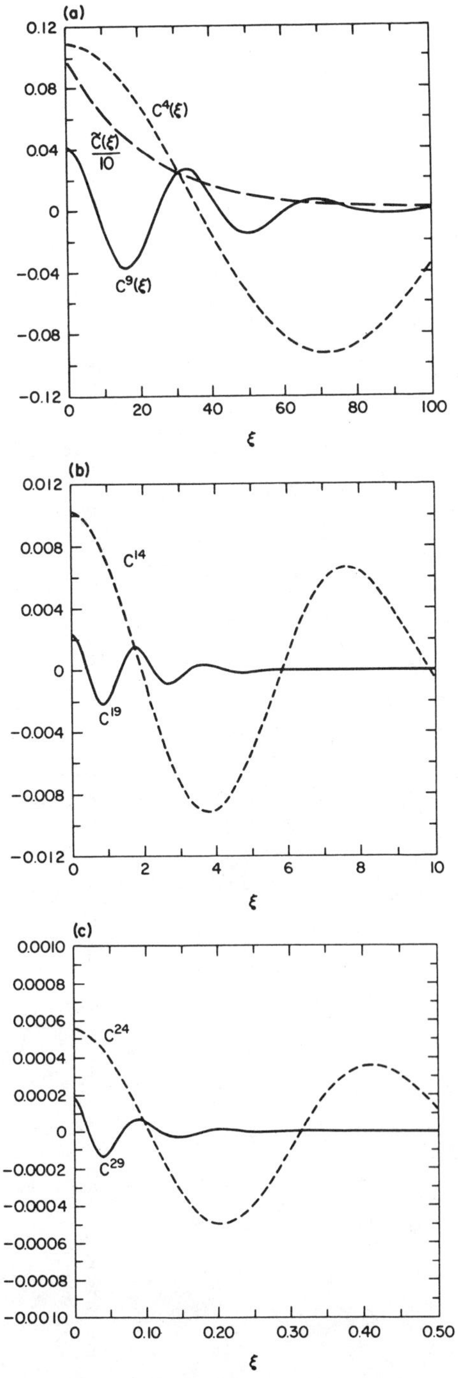

Fig. 4. Covariance functions versus lag distance $\xi = |x - y|$ for the example of section 6. Note the changes of scales between the panels.

With this approximation the matrix $\tilde{R}$ defines the fundamental information about the spectral deviation ν which is available from M realizations of $\phi(\mathbf{x})$, and it is the basic descriptor of the usefulness of the array for spectral estimation. $\tilde{R}$ is intrinsically positive definite. Any linear function of the spectral parameters,

$$\alpha = \sum_s \alpha_s \nu_s \quad (60)$$

can be estimated with an uncertainty

$$\langle \alpha'^2 \rangle = \sum_{s,t} \alpha_s (\tilde{R}^{-1})_{st} \alpha_t \quad (61)$$

For a given array, $\tilde{R}$ is proportional to the number of realizations M in the data sample, implying that, other things being equal, the uncertainty of every estimate is proportional to $M^{-1/2}$. If the array were equally spaced and long enough for end effects to be neglected, $\tilde{R}$ would be almost diagonal (in fact, tridiagonal) for the partly overlapping B spline basis features (49), corresponding to almost independent estimates for different ν_s: to the extent that $\tilde{R}$ is not diagonal, the uncertainties ν_s' are correlated for different s. The individual elements $\tilde{R}_{st}$ are all either positive or zero, and their sum is half the total number of degrees of freedom (measurements) available in the original data set,

$$\sum_{s,t} \tilde{R}_{st} = \tfrac{1}{2} NM \quad (62)$$

This result is a consequence of $\sum_s C^s = \tilde{C}$, which itself follows from (44). $\tilde{R}$ is given in full in Table 2 for the example under discussion.

However, in spite of the properties listed above, as soon as we attempt to use $\tilde{R}$ in (59), a difficulty arises. $\tilde{R}$ is so nearly singular that single-precision arithmetic on the computer fails to yield a reliable inverse. This implies that the optimal estimate $\hat{\nu}$ is so uncertain as to be quite useless.

c. *Principal Components*

To trace the origin of this difficulty, we list the S eigenvalues r_p of $\tilde{R}_{st}$ in Table 3 and plot some of the eigenvectors $\mathcal{G}_s^p$ in Figure 5. The latter we call principal components. They are independently measurable features (directions in the space of possible relative spectra ν), ranked with the best determined (those with largest r_p) first. Together they form an alternative basis set for the complete range of possibilities $\{F^s(\kappa)\}$. Their structure is thus an ordered indicator of the information available from the observational array, and an analysis in terms of them is in some sense a natural way to proceed.

If we adopt the normalization

$$\sum_s \mathcal{G}_s^p \mathcal{G}_s^q = \delta_{pq} \quad (63)$$

TABLE 2. $\tilde{R}_{st}$ for the Example of Section 6

	t																													
s	1	2	3	4	5	6	7	8	9	10	11	12	13	14	15	16	17	18	19	20	21	22	23	24	25	26	27	28	29	30	31
1	121																														
2	97	80																													
3	66	57	45																												
4	42	40	35	33																											
5	27	29	32	34	41																										
6	14	20	27	34	46	57																									
7	4	11	20	29	43	60	76																								
8	3	6	10	15	28	49	80	111																							
9	6	4	3	4	10	26	60	113	162																						
10	4	3	3	3	4	8	23	64	148	34																					
11	2	2	3	3	4	4	6	17	65	188	312																				
12	2	2	1	2	2	4	6	8	14	60	221	383																			
13	1	1	1	1	2	3	4	6	9	14	56	240	420																		
14	1	1	1	1	2	2	3	5	7	11	17	55	236	404																	
15	1	1	1	1	1	2	3	4	5	7	11	17	55	242	420																
16	0	0	0	0	1	1	2	3	5	8	14	27	50	93	242	352															
17	0	0	0	0	1	1	2	3	5	10	17	25	29	34	64	193	314														
18	0	0	0	0	0	1	1	2	4	7	11	17	21	20	26	71	205	332													
19	0	0	0	0	0	0	1	1	2	4	7	14	23	32	37	33	59	206	343												
20	0	0	0	0	0	0	0	1	2	4	7	11	15	19	20	18	19	63	219	369											
21	0	0	0	0	0	0	0	1	1	2	4	7	11	15	20	23	26	40	81	234	392										
22	0	0	0	0	0	0	0	0	1	2	3	6	9	12	15	17	17	21	26	58	232	418									
23	0	0	0	0	0	0	0	0	0	1	2	4	7	9	11	13	13	17	20	21	56	241	431								
24	0	0	0	0	0	0	0	0	0	1	2	3	5	6	8	9	10	16	22	25	25	48	216	383							
25	0	0	0	0	0	0	0	0	0	0	1	2	4	5	6	7	8	11	13	13	13	20	62	266	404						
26	0	0	0	0	0	0	0	0	0	0	1	1	2	3	4	5	6	8	12	15	25	52	107	146	116	64					
27	0	0	0	0	0	0	0	0	0	0	0	1	2	3	4	4	5	8	12	18	27	35	32	15	5	9	4				
28	0	0	0	0	0	0	0	0	0	0	0	1	1	2	3	3	4	6	9	13	19	28	36	36	28	16	3	5			
29	0	0	0	0	0	0	0	0	0	0	0	1	1	2	2	3	3	5	7	9	14	24	43	64	62	28	4	7	13		
30	0	0	0	0	0	0	0	0	0	0	0	1	1	2	2	3	3	5	8	11	16	25	38	48	43	21	3	6	10	7	
31	0	0	0	0	0	0	0	0	0	0	0	1	1	2	2	3	3	5	7	10	16	25	38	50	46	22	3	6	10	8	8

The format here is such that only the lower triangular part of this symmetric matrix is listed (i.e., $\tilde{R}_{st}$ for $t > s$ are deleted). The numbers are $100 \times \tilde{R}_{st}$, rounded to the nearest integer.

TABLE 3. Eigenvalues of $\tilde{R}_{st}$ and Estimated Coefficients of the Principal Components

p	r_p	$r_p^{1/2}\hat{\mu}_p$
1	11.5	−2.80
2	9.5	0.84
3	7.5	−4.22
4	6.8	−0.67
5	5.8	−1.90
6	5.0	−0.43
7	4.3	−1.95
8	3.4	−1.83
9	3.0	0.30
10	2.6	−0.67
11	2.1	0.48
12	1.7	−0.81
13	1.3	−0.11
14	0.95	0.63
15	0.72	−0.60
16	0.52	0.78
17	0.40	0.23
18	0.26	0.014
19	0.16	−0.73
20	0.03	−1.22
21	3×10^{-3}	0.17
22	1×10^{-4}	−1.43
23	3×10^{-5}	0.59
24	2×10^{-7}	−0.20
25	5×10^{-8}	1.73
26	1×10^{-8}	1.07
27	1×10^{-9}	−0.73
28	3×10^{-11}	0.02
29	3×10^{-11}	0.02
30	2×10^{-11}	−0.002
31	2×10^{-11}	0.02

then it follows from the well-known decomposition theorem for real symmetric matrices that

$$\tilde{R}_{st} = \sum_{p=1}^{S} r_p \mathcal{G}_s^p \mathcal{G}_t^p \tag{64}$$

and

$$(\tilde{R}^{-1})_{st} = \sum_{p=1}^{S} \frac{1}{r_p} \mathcal{G}_s^p \mathcal{G}_t^p \tag{65}$$

The optimal estimate of the coefficient associated with the pth component is

$$\hat{\mu}_p = \sum_{s=1}^{S} \mathcal{G}_s^p \hat{\nu}_s \tag{66}$$

and this coefficient has sampling uncertainty uncorrelated with that for any $q \neq q$ and has a standard deviation $1/r_p^{1/2}$. Given a particular set of observations and hence $\{\hat{\nu}_s\}$, the relative spectral deviation ν may be modeled by

$$\nu_s = \sum_{p=1}^{P} \left(\hat{\mu}_p + \frac{1}{r_p^{1/2}} \gamma_p\right) \mathcal{G}_s^p \tag{67}$$

where the γ_p are independent random variables, each normally distributed with zero mean and unit variance. Equation (67) defines an ensemble of possible deviations ν, each equally consistent with the observations. This ensemble has as mean and mode the optimal estimate $\hat{\nu}$ but scatter equal to that implied by the sampling errors in the observations, assuming that the latter are appropriate to a Gaussian process with spectrum the first guess.

Inspection of Table 3 and Figure 5 exposes the presence of very small eigenvalues which give difficulty with the inversion of (56). (In fact, r_p for p = 28–31 are so small as to be computationally suspect, as therefore are the corresponding $\mathcal{G}^p$.) Components 20–31 are primarily concerned with distinguishing wave numbers in the regions $\kappa > \pi/\Delta x_{\min} = 19$ and $\kappa < \pi/\Delta x_{\max} = 0.11$. From Figure 5 and Table 1 we see that these components have their greatest amplitude outside the range $\sigma = (6.4, 24.6)$(i.e., $s \approx (7, 25)$ or $\kappa \approx (\pi/\Delta x_{\min}, \pi/\Delta x_{\min})$), which is well sampled by the array, and such distinctions are essentially unmeasurable. Since the amplitudes of these principal components within the well-sampled range are relatively small, the fact that their coefficients are undetermined is more a procedural problem than a substantive one. Components 1–19, on the other hand, are individually fairly well measured, adding successively more detail in the sampled range. Several do extend with small amplitude outside, raising the prospect of aliasing, in the form of a misidentification of spectral deviations which really occur outside the sampled range as if they were located within, but with small magnitudes outside the problem does not appear to be too serious. However, closer inspection does reveal a further problem. From (58) and (65) these first 19 components, taken together, contribute an uncertainty to pointwise estimates of ν_s equal to

$$\sum_{p=1}^{19} \left\{\frac{1}{r_p} (\mathcal{G}_s^p)^2\right\}^{1/2}$$

which throughout the well-sampled range is approximately unity (Figure 7). This uncertainty is too large for such estimates to be of much use. Thus the resolution being sought in the basis set of features is too fine for the number M of realizations available. There are too many coefficients to be estimated for the aggregate to yield acceptable precision.

d. *Truncation*

These problems are not surprising, since the original scope of the analysis was deliberately set to exceed the estimated capacity of the array and the number of realizations. This was in order to guard against inadvertent exclusion of major features which are required by the data. No finite number of tests can eliminate all such possibilities, but one is provided by the normalized coefficients $r_p^{1/2}\hat{\mu}_p$ in Table 3. If the process ϕ is Gaussian and if the spectrum is indeed equal to the first guess, these should for each p be a sample of a normal distribution with zero mean and unit variance. Clearly, coefficients 1 and 3 are probably different from zero, but otherwise there is no compelling evidence that the true spectrum deviates from the first guess in any of the attributes permitted by the model family. (As was noted above, the values for p = 28–31 are computationally unreliable because of too small r_p.) Thus there is no inconsistency in restricting the number of principal components under consideration to, say, the first six, simply ignoring the remainder. This implies adopting a subspace of the complete set of possible modes of spectral variation as 'the measurable subspace.' Any vector ν describing such variation may be decomposed into a measurable part (its projection onto the measurable subspace) and an unmeasurable remainder, which is statistically independent (for sampling errors associated with this particular array and first-guess spectrum). The presentation of estimates and uncertainties is confined to the measurable part, though an understanding of the information content of the analysis requires also a characterization of the remainder (in particular, the criterion for dividing the total space in this way).

Following this approach and truncating at $P = 6$, Figure 6

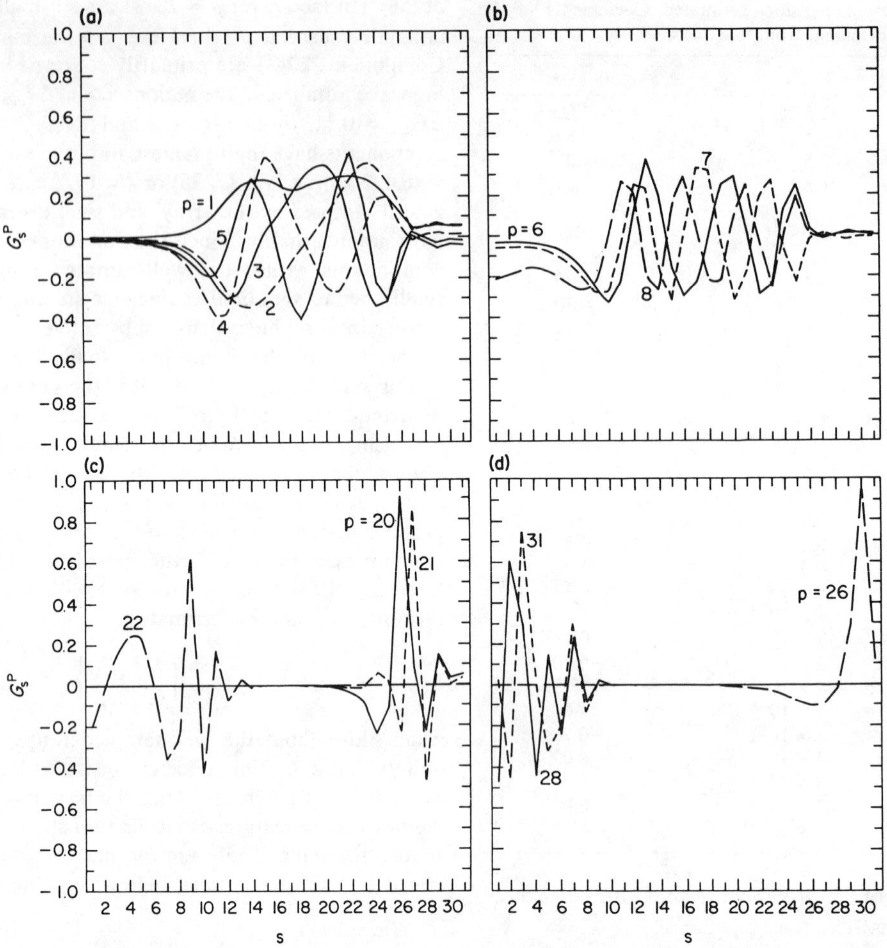

Fig. 5. Principal components $\mathcal{G}_s^p$ (eigenvectors of $\tilde{R}_{st}$) for several values of p for the example of section 6.

shows the best estimate $\hat{\nu}^+$ (solid curve), together with four alternative estimates generated from (67), each of which is equally consistent with the observations. The mean of this ensemble of alternatives is $\hat{\nu}^+$, and the scatter equals the uncertainty associated with sampling error in the observations. The pointwise standard deviations, though not the interrelations between uncertainties at different points, can be summarized in a single curve (Figure 7). The measurable subspace is best characterized by the ensemble portrayed in Figure 6 or, alternatively, by the basis set consisting of the curves for $p = 1, \cdots, 6$ in Figure 5. The unmeasurable part of any spectral deviation ν is the orthogonal complement of the measurable part, under the norm

$$\| \nu \| = \sum_{s=1}^{S} \nu_s^2 \tag{68}$$

Thus any vector $\{\nu_s'\}$ for which

$$\sum_{s=1}^{S} \nu_s' \mathcal{G}_s^p = 0 \qquad p = 1, \cdots, 6 \tag{69}$$

is completely unmeasurable. These conditions quantify the intuitive concept of 'fine structure within the well-sampled wave number range, as well as all structure outside it' being unmeasurable. Such unmeasurable aspects of the spectrum are excluded from the presentation in Figure 6.

In spite of the substantial reduction in the number of degrees of freedom associated with truncation the estimate $\hat{\nu}^+$ in Figure 6 is noisy and lacking in precision. It is, however, based on an optimal analysis, in the sense that within the measurable subspace defined by the first six principal components, the uncertainties are as small as or smaller than those from any other linear, unbiased estimation procedure which starts with the raw covariances. It is perhaps sobering that 25

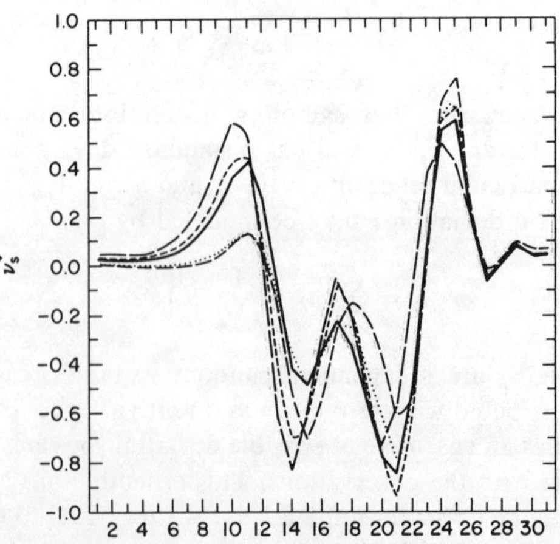

Fig. 6. Estimate of the spectral coefficients $\{\nu_s^+\}$ (solid curve) from 25 realizations on the array of Figure 2 using the truncated set of first-stage principal components. Also plotted are four alternative $\{\nu_s^+\}$ (dashed and dotted curves) generated from (67) with a truncation at $P = 6$.

realizations from 20 observation points do not really provide very much information, even when the observation points are efficiently distributed. (We will see in section 8 that this is indeed the case here.) The estimate $\hat{\nu}^+$ can be refined somewhat, as will be discussed in section 7, but the basic character cannot be significantly improved.

e. *Window Functions*

Another perspective on the characteristics of the combined measurement array and analysis procedure comes from the window function $W(\kappa, \kappa')$ defined in (11), here interpreted as the matrix of coefficients of the basis functions $F^s(\kappa)$. Thus the average of the estimates $\hat{\nu}^+$ over the ensemble of possible data samples is related to the true relative deviation ν by

$$\langle \hat{\nu}_s^+ \rangle = \sum_{t=1}^{S} W_{st}\nu_t \tag{70}$$

where

$$W_{st} = \sum_{p=1}^{P} \mathcal{G}_s^p \mathcal{G}_t^p \tag{71}$$

Several window functions for different values of s are shown in Figure 8. It should be noted that because of the truncation at $P = 6$ the estimator $\hat{\nu}^+$ is no longer completely unbiased; that is, the window function W_{st} is not precisely the Kronecker delta function δ_{st}. Instead, at points like $s = 14$ or $s = 19$ within the well-sampled range (i.e., $7 \lesssim s \lesssim 25$), $\langle \hat{\nu}_s^+ \rangle$ is sensitive to the value of ν_t in a surrounding band of finite width, with a reasonably small overshoot at greater separations. This is a reasonable, though not ideal, sensitivity characteristic for a measurement system. However, outside the well-sampled range, such as at $s = 4$ or 29, $\langle \hat{\nu}_s^+ \rangle$ bears no relationship to the local values of ν_t, the sensitivities being almost identical to those at $s = 9$ or 24, respectively, except for a substantial reduction in magnitude. This is a clear case of aliasing, and its correct identification is critical to the interpretation of an estimate $\hat{\nu}_s^+$ outside or close to the ends of the well-sampled range.

Another important interpretation can also be given to the curves in Figure 8. Because of the symmetry of the expression

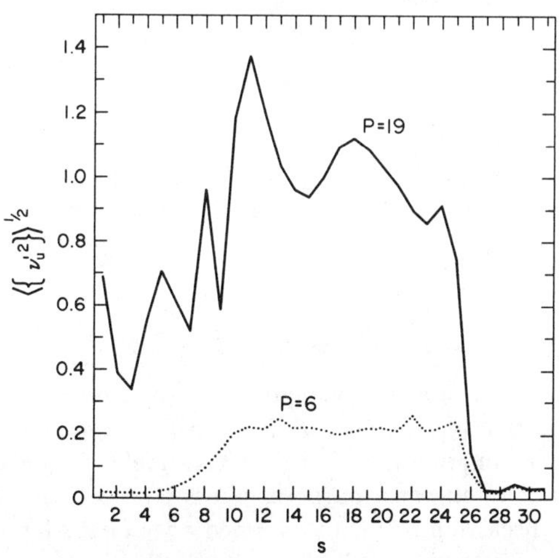

Fig. 7. Standard deviation of spectral estimates associated with truncated sets of the principal components, $\{\langle \nu_s'^2 \rangle\}^{1/2} = \sum_{p=1}^{P} (1/r_p)(\mathcal{G}_s^p)^2$ for $P = 6$ and 19.

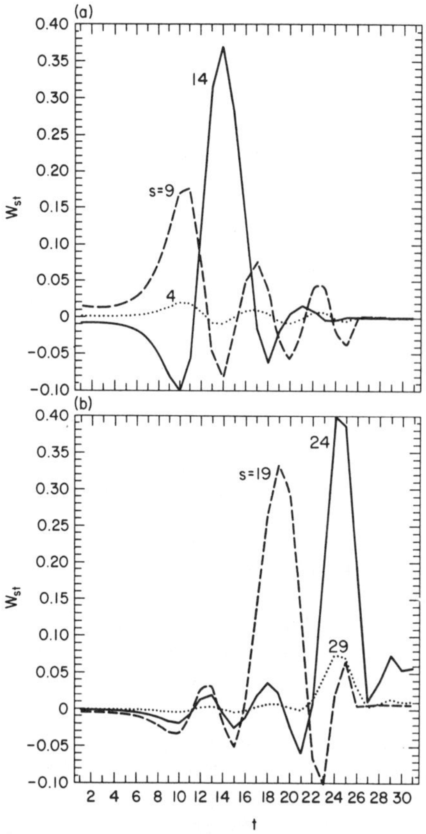

Fig. 8. Window functions for the best judgment first-stage spectrum (equation (71)).

(71) the W_{st} displays the influence on $\hat{\nu}_t^+$ of a discrete delta function at point s,

$$\nu_s = \delta_{ts}$$

Thus they indicate the degree of exposure of the estimate $\hat{\nu}^+$ to unexpectedly large power in the spectrum at both very large and very small wave numbers. For example, if the true spectrum were equal to the first guess everywhere except for the wave numbers covered by $F^t(\kappa)$ for $t = 4$, in which band it was $\bar{E}(\kappa)\{1 + F^4(\kappa)\}$, there would be a perceptible impact on the expected value of ν_s^+ throughout the range $s = 8$ to $s = 26$. However, the amplitude of this impact is everywhere less than 0.02, and a glance at Figure 7 shows that for a sample of 25 realizations it is quite undetectable within the noise. The power in this wave number band would have to be 10 times the first guess before a particular estimate $\hat{\nu}^+$ was substantially misleading in this regard.

7. THE SECOND-STAGE FIT

a. *A Priori and Measurement Norms*

Following the first-stage survey of the information available from the observational array and the crude spectral estimates obtained by arbitrarily truncating the sequence of eigenfunctions, we now consider how the procedure can be refined. The major arbitrary decisions made by the investigator in the first-stage analysis were the selections of the first-guess spectrum $\bar{E}(\kappa)$ and the attention function $\sigma(\kappa)$, the choice of $S = 31$ as the number of basis elements in the model of all possible relative spectra (i.e., the choice of resolution), and the truncation to $P = 6$ principal components. We will discuss the sensitivity of the outcome to changes in these decisions in the context of a formalism based on Bayesian inference. However, the inferential formalism should not be interpreted too literally: it allows a tuning of the analysis between questions asked and information available rather than a strict progression from inviolable, self-evident premises to a unique conclusion.

As an introduction to these ideas, consider the status of the first-stage analysis at the end of section 6*b*. The observations have been summarized by an unbiased linear operator yielding a set $\{\hat{\nu}_s\}$ of estimates of the coefficients defining the spectral power in discrete wave number bands. The sampling errors in these estimates are substantial and correlated, but the covariances between them are known: the inverse of the covariance matrix is $\tilde{R}_{st}$. Strictly speaking, the latter is only a guess, based upon $\tilde{E}$, but in the spirit of improving upon a significant amount of information available prior to the investigation we will continue to suppose for the time being that it is an adequate estimate. Also, it is the vector $\sum_t \tilde{R}_{st}\hat{\nu}_t$ that is readily computable rather than $\hat{\nu}$ itself. Yet it has become clear that to achieve useful results, further assumptions must be brought to the analysis.

An alternative to the subsequent course taken in section 6 is to postulate that certain values or directions for the spectral deviation $\hat{\nu}$ are a priori more likely than others and are to be preferred in the absence of compelling evidence to the contrary from the observations. A crude but useful statement of the investigator's prior expectations might be that deviations $\{\hat{\nu}_s\}$ are randomly distributed with mean zero and standard deviation unity, independently for each s. Thus in any wave number band, very large changes of E from $\tilde{E}$, of magnitude several times that of $\tilde{E}$ itself, are viewed as inherently unlikely. This is an expression of modest confidence in the first guess. We now regard the generalized least squares estimates $\hat{\nu}_s$ as measurements of the true values ν_s, albeit noisy ones, and use the approach to optimal interpolation described in section 2 to separate the signal from the noise as best we can. This requires knowledge of the covariance matrix between the observations, which in this case is the sum of the noise covariance $\tilde{R}^{-1}$ and the covariance matrix of the signal (the ν_s), which has just been postulated to be the unit matrix I. In addition, we need the covariance between the signal to be estimated, in this case ν_s, and the observations $\nu_t + \nu_t' = \hat{\nu}_t$ at all points t. Since the sampling noise ν' is approximately independent of ν, this is the a priori covariance between ν_s and ν_t, again simply the unit matrix. From (1) and (2) the least mean squares error estimator is therefore

$$\hat{\nu}_t^* = \sum_s [(I + \tilde{R}^{-1})^{-1}]_{ts}\hat{\nu}_s \tag{72}$$

Since

$$(I + \tilde{R}^{-1})^{-1} = (I + \tilde{R})^{-1}\tilde{R}$$

this procedure modifying $\hat{\nu}$ to $\hat{\nu}^*$ is equivalent to adding the unit matrix to $\tilde{R}$ on the left-hand side of (56). Inversion of $\tilde{R}$ itself is no longer required, and the solution of the modified equation is well defined. The principal components are unaltered, but the estimated coefficients are reduced in magnitude according to

$$\hat{\mu}_p^* = \frac{r_p}{1 + r_p}\hat{\mu}_p \tag{73}$$

For the well-measured components with $r_p \gg 1$ the effect of the reduction is negligible, but for the poorly measured ones with $r_p \ll 1$ the observations are given little weight relative to the prior confidence in the first guess. By analogy with (3) the residual error in the estimate $\hat{\nu}_t^*$ has covariance matrix

$$\langle \nu_s^{*\prime}\nu_t^{*\prime}\rangle = I - (I + \tilde{R}^{-1})^{-1} = (I + \tilde{R})^{-1} \tag{74}$$

To reproduce this in representative realizations, the denominator in (67) should be replaced by $(1 + r_p)^{1/2}$. This has the effect of strictly bounding the residual error, even for components with very small r_p, where the level of error is as if the observations were ignored completely. For the well-measured components, on the other hand, the residual error is determined by the sampling error in the observations alone, and for those components the estimator $\hat{\nu}^*$ is essentially unbiased. For the remainder there is a systematic bias toward the prior expected value $\hat{\nu} = 0$ (i.e., toward the first guess).

The qualitative effect of replacing $\hat{\nu}$ by $\hat{\nu}^*$ is clearly similar to that of truncating the sequence of principal components at the point where r_p is approximately unity. Components later in the sequence are given little, as opposed to zero, weight, whereas ones earlier are treated in an approximately, as opposed to completely, unbiased manner. The cutoff is a natural outcome of the a priori assumptions rather than an arbitrary artifact. Nonetheless, it is apparent that further modifications in the formalism are desirable. The criterion $r_p = 1$ would place the cutoff point at $P = 13$ (see Table 3), yet it was shown in section 6 that the more stringent $P = 6$ provides estimates which are still too noisy for many uses. Furthermore, the cutoff created by replacing $1/r_p$ by $1/(1 + r_p)$ is far from sharp, leading to still larger uncertainties in the estimates. More fundamentally, the postulated prior expectations, specifically

$$\tilde{\nu}_s = 0 \qquad \widetilde{\nu_s\nu_t} = \delta_{st} \tag{75}$$

are reasonable and natural only after the basic spectral model (43) has been confidently specified. Since this specification was itself somewhat arbitrarily made, a more general framework independent of any particular basis set would be preferable.

Such a framework may conveniently be derived as an application of Bayes' theorem, though it need not always be interpreted that way. It is also a device for rendering unambiguous the solution of an underdetermined inverse problem (e.g., the inversion of (14)). Suppose that the investigator is willing to say a priori that he expects derivations ν which may be modeled by

$$\nu_s = \sum_{u=1}^{S} \lambda_u g_s^u \tag{76}$$

where for each u the g_s^u are given functions of s and the $\{\lambda_u\}$ are random variables independently distributed with mean zero and variance unity. Such a model implies that

$$\tilde{\nu}_s = 0$$

$$\widetilde{\nu_s\nu_t} = \sum_u g_s^u g_t^u = T_{st} \tag{77}$$

where the tilde indicates averaging over prior possibilities before inspection of any data. T_{st} is a positive semidefinite matrix. The assumption that there are S variables $\{\lambda_u\}$ is convenient for computational purposes but is not necessary. In the previous discussion we tacitly assumed g_s^u to be the Kronecker delta function, so T was the unit matrix. Writing

$$g^u(\kappa) = \sum_s g_s^u F^s(\kappa) \tag{78}$$

(76) may also be written

$$\nu(\kappa) = \sum_s \nu_s F^s(\kappa) = \sum_u \lambda_u g^u(\kappa) \tag{79}$$

Thus the functions $g^u(\kappa)$ are a priori equiprobable, independent features defining the possible relative spectral derivations from the first guess $\tilde{E}(\kappa)$.

The vector $\boldsymbol{\nu}$ is thus joint normally distributed with an a priori probability density

$$\rho(\boldsymbol{\nu}) = \alpha \exp\left\{-\tfrac{1}{2} \sum_{s,t} \nu_s (T^{-1})_{st} \nu_t\right\}$$

which may be written symbolically as

$$\rho(\boldsymbol{\nu}) = \alpha \exp\{-\tfrac{1}{2}\boldsymbol{\nu}^T T^{-1} \boldsymbol{\nu}\} \tag{80}$$

Here $\boldsymbol{\nu}^T$ is the transpose of the vector $\boldsymbol{\nu}$, α is a normalizing constant (which need not concern us further and is in general different in each of the following equations in which it appears), and we are supposing for the moment that the inverse T^{-1} exists. If the magnitude of a particular feature $g^u(\kappa)$ is everywhere infinitesimally small, the implication is that spectral deviations with significant magnitude having that structure are extremely unlikely; that is, the a priori norm $\boldsymbol{\nu}^T T^{-1} \boldsymbol{\nu} = \boldsymbol{\lambda}^T \boldsymbol{\lambda}$ is extremely large. The analysis of the sampling errors in optimal least squares estimation shows that if the true spectral deviation were $\boldsymbol{\nu}$, the conditional probability of obtaining an estimate $\hat{\boldsymbol{\nu}}$ would be approximately

$$\rho(\hat{\boldsymbol{\nu}}|\boldsymbol{\nu}) = \alpha \exp\left\{-\tfrac{1}{2} \sum_{s,t} (\hat{\nu}_s - \nu_s)\tilde{R}_{st}(\hat{\nu}_t - \nu_t)\right\}$$

$$= \alpha \exp\{-\tfrac{1}{2}(\hat{\boldsymbol{\nu}} - \boldsymbol{\nu})^T \tilde{R}(\hat{\boldsymbol{\nu}} - \boldsymbol{\nu})\} \tag{81}$$

This supposes that the whole process $\phi(\mathbf{x})$ is Gaussian, though if the fourth-order cumulants vanish, it should be a good approximation under more general circumstances. Bayes' theorem [*Jeffreys*, 1961] asserts that the a posteriori distribution for $\boldsymbol{\nu}$ (i.e., the probability distribution given knowledge of the data $\hat{\boldsymbol{\nu}}$) is

$$\rho(\boldsymbol{\nu}|\hat{\boldsymbol{\nu}}) = \rho(\hat{\boldsymbol{\nu}}|\boldsymbol{\nu})\rho(\boldsymbol{\nu})/\rho(\hat{\boldsymbol{\nu}}) \tag{82}$$

which, since $\rho(\hat{\boldsymbol{\nu}})$ is independent of $\boldsymbol{\nu}$, gives

$$\rho(\boldsymbol{\nu}|\hat{\boldsymbol{\nu}}) = \alpha \exp\{-\tfrac{1}{2}[(\boldsymbol{\nu} - \hat{\boldsymbol{\nu}})^T \tilde{R}(\boldsymbol{\nu} - \hat{\boldsymbol{\nu}}) + \boldsymbol{\nu}^T T^{-1} \boldsymbol{\nu}]\} \tag{83}$$

The exponent in (83) is the sum of two quadratic forms, each positive definite, of which the first, the measurement norm, would give probable values for $\boldsymbol{\nu}$ clustered around the unbiased estimator $\hat{\boldsymbol{\nu}}$, and the second, the a priori norm, around the a priori mean $\hat{\boldsymbol{\nu}} = 0$. In combination the $\boldsymbol{\nu}$ dependence of the exponent may be written

$$-\tfrac{1}{2}(\boldsymbol{\nu} - \hat{\boldsymbol{\nu}}^*)^T[\tilde{R} + T^{-1}](\boldsymbol{\nu} - \hat{\boldsymbol{\nu}}^*) \tag{84}$$

where

$$\hat{\boldsymbol{\nu}}^* = [\tilde{R} + T^{-1}]^{-1}\tilde{R}\,\hat{\boldsymbol{\nu}} \tag{85}$$

Thus the modal estimate for $\boldsymbol{\nu}$ is $\hat{\boldsymbol{\nu}}^*$, which, in the event that $(T)^{-1}$ is the unit matrix I, is just as discussed in (72). The uncertainty in this estimate is determined by

$$\langle(\boldsymbol{\nu} - \hat{\boldsymbol{\nu}}^*)(\boldsymbol{\nu} - \hat{\boldsymbol{\nu}}^*)^T\rangle = [\tilde{R} + T^{-1}]^{-1} \tag{86}$$

the inverse of the sum of the matrices of the a priori and measurement norms.

These expressions are most easily interpreted and computed in terms of the underlying variables λ_u from which $\boldsymbol{\nu}$ may be derived according to (76) and (79). The $\boldsymbol{\lambda}$ are distributed a priori with mean 0 and covariance I. Given knowledge of the data, however, the distribution is

$$\rho(\boldsymbol{\lambda}|\hat{\boldsymbol{\nu}}) = \alpha \exp\left\{-\tfrac{1}{2}\left[\sum_{u,v} (\lambda_u - \hat{\lambda}_u)R_{uv}(\lambda_v - \hat{\lambda}_v) + \sum_u \lambda_u^2\right]\right\} \tag{87}$$

$$\rho(\boldsymbol{\lambda}|\hat{\boldsymbol{\nu}}) = \alpha \exp\left\{-\tfrac{1}{2} \sum_{u,v} (\lambda_u - \hat{\lambda}_u^*)(R + I)_{uv}(\lambda_v - \hat{\lambda}_v^*)\right\} \tag{88}$$

where the quantities $\hat{\boldsymbol{\lambda}}$, $\hat{\boldsymbol{\lambda}}^*$ and the matrix R_{uv} are defined in (91)–(94) below. Equation (87) shows that again the probable values of $\boldsymbol{\lambda}$ are determined by the sum of two norms, the a priori norm with matrix I and center point $\boldsymbol{\lambda} = 0$ and the measurement norm with matrix R and center point $\boldsymbol{\lambda} = \hat{\boldsymbol{\lambda}}$. Combined in (88), the optimal estimate is $\hat{\boldsymbol{\lambda}}^*$ with residual error covariance $(R + I)^{-1}$. The principal components, regarded as a basis set for the λ_u, are formed by the eigenvalues r_p and eigenvectors θ_u^p of the information matrix R, but the associated spectral deviations ν_s and $\nu_s F^s(\kappa)$ have the forms

$$\mathcal{G}_s^p = \sum_u \theta_u^p g_s^u \tag{89}$$

$$\mathcal{G}^p(\kappa) = \sum_u \theta_u^p g^u(\kappa) \tag{90}$$

The connection between $\hat{\boldsymbol{\lambda}}^*$ and $\hat{\boldsymbol{\lambda}}$ is isomorphic to that described in (72)–(74).

The information matrix R, introduced in (87), is defined by

$$R_{uv} = \sum_{s,t} g_s^u \tilde{R}_{st} g_t^v = \iint g^u(\kappa)H(\kappa, \kappa')g^v(\kappa')\,d\kappa\,d\kappa' \tag{91}$$

where

$$H(\kappa, \kappa') = \frac{M}{2} \sum_{\substack{i,j \\ i',j'}} \tilde{E}(\kappa) \cos \kappa(x_i - x_j)(\tilde{C}^{-1})_{ii'}(\tilde{C}^{-1})_{jj'} \cdot \cos \kappa'(x_{i'} - x_{j'})\tilde{E}(\kappa') \tag{92}$$

R describes the increase in information which is being adduced from the observations. According to (91), it depends on both the a priori equiprobable features $g^u(\kappa)$ and the basic matrix $H(\kappa, \kappa')$. The latter summarizes completely the information available from the observational array, in a form independent of any features, provided only the first guess is a reasonable one. It is closely approximated by the first-stage information matrix $\tilde{R}_{st}$ introduced in section 6 if the indices s and t are taken to represent nonoverlapping wave number bands which span the full wave number range. Note that $H(\kappa, \kappa')$ is intrinsically positive for all κ, κ'.

From (76) and (91) the unbiased estimator $\hat{\boldsymbol{\lambda}}$ and the optimal one $\hat{\boldsymbol{\lambda}}^*$ are defined by the equations

$$\sum_v R_{uv}\hat{\lambda}_v = \sum_{s,t} g_s^u \tilde{R}_{st}\hat{\nu}_t \tag{93}$$

and

$$\hat{\lambda}_u^* = \sum_{u',v} (R + I)_{uu'}^{-1} R_{u'v}\hat{\lambda}_v \tag{94}$$

Since some of the features $g^u(\kappa)$ may be small or even identically zero, R is only positive semidefinite, and the inversion of (93) to yield $\hat{\boldsymbol{\lambda}}$ is not always well behaved. However, combining (93) with (94), it will be seen that $\hat{\boldsymbol{\lambda}}^*$ is always completely determinate and is readily computed using standard matrix techniques. Once $\boldsymbol{\lambda}$ is known, $\nu(\kappa)$ follows from (79), and it is apparent that the only role of the basis features $\{F^s(\kappa)\}$ is to provide a finite approximation to what should be thought of as an infinite dimensional space of spectral deviations. The number of a priori nonzero features $g^u(\kappa)$, on the other hand (and hence the dimension of the matrix R_{uv}), is strictly finite.

The analysis best proceeds in terms of the principal components $\{\mathcal{G}^p(\kappa)\}$ defined by (90), which provide a basis set of features orthogonal with respect to both the a priori norm T^{-1} and the measurement norm H. They are the functions $\nu(\kappa)$ which make stationary the ratio

$$r = \boldsymbol{\nu}^T H \boldsymbol{\nu} / \boldsymbol{\nu}^T T^{-1} \boldsymbol{\nu} \tag{95}$$

They are normalized with respect to the norm

$$\boldsymbol{\nu}^T T^{-1} \boldsymbol{\nu} = \boldsymbol{\lambda}^T \boldsymbol{\lambda} \tag{96}$$

which selects from among the infinite class of conceivable spectral deviations a finite set of possible ones and sets a priori a probable range of amplitudes for each direction in this linear space. For general spectral deviations $\boldsymbol{\nu}(\kappa)$ we define

$$T(\kappa, \kappa') = \sum_u g^u(\kappa) g^u(\kappa') \tag{97}$$

as a counterpart to (77). Because of the definition of ν from $\boldsymbol{\lambda}$ according to (79) the nonexistence of T^{-1} as a general inverse of the covariance matrix T need cause no concern. Within this set of possible spectral deviations the measurement norm further restricts the range of probable values.

The eigenvalue r_p measures the amount of information available from the observations, relative to that available a priori, for the direction $\mathcal{G}^p(\kappa)$. When r_p is large in comparison to unity, this principal component is well measured; when it is small, the measurements are contributing little to the final estimate $\hat{\boldsymbol{\nu}}^*$.

b. *Tuning*

Armed with this formalism, we now reconsider the arbitrary choices made by the investigator during the first-stage analysis, 'tuning' or adjusting the functions $g^u(\kappa)$ to achieve a satisfactory balance in the final estimate between precision and resolution, with minimal contamination from wave numbers outside the well-sampled range. At stake are the number and functional forms of the features defining the measurable subspace (i.e., the principal components corresponding to eigenvalues greater than unity). Besides 'naturalness' of the principal components it is also desirable that the cutoff be sharp so that they fall clearly into two classes, those with $r_p \gg 1$ and the remainder with $r_p \ll 1$. The authors have no single objective criterion for a satisfactory result in this tuning process, and different investigators may well prefer different choices. It should be regarded as a process of compromise between the original interests and objectives of the investigator and the realities of the capability of the observational array. To preserve the integrity of the inferential process, it should be completed before any inspection of the observations themselves, indeed ideally before the array design is finally settled. The purpose of the first-stage analysis is then merely to explore whether the data are significantly inconsistent with some of the more dubious of the assumptions made in the a priori statement.

A pure Bayesian might reject such adjustment of prior expectations as inadmissible. Indeed, regarding at this stage a class of spectral deviations as inherently improbable does not mean that they are not present, just that their amplitude cannot be determined along with other, more important features and they are being excluded from primary consideration. Although computationally desirable, this exclusion may also be physically meaningless, and careful analysis is called for as to how sensitive the final outcome would be to their presence in reasonable amplitude and as to whether there is evidence of their presence in such large amplitude that exclusion would be seriously misleading.

These philosophical questions notwithstanding, we now manipulate the a priori norm to obtain the desired naturalness of the principal components. We will postpone until the end any considerations of adjusting $\tilde{E}(\kappa)$ or the mean $\langle \phi(x) \rangle$. We will also continue to define the $g^u(\kappa)$ in terms of overlapping basis elements on a revised attention function $\sigma(\kappa)$, the amplitude of such elements and the degree of overlap between them being the principal control on the resolution in the measurable subspace. This structure is of course more specialized than is required by (79), and different implementations of this approach to spectral estimation might well differ in this respect. Three manipulations of the first-stage analysis are particularly useful: redistribution, smoothing, and scaling. Redistribution is equivalent to changing the shape of the attention function $\sigma(\kappa)$. Smoothing decreases the effective resolution by increasing the interval in σ in which independent spectral deviations are expected a priori. Scaling alters the magnitude of such deviations and the position of the cutoff $r_p = 1$ which bounds the measurable subspace. The initial choices for these quantities were reasonably arbitrary. In the tuning process they are adjusted in light of experience with the analysis.

Redistributed features $f^u(\kappa)$, $u = 1, \cdots, S$, are generated as linear combinations of the $F^s(\kappa)$:

$$f^u(\kappa) = \sum_s f_s^{\,u} F^s(\kappa) \tag{98}$$

where the coefficients $f_s^{\,u}$ are derived from a transformation $\mathcal{f}(\sigma)$ of the attention function. The latter is a monotonically increasing function of σ, satisfying

$$\mathcal{f}(0) = 0 \qquad \mathcal{f}(S) = S \tag{99}$$

and the derivative $d\mathcal{f}/d\sigma$ is the new weight, relative to the old one, attached to wave numbers around σ. It is convenient to define $\mathcal{f}(\sigma)$ as a continuous function by quadrature from the discrete set of values of $d\mathcal{f}/d\sigma$ at the half-integer points $\sigma = s - \frac{1}{2}$ which correspond to the central wave numbers of the features $F^s(\kappa)$. Thus

$$\mathcal{f}(\sigma) = \sigma \left.\frac{d\mathcal{f}}{d\sigma}\right|_{1/2} \qquad 0 \le \sigma \le 1$$

$$\mathcal{f}(\sigma) = \sum_{s=1}^{l-1} \left.\frac{d\mathcal{f}}{d\sigma}\right|_{s-(1/2)} + (\sigma - l + 1) \left.\frac{d\mathcal{f}}{d\sigma}\right|_{l-(1/2)} \qquad l - 1 \le \sigma \le l \tag{100}$$

for each of $l = 2, \cdots, S$. The weights $(d\mathcal{f}/d\sigma)|_{s-(1/2)}$ may take any nonnegative values set by the investigator which are consistent with the normalization (99); this condition can be alternatively expressed as

$$\sum_{s=1}^{S} \left.\frac{d\mathcal{f}}{d\sigma}\right|_{s-(1/2)} = S \tag{101}$$

The redistribution coefficients f_s^u are derived from $\mathscr{f}(\sigma)$ by inverse interpolation at integer values of u to obtain σ_u such that

$$\mathscr{f}(\sigma_u) = u \tag{102}$$

Then if $[\sigma]$ is used to denote the largest integer less than or equal to σ and if

$$\begin{aligned} h_s(\sigma) &= 1 & s &\leq [\sigma] \\ h_s(\sigma) &= s - \sigma & s &= [\sigma] + 1 \\ h_s(\sigma) &= 0 & s &> [\sigma] + 1 \end{aligned} \tag{103}$$

an appropriate definition for the f_s^u is

$$\begin{aligned} f_s^u &= h_s(\sigma_1) & u &= 1 \\ f_s^u &= h_s(\sigma_u) - h_s(\sigma_{u-1}) & u &= 2, \cdots, S \end{aligned} \tag{104}$$

These manipulations lead to a new set of basis features $f^u(\kappa)$, which are nonnegative, continuously differentiable, and localized in the wave number band corresponding to $\sigma_{u-1} - 1 \leq \sigma \leq \sigma_u + 1$. From (98), (103), and (104) it can be shown that

$$\sum_u f_s^u = 1 \qquad \sum_u f^u(\kappa) = 1 \tag{105}$$

for all s and κ, preserving the normalization (44). If $\mathscr{f}(\sigma)$ is the identity function σ, then f_s^u is the identity matrix δ_{us}. The rows and columns of the redistributed information matrix

$$\begin{aligned} \tilde{\mathbf{R}}_{uv} &= \sum_{s,t} f_s^u \tilde{R}_{st} f_t^v \\ &= \iint f^u(\kappa) H(\kappa, \kappa') f^v(\kappa')\, d\kappa\, d\kappa' \end{aligned} \tag{106}$$

are weighted combinations of adjacent rows and columns of $\tilde{R}_{st}$ and correspond roughly to the amount of the positive 'substance' $H(\kappa, \kappa')$ in the wave number ranges κ, κ' corresponding to (σ_{u-1}, σ_u), (σ_{v-1}, σ_v).

The primary effect of redistribution on the principal components is to emphasize or deemphasize, at the discretion of the investigator, selected regions in wave number space. From the first-stage analysis of the example in section 6 it would seem desirable to concentrate attention in the well-sampled range $s \approx 7$ to $s \approx 25$, distinguishing only the general level of spectral deviations outside it. A transformation which accomplishes this is

$$\left.\frac{d\mathscr{f}}{d\sigma}\right|_{s-(1/2)} = \frac{2S}{NM} \sum_t \tilde{R}_{st} \tag{107}$$

though other formulations can achieve similar results. This function and its integral approximation (100) are plotted in Figure 9. Note its larger values for intermediate s, indicating relatively greater attention at these wave numbers. The effect of (107) is to repartition the substance of the information matrix so that the sum $\sum_v \tilde{\mathbf{R}}_{uv}$ of elements in each row of the new matrix is the same, $\frac{1}{2}NM/S$, with a corresponding condition on the columns, because of the symmetry of $\tilde{\mathbf{R}}$. This condition in turn implies that $\frac{1}{2}NM/S$ is an eigenvalue, with $\theta_u^p = \text{const}$ as the corresponding eigenvector and $\mathscr{G}^p(\kappa) = \text{const}$ as the principal component. Because all $\tilde{\mathbf{R}}_{uv}$ are nonnegative, this is also the largest eigenvalue (i.e., $p = 1$). Thus the first independently measurable feature is a change in magnitude of the

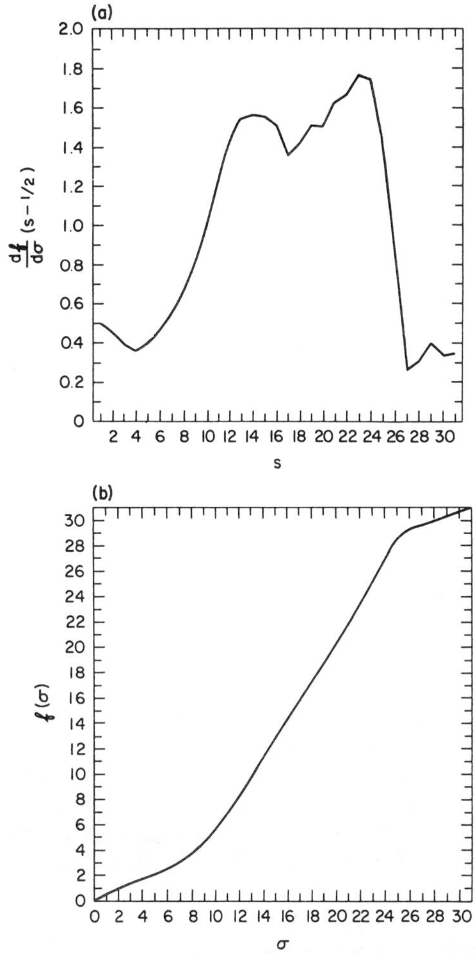

Fig. 9. Altered attention function for redistribution: (*a*) $(d\mathscr{f}/d\sigma)(s - \frac{1}{2})$ from (107) and (*b*) $\mathscr{f}(\sigma)$ from (100).

first-guess spectrum without change in shape. All other eigenvectors θ are orthogonal to this; that is, they satisfy

$$\sum_u \theta_u^q = 0 \tag{108}$$

showing that the redistributed attention function $\mathscr{f}[\sigma(\kappa)]$ is in certain respects a natural measure for this particular array.

In smoothing, the elementary features are replaced by broader, overlapping features constructed by successive multiplication by the three-point smoothing matrix

$$\begin{aligned} B_{vu} &= \tfrac{3}{4} & u &= v & u &= 1 \text{ or } S \\ B_{vu} &= \tfrac{1}{2} & u &= v & u &\neq 1 \text{ or } S \\ B_{vu} &= \tfrac{1}{4} & u &= v - 1 \text{ or } u = v + 1 \\ B_{vu} &= 0 & &\text{otherwise} \end{aligned} \tag{109}$$

Smoothed features can either be constructed from the $F^s(\kappa)$ directly or, as we shall do here, from the redistributed features $f^u(\kappa)$. Thus after redistribution and smoothing the basis features are

$$g^v(\kappa) = \sum_u b_u^v f^u(\kappa) \tag{110}$$

where

$$b_u^v = (B^n)_{vu} \tag{111}$$

TABLE 4. Eigenvalues of **R** After Tuning

p	r_p
1	121.3
2	89.3
3	41.8
4	12.6
5	2.6
6	0.4
7	4×10^{-2}
8	2×10^{-3}
9	1×10^{-4}
10	2×10^{-6}
11	4×10^{-8}
12	4×10^{-10}

for some positive integer n. For $n \gg 1$, $b_u{}^v$ approximates a normal distribution in u, centered on v, with unit integral and standard deviation $(\frac{1}{2}n)^{1/2}$, reflected symmetrically about the endpoints of the wave number range, $u - \frac{1}{2} = 0$ and $u - \frac{1}{2} = S$. The corresponding operation on the information matrix $\tilde{\mathbf{R}}_{uv}$ is a diffusion of the nonnegative substance in the domain $1 \leq u$, $v \leq S$, with impermeable walls at u, $v = \frac{1}{2}$ and $S + \frac{1}{2}$. Note that (109) and (111) imply $\sum_v b_u{}^v = 1$, which assures that the features g^u from (110) will satisfy a normalization condition analogous to (44) and (105). Thus both rows and columns are smoother functions of u and v after multiplication by b. The principal effect of this manipulation is to greatly reduce the eigenvalues associated with eigenfunctions or principal components that have significant fine structure in u on a half wavelength less than $\Delta\mathscr{f}_0 \sim n^{1/2}$. This sharpens the cutoff in the decreasing sequence of eigenvalues r_p and, taken in conjunction with the scaling described below, is the primary tool whereby the investigator may adjust the number of principal components within the measurable subspace. For given random coefficients λ_u in (79), increasing the value of n in (111) implies smoother spectral deviations and thus decreases the resolution of deviations deemed a priori to be likely.

Scaling consists of multiplying each a priori equiprobable feature by a constant α. Thus for any given spectral deviation the a priori norm is multiplied by α^2, and for the first time in this analysis the normalization constraint (44) is not preserved. The information matrix and eigenvalues are multiplied by α^2, while the principal components $\mathscr{G}^p$ are unchanged in form, though multiplied by the factor α. In (73), $\hat{\mu}_p$ is multiplied by $1/\alpha$, whereas $r_p/(r_p + 1)$ becomes $\alpha^2 r_p/(\alpha^2 r_p + 1)$. Thus for the well-measured components the spectral deviations $\hat{\mu}_p{}^* \mathscr{G}^p$ are not greatly affected by scaling, particularly if α and n are made large enough so that $\alpha^2 r_p$ is decreasing rapidly with increasing p near the cutoff of the measurable subspace (i.e., $\alpha^2 r_p \sim 1$). The choice

$$\alpha = (2\pi n)^{1/4} \tag{112}$$

yields an a priori covariance matrix which, away from the ends $\mathscr{f}(\sigma) = 0$ and S, may be represented approximately by

$$T_{st} = \exp\left\{-\frac{1}{2n}[\mathscr{f}(s - \tfrac{1}{2}) - \mathscr{f}(t - \tfrac{1}{2})]^2\right\} \tag{113}$$

This has local a priori variance $(\overline{\nu_s{}'^2})$ unity, but adjacent wave number bands are highly correlated, the correlation coefficient falling to e^{-1} when they are separated by $\Delta\mathscr{f} = (2n)^{1/2}$.

c. *Example of a Second-Stage Fit*

We now apply this tuning process to the example of section 6. The abscissa u, based upon the redefined attention function $\mathscr{f}[\sigma(\kappa)]$, represents the revised basis features $f^u(\kappa)$ defined in (98). Because of the restricted width of these features in wave number and the normalization condition (105) a graphical presentation is still appropriate; however, the scale relating u to κ needs to be altered. The fitting procedure is defined by (84)–(86) and (93)–(96), with the following substitution (note (98)–(104)):

$$\nu_s \rightarrow \nu_u$$
$$\tilde{R}_{st} \rightarrow \tilde{\mathbf{R}}_{uv} \tag{114}$$
$$T_{st} \rightarrow \mathbf{T}_{uv} = \overline{\nu_u \nu_v}$$

where

$$\nu_s = \sum_u f_s{}^u \nu_u$$

and

$$T_{st} = \sum_{u,v} f_s{}^u \mathbf{T}_{uv} f_t{}^v \tag{115}$$

The a priori independent features

$$g^u(\kappa) = \sum_v g_v{}^u f^v(\kappa)$$

are defined by a transformation matrix $g_v{}^u$, which is related to the covariance matrix **T** as in (77). This transformation matrix represents smoothing and scaling; hence

$$g_v{}^u = \alpha b_v{}^u \tag{116}$$

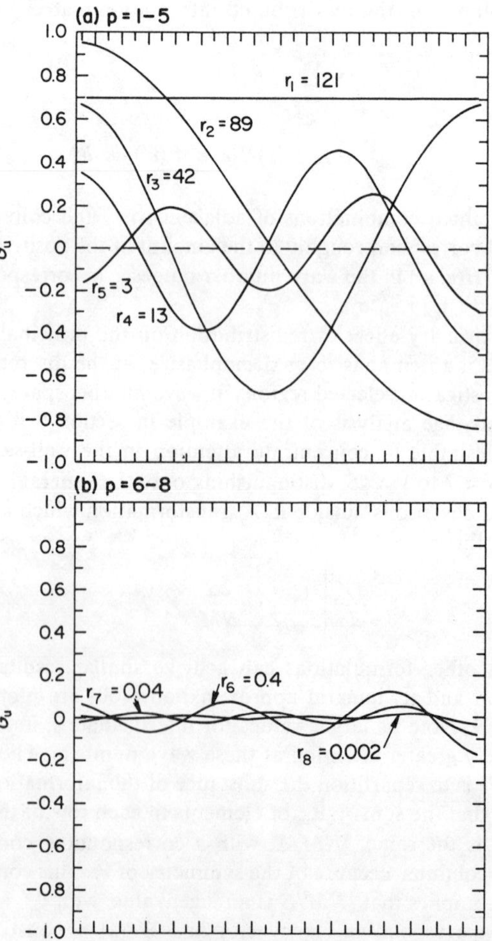

Fig. 10. First eight principal components for the second-stage fit. Figure 10*a* contains all those with $r_p > 1$.

In this case we choose $n = 36$ in (111); hence $\alpha = 3.88$ from (112). The eigenvalues and the first eight principal components are shown in Table 4 and Figure 10. Note that the largest eigenvalue is $r_p = 121$, implying an a posteriori uncertainty in $\hat{\mu}_1{}^*$ of less than 10%, so that the overall magnitude of the spectral shape $\hat{E}(\kappa)$ is known to high precision, while changes in that shape are less well determined. The remaining principal components approximate cosine functions of u, of order increasing directly with p. Note that these are not the eigenvectors $\boldsymbol{\Theta}^p$ of the information matrix $\mathbf{R}$, but because of the factor $g_s{}^u$ in (89) are smoothed and scaled versions of these eigenvectors. The eigenvalues decrease rapidly with increasing order, and only the first five fall in the measurable subspace. The optimal estimate $\hat{\nu}^*$ and a number of alternative estimates, equally consistent with the a posteriori mean $\hat{\nu}^*$ and the residual error covariance

$$\langle \nu_u' \nu_v' \rangle = \sum_{p=1}^{S} \frac{1}{1 + r_p} \mathscr{G}_u{}^p \mathscr{G}_v{}^p \tag{117}$$

are shown in Figure 11. The pointwise uncertainties $\langle \{\nu_u'^2\} \rangle^{1/2}$ associated with this ensemble are plotted in Figure 12 where, except near the ends of their range in u, they are approximately uniform at around $\pm 15\%$ precision.

Note that the factor $r_p/(r_p + 1)$ in (73) is nearly unity for the first five principal components and close to zero thereafter. Furthermore, the amplitude of $\mathscr{G}^p$ falls off rapidly for all features beyond $p = 5$, a consequence of the high degree of smoothing. Thus the optimal estimate $\hat{\nu}^*$ is essentially the same as an unbiased estimate $\hat{\nu}^+$ restricted to lie in the measurable subspace spanned by the first five second-stage principal components, and the residual error covariance (117) is dominated by uncertainties in these components. Hence

$$\langle \nu_u' \nu_v' \rangle \approx \sum_{p=1}^{5} \frac{1}{r_p} \mathscr{G}_u{}^p \mathscr{G}_v{}^p \tag{118}$$

Formal truncation of the set of features under consideration is thus unnecessary within this formalism. Since the principal components are themselves linear combinations of the original basis set $\{F^s(\kappa)\}$ and since least mean squares error fitting has been employed at each stage, it follows from the remark at the end of section 5 that $\hat{\nu}^*$ is optimal; that is, no other unbiased linear estimator operating on the same raw covariances within the same subspace of features can have smaller uncertainties. Note that according to (118), within this subspace the amplitudes (as opposed to the shapes) of the basis features are irrelevant, since a scaling factor α, multiplying both r_p and $\mathscr{G}_u^p$ $\mathscr{G}_v^p$ by its square α^2, would cancel in its effects.

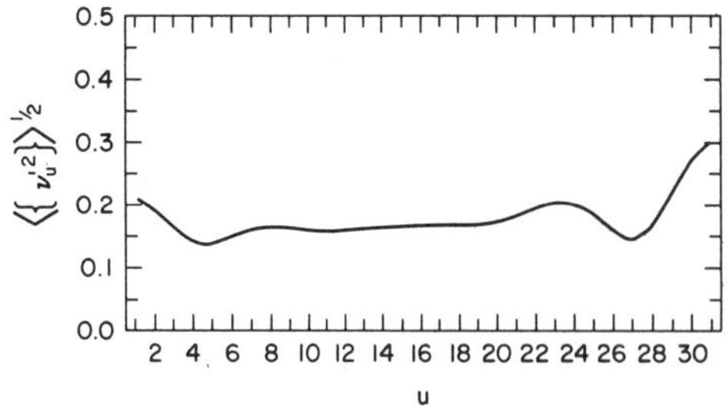

Fig. 12. Standard deviation of spectral estimates (see (110)) for the second-stage fit.

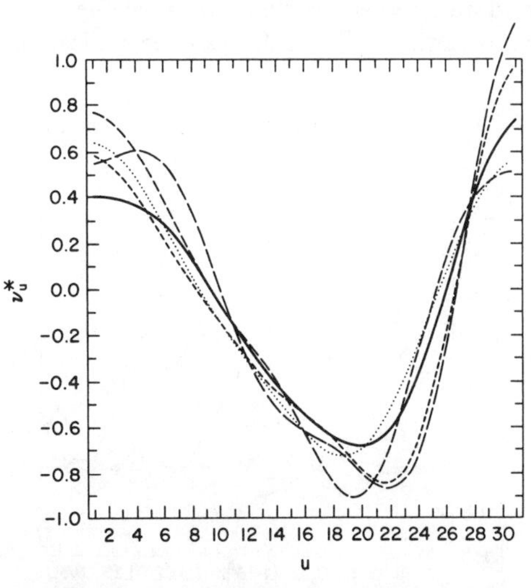

Fig. 11. Second-stage spectrum $\hat{\nu}_u{}^*$ (solid curve) together with four possible alternatives $\nu_u{}^*$ (dashed and dotted curves). These alternatives are defined by a variant of (67) appropriate to the second-stage estimate, namely, $\nu_u{}^* = \sum_p \{\hat{\mu}_p{}^* + [\lambda_p/(1 + r_p)]^{1/2}\} \mathscr{G}_u{}^p$.

Comparison of Figures 11 and 12 with Figures 6 and 7 shows that tuning has accomplished a substantial improvement in the analysis. The uncertainties in the best estimate are reduced in the second-stage analysis, and the independently measurable features are very natural ones. This naturalness appears to be associated with a good match between the array design and the logarithmic structure of the attention function. We will return at the end of this section to some general arguments which suggest why this might be so. Figure 10 and Table 5 show that a posteriori uncertainties are approximately independent in five distinct wave number bands, roughly of equal width in u, the rms amount in each being about $\pm 16\%$ of the first-guess spectrum, rising to ± 20–25% at the ends of the range. Such precision approaches that achievable for a uniformly spaced array with the same total number (500) of observations, namely, five independent estimates each to $\pm 14\%$. However, with a uniform array such estimates apply to spectral bands of equal width in κ, whereas here they are approximately of equal width in log κ, though more precisely of equal width in $\not{f}\,[\sigma(\kappa)]$.

d. *Window Functions*

Spectral windows for the second-stage fitting procedures can be defined analogously to those in (70)–(71):

$$\langle \nu_u{}^* \rangle = \sum_{v=1}^{S}{}' \, W_{uv} \langle \nu \rangle_v \tag{119}$$

where

$$W_{uv} = \sum_{v'=1}^{S} \frac{1}{1 + r_p} \mathscr{G}_u{}^p \mathscr{G}_{v'}{}^p \tilde{\mathbf{R}}_{v'v} \tag{120}$$

Examples of these window functions are plotted in Figure 13. Each curve shows the domain of sensitivity of the time spectral deviation in the estimate at the labeled value of u. Note that the area under each curve is unity, a consequence of the requirement that

$$\sum_v W_{uv} \mathscr{G}_v{}^p = \frac{r_p}{1 + r_p} \mathscr{G}_u{}^p \text{ for each } p$$

coupled with the statements that $r_1 \gg 1$ and $\mathscr{G}_u{}^1 = \text{const}$. The overall shape of these functions is quite satisfactory in that they represent local averages over adjacent redistributed wave numbers v in a band centered on wave number u of half width $\sim n^{1/2}$. The small negative lobes on the flanks of each window

TABLE 5. Residual Error Covariances for the Second-Stage Fit (Selected u and v)

v/u	2	8	15	22	29
2	0.036	−0.005	0.002	−0.001	0.000
8	−0.005	0.027	−0.006	0.002	−0.001
15	0.002	−0.006	0.027	−0.007	0.003
22	−0.001	0.002	−0.007	0.037	−0.020
29	0.000	−0.001	0.003	−0.020	0.047

Data are calculated from (116).

are associated with the sharp cutoff after $p = 5$; they could be largely eliminated by a smaller choice of the scaling parameter α such that $r_p < 1$ for all p (but at the price of a substantial bias toward zero in the estimated magnitude of all spectral deviations).

e. Some Consistency Tests

Before adopting the tuned $\hat{\nu}^*$ as the final best estimate of the spectral deviations it is necessary to test further whether the implied a posteriori probability distribution can be shown to be significantly inconsistent with the observational data. For this we need to consider possibilities which fall completely outside the framework in which the first-stage analysis was undertaken (i.e., the spectral model (43)).

The first question relates to the use of the first-guess spectrum $\tilde{E}(\kappa)$ to estimate the sampling errors in the raw covariances. Since $\hat{\nu}_u^*$ is everywhere substantially less than unity, this seems justified after the fact, but it is worth considering the issue explicitly. In particular, it often happens that the degree of confidence an investigator has a priori in the shape of $\tilde{E}(\kappa)$ may greatly exceed that in the magnitude of the constant multiplying this shape. Were $\tilde{E}(\kappa)$ to be set at a level, say, of one third of the true spectrum, but with roughly the correct shape, the unbiased spectral estimate $\tilde{E}(1 + \hat{\nu})$ would still be essentially correct, but the modified estimate $\tilde{E}(1 + \hat{\nu}^*)$ would tend to be the wrong value in poorly measured wave number regions. A more serious disadvantage would be that all estimates of rms uncertainty would be too small by a factor of 3. Under these circumstances it is desirable to estimate at the outset (i.e., at the beginning of the first-stage analysis) the average over wave number of the relative spectral deviation; if this is larger than a small fraction of unity, it should be used to rescale $\tilde{E}(\kappa)$. An optimal estimator for $1/S \sum_s \nu_s$ is

$$\sum_{s,t} \tilde{R}_{st}\hat{\nu}_t \Big/ \sum_{s,t} \tilde{R}_{st} \tag{121}$$

neglecting the differences between $\tilde{R}$ and R. In our example, (121) has the value −0.070, while one standard deviation in the estimate is $(2/NM)^{1/2} = 0.063$; hence no rescaling of $\tilde{E}(\kappa)$ is warranted.

A second question relates to the assumption that the mean $\langle\phi(x)\rangle$ is zero. To test this, we assume, for example, a linear trend

$$\langle\phi(x)\rangle = \alpha + \beta x \tag{122}$$

and check whether estimate values $\hat{\alpha}$ and $\hat{\beta}$ lie outside the 95% confidence interval based upon the null hypothesis $\langle\alpha\rangle = \langle\beta\rangle = 0$, where sampling uncertainties are calculated from the spectrum $\tilde{E}(\kappa)(1 + \hat{\nu}^*)$. We note that

$$\langle((\langle\phi_i\rangle^M - \langle\phi_i\rangle)(\langle\phi_j\rangle^M - \langle\phi_j\rangle)\rangle = \frac{1}{M} C_{ij} \tag{123}$$

so that the optimal generalized least squares fit $(\hat{\alpha}, \hat{\beta})$ results from minimizing

$$\frac{1}{2} M \sum_{i,j} (\langle\phi_i\rangle^M - \alpha - \beta x_i)(C^{-1})_{ij}(\langle\phi_j\rangle^M - \alpha - \beta x_j)$$

This defines two linear equations for $\hat{\alpha}$ and $\hat{\beta}$, namely,

$$\begin{aligned} R_{\alpha\alpha}\,\hat{\alpha} + R_{\alpha\beta}\,\hat{\beta} &= V_\alpha \\ R_{\beta\alpha}\,\hat{\alpha} + R_{\beta\beta}\,\hat{\beta} &= V_\beta \end{aligned} \tag{124}$$

where

$$\begin{aligned} R_{\alpha\alpha} &= M \sum_{i,j} (C^{-1})_{ij} \\ R_{\alpha\beta} = R_{\beta\alpha} &= M \sum_{i,j} x_i(C^{-1})_{ij} \\ R_{\beta\beta} &= M \sum_{i,j} x_i(C^{-1})_{ij}\, x_j \\ V_\alpha &= M \sum_{i,j} \langle\phi_i\rangle^M (C^{-1})_{ij} \\ V_\beta &= M \sum_{i,j} \langle\phi_i\rangle^M (C^{-1})_{ij}\, x_j \end{aligned} \tag{125}$$

According to (40) the covariance of the uncertainties in the estimates $\hat{\alpha}$ and $\hat{\beta}$ is the matrix R^{-1}. According to the null hypothesis we can therefore define independently measurable polynomials of degree 0 and 1 with respective coefficients λ_1 and λ_2 distributed with zero mean and unit variance,

$$\langle\phi(x)\rangle = \lambda_1 \left\{\frac{1}{R_{\alpha\alpha}^{1/2}}\right\} + \lambda_2 \left\{\frac{-R_{\alpha\beta}/R_{\alpha\alpha}^{1/2} + R_{\alpha\alpha}^{1/2}x}{(R_{\alpha\alpha}R_{\beta\beta} - R_{\alpha\beta}^2)^{1/2}}\right\} \tag{126}$$

For this example the quantities (125) were calculated from $\hat{\nu}^*$ in Figure 11, Fourier-transformed to give C_{ij}. The resulting estimates were $\hat{\lambda}_1 = 1.3$ and $\hat{\lambda}_2 = 0.2$, neither of which lies outside the 95% confidence limits (i.e., ±1.96) for a two-tailed, normal distribution with zero mean and unit variance. We therefore conclude that the data are consistent with the model assumption of $\langle\phi\rangle = 0$.

The third question is a partial test of whether $\phi(\mathbf{x})$ is non-Gaussian. In particular, if the M available realizations are di-

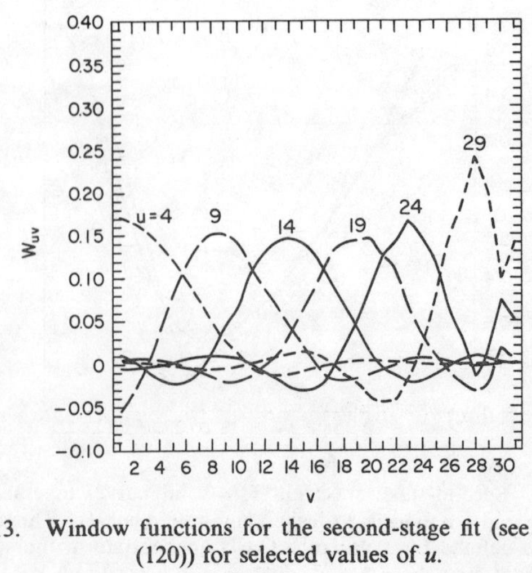

Fig. 13. Window functions for the second-stage fit (see (119) and (120)) for selected values of u.

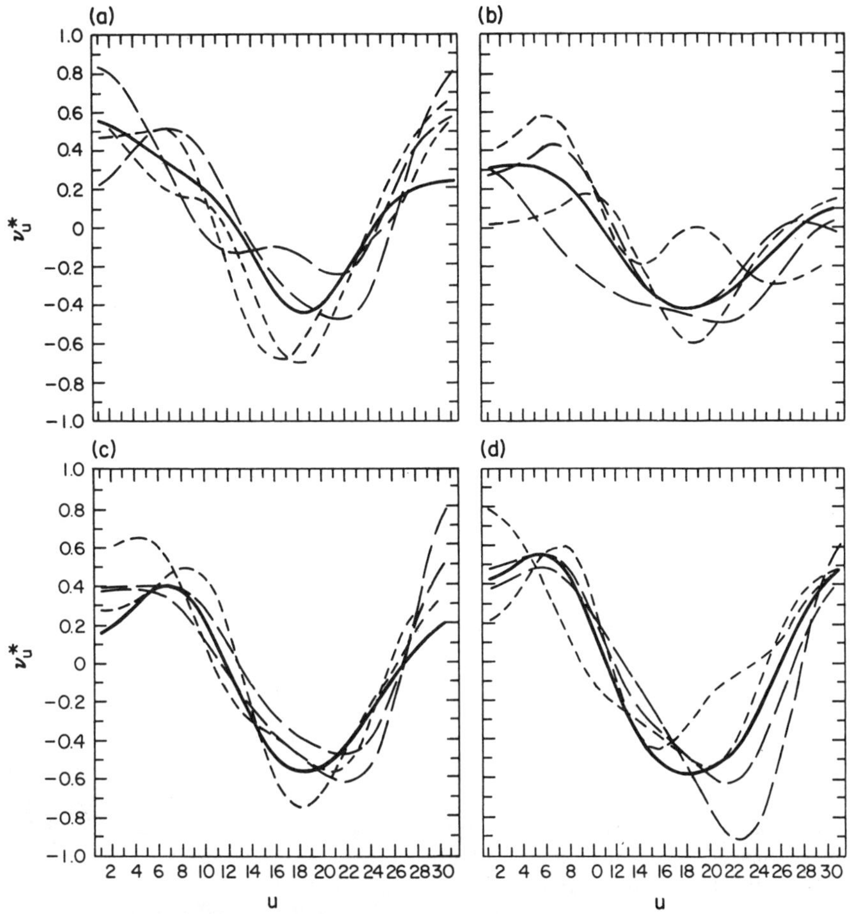

Fig. 14. Four sets of best estimates $\hat{\nu}_u^*$ and possible alternatives ν_u^* for independent experiments with $M = 25$ realizations. Each of these panels is directly comparable to Figure 11.

vided into subgroups, do the estimates $\hat{\mu}_p$ calculated from each subgroup separately scatter more than the assumption of zero fourth-order cumulants would indicate is likely? It must be recognized that a good estimate of a fourth-order moment requires more data than a good estimate of a second-order moment (i.e., of the spectrum). Accordingly, rather than subdivide the original sample, we have claimed the luxury afforded by artificially generated data and conducted four additional experiments, each with 25 realizations randomly produced from the original spectrum. For each of the five measurable principal components the standard deviation of the normalized coefficients $r_p^{1/2}\hat{\mu}_p$ was estimated from the scatter σ_p within the group of five experiments. According to the null hypothesis, $4\sigma_p^2$ should be distributed as chi-square with $5 - 1 = 4$ degrees of freedom. Furthermore, the sum of the values of $4\sigma_p^2$ over the $P = 5$ features should be distributed approximately as chi-square with $5 \times 4 = 20$ degrees of freedom, which has 95% confidence limits (9.6, 34.2). Since $\sum_p 4\sigma_p^2$ for this group of experiments was in fact 11.5, it may be concluded that the data are not inconsistent with the Gaussian assumption in at least this aspect.

f. Final Assessment

A by-product of this extension of the original data set is a direct comparison of the scatter in the five independent estimates of $\hat{\boldsymbol{\nu}}^*$ with the estimates of the uncertainty in the first $\hat{\boldsymbol{\nu}}^*$ as displayed in Figures 11 and 12. Figure 14 shows the results from each of the four additional experiments in the same format as Figure 11. These plots were a salutary reminder of how misleading impressions gained from small samples can be; in particular, the character of the a posteriori alternatives $\hat{\boldsymbol{\nu}}^* + \boldsymbol{\nu}'$ seems quite different in the different plots, even though all of the $\boldsymbol{\nu}'$ are drawn from the same statistical population (i.e., that defined in the caption of Figure 11).

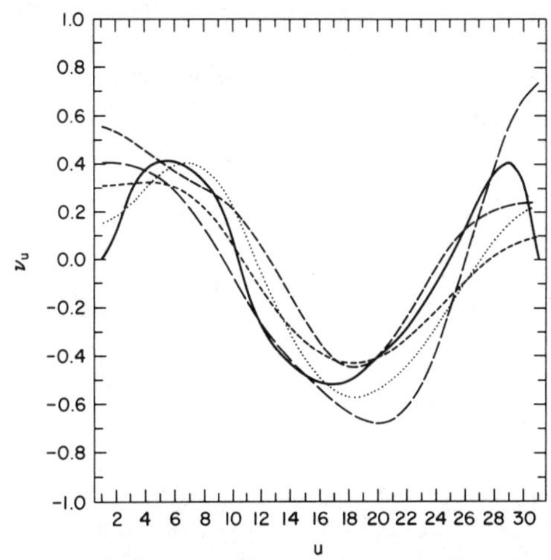

Fig. 15. The true spectrum $\langle \nu_u \rangle$ (solid curve) together with four second-stage estimates $\hat{\nu}_u^*$, each generated from 25 independent data realizations on the array in Figure 2. One of the $\hat{\nu}_u^*$ is equivalent to the solid curve in Figure 11; the other three are equivalent to those in Figures 14*a*–14*c*.

Finally, Figure 15 shows a collection of the first four $\hat{\nu}^*$ with the true spectral deviation $\langle \nu \rangle$ imposed (i.e., the ν used to artificially generate the data for our example). Together Figures 14 and 15 remind one of the amount of information truly available from a data set consisting of 500 observations.

8. ARRAY DESIGN

So far the discussion has centered on the question of, given a set of observations from an array $\{x_i\}$, what can be inferred about the spectrum of the process ϕ? We now turn to the issue of, given N measuring instruments and given a set of questions about the spectrum for which answers are required, how should the points $\{x_i\}$ be distributed so as to maximize the information gain from the observations?

As before, we suppose that the first guess spectrum $\tilde{E}(\kappa)$ is a reasonable guide as to what is likely to be observed. Attention is concentrated on a given set of features $\{g^p(\kappa)\}, p = 1, \cdots, P$, according to which the spectrum may deviate from $\tilde{E}(\kappa)$. As in section 7, the a priori expectations of the investigator about relative deviations are summarized as in (76):

$$\nu(\kappa) = \sum_1^P \lambda_p g^p(\kappa) \tag{127}$$

where the λ_p are independent, normally distributed random variables with zero mean and unit variance. If, following the observations, the a posteriori uncertainties are given by

$$\langle \lambda_p' \lambda_q' \rangle = [(I + R)^{-1}]_{pq} \tag{128}$$

then according to standard theory [*Middleton,* 1960] the gain in information $\Delta\mathscr{I}$ from the observations is the change in the mean value of the logarithm of the probability density, namely,

$$\begin{aligned} \Delta\mathscr{I} &= \tfrac{1}{2} \{\log \det (I + R) - \log \det I\} \\ &= \sum_1^P \tfrac{1}{2} \log (1 + r_p) \end{aligned} \tag{129}$$

where det $(I + R)$ is the determinant of $I + R$. The largest contributions to $\Delta\mathscr{I}$ come from these principal components for which $r_p \gg 1$, which are determined more by observation than initial assumption. Poorly measured components contribute little new information, since $\log (1 + r_p) \approx 0$ if $r_p \ll 1$.

The optimization problem is to adjust the positions $\{x_i\}$ of observation points so that $\Delta\mathscr{I}$ in (129) is maximized. This can be achieved numerically, noting that for each x_i,

$$\begin{aligned} \frac{\partial}{\partial x_i} \Delta\mathscr{I} &= \tfrac{1}{2} \operatorname{Tr} \left\{ (I + R)^{-1} \frac{\partial R}{\partial x_i} \right\} \\ &= \sum_{p,q=1}^P [(I + R)^{-1}]_{pq} \left\{ \sum_{j,i',j'}^N \left[\frac{\partial C_{ij}{}^p}{\partial x_i} (\tilde{C}^{-1})_{ii'} (\tilde{C}^{-1})_{jj'} C_{i'j'}{}^q \right.\right. \\ &\quad \left.\left. - \sum_{i'',j''} C_{j''j}{}^p (\tilde{C}^{-1})_{j''i} \frac{\partial \tilde{C}_{ii''}}{\partial x_i} (\tilde{C}^{-1})_{i''i'} (\tilde{C}^{-1})_{jj'} \tilde{C}_{i'j'}{}^q \right] \right\} \end{aligned} \tag{130}$$

which must vanish at a stationary value. For moderate values of N and P, values of $\Delta\mathscr{I}$ and $(\partial/\partial x_i)(\Delta\mathscr{I})$ can readily be computed for any initial $\{x_i\}$. Efficient iterative routines [e.g., *Fletcher and Powell,* 1963] then modify $\{x_i\}$ until a local maximum of $\Delta\mathscr{I}$ is obtained. There exists no certain technique for finding the global maximum.

This optimum array design depends upon the number N of array points, the number M of independent realizations to be observed, the first-guess spectrum $\tilde{E}(\kappa)$, and the features $g^p(\kappa)$. It does not, however, depend on having any data. This algorithm therefore provides an opportunity for refining the properties of the array before undertaking a costly field experiment. Of course, in a real situation other considerations besides maximizing $\Delta\mathscr{I}$ need to be taken into account, such as logistics, redundancy, and the possibility that the first guess may be badly wrong, but once the computer program is operational, it is straightforward to consider the merits of many alternative array designs, including their robustness under missing data points and large spectral deviations. The approach is readily extended to arrays in two dimensions. The optimal design provides a standard against which the loss of information due to other compromises can be measured. It should not be regarded as a uniquely decisive criterion but rather a useful tool among a variety of computer analyses and simulations.

The array shown in Figure 2, which was the basis for the illustrative example of sections 6 and 7, was in fact calculated as an optimal 20-point array with a first-guess spectrum proportional to κ^{-2}, and eight features which cover the range from $\kappa = 0.1$ to $\kappa = 25$ with a logarithmic attention function, when the number of realizations is large enough for all eight features to be well measurable. From this and several other examples the authors have inferred empirically [*Bretherton and McWilliams,* 1979] some general characteristics of nearly optimal arrays in one and two dimensions. This set of features $g^p(\kappa)$ is qualitatively similar to that of the second-stage fit in section 7 but not identical. One could, of course, go through the exercise of optimal array design for the second-stage spectral model. We have not done this but are convinced that the array would differ little from that in Figure 2, since, after all, the whole of the tuning process was directed to matching $\{g^p\}$ with $\{x_i\}$.

The algorithm (130) provides the best array design for determining the coefficients for a given set of features $g^p(\kappa)$. If these are the a priori equiprobable independent modes of spectral deviation, then the information gain from the measurements is a maximum. Once the array has been determined, the analysis may (and should) be based on a different set of a priori probabilities to investigate its sensitivity to the unexpected and the robustness of its conclusions under failure of the assumptions. In effect, the first-stage fitting and various consistency checks did just this for the array design exemplified in our previous example. The shape of the window functions after final tuning is a summary indicator of the quality of the end result. If they are well localized into wave number bands and are reasonably uniform over such bands, the implications of a measurement are straightforward and easily understood. If they have substantial fine structure and negative lobes extending well outside the primary band, serious confusion or aliasing between wave numbers may result. Judgments about what is substantial must take into account the random sampling error in any given experiment (a function of the number of realizations available) and the degree to which the first guess really represents the order of magnitude of the variance likely to be found in any given wave number band (so that W_{uv} fairly describes the risk of contamination by variance at wave number v into the estimate of wave number u).

After tuning, the analysis in section 7 seems quite satisfactory, but there remains the question of why it works so well. This is a very fundamental issue in this whole approach to spectral estimation, and the authors must admit that they do not fully understand the answer. The point is that redistribution and smoothing adjust the resolution available in the analysis, but they do not really alter the investigator's perception of prior probabilities, which remains one of preparedness for (though not necessarily expectation of) fine structure in the spectrum on all wave numbers with a wide range of magnitudes. Tuning is a recognition that such detail is largely unmeasurable, but it does not follow automatically that it will not contaminate coarse resolution estimates. The sensitivity of the independently measurable principal components to the statement of prior probabilities is a reminder of this. Yet in this case the sensitivity appears to be less than might at first sight be expected; the first-stage eigenfunctions in Figure 5 are remarkably well ordered, so that fine structure, appropriately defined, is independently measurable from coarse structure. This is a property of the basic matrix $H(\kappa, \kappa')$ which deserves explanation.

Partial insight comes from consideration of a white noise first-guess spectrum, so that

$$\tilde{C}_{ij} = C\,\delta_{ij} \tag{131}$$

for some constant C. Now consider two features, each concentrated in the same wave number band around κ_0, one of coarse resolution

$$F^1(\kappa) = \exp\,[-\tfrac{1}{2}a^2(\kappa - \kappa_0)^2]$$

and the other of fine resolution

$$F^2(\kappa) = \exp\,[-\tfrac{1}{2}a^2(\kappa - \kappa_0)^2]\cos\,(ma\kappa)$$

where m is a large integer. The corresponding covariances

$$C^1(\xi) = \alpha \exp\,[-\tfrac{1}{2}\xi^2/a^2]\cos\kappa_0\xi$$

and

$$C^2(\xi) = \tfrac{1}{2}\alpha\left\{\exp\left[-\frac{1}{2}\frac{(\xi - ma)^2}{a^2}\right] + \exp\left[-\frac{1}{2}\frac{(\xi + ma)^2}{a^2}\right]\right\}\cos\kappa_0\xi$$

occupy essentially disjoint regions of the ξ axis (α is again a generic normalization constant). Hence

$$R_{12} = \int F^1(\kappa)\,F^2(\kappa')\,H(\kappa, \kappa')\,d\kappa\,d\kappa'$$

$$= \frac{1}{2}\sum_{\substack{i,j\\ i',j'}} C_{ij}{}^1(\tilde{C}^{-1})_{ii'}(\tilde{C}^{-1})_{jj'}C_{i'j'}{}^2 \tag{132}$$

$$= \frac{\alpha^2}{2C^2}\sum_{i,j} C^1(\xi_i - \xi_j)C^2(\xi_i - \xi_j) \approx 0 \tag{133}$$

whereas

$$R_{11} = \frac{\alpha^2}{2C^2}\sum_{i,j}\{C^1(\xi_i - \xi_j)\}^2$$

$$R_{22} = \frac{\alpha^2}{2C^2}\sum_{i,j}\{C^2(\xi_i - \xi_j)\}^2 \tag{134}$$

Thus

$$R_{12}{}^2 \ll R_{11}R_{22} \tag{135}$$

and measurements of the coefficients of $F^1(\kappa)$ and $F^2(\kappa)$ are approximately independent. This reasoning is special to a white noise first guess. For more general but smooth $\tilde{E}(\kappa)$ the covariances associated with fine and coarse resolution features in the same wave number band will always occupy different regions of the ξ axis. However, the inverse of the matrix $\tilde{C}_{ij}$ is not in general readily characterizable, though for κ^{-2} it can be shown that the inverse is tridiagonal. From numerical experiments we have found that $(\tilde{C}^{-1})_{ii'}$ does normally appear to approach zero fairly rapidly as $|\xi_i - \xi_{i'}|$ increases when $\tilde{E}(\kappa)$ has a power law form with a negative exponent, limiting the extent to which $(\tilde{C}^{-1})_{ii'}$ and $(\tilde{C}^{-1})_{jj'}$ in (132) can couple widely different regions of the ξ axis. Thus (135) holds to some degree under a broad class of circumstances. That is why a separation into measurable and unmeasurable subspaces on the basis of resolution results in window functions which are fairly smooth (though not necessarily well localized).

9. DISCUSSION

In conclusion, we mention briefly a number of other considerations which may be of interest, in part by continuing the discussion of section 3 in the light of our results. Given the raw covariances $\langle C\rangle^M$, inference of the true spectrum (i.e., solution of (14)) is an example of 'the inverse problem' [*Backus and Gilbert*, 1970], involving solution of an underdetermined set of linear equations. Like them, we have adduced the structure of a Hilbert space (the a priori norm) to resolve the indeterminacy, but it is the additive combination of this norm with that associated with measurement uncertainties which provides the power of this analysis. This a priori norm is to a considerable extent arbitrary, but the basic matrix H summarizes in an optimal manner all the information available from the observations, and we know that no other linear method of reducing the data can result in smaller uncertainties. In practice, we also require the assumption that the first guess is a reasonable approximation to the true spectrum, so that H may be approximated by $\tilde{H}$, a calculable quantity, as in (92).

In section 3, methods of spectral estimation were classified into a priori, data adaptive, and model fitting. The present analysis provides a framework for intercomparing and evaluating all three approaches. For each method a single data sample yields a definite estimate $\mathscr{E}(\kappa)$. The a priori approaches imply that, averaged over a large number of samples, the mean estimate is related to the true spectrum E by (11), involving an explicit window function $W(\kappa, \kappa')$ (not necessarily symmetric in κ and κ'). Likewise, since the covariances of sampling errors in the $\langle C\rangle^M$ are known, there is an explicit noise covariance

$$\langle[\mathscr{E}(\kappa) - \bar{\mathscr{E}}(\kappa)][\mathscr{E}(\kappa') - \bar{\mathscr{E}}(\kappa')]\rangle = U_{\kappa\kappa'} \tag{136}$$

For the data adaptive and model-fitting methods these relationships are less clear, but when the sampling uncertainties are relatively small (this is always true for large enough M) and the first guess is a good one, we can linearize about the mean or first-guess situation, define the window function by the sensitivity of $\mathscr{E}$ to small changes in E, using coefficients $\alpha_{ij}(\kappa)$ computed for the first-guess conditions, and estimate the noise covariance likewise. Now the inference of the true spectrum from $\mathscr{E}(\kappa)$ is also an inverse problem, requiring the removal of any biases implied by $W(\kappa, \kappa')$, but we can obtain an optimal unbiased estimate $\hat{\nu}$ of the spectral deviation by minimizing

$$\frac{1}{2}\int\int\left\{\mathscr{E}(\kappa)-\int W(\kappa,\kappa'')\tilde{E}(\kappa'')\nu(\kappa'')\,d\kappa''\right\}(U^{-1})_{\kappa\kappa'}\cdot\left\{\mathscr{E}(\kappa')-\int W(\kappa',\kappa'')\tilde{E}(\kappa'')\nu(\kappa'')\,d\kappa''\right\}d\kappa\,d\kappa' \quad (137)$$

This yields a measurement norm for $\mathscr{E}(\kappa)$ defined by the matrix

$$\mathscr{H}(\kappa,\kappa')=\tilde{E}(\kappa)\left\{\int\int(U^{-1})_{\kappa''\kappa'''}W(\kappa'',\kappa)\cdot W(\kappa''',\kappa')\,d\kappa''\,d\kappa'''\right\}\tilde{E}(\kappa') \quad (138)$$

In passing by linear transformations from one unbiased estimator to another, information can only be lost, not gained. If optimal generalized least squares fits are used at each stage, all information is retained. Therefore, subject to the approximation of replacing C^{-1} by $\tilde{C}^{-1}$, the norm constructed from $\mathscr{H}$ (i.e., $\int\int \nu\mathscr{H}\nu$) can be no larger than that from H, and comparison of the two provides a measure of the efficiency of the estimator $\mathscr{E}$ with regard to a particular spectral feature $\nu(\kappa)$. Note that if $\mathscr{E}(\kappa)$ is smoothed or otherwise subject to a projection operator, the same holds for the κ dependence of $W(\kappa, \kappa')$ and $U_{\kappa\kappa'}$. The inverse U^{-1} should be computed within this more limited subspace of possibilities.

Analysis of the various methods mentioned in section 3 shows that some at least are efficient, in the sense that as $M \rightarrow \infty$, they tend asymptotically to yield the same information as optimal generalized least squares. The maximum likelihood method of *Capon* [1969] (see (15)) appears to be in this class, as does the model fitting of *Jones* [1977] (see (20)), provided the model chosen is general enough to include all the features which will ultimately be admitted as likely by the a priori norm. Optimal interpolation for $\phi(\mathbf{x})$, using the first-guess covariance function $\tilde{C}(\mathbf{x})$ (or more precisely the true covariance C), followed by Fourier transformation, yields a biased estimate $\mathscr{E}$, but one which contains the information to obtain an unbiased efficient estimate $\hat{\nu}$. If the initial interpolation is not optimal, the resulting spectral estimation is not efficient. The bin method is in general not efficient. Within the approach of model fitting, the concept of asymptotic efficiency as $M \rightarrow \infty$ [*Dzhaparidze and Yaglom,* 1974] gives precise meaning to these intercomparisons.

Another area of interest is aliasing. From a given array of $\{x_i\}$ of N data points, only $\frac{1}{2}N(N+1)$ distinct raw covariances can be formed, and the basic matrix H has at most this dimensionality. As the number of realizations M tends to infinity, the measurable subspace expands, but to a definite upper limit. Any feature ν for which the corresponding covariances vanish for all pairs (x_i, x_j) is intrinsically unmeasurable; it belongs to an infinite dimensional subspace known as the intrinsic alias. For finite M the effective alias of unmeasurable features certainly includes the intrinsic alias and will normally be larger (in the presentation of section 6 we restricted attention to its projection onto a finite basis set of functions $\{F^r\}$). It is not generally possible to state which aspects of array design lead to what characteristics of the effective alias in as simple a manner as that associated with the Nyquist wave number in uniformly spaced arrays (see section 4).

Finally, it must be admitted that computations for these analyses tend to be lengthy, increasing as N^3S^2 as the array becomes longer and greater resolution is called for. If an array is divided into segments sufficiently far apart to provide statistically independent samples of $\phi(x)$, the contributions of the distinct segments to the basic H matrix are simply additive. This suggests that for long arrays some approximations may be available to provide significant computational shortcuts, but a satisfactory understanding of these possibilities is not presently available.

One major outstanding research task is to test in practice how robust are the methods advocated here when the process ϕ is significantly non-Gaussian. Computer simulations should provide useful insights.

Acknowledgments. The authors greatly profited from discussions of this work with John Tukey. They also appreciate the computational assistance provided by Julianna Chow.

REFERENCES

Backus, G., and F. Gilbert, Uniqueness in the inversion of gross earth data, *Phil. Trans. Roy. Soc. London, Ser. A, 226,* 123–192, 1970.

Bretherton, F. P., and J. C. McWilliams, Spectral estimation from irregular arrays, *Tech. Note NCAR/TN-138+STR,* Nat. Center for Atmos. Res., Boulder, Colo., 1979.

Bretherton, F. P., R. E. Davis, and C. B. Fandry, A technique for objective analysis and design of oceanographic experiments applied to Mode-73, *Deep Sea Res., 23,* 559–582, 1976.

Capon, J., High resolution frequency-wavenumber spectrum analysis, *Proc. IEEE, 57,* 1408–1418, 1969.

Cooley, J., and J. Tukey, An algorithm for the machine calculations of complex Fourier series, *Math. Comput., 19,* 297–301, 1965.

Davis, R., and L. Regier, Methods for estimating directional wave spectra from multi-element arrays, *J. Mar. Res., 35,* 453–477, 1977.

Dzhaparidze, K. O., and A. M. Yaglom, Asymptotically efficient estimation of the spectrum parameters of stationary stochastic processes, in *Proceedings of the Prague Symposium on Asymptotic Statistics,* Charles University Press, Prague, 1974.

Fletcher, R., and M. Powell, A rapidly convergent descent method for minimization, *Comput. J., 6,* 163–168, 1963.

Gandin, L. S., *Objective Analysis of Meteorological Fields,* translated from Russian, No. 1373, p. 242, Israel Program for Scientific Translations, Jerusalem, 1965.

Green, P. E., R. A. Frosch, and C. F. Romney, Principles of an experimental large aperture seismic array (LASA), *Proc. IEEE, 53,* 1821, 1933, 1965.

Jeffreys, H., *Theory of Probability,* 459 pp., Oxford University Press, New York, 1961.

Jones, R., Spectrum estimation with unequally spaced data, in *5th Conference on Probability and Statistics (November 15-18, 1977, Las Vegas, Nevada),* pp. 277–282, American Meteorological Society, Boston, Mass., 1977.

Julian, P. R., and A. K. Cline, The direct estimation of spatial wavenumber spectra of atmospheric variables, *J. Atmos. Sci., 31,* 1526–1539, 1974.

Liebelt, P. B., *An Introduction to Optimal Estimation,* 273 pp., Addison-Wesley, Reading, Mass., 1967.

Middleton, D., *An Introduction to Statistical Communication Theory,* 1140 pp., McGraw-Hill, New York, 1960.

Saltzman, B., and A. Fletcher, Spectral statistics of the wind at 500 mb, *J. Atmos. Sci., 19,* 195–204, 1962.

Thompson, R., Spectral estimation from irregularly spaced data, *IEEE Trans. Geosci. Electron., GE-9,* 107–119, 1971.

(Received March 18, 1980;
accepted April 24, 1980.)

REVIEWS OF GEOPHYSICS AND SPACE PHYSICS, VOL. 20, NO. 2, PAGES 219–232, MAY 1982

Generalized Nonlinear Inverse Problems Solved Using the Least Squares Criterion

ALBERT TARANTOLA AND BERNARD VALETTE

Laboratoire de Sismologie (CNRS), Institut de Physique du Globe de Paris
Université Pierre et Marie Curie, 75230 Paris Cedex 05, France

We attempt to give a general definition of the nonlinear least squares inverse problem. First, we examine the discrete problem (finite number of data and unknowns), setting the problem in its fully nonlinear form. Second, we examine the general case where some data and/or unknowns may be functions of a continuous variable and where the form of the theoretical relationship between data and unknowns may be general (in particular, nonlinear integrodifferential equations). As particular cases of our nonlinear algorithm we find linear solutions well known in geophysics, like Jackson's (1979) solution for discrete problems or Backus and Gilbert's (1970) solution for continuous problems.

CONTENTS

1. INTRODUCTION

The aim of physical sciences is to discover the minimal set of parameters which completely describe physical systems and the laws relating the values of these parameters to the results of any set of measurements on the system. A coherent set of such laws is named a physical theory. To the extent that the values of the parameters can only be obtained as a results of measurements, one may equivalently consider that physical theories impose some relationships between the results of some measurements.

Theoretical relationships are often functional relationships, exactly relating the values of the parameters to the results of the measurements. Sometimes, theoretical relationships are probabilistic, as in geophysics when some property of the earth is statistically described, or as in quantum mechanics, where the probabilistic description of the results of the measurements is essential to the theory.

If, given some information on the values of the set of parameters, we try to use a theoretical relationship in order to obtain information on the values of some measurable quantities, we are solving, by definition, a 'direct (or forward) problem.' If, given some information on the values of some measured quantities, we try to use a theoretical relationship in order to obtain information on the values of the set of parameters, then we are solving an 'inverse problem.' For a direct problem the values of the parameters are 'data,' and the values of some observable quantities are 'unknowns.' For an inverse problem the data are the results of some measurements, and the unknowns are the values of parameters. We will see later that actual problems are in fact mixed problems.

One of the difficulties arising in the solution of some problems is the instability (a small change in the inputs of the problem produces a physically unacceptable large change in the outputs). This difficulty appears in direct as well as in inverse problems (see, for example, *Tikhonov* [1976]).

Inverse problems may have a more essential difficulty: nonuniqueness. There are two reasons for nonuniqueness. In some problems the nonuniqueness comes from the fact that the data are discrete; if the data were dense, the solution would be unique (see, for example, *Backus and Gilbert* [1970]). In other problems, nonuniqueness may be deeper, as, for example, in the inverse problem of obtaining the density structure of a region of the earth from the measurements of the local gravitational field: Gauss' theorem states that an infinity of different density configurations give identical gravitational fields.

The classical formulation of a problem (direct or inverse) may be stated as follows: (1) We have a certain amount of information on the values of our data set, for example, some confidence intervals. (2) We have some theoretical relationships relating data and unknowns. (3) We assume a total ignorance of the values of our unknowns; that is, the sole information must come from the data set. (4) Which are the values of the unknowns?

Such a problem may be ill posed, and the solution may be nonunique.

For maximum generality the problem should be formulated as follows:

1. We have a certain state of information on the values of our data set.

2. We have also a certain state of information on the values of our unknowns (eventually the state of null information).

3. We have a certain state of information on the theoretical relationships relating data and unknowns.

4. Which is the final state of information on the values of the unknowns, resulting from the combination of the three preceding states of information?

Posed in terms of states of information, all problems are well posed, and the uniqueness of the final state of information may be warranted. We have given some preliminary results of such an approach [*Tarantola and Valette,* 1982] and have shown that in the particular case where the states of information can be described by Gaussian probability density functions, the least squares formalism is obtained.

The main purpose of this paper will then be to state clearly the nonlinear least squares approach to the generalized inverse problem and to give practical procedures to solve it.

For the linear problem, generalized least squares solutions are today well known. *Franklin* [1970] gave a very general solution, valid for discrete as well as for continuous problems, and *Jackson* [1979] discussed the use of a priori information to resolve nonuniqueness in geophysical discrete inverse problems.

In contrast, the nonlinear generalized least squares problem has not received much attention, and the usual way of solving such a problem is by iteration of a linearized problem, but as we will show in this paper, this procedure may give wrong solutions.

In section 2 we will study the discrete problem, and in section 3 we will deal with problems involving functions of a continuous variable. As a particular case of our solution for continuous problems we find the *Backus and Gilbert* [1970] solution. In section 4 we give some numerical illustrations of our algorithms.

2. The Discrete Problem

a. Notations

We will mainly follow the notations used by *Tarantola and Valette* [1982]. We will refer to that work as paper 1.

Let $\mathcal{S}$ be a physical system in a large sense. By large sense we mean that we consider that $\mathcal{S}$ is composed of a physical system in the usual sense plus all the measuring instruments. We say that the system $\mathcal{S}$ is 'parameterizable' if any state of $\mathcal{S}$ may be described using some functions and some discrete parameters. This means that we limit ourselves to the quantitative aspects of $\mathcal{S}$. If any state of $\mathcal{S}$ may be described using a finite set of discrete parameters, we say that $\mathcal{S}$ is a 'discrete system.' In this section of the paper we will focus our attention on such discrete systems.

Let $\mathbf{X} = \{X^1, \cdots, X^m\}$ be the finite set of parameters needed to describe the system, and let us use lowercase letters to note any particular value of the parameter set: $\mathbf{x} = \{x^1, \cdots, x^m\}$. Since in $\mathcal{S}$ we also include the measuring instruments, the parameter set contains all the data and the unknowns of the problem. Let $\mathscr{E}^m$ be the m-dimensioned space where the parameters $\mathbf{X}$ take their values; then $\mathbf{x}$ is a point of $\mathscr{E}^m$ and will be called a 'state' of $\mathcal{S}$.

Physical theories impose constraints between the possible values of the parameters. In the simplest case these constraints take a functional form

$$\begin{aligned} f^1(x^1, \cdots, x^m) &= 0 \\ &\vdots \\ f^r(x^1, \cdots, x^m) &= 0 \end{aligned} \tag{1}$$

which may be written

$$\mathbf{f}(\mathbf{x}) = 0 \tag{2}$$

for short. In most usual cases one can naturally define a partition of the set of parameters as

$$\mathbf{X} = \begin{bmatrix} X^1 \\ \vdots \\ X^m \end{bmatrix} = \begin{bmatrix} D^1 \\ \vdots \\ D^r \\ P^1 \\ \vdots \\ P^s \end{bmatrix} = \begin{bmatrix} \mathbf{D} \\ \mathbf{P} \end{bmatrix} \tag{3}$$

In such a way that equations (1) simplify to

$$\begin{aligned} d^1 &= g^1(p^1, \cdots, p^s) \\ &\vdots \\ d^r &= g^r(p^1, \cdots, p^s) \end{aligned} \tag{4}$$

or, for short,

$$\mathbf{d} = \mathbf{g}(\mathbf{p}) \tag{5}$$

In the traditional terminology for inverse problems the set $\mathbf{D}$ is named the set of data, and the set $\mathbf{P}$ is named the set of unknowns, but this terminology may be misleading. For example, in a problem of earthquake hypocenter location the left-hand side of (5) consists of the arrival times of seismic waves at stations, the coordinates of seismic stations being on the right-hand side. But the coordinates of the stations are the results of some direct measurements, in exactly the same way as the arrival times of waves; there is thus no reason for reserving the term 'data' for the arrival times (except in the particular case where uncertainties on station coordinates have to be neglected). Since we will use the traditional terminology in this paper, this remark must be kept in mind.

b. The Inputs of the Problem

Let us consider a given parameter X^α. Two possibilities arise: either X^α is a directly measurable parameter or it is not. This partition of the set of parameters $\mathbf{X}$ is certainly more intrinsic that the one based on the form of (5), as discussed in section 2*a*.

If X^α is a directly measurable parameter, and if we have measured it, we assume in this paper that the result of the measurement has a Gaussian form; that is, it may conveniently be described using the expected value ${x_0}^\alpha$, the variance, and the covariances with other measurements.

If X^α is not a directly measurable parameter (it is an unknown), we will assume, in order to solve otherwise underdetermined problems, that we have some a priori knowledge and that this knowledge may also be expressed in a Gaussian form. If the a priori knowledge about a parameter is weak, the corresponding variance will be large, or even infinite.

This a priori information may come from different sources. For example, if the parameter set is describing the properties of the earth, some a priori information may be obtained from models of planet formation. Or the a priori information for a given problem may result from a posteriori information of a

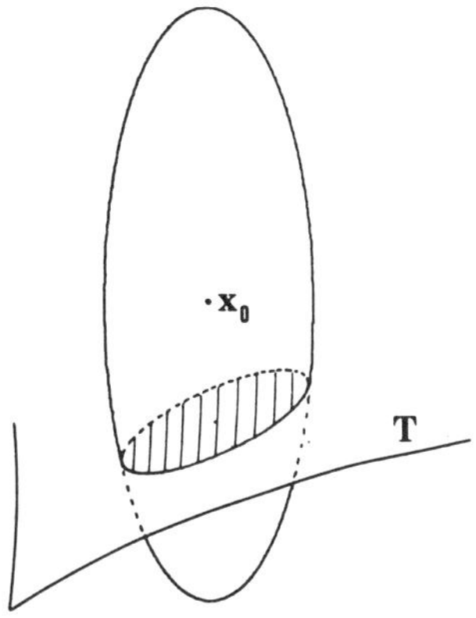

Fig. 1. Schematic illustration of the nonlinear theoretical manifold, the a priori expected point, the a priori ellipsoid of errors, and the 'induced probability' (shaded area) on the theoretical manifold.

previous inverse problem run with a different data set. More often the a priori information will simply be obtained by putting 'reasonable' error bars around a 'reasonable' central value. See *Jackson* [1979] for further discussion.

As is well known, and as we will see later, the definition of this a priori information, even with weak constraints (large variances), provides stability and uniqueness to the inversion.

In conclusion, in the least squares approach we assume that all a priori information on the parameter set (measurable as well as nonmeasurable parameters) takes the form of a vector of expected values $\mathbf{x}_0$ and a covariance matrix C_0.

When the theoretical equation (2) simplifies to (5) and the partition (3) is defined, the a priori information ($\mathbf{x}$, C_0) takes the partitioned form:

$$\mathbf{x}_0 = \begin{bmatrix} \mathbf{d}_0 \\ \mathbf{p}_0 \end{bmatrix} \qquad C_0 = \begin{bmatrix} C_{d_0d_0} & C_{d_0p_0} \\ C_{p_0d_0} & C_{p_0p_0} \end{bmatrix} \tag{6}$$

where in most useful applications the matrix $C_{d_0p_0} = (C_{p_0d_0})^T$ vanishes.

c. The Least Squares Problem

Let $\mathbf{x}_0$ be the vector of the a priori expected values, and C_0 be the a priori covariance matrix, as defined in section 2*b*. We have assumed that the a priori information has a Gaussian form. Then $\mathbf{x}_0$ and C_0 define in the parameter space $\mathscr{E}^m$ a Gaussian probability density function

$$\rho(\mathbf{x}) = \text{const} \cdot \exp\{-\tfrac{1}{2}(\mathbf{x} - \mathbf{x}_0)^T \cdot C_0^{-1} \cdot (\mathbf{x} - \mathbf{x}_0)\} \tag{7}$$

The nonlinear theoretical equation considered in (2),

$$\mathbf{f}(\mathbf{x}) = 0 \tag{8}$$

defines in $\mathscr{E}^m$ a certain nonlinear manifold (subspace) $\mathscr{T}$. The a priori density function $\rho(\mathbf{x})$ induces on the theoretical manifold $\mathscr{T}$ a probability density function which could naturally be taken as the a posteriori density function (see Figure 1).

If (8) is linear, $\mathscr{T}$ will be a linear manifold, and it has been demonstrated in paper 1 that the induced probability density function will be Gaussian. If (8) is not linear, the induced probability density function will generally not be Gaussian.

We state then the least squares problem as the search of the point $\hat{\mathbf{x}}$ of the theoretical manifold for which the induced density of probability is maximum.

In order to find this maximum we must minimize the argument of the exponential in (7). The problem is thus equivalent to searching for the point $\hat{\mathbf{x}}$ verifying the set of equations

$$\mathbf{f}(\hat{\mathbf{x}}) = 0 \tag{9}$$

$$s(\hat{\mathbf{x}}) = (\hat{\mathbf{x}} - \mathbf{x}_0)^T \cdot C_0^{-1} \cdot (\hat{\mathbf{x}} - \mathbf{x}_0) \quad \text{minimum over } \mathscr{T} \tag{10}$$

Equations (9) and (10) are our definition of the least squares problem.

We have assumed that the a priori information has a Gaussian form. If this is not the case and if we assume that the probabilistic distributions under consideration have an expected value and a covariance matrix, it is usual to employ the least squares criterion to define a 'least squares estimator' and to prove that this estimator has some properties (variance minimum, $\cdots$). See, for example, *Rao* [1973].

In fact, if the probabilistic distributions under consideration are clearly not Gaussian (long tailed, asymmetric, multimoded, laterally bounded, etc.), it is well known that the least squares criterion gives unacceptable results. For such cases the general techniques described in paper 1 should be used.

Equation (10) makes sense only if C_0 is regular. A covariance matrix becomes singular if there are null variances or if there are perfect correlations between parameters. In these cases the multidimensional ellipsoid representing C_0 degenerates into an ellipsoid of lower dimension. It is always possible to restrict the definition of the a priori probability density function to the corresponding subspace. We will see later that the algorithms leading to the solution of (9) and (10) are well defined even if C_0 is singular.

If a point $\hat{\mathbf{x}}$ verifies the set of equations (9) and (10), then it must verify the set

$$\mathbf{f}(\hat{\mathbf{x}}) = 0 \tag{11}$$

$$s(\hat{\mathbf{x}}) \text{ stationary over } \mathscr{T} \tag{12}$$

As is well known in all problems of searching for minima, the set of equations (11) and (12) may contain, in addition to the solution of (9) and (10), local minima, saddle points, maxima, etc. We will discuss this point later.

We show in the appendix that (11) and (12) are equivalent to the implicit equation

$$\hat{\mathbf{x}} = \mathbf{x}_0 + C_0 \cdot F^T \cdot (F \cdot C_0 \cdot F^T)^{-1} \cdot \{F \cdot (\hat{\mathbf{x}} - \mathbf{x}_0) - \mathbf{f}(\hat{\mathbf{x}})\} \tag{13}$$

where F is the matrix of partial derivatives:

$$F^{ik} = \partial f^i / \partial x^k \tag{14}$$

taken at the point $\hat{\mathbf{x}}$.

Equation (13) must of course be solved using an iterative procedure. If we assume that the elements F^{ik} of the matrix F are continuous functions of $\mathbf{x}$, the simplest procedure is obtained using a fixed point method:

$$\hat{\mathbf{x}}_{k+1} = \mathbf{x}_0 + C_0 \cdot F_k{}^T \cdot (F_k \cdot C_0 \cdot F_k{}^T)^{-1} \cdot \{F_k \cdot (\hat{\mathbf{x}}_k - \mathbf{x}_0) - \mathbf{f}(\hat{\mathbf{x}}_k)\} \tag{15}$$

where the derivatives are taken at the current point $\hat{\mathbf{x}}_k$.

In the rest of this paper the algorithm defined by (15) will be called the algorithm of total inversion (T.I.).

The iterative process (15) must be started at an arbitrary point $\hat{\mathbf{x}}_0$. A reasonable (although not obligatory) choice is to take the a priori point $\mathbf{x}_0$ as the starting point, that is, $\hat{\mathbf{x}}_0 = \mathbf{x}_0$.

If s has only one stationary point over $\mathcal{T}$, then it is easy to see that at this point, s is necessarily a minimum and that the T.I. algorithm converges always to the point $\hat{\mathbf{x}}$ which minimizes s (independently of the chosen starting point $\hat{\mathbf{x}}_0$). If we are not sure that s has only one stationary point, then we must check the possible existence of secondary minima or other stationary points (for instance, by starting the iterative procedure at different starting points $\hat{\mathbf{x}}_0 \neq \mathbf{x}_0$).

The total inversion algorithm is, in fact, some kind of generalization of Newton's algorithm for the search of the zeros of a nonlinear function $f(x)$. As is the case for Newton's algorithm, the total inversion algorithm will only converge in those problems where the nonlinearity is not too severe.

In ordinary iterative procedures for inverse problems, iterations are stopped when two successive solutions are close enough. In the total inversion algorithm we may also use this criterion, but because the solution $\hat{\mathbf{x}}$ must verify $\mathbf{f}(\hat{\mathbf{x}}) = 0$ and because $\mathbf{f}(\hat{\mathbf{x}}_k)$ is computed at each iteration, we can alternatively choose to stop the iterative procedure when the values $\mathbf{f}(\hat{\mathbf{x}}_k)$ are close enough to zero.

In order to give meaning to (13) and (15) we must assume that all the matrices of the type $F \cdot C_0 \cdot F^T$ are nonsingular. This is true in a great variety of circumstances. For example, from a practical point of view, a useful set of sufficient (although not necessary) conditions will be (1) that C_0 has neither null variances nor perfect correlations and (2) that theoretical equations (2) take the explicit form (5).

By hypothesis 1, C_0 is a positive definite matrix. By hypothesis 2 the matrix F takes the partitioned form

$$F = [I \quad -G] \tag{16}$$

where I is the identity matrix and G is a matrix of partial derivatives. Then F has a maximum rank, and $F \cdot C_0 \cdot F^T$ is then positive definite and thus regular.

In fact, the matrix $F \cdot C_0 \cdot F^T$ will only be singular in some very pathological cases. Nevertheless, in actual problems the matrix $F \cdot C_0 \cdot F^T$ may be numerically singular. This point will be discussed in section 2*d*.

In the iteration of algorithm (15) we must compute the vector $\mathbf{V} = F_k \cdot (\hat{\mathbf{x}}_k - \mathbf{x}_0) - \mathbf{f}(\hat{\mathbf{x}}_k)$. We must also compute the matrix $M = (F_k \cdot C_0 \cdot F_k^T)$ and then compute the vector $\mathbf{V}' = M^{-1} \cdot \mathbf{V}$. It is well known in numerical analysis that the computation of the vector $\mathbf{V}'$ does not require the effective inversion of the matrix M. This in an important point when the dimension of M is large, because the inversion is very time consuming.

d. Case $\mathbf{d} = \mathbf{g}(\mathbf{p})$

Let us assume that the parameters $\mathbf{X}$ may be divided into the data set $\mathbf{D}$ and the parameter set $\mathbf{P}$ in such a way that the theoretical equation $\mathbf{f}(\mathbf{x}) = 0$ simplifies to

$$\mathbf{f}(\mathbf{x}) = \mathbf{d} - \mathbf{g}(\mathbf{p}) = 0 \tag{17}$$

The matrix F defined by (14) takes then the partitioned form

$$F = [I \quad -G] \tag{18}$$

where G is the matrix of partial derivatives.

$$G^{i\alpha} = \partial g^i / \partial p^\alpha \tag{19}$$

Using (6), (17), and (18), the solution (15) gives the corresponding algorithms allowing the computation of $\hat{\mathbf{d}}$ and $\hat{\mathbf{p}}$. As is shown in the appendix, we obtain easily

$$\begin{aligned}\hat{\mathbf{p}}_{k+1} = \mathbf{p}_0 &+ [C_{p_0p_0} \cdot G_k^T - C_{p_0d_0}] \\ &\cdot [C_{d_0d_0} - C_{d_0p_0} G_k^T - G_k \cdot C_{p_0d_0} + G_k \cdot C_{p_0p_0} \cdot G_k^T]^{-1} \\ &\cdot \{\mathbf{d}_0 - \mathbf{g}(\hat{\mathbf{p}}_k) + G_k \cdot (\hat{\mathbf{p}}_k - \mathbf{p}_0)\}\end{aligned} \tag{20}$$

The corresponding algorithm for $\hat{\mathbf{d}}_{k+1}$ may then be written (see appendix) as

$$\hat{\mathbf{d}}_{k+1} = \mathbf{g}(\hat{\mathbf{p}}_k) + G_k \cdot (\hat{\mathbf{p}}_{k+1} - \hat{\mathbf{p}}_k) \tag{21}$$

Formula (20) generalizes to the nonlinear case the solution of *Franklin* [1970] for the linear problem.

Uncertainties in $\mathbf{d}_0$ are often independent of uncertainties in $\mathbf{p}_0$. Correspondingly,

$$C_{d_0p_0} = (C_{p_0d_0})^T = 0 \tag{22}$$

In that case, (20) becomes simpler and may be written in three different ways (see appendix):

$$\begin{aligned}\hat{\mathbf{p}}_{k+1} = \mathbf{p}_0 &+ C_{p_0p_0} \cdot G_k^T \cdot (C_{d_0d_0} + G_k \cdot C_{p_0p_0} \cdot G_k^T)^{-1} \\ &\cdot [\mathbf{d}_0 - \mathbf{g}(\hat{\mathbf{p}}_k) + G_k \cdot (\hat{\mathbf{p}}_k - \mathbf{p}_0)]\end{aligned} \tag{23}$$

$$\begin{aligned}\hat{\mathbf{p}}_{k+1} = \mathbf{p}_0 &+ (G_k^T \cdot C_{d_0d_0}^{-1} \cdot G_k + C_{p_0p_0}^{-1})^{-1} \cdot G_k^T \cdot C_{d_0d_0}^{-1} \\ &\cdot [\mathbf{d}_0 - \mathbf{g}(\hat{\mathbf{p}}_k) + G_k \cdot (\hat{\mathbf{p}}_k - \mathbf{p}_0)]\end{aligned} \tag{24}$$

$$\begin{aligned}\hat{\mathbf{p}}_{k+1} = \hat{\mathbf{p}}_k &+ (G_k^T \cdot C_{d_0d_0}^{-1} \cdot G_k + C_{p_0p_0}^{-1})^{-1} \cdot \{G_k^T \cdot C_{d_0d_0}^{-1} \\ &\cdot [\mathbf{d}_0 - \mathbf{g}(\hat{\mathbf{p}}_k)] - C_{p_0p_0}^{-1} \cdot (\hat{\mathbf{p}}_k - \mathbf{p}_0)\}\end{aligned} \tag{25}$$

For different configurations of matrices $C_{d_0d_0}$, $C_{p_0p_0}$, and G, these three algorithms, although mathematically equivalent, may be very different in time consumption.

It is easy to see that if the matrices $C_{d_0d_0}$ and $C_{p_0p_0}$ are regular, all the inverse matrices in (23)–(25) are also regular.

It may happen in actual problems that although mathematically regular, the algorithm may become numerically singular or unstable. This will always mean that the a priori information does not constrain the solution enough. A reduction of the variances in $C_{p_0p_0}$ will stabilize the solution, in the same way that dropping 'small eigenvalues' stablizes the solution of *Wiggins* [1972], which is based on the Lancsöz decomposition. But we must emphasize that since variances in $C_{p_0p_0}$ describe our a priori knowledge on parameters, reducing these variances in order to stabilize the solution means that we are introducing a priori information that, in fact, we do not possess. This fact may be hidden in formalisms where the a priori information on parameters is not explicitly stated.

Nevertheless, in most actual geophysical problems, variances in $C_{p_0p_0}$ representing the actual a priori knowledge (confidence in $\mathbf{p}_0$) allow a stable inversion of the data set, even when the number of parameters is, by many orders of magnitude, larger than the number of data (see section 2*f* for the measure of the extent to which the data set resolves the value of a given parameter).

The case where the a priori constraints on parameters are infinitely weak is obtained using a covariance matrix $C_{p_0p_0}$ of the form $C_{p_0p_0} = \sigma^2 \cdot I$ in the limit where $\sigma^2 \to \infty$. Two particular cases give simple solutions. For obvious reasons a problem where the matrix $G_k^T \cdot G_k$ is regular will be named purely overdetermined, whereas a problem where the matrix $G_k \cdot G_k^T$ is regular will be named purely underdetermined.

For a purely overdetermined problem the limit $\sigma^2 \to \infty$ gives, using (25),

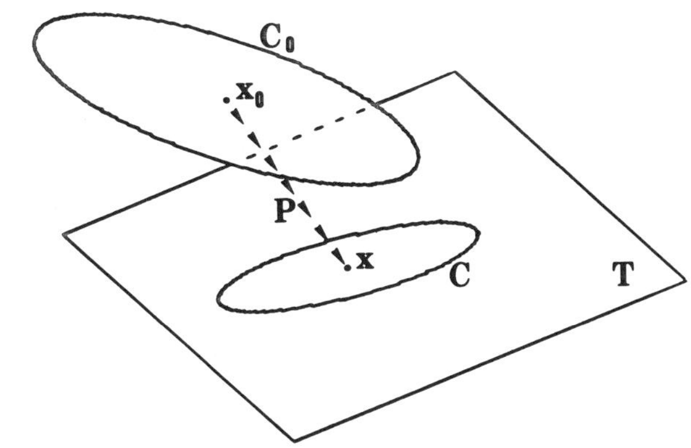

Fig. 2. The projector $P = I - C_0 \cdot F^T(F \cdot C_0 \cdot F^T)^{-1} \cdot F$ is the orthogonal (relative to C_0^{-1}) projector over the linear theoretical variety. It projects the a priori vector $\mathbf{x}_0$ as well as the a priori covariance tensor C_0, giving $\hat{\mathbf{x}}$ and C, respectively.

$$\hat{\mathbf{p}}_{k+1} = \hat{\mathbf{p}}_k + (G_k{}^T \cdot C_{d_0d_0}{}^{-1} \cdot G_k)^{-1} \cdot G_k{}^T \cdot C_{d_0d_0}{}^{-1} \cdot \{\mathbf{d}_0 - \mathbf{g}(\hat{\mathbf{p}}_k)\} \tag{26}$$

which corresponds to the solution of the classical nonlinear least squares problem.

For a purely underdetermined problem the limit $\sigma^2 \to \infty$ gives, using (23),

$$\hat{\mathbf{p}}_{k+1} = \mathbf{p}_0 + G_k{}^T \cdot (G_k \cdot G_k{}^T)^{-1} \cdot \{\mathbf{d}_0 - \mathbf{g}(\hat{\mathbf{p}}_k) - G_k \cdot (\mathbf{p}_0 - \mathbf{p}_k)\} \tag{27}$$

Equation (26) shows that the solution of a purely overdetermined problem is independent of the a priori value $\mathbf{p}_0$.

Equation (27) shows that the solution of a purely underdetermined problem is independent of the values of observational errors $C_{d_0d_0}$. If in the limit $k \to \infty$ we left-multiply (27) by G, we see that the solution $\hat{\mathbf{p}} = \hat{\mathbf{p}}_\infty$ of a purely underdetermined problem fits exactly the observed values of data: $\hat{\mathbf{d}} = \mathbf{g}(\hat{\mathbf{p}}) = \mathbf{d}_0$.

e. *Linear Problem*

By linear problem we mean that the general equations $\mathbf{f}(\mathbf{x}) = 0$ take the linear form

$$\mathbf{f}(\mathbf{x}) = F \cdot \mathbf{x} = 0 \tag{28}$$

where F is a matrix of perfectly known elements (if some of the elements are not perfectly known, they must be considered as parameters, and then the problem is not linear).

Using (28), (13) simplifies to the explicit solution

$$\hat{\mathbf{x}} = \mathbf{x}_0 - C_0 \cdot F^T \cdot (F \cdot C_0 \cdot F^T)^{-1} \cdot F \cdot \mathbf{x}_0 \tag{29}$$

As was pointed out in section 2*c*, if the probability density function describing the a priori information is Gaussian, the a posteriori density function will also be Gaussian; thus solution (29) will be the expected value. The a posteriori covariance matrix may also be computed, leading to (see appendix)

$$C = C_0 - C_0 \cdot F^T \cdot (F \cdot C_0 \cdot F^T)^{-1} \cdot F \cdot C_0 \tag{30}$$

If we define the linear operators

$$Q = C_0 \cdot F^T \cdot (F \cdot C_0 \cdot F^T)^{-1} \cdot F \qquad P = I - Q \tag{31}$$

it may easily be shown that they have the following properties:

$$\begin{aligned} P + Q &= I & P \cdot Q &= Q \cdot P = 0 \\ P^2 &= P & Q^2 &= Q \\ P \cdot C_0 &= C_0 \cdot P^T & Q \cdot C_0 &= C_0 \cdot Q^T \end{aligned} \tag{32}$$

which imply that P and Q are projectors and that they are orthogonal relative to $C_0{}^{-1}$.

Equations (29) and (30) then take the beautiful form

$$\hat{\mathbf{x}} = P \cdot \mathbf{x}_0 \tag{33}$$

$$C = P \cdot C_0 \tag{34}$$

These equations show that the same projector P projects the a priori value $\mathbf{x}_0$ to give the a posteriori value $\mathbf{x}$ and projects the a priori covariance matrix C_0 to give the a posteriori covariance matrix C (see Figure 2).

Equations (33) and (34) have been obtained in paper 1 for a Gaussian problem, without the use of the least squares criterion.

Let us now turn to the usual linear problem where the set of parameters $\mathbf{X}$ is divided into the data set $\mathbf{D}$ and the parameter set $\mathbf{P}$ and where the theoretical equation (28) simplifies to

$$F \cdot \mathbf{x} = [I \quad -G] \cdot \begin{bmatrix} \mathbf{d} \\ \mathbf{p} \end{bmatrix} = \mathbf{d} - G \cdot \mathbf{p} = 0 \tag{35}$$

that is,

$$\mathbf{d} = G \cdot \mathbf{p} \tag{36}$$

where G is a matrix of perfectly known elements.

Using (35) and equations (6), the solution of this problem is readily obtained from (33) and (34). For the parameter components, (33) gives

$$\hat{\mathbf{p}} = \mathbf{p}_0 + (C_{p_0p_0} \cdot G^T - C_{p_0d_0}) \cdot (C_{d_0d_0} - C_{d_0p_0} \cdot G^T - G \cdot C_{p_0d_0} + G \cdot C_{p_0p_0} \cdot G^T)^{-1} \cdot (\mathbf{d}_0 - G \cdot \mathbf{p}_0) \tag{37}$$

For the data components the expression obtained from (33) can be put in the form (see appendix for details)

$$\hat{\mathbf{d}} = G \cdot \hat{\mathbf{p}} \tag{38}$$

where $\hat{\mathbf{p}}$ is the solution (37). This simply means that the least squares solution exactly verifies the theoretical equation.

Equation (34) gives, for the parameter components of the a posteriori covariance matrix,

$$C_{\hat{p}\hat{p}} = C_{p_0p_0} - (C_{p_0p_0} \cdot G^T - C_{p_0d_0}) \cdot (C_{d_0d_0} - C_{d_0p_0} \cdot G^T - G \cdot C_{p_0d_0} + G \cdot C_{p_0p_0} \cdot G^T)^{-1} \cdot (G \cdot C_{p_0p_0} - C_{d_0p_0}) \tag{39}$$

while the other components may be put in the form (see appendix)

$$\begin{aligned} C_{\hat{d}\hat{p}} &= G \cdot C_{\hat{p}\hat{p}} \\ C_{\hat{p}\hat{d}} &= C_{\hat{p}\hat{p}} \cdot G^T \\ C_{\hat{d}\hat{d}} &= G \cdot C_{\hat{p}\hat{p}} \cdot G^T \end{aligned} \tag{40}$$

in accordance with (38).

In the particular case $\mathbf{p}_0 = \mathbf{0}$, (37) and (39) coincide with the solution of *Franklin* [1970].

If uncertainties in $\mathbf{d}_0$ are uncorrelated with uncertainties in $\mathbf{p}_0$, then $C_{d_0p_0} = (C_{p_0d_0})^T = 0$, and (37) simplifies to

$$\hat{\mathbf{p}} = \mathbf{p}_0 - C_{p_0p_0} \cdot G^T \cdot (C_{d_0d_0} + G \cdot C_{p_0p_0} \cdot G^T)^{-1} \cdot (\mathbf{d}_0 - G \cdot \mathbf{p}_0) \tag{41}$$

while the a posteriori covariance matrix becomes

$$C_{\hat{p}\hat{p}} = C_{p_0p_0} - C_{p_0p_0} \cdot G^T \cdot (C_{d_0d_0} + G \cdot C_{p_0p_0} \cdot G^T)^{-1} \cdot G \cdot C_{p_0p_0} \tag{42}$$

Using matricial identities, (41) and (42) may be written in a different form (see appendix):

$$\hat{\mathbf{p}} = \mathbf{p}_0 + (G^T \cdot C_{d_0d_0}{}^{-1} \cdot G + C_{p_0p_0}{}^{-1})^{-1} \cdot G^T \cdot C_{d_0d_0}{}^{-1} \cdot (\mathbf{d}_0 - G \cdot \mathbf{p}_0) \tag{43}$$

$$C_{\hat{p}\hat{p}} = (G^T \cdot C_{d_0d_0}{}^{-1} \cdot G + C_{p_0p_0}{}^{-1})^{-1} \tag{44}$$

Equations (41) and (42) coincide with the solution given by *Jackson* [1979].

The case where the a priori constraints on parameters are infinitely weak has been discussed in section 2*d*. For a purely underdetermined problem we obtain, using (41),

$$\hat{\mathbf{p}} = \mathbf{p}_0 + G^T \cdot (G \cdot G^T)^{-1} \cdot (\mathbf{d}_0 - G \cdot \mathbf{p}_0) \tag{45}$$

and, using (42),

$$C_{\hat{p}\hat{p}} = \sigma^2\{I - G^T \cdot (G \cdot G^T)^{-1} \cdot G\} \tag{46}$$

For a purely overdetermined problem we obtain, using (43),

$$\hat{\mathbf{p}} = (G^T \cdot C_{d_0d_0}{}^{-1} \cdot G)^{-1} \cdot G^T \cdot C_{d_0d_0}{}^{-1} \cdot \mathbf{d}_0 \tag{47}$$

and, using (44),

$$C_{\hat{p}\hat{p}} = (G^T \cdot C_{d_0d_0}{}^{-1} \cdot G)^{-1} \tag{48}$$

Equation (45) shows that the solution of a purely underdetermined problem is independent of $C_{d_0d_0}$ and fits exactly the observed values of the data ($\hat{\mathbf{d}} = G \cdot \hat{\mathbf{p}} = \mathbf{d}_0$).

Equation (47) shows that the solution of a purely overdetermined problem is independent of $\mathbf{p}_0$. This solution is well known and is traditionally called the 'normal solution.'

Of course, if G is a square matrix and if it is regular, then (45) and (47) reduce to the Kramer solution: $\hat{\mathbf{p}} = G^{-1} \cdot \mathbf{d}_0$.

f. Remarks

1. The usual approach to solving the nonlinear problem is through iteration of a linearized problem. If the data set **D** overdetermines the problem sufficiently so that all a priori information on the parameter set **P** can be neglected, then the iteration of a linearized problem always leads to the correct solution. If the data set does not overdetermine the problem, there is a common mistake which leads to a wrong solution. Let us examine this problem in some detail in the usual case where $C_{d_0p_0} = (C_{p_0d_0})^T = 0$.

Our solution to the nonlinear problem was (equation (25))

$$\hat{\mathbf{p}}_{k+1} = \hat{\mathbf{p}}_k + (G_k{}^T \cdot C_{d_0d_0}{}^{-1} \cdot G_k + C_{p_0p_0}{}^{-1})^{-1} \cdot \{G_k{}^T \cdot C_{d_0d_0}{}^{-1} \cdot [\mathbf{d}_0 - \mathbf{g}(\hat{\mathbf{p}}_k)] + C_{p_0p_0}{}^{-1} \cdot [\mathbf{p}_0 - \hat{\mathbf{p}}_k]\} \tag{49}$$

which has been shown to be equivalent to

$$\hat{\mathbf{p}}_{k+1} = \mathbf{p}_0 + (G_k{}^T \cdot C_{d_0d_0}{}^{-1} \cdot G_k + C_{p_0p_0}{}^{-1})^{-1} \cdot G_k{}^T \cdot C_{d_0d_0}{}^{-1} \cdot \{\mathbf{d}_0 - \mathbf{g}(\hat{\mathbf{p}}_k) + G \cdot (\hat{\mathbf{p}}_k - \mathbf{p}_0)\} \tag{50}$$

The solution to the linear problem was (equation (43))

$$\hat{\mathbf{p}} = \mathbf{p}_0 + (G^T \cdot C_{d_0d_0}{}^{-1} \cdot G + C_{p_0p_0}{}^{-1})^{-1} \cdot G^T \cdot C_{d_0d_0}{}^{-1} \cdot (\mathbf{d}_0 - G \cdot \mathbf{p}_0) \tag{51}$$

If we want our approach to be consistent, we must force the general (nonlinear) solution to give the linear solution as a particular case. It is easy to see that for a linear problem the algorithm (50) reduces to (51).

Let us show that this is not the case in the usual approach. Linearizing a problem means replacing the nonlinear equation $\mathbf{d} = \mathbf{g}(\mathbf{p})$ by its first-order development around a point $\hat{\mathbf{p}}_k$:

$$\mathbf{d} = \mathbf{g}(\hat{\mathbf{p}}_k) + G_k \cdot (\mathbf{p} - \hat{\mathbf{p}}_k) \tag{52}$$

If we call the values

$$\Delta\hat{\mathbf{d}}_k = \mathbf{d}_0 - \mathbf{g}(\hat{\mathbf{p}}_k) \tag{53}$$

'residuals' and we call the values

$$\Delta\hat{\mathbf{p}}_{k+1} = \hat{\mathbf{p}}_{k+1} - \hat{\mathbf{p}}_k \tag{54}$$

'corrections', then the linearized least squares problem consists of the search for the values $\Delta\mathbf{p}_{k+1}$ minimizing the sum

$$s = (G_k \cdot \Delta\hat{\mathbf{p}}_{k+1} - \Delta\hat{\mathbf{d}}_k)^T \cdot C_{d_0d_0}{}^{-1} \cdot (G_k \cdot \Delta\hat{\mathbf{p}}_{k+1} - \Delta\hat{\mathbf{d}}_k) \tag{55}$$

if the problem is overdetermined enough. For an underdetermined problem it is usually required that each successive correction $\Delta\hat{\mathbf{p}}_{k+1}$ be as small as possible, and the sum (55) is replaced by

$$s' = s + (\Delta\hat{\mathbf{p}}_{k+1})^T \cdot C_{p_0p_0}{}^{-1} \cdot (\Delta\hat{\mathbf{p}}_{k+1}) \tag{56}$$

The corresponding solution is then easily found to be

$$\Delta\hat{\mathbf{p}}_{k+1} = (G_k{}^T \cdot C_{d_0d_0}{}^{-1} \cdot G_k + C_{p_0p_0}{}^{-1})^{-1} \cdot G_k{}^T \cdot C_{d_0d_0}{}^{-1} \cdot \Delta\hat{\mathbf{d}}_k \tag{57}$$

Using (53) and (54) we see that (57) leads to the algorithm

$$\hat{\mathbf{p}}_{k+1} = \hat{\mathbf{p}}_k + (G_k{}^T \cdot C_{d_0d_0}{}^{-1} \cdot G_k + C_{p_0p_0}{}^{-1})^{-1} \cdot G_k{}^T \cdot C_{d_0d_0}{}^{-1} \cdot \{\mathbf{d}_0 - \mathbf{g}(\hat{\mathbf{p}}_k)\} \tag{58}$$

By comparison of (58) with (49) we see that in (58) the term $C_{p_0p_0}{}^{-1} \cdot [\mathbf{p}_0 - \hat{\mathbf{p}}_k]$ is missing. Thus the linear solution (51) cannot be obtained as a particular case of (58), which clearly means that (58) is wrong.

The mistake is to require each successive partial correction $\hat{\mathbf{p}}_{k+1} - \hat{\mathbf{p}}_k$ to be as small as possible. The right condition is, in fact, to require each successive total correction $\hat{\mathbf{p}}_{k+1} - \mathbf{p}_0$ to be as small as possible. It is then easy to see that the right equation (49) is obtained.

In any case the fully nonlinear approach developed in sections 2*c* and 2*d* is simpler and more natural than the approach consisting of the linearization of a nonlinear theory.

2. We have only been able to define the a posteriori covariance matrix for the linear case. For a strongly nonlinear problem the a posteriori errors may be far from Gaussian, and even if the covariance matrix could be computed, it would not be of great interest. If the nonlinearity is weak, then the a posteriori covariance matrix can be approximately computed using the formula obtained for the linear case.

3. Let us recall the formula (30) giving the a posteriori covariance matrix for a linear problem:

$$C = C_0 - C_0 \cdot F^T \cdot (F \cdot C_0 \cdot F^T)^{-1} \cdot F \cdot C_0 \tag{59}$$

Since the second right-hand term is clearly positive semidefinite, its diagonal terms are not negative, so the a posteriori variances are small or equal to the a priori ones. Furthermore, a parameter will be completely unresolved if it does not appear, in fact, in the equations $\mathbf{f}(\mathbf{x}) = 0$ and if no correlation is introduced by the a priori covariances between this parameter and other parameters. The corresponding column of F will then be null, and the corresponding row of C_0 will only have one nonnull element, the diagonal one. It is then easy to see that the corresponding diagonal term of the second right-hand term of (59) will be null. This implies that the a posteriori variance will equal the a priori one.

We have thus demonstrated that in the total inversion approach, the variances have the following properties:

In general,

$$\text{(a posteriori variance)} \leq \text{(a priori variance)}$$

For a nonresolved parameter,

$$\text{(a posteriori variance)} = \text{(a priori variance)}$$

The more the a posteriori variance differs from the a priori variance, the more we have increased the amount of knowledge on the value of the parameter.

We see thus that in the total inversion approach, the classical analysis of variance contains the analysis of resolution.

4. In this paper we have assumed that our theory allows an exact computation of the direct problem. Sometimes our theories contain some approximations and allow only an approximate computation of the values of the data. Let C_T be the covariance matrix of theoretical errors. In paper 1 we have demonstrated that to take into account these theoretical errors we must simply replace $C_{d_0d_0}$ by $C_{d_0d_0} + C_T$ in all formulas of sections 2*d* and 2*e*.

3. THE CONTINUOUS PROBLEM

a. Notations

In this section we will formally extend the results of section 2 to the case where some of the data and/or unknows are functions of a continuous variable.

Let us start with one linear problem involving only one data function and one unknown function:

$$d(s) = \int g(s, r) \cdot p(r)\, dr \tag{60}$$

In order to have compact notations, (60) is usually written in vectorial notation:

$$\mathbf{d} = G \cdot \mathbf{p} \tag{61}$$

where, as is easily seen from (60), G is a linear operator. The function $g(s, r)$ is then called the kernel of G.

In some problems, instead of having two functions $d(s)$ and $p(r)$ we may have one function and one discrete vector. For example, in a problem with discrete data, (60) becomes

$$d^i = \int g(s^i, r) \cdot p(r)\, dr \tag{62}$$

while (61) keeps the same form.

We will admit that the linear operator G may be of the most general form. In particular, we accept for G differential operators.

It must be pointed out that if we accept distributions as kernels, the differential equation

$$d(s) = (dp(r)/dr)_{r=s} \tag{63}$$

may be written as

$$d(s) = \int [-\delta'(s - r)] \cdot p(r)\, dr \tag{64}$$

where δ' is the derivative of the Dirac distribution. With this convention, (60) may represent integral as well as differential equations.

A slightly more general equation than (61) is

$$F^1 \cdot \mathbf{x}^1 = F^2 \cdot \mathbf{x}^2 \tag{65}$$

where F^1 and F^2 are linear operators and $\mathbf{x}^1$ and $\mathbf{x}^2$ are continuous or discrete vectors. In actual problems in geophysics we must deal with more than one data vector (for example, some seismograms or magnetic records) and with more than one parameter vector (for example, some functions describing the earth). Let us write $\mathbf{x}^1, \cdots, \mathbf{x}^m$ for all the continuous or discrete vectors we need to describe our system. A general linear relationship between these vectors is written as

$$\begin{matrix} F^{11}\,\mathbf{x}^1 & + & \cdots & + & F^{1m}\,\mathbf{x}^m & = & 0 \\ \vdots & & \vdots & & \vdots & & \vdots \\ F^{r1}\,\mathbf{x}^1 & + & \cdots & + & F^{rm}\,\mathbf{x}^m & = & 0 \end{matrix} \tag{66}$$

Let us define

$$\mathbf{X} = \begin{bmatrix} \mathbf{X}^1 \\ \vdots \\ \mathbf{X}^m \end{bmatrix} \tag{67}$$

If $\mathbf{X}^j$ belongs to a Hilbert space H^j, the vector $\mathbf{X}$ belongs to the real Hilbert space $H = H^1 \times \cdots \times H^m$. If we also define the linear operator

$$F = \begin{bmatrix} F^{11} & \cdots & F^{1m} \\ \vdots & & \vdots \\ F^{r1} & \cdots & F^{rm} \end{bmatrix} \tag{68}$$

then the linear equations (66) are simply written as

$$F \cdot \mathbf{x} = 0 \tag{69}$$

Since we also want to deal with nonlinear problems, we must consider a general nonlinear relationship of the form

$$\mathbf{f}(\mathbf{x}) = 0 \tag{70}$$

where $\mathbf{f}$ is any nonlinear integrodifferential operator acting on $\mathbf{x}$. We will later assume that $\mathbf{f}$ is differentiable. Its derivative is, by definition, a linear operator and will have the form (68).

b. Results of Measurements and a Priori Information on Parameters

Let $\mathbf{X}^i$ be one of the elements of $\mathbf{X}$. If $\mathbf{X}^i$ is a discrete vector, as was discussed in section 2*b*, we use a vector of expected values and a covariance matrix to describe the results of our measurements, as well as the a priori information on nondirectly measurable parameters.

If $\mathbf{X}$ is a continuous vector, that is, a function $s \rightarrow X(s)$ of a continuous variable s, we must use the concepts of the theory of random functions. A random function is defined as a function which, for each value of the variable, is a random variable. The expected value $X_0(s)$ of the random function is defined as the (nonrandom) function whose value at each point s equals the expected value of the random variable $X(s)$. The covariance function $C_0(s, s')$ is defined as the two-variable (nonrandom) function whose value at the point (s, s') equals the covariance between the random variables at the points s and s'. It is well known (see, for example, *Pugachev* [1965]) that covariance functions have almost all the properties of covariance matrices: they are symmetric, positive semidefinite, etc.

Covariance matrices and covariance functions naturally define linear operators which are named covariance operators.

For the sake of generality we will not assume that errors between two vectors $\mathbf{X}^j$ and $\mathbf{X}^k$ must be uncorrelated; we will thus also consider cross-covariance operators C^{jk}. To make our notations compact, we will define the matrix of covariance operators as

$$C_0 = \begin{bmatrix} C^{11} & \cdots & C^{1m} \\ \vdots & & \vdots \\ C^{m1} & \cdots & C^{mm} \end{bmatrix} \tag{71}$$

We will assume that C_0 is a positive definite operator.

Matrix C_0 and vector

$$\mathbf{x}_0 = \begin{bmatrix} (\mathbf{x}^1)_0 \\ \vdots \\ (\mathbf{x}^m)_0 \end{bmatrix} \tag{72}$$

display the results of measurements, the a priori information, and our confidence in these a priori estimators.

c. *The General Nonlinear Least Squares Problem*

Let a system be described through a set $\mathbf{X}$ of continuous and/or discrete vectors. Let $\mathbf{x}_0$ be the a priori value of $\mathbf{X}$, and let C_0 be the corresponding covariance operator. Let a physical theory impose a nonlinear relationship of the form

$$\mathbf{f}(\mathbf{x}) = 0 \tag{73}$$

on the possible values of $\mathbf{X}$, where $\mathbf{f}$ is any nonlinear differentiable operator acting on $\mathbf{X}$. There is no reason for $\mathbf{x}_0$ to verify (73). Since C_0 is a positive definite operator, its inverse C_0^{-1} can be defined, and the least squares problem may then be stated as the search for the point $\hat{\mathbf{x}}$ minimizing

$$s(\mathbf{x}) = [(\hat{\mathbf{x}} - \mathbf{x}_0), C_0^{-1}(\hat{\mathbf{x}} - \mathbf{x}_0)] \tag{74}$$

among the points verifying (73), where [,] represents the scalar product of H.

The problem defined by (73) and (74) is formally the same problem as the one defined in section 2c, so the solution given by (13) can be formally extended to give the solution of the present, more general problem. The solution of (73) and (74) will then verify

$$\hat{\mathbf{x}} = \mathbf{x}_0 + C_0 \cdot F^* \cdot (F \cdot C_0 \cdot F^*)^{-1} \cdot \{F \cdot (\hat{\mathbf{x}} - \mathbf{x}_0) - f(\hat{\mathbf{x}})\} \tag{75}$$

where the linear operator F is the derivative of the nonlinear operator $\mathbf{f}$ (having structure identical with that of the operator defined in (68)) and where F^* is its adjoint.

The solution of (75) may be obtained using a fixed point method:

$$\hat{\mathbf{x}}_{k+1} = \mathbf{x}_0 + C_0 \cdot F_k^* \cdot (F_k \cdot C_0 \cdot F_k^*)^{-1} \cdot \{F_k \cdot (\hat{\mathbf{x}}_k - \mathbf{x}_0) - f(\hat{\mathbf{x}}_k)\} \tag{76}$$

d. *Remarks*

1. If the problem is discrete, the linear operator C_0 is a matrix, and its inverse C_0^{-1} can act on any vector $(\hat{\mathbf{x}} - \mathbf{x}_0)$, so (74) has always a sense. If the problem involves some functions of a continuous variable, the operator C_0 is not necessarily surjective (the image of C_0 is not the entire Hilbert space). To give a meaning to (74), we must assume that $(\mathbf{x} - \mathbf{x}_0)$ belongs to the image of C_0. It is easy to see from (75) that the solution furnished by the algorithm (76) verifies this constraint. From a practical point of view, this means that the definition of C_0 defines also the space of possible solutions. For example, covariance operators are, in general, smoothing operators; if this is the case, then the difference between the computed solution and the a priori point, $(\hat{\mathbf{x}} - \mathbf{x}_0)$, will be smooth (see section 3*f*).

2. Since (75) is a transcription of (13), the particular cases studied in sections 3*d* and 3*e* are automatically generalized to the case where some of the vectors are continuous, and they do not have to be studied here.

e. *The Backus and Gilbert Problem*

Backus and Gilbert [1970] have examined the problem of the inversion of a finite set of discrete data, d^i, when the unknown is a function $p(r)$.

We will first state the problem in a general way and later make the particular Backus and Gilbert assumptions. If a problem involves a discrete vector $\mathbf{d}$ and a continuous vector $\mathbf{p}$, the vector $\mathbf{x}$ defined in (67) takes the form

$$\mathbf{x} = \begin{bmatrix} \mathbf{d} \\ \mathbf{p} \end{bmatrix} \tag{77}$$

We will assume that the theoretical equation is nonlinear and takes the explicit form

$$\mathbf{f}(\mathbf{x}) = \mathbf{d} - \mathbf{g}(\mathbf{p}) = 0 \tag{78}$$

With our assumptions, $\mathbf{g}$ is a vector of ordinary nonlinear functionals

$$d^i = g^i[p(r)] \tag{79}$$

The results of the measurements will be described by the discrete vector of expected values $\mathbf{d}_0$ and the covariance matrix $C_{d_0d_0}$. If we have some a priori knowledge on the function $p(r)$, let us describe it using the expected value $p_0(r)$ and the covariance function $C_{p_0p_0}(r, r')$. If we assume null covariances between $\mathbf{d}_0$ and $\mathbf{p}_0$, the covariance operator defined by (71) takes the form

$$C_0 = \begin{bmatrix} C_{d_0d_0} & 0 \\ 0 & C_{p_0p_0} \end{bmatrix} \tag{80}$$

With these hypotheses the general algorithm (76) leads to the equivalent of (23):

$$\mathbf{p}_{k+1} = \mathbf{p}_0 + C_{p_0p_0} \cdot G_k^* \cdot (C_{d_0d_0} + G_k \cdot C_{p_0p_0} \cdot G_k^*)^{-1} \cdot \{\mathbf{d}_0 - \mathbf{g}(\hat{\mathbf{p}}_k) + G_k \cdot (\hat{\mathbf{p}}_k - \mathbf{p}_0)\} \tag{81}$$

Explicitly,

$$\hat{p}_{k+1}(r) = p_0(r) + \int dr' \sum_i \sum_j C_{p_0p_0}(r, r') \cdot G_k^i(r') \cdot (S^{-1})^{ij} \cdot \{d_0^j - g^j(\hat{\mathbf{p}}_k) + \int dr'' \cdot G_k^j(r'') \cdot [\hat{p}_k(r) - p_0(r)]\} \tag{82}$$

where the matrix S_k is given by

$$S_k^{ij} = (C_{d_0d_0})^{ij} + \int dr' \int dr'' \, G_k^i(r') \cdot C_{p_0p_0}(r', r'') \cdot G_k^j(r'') \tag{83}$$

and $G_k^i(r)$ is the derivative of the nonlinear functional $g^i(\mathbf{p})$ taken at $\mathbf{p} = \hat{\mathbf{p}}_k$.

For a linear problem, (79) becomes

$$d^i = \int dr \cdot G^i(r) \cdot p(r) \quad (84)$$

Equation (81) then gives the explicit solution

$$\hat{\mathbf{p}} = \mathbf{p}_0 + C_{p_0p_0} \cdot G^* \cdot (C_{d_0d_0} + G \cdot C_{p_0p_0} \cdot G^*)^{-1} \cdot \{\mathbf{d}_0 - G \cdot \mathbf{p}_0\} \quad (85)$$

Explicitly,

$$\hat{p}(r) = p_0(r) + \int dr' \sum_i \sum_j C_{p_0p_0}(r, r') \cdot G^i(r') \cdot (S^{-1})^{ij} \cdot \left\{ d_0^j - \int dr'' \cdot G^j(r'') \cdot p_0(r'') \right\} \quad (86)$$

where the matrix S is given by

$$S^{ij} = (C_{d_0d_0})^{ij} + \int dr' \int dr''\, G^i(r') \cdot C_{p_0p_0}(r', r'') \cdot G^j(r'') \quad (87)$$

In this linear case we can compute exactly the a posteriori covariance function of $\hat{p}(r)$. By analogy with (43) we have

$$C_{\hat{p}\hat{p}} = C_{p_0p_0} - C_{p_0p_0} \cdot G^* \cdot (C_{d_0d_0} + G \cdot C_{p_0p_0} \cdot G^*)^{-1} \cdot G \cdot C_{p_0p_0} \quad (88)$$

Explicitly,

$$C_{\hat{p}\hat{p}}(r, r') = C_{p_0p_0}(r, r') - A(r, r') \quad (89)$$

where

$$A(r, r') = \int dr'' \sum_i \sum_j \int dr''' \cdot C_{p_0p_0}(r, r'') \cdot G^i(r'') \cdot (S^{-1})^{ij} \cdot G^j(r''') \cdot C_{p_0p_0}(r''', r') \quad (90)$$

The closer the function $A(r, r')$ approaches the function $C_{p_0p_0}(r, r')$, the closer the a posteriori covariance function approaches zero, that is, the better the data set resolves the function $p(r)$.

In order to obtain the Backus and Gilbert solution of the problem we must consider the particular case where the data are assumed to be error free.

$$C_{d_0d_0} = 0 \quad (91)$$

and the confidence we have in our a priori information $p_0(r)$ tends to vanish; that is, the a priori covariance function $C_{p_0p_0}(r, r')$ has infinite variances and null covariances:

$$C_{p_0p_0}(r, r') = k \cdot \delta(r - r') \quad (92)$$

where δ is the Dirac distribution and k is a constant. The linear operator whose kernel is given by (92) is proportional to the identity operator

$$C_{p_0p_0} = k \cdot I \quad (93)$$

Using (91) and (93), (85) and (88) become

$$\hat{\mathbf{p}} = \mathbf{p}_0 + G^*(G \cdot G^*)^{-1} \cdot (\mathbf{d}_0 - G \cdot \mathbf{p}_0) \quad (94)$$

$$C_{\hat{p}\hat{p}} = k \cdot [I - G^* \cdot (G \cdot G^*)^{-1} G] \quad (95)$$

Explicitly,

$$\hat{p}(r) = p_0(r) + \sum_i \sum_j G^i(r) \cdot (S^{-1})^{ij} \cdot \left\{ d_0^j - \int dr' \cdot G^j(r') \cdot p_0(r') \right\} \quad (96)$$

$$C_{\hat{p}\hat{p}}(r, r') = k \cdot [\delta(r - r') - A(r, r')] \quad (97)$$

where

$$S^{ij} = \int dr \cdot G^i(r) \cdot G^j(r) \quad (98)$$

$$A(r, r') = \sum_i \sum_j G^i(r) \cdot (S^{-1})^{ij} \cdot G^j(r') \quad (99)$$

By left-multiplying (94) by G, we see that our solution exactly fits the observed data set ($\hat{\mathbf{d}} = G \cdot \hat{\mathbf{p}} = \mathbf{d}_0$).

If we put $p_0(r) \equiv 0$ in (96), we obtain the Backus and Gilbert solution of the problem. Let us explain the existence of $\mathbf{p}_0(r)$ in (94) from the Backus and Gilbert point of view. With our notations their solution is written as

$$\hat{\mathbf{p}}_{BG} = G^* \cdot (G \cdot G^*)^{-1} \cdot \mathbf{d}_0 \quad (100)$$

Backus and Gilbert argue that we can add to (100) any function $p'(r)$ which has no effect on the values of computed data, that is, such that

$$\int dr \cdot G^i(r) \cdot p'(r) = 0 \quad (101)$$

that is,

$$G \cdot \mathbf{p}' = 0 \quad (102)$$

Our solution (94) differs from (100) by the additive term

$$\mathbf{p}' = \mathbf{p}_0 - G^* \cdot (G \cdot G^*)^{-1} \cdot G \cdot \mathbf{p}_0 \quad (103)$$

By left-multiplying this term by G we see that the addition of $\mathbf{p}'$ to $\hat{\mathbf{p}}_{BG}$ has no effect on the values of computed data, so the solution (94) verifies the Backus and Gilbert requirements. From our point of view, the Backus and Gilbert solution corresponds to the particular choice $\mathbf{p}_0 \equiv 0$.

The function (99) is named the 'resolving kernel' in the Backus and Gilbert paper, and they show that the closer $A(r, r')$ approaches a Dirac function $\delta(r - r')$, the better the solution is resolved by the data set. Equation (97) gives the probabilistic interpretation of their 'δ-ness' criterion: If $A(r, r')$ approaches $\delta(r - r')$, the a posteriori covariance function $C_{\hat{p}\hat{p}}(r, r')$ tends to vanish; that is, the solution tends to be perfectly resolved. Equation (89) shows the generalization of the δ-ness criterion to the case where the a priori information is used.

We have then shown that our solution contains the Backus and Gilbert solution and that we generalize this solution to the case where the data may have a general Gaussian error distribution, where we may take into account a priori assumptions on the unknown function and where the theoretical relationship between the data and the unknown function is nonlinear.

4. THREE NUMERICAL ILLUSTRATIONS

In this section we examine three problems that cannot naturally be solved using traditional approaches. These three problems are as follows.

1. The computation of a regional stress tensor from the measurements of the strike and dip of faults and the direction and sense of striae on these faults. This is a discrete

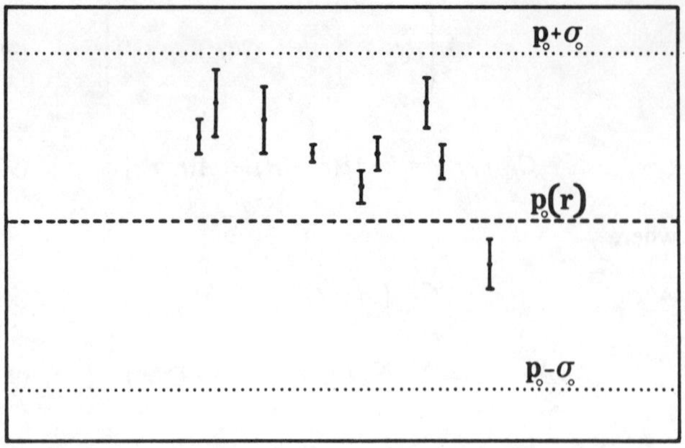

Fig. 3. Data used for the problem of estimation of a curve, given some of its points, and error bars. We also show the a priori value $p_0(r)$ and its a priori error bar.

problem, and its interest arises from the fact that the theoretical nonlinear equations do not take the form $\mathbf{d} = \mathbf{g}(\mathbf{p})$ but the more general form $\mathbf{f}(\mathbf{x}) = 0$.

2. The estimation of a curve given some of its points, or given the values of its derivative at some points. This is a linear and highly underdetermined problem. We will show that we can naturally solve it using the a priori hypothesis that the corresponding curve is smooth.

3. The solution of a gravimetric nonlinear problem with an unknown function and a discrete set of data.

a. Computation of a Regional Stress Tensor

To describe the orientation of a fault plane and striae, we use three angles: strike (d), dip (p), and slip (i). The components of the unit normal, $\mathbf{n}$, and unit striae, $\mathbf{s}$, are then easily computed as functions of d, p, and i. Let T be the regional deviatoric stress tensor. It is shown by *Angelier et al.* [1982] that if we assume that slickenslides on faults are caused by the existence of a regional stress tensor T, then on each fault we must have

$$\mathbf{s} \cdot T \cdot \mathbf{n} - [\|T \cdot \mathbf{n}\|^2 - (\mathbf{n} \cdot T \cdot \mathbf{n})^2]^{1/2} = 0 \qquad (104)$$

We see then that each set of three measured quantities (d, p, i) leads to one theoretical nonlinear equation of the form $\mathbf{f}(\mathbf{x}) = 0$. Algorithms which assume that the theoretical equations have the form $\mathbf{d} = \mathbf{g}(\mathbf{p})$, where $\mathbf{d}$ is the data set, cannot solve this problem in a natural way. See the paper by Angelier et al. for more details and numerical results.

b. Estimation of a Curve Given Some Points

This is a very common problem in all branches of physics. Let d be a quantity that physically depends on a quantity r; that is, we assume that d and r are functionally related:

$$d = p(r) \qquad (105)$$

Let us assume that we have measured the value of d for some values of r and that we have obtained the results shown in Figure 3. Our aim is to give a 'reasonable' expression for $p(r)$ that fits the experimental points 'reasonably well.'

Our unknown is the function $p(r)$, and our data are the points d^i. The theoretical equation is

$$d^i = p(r^i) \qquad (106)$$

which can also be written as

$$d^i = \int \delta(r - r^i) \cdot p(r)\, dr \qquad (107)$$

Equation (107) 'shows that our problem is linear with kernel

$$G^i(r) = \delta(r^i - r) \qquad (108)$$

The a priori information on the problem is as follows: (1) The results of the measurements are described using the observed values $d_0{}^i$ of Figure 3 and their standard deviations $\sigma_0{}^i$. (2) Since the problem is highly underdetermined, we must use some kind of a priori information. Let us assume that we have some reasons to force the solution $p(r)$ to be not too far from $p_0(r)$. Let σ be the confidence we have on this a priori value. This means that we expect the dispersion of $p(r)$ around $p_0(r)$ to be of the order of σ. Finally, we will introduce the assumption that $p(r)$ is smooth. Since $p_0(r)$ is smooth by definition, the simplest way to impose the smoothness of $p(r)$ is to impose that if, at a given point r, the value of $p(r)$ has a deviation $p(r) - p_0(r)$ of given sign and magnitude, we want, at a neighboring point r', the deviation $p(r') - p_0(r')$ to have the same sign and similar magnitude. In other words, we want to impose a priori nonnull covariances between points r and r'. Many choices of covariance functions may correspond to the physics of each problem. Let us take here a covariance function of the form

$$C_{p_0p_0}(r, r') = \sigma^2 \exp\left\{-\frac{1}{2}\frac{(r - r')^2}{\Delta^2}\right\} \qquad (109)$$

which means that the variance at the point r, $C_{p_0p_0}(r, r)$, equals σ and that the correlation length between errors is Δ.

Since we have precisely defined the a priori values $\mathbf{d}_0$, $C_{d_0d_0}$, $\mathbf{p}_0$, and $C_{p_0p_0}$ and the theoretical equation (107), the solution of our problem is readily obtained from (86). Using (108), (86) gives

$$\hat{p}(r) = p_0(r) + \sum_i \sum_j C_{p_0p_0}(r, r^i) \cdot (S^{-1})^{ij} \cdot [d_0{}^j - p_0(r^j)] \qquad (110)$$

where

$$S^{ij} = (C_{d_0d_0})^{ij} + C_{p_0p_0}(r^i, r^j) \qquad (111)$$

The a posteriori covariance function is obtained from (89):

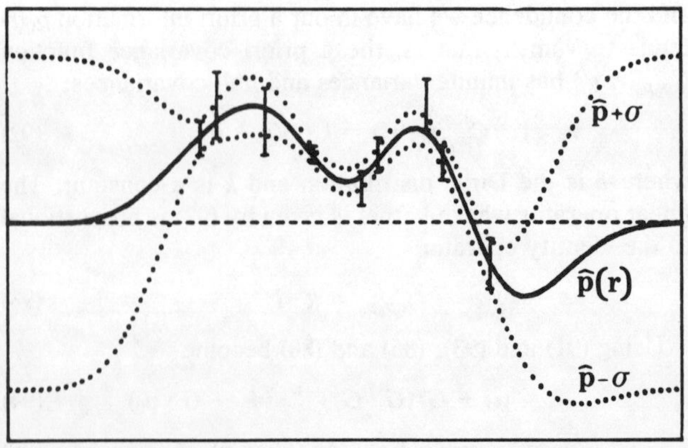

Fig. 4. Solution obtained using our algorithm (solid curve). We also show the standard error at each point $[C_{\hat{p}\hat{p}}(r, r)]^{1/2}$). Note that far from the data the standard error tends to equal the a priori standard error.

$$C_{\hat{p}\hat{p}}(r, r') = C_{p_0p_0}(r, r') - \sum_i \sum_j C_{p_0p_0}(r, r^i) \cdot (S^{-1})^{ij} \cdot C_{p_0p_0}(r^j, r') \quad (112)$$

In Figure 4 we show the solution $\hat{p}(r)$ given by (110), and we also show the a posteriori standard deviation $[C_{\hat{p}\hat{p}}(r, r)]^{1/2}$.

Figure 5 shows the covariance function $C_{\hat{p}\hat{p}}(r, r')$ for a particular value of r'.

If instead of giving the value of the function at some points we give the value of its derivative, the problem is very similar. Equations (106), (107), and (108) should be written as

$$d^i = p'(r^i) \quad (113)$$

$$d^i = \int (-\delta'(r^i - r)) \cdot p(r)\, dr \quad (114)$$

$$G^i(r) = -\delta'(r^i - r) \quad (115)$$

respectively, where δ' is the derivative of the Dirac distribution.

Equations (110), (111), and (112) then become

$$\hat{p}(r) = p_0(r) + \sum_i \sum_j \left(\frac{\partial C_{p_0p_0}(r, r')}{\partial r'} \right)_{r'=r^i} \cdot (S^{-1})^{ij} \cdot [d_0^j - p_0'(r^j)] \quad (116)$$

$$S^{ij} = (C_{d_0d_0})^{ij} + \left(\frac{\partial^2 C_{p_0p_0}(r, r')}{\partial r\, \partial r'} \right)_{r=r^i, r'=r^j} \quad (117)$$

$$C_{\hat{p}\hat{p}}(r, r') = C_{p_0p_0}(r, r') - \sum_i \sum_j \left(\frac{\partial^2 C_{p_0p_0}(r, r'')}{\partial r''} \right)_{r''=r^i} \cdot (S^{-1})^{ij} \cdot \left(\frac{\partial^2 C_{p_0p_0}(r''', r')}{\partial r'''} \right)_{r'''=r^j} \quad (118)$$

respectively.

The curve $\hat{p}(r)$ is a smooth curve whose derivative at the points r^j approaches d_0^j.

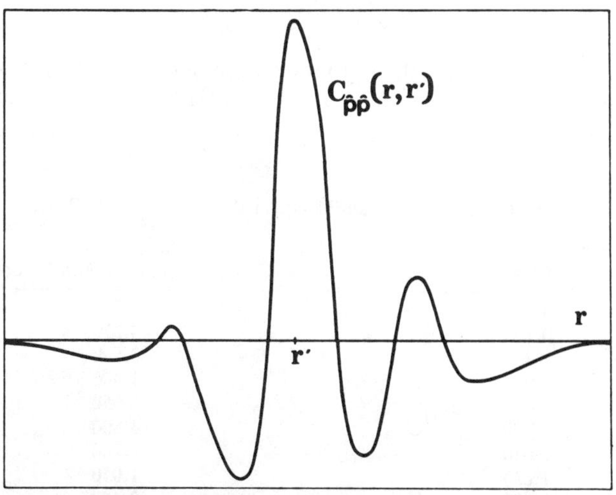

Fig. 5. The covariance curve $C_{\hat{p}\hat{p}}(r, r')$ computed at an arbitrary point r'. This curve generalizes the 'resolving kernel' of *Backus and Gilbert* [1970].

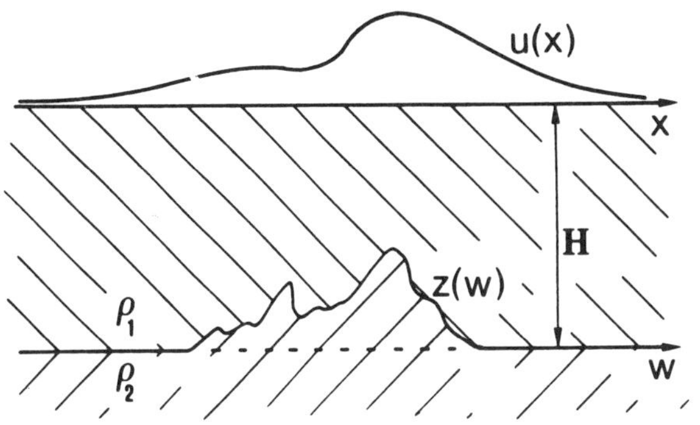

Fig. 6. The gravitational inverse problem of obtaining the shape of a frontier between two media of different density, using as data the anomaly at the surface.

c. Nonlinear Problem With Discrete Data and a Function as Unknown

Our last example has been borrowed from *Tikhonov* [1976], who shows it as an example of an ill-posed nonlinear problem.

Let us assume a layer of uniform density over a half space of uniform but different density. If between two points a and b the interface is not planar but has a shape $z(w)$ (see Figure 6), the vertical component of gravity and the free surface will have an anomaly $u(x)$. Assuming a two-dimensional problem, it may be shown that

$$u(x) = \int_a^b \log \frac{(x - w)^2 + H^2}{(x - w)^2 + [H - z(w)]^2}\, dw \quad (119)$$

We want to solve the inverse problem of estimating the function $z(w)$ from a finite set of measurements of the anomaly

$$d_0^{\,i} = u(x^i) \quad (120)$$

The general solution of the problem is given by (81) or (82). Let us explain the computation of the kernel $G_k^{\,i}(w)$ of the nonlinear functional (119).

Equation (119) defines a nonlinear operator which may be written as

$$\mathbf{u} = \mathbf{g}(\mathbf{z}) \quad (121)$$

Let $\boldsymbol{\varepsilon}$ be a function of w. It can be shown that in the limit when $\boldsymbol{\varepsilon} \to 0$, the operator which associates to $\boldsymbol{\varepsilon}$, the function

$$\mathbf{g}(\mathbf{z}_k + \boldsymbol{\varepsilon}) - \mathbf{g}(\mathbf{z}_k) \quad (122)$$

is a linear operator. It is by definition the derivative of $\mathbf{g}(\mathbf{z})$ at the point $\mathbf{z}_k$. If we note $G_k^{\,i}(w)$, the kernel of the derivative, we have then

$$\int_a^b G_k^{\,i}(w)\, \varepsilon(w)\, dw = \int_a^b \left\{ \log \left(\frac{(x^i - w)^2 + H^2}{(x^i - w)^2 + \{H - [z_k(w) + \varepsilon(w)]\}^2} \right) - \log \left(\frac{(x^i - w)^2 + H^2}{(x^i - w)^2 + [H - z_k(w)]^2} \right) \right\} dw \quad (123)$$

in the limit $\boldsymbol{\varepsilon} \to 0$. The computation of the limit on the right-hand side is carried out formally in exactly the same way as

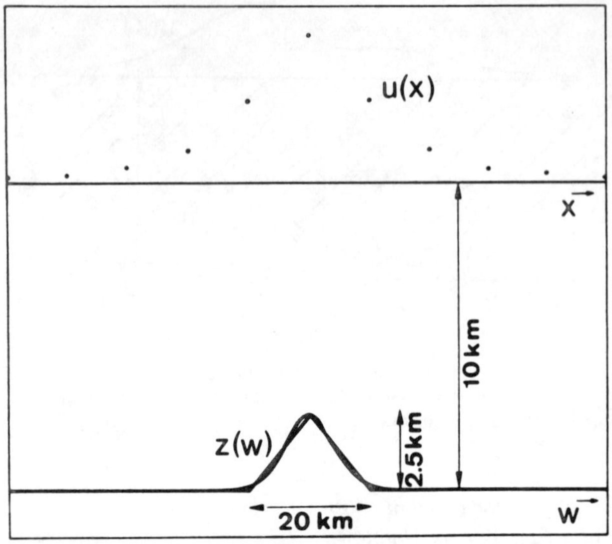

Fig. 7. Synthetic data (top) generated from a 'true model' (triangle at bottom). The smooth curve near the triangle corresponds to the results of the inversion of the synthetic data (which have been contaminated with errors, as shown in Table 1).

for the computation of the classical derivative of the expression

$$\log \frac{(x^i - w)^2 + H^2}{(x^i - w)^2 + (H - z)^2} \tag{124}$$

with respect to the variable z. We obtain then

$$G_k^{\,i}(w) = \frac{2(H - z(w))}{(x^i - w)^2 + [H - z(w)]^2} \tag{125}$$

We have generated a set of synthetic data as described in Figure 7 and Table 1. This data set has been contaminated with arbitrary errors, and it has then been inverted in order to evaluate $z(w)$. The following assumptions have been made:

1. The standard error of the data is 0.10 units ($C_{d_0d_0} = 0.01 \cdot I$).
2. The a priori estimate of $z(w)$ is $z_0(w) \equiv 0$ with an error bar of 5 km.
3. We do not want any oscillations of the solution with wavelength shorter than 1 km.

From assumptions 2 and 3 it follows that an adequate a priori covariance function may be

$$C_{p_0p_0}(w, w') = \sigma^2 \exp\left\{-\frac{1}{2}\frac{(w - w')^2}{\Delta^2}\right\} \tag{126}$$

with $\sigma = 5$ km and $\Delta = 1$ km.

The result is obtained by application of the algorithm (82):

$$\hat{z}_{k+1}(w) = \int dw' \sum_i \sum_j C_{p_0p_0}(w, w') \cdot G_k^{\,i}(w') \cdot (S^{-1})^{ij}$$

$$\cdot \left\{ d_0^{\,j} - g^j(\hat{\mathbf{z}}_k) + \int dw'' G_k^{\,j}(w'') \cdot z_k(w'') \right\} \tag{127}$$

and is shown in Figure 7.

The number of iterations needed depends strongly on the starting point. For the starting point $\hat{z}_0(w) \equiv z_0(w) \equiv 0$ an accuracy of a few percent is obtained in two iterations. Figure 8 shows the results of the first two iterations for a remote starting point. The final result is independent of the starting point $\hat{z}_0(w)$ (but of course depends on the a priori point $z_0(w)$, which has always been taken as null in this example.

The result of the inversion looks similar to the true value of Figure 8, but the sharp slope discontinuities of the true solution have been smoothed.

The integrations involved in the algorithm (82) have been numerically performed using a grid of 100 points. The reduction of the number of points to 50 does not alter the solution significantly.

5. CONCLUSION

We have shown that the least squares problem admits a general definition, valid for discrete as well as for continuous problems, for overdetermined as well as for underdetermined problems and for linear as well as for nonlinear problems. We have shown that the use of the concept of an a priori covariance matrix (or function) allows one to obtain stable solutions for otherwise unstable problems, or smooth solutions when required. Our general solution (76) solves the simpler problems of least squares adjustments, as well as problems involving any number of integrodifferential equations. The convergence of the algorithm will only be ensured if the nonlinearity is not too strong.

APPENDIX

Let us first demonstrate the equivalence of (13) and (11) and (12).

We will call $\mathcal{T}$ the nonlinear manifold defined by the equation $\mathbf{f}(\mathbf{x}) = 0$ (the theoretical manifold). We assume $\mathbf{f}$ to be differentiable. The matrix of partial (Fréchet) derivatives,

$$F^{ik} = \partial f^i / \partial x^k \tag{A1}$$

defines a linear application which is named the tangent linear application. We will assume F to be of maximum rank. Let S be the tangent linear application of s:

$$S^k = \partial s / \partial x^k \tag{A2}$$

Let $\hat{\mathbf{x}}$ be a solution of (11) and (12). Then s is stationary at $\hat{\mathbf{x}}$; that is, the tangent linear application S is null over the tangent linear manifold to $\mathcal{T}$ at $\hat{\mathbf{x}}$. We easily obtain

$$S = 2(\hat{\mathbf{x}} - \mathbf{x}_0)^T \cdot C_0^{-1} \tag{A3}$$

Since a vector $\mathbf{V}$ belongs to the tangent linear manifold to $\mathcal{T}$ at $\hat{\mathbf{x}}$ if and only if $F \cdot \mathbf{V} = 0$, (11) and (12) are equivalent to

TABLE 1. Values of Data Used for the Gravitational Problem

Error Free Values of Data	Values of Data Contaminated With Errors
0.181	0.200
0.280	0.250
0.487	0.500
1.023	1.000
2.676	2.650
4.770	4.800
2.676	2.700
1.023	1.050
0.487	0.450
0.280	0.300
0.181	0.150

$$\mathbf{f}(\hat{\mathbf{x}}) = 0 \tag{A4}$$

$$F \cdot \mathbf{v} = 0 \Rightarrow (\hat{\mathbf{x}} - \mathbf{x}_0)^T \cdot C_0^{-1} \cdot \mathbf{v} = 0 \tag{A5}$$

Since F is of maximum rank, (A5) implies the existence of a vector of Lagrange parameters, L, such that $(\hat{\mathbf{x}} - \mathbf{x}_0)^T \cdot C_0^{-1} = \mathbf{L}^T \cdot F$. Equations (A4) and (A5) are then equivalent to

$$\mathbf{f}(\hat{\mathbf{x}}) = 0 \tag{A6}$$

$$\exists\, \mathbf{L} : (\hat{\mathbf{x}} - \mathbf{x}_0) = C_0 \cdot F^T \cdot \mathbf{L} \tag{A7}$$

By left-multiplying (A7) by F we obtain

$$F \cdot (\hat{\mathbf{x}} - \mathbf{x}_0) = (F \cdot C_0 \cdot F^T) \cdot \mathbf{L} \tag{A8}$$

and since C_0 is positive definite and F is of maximum rank,

$$\mathbf{L} = (F \cdot C_0 \cdot F^T)^{-1} \cdot F \cdot (\hat{\mathbf{x}} - \mathbf{x}_0) \tag{A9}$$

Equations (A6) and (A7) are then equivalent to the set of equations

$$\mathbf{f}(\hat{\mathbf{x}}) = 0 \tag{A10}$$

$$(\hat{\mathbf{x}} - \mathbf{x}_0) = C_0 \cdot F^T \cdot (F \cdot C_0 \cdot F^T)^{-1} \cdot F \cdot (\hat{\mathbf{x}} - \mathbf{x}_0) \tag{A11}$$

which are equivalent to the single equation

$$(\hat{\mathbf{x}} - \mathbf{x}_0) = C_0 \cdot F^T \cdot (F \cdot C_0 \cdot F^T)^{-1} \cdot \{F \cdot (\hat{\mathbf{x}} - \mathbf{x}_0) - \mathbf{f}(\hat{\mathbf{x}})\} \tag{A12}$$

It is obvious that (A10) and (A11) imply (A12). The reciprocal is easily demonstrated by left-multiplying (A12) by F.

Let us now turn to the demonstration of equations (20) and (21).

Using (6), (17), and (18), we successively obtain

$$\begin{aligned} F_k \cdot (\mathbf{x}_k - \mathbf{x}_0) - \mathbf{f}(\mathbf{x}_k) &= [I \quad -G_k] \cdot \begin{bmatrix} \hat{\mathbf{d}}_k - \mathbf{d}_0 \\ \hat{\mathbf{p}}_k - \mathbf{p}_0 \end{bmatrix} \\ &\quad - [\hat{\mathbf{d}}_k - \mathbf{g}(\hat{\mathbf{p}}_k)] \\ &= -\{\mathbf{d}_0 - \mathbf{g}(\hat{\mathbf{p}}_k) + G_k \cdot (\hat{\mathbf{p}}_k - \mathbf{p}_0)\} \end{aligned} \tag{A13}$$

$$\begin{aligned} C_0 \cdot F_k^T &= \begin{bmatrix} C_{d_0d_0} & C_{d_0p_0} \\ C_{p_0d_0} & C_{p_0p_0} \end{bmatrix} \cdot \begin{bmatrix} I \\ -G_k^T \end{bmatrix} \\ &= \begin{bmatrix} C_{d_0d_0} - C_{d_0p_0} \cdot G_k^T \\ C_{p_0d_0} - C_{p_0p_0} \cdot G_k^T \end{bmatrix} \end{aligned} \tag{A14}$$

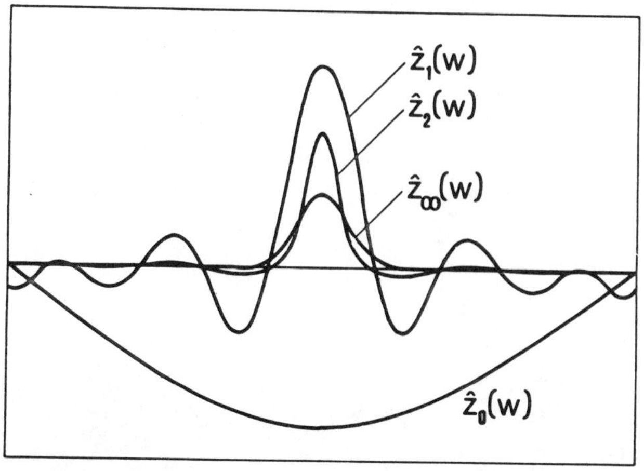

Fig. 8. The problem is nonlinear, and this figure illustrates the convergence of the algorithm when the 'starting point' is far from the 'true solution.' The final solution is rigorously independent of the starting point.

$$\begin{aligned} F_k \cdot C_0 \cdot F_k^T &= C_{d_0d_0} - C_{d_0p_0} \cdot G_k^T \\ &\quad - G_k \cdot C_{p_0d_0} + G_k \cdot C_{p_0p_0} \cdot G_k^T \end{aligned} \tag{A15}$$

then (15) gives

$$\begin{aligned} \begin{bmatrix} \hat{\mathbf{d}}_{k+1} \\ \hat{\mathbf{p}}_{k+1} \end{bmatrix} &= \begin{bmatrix} \mathbf{d}_0 \\ \mathbf{p}_0 \end{bmatrix} + \begin{bmatrix} C_{d_0p_0} \cdot G_k^T - C_{d_0d_0} \\ C_{p_0p_0} \cdot G_k^T - C_{p_0d_0} \end{bmatrix} \\ &\quad \cdot [C_{d_0d_0} - C_{d_0p_0} \cdot G_k^T - G_k \cdot C_{p_0d_0} + G_k \cdot C_{p_0p_0} \cdot G_k^T]^{-1} \\ &\quad \cdot [\mathbf{d}_0 - \mathbf{g}(\hat{\mathbf{p}}_k) + G_k \cdot (\hat{\mathbf{p}}_k - \mathbf{p}_0)] \end{aligned} \tag{A16}$$

The solution for $\hat{\mathbf{p}}_{k+1}$ coincides with (20). To obtain (21), we must first demonstrate a matricial identity. From the trivial equality

$$\begin{aligned} (C_{d_0d_0} - C_{d_0p_0} \cdot G^T) &= (C_{d_0d_0} - C_{d_0p_0} \cdot G^T - G \cdot C_{p_0d_0} \\ &\quad + G \cdot C_{p_0p_0} \cdot G^T) - G \cdot (C_{p_0p_0} \cdot G^T - C_{p_0d_0}) \end{aligned} \tag{A17}$$

we deduce, by right-multiplication with the appropriate matrix,

$$\begin{aligned} &(C_{d_0d_0} - C_{d_0p_0} \cdot G^T) \cdot (C_{d_0d_0} - C_{d_0p_0} \cdot G^T - G \cdot C_{p_0d_0} \\ &\quad + G \cdot C_{p_0p_0} \cdot G^T)^{-1} = I - G \cdot (C_{p_0p_0} \cdot G^T - C_{p_0d_0}) \\ &\quad \cdot (C_{d_0d_0} - C_{d_0p_0} \cdot G^T - G \cdot C_{p_0d_0} + G \cdot C_{p_0p_0} \cdot G^T)^{-1} \end{aligned} \tag{A18}$$

Using (A18), the solution for $\hat{\mathbf{d}}_{k+1}$, as given by (A16), may be written as

$$\begin{aligned} \hat{\mathbf{d}}_{k+1} &= \mathbf{d}_0 - [I - G_k \cdot (C_{p_0p_0} \cdot G_k^T - C_{p_0d_0}) \\ &\quad \cdot (C_{d_0d_0} - C_{d_0p_0} \cdot G_k^T - G_k \cdot C_{p_0d_0} + G_k \cdot C_{p_0p_0} \cdot G_k^T)^{-1}] \\ &\quad \cdot [\mathbf{d}_0 - \mathbf{g}(\hat{\mathbf{p}}_k) + G_k \cdot (\hat{\mathbf{p}}_k - \mathbf{p}_0)] \\ &= \mathbf{g}(\hat{\mathbf{p}}_k) - G_k \cdot \hat{\mathbf{p}}_k + G_k \cdot \hat{\mathbf{p}}_{k+1} \end{aligned} \tag{A19}$$

which demonstrates (21).

To demonstrate (24) and (25), we will use the matricial identities

$$\begin{aligned} &C_{p_0p_0} \cdot G^T \cdot (C_{d_0d_0} + G \cdot C_{p_0p_0} \cdot G^T)^{-1} \\ &\quad = (G^T \cdot C_{d_0d_0}^{-1} \cdot G + C_{p_0p_0})^{-1} \cdot G^T \cdot C_{d_0d_0}^{-1} \end{aligned} \tag{A20}$$

$$\begin{aligned} &(G^T \cdot C_{d_0d_0}^{-1} \cdot G + C_{p_0p_0}^{-1})^{-1} = C_{p_0p_0} \\ &\quad - C_{p_0p_0} \cdot G^T \cdot (C_{d_0d_0} + G \cdot C_{p_0p_0} \cdot G^T)^{-1} \cdot G \cdot C_{p_0p_0} \end{aligned} \tag{A21}$$

which have been demonstrated in paper 1.

Equation (24) is deduced from (23) using (A20). It is obvious that (23) can be written as

$$\begin{aligned} \hat{\mathbf{p}}_{k+1} &= \hat{\mathbf{p}}_k + C_{p_0p_0} G_k^T \cdot (C_{d_0d_0} + G_k \cdot C_{p_0p_0} \cdot G_k^T)^{-1} \\ &\quad \cdot [\mathbf{d}_0 - \mathbf{g}(\hat{\mathbf{p}}_k)] - [C_{p_0p_0} - C_{p_0p_0} \cdot G_k^T \\ &\quad \cdot (C_{d_0d_0} + G_k \cdot C_{p_0p_0} \cdot G_k^T)^{-1} \cdot G_k \cdot C_{p_0p_0}] \\ &\quad \cdot C_{p_0p_0}^{-1} \cdot (\hat{\mathbf{p}}_k - \mathbf{p}_0) \end{aligned} \tag{A22}$$

Equation (25) is deduced from (A22) using (A20) and (A21).

A proper derivation of (30) has been made in paper 1, where C is obtained as the covariance matrix of the a posteriori probability density function. A formal derivation of (30) is obtained if we consider (29) as a relation between random variables. By definition of covariance matrix,

$$C = E\{[\hat{\mathbf{x}} - E(\hat{\mathbf{x}})][\hat{\mathbf{x}} - E(\hat{\mathbf{x}})]^T\} \tag{A23}$$

where E stands for mathematical expectation. Using the notations of (31), (32), and (33), we successively obtain

$$\begin{aligned} C &= E\{[P\mathbf{x}_0 - E(P\mathbf{x}_0)][P\mathbf{x}_0 - E(P\mathbf{x}_0)]^T\} \\ &= P \cdot E\{[\mathbf{x}_0 - E(\mathbf{x}_0)][\mathbf{x}_0 - E(\mathbf{x}_0)]^T\}P^T \\ &= P \cdot C_0 \cdot P^T = P^2 \cdot C_0 = P \cdot C_0 \end{aligned} \tag{A24}$$

which demonstrates (34) and (30).

The proof of (37) and (38) is a particular case of that of (20) and (21). Using (6), (35), and (36), (33) becomes

$$\begin{aligned} \begin{bmatrix} \hat{\mathbf{d}} \\ \hat{\mathbf{p}} \end{bmatrix} &= \begin{bmatrix} \mathbf{d}_0 \\ \mathbf{p}_0 \end{bmatrix} + \begin{bmatrix} C_{d_0p_0} \cdot G^T - C_{d_0d_0} \\ C_{p_0p_0} \cdot G^T - C_{p_0d_0} \end{bmatrix} \\ &\quad \cdot [C_{d_0d_0} - C_{d_0p_0} \cdot G^T - G \cdot C_{p_0d_0} \\ &\quad + G \cdot C_{p_0p_0} \cdot G^T]^{-1}\, [\mathbf{d}_0 - G \cdot \mathbf{p}_0] \end{aligned} \tag{A25}$$

The solution for $\hat{\mathbf{p}}$ coincides with (37). Using (A18), the solution for $\hat{\mathbf{d}}$ may be written as

$$\begin{aligned} \hat{\mathbf{d}} &= \mathbf{d}_0 - [I - G \cdot (C_{p_0p_0} \cdot G^T - C_{p_0d_0}) \cdot (C_{d_0d_0} - C_{d_0p_0} \cdot G^T \\ &\quad - G \cdot C_{p_0d_0} + G \cdot C_{p_0p_0} \cdot G^T)^{-1} \\ &\quad \cdot (\mathbf{d}_0 - G \cdot \mathbf{p}_0) = G \cdot \hat{\mathbf{p}} \end{aligned} \tag{A26}$$

which demonstrates (38).

Using (6) and (35), (34) becomes

$$\begin{aligned} \begin{bmatrix} C_{\hat{d}\hat{d}} & C_{\hat{d}\hat{p}} \\ C_{\hat{p}\hat{d}} & C_{\hat{p}\hat{p}} \end{bmatrix} &= \begin{bmatrix} C_{d_0d_0} & C_{d_0p_0} \\ C_{p_0d_0} & C_{p_0p_0} \end{bmatrix} - \begin{bmatrix} C_{d_0d_0} - C_{d_0p_0} \cdot G^T \\ C_{p_0d_0} - C_{p_0p_0} \cdot G^T \end{bmatrix} \\ &\quad \cdot [C_{d_0d_0} - C_{d_0p_0} \cdot G^T - G \cdot C_{p_0d_0} + G \cdot C_{p_0p_0} \cdot G^T]^{-1} \\ &\quad \cdot [C_{d_0d_0} - G \cdot C_{p_0d_0} \qquad C_{d_0p_0} - G \cdot C_{p_0p_0}] \end{aligned} \tag{A27}$$

For $C_{\hat{p}\hat{p}}$ we directly obtain from (A27) the expression (39). For $C_{\hat{d}\hat{p}}$ we obtain successively

$$\begin{aligned} C_{\hat{d}\hat{p}} &= C_{d_0p_0} - (C_{d_0d_0} - C_{d_0p_0} \cdot G^T)[\ \ ]^{-1}(C_{d_0p_0} - G \cdot C_{p_0p_0}) \\ &= C_{d_0p_0} - [I - G \cdot (C_{p_0p_0} \cdot G^T - C_{p_0d_0})[\ \ ]^{-1}] \\ &\quad \cdot (C_{d_0p_0} - G \cdot C_{p_0p_0}) = G \cdot C_{\hat{p}\hat{p}} \end{aligned} \tag{A28}$$

where we have used (A18). For $C_{\hat{d}\hat{d}}$ we obtain successively from (A27), using (A18),

$$\begin{aligned} C_{\hat{d}\hat{d}} &= C_{d_0d_0} - (C_{d_0d_0} - C_{d_0p_0} \cdot G^T)[\ \ ]^{-1}(C_{d_0d_0} - G \cdot C_{p_0d_0}) \\ &= C_{d_0d_0} - [I - G \cdot (C_{p_0p_0} \cdot G^T - C_{p_0d_0})[\ \ ]^{-1}] \\ &\quad \cdot (C_{d_0d_0} - G \cdot C_{p_0d_0}) \\ &= G \cdot C_{p_0d_0} + G \cdot (C_{p_0p_0} \cdot G^T - C_{p_0d_0})[\ \ ]^{-1} \\ &\quad \cdot (C_{d_0d_0} - G \cdot C_{p_0d_0}) \\ &= G \cdot C_{p_0d_0} + G \cdot (C_{p_0p_0} \cdot G^T - C_{p_0d_0})[I - [\ \ ]^{-1} \\ &\quad \cdot (G \cdot C_{p_0p_0} - C_{d_0p_0}) \cdot G^T] \\ &= G \cdot C_{p_0p_0} \cdot G^T - G \cdot (C_{p_0p_0} \cdot G^T - C_{p_0d_0})[\ \ ]^{-1} \\ &\quad \cdot (G \cdot C_{p_0p_0} - C_{d_0p_0}) \cdot G^T \\ &= G \cdot C_{pp} \cdot G^T \end{aligned} \tag{A29}$$

Equations (A28) and (A29) demonstrate (40).

Equations (43) and (44) are obtained from (41) and (42) using (A20) and (A21).

Acknowledgments. We wish to acknowledge for helpful suggestions and discussions our colleagues G. Jobert, A. Nercessian, J. L. Piednoir, A. Provost, R. Madariaga, J. C. De Bremaecker, and J. F. Minster. This work has partially been supported by the Recherche Coopérative sur Programme 264, Etude Interdisciplinaire des Problèmes Inverses. Contribution I.P.G.P. 550.

REFERENCES

Angelier, J., A. Tarantola, S. Manoussis, and B. Valette, Inversion of field data in fault tectonics to obtain the regional stress tensor, *Geophys. J. R. Astron. Soc., 69,* in press, 1982.

Backus, G., and F. Gilbert, Uniqueness in the inversion of inaccurate gross earth data, *Philos. Trans. R. Soc. London, Ser. A, 266,* 123–192, 1970.

Franklin, J. N., Well-posed stochastic extensions of ill-posed linear problems, *J. Math. Analy. Appl., 31,* 682–716, 1970.

Jackson, D. D., The use of a priori data to resolve non-uniqueness in linear inversion, *Geophys. J. R. Astron. Soc., 57,* 137–157, 1979.

Pugachev, V. S., *Theory of Random Functions,* Pergamon, New York, 1965.

Rao, C. R., *Linear Statistical Inference and Its Applications,* John Wiley, New York, 1973.

Tarantola, A., and B. Valette, Inverse problems: Quest of information, *J. Geophys.,* in press, 1982.

Tikhonov, A., and V. Arsénine, Méthodes de résolution de problèmes mal posés, Editions MIR, Moscow, 1976.

Wiggins, R. A., The general linear inverse problem, *Rev. Geophys. Space Phys., 10*(1), 251–285, 1972.

(Received August 15, 1981;
accepted December 8, 1981.)

REVIEWS OF GEOPHYSICS AND SPACE PHYSICS, VOL. 19, NO. 2, PAGES 299–306, MAY 1981

A Characterization of the Inverse Gravimetric Source Problem Through Extremal Measures

GOTTFRIED ANGER

Department of Mathematics, Martin Luther University
4010 Halle (Saale), German Democratic Republic

Most problems in applications are inverse source problems or identification problems. This review deals with a systematic treatment of the nonnegative sources (measures) relative to the Laplace equation and the heat conduction equation producing the same potential outside a domain $\Omega \subset \mathbf{R}^3$. This set is convex and weakly compact. Therefore the basic theorems on convex sets can be applied to such problems. Sets of extremal measures are characterized by geometric conditions. Almost nothing is known about the structure of the sources which produce the earth's gravitational field. Therefore these investigations may be of interest for the inverse problem in gravimetry. The following question remains unanswered in this review: to find additional conditions to select uniquely a measure producing the earth's gravitational field. This question is unanswered in most inverse source problems.

1. Introduction

Inverse problems in geophysics are known to be among the most difficult of mathematical problems. Frequently, particular problems in geophysics can be solved by methods of classical mathematics (the Riemann integral, series expansions using special functions, etc.). Such methods form the basis for most previous work in this field. In the earth sciences many parameters can be determined very accurately, for example, measurements of gravity. On the other hand, almost nothing is known about the structure of the sources which produce the earth's gravitational field or magnetic field. Similar questions remain unanswered with respect to heat sources, the sources of electrostatic or electromagnetic fields, or special sources in continuum mechanics and solid-state physics (sources with respect to the biharmonic equation).

Furthermore, there is almost a complete lack of a theory which would allow the determination of individual coefficients (material parameters) of a differential equation from measured data (identification problems) [e.g., *Anger,* 1979*a*; *Kluge,* 1979; *Nashed,* 1980]. Such a theory should yield indications about the physical quantities which should be measured in very complicated systems.

Such observations justify an attempt at a systematic treatment of inverse source problems, since a large part of physics is a theory of fields.

A systematic study of inverse source problems (for the Laplace equation, see *Anger* [1961, 1976, 1977, 1978, 1979*a*, 1980]; the equation of heat conduction, *Anger* [1979*b*], *Anger and Czerner* [1979], *Czerner* [1980], and *Cannon and DuChateau* [1980]; the Helmholtz equation, *Bleistein and Cohen* [1978]; the wave equation, *Juhl* [1979]; the biharmonic equation, *Michel* [1979] and *Schulze and Wildenhain* [1977], etc.) would presuppose a knowledge of the results of modern potential theory [e.g., *Helms,* 1969; *Schulze and Wildenhain,* 1977], partial differential equations [e.g., *Trèves,* 1975; *Schulze and Wildenhain,* 1977], and functional analysis [e.g., *Anger,* 1980; *Choquet,* 1969; *Schulze and Wildenhain,* 1977].

If one examines the historical development of such problems, one finds that the works of C. F. Gauss occupy a central position [e.g., *Garland,* 1979; *Gauss,* 1877]. Gauss knew that his investigations would have to be made mathematically rigorous (development of the concept of measure). It required more than 120 years to develop the complete foundations and to apply them to the study of inverse problems. It is very desirable to close the gap which exists between highly developed measurement techniques and the virtually nonexistent methods of characterizing the sources.

In physics a source is usually introduced as the limit of a sum of finitely many point measures (a sum of Dirac measures). If one requires that these sums converge for every continuous function, then one immediately sees connections to functional analysis [*Anger,* 1980], according to which a measure is a continuous linear form on the space of continuous functions. Using the concept of measure as a starting point, the potential relative to the Newton kernel or, more generally, with respect to the fundamental solution of a differential equation can be introduced [*Anger,* 1980; *Schulze and Wildenhain,* 1977]. The inverse transformation from potential to a measure can also be found with the help of continuous potentials and the space of continuous functions [*Anger,* 1961, 1980; *Schulze and Wildenhain,* 1977]. In this way there arises a close relationship between potentials and the space of continuous functions. The results of linear algebra and functional analysis are applicable for certain questions in potential theory. The determination of the relationship between physical concepts on the one hand and mathematical concepts on the other hand is among the most difficult questions in mathematics.

Let Ω be a domain of the Euclidean space $\mathbf{R}^3$ and μ be a positive mass (measure) on $\bar{\Omega} = \Omega \cup \partial\Omega$, where $\partial\Omega$ is the boundary of Ω. Further, let $C(\bar{\Omega})$ be the space of continuous functions defined on $\bar{\Omega}$ and $C^*(\bar{\Omega})$ be the space of all measures on $\bar{\Omega}$. As a first step we introduce the set $\mathscr{B}(\mu)$ of all positive measures on $\bar{\Omega}$ which produce the same potential outside $\bar{\Omega}$. This set is convex and weakly compact. The extremal elements of convex sets are of special interest. Using the famous Krein-Milman-Choquet theorem [e.g., *Choquet,* 1969], one can represent each measure $\nu \in \mathscr{B}(\mu)$ with the aid of an integral over the extremal measures. In physics such integral representations are used for concrete problems. We can characterize the extremal measures of $\mathscr{B}(\delta_{x^0})$ in $\mathbf{R}^2$ by geometrical conditions [*Anger,* 1977, 1980; *Anger and Czerner,* 1979; *Czerner,* 1980], where δ_{x^0} is the Dirac measure (see formula (17)). There exist two classes of extremal measures of $\mathscr{B}(\delta_{x^0})$. The second class is of special interest, because these measures must have a concrete physical interpretation. Let μ be a positive measure satisfying supp $\mu = \bar{\Omega}$, and let ν_0 be an element of $\mathscr{B}(\mu)$. If supp ν_0 contains an interior point, then ν_0 is no extremal element of $\mathscr{B}(\mu)$. The earth's gravitational potential $G_L\nu_0$ satisfies this condition.

Let $A: C(\bar{\Omega}) \leftrightarrow C(\bar{\Omega})$ be the mapping which describes the solution of the Dirichlet problem for the Laplace equation. The adjoint mapping $A^*: C^*(\bar{\Omega}) \leftrightarrow C^*(\bar{\Omega})$ of A can be interpreted as the balayage principle (sweeping out process) of Poincaré. The relation $\mathscr{B}(\mu) \subset (A^*)^{-1}(\mu)$ holds, where $(A^*)^{-1}$ is the inverse of A^*. The same relations hold for the equation of heat conduction. Further work must be done in these investigations. There are relations relative to the design of antennas [*Bleistein and Cohen,* 1978], the inverse source problem of the wave equation [*Juhl,* 1979], and the inverse source problem of solid-state physics [*Michel,* 1979]. Further, there exist relations to probability [*Devaney,* 1979; *Karr and Pittenger,* 1977]. A complete version of these ideas is given by *Anger* [1980].

2. The Inverse Gravimetric Problem

We denote by $\mathbf{R}^3$ the three-dimensional Euclidean space, by $x = (x_1, x_2, x_3)$, $y = (y_1, y_2, y_3)$ points of $\mathbf{R}^3$, and by

$$|x - y| := [(x_1 - y_1)^2 + (x_2 - y_2)^2 + (x_3 - y_3)^2]^{1/2}$$

the distance between the points x and y. Further, we denote by $\mathbf{R}$ the set of real numbers, by $\bar{\mathbf{R}}$ the set $\{x: -\infty \leqq x \leqq +\infty\}$, and by $\mathbf{R}_+$ the set of all real numbers $x \geqq 0$. We often consider open sets $\Omega \subset \mathbf{R}^3$ and compact sets $K \subset \mathbf{R}^3$. In $\mathbf{R}^3$ a compact set is equivalent to a closed and bounded set. We denote by $\partial\Omega$ the boundary of Ω and by $\bar{\Omega} = \Omega \cup \partial\Omega$ the closure of Ω. Let

$$B(x^0, r) := \{x: |x - x^0| < r\}$$

be an open ball, and

$$K(x^0, r) := \{x: |x - x^0| \leqq r\}$$

a closed ball. Let $f: \mathbf{R}^3 \leftrightarrow \mathbf{R}$ be a function defined on $\mathbf{R}^3$. Then $\text{rest}_K f$ or $f|_K$ denotes the restriction of f on K, that is, $\text{rest}_K f$ is a function on K such that $\text{rest}_K f(x) = f(x)$ for all $x \in K$.

Our notations also hold in the n-dimensional Euclidean space $\mathbf{R}^n$. Then $x = (x_1, \cdots, x_n)$ and $y = (y_1, \cdots, y_n)$ are points of $\mathbf{R}^n$. Most definitions and theorems of this article hold in $\mathbf{R}^n$. Further information about (modern) potential theory is given by *Anger* [1961, 1976, 1980] and *Schulze and Wildenhain* [1977].

The inverse gravimetric problem consists in determining a potential u_ρ having its mass distribution $\rho \geqq 0$ on $\bar{\Omega}$ such that

$$|\text{grad } u_\rho|\Big|_{\partial\Omega} = h \qquad (1)$$

Let

$$-P(D)u := \Delta_3 u = \frac{\partial^2 u}{\partial x_1^2} + \frac{\partial^2 u}{\partial x_2^2} + \frac{\partial^2 u}{\partial x_3^2} \qquad (2)$$

be the Laplace operator and

$$G_L(x, y) := \frac{1}{4\pi} \frac{1}{|x - y|} \qquad (3)$$

the fundamental solution of $-\Delta_3 u$. In $\mathbf{R}^2$ the fundamental solution has the form

$$G_L(x, y) := \frac{1}{2\pi} \log \frac{1}{|x - y|} \qquad (4)$$

and in $\mathbf{R}^1$ the form

$$G_L(x, y) := (x - y)\Theta(x - y) + c_1(x - y) + c_2 \qquad (5)$$

where c_1 and c_2 are constants and Θ is the Heaviside function, that is, $\Theta(z) = 1$ for $z > 0$ and $\Theta(z) = 0$ for $z \leqq 0$. The potential u_ρ of a volume distribution ρ is defined by

$$u_\rho(x) := \int_\Omega G_L(x, y)\rho(y)\, dy \qquad dy = dy_1\, dy_2\, dy_3 \qquad (6)$$

the potential of a single layer by

$$u_\sigma(x) := \int_{\partial\Omega} G_L(x, y)\sigma(y)\, dS(y) \qquad (7)$$

where dS is the surface element, and the potential of the Dirac measure δ_{x^0} (unit mass concentrated at $x^0 \in \mathbf{R}^3$) by

$$u_{x^0}(x) := G_L(x, x^0) \qquad (8)$$

The following result is well known: for each linear differential equation (with constant coefficients) there exists a fundamental solution G [see *Schulze and Wildenhain,* 1977; *Trèves,* 1975]. If G is a locally integrable fundamental solution, the following equation can be used for introducing a fundamental solution G of the linear differential operator $P(x, D)$ defined on $\mathbf{R}^3$

$$\int G(x, y)P^*(x, D)\varphi(x)\, dx = \varphi(y) \qquad (9)$$

for all $\varphi \in C_0^\infty(\mathbf{R}^3)$ and all $y \in \mathbf{R}^3$, where $P^*(x, D)$ is the adjoint differential operator and $C_0^\infty(\mathbf{R}^3)$ the space of all continuous function $\varphi: \mathbf{R}^3 \leftrightarrow \mathbf{R}$ having compact support

$$\text{supp } \varphi := \overline{\{x: \varphi(x) \neq 0\}}$$

and continuous derivatives of arbitrary order. In the case of the Laplace operator, $P^*(x, D)\varphi$ is equal to $-\Delta_3\varphi$. Using Green's theorem, one can prove that the relation (9) holds for G_L.

From (9) the following important theorem follows.

Theorem 1: There exist at least two positive densities ρ^+ and ρ^- on $\bar{\Omega}$ such that the relation

$$u_{\rho^+}(x) = u_{\rho^-}(x) \qquad (10)$$

holds for all $x \notin \bar{\Omega}$.

Proof: Let $\rho(x) := P(x, D)\varphi(x)$ and

$$\rho^+(x) = \max\, [P(x, D)\varphi(x), 0]$$

$$\rho^-(x) := -\min\, [P(x, D)\varphi(x), 0]$$

where ρ^+ is the positive part and ρ^- the negative part of ρ, that is, $\rho = \rho^+ - \rho^-$. If supp $\varphi \subset \Omega$, then $\varphi(y) = 0$ for all $y \notin \bar{\Omega}$. For these y the relation

$$0 = \int_\Omega G(x, y)P(x, D)\varphi(x)\, dx = \int_\Omega G(x, y)\rho^+(x)\, dx - \int_\Omega G(x, y)\rho^-(x)\, dx$$

holds. Therefore we get

$$u_{\rho^+}(y) = \int_\Omega G_L(x, y)\rho^+(x)\, dx = \int_\Omega G_L(x, y)\rho^-(x)\, dx = u_{\rho^-}(y)$$

for all $y \notin \bar{\Omega}$.

Remark 1: The relation (10) holds for each linear differential operator $P(x, D)$ having a locally integrable fundamental solution.

Remark 2: From theorem 1 it follows that ρ cannot be determined uniquely from the values $u_\rho|_{\partial\Omega}$ on the boundary $\partial\Omega$. As a first step we introduce the set $\mathscr{B}(\mu)$ of all positive mass distributions on $\bar{\Omega}$ producing the same potential outside $\bar{\Omega}$ [see *Anger,* 1976, 1977, 1978, 1979*a*, 1979*b*, 1980; *Anger and Czerner,* 1979].

Remark 3: From remark 2 it follows that if a solution u_ρ of the inverse gravimetric problem exists, then there exist different mass distributions on $\bar{\Omega}$ producing the same potential outside $\bar{\Omega}$. Therefore we have to introduce additional conditions to select uniquely a mass distribution solving the inverse problem.

In our theory we are using the space $C(K)$ of all continuous functions defined on $K \subset \mathbf{R}^3$. Especially, K can be the closure of a bounded domain Ω or its boundary $\partial\Omega$. On $C(K)$ we introduce the norm

$$\| f \| := \max_{x \in K} |f(x)| \tag{11}$$

The relation $\| f - f_k \| \to 0$ for $k \to \infty$ is equivalent to uniform convergence of the sequence (f_k) to f.

We write the mass distributions, producing the potentials (6) and (7), in integral form. The integrals are Riemann integrals:

$$\mu_\rho(f) := \int_\Omega f(y)\rho(y)\, dy \qquad f \in C(\bar{\Omega}) \tag{12}$$

$$\mu_\sigma(f) := \int_{\partial\Omega} f(y)\sigma(y)\, dS(y) \qquad f \in C(\partial\Omega) \tag{13}$$

If $f(y) = G_L(x, y)$, $x \notin \bar{\Omega}$, then we get the potential (6) and for $x \notin \partial\Omega$ the potential (7).

Let $\mu(f)$ be $\mu_\rho(f)$ or $\mu_\sigma(f)$. Then the following property holds for $\mu: C(\bar{\Omega}) \mapsto \mathbf{R}$ (mapping from $C(\bar{\Omega})$ in $\mathbf{R}$):

$$\mu(c_1 f_1 + c_2 f_2) = c_1\mu(f_1) + c_2\mu(f_2) \tag{14}$$

where $c_1, c_2 \in \mathbf{R}$ and $f_1, f_2 \in C(\bar{\Omega})$. Further, we can prove the inequality

$$|\mu(f)| \leqq M \| f \| \tag{15}$$

If $\rho \geqq 0$ and $\sigma \geqq 0$, then

$$\mu(f) \geqq 0 \text{ for } f \geqq 0 \tag{16}$$

The special measures μ_ρ and μ_σ are linear forms (relation (14)) which are continuous (relation (15)). If $\rho \geqq 0$ and $\sigma \geqq 0$, then μ_ρ and μ_σ are positive linear forms (relations (14) and (16)). Each positive linear form μ on $C(K)$ is continuous, that is, satisfies (15) [see *Anger,* 1980; *Choquet,* 1969].

The dual space $C^*(K)$ of $C(K)$ consists of all continuous linear forms on $C(K)$. Therefore

$$\mu_\rho \in C^*(\bar{\Omega}) \qquad \mu_\sigma \in C^*(\bar{\Omega})$$

Further, the Dirac measure δ_{x^0}, $x^0 \in \bar{\Omega}$, is an element of $C^*(\bar{\Omega})$, where

$$\delta_{x^0}(f) := f(x^0) \tag{17}$$

In our theory it is more convenient to use linear forms than mass distributions defined on subsets of $\bar{\Omega}$. In this case we can apply the concepts of linear algebra and functional analysis to $C(\bar{\Omega})$ and $C^*(\bar{\Omega})$.

We consider the following special mass distributions (averages) in $\mathbf{R}^3$:

$$\delta_{x^0,r}(f) := \frac{1}{4\pi r^2}\int_{\partial B(x^0,r)} f(y)\, dS(y) \tag{18}$$

$$m_{x^0,R}(f) := \frac{3}{4\pi R^3}\int_{B(x^0,R)} f(y)\, dy \tag{19}$$

They produce the following special potentials:

$$\begin{aligned} G_L\,\delta_{x^0,r}(x) &= G_L(x, x^0) && |x - x^0| > r \\ G_L\,\delta_{x^0,r}(x) &= \frac{1}{4\pi r} && |x - x^0| \leqq r \end{aligned} \tag{20}$$

$$\begin{aligned} G_L m_{x^0,R}(x) &= G_L(x, x^0) && |x - x^0| > R \\ G_L m_{x^0,R}(x) &= \frac{1}{4\pi R^3}\left(\tfrac{3}{2}R^2 - \tfrac{1}{2}|x - x^0|^2\right) && |x - x^0| \leqq R \end{aligned} \tag{21}$$

The mass distribution $m_{x^0,R}$ can be calculated by using the mass distributions $\delta_{x^0,r}$, $r \leqq R$:

$$\begin{aligned} m_{x^0,R}(f) &= \frac{3}{4\pi R^3}\int_{B(x^0,R)} f(y)\, dy \\ &= \int_{r=0}^{R}\left[\frac{1}{4\pi r^2}\int_{\partial B(x^0,r)} f(y)\, dS(y)\right] dm(r) \\ &= \int_{r=0}^{R} \delta_{x^0,r}(f)\, dm(r) \end{aligned} \tag{22}$$

where

$$dm(r) = \frac{3r^2}{R^3}\, dr \qquad \int dm(r) = \int_{r=0}^{R} \frac{3r^2}{R^3}\, dr = 1 \tag{23}$$

The mass distributions $\delta_{x^0,r}$, $r \leqq R$, and $m_{x^0,R}$ produce the same potential outside $\bar{\Omega}$. Each positive mass distribution m depending only on r and satisfying $\int dm(r) = 1$, for instance, $dm(r) = \rho(r)\, dr$, produces the same potential as $\delta_{x^0,r}$. To prove this, we use the relation

$$\nu(f) := \int \delta_{x^0,r}(f)\, dm(r) \tag{24}$$

Therefore we introduce the set $\mathscr{B}(\delta_{x^0})$ of all positive mass distributions on $\bar{\Omega}$ (i.e., positive linear forms on $C(\bar{\Omega})$) producing the same potential outside $\bar{\Omega}$. Later we prove that if μ_1, μ_2 are elements of $\mathscr{B}(\delta_{x^0})$, then

$$\mu := t\mu_1 + (1 - t)\mu_2 \qquad 0 < t < 1 \tag{25}$$

is also an element of $\mathscr{B}(\delta_{x^0})$; that is, $\mathscr{B}(\delta_{x^0})$ is a convex set. If an element of a convex set cannot be represented in the form (25), it is called an extremal element. *Anger* [1976] and *Schulze and Wildenhain* [1977] prove that measures of the special form $\delta_{x^0,r}$ are extremal elements of $\mathscr{B}(\delta_{x^0})$. There are two classes of extremal elements.

The mass distribution $m_{x^0,R} \in \mathscr{B}(\delta_{x^0})$ can be calculated by using an integration procedure (22), where the elements $\delta_{x^0,r}$ in (22) are extremal elements of $\mathscr{B}(\delta_{x^0})$. This fact holds for each measure $\nu \in \mathscr{B}(\delta_{x^0})$ by using extremal elements of $\mathscr{B}(\delta_{x^0})$ (Krein-Milman-Choquet theorem [e.g., *Choquet,* 1969; *Schulze and Wildenhain,* 1977]). The Krein-Milman-Choquet theorem is a generalization of the concept 'center of gravity' from the $\mathbf{R}^3$ to a Banach space [e.g., *Anger,* 1976].

3. Measures and Potentials

For developing a theory for inverse (source) problems we have to apply the space $C(K)$ and its dual space $C^*(K)$. *Riesz* [1909] proved that each continuous linear form $L \in C^*(K)$ can be represented by a measure on K and reverse a measure on K produces a continuous linear form on $C(K)$ [e.g., *Choquet,* 1969]. Therefore we can introduce a measure as a continuous linear form on $C(K)$ [*Anger,* 1980].

The Riemann integral (12) can be calculated approximately by the following finite sums:

$$\mu_N(f) = \sum_{k=1}^{N} c_k f(\xi^k) \qquad \xi^k \in \bar{\Omega} \tag{26}$$

Using the Dirac measures δ_{ξ^k}, we can write (26) in the form

$$\mu_N(f) = \sum_{k=1}^{N} c_k \delta_{\xi^k}(f) \tag{26'}$$

The measure μ_N is a finite sum of Dirac measures. It is well known that $\mu_N(f)$ converges to $\mu_\rho(f)$ for all $f \in C(\bar{\Omega})$. Further, each measure μ on K can be approximated by finite sums μ_N such that the limit exists for all $f \in C(K)$ (weak convergence). Because the Dirac measures δ_ξ, $\xi \in K$, are the extremal elements of the unit sphere in $C^*(K)$, the approximation (26′) is a sum (integral) using extremal elements [e.g., *Choquet,* 1969]. The formulas (22) and (24) of section 2 have the same structure as (26′) on the corresponding convex sets.

In physics the notion of a mass is introduced as a limit of finite sums of Dirac measures. The proper notion of convergence on the set of mass distributions is the weak convergence for measures (linear forms), which we have to add to the physical notion of masses [e.g., *Anger,* 1980; *Choquet,* 1969; *Schulze and Wildenhain,* 1977].

Let $\mu: C(K) \leftrightarrow \mathbf{R}$ be a positive measure, that is, $\mu(f) \geqq 0$ for $f \geqq 0$. For $\mu(f)$ we often write $\int f\, d\mu$ or $\int f(y)\, d\mu(y)$. The value $\mu(f) = \int f\, d\mu$ is called the integral of the measure relative to $f \in C(K)$. If $x \notin K$, then $y \leftrightarrow f(y) = G_L(x, y)$ is a continuous function, and the potential

$$G_L\mu(x) := \int G_L(x, y)\, d\mu(y) \tag{27}$$

is defined. If $x \in K$, then G_L can be approximated on the product space $\mathbf{R}^3 \times \mathbf{R}^3$ by an increasing sequence of continuous kernels $G_k \geqq 0$. The value $G_k\mu(x)$ is defined in the same manner as (27). From $\mu(f) \geqq 0$ it follows that for $G_{k+1} - G_k \geqq 0$ the relation $\mu(G_{k+1} - G_k) \geqq 0$ holds. Therefore the limit

$$G_L\mu(x) := \lim_{k\to\infty} \int G_k(x, y)\, d\mu(y) \tag{28}$$

is finite or infinite. It can be proved that the potentials

$$x \leftrightarrow G_k\mu(x)$$

are continuous on $\mathbf{R}^3$ [*Anger,* 1980]. $G_L\mu$ is as limit of an increasing sequence of continuous functions, a so-called lower semicontinuous function. A lower semicontinuous function can be defined as a limit of an increasing sequence of continuous functions (equivalent definition).

If $\mu = \mu^+ - \mu^-$, where μ^+ and μ^- are positive measures, then

$$G_L\mu(x) := G_L\mu^+(x) - G_L\mu^-(x) \tag{29}$$

provided that $G_L\mu^+(x)$ and $G_L\mu^-(x)$ are not infinite simultaneously.

The definition of the special potentials (6) and (7) of section 2 is equivalent to the definition (28) and (29) in these special cases [*Anger,* 1980].

The potentials (6) and (7) are continuous if ρ and σ are, for instance, continuous. The potential (29) of a measure μ is the difference of two lower semicontinuous functions.

4. A New Transformation in Potential Theory

There are enough measures λ on $K = \bar{\Omega}$ (or $K = \partial\Omega$) producing a continuous potential [see *Anger,* 1980, equations (1.8) and (1.10); *Schulze and Wildenhain,* 1977]. Then the following important relation holds

$$\mu(G_L\lambda) = \int G_L\lambda\, d\mu = \int[\int G_L(x, y)\, d\lambda(y)]\, d\mu(x)$$
$$= \int[\int G_L(x, y)\, d\mu(x)]\, d\lambda(y) = \int G_L\mu\, d\lambda \tag{30}$$

The set $D(K)$ of all continuous functions $f \in C(K)$ which can be represented as a potential $G_L\lambda$ of a measure λ on K is dense in $C(K)$ if the boundary ∂K of K is, for instance, smooth. This condition is fulfilled in applications. If the potential $G_L\mu$ is known, then we can calculate the value $\mu(G_L\lambda)$ by using the relation (30). Instead of the potential $G_L\mu$ we consider the measure μ producing the potential $G_L\mu$. The measure μ is an element of the space $C^*(K)$. The correspondence between $G_L\mu$ and μ is one to one. One can find the proofs in the work of *Anger* [1961, 1980] and *Schulze and Wildenhain* [1977]. To use the transformation (30), we have to introduce exceptional sets, which are called sets of capacity zero.

As an example of the transformation (30), let us consider a sequence of potentials $G_L\mu_k$, $k = 1, 2, \cdots$. In general, the potentials $G_L\mu_k$ are unbounded and discontinuous functions, their corresponding measures being elements of the dual space $C^*(K)$. It is easier to work with the measures μ_k than with the potentials $G_L\mu_k$. For such problems we can apply the (modern) results of algebra and functional analysis.

5. A Systematic Approach for Studying Inverse Problems

In section 2 we proved that there exist different measures on $K = K(x^0, r)$ producing the same potential outside $K(x^0, r)$. The same result holds for a bounded domain $\Omega \subset \mathbf{R}^3$. Therefore we introduce for a positive measure μ on $\bar{\Omega}$ (i.e., the support of μ is a subset of $\bar{\Omega}$) the set $\mathscr{B}(\mu)$ of all positive measures ν on $\bar{\Omega}$ producing the same potential outside $\bar{\Omega}$.

If $\nu \in \mathscr{B}(\mu)$, the relation

$$\int u\, d\nu = \int u\, d\mu \tag{31}$$

holds for all $u \in C(\bar{\Omega})$ harmonic on Ω, that is, $\Delta_3 u = 0$ on Ω [see *Anger,* 1961, 1980; *Schulze and Wildenhain,* 1977]. From this equation it follows that the masses of ν and μ are equal, because $u = 1$ is harmonic:

$$\int d\nu = \int d\mu \tag{31'}$$

A sequence of measures $\nu_k \in C^*(\bar{\Omega})$ converges weakly to $\nu \in C^*(\bar{\Omega})$ if

$$\nu_k(f) \to \nu(f) \quad \text{for all} \quad f \in C(\bar{\Omega}) \tag{32}$$

Because $\mathscr{B}(\mu) \subset C^*(\bar{\Omega})$, we can introduce the weak convergence on $\mathscr{B}(\mu)$. The set $\mathscr{B}(\mu)$ is convex; that is, from $\mu_1, \mu_2 \in \mathscr{B}(\mu)$ it follows that

$$\nu := t\mu_1 + (1 - t)\mu_2 \qquad 0 \leqq t \leqq 1 \tag{33}$$

is an element of $\mathscr{B}(\mu)$. To prove this, we remark that ν is a positive measure on $\bar{\Omega}$ and for $x \notin \bar{\Omega}$ the relation

$$G_L\nu(x) = tG_L\mu_1(x) + (1-t)G_L\mu_2(x)$$
$$= tG_L\mu(x) + (1-t)G_L\mu(x) = G_L\mu(x)$$

holds. Further, $\mathscr{B}(\mu)$ is a (weakly) compact set [*Anger,* 1976, 1980; *Schulze and Wildenhain,* 1977]. Therefore the following theorem holds.

Theorem 1: $\mathscr{B}(\mu)$ is a convex and weakly compact set.

For convex and compact subsets B of a Banach space X a rich theory exists [*Anger,* 1976; *Choquet,* 1969]. Each point $x^0 \in B$ can be represented with the aid of extremal points of B using an integral as in relation (22) or (24) of section 2. Special extremal points of $\mathscr{B}(\mu)$ were calculated by *Anger* [1976, 1978, 1980] and *Anger and Czerner* [1979].

In the following discussion we shall study the set ex $\mathscr{B}(\delta_{x^0})$ of all extremal elements of $\mathscr{B}(\delta_{x^0})$. If $\nu \in \mathscr{B}(\delta_{x^0})$, then

$$G_L\nu(x) = G_L(x, x^0) \text{ for all } x \notin \bar{\Omega} \tag{34}$$

The measures $\nu \in \mathscr{B}(\delta_{x^0})$ produce the fundamental solution $G_L(x,x^0)$ outside $\bar{\Omega}$.

It is possible to characterize the extremal elements of $\mathscr{B}(\mu)$ by a general approximation property. Let L_ν^1 be the space of all ν-integrable functions. The norm on L_ν^1 is given by

$$\| f \| := \int |f(x)|\, d\nu(x) \tag{35}$$

Let

$$H_G(\bar{\Omega}) := \left\{ f \in C(\bar{\Omega}) : f(y) = \sum_{k=1}^{N} d_k G_L(y, x^k), x^k \in \bar{\Omega} \right\} \tag{36}$$

The following theorem is important for studying the extremal points of $\mathscr{B}(\delta_{x^0})$ by geometrical properties [see *Anger,* 1977, 1980; *Schulze and Wildenhain,* 1977].

Theorem 2: A measure $\nu \in \mathscr{B}(\mu)$ is extremal if and only if $H_G(\bar{\Omega})$ is dense in L_ν^1.

Let

$$d\mu_x^{\Omega}(y) := \frac{\partial G_L^{\Omega}(x, y)}{\partial n_y}\, dS(y) \tag{37}$$

be the so-called harmonic measure of Ω, where G_L^{Ω} is Green's function of Ω. The harmonic measure solves the Dirichlet problem for the boundary values f:

$$u_f(x) = \int_{\partial\Omega} f(y) \frac{\partial G_L^{\Omega}(x, y)}{\partial n_y}\, dS(y) =: \int f(y)\, d\mu_x^{\Omega}(y) \tag{38}$$

The existence of the harmonic measure μ_x^{Ω} can be proved for a bounded domain [e.g., *Helms,* 1969; *Trèves,* 1975] by using the so-called Perron-Wiener-Brelot method.

Special measures of ex $\mathscr{B}(\delta_{x^0})$ were constructed in section 2. They have the form

$$\delta_{x^0,r} = \mu_{x^0}^{\Omega} \qquad \Omega = B(x^0, r) \tag{39}$$

Theorem 3: Let $\Omega' \subset \Omega$ be a domain satisfying $\partial\Omega' = \partial\bar{\Omega}'$ and $x^0 \in \bar{\Omega}$. Then the harmonic measure $\mu_{x^0}^{\Omega'}$ is an extremal element of $\mathscr{B}(\delta_{x^0})$.

A proof of this theorem is given by *Anger* [1980] and *Schulze and Wildenhain* [1977].

There exists a second class of extremal measures for $\mathscr{B}(\delta_{x^0})$. For simplicity we only consider the case $\Omega = B(x^0, r) \subset \mathbf{R}^2$. Let $K \subset B(x^0, r)$ be a compact set and

$$\nu = \nu_K + \gamma \in \mathscr{B}(\delta_{x^0}) \tag{40}$$

where ν_K is concentrated on K and γ on $\partial B(x^0, r)$.

For proving the following theorems we have to use the so-called balayage principle (sweeping out process [e.g., *Anger,* 1961, 1980; *Helms,* 1969]) relative to the Laplace equation. This principle also holds for the equation of heat conduction. Therefore most of our results hold for this equation [*Anger,* 1979*b*; *Anger and Czerner,* 1979].

Balayage principle: To each measure ν on $\bar{\Omega}$ there exists a measure μ' on $\partial\Omega$ such that (31) is satisfied.

Let us consider (40) for harmonic functions $u \in C(\bar{\Omega})$. Then

$$\nu_K(u) + \gamma(u) = \mu(u) \tag{41}$$

The swept out measure of ν is denoted by $\Pi_\Omega\nu$. In the case $\Omega = B(x^0, r) \subset \mathbf{R}^2$ it is given by

$$\Pi_\Omega\nu(u) = \int_0^{2\pi} \left[\int P(x, y)\, d\nu(x) \right] u(y)\, d\varphi(y) \tag{42}$$

where $y = y(\varphi) = (r, \varphi) \in \partial B(x^0, r)$.

$$P(x, y) = \frac{1}{2\pi} \frac{|y|^2 - |x|^2}{|x - y|^2} \tag{43}$$

is the kernel of the Poisson integral. The relation

$$d\mu_x^{\Omega}(y) = \frac{\partial G_L(x, y)}{\partial n_y} d\varphi(y) = P(x, y)\, d\varphi(y) \tag{44}$$

holds. Using the balayage principle, it follows from (41) that ($\mu = \delta_{x^0}$)

$$\gamma(u) = \Pi_\Omega\mu(u) - \Pi_\Omega\nu_K(u)$$
$$= \int_0^{2\pi} \left[\frac{1}{2\pi} - \int P(x, y)\, d\nu_K(x) \right] u(y)\, d\varphi(y)$$
$$= \int_0^{2\pi} g(\varphi) u(\varphi)\, d\varphi \tag{45}$$

where

$$g(\varphi) := \frac{1}{2\pi} - \int P(x, y)\, d\nu_K(x) \qquad y = (r, \varphi) \tag{46}$$

Special extremal elements $\nu = \nu_K + \gamma \in \mathscr{B}(\delta_{x^0})$ are as follows:

Element 1

$$z^0 \in B(x^0, r') \subset B(x^0, r) \qquad K = \{z^0\} \qquad \nu_K = a\delta_{z^0}$$

$\nu = \nu_K + \gamma$ is extremal if

$$a = \frac{r' - r_1}{r' + r_1} \qquad r_1 = |z^0 - x^0|$$

Element 2

$$z^{(1)}\, z^{(2)} \in B(x^0, r') \subset B(x^0, r) \qquad |z^{(1)} - x^0| = |z^{(2)} - x^0| = r_1$$
$$z^{(1)} = (r_1, 0) \qquad z^{(2)} = (-r_1, 0) \qquad \nu_K = a\delta_{z^{(1)}} + a\delta_{z^{(2)}}$$

$\nu = \nu_K + \gamma$ is extremal if

$$a = \frac{(r' + r_1)^4 (r' - r_1)^2 - (r' - r_1)^4 (r' + r_1)^2}{[(r' + r_1)^4 - (r' - r_1)^4](r'^2 - r_1^2)}$$

To characterize extremal elements of the second class we have to prove the following theorems [*Anger*, 1980; *Anger and Czerner*, 1979] relative to a domain $\Omega' = B(x^0, r') \subset B(x^0, r)$.

Theorem 4: Let $g:\partial\Omega' \leftrightarrow \mathbf{R}$ be positive. Then there exists a constant $d > 0$ such that the inequality

$$g(\varphi) \geqq d \int P(x, y)\, d\nu_K(x) \qquad y = (r, \varphi) \tag{47}$$

holds on $\partial\Omega'$.

From theorem 4 follows theorem 5.

Theorem 5: Let $g:\partial\Omega' \leftrightarrow \mathbf{R}$ be positive and $u_k \in H_G(\bar{\Omega})$, satisfying

$$\int |u_k(y)|\, d\gamma(y) = \int_0^{2\pi} |u_k(y)| g(\varphi)\, d\varphi \to 0 \quad \text{for} \quad k \to \infty$$

Then

$$\int |u_k(x)|\, d\nu_K(x) \to 0 \qquad |u_k(x^0)| \to 0 \qquad \text{for } k \to \infty$$

A consequence of theorems 4 and 5 is theorem 6.

Theorem 6: Let $g:\partial\Omega' \leftrightarrow \mathbf{R}$ be nonnegative and $\nu = \nu_K + \gamma$ an extremal element of $\mathscr{B}(\delta_{x^0})$. If $g > 0$, then $\nu_K = 0$; if $\nu_K \neq 0$, then g must have at least one root.

Let g have a root in φ^0 and $g(\varphi) > 0$ for $\varphi \neq \varphi^0$. Then $g(\varphi) \geqq c > 0$ outside an interval $I_\epsilon := [\varphi^0 - \epsilon, \varphi^0 + \epsilon]$, and

$$\int |u_k(y)| g(\varphi)\, d\varphi \geqq c \int_{I_\epsilon'} |u_k(y)|\, d\varphi$$

where $I_\epsilon' := \partial B(x^0, r) \backslash I_\epsilon$. If

$$\int |u_k(y)|\, d\gamma(y) \to 0$$

then the integrals

$$\int_{I_\epsilon'} |u_k(y)|\, d\varphi \geqq 0$$

also tend to zero; at φ^0 the functions $u_k(y^0)$ can tend to infinity while the product

$$|u_k(y^0)| g(\varphi^0)$$

remains finite. The following theorem gives an answer relative to this fact.

Theorem 7: Let $g:\partial\Omega' \leftrightarrow \mathbf{R}$ be the function defined in (46). If g has one root in $y^0 = (r, \varphi^0)$, then there exists a sequence of functions $u_k \in H_G(\bar{\Omega})$ such that

$$\int_0^{2\pi} g(\varphi) |u_k(y)|\, d\varphi \to 0 \quad \text{for} \quad k \to \infty$$

and

$$u_k(x) \to d_1 P(x, y^0) + d_2 \frac{\partial P(x, y^0)}{\partial\varphi} \tag{48}$$

where d_1 and d_2 are arbitrary constants.

Remark 1: The function $g:\partial B(x^0, r') \leftrightarrow \mathbf{R}$ is analytic. Therefore there exist at most m roots in $\varphi^{(1)}, \cdots, \varphi^{(m)}$. In this case there exists a sequence of functions v_k such that

$$v_k(x) \to d_{11} P(x, y^{(1)}) + \cdots + d_{1m} P(x, y^{(m)}) + d_{21} \frac{\partial P(x, y^{(1)})}{\partial\varphi} + \cdots + d_{2m} \frac{\partial P(x, y^{(m)})}{\partial\phi} \tag{49}$$

Using theorem 7 and remark 1, we can prove theorem 8 [*Anger*, 1977, 1980; *Anger and Czerner*, 1979; *Czerner*, 1980].

Theorem 8: Let $\nu = \nu_K + \gamma \in \mathscr{B}(\delta_{x^0})$ be an extremal measure, $\nu_K \neq 0$, and $g(\varphi) = 0$ in $\varphi^{(1)}, \cdots, \varphi^{(m)}$. Then the support of ν_K consists at most of $2m$ points $x^{(1)}, \cdots, x^{(2m)}$, that is, $K = \{x^{(1)}, \cdots, x^{(2m)}\}$.

Remark 2: If $g \geqq 0$ is positive, the boundary $\partial\Omega$ is a 'barrier' for the functions $v \in H_G(\bar{\Omega})$. If g is zero in $\varphi^{(1)}, \cdots, \varphi^{(m)}$, we can penetrate into Ω using functions $v \in H_G(\bar{\Omega})$. In Ω we get the functions of the right-hand side of (49). This fact can be important in physics, for instance, in astrophysics or diffusion. We do not know the physical interpretation.

In the preceding theorems we studied the structure of the convex set $\mathscr{B}(\delta_0)$, where $\delta_0(f) = f(0)$ is the Dirac measure. Theorem 9 gives an answer for positive measures whose support contains an interior point. The mass distribution μ of the earth has such a property.

Theorem 9: Let $\Omega \subset \mathbf{R}^n$, $n = 2, 3$, be a bounded domain and μ a positive measure satisfying supp $\mu = \bar{\Omega}$. A measure $\nu_0 \in \mathscr{B}(\mu)$ having the form $\nu_0 = \nu_1 + \nu_2$, where $d\nu_1(y) = \rho(y)\, dy$, $0 < c \leqq \rho$, and supp $\nu_1 = \bar{\Omega}$ cannot be an extremal element of $\mathscr{B}(\mu)$.

Proof: By assumption there exists a ball $K(x^0, r) \subset \text{supp } \nu_0$. Let $B := \{x:(3r/4) < |x^0 - x| < r\}$ and $K(x, r') \subset B$ a further ball. From Gauss' theorem of the arithmetic mean it follows that harmonic functions satisfy the relation

$$u(x) = \frac{1}{V_{r'}} \int_{K(x,r')} u(y)\, dy \tag{50}$$

where $V_{r'}$ is the volume of the ball. Using (50), we get

$$|u(x)| \leqq \frac{1}{cV_r} \int_{K(x,r')} |cu(y)|\, dy \leqq C \int_{K(x,r')} |\rho(y) u(y)|\, dy \tag{51}$$

and from this inequality,

$$|u(x)| \leqq C \int_{K(x,r')} \rho(y) |u(y)|\, dy = C \int_B \rho(y) |u(y)|\, dy \tag{52}$$

A measure $\nu_0 \in \mathscr{B}(\mu)$ is extremal if and only if for $f \in L\nu_0^1$ and each $\epsilon > 0$ there exists a $u_k \in H_G(\bar{\Omega})$ such that (see theorem 2)

$$\|f - u_k\| = \int |f - u_k|\, d\nu_0 < \epsilon \tag{53}$$

From

$$\int |f - u_k|\, d\nu_0 = \int_\Omega |f - u_k| \rho\, dy + \int |f - u_k|\, d\nu_2 < \epsilon$$

it follows that

$$\int_\Omega |f(y) - u_k(y)| \rho(y)\, dy < \epsilon \tag{54}$$

Let $f(y) = 0$ on B and $f(y) = 1$ on $K(x^0, r/4)$. From (54) we get

$$\int_B |u_k(y)| \rho(y)\, dy < \epsilon \qquad \int_{K(x^0, r/4)} |1 - u_k(y)| \rho(y)\, dy < \epsilon \tag{55}$$

and from (52) and (54),

$$|u_k(x)| \leqq C \int_B |u_k(y)| \rho(y)\, dy < \epsilon$$

that is

$$\lim_{k\to\infty} u_k(x) = 0 \quad \text{on} \quad B'$$

where

$$B' := \{x : (3r/4) + \epsilon' < |x^0 - x| < r - \epsilon'\} \qquad \epsilon' > 0$$

is sufficiently small. Using the maximum-minimum principle of the harmonic functions, we get

$$\lim_{k\to\infty} u_k(x) = 0 \quad \text{on} \quad K(x^0, r - \epsilon') \tag{56}$$

uniformly. On the other hand it follows from (55) that

$$\int_{K(x^0,r/4)} \rho(y)\,dy = \lim_{k\to\infty} \int_{K(x^0,r/4)} u_k(y)\rho(y)\,dy \tag{57}$$

Using (56), we get zero on the right-hand side of (57), which is a contradiction. Therefore the measure ν_0 cannot be extremal.

Remark 3: Newton's gravitational law is one of the most important laws in physics. Many scientists have studied the gravitational field. From a physical point of view one can find such considerations in the work of *Pick et al.* [1973] and *Goncharskiy et al.* [1978]. Concrete formulas for calculating potentials are given by *Butkovskiy* [1978].

Remark 4: The earth's gravity field plays an important role in determining the shape of the earth. This nonlinear free boundary problem was studied by *Hörmander* [1976] (see also *Pick et al.* [1973]).

Remark 5: The solution u of the Dirichlet problem defines a mapping

$$A : C(\bar{\Omega}) \leftrightarrow C(\bar{\Omega}) \qquad Au(x) := \int u(y)\,d\mu_x^{\Omega}(y)$$

which is a projection. Let A^* be the adjoint mapping defined on $C^*(\bar{\Omega})$ by $(Af, \mu) = (f, A^*\mu)$, where $(f, \mu) := \int f\,d\mu$, $f \in C(\bar{\Omega})$, $\mu \in C^*(\bar{\Omega})$. Let $(A^*)^{-1}$ be the inverse of A^*, and let μ be a measure on $\partial\Omega$. Then the relation $\mathscr{B}(\mu) \subset (A^*)^{-1}(\mu)$ holds [*Anger*, 1978, 1980]. A^* describes the balayage principle.

6. An Inverse Problem for the Equation of Heat Conduction

The preceding results also hold for the equation of heat conduction

$$\frac{\partial u}{\partial t} - \frac{\partial^2 u}{\partial x^2} = 0 \tag{58}$$

or more general in $\mathbf{R}^3$,

$$\frac{\partial u}{\partial t} - \frac{\partial^2 u}{\partial x_1^2} - \frac{\partial^2 u}{\partial x_2^2} - \frac{\partial^2 u}{\partial x_3^2} = 0 \tag{59}$$

The adjoint equation of heat conduction is defined by

$$-\frac{\partial u}{\partial t} - \frac{\partial^2 u}{\partial x^2} = 0 \tag{60}$$

The fundamental solution of (58) has the form

$$G_H(x, t; z, \tau) := \frac{1}{2[\pi(t-\tau)]^{1/2}} \exp \frac{(x-z)^2}{4(t-\tau)} \qquad t > \tau \tag{61}$$

$G_H(x, t; z, \tau) = 0$ otherwise.

Let

$$\begin{aligned} \Omega_T &:= \{(x, t) \in \mathbf{R}^2 : 0 < t < T\} \\ \partial\Omega_T &:= \{(x, t) \in \mathbf{R}^2 : t = T\} \end{aligned} \tag{62}$$

The solution of the Cauchy problem relative to $t = \tau$ is given by

$$\begin{aligned} u(x, t) &= \int_{-\infty}^{+\infty} u(z, \tau) G_H(x, t; z, \tau)\,dz \\ &=: \int u(Z)\,d\mu_X(Z) \end{aligned} \tag{63}$$

where $X = (x, t)$ and $Z = (z, \tau)$, $t > \tau$. The solution of the Cauchy problem for the adjoint equation has the form

$$\begin{aligned} u(x, t) &= \int_{-\infty}^{+\infty} u(z, \tau) G_H(z, \tau; x, t)\,dz \\ &=: \int u(Z)\,d\mu_X^*(Z) \qquad t < \tau \end{aligned} \tag{64}$$

For studying an inverse problem we introduce the set $\mathscr{B}(\delta_{X^0})$ of all positive measures ν on $\bar{\Omega}_T$ such that $G_H\nu$ and $G_H\delta_{X^0}$, $X^0 = (0, 0)$, are equal for $t > T$.

It is possible to introduce a balayage principle relative to the Cauchy problem and relative to the adjoint Cauchy problem [*Anger*, 1979*b*; *Anger and Czerner*, 1979]. The set $\mathscr{B}(\delta_{X^0})$ is a convex and weakly compact set. We can study the extremal elements of $\mathscr{B}(\delta_{X^0})$ and so on. The results are similar to those of section 5. The structure of the extremal elements of $\mathscr{B}(\delta_{X^0})$ is given by *Anger and Czerner* [1979]. *Czerner* [1980] gives the complete proofs.

7. Concluding Remarks

The inverse gravimetric problem and the inverse magnetic problem [*Anger*, 1979*a*; *Garland*, 1979; *Phillips and Lambeck*, 1980; *Strakhov*, 1980*b*; *Zhdanov*, 1980] are famous unsolved problems in mathematics and its applications. The right-hand side (sources) of a given differential equation is to be determined.

As a first step many authors consider the inverse problem relative to a potential (and not relative to the gradient on a surface). Historical remarks can be found in the work of *Anger* [1979*a*], *Bleistein and Cohen* [1978], *Garland* [1979], and *Strakhov* [1980*a*, *b*]. In this form there are relations to other problems, for instance, to distillation processes [*Karr and Pittenger*, 1977], to an inverse problem in heat conduction [*Anger*, 1979*b*; *Anger and Czerner*, 1979; *Czerner*, 1980], and to an inverse problem in acoustics and electromagnetics [*Bleistein and Cohen*, 1978]. A similar inverse problem for the biharmonic equation (solid-state physics) can be found in the work of *Michel* [1979] and *Phillips and Lambeck* [1980].

There is no systematic approach to solving such problems. Therefore the author introduced the set $\mathscr{B}(\mu)$ as a first step. In this form the results of modern mathematics can be applied. Much more work must be done to solve the inverse gravimetric problem and so on. Additional conditions on $\mathscr{B}(\mu)$ must be introduced in these considerations [*Anger*, 1980; *Phillips and Lambeck*, 1980].

The paper by *Anger* [1979*a*] is a survey of inverse and improperly posed problems in differential equations and contains 104 references. *Anger*'s [1980] lecture on modern potential theory and inverse problems contains 75 references on inverse and improperly posed problems. A more complete version of this lecture will be published as a book by Akademie-Verlag, Berlin. The author would appreciate comments and suggestions concerning the present review.

Note added in proof. The author would also like to call attention to recent publications by *Baltes* [1978], *Lavrent'ev et al.* [1980], and *Sabatier* [1980].

REFERENCES

Anger, G., Die Entwicklung der Potentialtheorie im Hinblick auf ihre grundlegenden Existenzsätze I, II, *Beitr. Dtsch. Math. Ver., 64,* 51–78, 100–134, 1961.

Anger, G., Die Rolle der modernen Potentialtheorie in der Theorie der inversen Aufgabenstellungen, *Gerlands Beitr. Geophys., 85,* 1–20, 1976.

Anger, G., Uniquely determined mass distributions in inverse problems, *Veroeff. Zentralinst. Phys. Erde Potsdam, 52,* 633–656, 1977.

Anger, G., Convex sets in inverse problems (in Russian), in *Trudy Vsesoyusnoy Konferentsii po Uravneniyam c Tsastnymi Proizvodnymi,* edited by P. S. Aleksandrov and O. A. Oleynik, pp. 23–30, Moscow University Press, Moscow, 1978.

Anger, G., Some remarks on inverse problems in differential equations, Elliptische Differentialgleichungen, Proceedings of Meeting Held at the University of Rostock, October 10–16, 1977, *Rostocker Math. Kolloq., Spec. Vol.,* 31-51, 1979*a*.

Anger, G., Ein inverses Problem der Wärmeleitungsgleichung, I, *Lect. Notes Math., 743,* 485–507, 1979*b*.

Anger, G., Lectures on potential theory and inverse problems, lectures held at the Department of Geophysics, Bergakademie Freiberg (Sachsen), during a summer school from June 6–16, 1978, in *Geodätische und Geophysikalische Reihe,* vol. 45, Nationalkomitee für Geodäsie und Geophysik bei der Akademie der Wissenschaften der DDR, Reihe 3, Berlin, 1980.

Anger, G., and R. Czerner, Solution of an inverse problem for the heat equation by methods of modern potential theory, in *Inverse and Improperly Posed Problems in Differential Equations,* edited by G. Anger, pp. 7–23, Akademie, Berlin, 1979.

Baltes, H. P. (Ed.), *Inverse Source Problems in Optics, Topics in Current Physics,* vol. 9, Springer, New York, 1978.

Bleistein, N., and J. K. Cohen, A survey of recent progress on inverse problems, *Seminar MS-R-7806,* Dep. of Math., Univ. of Denver, Denver, Colo., 1978.

Butkovskiy, A. G., *Characteristics of Systems With Distributed Parameters* (in Russian), Nauka, Moscow, 1978.

Cannon, J. R., and P. DuChateau, An inverse problem for an unknown source in a heat equation, *J. Math. Anal. Appl., 45,* 465–485, 1980.

Choquet, G., *Lectures on Analysis,* vol. I, II, III, Benjamin, New York, 1969.

Czerner, R., Charakterisierung von konvexen Mengen beim inversen Problem der Wärmeleitungsgleichung, thesis, Martin-Luther-Universität Halle-Wittenberg, Halle (Saale), German Democratic Republic, 1980.

Devaney, A. I., The inverse problem for random sources, *J. Math. Phys., 20,* 1687–1691, 1979.

Garland, G. D., The contributions of Carl Friedrich Gauss to geomagnetism, *Hist. Mat., 6,* 5–29, 1979.

Gauss, C. F., Allgemeine Lehrsätze in Beziehung auf die im verkehrten Verhältnisse des Quadrates der Entfernung wirkenden Anziehungs- und Abstossungskräfte, in *Gauss' Werke,* vol. 5, pp. 194–242, Koenigliche Gesellschaft der Wissenschaften zu Goettingen, Göttingen, 1877.

Goncharskiy, A. V., A. M. Cherepashchuk, and A. G. Yagola, *Numerical Methods for the Solution of Inverse Problems in Astrophysics* (in Russian), Nauka, Moscow, 1978.

Helms, L. L., *Introduction to Potential Theory,* Wiley-Interscience, New York, 1969.

Hörmander, L., The boundary problems of physical geodesy, *Arch. Ration. Mech. Anal., 62,* 1–52, 1976.

Juhl, A., The inverse problem for the wave equation, Elliptische Differentialgleichungen, Proceedings of Meeting Held at the University of Rostock, October 10–16, 1977, *Rostocker Math. Kolloq., Spec. Vol.,* 125–131, 1979.

Karr, A. F., and O. A. Pittenger, An inverse balayage problem for Brownian motion, *Tech. Rep. 277,* Dep. of Math. Sci., Johns Hopkins Univ., Baltimore, Md., 1977.

Kluge, R., On some inverse problems in variational and quasi-variational inequalities, in *Inverse and Improperly Posed Problems in Differential Equations,* edited by G. Anger, pp. 141–149, Akademie, Berlin, 1979.

Lavrent'ev, M. M., V. G. Romanov, and S. P. Shishatskiy, Improperly Posed Problems in Mathematical Physics and Analysis (in Russian), Nauka, Moscow, 1980.

Michel, B., About inverse problems in continuum mechanics and solid state physics, in *Inverse and Improperly Posed Problems in Differential Equations,* edited by G. Anger, pp. 171–180, Akademie, Berlin, 1979.

Nashed, M. Z., *Ill-Posed Problems: Theory and Practice* vol. I, II, D. Reidel, Hingham, Mass., 1980.

Phillips, R. J., and K. Lambeck, Gravity fields of the terrestrial planets: Long-wavelength anomalies and tectonics, *Rev. Geophys. Space Phys., 18,* 27–76, 1980.

Pick, M., J. Picha, and V. Vyskočil, *Theory of the Earth's Gravity Field,* Academia, Prague, 1973.

Riesz, F., Sur les opérations linéaires, *C. R. Hebd. Seances Acad. Sci., 149,* 974–977, 1909.

Sabatier, P. C. (Ed.), *Inverse Problems, Non Linear Evolution, Proceedings of the Workshop R.C.P. 264, Montpellier, October 1979,* Éditions du Centre National de la Recherche Scientifique, Paris, 1980.

Schulze, B.-W., and G. Wildenhain, *Methoden der Potentialtheorie für elliptische Differentialgleichungen beliebiger Ordnung,* Birkhäuser, Boston, Mass., 1977.

Strakhov. V. N., Equivalence in gravity inversion problems and possible implications for interpreting gravity anomalies, I. (in Russian), *Izv. Akad. Nauk SSSR Fis. Zemli,* No. 2, 44–64, 1980*a*.

Strakhov, V. N., Unsolved plane problems in mathematical theory of gravity and magnetometry (in Russian), *Izv. Akad. Nauk SSSR Fis. Zemli,* No. 8, 3–28, 1980*b*.

Trèves, F., *Basic Linear Partial Differential Equations,* Academic, New York, 1975.

Zhdanov, M. S., Cauchy integral analogues for the separation and continuation of electromagnetic fields within conducting matter, *Geophys. Surv., 4,* 115–136, 1980.

(Received August 29, 1980;
accepted November 17, 1980.)

Comparison of Some Methods for the Detailed Representation of the Earth's Gravity Field

C. C. TSCHERNING

Geodaetisk Institut, DK-2920 Charlottenlund, Denmark

Currently, two types of methods are used for the detailed representation of the earth's gravity potential, namely, integral formulae and approximation techniques. The theoretical advantages and disadvantages of the methods (or of a combination of the methods) are discussed, and the results obtained are compared. If the methods are judged according to their general applicability, flexibility, and possibilities for obtaining error estimates, then approximation techniques like least squares or minimum norm collocation must be preferred. Integral formulae techniques are, on the other hand, well suited in areas with a homogeneous data coverage.

1. INTRODUCTION

Let W denote the gravitational potential of the earth and U a suitable approximation to W, for example, defined as a linear combination of low-order solid spherical harmonic plus the rotational potential. We will not presuppose that the anomalous potential $T = W - U$ is harmonic outside the surface of the earth. Neither will we presuppose that the usual regularity conditions at infinity are fulfilled. This means that we do not presuppose the precise knowledge of the rotational axis, the gravity center, or the total mass of the earth.

We will keep in mind the fact that W is produced by a volume distribution of mass and the rotation of the earth. This means that T not only is smooth in the open set Ω outside the surface of the earth but also behaves well at the boundary ω [*Sanso',* 1980].

A representation of the earth's gravity field will in this paper be any numerical procedure capable of computing an approximation $\tilde{T}$ to the anomalous potential or to the value of some functional F (linear or nonlinear) applied on T, $\tilde{F}T$. (Sometimes an approximation $\tilde{T}$ will be constructed, in which case we may be able to compute $F(\tilde{T})$.) The main point is that we do not make a distinction between T and the values of functionals applied on T.

The functionals, which we will actually regard, will all be linearized and will be denoted by L, contingently with a subscript, for example, L_i. The functionals will be of the kind which, when applied on T, give linear combinations of some of the following quantities:

Height anomaly

$$\zeta = T/\gamma \tag{1}$$

Gravity anomaly

$$\Delta g = -\frac{\partial T}{\partial r} - \frac{2}{r}T \tag{2}$$

Meridian and prime vertical components of the deflection of the vertical

$$\xi = -\frac{1}{\gamma \cdot r}\frac{\partial T}{\partial \varphi} \tag{3}$$

$$\eta = -\frac{1}{\gamma \cdot r \cdot \cos\varphi}\frac{\partial T}{\partial \lambda} \tag{4}$$

Gravity gradients

$$\partial^2 T/\partial x_i\, \partial x_j$$

where the quantities are evaluated at a point P, with latitude φ, longitude λ, radial distance r, and the normal gravity $\gamma = |\nabla U|$, x_i being the coordinates of a suitable local or global Cartesian coordinate system.

Gravity field representations are used for many different purposes, and some types of representations are suitable for one purpose but not necessarily for the others. A representation of $\tilde{T}$ as a series in solid spherical harmonics is suitable for the analysis of certain satellite orbit perturbations but is certainly not suitable for the computation of a precise gravity value needed at a leveling point.

Let us mention different areas where a precise, detailed representation of the gravity field is needed: (1) three-dimensional adjustment of geodetic networks or 'reduction' of observations for a two-dimensional adjustment (standard deviations σ needed: $\sigma(\tilde{\zeta}) \approx \pm 0.5$ m, $\sigma(\tilde{\xi}) = \sigma(\tilde{\eta}) \approx \pm 1''$), (2) computation of orthometric heights H from ellipsoidal heights h ($\sigma(\tilde{H} - h) = \sigma(\tilde{\zeta}) < 1$ m for mapping purposes and <0.05 m for geodynamic purposes), (3) computation of potential differences in a leveling network from observed height differences ($\sigma(\tilde{\Delta g}) \approx \pm 1$ mGal), (4) determination of a reference surface for oceanography ($\sigma(\tilde{\zeta}) \approx \pm 0.1 - \pm 0.01$ m), (5) computation of gravity vector components for inertial navigation and surveying and support of advanced military weapon systems ($\sigma(\tilde{\xi}) = \sigma(\tilde{\eta}) \approx \pm 0.5''$, $\sigma(\tilde{\Delta g}) \approx \pm 1$ mGal), and (6) computation of the density distribution inside the earth (prospecting): ($\sigma(\tilde{\Delta g}) \approx \pm 10 - \pm 0.1$ mGal, depending on actual purpose).

One method of obtaining a quantity $L(T)$ is, naturally, to observe the quantity. If this is not feasible or possible, a representation $\tilde{T}$ (or $L(\tilde{T})$) must be constructed by using our knowledge of the gravity field, that is, the many different types of observations. These do not consist only of the observed values of ζ, Δg, ξ, η, etc., but also of the coordinates of the points in which they have been observed. Furthermore, angles between points, observed by using a theodolite, contain gravity field information, because the scale readings are carried out with the instrument axis near to the plumb line [see *Eeg and Krarup,* 1975].

The purpose of this paper is to compare different representation techniques. This will be done not only on the basis of actual results obtained by using the different techniques but also on the basis of whether the methods can handle the type of information we actually have: we do not know station positions precisely; the position of the coordinate systems with respect to the earth's center of mass and rotational axis is not yet known exactly, and the total mass is given approximately.

In section 2 we have tried to review the different methods in a systematic manner, and in section 3 we confront the methods with three types of questions concerning (1) theoretical properties, (2) flexibility and reliability, and (3) computational properties. Finally, in section 4 some of the results actually obtained are reviewed.

2. METHODS FOR GRAVITY FIELD REPRESENTATION

A General Overview of the Methods

A wide range of techniques are used, or have been proposed, for the representation of the gravity field. Following *Moritz* [1978*a*], we may regard the methods as belonging to one of two approaches to physical geodesy: the model approach or the operational approach.

In the model approach a mathematical model is set up, which from a given set of data will determine the required quantity. The problem is then to obtain the data.

The operational approach starts with the available data and asks how these data can be combined in the best way in order to obtain the required quantity.

Now these approaches are in a certain sense dual: the justification for large-scale observation campaigns of gravity was that the data could be used for the solution of the geodetic boundary value problem. On the other hand, data have often been collected for some specific reason (like rock density values for geological investigations), and the problem has then been how to use this kind of data for gravity field representation.

Therefore the model approach will sometimes fit the real-world situation, at least to a certain degree. Mostly, it will not fit, but in reality we will see that the two approaches go hand in hand. In fact, a model approach, which has no restrictions whatsoever in the use of data, is equivalent to an operational approach, and an operational approach may (partially) consist of finding such 'corrections' to or 'reductions' of the observations that they will suit the model.

The Model Approach (Not an Exhaustive Review)

In the model approach, known relations between T and different types of observations are used. Presupposing harmonicity of T in Ω and regularity at infinity, $\tilde{T}$ will be the solution to different boundary value problems. The solutions manifest themselves as integral formulae: Stokes' integral, Poisson's integral (with data given on a sphere), or the proposed integral formula solution to Molodensky's problem. Application of the respective linear functionals on these formulae gives corresponding integral formulae for these functionals, known, for example, as Vening Meinesz integral formulae.

The general expression for $\tilde{T}$ is

$$\tilde{T}(P) = \int_{\omega} K(P, Q) \left[D(Q) - \sum_{i=1}^{n} C_i(Q) \right] d\omega_Q \qquad (5)$$

where P is a point in Ω, Q is a point on the boundary ω, $K(P, Q)$ is a kernel function, $D(Q)$ is the data, and the $C_i(Q)$ are certain correction terms (compare, for example, *Heiskanen and Moritz* [1967, equation (8-50)]). Note that the data necessary for the numerical evaluation of (5) must be available in a regular grid around P (or around P's projection on ω, if P is a point in free space). The grid values are generally determined from irregularly spaced data by methods developed in the operational approach.

The technique of astrogeodetic geoid determination also belongs to the model approach. The model is very simple, and harmonicity or regularity of T is not presupposed. The anomalous potential restricted to the reference ellipsoid is regarded as a function of latitude and longitude only. The deflections of the vertical are then associated in a simple manner with T through the horizontal derivatives (compare (3) and (4)). $\tilde{T}$ (or $\zeta = \tilde{T}/\gamma$) is then determined through numerical integration. This model requires, however, that the deflections of the vertical in some way be reduced to the ellipsoid (or geoid). The method has the drawback that the two-dimensional numerical integration may break down if only one component of the deflection of the vertical is available. In this situation, only geoid profiles can be determined, or the one-component stations are left out! (See, for example, *Wenzel* [1978, Figure 1]).

The 'worst' example of the model approach is probably the harmonic sampling functions of *Giacaglia and Lundquist* [1972]. This method requires that the values of ζ or Δg be available in a specific, predefined set of points in space. The approximation $\tilde{T}$ is then given as a linear combination of harmonic functions with the observed values as coefficients. (Because of the global character of the sampling functions they are not well suited for the detailed representation of the gravity field.)

The Operational Approach

While the model approach, until quite recently, has been used exclusively when dealing with the gravity field of the earth, the operational approach (namely, least squares adjustment) has always dominated geometrical geodesy. This is obviously due to the simple relation between model and data. First, through recent developments in mathematics and statistics it has been possible to generalize least squares techniques from finite dimensional spaces to spaces of infinite dimension.

We will now presuppose that T can be expressed as the sum of a harmonic and a regular function (which we will also denote by T) and some correction functions $A \cdot X$, where X is a vector of constants describing corrections to the reference system orientation and position with respect to the earth's rotation axis and gravity center, for example. (A modeling of time-dependent changes can also be made here.) We may also introduce other parameters, such as station position corrections. The important point here is that we only deal with a finite number of parameters.

Here T will be an element of a reproducing kernel Hilbert space of harmonic functions (because it is the potential of a difference between volume density distributions), and X will be an element of a finite dimensional real vector space, which we may furnish with a norm, for example, inherited from the statistical distribution of the observation errors. This means that our task is the determination of a point in an infinite dimensional Hilbert space spanned by the harmonic functions and the parameters X. We will suppose that there are a total of k parameters, X_j, $j = 1, \cdots, k$.

The observations D_i will be related to T and X through linearized functionals

$$D_i = L_i(T) + L_i(A) \cdot X + \text{observational error} \qquad i = 1, \cdots, n \qquad (6)$$

We are hence faced with an improperly posed problem, namely, the determination of the T and X in all infinitely many coordinates.

Different techniques can be used in order to find a solution; however, only the following two seem to have been used in geodesy: (1) a change of the space and/or topology and (2) the use of a regularization operator.

Moritz [1978*a*] also lists the use of probabilistic methods. However, the methods, which are used for gravity field representation and are called probabilistic, do not seem to me to be truly probabilistic. (The proposed methods are under all circumstances equivalent to regularization methods, because the covariance functions used are reproducing kernels belonging to some Hilbert space of harmonic functions.)

A large number of methods have been based on a change in the solution space. A finite set of functions h_i, $i = 1, \cdots, m$ are selected ($m + k \leq n$), and the approximation $\tilde{T}$ is obtained as a linear combination

$$\tilde{T}(P) = \sum_{i=1}^{m} a_i h_i \tag{7}$$

The observation equations

$$D_i = L_i(\tilde{T}) + L_i(A) \cdot X = \sum_{j=1}^{m} a_j L_i h_j + L_i(A) \cdot X \tag{8}$$

form, for $m + k < n$, an overdetermined system, which can be solved by using traditional least squares techniques, using the variance-covariance matrix of the observation errors $\{P_{ij}\}^{-1}$. In case the resulting system of normal equations is not of full rank, or in case the solution oscillates 'too much,' a regularization procedure may be used. Let $\| \quad \|$ be a norm on the $m + k$ dimensional space spanned by the functions h_i and the parameters, with

$$\| (\tilde{T}, X) \|^2 = \| \tilde{T} \|_0^2 + X^T Q X \tag{9}$$

where $\| \quad \|_0$ is the norm in the Hilbert space and Q is a positive definite $k \times k$ matrix. A solution is then obtained by minimizing

$$\lambda \cdot \| (\tilde{T}, X) \|^2 + \{D_j - L_j(\tilde{T}) - L_j(A) \cdot X\}^T \{P_{ij}\} \{D_i - L_i(T) - L_i(A) \cdot X\} \tag{10}$$

where $\lambda > 0$. The result is a new set of normal equations. (This is the procedure used, for example, for the construction of the satellite solutions GEM 9 and 10 [see *Lerch et al.*, 1979].

In the case $k = 0$ and $m = n$ the coefficients $\{a_i\}$ are determined as solutions to the system of equations

$$L_j(\tilde{T}) = D_j = \sum_{i=1}^{n} a_i L_j h_i = \{a_i\}^T \{L_j h_i\} \tag{11}$$

This set of equations may be solved when the functions h_i and the linear functionals L_i (regarded as elements of the space dual to the space spanned by $\{h_j\}$) are linearly independent. This technique is called 'pure collocation,' because an exact agreement between $\tilde{T}$ and the data is enforced.

If $m + k < n$, we are again faced with an improperly posed problem, which is generally solved by using a regularization technique (see discussion below).

If the main goal is not to determine $\tilde{T}$ but some operator F applied on $\tilde{T}$, the same procedure can be used. F can, for example, be the restriction operator which restricts T to a set R, for example, the surface of the earth, an ellipsoid, or a sphere. In these cases, T_R is a function of two variables, which are generally taken to be the latitude and the longitude. This requires that the data refer to points in R. This fact is sometimes overlooked and results in a misleading terminology (such as 'inverse Stokes operator,' the simple, correct version of which is (2); see also *Nash and Jordan* [1978], who never take into account the topography of the earth).

The different methods vary mostly according to the kind of base functions h_i which are used. Preferably, harmonic functions are used, such as the potentials of point masses or surface densities or harmonic kernel functions [*Lelgemann*, 1978*a*]. These functions can be chosen so that they are well suited for local, detailed gravity field approximation. A suitable choice of base functions can make $\tilde{T}$ identical to a solution obtained by using regularization.

Also, nonharmonic functions such as spline functions and traditional polynomials are used. So-called finite elements (for example, a set of polynomials, each valid for disjoint sets in $\mathbf{R}^3$) have also been proposed [*Junkins and Engels*, 1979], and *Batdorf* [1970] has considered some nearly harmonic 'B functions.'

Such methods require that the harmonicity of T also be represented as an observation, a fact which sometimes seems to have been forgotten. In the same way, additional conditions have to be introduced, when using potentials of point masses, in order not to violate the condition that T be regular at infinity (see, for example, *Blaha* [1977]).

The 'method of regularization' consists of minimizing linear combinations of norms (e.g., those given by (10)), where $\tilde{T}$ is now regarded as an element of an infinite dimensional Hilbert space in which T is an element. In this case the minimization of the quantity in (10) will furnish us with an n-dimensional subspace spanned by harmonic functions. These functions are equal to the linear functional L_i applied on the reproducing kernel $K(P, Q)$, that is,

$$\tilde{T}(Q) = \sum_{i=1}^{n} a_i L_i K(\cdot, Q) \tag{12}$$

(where the dot indicates that L_i has been applied on $K(P, Q)$ with respect to the variable P). The coefficients can be determined by pure collocation, (11) for $k = 0$. This technique is known as minimum norm collocation [cf. *Tscherning*, 1978*b*], and the normal equations become

$$\{L_i L_j K(\cdot, \cdot) + P_{ij}'\} \{a_j\} = \{D_i\} \tag{13}$$

which are symmetric ($\{P_{ij}'\} = \{P_{ij}\}^{-1}$). For $k > 0$, see, for example, *Tscherning* [1978*a*].

Using regularization, we are faced with the problem of selecting a suitable Hilbert space, that is, a set of functions and a quadratic norm. This choice will determine the reproducing kernel and vice versa.

Until recently, the only reproducing kernels which were explicitly known and computable were for a Hilbert space consisting of functions harmonic outside a sphere. So in order to use regularization one had to change solution space first.

These kernels could in many cases be represented as closed expressions, which greatly facilitated the computations [see *Tscherning and Rapp*, 1974; *Tscherning*, 1976*a*; *Heller and Jordan*, 1979]. The kernel proposed by *Krarup* [1978] does not require a change in solution space, but integration along the true earth's surface is necessary.

Hence numerical reasons have until now made it necessary to change solution space. (This is quite a serious operation [e.g., *Sjöberg*, 1978, Figure 6].) However, the change still permits us to associate the linear functionals with their true position in space. On the other hand, it is customary to work in spherical approximation, and the error introduced thereby is small but not negligible.

We lose, however, the important property that $\tilde{T}$ is an element of the same Hilbert space as T. If they are elements of the same space, then maximal error estimates can be computed [see *Dermanis,* 1976; *Krarup,* 1978; *Tscherning,* 1977].

Solutions optimal in a least squares sense may be obtained by choosing a norm which is derived from T itself. This is the well-known method of least squares collocation. The only implemented version of the method requires that we immediately change solution space by turning to spherical approximation. Here the earth is bounded by a sphere with radius R, and T can be expressed as a series in fully normalized solid spherical harmonics with coefficients $\bar{C}_{ij}$, $\bar{S}_{ij}$, $i = 0, \cdots, \infty, 0 \leq j \leq i$. Then the empirical covariance function is $K_E(P, Q)$, where

$$K_E(P, Q) = \sum_{n=0}^{\infty} \sigma_n \left(\frac{R^2}{r_P r_Q} \right)^{n+1} P_n(\cos \psi) \tag{14}$$

$$\sigma_n = \sum_{j=0}^{n} (\bar{C}_{ij}^2 + \bar{S}_{ij}^2) \tag{15}$$

Here ψ is the spherical distance between P and Q; r_P and r_Q are the radial distances of P and Q, respectively; and P_n is the nth Legendre polynomial. The σ_n are the (potential) degree variances. We can use $K_E(P, Q)$ as our reproducing kernel, thereby forcing $\tilde{T}$ to be an element of the corresponding Hilbert space. The use of the technique requires the estimation of the degree variances σ_n. This is a difficult process [see *Tscherning and Rapp,* 1974; *Moritz,* 1977; *Jekeli,* 1978; *Rapp,* 1979*b*].

The least squares principle, the use of which results in (14) and (15), is of a global character [see *Heiskanen and Moritz,* 1967, chapter 7]. For this reason some authors have rejected the use of least squares collocation for the local, detailed approximation of T [e.g., *Kearsley,* 1976]. However, one must keep in mind that the use of a least squares principle, resulting in the isotropic covariance function (14), does not imply that T must fulfill any isotropy conditions.

If a high-order reference field U is used, the 'global' least squares principle may be adapted for local use. A sufficiently high order reference field has the significance that the gravity field outside the regarded area is of minor importance when constructing $\tilde{T}$ for the area. Hence the gravity field outside the area can be regarded as having the same variations globally as locally. The global sampling process used when estimating $K_E(P, Q)$ may hence be restricted to the local area, as was done by *Tscherning* [1974, section 2.3]. In this manner the least squares collocation method may be regarded as being optimal in a least squares sense for a certain local area.

The method of least squares collocation will in theory enable the estimation of mean square errors. The theoretical interpretation of this type of error is somewhat doubtful, as there is no explicit probability distribution given for the random or stochastic variables.

The techniques described above may, as was explained above for the model approach, be used for the restriction of T to a two-dimensional surface in $\mathbf{R}^3$ or for the approximation of another operator applied on T. The restriction T_R may be regarded as a function of (φ, λ) or of some plane set of coordinates (X, Y). The two techniques (change in solution space and regularization) are then adapted to this situation. The change in solution space will result in the choice of a finite set of functions of two variables, for example, surface harmonics, polynomials in two variables, multiquadratic functions, or the binary sample functions used by *Brown* [1975]. Regularization techniques will appear through the use of reproducing kernels and covariance functions depending only on the spherical distance ψ and not on r_P and r_Q as in (14) (see, for example, *Nash and Jordan* [1978] and *Hein and Lenze* [1979]).

As was mentioned above, this type of procedure has the drawback that data must refer to the selected reference surface. This problem is considerably reduced if the surface of the earth or a sphere outside the earth is used and not a sphere inside the earth.

Combined Methods

Techniques which combine the model approach and the operational approach are in frequent use. As was mentioned above, the integral formula (5) requires that data be available in a regular grid. Here two-dimensional operational approaches are followed in order to obtain the gridded data (see, for example, *Sünkel* [1977] and *C. R. Schwarz* [1978]).

The integral equation (5) must in principle be evaluated over the whole surface ω. In order to save computational time the integration is frequently only carried out to a certain distance (depending on the reference potential U used). The error arising will be approximately constant in a small area. A comparison with contingently observed values of the geoid undulations or deflections of the vertical may then (for example, through a least squares process) determine these 'systematic errors.' An example of this type of technique is the method of astrogravimetric geoid determination (see, for example, *Heiskanen and Moritz* [1967, p. 203] and *Merry and Vaníček* [1974]).

Least squares or collocation techniques require the solution of systems of linear equations. Here the number of equations can be reduced by first computing an approximation $\tilde{T}_1$ exclusively from data given in a local area ω_1. As a second step, $T_2 = T - \tilde{T}_1$ may be determined by using an integral formulae method, where the integration will only have to be performed in the area outside ω_1, because $T - \tilde{T}_1$ is zero in ω_1. This technique is developed by *Moritz* [1976], *Lachapelle* [1977], and *Lachapelle and Tscherning* [1978].

The prediction methods developed by Bjerhammar [e.g., *Bjerhammar,* 1978] can to a certain extent also be regarded as combined methods. A discretized integral equation is used for the determination of quantities (e.g., the so-called reduced gravity anomalies Δg^*) associated with points or surface elements on a sphere inside the earth. A harmonic function is then determined outside the sphere by using an integral formula. By a suitable choice of 'carrier points' and values we may arrive at techniques identical to the collocation methods, which use reproducing kernel Hilbert spaces of functions harmonic down to an internal sphere [see, *Sjöberg,* 1978].

3. COMPARISON OF THE METHODS

Realizing which kind of treatment our poor T is given by the different methods, one may be a little suspicious about the performance of some of the methods. Do they really work?

All the methods mentioned have been tried in practice with some success.

However, it is one thing to test a method in a scientific environment, where users really know what they are doing. It is quite another in a production environment, where the methods really are applied. They must be as foolproof as possible. This again means that different user groups will favor different techniques. The theoretically minded will favor methods which do not use reductions of data and which have been proved mathematically correct. The practician may prefer methods which, for a small cost, deliver a precise result within prespecified limits and may not worry too much about the method used.

In Table 1 an overview of the methods and the types of observations they can use is given. In Table 2 we have summarized the answers to 22 questions which have been 'asked' of every method. The questions have been asked so that a 'yes' indicates something positive about a method. In a few cases some of the questions are not relevant for judging a method. In this case no answer is given.

We have asked three types of questions: (1) about theoretical properties, (2) about the flexibility and reliability, and (3) about computational properties.

The most important theoretical question to be raised is whether or not a method converges. If it does not converge, there is a limit to the precision which can be achieved by using the method. Convergence has been proved for most of the methods which are based on the operational approach [see *Moritz*, 1976; *Tscherning*, 1978*a*; *Sanso' and Tscherning*, 1980; *Krarup*, 1980; *Keller*, 1978; *Sjöberg*, 1979; *Bjerhammar and Svensson*, 1979]. The reasons for a nonconvergence appear clearly in Table 2. The methods which require spherical approximation or the reduction of data to a sphere or ellipsoid will not converge. It should also be noted that a set of harmonicity which is bigger than Ω (or even nonharmonicity) does not exclude convergence. Also, the condition $\Delta\tilde{T} = 0$ may be better and better approximated when the data density increases.

Naturally, it is a condition for convergence that station positions are improved simultaneously with an improvement in $\tilde{T}$. This relationship is also clearly seen in the table.

For large-scale and/or routine computations the flexibility and reliability of a method is very important. There is here again a remarkable difference between the 'model approach' methods and the 'operational approach' methods, a difference which nearly makes it justifiable to call the model approach methods inoperable. There are, however, situations where all data are nicely homogeneously distributed, have insignificant errors, and are available to such an extent that a required precision can be achieved. We should also keep in mind that error estimates may be computed (by comparison with an external standard) in cases where they cannot be computed as a part of the computational process (see, for example, *Strange and Fury* [1979]). However, no method can really give 'true'

TABLE 1. Methods for Gravity Field Representation

Method	Short Name	Permitted Data Types
Model Approach		
Integral formulae		(all data gridded)
Poisson's integral	Pois. I	geoid heights on sphere
Stokes' integral	Stok. I	Δg
Molodensky series	Mol. Se.	Δg on ω
Numerical integration		
Astrogeodetic leveling	As. Ge. Le.	(ξ, η) pairs of ξ or η in profiles on ellipsoid
Harmonic sampling functions	Harm. Sa. Fct.	T or Δg in fixed points on sphere
Operational Approach		
Change in solution space		
Least squares approximation in three dimensions ($n \geq m$)		
Harmonic base functions	LSAH3D	all data types
Nonharmonic base functions	LSA3D	all data types
Least squares approximation in two dimensions ($n \geq m$)	LSA2D	data which can be 'reduced' to the same two-dimensional surface
Regularization methods		
Least squares collocation	LSCOL	all data types
Minimum norm collocation	MNCOL	all data types
Combined Methods		
Regularization, then integration		
Least squares collocation, then	LSCOL +	all data types in the inner area
Stokes integral	Stok. I	Δg on sphere, outside
Change in solution space, then integration		
Least squares approximation in two dimensions, then	LSA2D +	data which can be reduced to two-dimensional surface
Poissons integral	Pois. I	
Integration, then change in solution space		
Astrogravimetric leveling	As. Gra. le	Δg on part of Ω, ξ, η, in area of interest, plus one ζ value

TABLE 2. A Comparison of the Different Methods

	Pois. I	Stok. I.	Mol. Se.	As. Ge. Le.	Harm. Sa. Fct.	LSAH3D	LSA3D	LSA2D	LSCOL	MNCOL	LSCOL ⊥ Stok. I	LSA2D + Pois. I	As. Gra. Le.
Theoretical Properties													
$\tilde{T}$ harmonic, regular at infinity	(+)	(+)	(+)		(+)	(+)	−	−	(+)	(+)	(+)	(+)	(+)
Set of harmonicity correct	−	−	+		−	(+)(1)	(−)		−	(+)(1)	−	−	(−)
Parameter estimation possible	−	−	−	−	−	+	+	(−)	+	+	(+)	−	(+)
Regularity at infinity not required	+	−	−	(+)	+	+	+		+	(+)(4)	−	+	+
Spherical approximation not needed	−	−	(+)	(+)	−	+	+	(+)(2)	−	+	−	+	(+)
Ultimate precision possible (convergence)	−	−	?	−	−	+	+	(−)(2)	(−)	+	(−)	+	(+)
'Reduced' data not needed	−	−	+	−	(−)	+	+	−	+	+	(+)	+	−
Flexibility and Reliability													
All functionals can be evaluated	+	+	+	only ζ	+	+	+	(−)	+	+	+	+	−
All data types can be used	−	−	−	−	−	+	+	(−)	+	+	+	(−)	−
Data quality (σ) taken into account	−	−	−	(−)	−	+	+	+	+	+	+	+	(−)
Error estimates computable	−	−	−	(−)	−	(+)	(+)	(−)	(+)	(+)	(+)	−	−
Global data coverage not required	(−)	(−)	(−)	+	(−)	+	+	+	+	+	(+)	−	+
Stable in cluster/downward continuation	(+)	(+)	(+)	(+)	(+)	(+)(3)	(+)(3)	(+)(3)	(+)(4)	(+)(4)	(+)(4)	(+)(3)	(+)
New data/correction easily introduced	−	−	−	(−)	−	(+)	(+)	(+)	+	+	+	−	−
Behavior in data holes known	(+)	(+)	(+)	+	(+)	+	+	+	(+)	+	+	+	+
Computational Properties													
Method easily implemented	(+)	(+)	(+)	+	+	+	+	+	(−)	+	(−)	+	(+)
Method can be fully automated	(+)	(+)	(+)	+	(−)	(+)	(+)	(+)	−	+	−	+	(−)
Data management uncomplicated	(+)	(+)	(+)	−	(+)	+	+	(+)	(+)	+	(−)	(+)	(−)
No solution of linear equations	+	+	+	+	−	−	−	−	−	−	−	−	(−)
Gridded data not needed	−	−	−		−	+	+	+	+	+	(+)	+	(−)
Fast method	(+)	(+)	−	+	(+)	(+)	+	+	(+)	(−)	+	(+)	+
Automatic selection of base function	+	+	+	(+)	+	(−)	(−)	(−)	(+)	+	(+)	+	(+)

The explanation of the symbols is as follows: absolutely yes, +; yes with modifications, (+); unknown, ?; no with modifications, (−); absolutely no, −; depends on choice of base functions, but correct set of harmonicity can be achieved, (1); depends on choice of two-dimensional surface, (2); depends on selection of base functions, (3); and depends on norm and/or set of harmonicity, (4). For explanation of method name abbreviations, see Table 1.

error estimates: In least squares collocation the empirical covariance function must be known, and in minimum norm collocation it is necessary to know the norm of T, $\| T \|$.

Solutions obtained by least squares or minimum norm collocation can be easily updated or corrected. If the reduced normal equations are stored, an updated value of an observation only means that the right-hand side must be reduced again and a back substitution must be carried out. A new observation just means that a new column is added, which must be reduced, etc. Least squares approximation requires a change in the unreduced normal equations when a new observation is added [see *Tscherning,* 1978*a*; *Wolf,* 1979].

The stability of the methods in situations where data are lacking, occur in clusters, or must be continued downward from satellite altitude depends very much on the selected base functions and the degree of regularization used. This is a field where more investigations must be carried out [see *Tscherning,* 1975; *Lelgemann,* 1978*b*; *Rummel,* 1979].

The computational complexities of the different methods are of the same magnitude. Methods which do not require the solution of linear equations will need gridded data, and the process of producing these data is comparable to the effort spend for solving systems of equations. The main reason for this is that a different grid must be used for each computational point.

We should also keep in mind that a solution $\tilde{T}$ computed by using collocation or an approximation technique will be valid for a certain (smaller or larger) area. A data grid usable for precise computations by an integral formula will be only partly valid for a set of different points. This is because of the strong singularities of the kernels used in the integral formulae, which require special precautions during the numerical evaluation. The methods may be compared with respect to computational speed. This is, however, somewhat dangerous, because it depends on the smartness of the programer writing the computer algorithms and other computer dependent factors. But it is obvious that the most complex methods will need longer computational time. However, the computational time used must be balanced against the precision needed for the result. A very precise $\tilde{T}$ must necessarily cost more than a less precise $\tilde{T}$.

The considerations made when selecting a specific method should ideally be of the cost-benefit type: How much computer time will be spent in order to obtain a certain precision? Whether a 'better' result could have been obtained for the same cost seems, however, never to have been taken into consideration. This is because the 'capital' investments for implementing one method are very great in terms of manpower. We must realize that there are very few geodetic organizations which are capable of designing, implementing, and maintaining a system of computer programs for gravity field representation.

Let us conclude this section by again reviewing Table 2. Here the operational approach methods like least squares approximation and collocation stand out with many pluses. This is also why many authors favor these methods [*K.-P. Schwarz and H. Sünkel,* 1978; *Moritz,* 1978*a*; *Lachapelle and Tscherning,* 1978; *Reinhart et al.,* 1978].

However, in many cases a homogeneous gravity coverage exists in areas with moderately varying topography. Altimeter-derived geoid undulation at sea also forms a nice homogeneous set. Hence integral formulae methods are feasible in many situations [*Strange and Fury,* 1979; *C. R. Schwarz,* 1978].

4. RESULTS FROM METHODS USED FOR THE REPRESENTATION OF THE DETAILED GRAVITY FIELD

In this section we will review some of the results obtained with a gravity field representation method for the computation or prediction of 'point values' of gravity anomalies and deflections of the vertical. Such results are presented by different investigators in the form of comparisons between (1) computed and observed values, (2) computed and observed values where all values have been artificially generated, and (3) values computed with different methods.

First, it is well known that the computational results will depend strongly on the local gravity field variations (see, for example, *Tscherning and Forsberg* [1978, Table 1]). So different methods must be compared in terms of

$$\rho = \frac{\text{root-mean-square variation (observed} - \text{computed)}}{\text{root-mean-square variation of the observations}}$$

Second, the result (ρ) will depend on the kind of data used and the spacing of the data. Through theoretical investigations we know that some of the methods are able, in principle, to give an unlimited precision if the necessary data are available. However, methods which involve the use of reduced data or spherical approximation may also, in practice, give fine results. This is primarily because the errors in the 'model' will in many cases be small in comparison to the errors in the observations. Furthermore, different 'corrections' can be applied as a weapon against the 'model errors.'

The dependence of ρ on the data spacing is shown in the numerical studies of *Tscherning* [1975, Figures 5–7] and *K.-P. Schwarz* [1976]. When the spacing s of the data is smaller than the first zero point of the used covariance function, ρ will decrease almost linearly toward zero for $s \rightarrow 0$. The decrease will depend not only on the spacing of the data in the vicinity of the point of computation but also on the extent of the data distribution.

Figures 1 and 2 show results obtained by comparing computed and observed deflections of the vertical and gravity anomalies. We clearly see the decrease of ρ for decreasing s and the significant improvement obtained by adding a small

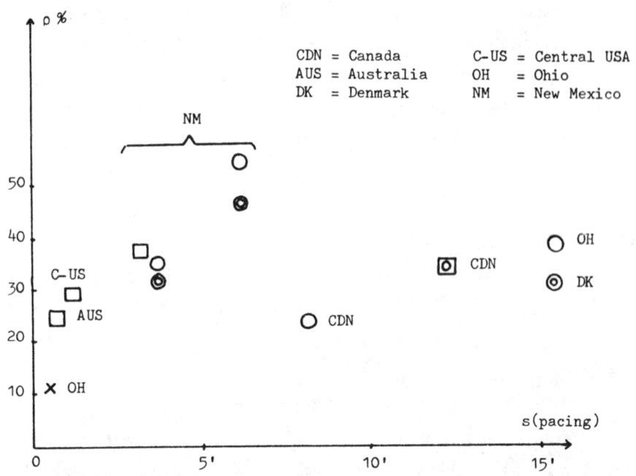

Fig. 1. Results obtained when computing deflections of the vertical: circles, using collocation with Δg only; double circles, using collocation with Δg and (ξ, η) spaced $\frac{1}{2}°$–$1°$ apart; squares, using Vening Meinesz formula; circles within squares, using collocation and Vening Meinesz formula combined; and crosses, using numerical integration of horizontal gravity gradients and (ξ, η) spaced $\frac{1}{2}°$ apart. The results are from different authors, collected by *Tscherning and Forsberg* [1978, Table 1], *Kearsley* [1976] and *C. R. Schwarz* [1978], and computations by the author in the New Mexico area.

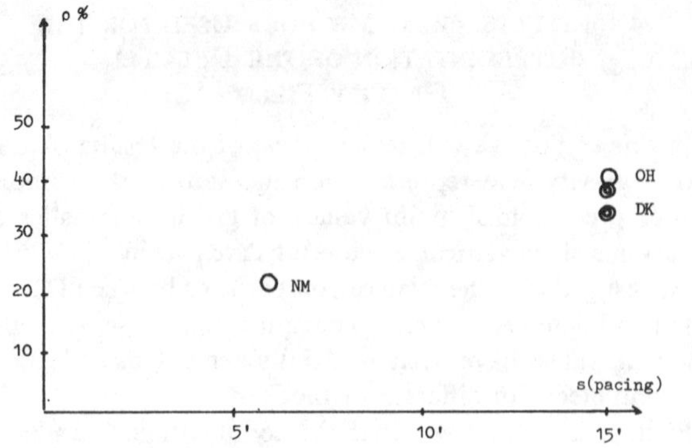

Fig. 2. Results obtained when computing gravity anomalies. For explanation of symbols, see Figure 1. These results were collected by *Lachapelle and Tscherning* [1978, Table 1].

number of deflections of the vertical (Figure 1). The results obtained in the same area with different methods (collocation and Vening Meinesz integral) are also of the same quality. But we also see that the decrease of ρ for decreasing s is different in New Mexico and in Canada. This is without doubt caused by the circumstance that the topography in the New Mexico area varies much more than the topography in the Canadian plains. Remember that we have taken into account the different variations of the gravity through the definition of ρ.

Height anomalies (or geoid undulations) have recently become available as observations. One type is obtained as the difference between the ellipsoidal height determined by Doppler satellite observations and the orthometric height. The other is obtained by setting the geoid height equal to the sea surface topography as determined by radar altimetry. These 'observational' values have standard deviations between 1 and 0.3 m.

The values have been compared with values obtained by different computational procedures (see, for example, *Rapp and Rummel* [1976], *Lachapelle* [1979], *Torge* [1980], *Monka et al.* [1978], *Balmino et al.* [1979], and *Chapman and Talwani* [1979]). Unfortunately, the authors have not given enough information that the results can be expressed in terms of ρ and s. The general conclusion is, however, that quite different methods give nearly the same results. Furthermore, the values have been used for mean gravity anomaly prediction and for the prediction of point values [see *Rapp*, 1978, 1979*a*].

Over the years, several investigations have been carried out by varying a specific method [*Tengström*, 1971; *Tscherning*, 1975; *Kearsley*, 1976; *Sjöberg*, 1978] or by comparing different methods [*Rapp and Agajelu*, 1975] while data spacing or type was not varied. Artificially generated data have also been used, mainly to test a specific method [see *K.-P. Schwarz*, 1978; *Sjöberg*, 1978]. These investigations confirm the impression that different methods will give results of approximately the same quality when the same set of data is used.

Some of the results have a special interest, because they show how much or how little the results depend on the 'changes in solution space' or the regularization used, that is, on the covariance function (see, for example, *Sjöberg* [1978] and *Rapp* [1979*a*]).

Finally, let us notice that very few investigations have been carried out in which a set of parameters and a local (detailed) gravity field approximation have been determined simultaneously [*Tscherning*, 1976*b*].

5. CONCLUSION

Our investigations of the different methods for gravity field representation show that the least squares approximation and collocation methods are conceptually superior to the integral formulae methods. These methods will, however, give results comparable to the operational methods in situations where only one kind of data is available and these data are sufficiently homogeneously distributed.

Today it is impossible to single out one method as the best, because the different methods have not been compared in a systematic way. We know only very little about how the different changes of the solution space and the different regularizations influence the precision of $\tilde{T}$, how data should be distributed, and which kinds of data should be used in order to obtain a given precision at the lowest cost. Answers to these important questions can only be found through international cooperation.

REFERENCES

Balmino, G., C. Brossier, A. Cazenave, and F. Nouel, Geoid of the Kerguelen Islands area determined from Geos 3 altimeter data, *J. Geophys. Res., 84*(B8), 3827–3831, 1979.

Batdorf, S. B., On a local representation of the fine structure of the earth's gravitational field, *J. Appl. Mech.*, 738–743, 1970.

Bjerhammar, A., A review of discrete methods in physical geodesy, in *Approximation Methods in Geodesy*, edited by H. Sünkel and H. Moritz, pp. 211–233, H. Wichmann, Karlsruhe, West Germany, 1978.

Bjerhammar, A., and L. Svensson, The convergence of harmonic reduction to an internal sphere, *Bull. Geod., 53*, 247–257, 1979.

Blaha, G., Refinement in the combined adjustment of satellite altimetry and gravity anomaly data, *Rep. AFGL-TR-77-0169*, Air Force Geophys. Lab., Bedford, Mass., 1977.

Brown, R. D., Geopotential modeling by binary sample functions, *Rep. CSC/TR-75/6006*, Comput. Sci. Corp., Silver Spring, Md., 1975.

Chapman, M. E., and M. Talwani, Comparison of gravimetric geoids with Geos 3 altimetric geoid, *J. Geophys. Res., 84*(B8), 3803–3816, 1979.

Dermanis, A., Probabilistic and deterministic aspects of linear estimation in geodesy, *Rep. 244*, Dep. of Geod. Sci., Ohio State Univ., Columbus, 1976.

Eeg, J., and T. Krarup, Integrated geodesy, *Method. Verfahren Math. Phys., 13*, 77–123, 1975.

Giacaglia, G. E. O., and C. A. Lundquist, Sampling functions for geophysics, *Smithson. Astrophys. Observ. Spec. Rep., 34*, 1972.

Hein, G. W., und K. Lenze, Zur Genauigkeit und Wirtschaftlichkeit verschieden Interpolations- und Prädiktionsmethoden, *Z. Vermess., 104*, 492–504, 1979.

Heiskanen, W. A., and H. Moritz, *Physical Geodesy*, W. H. Freeman, San Francisco, 1967.

Heller, W. G., and S. K. Jordan, Attenuated white noise statistical gravity model, *J. Geophys., Res., 84*(B9), 4680–4688, 1979.

Jekeli, C., An investigation of two models for the degree variances of global covariance functions, *Rep. 275*, Dep. of Geod. Sci., Ohio State Univ., Columbus, 1978.

Junkins, J. L., and R. C. Engels, The finite-element approach in gravity modelling, *Manuscripta Geod., 4*, 185–206, 1979.

Kearsley, A. H. W., The computation of deflections of the vertical from gravity anomalies, *UNISURV*, S-15, 1976.

Keller, W., Zum Konvergenzverhalten des Kollokationsverfahrens, *Vermessungstechnik, 26*, 420–422, 1978.

Krarup, T., Some remarks about collocation, in *Approximation Methods in Geodesy*, edited by H. Moritz and H. Sünkel, pp. 193–209, H. Wichmann, Karlsruhe, West Germany, 1978.

Krarup, T., A convergence problem in collocation theory, *Boll. Geod. Sci. Affini*, in press, 1980.

Lachapelle, G., Estimation of disturbing potential components using a combined integral formula and collocation approach, *Manuscripta Geod., 2*, 233–262, 1977.

Lachapelle, G., Comparison of Doppler-derived and gravimetric geoid undulations in North America, in *Proceedings of the Second International Geodetic Symposium on Satellite Doppler Positioning,* pp. 653–670, Applied Research Laboratories, University of Texas, 1979.

Lachapelle, G., and C. C. Tscherning, Use of collocation for predicting geoid undulations and related quantities over large areas, in *Proceedings of the International Symposium on the Geoid in Europe and Mediterranean Area,* pp. 165–182, Società Italiana di Fotogrammetria e Topografia, Ancona, Italy, 1978.

Lelgemann, D., Ein Verfahren zur astro-gravimetrischen Geoidbestimmung, *Publ. Dtsch. Geod. Komm., Reihe C, 247,* 1978*a*.

Lelgemann, D., The astro-gravimetric computation of the geoid in the Federal Republic of Germany, *Proceedings of the International Symposium on the Geoid in Europe and Mediterranean Area,* pp. 165–182, Società Italiana di Fotogrammetria e Topografia, Ancona, Italy, 1978*b*.

Lerch, F. J., S. M. Klosko, R. E. Laubscher, and C. A. Wagner, Gravity model improvement using Geos 3 (GEM 9 and 10), *J. Geophys. Res., 84*(B8), 3897–3916, 1979.

Merry, C. L., and P. Vaníček, A technique for determining the geoid from a combination of astro-geodetic and gravimetric deflections, *Can. Surv., 28,* 549–554, 1974.

Monka, F. M., W. Torge, G. Weber, and H.-G. Wenzel, Gravimetric, altimetric and astrogeodetic geoid determinations in the North Sea region, *Wiss. Arb. Lehrstühle Geod. Photogramm. Kartogr. Tech. Univ. Hannover, 80,* 1978.

Moritz, H., Integral formulas and collocation, *Manuscripta Geod., 1,* 1–40, 1976.

Moritz, H., On the computation of a global covariance model, *Rep. 255,* Dep. of Geod. Sci., Ohio State Univ., Columbus, 1977.

Moritz, H., The operational approach to physical geodesy, *Rep. 277,* Dep. of Geod. Sci., Ohio State Univ., Columbus, 1978*a*.

Nash, R. A., Jr., and S. K. Jordan, Statistical geodesy—An Engineering perspective, *Proc. IEEE, 66,* 532–550, 1978.

Rapp, R. H., Gravity anomaly and geoid undulation results in local areas from Geos-3 altimeter data (abstract), *Eos Trans. AGU, 59,* 259, 1978.

Rapp, R. H., Global anomaly and undulation recovery using Geos 3 altimeter data, *Rep. 285,* Dep. of Geod. Sci., Ohio State Univ., Columbus, 1979*a*.

Rapp, R. H., Potential coefficient and anomaly degree variance modelling revisited, *Rep. 293,* Dep. of Geod. Sci., Ohio State Univ., Columbus, 1979*b*.

Rapp, R. H., and S. I. Agajelu, Comparison of upward continued anomalies by the Poisson integral and by collocation, *Rep. 227,* Dep. of Geod. Sci., Ohio State Univ., Columbus, 1975.

Rapp, R. H., and R. Rummel, Comparison of Doppler derived undulations with gravimetric undulations considering the zero-order undulation of the geoid, in *Proceedings of the International Geodetic Symposium Satellite Doppler Positioning,* pp. 389–398, Physical Science Laboratory, New Mexico State University, Las Cruces, 1976.

Reinhart, E., G. Boedecker, and R. Rummel, A note on the precise geoid determination in the alps from the geodetic and geodynamic point of view, in *Proceedings of the International Symposium on the Geoid in Europe and Mediterranean Area,* pp. 207–222, Società Italiana di Fotogrammetria e Topografia, Ancona, Italy, 1978.

Rummel, R., Determination of short-wavelength components of the gravity field from satellite-to-satellite tracking or satellite gradiometry—An attempt to an identification of problem areas, *Manuscripta Geod., 4,* 107–148, 1979.

Sanso', F., Internal collocation, *Mem. Accad. Naz. Lincei, Ser. VIII, XVI,* 1980.

Sanso', F., and C. C. Tscherning, Notes on convergence problems in collocation theory, *Boll. Geod. Sci. Affini, 39,* 221–252, 1980.

Schwarz, C. R., Deflection computations for network adjustment in the United States, in *Proceedings of the Second International Symposium on Problems Related to the Redefinition of North American Geodetic Networks,* pp. 91–102, U.S. Department of Commerce, Washington, D. C., 1978.

Schwarz, K.-P., Geodetic accuracies obtainable from measurements of first and second order gravitational gradients, *Rep. 242,* Dep. of Geod. Sci., Ohio State Univ., Columbus, 1976.

Schwarz, K.-P., Airborne inertial systems for gravity determination in ocean areas, in *Proceedings 1st. International Symposium on Inertial Technology for Surveying and Geodesy,* pp. 351–360, Canadian Institute of Surveying, Ottawa, Ont., 1978.

Schwarz, K.-P., and H. Sünkel, Vorarbeiten zur Geoidbestimmung in Österreich, *Oesterr. Z. Vermess. Photogramm., 66,* 57–70, 1978.

Sjöberg, L., A comparison of Bjerhammar methods and collocation in physical geodesy, *Rep. 273,* Dep. of Geod. Sci., Ohio State Univ., Columbus, 1978.

Sjöberg, L., On the existence of solutions for the methods of Bjerhammar in the continuous case, *Bull. Geod., 53,* 227–230, 1979.

Strange, W. E., and R. J. Fury, Computation of deflections of the vertical in support of the readjustment of the North American datum, in *Proceedings of the International Symposium on Optimization of Design and Computation of Control Networks,* edited by F. Halmos and J. Somogyi, pp. 695–706, Akademiai Kiado, Budapest, 1979.

Sünkel, H., Die darstellung Geodätischer Integralformeln durch Bikubische Spline-Funktionen, *Mitt. Geod. Inst. Tech. Univ. Graz, 28,* 1977.

Tengström, E., Report of IAG SSG no. 5.16, paper presented at IUGG/IAG General Assembly, Moscow, Int. Union of Geod. and Geophys., 1971.

Torge, W., Regional gravimetric geoid calculations in the North Sea test area, *Mar. Geod., 3,* 257–271, 1980.

Tscherning, C. C., A Fortran IV program for the determination of the anomalous potential using stepwise least squares collocation, *Rep. 212,* Dep. of Geod. Sci., Ohio State Univ., Columbus, 1974.

Tscherning, C. C., Application of collocation for the planning of gravity surveys, *Bull. Geod., 116,* 183–198, 1975.

Tscherning, C. C., Covariance expressions for second and lower order derivatives of the anomalous potential, *Rep. 225,* Dep. of Geod. Sci., Ohio State Univ., Columbus, 1976*a*.

Tscherning, C. C., Determination of datum-shift parameters using least squares collocation, *Boll. Geod. Sci. Affini, 35,* 173–183, 1976*b*.

Tscherning, C. C., A note on the choice of norm when using collocation for the computation of approximations to the anomalous potential, *Bull. Geod., 51,* 137–147, 1977.

Tscherning, C. C., A users guide to geopotential approximation by stepwise collocation on the RC 4000-computer, *Publ. 53,* Dan. Geod. Inst., Copenhagen, 1978*a*.

Tscherning, C. C., Introduction to functional analysis with a view to its application in approximation theory, in *Approximation Methods in Geodesy,* edited by H. Moritz and H. Sünkel, pp. 157–192, H. Wichmann, Karlsruhe, West Germany, 1978*b*.

Tscherning, C. C., and R. Forsberg, Prediction of deflections of the vertical, in *Proceedings of the Second International Symposium on Problems Related to the Redefinition of North American Geodetic Networks,* pp. 117–134, U.S. Department of Commerce, Washington, D. C., 1978.

Tscherning, C. C., and R. H. Rapp, Closed covariance expressions for gravity anomalies, geoid undulations and deflections of the vertical implied by anomaly degree variance models, *Rep. 208,* Dep. of Geod. Sci., Ohio State Univ., Columbus, 1974.

Wenzel, H. G., An astrogeodetic geoid around the North Sea (AGNS1), *Wiss. Arb. Lehrstühle Geod. Photogramm. Kartogr. Tech. Univ. Hannover, 80,* 23–35, 1978.

Wolf, H., Kollokation mit Hilf des Gausschen Algoritmus, *Z. Vermess., 103,* 13–19, 1979.

(Received August 6, 1980;
accepted October 14, 1980.)

REVIEWS OF GEOPHYSICS AND SPACE PHYSICS, VOL. 19, NO. 2, PAGES 407–414, AUGUST 1981

Local and Global Gravity Field Representation

E. GROTEN

Institute of Physical Geodesy, Technical University, Darmstadt, Federal Republic of Germany

Gravity fields applied in geodesy are usually built up by (1) a global satellite (low harmonics) part obtained from altimetry and/or differential orbit improvement (in the future, satellite to satellite tracking techniques will also contribute), (2) a regional part obtained from conventional terrestrial material, and (3) a local part of ±1-mGal accuracy. Whenever four-dimensional considerations are of interest, ±0.001-mGal accuracy is necessary. The combination of these parts is considered; attention is also paid to the zero-degree term in view of very accurate modern absolute measurements of gravity. Integral equation solutions dominated by kernel functions such as l^{-n} as well as collocation formulas lend themselves to useful combinations of high-precision regional and low-precision global parts where many difficulties arise in practical applications. But no ideal method is available in order to overcome deficient knowledge of the gravity field. Detailed applications of Stokes- and Poisson-type solutions are discussed. In spite of the fact that a lot of local information is averaged out because of its random behavior, valuable information on systematic effects gets lost unless local parameters are incorporated. Model considerations are based on Kaula's rule of thumb for high harmonics.

1. Introduction

It is the aim of this paper to critically review and comment on local and global gravity field representations. In view of the very many types of representations it is necessary to concentrate on a few well-selected aspects. Nevertheless, some specific points had to be picked out which are only loosely connected with each other. More general reviews were given by *Kaula* [1963] and *Rapp* [1975]; specific statistical aspects of the global problem have been discussed by *Kaula* [1967]. The fundamental ideas of least squares collocation were outlined by *Moritz* [1978].

As the topic is extremely extended and vast, the review is basically conceptual. Global mean and point value representation, harmonic series expansion, transformation of gravity to potential, and first and second derivatives are considered on one hand; on the other hand, local field representation is briefly inspected. Remembering, moreover, that two thirds of the surface of the earth is covered by oceans and knowing the geoid in continental areas with sufficient accuracy, we have to reconsider the classical concept of representing the earth's gravity field in terms of free air gravity.

2. Poisson-Type Solution

The following considerations represent a supplement of earlier studies [*Groten*, 1975, 1976, 1978*a*; *Brennecke et al.*, 1976*a*; *Groten and Joehemczyk*, 1978]; *Koch* [1971] was the first to consider some theoretical aspects of a *Hotine* [1969] function associated with an oblique boundary value problem; *Moritz* [1974] used the classical version for the evaluation of the separation of the sea surface from the geoid; *Jekeli* [1979] and others did very detailed investigations related to practical applications of Hotine's function; the 'unmodified' version of derivatives of Hotine's function were treated by *Groten* [1975, 1976], 'modified' versions were discussed, for example, by *Jekeli* [1979] and others. By modified we mean a Hotine function where the harmonics of degrees zero and one are omitted.

With the feasibility of global positioning system–type satellite tracking having an accuracy of the order of a few centimeters (P. Bender, private communication) and resolution of satellite altimeters of the order of ±10 cm we can anticipate the earth's surface S to be known with decimeter accuracy or better; the corresponding accuracy for gravity is better than ±0.5 mGal; this estimate is based on the assumption that any vertical translocation δS of S produces a corresponding change of gravity at the same point with limits given, in general, by a Bouguer coefficient of 0.11 mGal/m and a free air coefficient of 0.31 mGal/m. If the ocean surface is supposed to coincide with the geoid, we get geoid heights N from altimetry at sea; whenever the separation of the ocean surface from the geoid is taken into account, we get gravity disturbances at the oceanic surface from altimeter data plus sea surface gravimetry; then Hotine's formula yields geoid undulations.

Thus it is now straightforward to replace the gravity anomaly in most cases by the gravity disturbance. Consequently, Stokes' integral can be replaced by the corresponding integral formula where gravity disturbances $-\partial T/\partial n = \delta g$ take the place of gravity anomalies Δg. *Hotine* [1969] was, to my knowledge, the first to discuss geodetic details of the corresponding Neumannian boundary value problem in terms of a closed kernel formula. Therefore I introduced some years ago the Hotine function as a name for

$$\bar{S} = \sum_{n=0}^{\infty} \frac{2n+1}{n+1} P_n(\cos\psi) \tag{1}$$

where

- T disturbing potential;
- P_n zonal harmonic of degree n;
- ψ 'spherical distance' between two points on the unit sphere.

The consequences of representing the gravity field by δg and various other aspects will be discussed below.

3. Modified Hotine Function

The treatment of the first and zero harmonics is directly related to the coordinate system origin, and the scale differs basically in the case of Hotine's approach from the Stokesian method. This well-known topic can be omitted if compatibility is achieved by introducing a modified Hotine function

$$\tilde{S} = \sum_{n=2}^{\infty} \frac{2n+1}{n+1} P_n(\cos\psi) \tag{2}$$

$$\tilde{S} = \bar{S} - P_0 - \tfrac{3}{2}P_1 = \bar{S} - \bar{S}_0 - \bar{S}_1 \qquad (3)$$

However, it should be realized that the application of the original function $\bar{S}$ in (1) implies a nongeocentric representation; by omitting first- and zeroth-degree harmonics in the Stokesian approach, scale corrections N_0 for geoid heights $N = \sum N_n$ have to be considered which are sometimes difficult to separate from first-order terms [*Schaab and Groten,* 1979]. So far the application of (1) instead of Stokes' kernel represents an alternative. Since linear model (collocation) approaches were often found to be more efficient than conventional integral equation and integral formula approaches, a few remarks on these approaches are in order:

Cross-regression approaches are given, in general, by [*Moritz,* 1980]

$$\mathbf{f}(P) = \mathbf{K}_{Pi}\mathbf{C}_{ij}^{-1}\mathbf{F}_j \qquad (4a)$$

whereas autoregression is based on

$$\mathbf{f}(P) = \mathbf{C}_{Pi}\mathbf{C}_{ij}^{-1}\mathbf{f}_j \qquad (4b)$$

where **K** denotes the cross-covariance matrix and **C** denotes the autocovariance matrix; **f** and **F** in (4*a*) denote the set of unknown and known data, respectively. Future applications might be based on combinations of linear model methods in connection with integral formulas and equations [*Groten,* 1979*a*]. In such combinations we can replace the Stokesian approximation of Molodensky's problem by a collocation approach and use Molodensky's corrections in terms of integral formulas (using, for example, the gravity gradient or terrain correction–type formulas) in order to get higher-order approximations. Computational advantages are one reason for applying such combinations.

If cross correlation of gravity disturbances with potential or deflections of the vertical are used instead of Stokes' and Vening-Meinesz's approaches, respectively, one recognizes the fact [*Groten,* 1979*a*] that the gravity anomalies Δg can be considered as high-pass-filtered gravity disturbances δg, that is,

$$\delta g_n = \frac{n+1}{n-1}\,\Delta g_n \qquad (5)$$

Consequently, the autocovariance function C of δg contains relatively big low-harmonics coefficients, since the coefficients a_n of

$$C(\delta g) = \sum_{n=0} a_n P_n(\cos\psi) \qquad (6)$$

are the degree variances of δg. Therefore the correlation length of δg is greater than the correlation length of Δg.

Nevertheless, the contribution of the remote zones is less in Hotine's integral than in Stokes' integral [*Groten,* 1978*a*; *Jekeli,* 1980]. This fact is still more pronounced in the case of first and second 'horizontal' derivatives, that is, when deflections of the vertical (ξ, η) and second derivatives of the potential are computed from δg. It is described below.

4. COATING APPROACH

As soon as Δg and T are known, it is straightforward to reconsider Helmert's old method, where a coating of density

$$\mu = \Delta g + (3T/2R) = 2\pi G\rho$$

(with G the Newtonian gravitational constant and ρ the single-layer density) is applied on a sphere of radius $r = R$; R denotes the mean radius of the earth. In the high-harmonics part, δg, Δg, and μ are practically identical because the high harmonics of T are, in general, of low amplitude; using, for example, *Kaula*'s [1966] rule of thumb, we immediately realize that fact. Moreover [*Heiskanen and Moritz,* 1967, pp. 97, 237], on the unit sphere s we have

$$T = \sum T_n = R\sum_{n=2}^{\infty}\frac{\Delta g_n}{n-1} = \frac{R}{4\pi}\iint \Delta g S(\psi)\,ds \qquad (7)$$

$$T = R\sum_{n=2}^{\infty}\frac{\delta g_n}{n+1} = \frac{R}{4\pi}\iint \delta g\tilde{S}(\psi)\,ds \qquad (8)$$

$$T = \frac{R^2}{2\pi}\iint \frac{\mu}{l}\,ds \quad \text{or} \quad T = G\iint_\sigma \frac{\rho}{l}\,d\sigma \qquad (9)$$

where σ is supposed to be a closed surface, $d\sigma$ being an element of σ. We can omit here the corrections according to Molodensky, because they are well known and do not play a role in this context; $S(\psi)$ is Stokes' function, and the kernel l^{-1} is the inverse of l, being the straight-line separation corresponding to the ψ on the unit sphere s.

5. IMPLICATIONS OF THE GRAVITY DISTURBANCE AND COATING APPROACHES

In the overwhelming part of the spectrum (omitting low harmonics of degrees $3 \leq n \leq 12$), δg, Δg, and μ are dominated by the contribution of Δg. When the computation of deflections is of interest, the kernel functions

$$\frac{\partial\bar{S}}{\partial\psi} = -\frac{\cos(\psi/2)}{2\sin^2(\psi/2)} + \frac{\cos(\psi/2)}{2[1+\csc(\psi/2)]\sin^2(\psi/2)} \qquad (10)$$

$$\frac{\partial\tilde{S}}{\partial\psi} = \frac{\partial}{\partial\psi}(\bar{S} - P_0 - \tfrac{3}{2}P_1) = \frac{\partial\bar{S}}{\partial\psi} + \tfrac{3}{2}\sin\psi \qquad (11)$$

$$\frac{\partial(1/l)}{\partial\psi} = \frac{\partial}{\partial\psi}[\tfrac{1}{2}\sin^{-1}(\psi/2)] = \frac{-\cos(\psi/2)}{4\sin^2(\psi/2)} \qquad (12)$$

are needed. For the computation of the second horizontal derivatives (from δg and μ) we need, analogously,

$$\frac{\partial^2\bar{S}}{\partial\psi^2} = -\frac{1}{4\sin(\psi/2)} + \frac{1}{2\sin^3(\psi/2)} + \frac{\sin^3(\psi/2) - 2\sin(\psi/2) - 1}{4[\sin^2(\psi/2) + \sin(\psi/2)]^2} \qquad (13)$$

$$\frac{\partial^2\tilde{S}}{\partial\psi^2} = \frac{\partial^2\bar{S}}{\partial\psi^2} + \tfrac{3}{2}\cos\psi \qquad (14)$$

$$\frac{\partial^2(1/l)}{\partial\psi^2} = -\frac{1}{8\sin(\psi/2)} + \frac{1}{4\sin^3(\psi/2)} \qquad (15)$$

where, of course, $\tilde{S}$ is always associated with

$$\delta g' = \sum_{n=2}^{\infty}\delta g_n \qquad (16)$$

and $\bar{S}$ with

$$\delta g = \sum_{n=0}^{\infty}\delta g_n \qquad (17)$$

It should, however, be emphasized that in contrast to the error integrals of Stokes' function S, as well as $\bar{S}$, $\tilde{S}$, and $1/l$, things are more involved when the derivatives are considered,

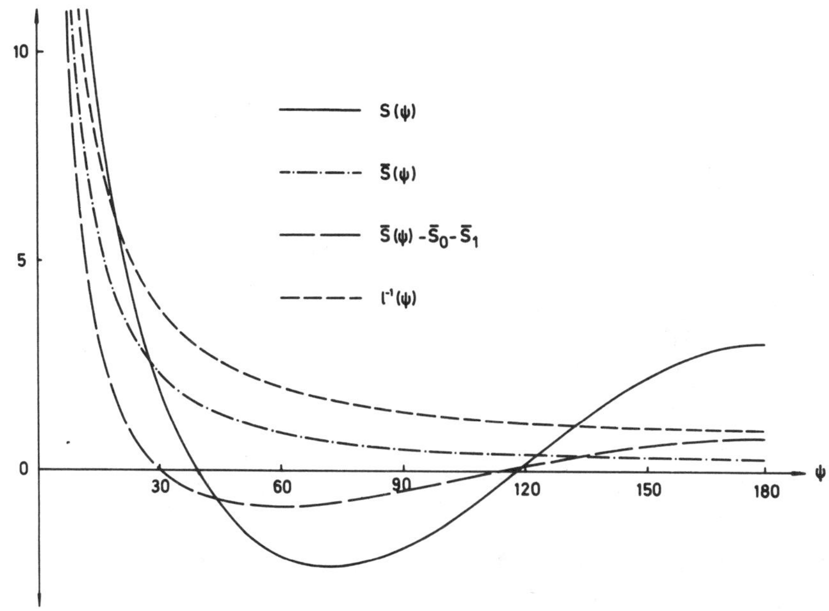

Fig. 1. Different kernel functions for computing the potential utilizing Δg, δg, and ρ.

because (ξ, η) is basically a vector quantity [*Grafarend and Offermanns,* 1975], so that the corresponding error functions (as defined by *Groten* [1979*a*]) are no longer simply proportional to the error integrals (as defined by *Groten and Moritz* [1964]). Moreover, the planar approximation is, in these cases, appropriate. These statements refer also analogously to second derivatives. Nevertheless, first and second derivatives (as shown, for example, in (21)) give an average insight into the relative behavior of truncation with increasing distance from the point of computation. Thus tendencies are revealed, whereas detailed numerical investigations had to start from explicit formulas.

When the integrals in (7), (8), and (9) are applied to a cap of radius ψ_0 around the computation point P instead of over the whole surface s, a truncation error arises which depends basically on the degree variances of Δg, δg, and μ, respectively (the alternative is given by adding a series of type $(R/2\gamma) \sum_{n+1}^{\infty} Q_n \Delta g_n$). Molodensky considered this truncation error in various papers; numerical studies for Δg were made later on by *Groten and Moritz* [1964]; analogous considerations for the kernel $1/l$ were made by *Groten* [1970] and for the kernel $\bar{S}(\psi)$ by *Groten* [1978*a*, 1979*a*]. The corresponding truncation coefficients $Q_n(\psi_0)$ for Δg and $q_n(\psi_0)$ for δg were considered by *Heiskanen and Moritz* [1967] and *Groten and Jochemczyk* [1978], respectively. More detailed investigations were made in various papers by the Ohio State University group [e.g., *Jekeli,* 1979].

The related integrals

$$\int_{\psi_0}^{\pi} [\bar{S}(\psi)]^2 \sin \psi \, d\psi \tag{18}$$

$$\int_{\psi_0}^{\pi} [\bar{S}(\psi)]^2 \sin \psi \, d\psi \tag{19}$$

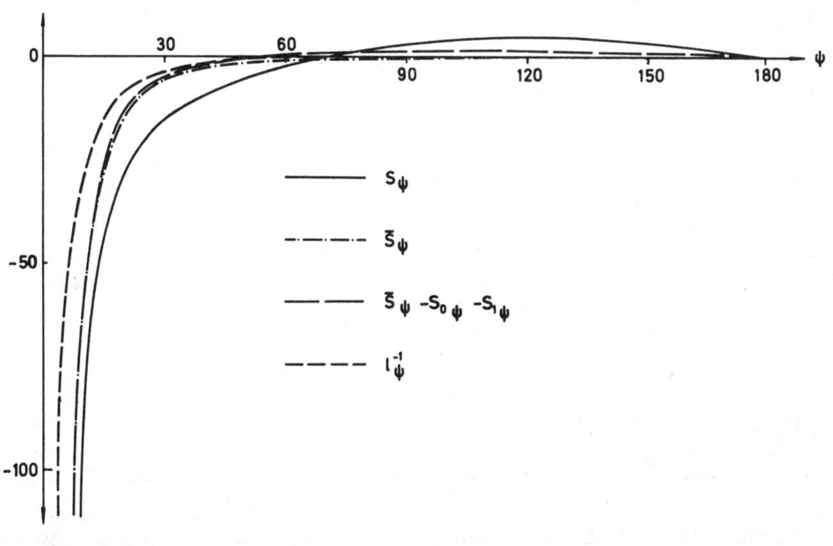

Fig. 2. Different kernel functions for computing deflections of the vertical utilizing Δg, δg, and ρ.

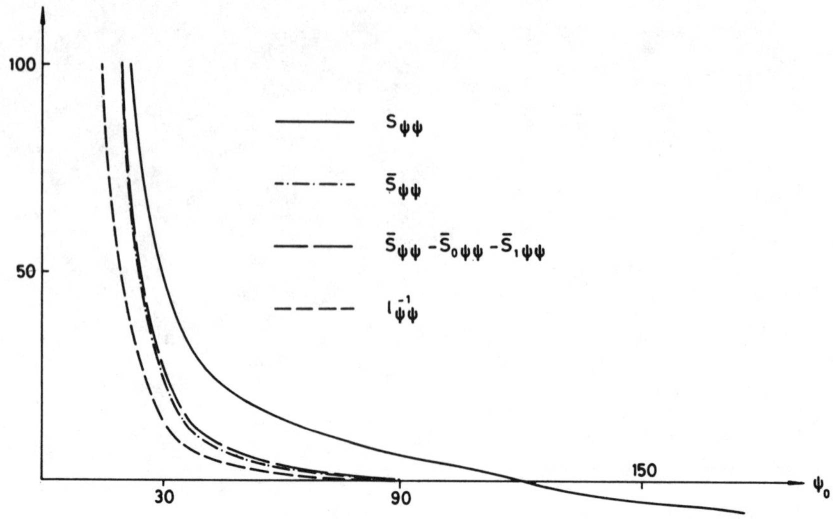

Fig. 3. Different kernel functions for computing the second horizontal derivation, that is, horizontal derivatives of ξ and η.

$$\int_{\psi_0}^{\pi} \left(\frac{1}{l}\right)^2 \sin\psi \, d\psi \tag{20}$$

for the computation of T as well as the corresponding expressions, such as

$$\int_{\psi_0}^{\pi} \left(\frac{\partial \bar{S}}{\partial \psi}\right)^2 \sin\psi \, d\psi \tag{21}$$

were denoted by 'truncation functions' by *Groten* [1979]. In order to illustrate these truncation functions for Δg, δg, and μ they are plotted in Figures 1–6. Since the degree variances of δg and Δg are almost of the same order of magnitude for $n > 13$ as that mentioned above, the truncation functions for (1) S, $\partial S/\partial\psi \equiv S_\psi$, and $\partial^2 S/\partial\psi^2 \equiv S_{\psi\psi}$, (2) $\bar{S}$, $\bar{S}_\psi$, and $\bar{S}_{\psi\psi}$ and (3) $\bar{\bar{S}}$, $\bar{\bar{S}}_\psi$, and $\bar{\bar{S}}_{\psi\psi}$ can be directly compared with each other in order to get an idea of the magnitude of the truncation error as a function of ψ_0. If we omit the harmonics for $n < 6$, the degree variances of δg are always less than twice as big as those of Δg.

In two forthcoming papers, one by Stock and the other by Groten, details referred to specific gravity fields are studied. The comparison of the truncation functions of $1/l$ and its derivatives is only a relative one; that is, the shape of $1/l$ and its derivatives indicates the relatively small influence of the remote zones. By applying factors such as $\bar{\sigma}_0(\Delta g)R^2/(8\pi\gamma^2)$ (with σ_0 indicating error variances, $\bar{\sigma}_0 = \sigma_0$ times the area of the block to which corresponding mean gravity is related) to the curves in Figure 4, etc., we get directly comparable, that is, normalized, truncation error effects for specific models.

Using the concept described by *Groten and Moritz* [1964] and *Heiskanen and Moritz* [1967], a comparison of the methods utilizing δg and Δg, respectively, is possible. By taking account of the variances of δg and Δg (found from (6) using $\psi = 0$) it is seen that δg methods are more efficient than Δg methods if the outer zone ψ_0 is totally neglected. Such results depend strongly on the numerical values of those variances which are still somewhat uncertain. On the other hand, the er-

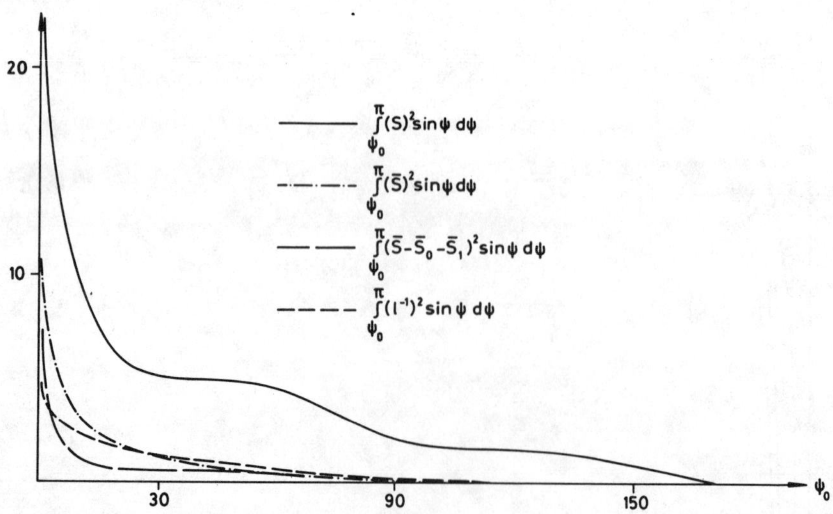

Fig. 4. Relative behavior of the error influence of remote zones for $\psi > \psi_0$ associated with specific kernel functions of Figure 1.

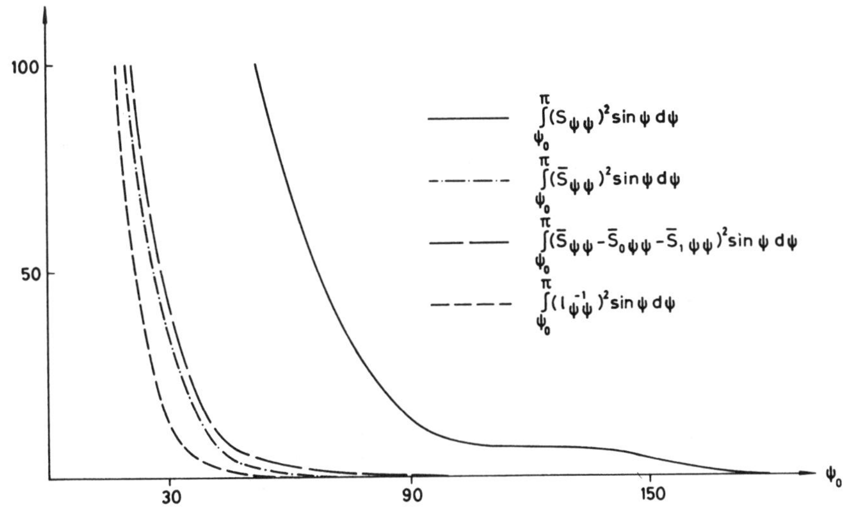

Fig. 5. Relative behavior of the error influence of remote zones for $\psi > \psi_0$ associated with the kernel function of Figure 2.

rors in N, ξ, η, etc., caused by the various errors of approximations and models of the earth's gravity field can be estimated by using error covariances $\sigma(\Delta g)$, $\sigma(\delta g)$, etc., for specific approximations.

Using different models, we end up with results which differ slightly from *Jekeli's* [1980] result; he concluded from his studies that Δg methods are often superior to δg methods. He omitted the case where gravity is neglected in the Stokes and Hotine integrals. However, this question is not of real practical impact. A more realistic question is related to the error of the gravity field. *Jekeli* [1980] pointed out that the δg method is more effective if the gravity field is represented by a well-selected truncated harmonic series using satellite results together with terrestrial data. His results are based on a series representation of the type $(R/2\gamma)\ \Sigma Q_n \Delta g_n$. This result is in basic agreement with the results obtained in the present investigation, being based on the standardization of quantities such as those in (18) and (19). This corresponds to the case where Δg and δg are perfectly known within a neighborhood zone of spherical radius ψ_0, and an approximation of it for $\psi > \psi_0$ is affected by an error of approximation. In other words, uncertainties in the gravity field still inherent in present approximations of its global part have a smaller influence in applications of Hotine's integral than in Stokes' approach. Using various approximation models it is realized that the latter statement holds in general. It is also applicable to the coating method.

As integral formula approaches can be considered as limiting cases of the collocation method described by (4*a*), the same result holds, of course, for collocation approaches. The inversion of $\mathbf{C}_{ij}$ in (4*a*) is only slightly more difficult in the case of δg; but the row matrix $\mathbf{K}_{Pi}$ is easier to handle in the case of δg. It should be noted that even though $|\Delta g| < |\delta g|$, in general, whenever the geoid height is positive (i.e., the geoid is above the ellipsoid), Δg is positive too. At sea, for example, in the Mediterranean area, Δg is negative, in general, and geoid heights are sometimes positive. In those cases, $|\delta g| < |\Delta g|$, in general. The general superiority of δg approaches is therefore straightforward.

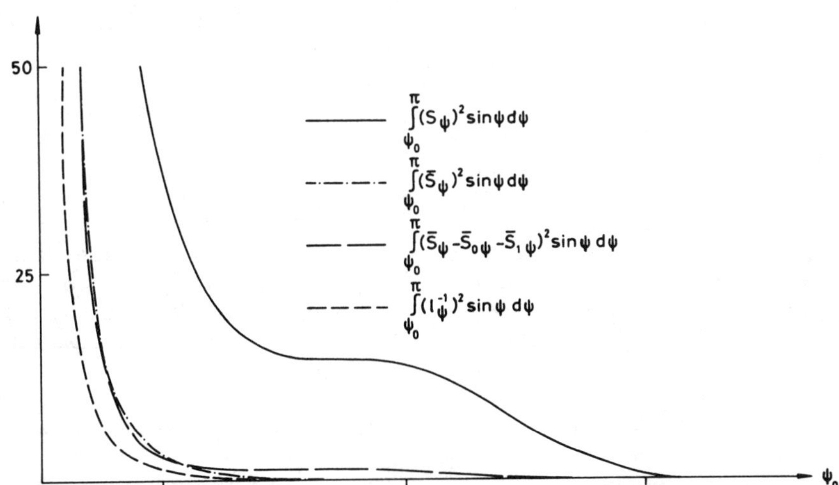

Fig. 6. Relative behavior of the error influence of remote zones for $\psi > \psi_0$ associated with the kernel functions of Figure 3; by normalizing these values using factors proportional to the error variances of specific gravity models we get error propagation for N, ξ, η, etc., from Figures 4–6.

6. SPHERICAL HARMONICS REPRESENTATION OF THE GLOBAL FIELD

a. *Mean Value Problem*

The most detailed investigations on this topic were made by R. H. Rapp and his group. One of the basic problems is the evaluation of spherical harmonics from discrete values on a sphere. *Katsambalos* [1979] has given an excellent review of the question of how to get these harmonics from mean anomalies, in which the various investigations by Pellinen as well as *Meissl*'s [1971], *Rapp*'s [1977], *Lowes*' [1979], and many other investigations are critically considered. Other important contributions are, for example, by *Pollack* [1973], *Sjöberg* [1979], and *Gaposchkin* [1979]. They finally led to many revisions of apparently well-established rules such as

$$s' \doteq 180°/n \tag{22a}$$

where n is the degree of the highest harmonic in the series expansion computed from mean anomalies in 'square' (or equal area) blocks of side length s' (expressed in degrees). *Jochemczyk* [1979] has recently investigated various types of equal area blocks; in order to preserve orthogonality he favors the use of equal area blocks of constant longitude difference. In that case the latitude difference is varied along the meridian in order to get blocks of the same size all over the earth.

However, since we are no longer interested in low-harmonic expansions using mean values of 5° × 5° or similar large blocks (where the mean values obtained from nonuniformly distributed data are distorted anyway) but rather in higher-harmonic expansions corresponding to representative values of relatively small blocks, it seems indeed more appropriate to use discrete values, the 'error of representativity' of which is relatively small in corresponding small blocks. Since the averaging process, in the case of mean values, does not inherently limit the number of spectral components in the averaged function, the use of mean anomalies becomes more problematic than was earlier anticipated. This is all the more true since neither averaging nor sampling can be completely undone [*Gaposchkin,* 1979] but rather approximate methods (such as those mentioned above) might partly eliminate the averaging effects.

Consequently, the autocovariance function of mean anomalies contains harmonics of almost any degree beyond

$$n = 180°/s' \tag{22b}$$

A finite upper limit of the sum (6) is impossible anyway whenever positive definiteness should be preserved. It therefore seems appropriate to abandon the use of mean anomalies in most cases.

On the other hand, high-frequency information on the spectrum above the Nyquist frequency cannot be obtained even from equidistant discrete data; we have to introduce hypotheses whenever positive definite functions of type (6) are empirically determined. Nevertheless, the use of mean values is more problematic.

b. *Representation of Three-Dimensional Functions*

Arnold [1979] and others have considered products of spherical harmonics S_n leading to sums of the form

$$S_n(\theta, \lambda)S_m(\theta, \lambda) = \sum_{i=0}^{\infty} y_i(n, m)\, S_i(\theta, \lambda) \tag{23}$$

With increasing numbers of harmonics such formulas become more and more important because they yield the possibility of representing a three-dimensional function in terms of a series of surface harmonics. For example, let the topography as well as the gravity field be given in terms of truncated harmonic series; then products of these two series are expanded as shown in (23), thus leading to a two-dimensional expansion representing the gravity field along the topographic surface.

7. THE PROBLEM OF HARMONICITY

With increasing accuracy of observations, attention has to be paid to minute effects which previously did not deserve attention: *Heikkinen* [1979] has recently raised the question of harmonicity. In geodesy we are used to dealing with the harmonic disturbing potential T related to the gravity anomalies $\Delta g = \sum \Delta g_n$. In both functions the nonharmonic parts, such as centrifugal potentials, are supposed to be eliminated. Unfortunately, the so-called Honkasalo correction has been applied to global gravity systems in order to take account of the stationary tidal deformation of zonal type. The application of this correction implies a slight nonharmonic part in the disturbing potential caused by the attraction of a homogeneous mass halo around the earth made up by the masses of the sun and moon; for details see *Groten* [1978*b*]. The perfect elimination of the tidal perturbation as proposed by *Heikkinen* [1979] is impossible because we do not know the corresponding Love number. It is not straightforward to substitute empirical deformation coefficients obtained from semidiurnal earth tide observations because the aforementioned stationary tidal effect is associated with a secular Love number which is not necessarily identical with the Love numbers obtained for diurnal or semidiurnal waves, even though there are arguments in favor of a relatively small difference between the secular and the tidal effective Love numbers. But the secular Love number cannot be observed; on the other hand, in spite of the relatively high rigidity of the earth's crust we cannot infer that the earth could react in an elastic but rather viscoelastic or Maxwellian way to small stresses such as stationary tidal stresses. However, such a global stress involving the whole earth has to be considered carefully in view of various arguments against these types of response. An optimal representation of the gravity field may therefore be obtained by eliminating the pure attractional part of the stationary tidal effect as described by *Groten* [1978*b*], where the 'induced' tidal potential (i.e., the small part of the stationary tidal effect) which is caused by the stationary deformation of the earth is not eliminated. The disturbing potential T, as well as Δg, obtained in this way refers to the actual earth's gravity field as found, for example, from satellite orbit analysis. This actual earth has a flattening which is different from the flattening of the earth to which Heikkinen refers, because the latter earth is not supposed to be affected by the stationary tidal deformation. The perfect transition from this actual flattening to the flattening involved in Heikkinen's concept is not possible because of the deficient knowledge of the secular Love number as described above. The representation of the field proposed in this paper leads to a harmonic potential, avoiding the uncertainties inherent in Heikkinen's proposal.

This uncertainty affects the external potential by a term of the form kV_2, k being a Love number and V_2 being the zonal tidal potential, whereas terrestrial gravity measurements at the (deformed) surface of the earth involve an additional term of the form hV_2, h being a second secular Love number which again is not precisely known; for details see also *Groten* [1978*b*].

Let us denote the sea surface corrected for periodic tidal deformation as 'mean sea level' (MSL); its stationary part almost coincides with a level surface. This surface reflects, to some extent, the secular Love numbers (h, k) associated with the 'stationary' tidal constituents. If we replace the secular Love numbers by any other set (h, k), such as the purely elastic Love numbers in the correction of gravimetric data, we end up with a corresponding more or less arbitrarily deformed geoid which deviates from the true geoid by a second zonal degree type of deformation. In order to avoid such an arbitrary deformation it is straightforward to set

$$h = k = 0$$

in the tidal correction for the zonal tides M_0 and S_0. This assumption corresponds to a rigid body behavior, in principle, and avoids any change of the earth's shape but eliminates only the nonharmonic part of the disturbing potential in geodetic applications.

It should be noted (even though this fact is only of academic interest) that the volume of the deformed earth is slightly different from the volume of the earth which is not subjected to tidal deformation. However, since the volume of the geoid (being different from the volume of the earth) is often not clearly distinguished from the volume of the actual earth—the difference being substantially bigger—we need not discuss such trifles.

8. A REMARK ON LOCAL REPRESENTATION

Statistical as well as nonstatistical representations of the gravity field have been discussed by many authors [e.g., *Brennecke et al.,* 1976*b*; *Grafarend,* 1976; *Hardy,* 1977; *Hein and Lenze,* 1977; *Lenze,* 1978; *Sünkel,* 1977] considering various types of representation (linear triangle interpolation, polynomial approximation, two-dimensional spline methods, collocation approaches, and multiquadric methods) together with specific advantages and disadvantages. Each of these seems to deserve attention, depending on specific goals, data types, etc.; none is absolutely superior, as our various numerical studies reveal.

As far as collocation is concerned, this is a consequence of section 6*a*: The transition from local to global autocovariance functions and vice versa is seen to be problematic because any autocovariance function obtained from a data set (with zero average) in a limited area of the sphere will not yield a well-determined number of terms of the sum (6) but rather a not clearly defined part of it. Formulas such as [*Groten,* 1979*b*]

$$\sum_{n_1}^{\infty} a_n P_n(\cos \psi)$$

where n_1 simply depends on the block size of the aforementioned limited area using rules such as (22*b*), are found to be unreliable, since (22*b*) was found to be no longer true.

A second consequence is that local variations of the given function must be carefully split off from the 'trend' or model and a residual function has to be used instead of the original function in order to achieve accuracy as high as possible by applying a (global) covariance function which really corresponds to the model to be dealt with. Finally, the original function can be restored from the residuals and the trend.

The importance of this approach becomes obvious when we are dealing with 'irregular' functions such as second derivatives of the potential in rough topography.

9. CONCLUSIONS

In view of uncertainties still inherent in present models of the earth's gravity field the selection of optimal kernel functions is necessary in integral transforms and convolution integrals such as the Stokes- and Vening-Meinesz-type formulas by which the influence of the remote zones, etc., is minimized. Similarly, collocation procedures should be applied using optimal data sets. A variety of methods are now available; there are several arguments in favor of using gravity disturbances, mainly at sea. If physical geodesy is supposed to provide precise information on N, ξ, η, etc., which can be utilized in the combination of high-precision results obtained from very long baseline interferometry and laser satellite techniques, carefully selected methods should be applied, depending on the specific goal. There is obviously no single ideal approach.

Acknowledgment. B. Stock prepared the computer programs and checked the numerical formulas.

REFERENCES

Arnold, K., On the determination of gravity anomalies at the topographical surface of the earth by cosmic methods, *Gerlands Beitr. Geophys., 88*(4), 265–267, 1979.

Brennecke, J., E. Groten, and G. Hein, Some problems of digital representations of geodetic functions and integral transforms, *Manuscr. Geod., 1,* 143–171, 1976*a*.

Brennecke, J., E. Groten, G. Hein, H. Schaab, and J. Schmitt, Lokaler Vergleich von astrogravimetrischen Ergebnissen mit anderen geodätischen Daten, Beiträge zur gravimetrische Geodäsie und Gezeitenforschung, *Rep. B217,* pp. 67–95, Dtsch. Geod. Komm., Munich, 1976*b*.

Gaposchkin, E. M., Averaging on the surface of a sphere, *Reprint Ser. 1169,* Center for Astrophys., Cambridge, Mass., 1979.

Grafarend, E., Geodetic applications of stochastic processes, *Phys. Earth Planet. Interiors, 12,* 151–179, 1976.

Grafarend, E., and G. Offermanns, Eine Lotabweichungskarte Westdeutschlands nach einem konsistenten Kolmogorow-Wiener Modell, *Rep. A82,* Dtsch. Geod. Komm., Munich, 1975.

Groten E., On the spherical harmonics series of the sea tide correction of earth tide observation, in *Report of Symposium on Coastal Geodesy,* edited by R. Sigl, pp. 537–547, International Association of Geodesy, Special Study Group 2.22, Munich, 1970.

Groten, E., On the determination and applications of gravity gradients in geodetic systems, *Boll. Geod. Sci. Affini, 34*(4), 357–395, 1975.

Groten, E., Some numerical investigations on combination solutions, *Rep. B217,* pp. 99–127, Dtsch. Geod. Komm., Munich, 1976.

Groten, E., Gravity gradients, Lecture Notes, 2nd course, International School of Advanced Geodesy, Erice, Italy, May 18 to June 2, 1978*a*.

Groten, E., *Geodesy and the Earth's Gravity Field,* vol. I, II, Dümmler, Bonn, 1978*b*.

Groten, E., Derivatives of geopotential obtained from combination solutions, paper presented at European Geophysical Society Meeting, Vienna, 1979*a*.

Groten, E., On the mathematical structure of the gravity field, in *Proceedings of the International Symposium on Optimization of Design and Computation of Control Networks, Sopron, 1977,* pp. 647–682, edited by F. Halmos and J. Somogyi, Akadémiai Kiadó, Budapest, 1979*b*.

Groten, E., and H. Jochemczyk, Erweiterung einer Molodenskii-Reihenent-wicklung für das Störpotential, *Rep. B233,* pp. 79–86, Dtsch. Geod. Komm., Munich, 1978.

Groten, E., and H. Moritz, On the accuracy of geoid heights and deflections of the vertical, *Rep. 38,* Inst. of Geod., Photogramm. and Cartogr., Ohio State Univ., Columbus, 1964.

Hardy, R. L., Least squares prediction, *Photogramm. Eng. Remote Sens., 43,* 475–492, 1977.

Heikkinen, M., On the Honkasalo term in tidal corrections to gravimetric observations, *Bull. Geod., 53*(3), 239–245, 1979.

Hein, G., and K. Lenze, Zur Genauigkeit und Wirtschaftlichkeit verschiedener Interpolations- und Prädiktionsmethoden, *Z. Vermess., 104*, 492–505, 1977.

Heiskanen, W. A., and H. Moritz, *Physical Geodesy*, W. H. Freeman, San Francisco, Calif., 1967.

Hotine, M., *Mathematical Geodesy, ESSA Monogr. 2*, U.S. Department of Commerce, Washington, D. C., 1969.

Jekeli, C., Global accuracy estimates of point and mean undulation differences obtained from gravity disturbances, gravity anomalies and potential coefficients, *Rep. 288*, Dep. of Geod. Sci., Ohio State Univ., Columbus, 1979.

Jekeli, C., Comparison of undulation difference accuracies using gravity anomalies and gravity disturbances, *Bull. Geod., 54*(2), 149–163, 1980.

Jochemczyk, H., Determination of spherical harmonics from mean anomalies, translated from German, report, Tech. Hochsch., Darmstadt, Federal Republic of Germany, 1979.

Katsambalos, K. E., The effect of the smoothing operator on potential coefficient determinations, *Rep. 287*, Dep. of Geod. Sci., Ohio State Univ., Columbus, 1979.

Kaula, W. M., Determination of the earth's gravitational field, *Rev. Geophys. Space Phys., 1*, 507–551, 1963.

Kaula, W. M., *Theory of Satellite Geodesy*, Blaisdell, Waltham, Mass., 1966.

Kaula, W. M., Theory of statistical analysis of data distributed over a sphere, *Rev. Geophys. Space Phys., 5*, 83–107, 1967.

Koch, K. R., Die geodätische Randwertaufgabe bei bekannter Erdoberfläche, *Z. Vermess., 6*, 218–224, 1971.

Lenze, K., Zum Problem der Isoliniendarstellung—Interpolation und computerunterstütztes Zeichnen, thesis, Tech. Hochsch., Darmstadt, Federal Republic of Germany, 1978.

Lowes, F. J., Comment on 'The relationship between mean anomaly block sizes ans spherical harmonic representations,' by Richard H. Rapp., *J. Geophys. Res., 84*(B9), 4781–4782, 1979.

Meissl, P., A study of covariance functions related to the earth's disturbing potential, *Rep. 151*, Dep. of Geod. Sci., Ohio State Univ., Columbus, April 1971.

Moritz, H., Precise gravimetric geodesy, *Rep. 219*, Dep. of Geod. Sci., Ohio State Univ., Columbus, 1974.

Moritz, H., Least squares collocation, *Rev. Geophys. Space Phys., 16*, 421–430, 1978.

Moritz, H., *Advanced Physical Geodesy*, Wichmann, Karlsruhe, Federal Republic of Germany, 1980.

Pollack, H. N., Spherical harmonic representation of the gravitational potential of a point mass, a spherical cap, and a spherical rectangle, *J. Geophys. Res., 78*(11), 1760–1768, 1973.

Rapp, R. H., The earth's gravity field, *Geophys. Surv., 2*, 193–216, 1975.

Rapp, R. H., The relationship between mean anomaly block sizes and spherical harmonic representation, *J. Geophys. Res., 82*(33), 5360–5364, 1977.

Schaab, H., and E. Groten, Comparison of geocentric origins of global systems from uniformly distributed data, *Bull. Geod., 53*(1), 11–16, 1979.

Sjöberg, L., The accuracy of geoid undulations by degree implied by mean gravity anomalies on a sphere, *J. Geophys. Res., 84*(B11), 6226–6230, 1979.

Sünkel, H., Die Darstellung geodätischer Intergralformeln durch bikubische Spline-Funktionen, *Mitt. Geod. Inst. Tech. Hochsch. Graz, 28*, 1977.

(Received October 27, 1980;
accepted November 24, 1980.)

REVIEWS OF GEOPHYSICS AND SPACE PHYSICS, VOL. 19, NO. 3, PAGES 415–420, AUGUST 1981

Stability Investigations of Various Representations of the Gravity Field

MICHAEL GERSTL

Deutsches Geodätisches Forschungsinstitut and Sonderforschungsbereich 78
D-8000 München 22, Federal Republic of Germany

REINER RUMMEL[1]

Bayerische Kommission für die Internationale Erdmessung, Bayerische Akademie der Wissenschaften
D-8000 München 22, Federal Republic of Germany

A number of problems in physical geodesy are improperly posed in the sense that the approximate solution is not continuously dependent on the given observations. A transformation of the minimum norm least squares estimates for different types of gravity representations to the spectral domain, via a singular value decomposition, allows an easy insight into the stability situation of these representations. For the density layer or point mass model, Bjerhammar's method, and the least squares collocation model the filter expressions are derived which control their stability. It is shown in which way the shape of the filters is influenced by the a priori variance of the measurement noise, the choice of the radius of the Bjerhammar sphere, and the degree variance model.

1. DEFINITION OF THE PROBLEM

The problem of representing the gravity field of the earth may be seen as consisting of two parts: (1) the computation of an interpolant $\hat{t}$ (circumflex for estimated) to a function t from a set of data $\tilde{b}_1(t), \cdots, \tilde{b}_n(t)$ (tilde for observed), the $\tilde{b}_j(t)$ being bounded linear functionals of t, and (2) the determination of an arbitrary function $\hat{g}$ related to the gravity field of the earth. As described by *Moritz* [1978], a possible representation of $\hat{t}$ is of the form

$$\hat{t}(P) = \sum_i \hat{a}_i E_i(P) \qquad i = 1, \cdots, m \tag{1}$$

where $E_1, \cdots, E_m$ is a chosen set of base functions and the $\hat{a}_i$ are the coefficients of the representation. The coefficients are derived from relations

$$\tilde{b}_j(t) \doteq \sum a_i \tilde{b}_j(E_i) \qquad j = 1, \cdots, n \tag{2}$$

or in matrix form,

$$\tilde{\mathbf{b}} \doteq \mathbf{Ba} \qquad \mathbf{B}_{ji} = \tilde{b}_j(E_i) \tag{2'}$$

When t itself is observed, $\tilde{b}$ becomes the evaluation functional at points $P_1, \cdots, P_n$ or $\tilde{b}_j(t) = \tilde{t}(P_j)$ and $\tilde{b}_j(E_i) = E_i(P_j)$. The estimation of $\hat{g}$ can be based on a mathematical model of type

$$t = Sg \tag{3}$$

with S an operator connecting t and g and an estimate of t coming from (1). Depending on the characteristics of the base functions $E_1, \cdots, E_m$, the matrix of the interpolation problem, (2'), may be ill-conditioned in such a manner that any increase of the order of approximation m or of the number of observations n causes a deterioration of the condition of $\mathbf{B}$. Such a situation is met when the representation in (2') approaches the mathematical characteristics of the underlying original problem, that is, when m and n tend to infinity. In such a case the approximate solution $\hat{\mathbf{a}}$ does not continuously depend on the given left-hand side $\tilde{\mathbf{b}}$. In other words, infinitesimally small variations in $\tilde{\mathbf{b}}$ may cause large variations in $\hat{\mathbf{a}}$. Such a problem is said to be 'improperly posed.' In the same sense, the determination of g may be improperly posed depending on the problem to be solved. Examples are the determination of an approximation to the gravity field on the surface of the earth from satellite or airborne data, the computation of gravity anomalies from altimeter data, and the determination of the vertical gradient of gravity from gravity anomalies. These types of problems are discussed in the geodetic literature [e.g., *Schwarz,* 1971, 1979; *Moritz,* 1966, 1975; *Rummel et al.,* 1979; *Schaffrin et al.,* 1977]. Desirable for the representation of the gravity field would be on one hand its global character and on the other hand that local changes in the given gravity information have only local influence on the representation. One specific representation, based on sampling functions [*Giacaglia and Lundquist,* 1972], where

$$E(P) = \sum_i \tilde{t}(P_i)\chi_i(P)$$

possesses both features if the sample functions χ_i are based on the system of spherical harmonics. But exactly this type of representation fails entirely because of stability problems, as described by *Balmino* [1973] or *Schmidt* [1978].

2. SPECTRAL DECOMPOSITION

When solving a problem improperly posed in the sense defined above, that is, being unstable, the instability usually shows up in the form of high-frequency oscillations of the approximating function. Therefore the best form to represent the stability characteristics of a problem is its transformation into the frequency domain, that is, the representation in an orthonormal system of eigenfunctions of the operator connecting the observed values with the estimate $\hat{g}$. An outline of the eigenfunction representation is given by *Rummel et al.* [1979].

In a general case the representation of t (equation (1)) is compared with a corresponding representation for $\hat{g}$:

[1] Now at Technische Hogeschool Delft, Afdeling der Geodesie, NL-2629 JA Delft, Netherlands.

$$\hat{g}(P) = \sum_k \hat{c}_k F_k(P) \qquad k = 1, \cdots, l \tag{4}$$

or in matrix form,

$$\hat{g}(P) = \mathbf{F}(P)^T \hat{\mathbf{c}} \tag{4'}$$

(As special choices are included, for example, E_k for F_k or $\hat{a}_k$ for $\hat{c}_k$.) Equations (1), (3), and (4) yield

$$\sum_i \hat{a}_i E_i(P) = \hat{t}(P) = \sum_k \hat{c}_k (SF_k)(P)$$

and for the coefficients $\hat{a}_i$,

$$\hat{a}_i = \langle E_i, \hat{t} \rangle = \sum_k \langle E_i, SF_k \rangle \hat{c}_k$$

($\langle\ ,\ \rangle$ being the inner product), or in matrix form,

$$\hat{\mathbf{a}} = \mathbf{S}\hat{\mathbf{c}} \qquad \mathbf{S}_{ik} = \langle E_i, SF_k \rangle$$

Finally, with (2′) we arrive at

$$\tilde{\mathbf{b}} = \mathbf{BS}\hat{\mathbf{c}} = \mathbf{A}\hat{\mathbf{c}}$$

The solution of this equation, $\hat{\mathbf{c}}$, determines the representation (4) of g. A singular value decomposition of **A** transforms the latter equation to

$$\tau = \Sigma \zeta \tag{5}$$

where $\Sigma = \mathbf{U}^T\mathbf{AV}$ is the singular value decomposition of A, $\tau = \mathbf{U}^T\tilde{\mathbf{b}}$ is the transformation of $\tilde{\mathbf{b}}$ to the spectral base, and $\zeta = \mathbf{V}^T\hat{\mathbf{c}}$ is the corresponding transformation of $\hat{\mathbf{c}}$. **U** and **V** contain the orthonormal systems of eigenvectors or eigenfunctions of $\mathbf{AA}^T$ and $\mathbf{A}^T\mathbf{A}$ with $\mathbf{U}^T\mathbf{U} = \mathbf{I}$ and $\mathbf{V}^T\mathbf{V} = \mathbf{I}$.

Σ is a $n \times l$ diagonal matrix with the singular values σ_i in decreasing order in its diagonal. For questions concerning such a transformation the reader is referred to *Rummel et al.* [1979]. For a specific singular value, (5) becomes

$$\tau_i = \sigma_i \zeta_i \tag{5'}$$

and for the unknown ζ_i one obtains from the minimum norm least squares solution

$$\zeta_i = \sigma_i^{-1}\tau_i \qquad \sigma_i > 0$$
$$\zeta_i = 0 \qquad \sigma_i = 0$$

If $\tilde{\tau}_i$ is derived from real observations and thus contains an error ε_i, for example, from observational noise, the solution transfers to

$$\hat{\zeta}_i = \frac{\tau_i}{\sigma_i} + \frac{\varepsilon_i}{\sigma_i} \tag{6}$$

For high frequencies, that is, small σ_i, the influence of the second term of (6) becomes more and more pronounced, causing the above mentioned high-frequency fluctuations or destroying the convergency of the series $\hat{g} = \sum_k \hat{c}_k F_k$ at all. A unique and stable solution to this type of problem is obtained, as described by *Rummel et al.* [1979], from the minimum condition

$$\min\left(\| \tilde{\mathbf{b}} - \mathbf{Ac} \|_{D^{-1}}^2 + \alpha \| \mathbf{c} \|_{C_{cc}^{-1}}^2 \right) \qquad \alpha > 0 \tag{7}$$

with $\mathbf{D} = \sigma_0^2 I$ the a priori variance-covariance matrix of the observations, α the stabilization factor, and $\mathbf{C}_{cc}^{-1}$ the positive-definite weight matrix for **c**.

The estimated $\hat{\mathbf{c}}$ becomes

$$\hat{\mathbf{c}} = \mathbf{C}_{cc}\mathbf{A}^T(\mathbf{AC}_{cc}A^T + \alpha\mathbf{D})^{-1}\tilde{\mathbf{b}} \tag{8}$$

and has with $\mathbf{C}_{cc} = \mathbf{RR}^T$ the spectral form

$$\hat{\zeta} = \Sigma^T(\Sigma\Sigma^T + \alpha\sigma_0^2\mathbf{I})^{-1}\tilde{\tau} \tag{9}$$

where Σ is now obtained from $\Sigma = \mathbf{U}^T\mathbf{ARV}$.

A single spectral component of the stabilized solution is therefore

$$\hat{\zeta}_i = \frac{\sigma_i}{\sigma_i^2 + \alpha\sigma_0^2}\tilde{\tau}_i \tag{10}$$

or with (6),

$$\hat{\zeta}_i = \frac{\sigma_i^2}{\sigma_i^2 + \alpha\sigma_0^2}\zeta_i + \frac{\sigma_i^2}{\sigma_i^2 + \alpha\sigma_0^2}\frac{\varepsilon_i}{\sigma_i} \tag{11}$$

The factor

$$f_i = \frac{\sigma_i^2}{\sigma_i^2 + \alpha\sigma_0^2} = \frac{1}{1 + (\alpha\sigma_0^2/\sigma_i^2)} \tag{12}$$

is the filter through which $\tilde{\tau}$ has to pass and which best eliminates the noise ε. Thus the filter reflects rather objectively the stability characteristics of a specific estimation problem. In the following discussion we will therefore try to derive the expression for the filter for various representations of the gravity field and analyze factors influencing the shape of these filter expressions.

3. REPRESENTATIONS

Spherical harmonics: For a representation of the gravity field in terms of spherical harmonics, (1) takes the form

$$\tilde{t}(P) = \mathrm{Re} \sum_{l=0}^{\infty} \sum_{m=-l}^{+l} t_{lm} Y_{lm}(P) \tag{13}$$

$$Y_{lm} = \frac{(-1)^m}{(2 - \delta_{0m})^{1/2}} \bar{P}_{lm}(\cos\theta)e^{im\lambda} \qquad m \geq 0$$
$$Y_{lm} = \frac{(-1)^m}{(2 - \delta_{0m})^{1/2}} \bar{P}_{l|m|}(\cos\theta)ie^{im\lambda} \qquad m \leq 0 \tag{14}$$

where the $\bar{P}_{lm}$ are the fully normalized associated Legendre functions. A prerequirement for a computation of meaningful coefficients t_{lm} from a solution of (2) is, however, a more or less homogeneous data distribution all over the earth. Even for a truncation of the series at a comparably low degree, for example, $l_{max} = 30$, the computational effort becomes tremendous, and rather large distortions occur in the estimated coefficients, as described by *Colombo* [1979, private communication, 1980]. For the long-wavelength representation, however, where the instabilities remain modest, this type of representation has enormous merits, as discussed by *Kaula* [1969].

Mean values over equal angular or equal area blocks: Most of the terrestrial gravity information is kept in the form of mean values of certain block sizes. The base functions in (1) are then a system of (discontinuous) step functions. Stability problems do not occur because of the local nature of the base functions. Disadvantages are, as shown by *Rapp* [1977], the discontinuous character of the representation, distortions due to aliasing, and the fact that there is no simple one-to-one correspondence to a representation in terms of spherical harmonics.

The representation of the gravity field in terms of spherical harmonics and in the form of mean block values can be regarded as a global and local limit. In the following discussion we will therefore concentrate on representations which allow a local as well as a global approximation to the earth's gravity field.

The comparison in terms of the filter equation (12) will be performed by means of an example, where we assume a given set of 'observations' $\tilde{t}_j, j = 1, \cdots, n$ from which a global approximation $\hat{t}$ to the disturbing potential will be derived. In the same way we could have discussed the determination of any other gravity parameter type from this set of observations, or any other type of observations, too. In agreement with physical reality the observations are assumed to contain an error e_j.

Point mass model [e.g., *Needham,* 1970; *Balmino,* 1973; *Reilly and Herbrechtsmeier,* 1978; *Blaha,* 1979]: The observed disturbing potential $\tilde{t}(P_j) = \tilde{t}_j$ is represented in terms of the dimensionless algebraic point masses μ at the points $Q_1, \cdots, Q_m$:

$$\tilde{t}_j = kM \sum_i \frac{\mu_i}{l_{ij}} + e_j \qquad j = 1, \cdots, n \tag{15}$$

where kM is the earth's mass times gravitational constant and l_{ij} is the spatial distance between P_j and Q_i, the Q_i usually being located on one or two spherical shells inside the earth's surface ($r_Q < r_P$). The kM/l_{ij}, $i = 1, \cdots, m$ correspond to the base functions $E_i(P_j)$ in (1). We may rewrite (15) in the matrix form

$$\tilde{\mathbf{t}} = \mathbf{E} \cdot \boldsymbol{\mu} + \mathbf{e}$$

Equation (8) yields the estimated point masses

$$\hat{\boldsymbol{\mu}} = \mathbf{C}_{\mu\mu}\mathbf{E}^T(\mathbf{E}\mathbf{C}_{\mu\mu}\mathbf{E}^T + \alpha\mathbf{D})^{-1}\tilde{\mathbf{t}} \tag{16}$$

and from them, in analogy to (1), the unknown approximation to t becomes, with (15),

$$\hat{t}(P) = \mathbf{E}(P, Q_i)^T\mathbf{C}_{\mu\mu}\mathbf{E}^T(\mathbf{E}\mathbf{C}_{\mu\mu}\mathbf{E}^T + \alpha\mathbf{D})^{-1}\tilde{\mathbf{t}} \tag{17}$$

The norm $||\boldsymbol{\mu}||_{C_{\mu\mu}^{-1}}{}^2$ in minimum condition (7) from which (16) results has to be defined through a proper choice of the weight matrix $\mathbf{C}_{\mu\mu}{}^{-1}$ or its inverse $\mathbf{C}_{\mu\mu}$. As a general expression for a matrix element ii' of $\mathbf{C}_{\mu\mu}$ we choose

$$(\mathbf{C}_{\mu\mu})_{ii'} = \sum_{l=0}^{\infty} c_l^{\mu} s^{l+2} P_l(\cos\psi_{ii'}) \tag{18}$$

where c_l^{μ} is the degree 'variance' of the point masses, $s = r_B{}^2/r_{Q_i}r_{Q_{i'}}$ is the ratio of the square of some internal (Bjerhammar) sphere to the geocentric radii of the mass points Q_i and $Q_{i'}$, and P_l is the Legendre polynomial.

An arbitrary element of the matrix product $\mathbf{E}\mathbf{C}_{\mu\mu}\mathbf{E}^T$ takes the form

$$(\mathbf{E}\mathbf{C}_{\mu\mu}\mathbf{E}^T)_{jk} = (kM)^2 \sum_i \sum_{i'} l_{ji}{}^{-1}(\mathbf{C}_{\mu\mu})_{ii'}l_{i'k}{}^{-1} \tag{19}$$

In terms of Legendre polynomials the expression for the inverse distance l_{ji} between P_j and Q_i is

$$l_{ji}{}^{-1} = \frac{1}{r_i}\sum_{l=0}^{\infty}\left(\frac{r_i}{r_j}\right)^{l+1} P_l(\cos\psi_{ij}) \tag{20}$$

where with the addition theorem

$$P_l(\cos\psi_{ij}) = \frac{1}{2l+1}\sum_{m=-l}^{+l} Y_{lm}{}^*(P_j)Y_{lm}(Q_i) \tag{21}$$

If all point masses are located on a geocentric sphere σ_Q with radius r_Q and the arrangement of point masses becomes globally dense (this type of limit is explained by *Moritz* [1975]), a single density layer model is obtained. The representation of the gravity field with a density layer model is discussed, for example, by *Koch and Witte* [1971] or *Morrison* [1976]. Then the elements of $\mathbf{E}\mathbf{C}_{\mu\mu}\mathbf{E}^T$ become, with (19)–(21),

$$\lim_{\substack{Q_i \text{ dense} \\ \text{on } \sigma_Q}} (\mathbf{E}\mathbf{C}_{\mu\mu}\mathbf{E}^T)_{jk} = \frac{(kM)^2}{r_Q{}^2}\sum_l \bar{c}_l^{\mu}s^{l+2}\left(\frac{r_Q{}^2}{r_{P_j}r_{P_k}}\right)^{l+1} \cdot \frac{1}{2l+1}P_l(\cos\psi_{jk}) \tag{22}$$

where

$$\bar{c}_l^{\mu} = \frac{c_l^{\mu}}{2l+1}$$

The corresponding limit in $\mathbf{E}(P, {}_i)^T\mathbf{C}_{\mu\mu}\mathbf{E}^T$ in (17) is obtained by replacing the fixed observation point P_k by P.

For this finite case with $n < \infty$ observations the shape of the filter has to be derived numerically from a singular value decomposition of (17). But the basic stability characteristics of a certain type of gravity representation can be deduced from a global limit, where also the arrangement of the given data points becomes globally dense. Then instead of a finite vector of observations $\tilde{\mathbf{t}}$ we assume to have given a global function $\tilde{t}$ still containing the observation error e. The function $\tilde{t}$ will be represented by its expansion into the infinite series of spherical harmonics:

$$\tilde{t}(P) = \text{Re}\sum_{lm}\tilde{\tau}_{lm}Y_{lm}(P)$$

In the same way the estimate $\hat{t}$ is represented by

$$\hat{t}(P') = \text{Re}\sum_{lm}\hat{\tau}_{lm}Y_{lm}(P')$$

and $\mathbf{E}\mathbf{C}_{\mu\mu}\mathbf{E}^T$ becomes

$$\lim_{\substack{Q_i \text{ dense} \\ \text{on } \sigma_Q}}\ \lim_{\substack{P_j \text{ dense} \\ \text{on } \sigma_P}} (\mathbf{E}\mathbf{C}_{\mu\mu}\mathbf{E}^T)_{jk} = \sum_{lm}\sigma_l{}^2 Y_{lm}{}^*(P)Y_{lm}(P')$$
$$= \sum_l (2l+1)\sigma_l{}^2 P_l(\cos\psi_{PP'}) \tag{23}$$

where for $\sigma_l{}^2$, (22) yields

$$\sigma_l{}^2 = \frac{(kM)^2}{r_Q{}^2}\bar{c}_l^{\mu}s^{l+2}\left(\frac{r_Q{}^2}{r_P{}^2}\right)^{l+1}(2l+1)^{-2} \tag{24}$$

For the global limit the expression for the filter (compare (12)) of the density layer or point mass model becomes

$$f_l = \left[1 + \frac{\alpha\sigma_0{}^2}{[(kM)^2/r_Q{}^2]\,\bar{c}_l^{\mu}s^{l+2}\,(r_Q{}^2/r_P{}^2)^{l+1}\,(2l+1)^{-2}}\right]^{-1} \tag{25}$$

In the case of $\mathbf{C}_{\mu\mu} = \mathbf{I}$, as is commonly assumed in point mass applications, the filter is

$$f_l = \left[1 + \frac{\alpha\sigma_0{}^2}{[(kM)^2/r_Q{}^2]\,(r_Q{}^2/r_P{}^2)^{l+1}\,(2l+1)^{-2}}\right]^{-1}$$

Bjerhammar's method [e.g., *Bjerhammar,* 1973, 1975; *Sjöberg,* 1975]: This method is based on a reduction of the gravity anomaly field to an internal sphere with radius r_Q using a discrete version of Poisson's integral for $r_P\Delta g(P)$:

$$r_P\Delta g(P) = \frac{r_Q(r_P^2 - r_Q^2)}{4\pi} \int_\sigma \frac{r_Q\Delta g^*(Q)}{l(P,Q)^3}\, d\sigma$$

with Δg^* a reduced gravity anomaly function at the internal sphere (σ being a unit sphere). In our example we have instead to employ a discrete form of Pizetti's extension of Stokes' equation:

$$\tilde{t}_j = r_Q \sum_i g_i^* d_{ij} + e_j \tag{26}$$

with $\tilde{t}_j$ the 'observed' disturbing potential, $j = 1, \cdots, n$; g_i^* the reduced gravity anomalies, $i = 1, \cdots, m$; and d_{ij} the discrete value of the extended Stokes kernel $S(P_j,Q_i)$.

Now the $r_Q d_{ij}$, $i = 1, \cdots, m$ correspond to the base functions $E_i(P_j)$ of (1). The matrix form of (26) becomes

$$\tilde{\mathbf{t}} = \mathbf{S}\mathbf{g}^* + \mathbf{e}$$

All further derivations for the determination of the global limit of the filter expression are repeated analogously to the point mass model. The results are (with the corresponding equation number for the point mass model in parentheses) as follows:

Estimated reduced gravity anomalies (16)

$$\hat{\mathbf{g}}^* = \mathbf{C}_{g^*g^*}\mathbf{S}^T(\mathbf{S}\mathbf{C}_{g^*g^*}\mathbf{S}^T + \alpha\mathbf{D})^{-1}\tilde{\mathbf{t}}$$

Estimated approximation to t (17)

$$\hat{t}(P) = \mathbf{S}(P,Q_i)^T\mathbf{C}_{g^*g^*}\mathbf{S}^T(\mathbf{S}\mathbf{C}_{g^*g^*}\mathbf{S}^T + \alpha\mathbf{D})^{-1}\tilde{\mathbf{t}}$$

Element of inverse weight matrix (18)

$$(\mathbf{C}_{g^*g^*})_{ii'} = \sum_{l=0}^{\infty} c_l^{g^*} s^{l+2} P_l(\cos\psi_{ii'})$$

Element of $\mathbf{S}\mathbf{C}_{g^*g^*}\mathbf{S}^T$ (19)

$$(\mathbf{S}\mathbf{C}_{g^*g^*}\mathbf{S}^T)_{jk} = r_Q^2 \sum_i \sum_{i'} d_{ji}(\mathbf{C}_{g^*g^*})_{ii'} d_{ki'}$$

Representation of the extended Stokes kernel in terms of Legendre polynomials (20)

$$d_{ji} = \sum_{l=2}^{\infty} \left(\frac{r_{Q_i}}{r_{P_j}}\right)^{l+1} \frac{2l+1}{l-1} P_l(\cos\psi_{ji})$$

Element of $\mathbf{S}\mathbf{C}_{g^*g^*}\mathbf{S}^T$ for globally dense g^* on σ_Q (22)

$$\lim_{\substack{Q_i \text{ dense} \\ \text{on } \sigma_Q}} (\mathbf{S}\mathbf{C}_{g^*g^*}\mathbf{S}^T)_{jk} = r_Q^2 \sum_l \bar{c}_l^{g^*} s^{l+2} \left(\frac{r_Q^2}{r_{P_j} r_{P_k}}\right)^{l+1} \frac{2l+1}{(l-1)^2} P_l(\cos\psi_{jk})$$

Limit for globally dense observations $\tilde{t}$ on σ_P (24)

$$\sigma_l^2 = r_P^2 \bar{c}_l^{g^*} s^{l+2} \left(\frac{r_Q^2}{r_P^2}\right)^{l+2} (l-1)^{-2}$$

Filter expression per degree of the generalized Bjerhammar method (25)

$$f_l = \left(1 + \frac{\alpha\sigma_0^2}{r_P^2 \bar{c}_l^{g^*} s^{l+2} (r_Q^2/r_P^2)^{l+2} (l-1)^{-2}}\right)^{-1} \tag{27}$$

Least squares collocation [*Krarup*, 1969; *Moritz*, 1972]: In collocation the observations $\tilde{t}(P_j) = \tilde{t}_j$ are represented through

$$\tilde{t}_j = \sum_{i=1}^{m} a_i K(P_j,Q_i) + e_j \qquad j = 1, \cdots, n \tag{28}$$

where, by definition, $m = n$ and $K(P,Q)$ is a reproducing kernel or a variance-covariance model for t.

The properties of $K(P,Q)$ are described, for example, by *Krarup* [1969] or *Tscherning* [1975]. It has the form

$$K(P,Q) = \sum_{l=0}^{\infty} c_l' s^{l+1} P_l(\cos\psi_{PQ}) \tag{29}$$

Since, analogous to (1),

$$\hat{t}(P) = \sum_i a_i K(P,Q_i)$$

the norm of $\hat{t}$ becomes, by the reproducing property,

$$\|\hat{t}\|^2 = \langle \hat{t},\hat{t}\rangle = \sum_{i,j} a_i \langle K(\ ,Q_i),K(\ ,Q_j)\rangle a_j$$

$$= \sum_{i,j} a_i K(Q_i,Q_j) a_j = \mathbf{a}^T\mathbf{K}\mathbf{a} = \|\mathbf{a}\|_{\mathbf{K}}^2 = \|\mathbf{a}\|_{\mathbf{C}_{aa}^{-1}}^2 \tag{30}$$

With (30) the equation for the estimated a can be transformed to

$$\hat{\mathbf{a}} = \mathbf{C}_{aa}\mathbf{K}^T(\mathbf{K}\mathbf{C}_{aa}\mathbf{K}^T + \alpha\mathbf{D})^{-1}\tilde{\mathbf{t}} = (\mathbf{K} + \alpha\mathbf{D})^{-1}\tilde{\mathbf{t}}$$

The resulting equations are as follows:

Estimated disturbing potential (17)

$$\hat{t}(P) = \mathbf{K}(P,\ Q_i)^T(\mathbf{K} + \alpha\mathbf{D})^{-1}\tilde{\mathbf{t}}$$

Limit for globally dense observations $\tilde{t}$ on σ_P (24)

$$\sigma_l^2 = \bar{c}_l' s^{l+1}$$

The filter per degree of the least squares collocation method (24)

$$f_l = \left(1 + \frac{\alpha\sigma_0^2}{\bar{c}_l' s^{l+1}}\right)^{-1} \tag{31}$$

4. DISCUSSION

The spectral relations between the spherical harmonic coefficients τ_{lm}, μ_{lm}, and g_{lm}^* of disturbing potential t, a global algebraic density layer μ, and the reduced gravity anomaly

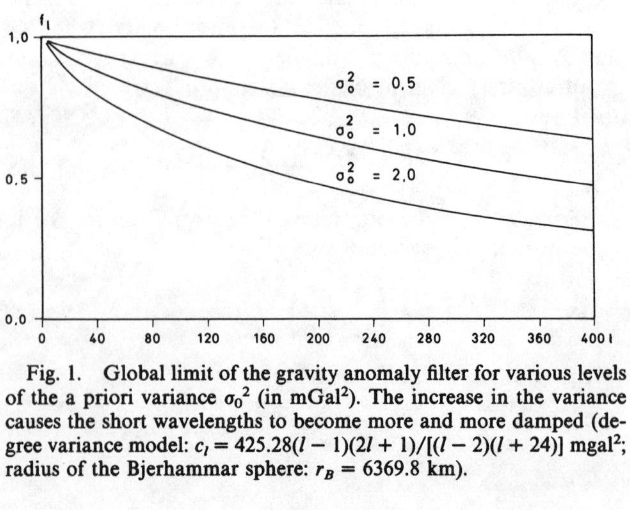

Fig. 1. Global limit of the gravity anomaly filter for various levels of the a priori variance σ_0^2 (in mGal²). The increase in the variance causes the short wavelengths to become more and more damped (degree variance model: $c_l = 425.28(l-1)(2l+1)/[(l-2)(l+24)]$ mgal²; radius of the Bjerhammar sphere: $r_B = 6369.8$ km).

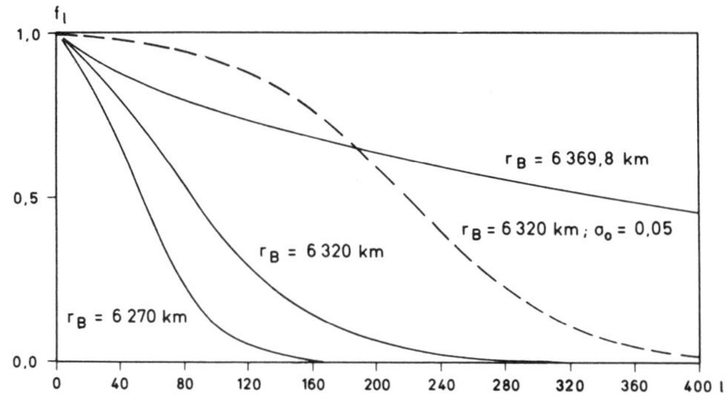

Fig. 2. Global limit of the gravity anomaly filter for various choices of the radius of the Bjerhammar sphere. The dashed line shows a constructed filter with an approximate low-pass characteristic (degree variance model, see Figure 1; a priori variance: $\sigma_0^2 = 1$ mgal^2).

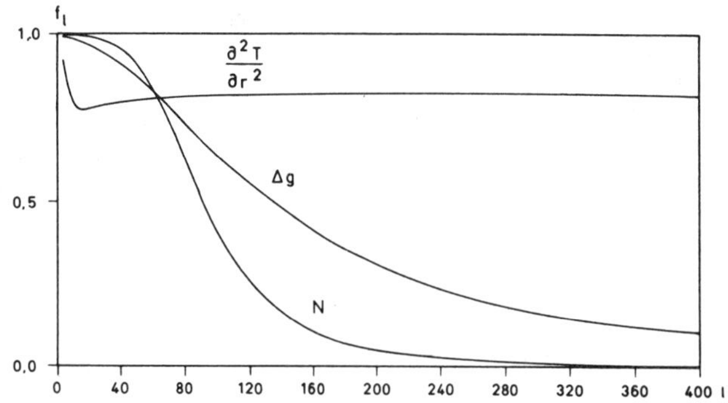

Fig. 4. Global limit of the filter for geoid heights N, gravity anomalies Δg, and second radial derivatives of the disturbing potential $\partial^2T/\partial r^2$ (degree variance model: $c_l = 425.28(l-1)/[(l-2)(l+24)]$ mgal^2 [*Tscherning and Rapp,* 1974]; radius of the Bjerhammar sphere: r_B = 6369.8 km; a priori standard deviation: $\sigma(N) = \pm 0.01$ m, $\sigma(\Delta g) = \pm 0.1$ mgal, $\sigma(\partial^2T/\partial r^2) = \pm 0.01$ Eötvös units ($= \pm 0.01 \times 10^{-9}$ s^{-2})).

function g^* are

$$\tau_{lm} = \frac{kM}{r_Q}\left(\frac{r_Q}{r_P}\right)^{l+1}\frac{1}{2l+1}\,\mu_{lm} \tag{32}$$

$$\tau_{lm} = r_P\left(\frac{r_Q}{r_P}\right)^{l+2}\frac{1}{l-1}\,g_{lm}^* \tag{33}$$

Thus if all three gravity representation models—point mass or density layer, Bjerhammar's method, and collocation—are based on the same weight model (reproducing kernel, variance-covariance model), their filter expressions become identical in the global limit. The similarities between these three types of representation are also discussed by *Sjöberg* [1978]. For an arbitrary gravity parameter x, assumed to be the observed quantity, with the spectral relation to t:

$$x_{lm} = \lambda_l \tau_{lm}$$

the general filter expression becomes, in the global limit,

$$f_l = \left(1 + \frac{\alpha\sigma_0^2}{\lambda_l^2 \bar{c}_l' s^{l+1}}\right)^{-1} \tag{34}$$

where $\alpha\sigma_0^2/\lambda_l^2\bar{c}_l's^{l+1}$ is the a priori noise to signal ratio per degree. This type of filter expression was derived for the Stokes equation by *Neyman* [1974] and *Moritz* [1975].

We see that the shape of the filter depends on (1) the a priori noise level (level of stabilization) $\alpha\sigma_0^2$, (2) the ratio $s = r_B^2/r_Pr_{P'}$ and therefore the chosen radius of the Bjerhammar sphere, (3) the chosen degree variance model c_l', and (4) the type of quantity that has been observed. These effects are demonstrated in Figures 1–4.

Figure 1 shows the variations of a filter for the gravity anomaly field due to variations of the assumed noise level (α = 1). The underlying degree variance model is a modification of *Tscherning and Rapp*'s [1974] model 4:

$$c_l = \frac{425.28(l-1)(2l+1)}{(l-2)(l+24)} \qquad s = 0.999617$$

In Figure 2 the filters for the same model and $\sigma_0^2 = 1$ mGal^2 are given for three different choices of the Bjerhammar sphere. Through the proper selection of $\alpha\sigma_o^2$ and r_B one may therefore construct an 'ideal' filter for a specific experiment. The dashed line shows an example for an arbitrary low-pass filter.

The influence of the degree variance models is demonstrated in Figure 3, which provides the filters based on the degree variance models A, B, and D* of *Tscherning* [1976, Table 7*a*, p. 39]. Discussions on the proper choice of the reproducing kernel or degree variance model are given, for example, by *Rapp* [1972], *Tscherning* [1973], or *Tscherning and Rapp* [1974].

The different shapes of the filter for the geoid height, gravity anomaly field, and second radial derivative of t are shown in Figure 4. It indicates especially that for the analysis of observed second derivatives a careful design of the weight model is a precondition.

Finally, it should be mentioned that the filter expression (34) allows the construction of a stabilized operator even for inverse problems, such as the downward continuation from satellite altitudes. This could become a very useful tool for the numerical treatment of problems where, on one hand, regionally very dense data allow its application and, on the other hand, the large amount of data circumvents the solution of the resulting large system of linear equations, for example, in least squares collocation.

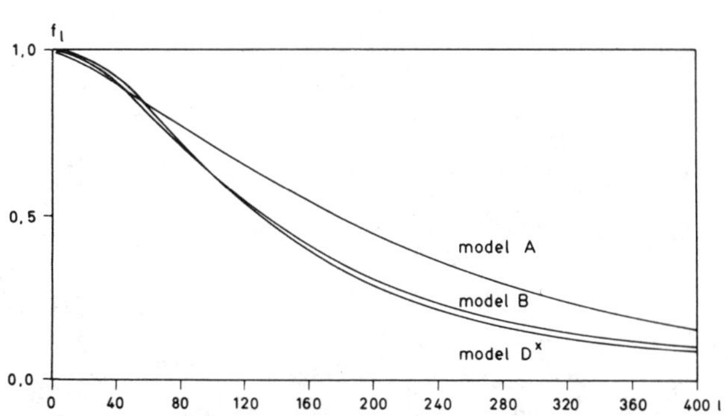

Fig. 3. Global limit of the degree variance filter for three different degree variance models published by *Tscherning* [1976, Table 7*a*, p. 39] (a priori variance: σ_0^2 = 0.01 mgal^2; radius of the Bjerhammar sphere: r_B = 6369.8 km).

Acknowledgment. This study has been carried out in the framework of the special research group on satellite geodesy (SFB 78) at the Technische Universität München.

REFERENCES

Balmino, G., The geopotential representation by point masses, in *Proceedings of the International Symposium on the Use of Artificial Satellites for Geodesy and Geodynamics,* edited by G. Veis, pp. 335–348, National Technical University, Athens, 1973.

Bjerhammar, A., On the discrete boundary value problem in physical geodesy, report, Div. of Geod., R. Inst. of Technol., Stockholm, 1973.

Bjerhammar, A., Discrete solutions of the boundary value problem in physical geodesy, *Tellus, 27*(2), 97–106, 1975.

Blaha, G., Improved determinations of the earth's gravity field, *Rep. AFGL-TR-79-0058,* Air Force Geophys. Lab., Bedford, Mass., 1979.

Colombo, O., Optimal estimation from data regularly sampled on a sphere with applications in geodesy, *Rep. 291,* Dep. of Geod. Sci., Ohio State Univ., Columbus, 1979.

Giacaglia, G. E. O., and C. A. Lundquist, Sampling functions for geophysics, Research in Space Science, Smithson. Astrophys. Lab. *Spec. Rep., 344,* 1972.

Kaula, W. M., The appropriate representation of the gravity field for satellite geodesy, paper presented at the 4th Symposium on Mathematical Geodesy, Bologna, Italy, 1969.

Koch, K. R., and B. U. Witte, Earth's gravity field represented by a simple layer potential from Doppler tracking of satellites, *J. Geophys. Res., 76,* 8471–8479, 1971.

Krarup, K., A contribution to the mathematical foundation of physical geodesy, *Medd. 44,* Geod. Inst., Copenhagen, 1969.

Moritz, H., Methods for downward continuation of gravity, *Rep. A-50,* Dtsch. Geod. Komm., Munich, 1966.

Moritz, H., Advanced least-squares methods, *Rep. 175,* Dep. of Geod. Sci., Ohio State Univ., Columbus, 1972.

Moritz, H., Integral formulas and collocation, *Rep. 234,* Dep. of Geod. Sci., Ohio State Univ., Columbus, 1975.

Moritz, H., Introduction to interpolation and approximation, in *Approximation Methods in Geodesy, Neue Folge,* vol. 10, edited by H. Moritz and H. Sünkel, pp. 1–45, Sammlung Wichmann, Karlsruhe, Federal Republic of Germany, 1978.

Morrison, F., Algorithms for computing the geopotential using a simple density layer, *J. Geophys. Res., 81*(26), 4933–4936, 1976.

Needham, P. E., The formation and evaluation of detailed geopotential models based on point masses, *Rep. 149,* Dep. of Geod. Sci., Ohio State Univ., Columbus, 1970.

Neyman, Yu. M., Probability modification of Stokes formula for calculating the height anomaly, *Geod. Mapp. Photogramm., 16,* 137–139, 1974.

Rapp, R. H., Geopotential coefficient behavior to high degree and geoid information by wavelength, *Rep. 180,* Dep. of Geod. Sci., Ohio State Univ., Columbus, 1972.

Rapp, R. H., The relationship between mean anomaly block sizes and spherical harmonic representations, *J. Geophys. Res., 82*(33), 5360–5364, 1977.

Reilly, J. P., and E. H. Herbrechtsmeier, A systematic approach to modeling the geopotential with point mass anomalies, *J. Geophys. Res., 83*(B2), 841–845, 1978.

Rummel, R., K.-P. Schwarz, and M. Gerstl, Least squares collocation and regularization, *Bull. Geod., 53*(4), 343–361, 1979.

Schaffrin, B., E. Heidenreich, and E. Grafarend, A representation of the standard gravity field, *Manuscr. Geod., 2*(2), 135–174, 1977.

Schmidt, H. F., Untersuchungen zur Anwendung von Sampling-Funktionen in der Geodäsie, *Rep. C-241,* Dtsch. Geod. Komm., Munich, 1978.

Schwarz, K.-P., Numerische Untersuchungen zur Schwerefortsetzung, *Rep. C-171,* Dtsch. Geod. Komm., Munich, 1971.

Schwarz, K.-P., Geodetic improperly posed problems and their regularization, *Boll. Geod. Sci. Affini, 38,* 389–416, 1979.

Sjöberg, L., On the discrete boundary value problem of physical geodesy with harmonic reduction to an internal sphere, report, Div. of Geod., R. Inst. of Technol., Stockholm, 1975.

Sjöberg, L., A comparison of Bjerhammar's methods and collocation in physical geodesy, *Rep. 273,* Dep. of Geod. Sci., Ohio State Univ., Columbus, 1978.

Tscherning, C. C., On the relation between the variation of the degree-variances and the variation of the anomalous potential, *Boll. Geod. Sci. Affini, 32,* 149–159, 1973.

Tscherning, C. C., Application of collocation determination of a local approximation to the anomalous potential of the earth using "exact"-astro-gravimetric collocation, in *Methoden und Verfahren der mathematischen Physik,* vol. 14/3, edited by B. Brosowski and E. Martensen, pp. 83–110, Bibliographisches Institut, Mannheim, Federal Republic of Germany, 1975.

Tscherning, C. C., Covariance expressions for second and lower order derivatives of the anomalous potential, *Rep. 225,* Dep. of Geod. Sci., Ohio State Univ., Columbus, 1976.

Tscherning, C. C., and R. H. Rapp, Closed covariance expressions for gravity anomalies, geoid undulations, and deflections of the vertical implied by anomaly degree variance models, *Rep. 208,* Dep. of Geod. Sci., Ohio State Univ., Columbus, 1974.

(Received November 17, 1980;
accepted February 13, 1981.)

REVIEWS OF GEOPHYSICS AND SPACE PHYSICS, VOL. 19, NO. 3, PAGES 421–436, AUGUST 1981

Sampling Function and Finite Element Method Representation of the Gravity Field

H. F. SCHMIDT

Deutsches Geodätisches Forschungsinstitut, Zentralleitung D-8000 Munich 2, Federal Republic of Germany

After introducing the fundamental method of interpolation and approximation of functions using sampling functions, some examples of one-dimensional models are demonstrated. Then a method of sampling functions for the representation and determination of the earth's gravity field—based on spherical harmonics and an arrangement of sampling points on a sphere—is described and discussed by investigating its mathematical implication. Additionally, this method is compared with the finite element method, with a view to the interpolation and modeling of the gravity field of the earth.

CONTENTS

A. INTRODUCTION

In geodesy the problem of describing, in an easy, manageable, and mathematically effective manner the detailed structure of the earth's gravity field still exists. To solve the problem, dense data distributions of high accuracy are necessary as well as appropriate analytical models for global or local processing. The latter are the subject of this paper.

One of the global models commonly used to represent the gravitational potential of the earth is [cf. *Heiskanen and Moritz,* 1967]

$$V(r, \Theta, \lambda) = \frac{kM}{r} \sum_{n=0}^{N} \sum_{m=0}^{n} \left(\frac{a}{r}\right)^n \{C_{nm} X_{nm}(\Theta, \lambda) + S_{nm} Y_{nm}(\Theta, \lambda)\} \tag{1a}$$

with the potential coefficients C_{nm}, S_{nm} and the base system of spherical harmonics

$$\begin{aligned} X_{nm}(\Theta, \lambda) &= P_{nm}(\Theta) \cdot \cos m\lambda \\ Y_{nm}(\Theta, \lambda) &= P_{nm}(\Theta) \cdot \sin m\lambda \end{aligned} \tag{1b}$$

$P_{nm}(\Theta)$ representing the associated Legendre functions.

However, for existing data of high accuracy and density (e.g., from altimetry) and for future data from satellite-to-satellite tracking and gradiometry [e.g., *Rummel,* 1979], this global model (1*a*), because of its serious numerical problems, cannot yield an equivalently accurate representation. To solve this problem, other models were suggested, based on, for example, mass points [e.g., *Balmino,* 1972], surface density layer [e.g., *Koch and Witte,* 1971; *Vinti,* 1971], 'carrier' points of gravity anomalies on the surface of an internal sphere [*Bjerhammar,* 1976], and sampling functions [*Lundquist et al.,* 1969; *Giacaglia and Lundquist,* 1972], described in detail in section D.

B. SAMPLING FUNCTION METHOD

Let the approximation $f(P)$ of a function $g(P)$ in a one-, two-, or three-dimensional space B containing the points P be given by a linear combination of base functions $G_j(P)$:

$$g(P) \approx f(P) = \sum_{j=1}^{n} a_j G_j(P) \tag{2}$$

The determination of the unknown parameters a_j is based in the simplest case on the assumption of error-free measurements $g(P_i) = f(P_i) = f_i$ at n points P_i, $i = 1, 2, \cdots, n$ and on the exact reproduction of the values f_i in P_i [*Moritz,* 1978] via

$$\sum_{j=1}^{n} a_j G_j(P_i) = f_i \tag{3a}$$

Written in matrix form and with $G_j(P_i) = G_{ji} = \mathbf{G}$, (3*a*) becomes

$$\mathbf{aG} = \mathbf{f} \tag{3b}$$

If the n equations are linearly independent, the unknowns a_j result from

$$\mathbf{a} = \mathbf{G}^{-1} f \tag{4}$$

Generally, these parameters a_j influence the whole area B; therefore they are called area parameters [*Heitz,* 1975]. An example of this type of parameter represents the potential coefficients C_{nm}, S_{nm} in the spherical harmonic model (1*a*).

A derivation for the sampling function model may be given—the departure from (7) is possible, too—if the matrix $\mathbf{G}$ in (3*a*) and (3*b*) is a unit matrix [*Moritz,* 1978]

$$\mathbf{G} = \mathbf{I} \tag{5}$$

Consequently, for the parameters a_j, (4) yields nothing else but the function values f_j (sample on the sampling point P_j), for example, the so-called point parameters

$$\mathbf{a} = \mathbf{f} \tag{6}$$

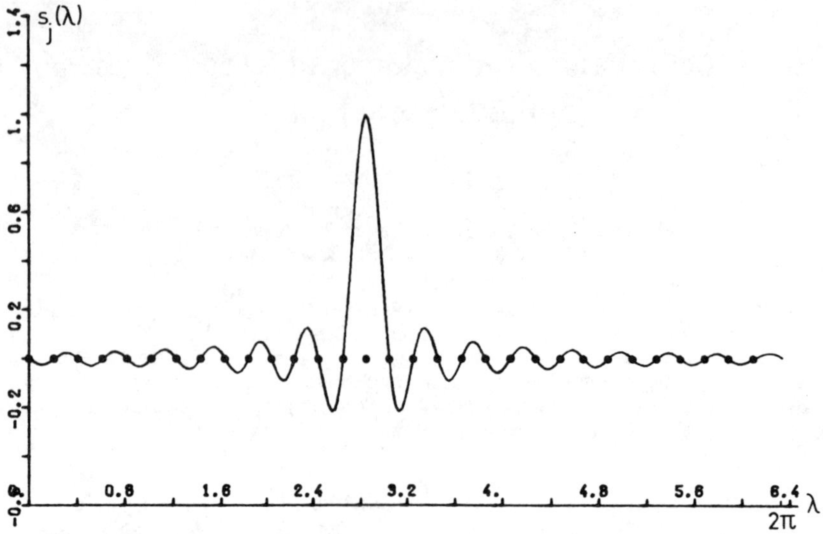

Fig. 1. Whittaker's cardinal function (14a) with the sampling points λ_j in (0, 2π) for n = 15, $w = \pi/n$, and j = 1.

Thus the base functions in (3a) should be called sampling functions $S_j(P)$, having the property

$$G_j(P_i) = S_j(P_i) = \delta_{ji} \qquad j, i = 1, 2 \cdots, n \tag{7}$$

where δ_{ji} is the Kronecker symbol.

According to (2) the sampling function approximation model obtains the form

$$f(P) = \sum_{j=1}^{n} f_j S_j(P) \tag{8}$$

The advantage of model (8)—as against the approximation (2) by means of area parameters, determined by (4)—is that for a certain set of sampling points the computation of the base of the approximation model (8), the sampling functions $S_j(P)$, is only necessary one time; that is, the base $S_j(P)$ is also fixed for changing sets of function values f_j.

1. *Determination of Sampling Functions*

The determination of a set of sampling functions $S_j(P)$ from an arbitrary base system $G_j(P)$, such as the one given in (2), can be realized by [*Moritz*, 1978]

$$S_j(P) = \sum_{i=1}^{n} G_{ij}^{-1} G_i(P) \tag{9a}$$

where the G_{ij}^{-1} are the elements of the inverse of G, because at the sampling points P_k, (9a) becomes

$$S_j(P_k) = \sum_{i=1}^{n} G_{ij}^{-1} G_{ki} = \delta_{kj} \tag{9b}$$

which therefore satisfies exactly the conditions of sampling functions (7).

2. *Prediction, Collocation, and the Sampling Function Method*

The purpose of prediction can be seen, on the basis of a data distribution of a function, in the interpolation as well as the extrapolation of this function. Because the approximation model (8) is also usable for interpolation purposes, a comparison of the prediction and, related to it, of the collocation method with the sampling function model seems to be worthwhile.

The principle of the prediction method may be explained as

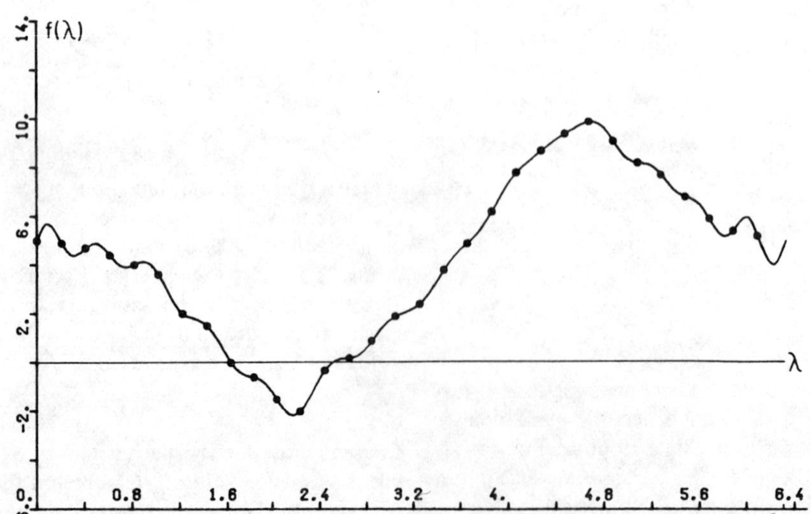

Fig. 2. Example of the interpolation (13) using the cardinal function (14a) (Figure 1).

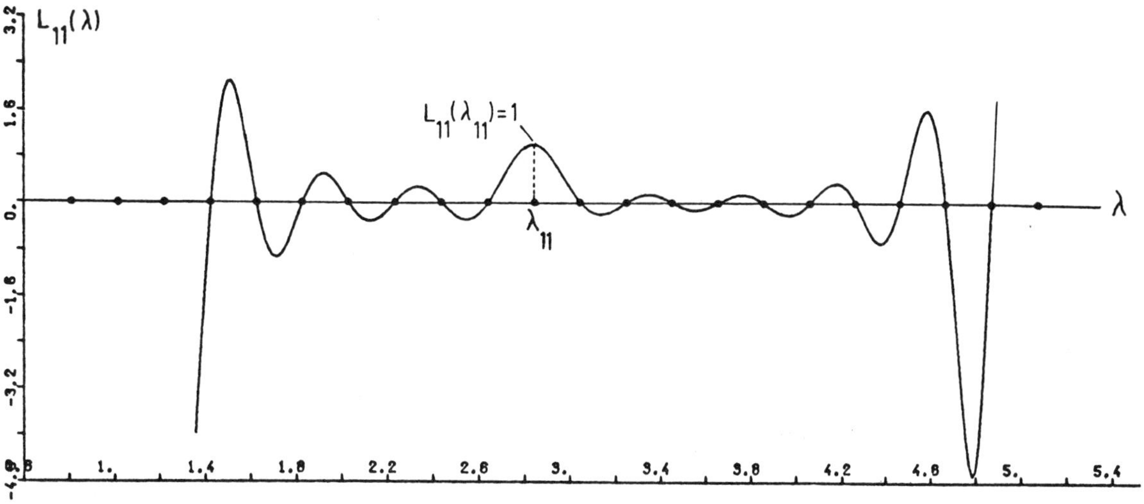

Fig. 3. Lagrange polynomial (14*b*) for $n = 15, j = 11$.

follows for the case in which the function $g(P)$ is the earth's disturbing potential $T(P)$: Since by a finite linear combination of the base functions $G_j(P)$ a corresponding approximation (2) can reproduce $T(P)$ only at n data points P_i, then the error

$$\varepsilon_p = T(P) - f(P) \tag{10a}$$

exists at the other points P. Now the prediction method minimizes the mean square interpolation error m_p, defined by

$$m_p^2 = M\{\varepsilon_p^2\} \tag{10b}$$

(M being a suitable average defined as a statistical aspectation value [e.g., *Heiskanen and Moritz,* 1967]) by the least squares method and gives as a result for the basis system in (2) according to N. Wiener and A. N. Kolmogorov [cf. *Moritz,* 1976*a*]

$$G_k(P) = C(P, P_k) \tag{11a}$$

with $C(P, P_k) = C_{pk}$, for example, the covariance function of the disturbing potential at the points P and P_k.

Therefore the prediction method is based on the statistical behavior of the gravity field characterized, for example, by the globally distributed gravity anomalies Δg. For an arbitrary point P the prediction method gives for a gravity anomaly Δg_p, on the basis of n existing 'samples' $\mathbf{x} = (\Delta g_1, \Delta g_2, \cdots, \Delta g_n)^T$ and the covariance matrices

$$\mathbf{C}_p = \begin{pmatrix} C_{p1} \\ C_{p2} \\ \vdots \\ C_{pn} \end{pmatrix} \qquad \mathbf{C} = \begin{pmatrix} C_{11} \cdots C_{1n} \\ C_{21} \cdots C_{2n} \\ \vdots \\ C_{n1} \cdots C_{nn} \end{pmatrix} \tag{11b}$$

the prediction value

$$\tilde{\Delta g}_p = \mathbf{C}_p^T\, \mathbf{C}^{-1}\mathbf{x} \tag{11c}$$

and is thus minimizing

$$\varepsilon_p = \Delta g_p - \tilde{\Delta g}_p$$

On the basis of the statistical principle of the prediction representation (11*c*) the approximation model of collocation additionally takes into consideration not only random noise but also systematic errors, as well as the possibility of using different types of observations [e.g., *Tscherning,* 1974]. The basic model for collocation is [*Moritz,* 1978]

$$\mathbf{x} = \mathbf{AX} + \mathbf{s} + \mathbf{n} \tag{12a}$$

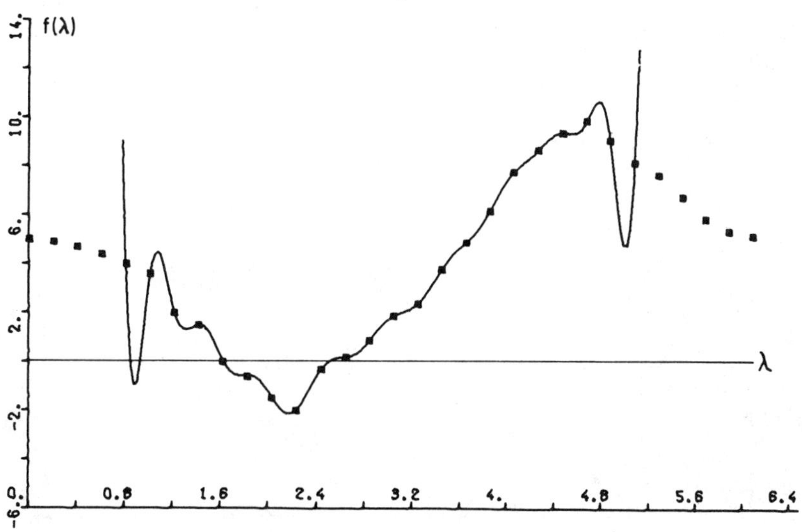

Fig. 4. Example of the interpolation (13) using the Lagrange polynomial (14*b*) (Figure 3).

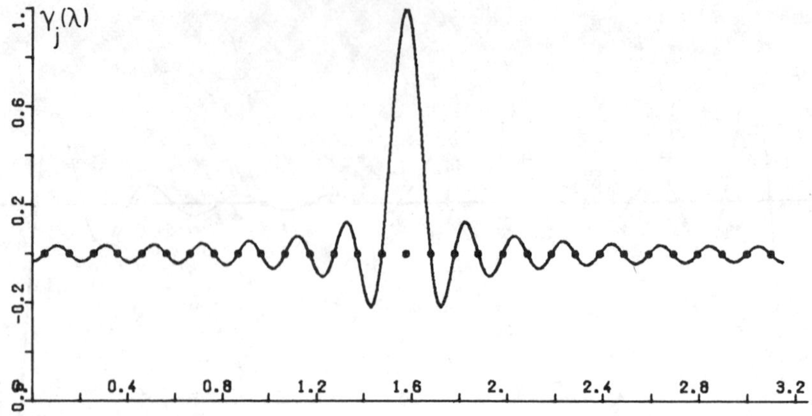

Fig. 5. Sampling function (15*b*) based on the Chebyshev polynomial (15*a*), $n = 30$, $j = 16$ in $[0, \pi]$.

with **x** the measured quantities, **s** the random 'signal' (representing, for example, the anomalous gravitational field like the Δg_p), **n** the random measuring error (noise), **X** the systematic nonrandom parameters, and **A** the given matrix of the coefficients. Using the least squares principle for

$$\mathbf{v}^T\mathbf{P}\mathbf{v} = \min$$

with

$$\mathbf{z} = \mathbf{s} + \mathbf{n} \quad \text{in} \quad \mathbf{v} = \begin{pmatrix} \mathbf{s} \\ \mathbf{z} \end{pmatrix}$$

and

$$Q = \begin{pmatrix} \mathbf{C}_{ss} & \mathbf{C}_{sx} \\ \mathbf{C}_{xs} & \mathbf{C}_{xx} \end{pmatrix} \quad \text{in} \quad \mathbf{P} = Q^{-1}$$

the result for estimating the signal at a point P is

$$s_p = \mathbf{C}_p^T \mathbf{C}_{xx}^{-1}(\mathbf{x} - \mathbf{A}\mathbf{X}) \qquad (12b)$$

Thus the collocation method can be used universally, unlike the sampling function model (8). As a numerical problem (the determination of a set of sampling functions (9*a*) needs, for a dense sampling point distribution, the solution of a large inverse matrix G_{ij}^{-1}) the collocation model (12*b*), as well as the prediction method (11*c*), also includes the computation of a possible large inverse matrix $\mathbf{C}_{xx}^{-1}$ and $\mathbf{C}^{-1}$, respectively, depending on the number of observations. But the sampling function determination is necessary only once for a fixed grid of sampling points, independent of the observation type.

C. Models of One-Dimensional Sampling Function Representation

The geodetic application of the one-dimensional sampling function approximation may be possible for observations along rectilinear profiles or along parallels of latitude and longitude. The characteristics of the sampling function representation are, according to (7) and (8), that at a certain set of sampling points $P_j = \lambda_j$, $j = 1, 2, \cdots, n$, the function values $f(\lambda_j) = f_j$ are given and that the interpolating curve (8), now

$$f(\lambda) = \sum_{j=1}^{n} f_j S_j(\lambda) \qquad (13)$$

passes exactly through these point parameters f_j. Since nothing is defined on how $f(\lambda)$ ought to behave between the sampling points, the door is open for a variety of methods with different interpolation behavior.

In the following discussion some examples of one-dimensional sampling functions usable as base functions $s_j(\lambda)$ are demonstrated:

a. Cardinal function. The cardinal function $s_j(\lambda)$ [*Whittaker*, 1915] for an equidistant set of sampling points $\lambda_j = jw$ (step width w), $j = 0, \pm 1, \pm 2, \cdots, \pm n$ (Figure 1) is given by

$$s_j(\lambda) = \frac{\sin[\pi(\lambda - \lambda_j)/w]}{\pi(\lambda - \lambda_j)/w} \qquad (14a)$$

which is referred to by *Moritz* [1976*a*] as the sin x/x method.

An example of the use of the cardinal function in interpolation is given in Figure 2.

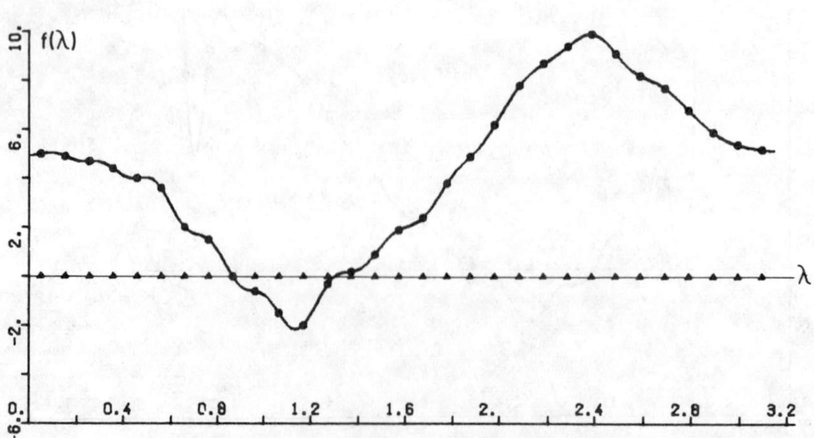

Fig. 6. Example of the interpolation (13) based on the sampling function (15*b*) (Figure 5).

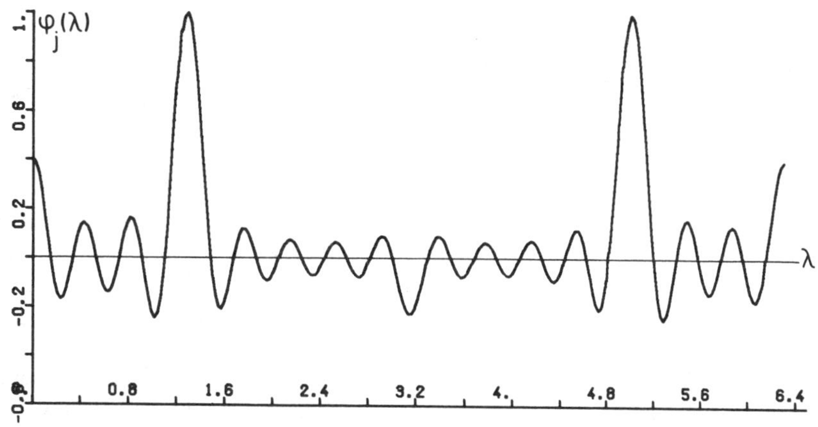

Fig. 7. Sampling function (16*b*) from Legendre polynomials, $N = 15$, $j = 7$ in an interval $[0, 2\pi]$.

b. Lagrange polynomial. The Langrange polynomial $L_j(\lambda)$ for the equidistant sampling point distribution λ_j, $j = 0, 1, 2, \cdots, n$ [e.g., *Giacaglia,* 1970] is given by

$$L_j(\lambda) = \frac{(\lambda - \lambda_0)(\lambda - \lambda_1) \cdots (\lambda - \lambda_{j-1})(\lambda - \lambda_{j+1}) \cdots (\lambda - \lambda_n)}{(\lambda_j - \lambda_0)(\lambda_j - \lambda_1) \cdots (\lambda_j - \lambda_{j-1})(\lambda_j - \lambda_{j+1}) \cdots (\lambda_j - \lambda_n)} \tag{14b}$$

At the boundaries of the definition interval the Lagrange polynomial exhibits too large amplitudes (cf. Figure 3), yielding, according to model (13), unrealistic interpolation values (Figure 4) there.

c. Chebyshev polynomial. The Chebyshev polynomial of first order in the interval $[-1, 1]$,

$$T_K(\lambda) = \cos k\lambda \qquad k = 0, 1, \cdots, n \tag{15a}$$

with the equidistant points

$$\lambda_K = \frac{\pi}{2} \cdot \frac{2k+1}{n+1}$$

was applied by *Brown* [1975] to define the sampling function (Figure 5)

$$\gamma_j(\lambda) = \frac{2}{n+1} \sum_{r=0}^{n} {}^*T_r(\lambda_j)T_r(\lambda) \qquad j = 0, 1, \cdots, n \tag{15b}$$

where the asterisk denotes a weight factor of 0.5 for the first term.

An example of the use of the Chebyshev polynomial in interpolation is given in Figure 6.

d. Legendre polynomial. The $(N + 1)$ Legendre polynomials [e.g., *Moritz,* 1978]

$$P_n(\cos\lambda) = -2^{-n} \sum_{r=0}^{\langle n/2 \rangle} \frac{(-1)^r}{r!} \frac{(2n-2r)!}{(n-r)!(n-2r)!} (\cos\lambda)^{n-2r} \tag{16a}$$

$$n = 0, 1, \cdots, N$$

where the angle brackets denote the greatest integer, give $(N + 1)$ sampling functions [*Brown,* 1975]

$$\phi_j(\lambda) = \left[\sum_{n=0}^{N} \frac{2n+1}{2} P_n(\cos\lambda_j) P_n(\cos\lambda) \right] \cdot \left[\sum_{n=0}^{N} \frac{2n+1}{2} P_n^2(\cos\lambda_j) \right]^{-1} \tag{16b}$$

for the sampling points λ_j, which are the zero points of the polynomial of order $(N + 1)$ determined numerically (see Figure 7).

An example of the use of the Legendre polynomial in interpolation is given in Figure 8.

e. Binary functions. In the same way as for the other examples in this section, 'binary' functions of the order ν

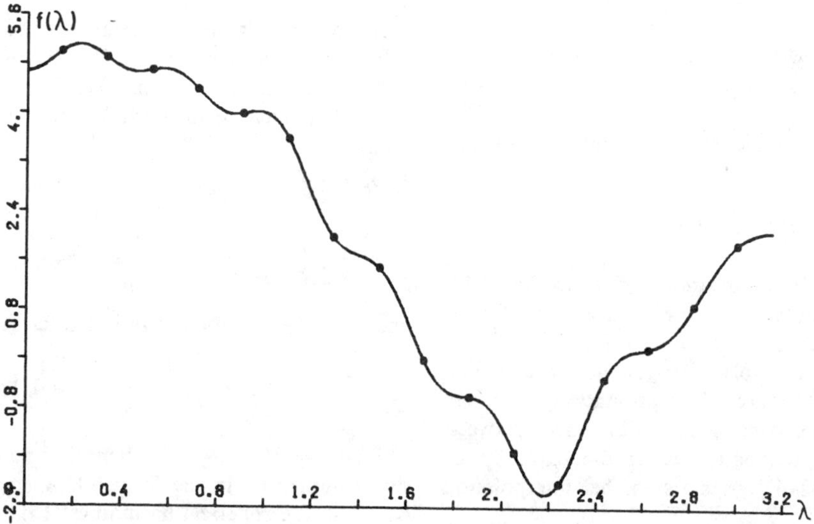

Fig. 8. Example of the interpolation (13) using the sampling function type (16*b*) (Figure 7).

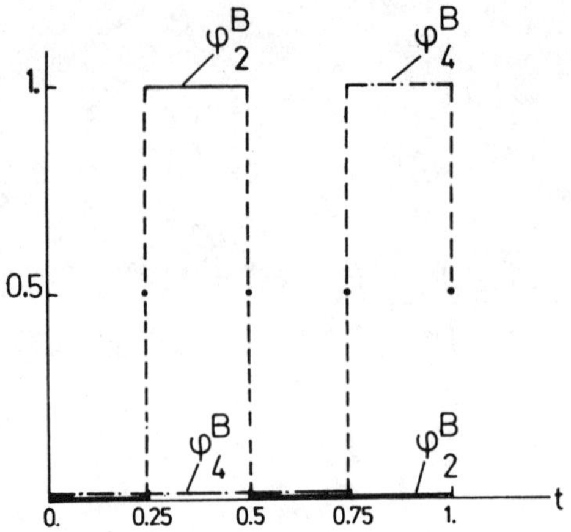

Fig. 9. Two examples of 'binary' functions (17) of order $\nu = 2$, for $k = 2, 4$ in [0, 1].

[*Brown*, 1975] also fulfill the conditions (7) for sampling functions $S_k(P_i) = \delta_{ki}$:

$$\phi_k^B(\lambda) = 1 \qquad \frac{k-1}{2^\nu} < \lambda < \frac{k}{2^\nu}$$
$$\phi_k^B = 0 \qquad 0 \leqq \lambda \leqq \frac{k-1}{2^\nu} \qquad \frac{k}{2^\nu} \leqq \lambda \leqq 1 \tag{17}$$

where $k = 1, 2, \cdots, 2^\nu$, in [0, 1], λ is

$$(\lambda_k - \Delta\lambda) < \lambda < (\lambda_k + \Delta\lambda)$$

with $\Delta\lambda = (\lambda_k - \lambda_{k-1})/2$.

At the points $(k-1)/2^\nu$ and $k/2^\nu$, (17) is defined by the value 0.5 (cf. Figure 9). Examples of the application of this function are given by *Brown* [1975].

f. Smooth interpolation. For the order n, *Giacaglia* [1970] defined $(2n + 1)$ sampling points λ_k in $[0, 2\pi]$ by

$$\lambda_k = \frac{2\pi}{2n+1} k \qquad k = 0, 1, \cdots, 2n \tag{18a}$$

and according to (18*a*) the trigonometric sampling functions (cf. Figures 10*a* and 10*b*) are

$$q_j^n(\lambda) = \frac{1}{2n+1}\left[1 + 2\sum_{k=1}^{n} \cos k(\lambda - \lambda_j)\right] \tag{18b}$$

with the normality

$$\sum_{j=0}^{2n} q_j^n(\lambda) = 1 \tag{18c}$$

Again the approximation model is, according to (8) and (13),

$$f(\lambda) = \sum_{j=0}^{2n} f(\lambda_j) q_j(\lambda) \tag{18d}$$

An example of the smooth interpolation is given in Figure 11. The model (18*d*) also works for a nonequidistant set of sampling functions [*Giacaglia*, 1970; *Schmidt*, 1978].

Although the second example of trigonometric sampling function interpolation (Figure 12) represents a relatively smooth curve, for intervals with greater differences between function values the approximation using the cubic spline function [e.g., *De Boor*, 1962] gives ı smoother interpolation (cf. Figure 12).

g. Dirac δ function. Let, as a classic example, the Dirac δ function [*Laugwitz*, 1967] be given by

$$\delta(\lambda) = 0 \qquad \lambda \neq 0$$
$$\int_{-\infty}^{+\infty} \delta(\lambda)\, d\lambda = 1 \tag{19a}$$

As the Fourier series of a Dirac distribution of order n, periodic in 2π, according to *Lanczos* [1957] gives

$$\delta_{2\pi}^n(\lambda) = \frac{1}{2\pi}\left(1 + 2\sum_{k=1}^{n} \cos k\lambda\right) \tag{19b}$$

it may be possible to replace the trigonometric sampling function (18*b*) for $j = 0$ by the relation

$$q_0^n(\lambda) = \frac{2\pi}{2n+1}\delta_{2\pi}^n(\lambda) \tag{20}$$

In connection with these trigonometric sampling functions (18*a*)–(18*d*), section D will demonstrate a similar sampling function model for the sphere.

D. Two-Dimensional Sampling Function Representation

A great part of the effort in geodesy aims at finding approximation possibilities for the two- and three-dimensional space.

In principle, the application of the one-dimensional sampling function model (section C) on crossing profiles is one possible method for the representation of two-dimensional areas. *Brown* [1975] has demonstrated the application of his binary function on a sphere. *Bjerhammar* [1976] uses the Dirac δ function (section C) to represent carrier points of gravity anomalies in the interior of the earth. *Hardy* [1972] has determined, on the basis of the so-called 'multiquadric analysis,' a representation of the world's topography and bathymetric irregular surface from a limited number of data points by summation of regular, that is, mathematically defined, surfaces such as the hyperboloid or hyperboloidlike surfaces in a spherical coordinate system. The regular surfaces are associated with particular coordinated control points on the irregular surface.

1. *Spherical Harmonic Sampling Function on a Sphere*

As an approximate equidistant grid of sampling points on a sphere for an arbitrary order N on $(N + 1)$ parallels of polar distance Θ_k, *Lundquist et al.* [1969] and *Giacaglia and Lundquist* [1972] have defined, for $k = 0, 1, \cdots, N$,

$$\Theta_k = \frac{2\pi}{2N+1} i \qquad i = \frac{k}{2} \qquad k = 0, 2, 4, \cdots$$
$$\Theta_k = \frac{2\pi}{2N+1}(N - i) \qquad i = \frac{k-1}{2} \qquad k = 1, 3, \cdots \tag{21a}$$

and on it the distribution in longitude (Figure 13)

$$\lambda_{kj} = \frac{2\pi}{2k+1} j \qquad j = 0, 1, \cdots, 2k \tag{21b}$$

For these $M = (N + 1)^2$ sampling points (Θ_k, λ_{kj})—successively numbered in the form (Θ_j, λ_j), $j = 1, 2, \cdots, M$—and based on the spherical harmonics (1*b*), $X_{nm}(\Theta, \lambda)$, $Y_{nm}(\Theta, \lambda)$,

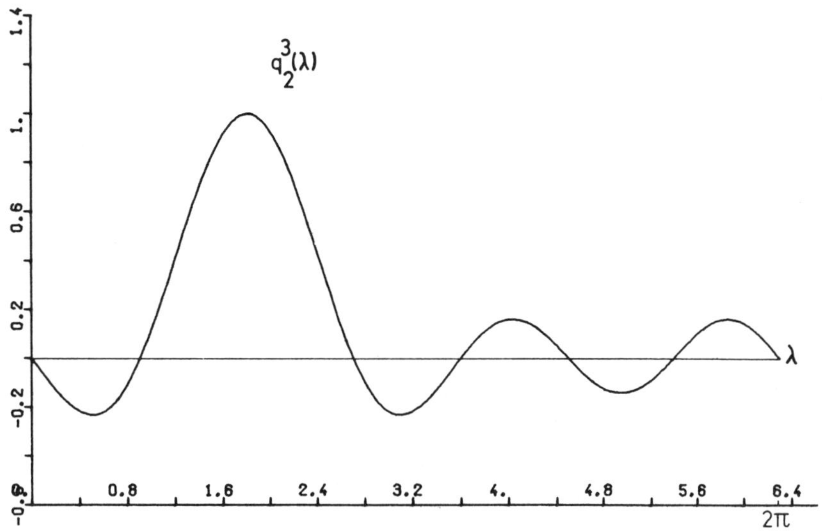

Fig. 10*a*. Trigonometric sampling function (18*b*) for the order $n = 3, j = 2$.

the corresponding sampling functions (Figure 14) are defined as

$$W_j(\Theta, \lambda) = \sum_{n=0}^{N} \sum_{m=0}^{m} [A_j^{nm} X_{nm}(\Theta, \lambda) + B_j^{nm} Y_{nm}(\Theta, \lambda)] \qquad (22)$$

$$j = 1, 2, \cdots, M$$

For the set of sampling points (21*a*) and (21*b*) the conditions (7) have to be fulfilled:

$$W_j(\Theta_i, \lambda_i) = \delta_{ji} \qquad j, i = 1, 2, \cdots, M \qquad (23)$$

Accordingly, for the property (18*c*) for the one-dimensional trigonometric sampling function the normality

$$\sum_{j=1}^{M} W_j(\Theta, \lambda) = 1 \qquad (24)$$

holds.

Beyond, along a parallel of latitude Θ_k with its $(2k + 1)$ sampling points (Θ_k, λ_{kj}), $j = 0, 1, \cdots, 2k$ (cf. (21*b*)) the sampling functions $W_{kj}(\Theta_k, \lambda_{kj})$ represent the one-dimensional trigonometric sampling function (18*b*) of the form [*Lundquist et al.*, 1969]

$$W_{kj}(\Theta_k, \lambda) = q_j^k(\lambda) = \frac{1}{2k+1}\left[1 + 2\sum_{i=1}^{k} \cos i(\lambda - \lambda_{kj})\right] \qquad (25)$$

The determination of the sampling function parameters A_j^{nm}, B_j^{nm} in (22) is possible according to the conditions (23). Writing for

$$(A_j^{00}, A_j^{10}, \cdots, A_j^{NN}, B_j^{11}, \cdots, B_j^{NN}) = (A_j^1, A_j^2, \cdots, A_j^M) = \mathbf{A}_j \qquad (26a)$$

$$[X_{00}(\Theta_j, \lambda_j), X_{10}(\Theta_j, \lambda_j), \cdots, X_{NN}(\Theta_j, \lambda_j),$$
$$Y_{11}(\Theta_j, \lambda_j), \cdots, Y_{NN}(\Theta_j, \lambda_j)] = (X_{1j}, X_{2j}, \cdots, X_{Mj}) = \mathbf{K}_j \qquad (26b)$$

and subsequently for the $M \times M$ matrix

$$(\mathbf{K}_1, \mathbf{K}_2, \cdots, \mathbf{K}_M)^T = \mathbf{K} \qquad (26c)$$

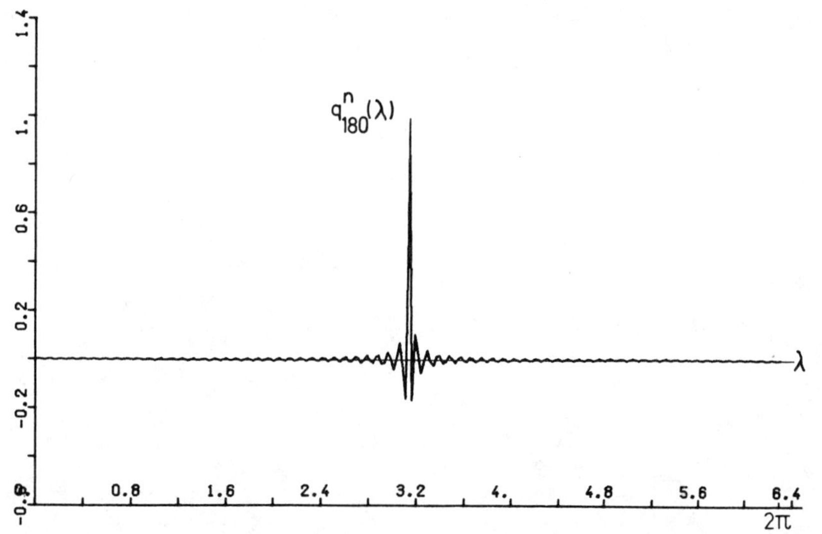

Fig. 10*b*. Trigonometric sampling function (18*b*) for the order $n = 180, j = 180$. The modeling of sampling function condition (7) is better than in Figure 10*a* because of the higher order *n*.

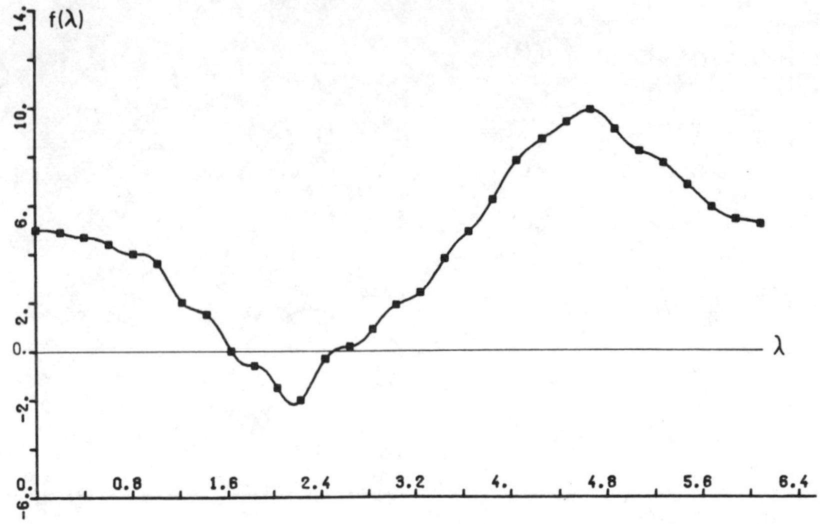

Fig. 11. Example of the interpolation (18*c*) by means of the trigonometric sampling function (18*b*) (Figures 10*a* and 10*b*).

and with

$$\underbrace{(0, 0, \cdots, 1, \cdots, 0)}_{j} = \mathbf{e}_j \tag{26d}$$

(22) and (23) yield the system of linear equations $\mathbf{KA}_j = \mathbf{e}_j$ and from that the solution

$$\mathbf{A}_j = \mathbf{K}^{-1}\,\mathbf{e}_j \tag{26e}$$

The determination of all $M \times M$ parameters $\mathbf{A}_j$, with $j = 1, 2, \cdots, M$, is equivalent to the computation of the inverse matrix of $\mathbf{K}$, because with

$$(\mathbf{A}_1, \mathbf{A}_2, \cdots, \mathbf{A}_M) = \mathbf{A} \tag{26f}$$

and the unit matrix $\mathbf{I}$ the solution becomes (from (26*e*))

$$\mathbf{A} = \mathbf{K}^{-1}\mathbf{I} \tag{26g}$$

However, since an analytical solution for the sampling function parameters A_j^{nm}, B_j^{nm} has not yet been obtained, the computation solution of the system of linear equations represents a difficult numerical problem for a high-order N, even for computers with large capacities. For example, for $N = 36$ it is already necessary to solve 1369 systems (26*e*), each time with 1369 unknowns $\mathbf{A}_j$ of the *j*th sampling function (22) or the inversion of a 1 874 161-element matrix (26*c*).

2. *Spherical Harmonic Sampling Function Models of the Gravity Field*

According to the general sampling function model (8) and by using the base of spherical harmonic sampling functions (22) as well as the given function values

$$f(\Theta_j, \lambda_j) = f_j$$

at the M sampling points (Θ_j, λ_j), from the definition (21*a*) and (21*b*) the approximation model becomes

$$f(P) = f(\Theta, \lambda) = \sum_{j=1}^{M} f_j W_j(\Theta, \lambda) \tag{27}$$

a. Examples for geodetic sampling function modeling. Once the base of the sampling functions $W_j(\Theta, \lambda)$ is computed, the model (27) can theoretically be used for arbitrary sets of function values f_j, $j = 1, 2, \cdots, M$ at the sampling

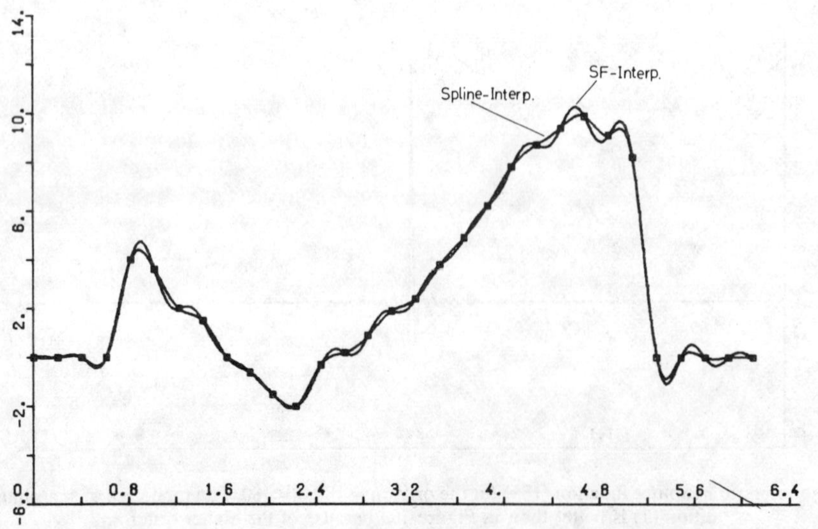

Fig. 12. Comparison of sampling function (SF) interpolation (18*d*) with cubic spline function model.

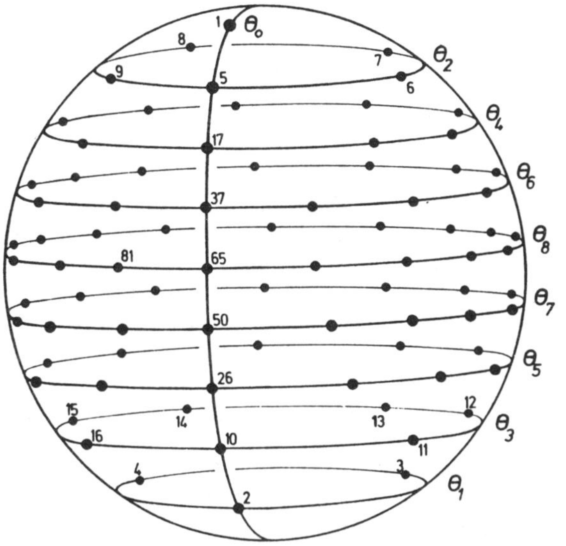

Fig. 13. Diagram of the sampling points (21*a*) and (21*b*), order $N = 8$.

points (Θ_j, λ_j), and since (27) is analytical, any functional of the earth's gravity field can easily be represented.

As a first example the global geoid can be represented by means of M geoid undulations $\Delta h(\Theta_j, \lambda_j) = \Delta h_j$ according to (27) in the form

$$\Delta h(\Theta, \lambda) = \sum_{j=1}^{M} \Delta h_j W_j(\Theta, \lambda)$$

$$= \sum_{j=1}^{M} \Delta h_j \sum_{n=0}^{N} \sum_{m=0}^{n} [A_j^{nm} X_{nm}(\Theta, \lambda) + B_j^{nm} Y_{nm}(\Theta, \lambda)] \qquad (28a)$$

Further geodetic models are possible by using sets of function values f_j from *Lundquist et al.* [1969] and *Schmidt* [1978]: (1) the geopotential V_j computed, for example, from (1*a*) on spheres in the outer space of the earth, (2) global surface densities $\kappa_j = \kappa(\Theta_j, \lambda_j)$, where

$$\kappa(\Theta, \lambda) = \sum_{j=1}^{M} \kappa_j W_j(\Theta, \lambda) \qquad (28b)$$

for use in the gravity potential (cf. Figure 15)

$$V(P) = k^2 \int\int_S \frac{dm}{l} = k^2 \int\int_S \frac{\kappa}{l}\, dS \qquad (28c)$$

(3) the components of the deflection of the vertical ξ_j, η_j, (4) the distribution of gravity anomalies $\Delta g_j = \Delta g(\Theta_j, \lambda_j)$ for a global representation, (5) a set of mean geoidal radii $r_j = r(\Theta_j, \lambda_j)$, derived (e.g., from altimeter measurements) for a global geoid representation, and (6) topographic values $G_j = G(\Theta_j, \lambda_j)$ for a global approximation of the surface of the earth.

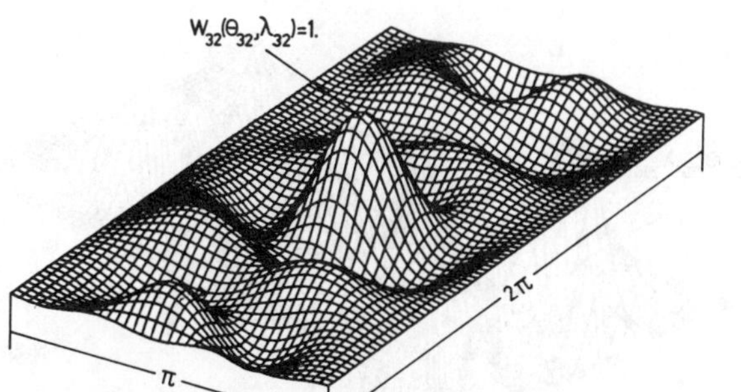

Fig. 14. Example of the spherical harmonic sampling function (22) on a sphere for $N = 5$, $j = 32$; Θ, λ are represented by a rectangle.

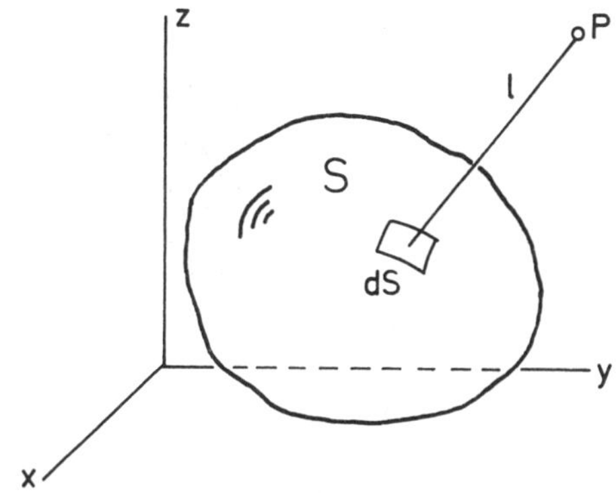

Fig. 15. Gravity potential (equation (28*c*)).

b. Determination of the potential coefficients C_{nm}, S_{nm} of the spherical harmonic gravity model. Since the sampling functions $W_j(\Theta, \lambda)$ in (27) and (28*a*) are defined as a spherical harmonic expansion (22), a very efficient application for the determination of the parameters of spherical harmonic models seems to be possible [*Giacaglia and Lundquist,* 1972]: Starting from the representation of a function $f(\Theta, \lambda)$ by the spherical harmonics (1*b*)

$$f(\Theta, \lambda) = \sum_{n=0}^{N} \sum_{m=0}^{n} [a_{nm} X_{nm}(\Theta, \lambda) + b_{nm} Y_{nm}(\Theta, \lambda)] \qquad (29)$$

and after comparison with the sampling function model (27)

$$f(\Theta, \lambda) = \sum_{j=1}^{M} f_j \left[\sum_{n=0}^{N} \sum_{m=0}^{n} (A_j^{nm} X_{nm}(\Theta, \lambda) + B_j^{nm} Y_{nm}(\Theta, \lambda) \right] \qquad (30)$$

the coefficients of (29) and (30) show the relation

$$\left\{ \begin{matrix} a_{nm} \\ b_{nm} \end{matrix} \right\} = \sum_{j=1}^{M} f_j \left\{ \begin{matrix} A_j^{nm} \\ B_j^{nm} \end{matrix} \right\} \qquad (31)$$

As a geodetic example the determination of the geopotential coefficients C_{nm}, S_{nm} in (1*a*) is demonstrated according to (29)–(31). The spherical harmonic expansion for geoid undulations is [*Heiskanen and Moritz,* 1967]

$$\Delta h(\Theta, \lambda) = \frac{k^2 M}{R\gamma} \sum_{n=0}^{N} \sum_{m=0}^{n} [\Delta C_{nm} X_{nm}(\Theta, \lambda) + \Delta S_{nm} Y_{nm}(\Theta, \lambda)] \qquad (32)$$

with $\Delta C_{nm} = C_{nm}$, $\Delta S_{nm} = S_{nm}$, except $\Delta C_{20} = C_{20} - C_{20}'$ and $\Delta C_{40} = C_{40} - C_{40}'$ with C_{40}', C_{20}' the potential coefficients of the normal gravity field model. The result of comparing (32) with the sampling function model (28*a*) is

$$\left\{ \begin{matrix} C_{nm} \\ S_{nm} \end{matrix} \right\} = \frac{R\gamma}{k^2 M} \sum_{j=1}^{M} \Delta h(\Theta_j, \lambda_j) \left\{ \begin{matrix} A_j^{nm} \\ B_j^{nm} \end{matrix} \right\} \qquad (33)$$

This means that the potential coefficients C_{nm}, S_{nm}, which are of great interest in geodesy, for example, in (1*a*), should be computable only by summation of the products of the given point parameters Δh_j and the given sampling function coefficients A_j^{nm}, B_j^{nm} (equation (26*g*)). Similar relations for the C_{nm}, S_{nm} can be derived theoretically, for example, for a set of gravity anomalies Δg_j, surface densities κ_j, geoid radii r_j [*Giacaglia and Lundquist,* 1972]. But real computations of the potential coefficients C_{nm}, S_{nm} using gravity anomalies Δg_j or

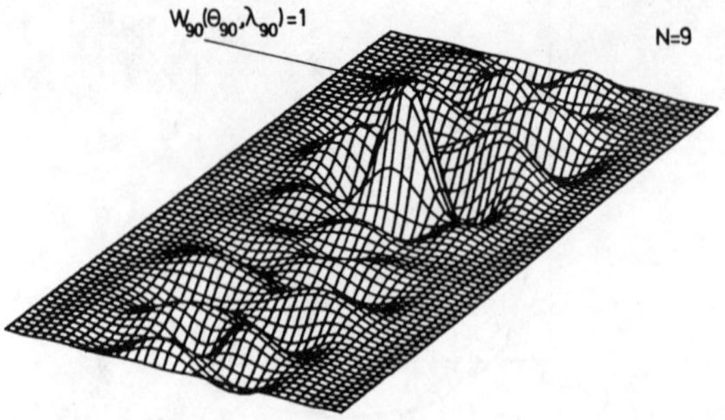

Fig. 16*a*. Sampling function $W_j(\Theta, \lambda)$ according to (22) for order $N = 9$, $j = 90$, still showing a relatively smooth surface outside the corresponding sampling point area.

geoid undulations Δh_j (equation (33)) have shown that the method (31) using (33) is not practically feasible for high-order N. The reason is given in section D3.

3. *Problems of the Spherical Harmonic Sampling Function Method*

It is not possible either to produce a smooth shape of the sampling function $W_j(\Theta, \lambda)$ as well as a smooth interpolation surface according to the general spherical harmonic sampling function model (27) or to compute plausible coefficients of a spherical harmonic expansion (31), such as the potential coefficients C_{nm}, S_{nm} from, for example, (33), especially for higher-order N (i.e., $N > 10$) [*Schmidt,* 1978]. The reason is that the sampling function approximation (27) really represents only a spherical harmonic expansion passing through a more or less arbitrary (i.e., only approximately equidistant) set of function values f_j, and as is well known, this model is not able to guarantee a smooth approximation surface and consequently, for example, a plausible computation of the C_{nm}, S_{nm}. To prove this fact, the following representation, namely, the sampling function model (27), can be given, using

$$[X_{00}(\Theta, \lambda), X_{10}(\Theta, \lambda), \cdots, Y_{NN}(\Theta, \lambda)] = \mathbf{K}_{nm} \quad (34a)$$

and (26*a*) in a first step as

$$f(\Theta, \lambda) = \sum_{j=1}^{M} f_j \mathbf{K}_{nm} \mathbf{A}_j$$

This gives, with

$$(f_1, f_2, \cdots, f_M) = \mathbf{f} \quad (34b)$$

and (26*f*),

$$f(\Theta, \lambda) = \mathbf{K}_{nm}\mathbf{A}\mathbf{f}$$

and finally, using the solution (26*g*) for **A**,

$$f(\Theta, \lambda) = \mathbf{K}_{nm}\mathbf{K}^{-1}\mathbf{f} \quad (34c)$$

The same result is obtained for an expansion into a series of spherical harmonics from a set of point parameters f_j or sampling function values. The general spherical harmonic expression (29), using

$$(a_{00}, a_{10}, \cdots, b_{NN})^T = \mathbf{a}_j \quad (35a)$$

and (34*a*), provides

$$f(\Theta, \lambda) = \mathbf{K}_{nm}\mathbf{a}_j \quad (35b)$$

The $M = (N + 1)^2$ parameters $\mathbf{a}_j$ in (35*b*) result from the systems of equations (29) by inserting the M f_j (equation (34*b*)) on the left side at the M sampling points (Θ_j, λ_j); that is, from

$$\mathbf{f} = \mathbf{K}\mathbf{a}_j$$

the parameters $\mathbf{a}_j$ become

$$\mathbf{a}_j = \mathbf{K}^{-1}\mathbf{f} \quad (35c)$$

Equation (35*c*) combined with (35*b*) yields

$$f(\Theta, \lambda) = \mathbf{K}_{nm}\mathbf{K}^{-1}\mathbf{f} \quad (35d)$$

the same solution as (34*c*); that is, the spherical harmonic sampling function model (27) and the spherical harmonic expansion (29), concerning the function values f_j, are identical. A smooth approximation model (27) cannot be expected.

The sampling function $W_j(\Theta, \lambda)$ itself is also, according to (22), nothing but a spherical harmonic expansion from the 'values' (26*d*). Therefore the $W_j(\Theta, \lambda)$, too, show unrealistic amplitudes for higher-order N, disagreeing with (23) and Figure 14 (see Figures 16*a* and 16*b*).

The relatively high order of $N = 36$ produced maximal values for the $W_j(\Theta, \lambda)$ of even ± 20, although the one-dimensional trigonometric sampling function (18*b*) is valid, according to (25), along the parallels of latitude Θ_k defined in (21*a*).

In fact, it is possible to accommodate the sampling function $W_j(\Theta, \lambda)$ to the 'ideal figure' (i.e., compliance with (23) and approximately zero values between the sampling points, as is indicated in Figure 14) by using additional conditions for points in the areas between the sampling points and by following a least squares adjustment procedure to compute the sampling function parameters A_j^{nm}, B_j^{nm}. But after that, neither the sampling function model (27) nor, for example, the determination of the geopotential coefficients C_{nm}, S_{nm} (equation (33)) has any practical relevance.

4. *Possible Improvement of the Sampling Function Model on a Sphere*

The reason for the failure of the sampling function model (27) defined by *Lundquist et al.* [1969] and *Giacaglia and Lundquist* [1972] is assumed to be that the sampling point distribution (21*a*) and (21*b*) is only approximately equidistant, although an equidistant grid of points is theoretically indispensable. But as is known, it is not possible, apart from a few cases, to distribute equidistantly any arbitrary number of points on a sphere.

Because of that, a more equidistant point distribution different from (21*a*), concerning the parallels of latitude Θ_k, was introduced, for example,

$$\Theta_k^{(\text{new})} = \Theta_k - \frac{2\pi}{M} \cdot 4 \cos [\Theta_k (1 - \cos^2 \Theta_k)] \quad (36)$$

for investigating the fundamental influence of such a point distribution. Indeed, by using (36), for example, the mean am-

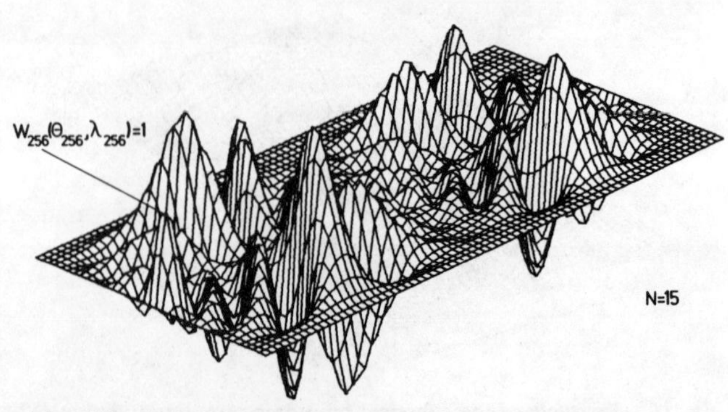

Fig. 16*b*. Spherical harmonic sampling function (22) for $N = 15$, $j = 256$. The higher order, compared with Figures 16*a* and 14 gives unrealistic values (i.e., $>+1$), disagreeing with conditions (23).

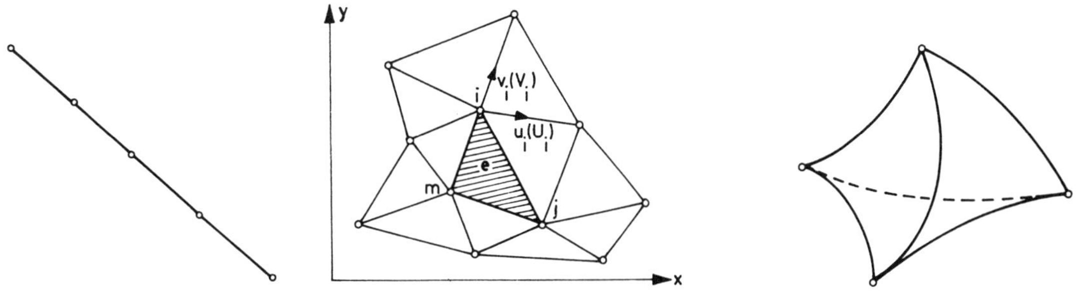

Fig. 17. Examples of finite elements: (left) one dimensional, (middle) two dimensional, and (right) three dimensional.

plitudes of the sampling functions $W_j(\Theta, \lambda)$ decrease (for an order $N = 15$ in Figure 16*b* the diminution was about 50%), so that a smoother approximation model (27) is to be expected.

Following this example, a more comprehensive investigation could be successful. Starting from (21*a*) and (21*b*), an iterative displacement of the sampling point in latitude as well as in longitude may result in a better approximate equidistant distribution to improve the proposed earth gravity models (section D2).

E. FINITE ELEMENT METHOD

The primary use of the finite element method consists of solving numerous stress and deformation problems in elasticity and structural mechanics. The method is based on the idea of dividing a complex structure into a certain number of relatively simple elements, so that a computation becomes possible. From the discrete element solution an attempt is made to construct a numerical result for the whole system, considering conditions of continuum and equilibrium [e.g., *Dankert,* 1977; *Collatz and Krabs,* 1977]. An important advantage over other numerical methods can be seen in the possibility of describing a great variety of structures by using the flexible modification of the elements in shape and dimension, such as that shown in the first applications in engineering statics. Gradually, this method was used to solve many technical and physical problems [*Zienkiewicz,* 1975], for example, those in engineering fields, such as aeronautical, automobile, mechanical, or shipbuilding applications [*Robinson,* 1973]. The theoretical fundamentals as well as a great number of practical examples are shown, to mention only a small selection, for example, by *Gallagher* [1975], *Hahn* [1975], *Huebner* [1975], *Schwarz* [1980], *Ural* [1973], and *Zienkiewicz* [1972, 1975].

1. *Establishment of the Finite Element Method for Geodetic Use*

In the wide field of geodesy there are also many possibilities for using the finite element method as an approximation model for pure interpolation (also as a replacement for the time-wasting integration modes) or perhaps, since the conventional spherical harmonic model (cf. (1*a*)) is unable to yield the high accuracy of partially available and future observations, to develop new models for detailed information in geodesy (e.g., topography, geoid, and gravity field) or in some geodetic applications of geophysics (e.g., crustal deformation, plate tectonics, and tidal models).

In geodetic and geophysical fields the method of finite elements has been applied (e.g., by *Richardson and Hopkins* [1978]) to the computation of gravity disturbance components using mean free-air gravity anomalies and finite elements of '1° × 1° × 9' km dimension. Further, *Szameitat* [1979] used elements of sphere segments to apply the Love theory to form an elastic earth model from tidal forces. A theoretical representation of the finite element method in interpolation theory for the application in the Stokes and Vening Meinesz integrals is demonstrated by *Moritz* [1978] and compared with spline approximation. For the representation of the gravity field, *Junkins and Engels* [1979] have developed a special finite element model for the computation of the gradients of the geopotential that is highly efficient because of its speed advantage, especially in computing satellite orbits [*Junkins,* 1977].

Thus it seems to be in order to pay attention to the development of the finite element method for gravity field representation. In the following discussion an attempt is therefore made to demonstrate briefly the fundamentals of the finite element method for an applied interpolation model and to show the connection with the sampling function approximation model (sections B–D).

2. *Basic Principle of the Finite Element Method*

In structural mechanics it is often necessary to determine the distribution or the field of stresses in an elastic continuum caused by internal and external forces which are everywhere in equilibrium, whereby the resulting displacement should be continuous. Applying the finite element method, this continuum is divided into a number of elements by means of assumed points, lines, or surfaces connected with each other by the variables at specified points called nodes or nodal points (cf. Figure 17).

The general mathematical formulation of the finite element method is explained for the two-dimensional case, following the notation of *Zienkiewicz* [1975]: The forces, for example, V_i, U_i, at the node i cause the displacements

$$v_i(V_i) = v_i(x, y), \; u_i(U_i) = u_i(x, y)$$

or, in vector notation,

$$\{\delta_i\} = \begin{Bmatrix} v_i(V_i) \\ u_i(U_i) \end{Bmatrix} \tag{37a}$$

For all nodes of the element e the displacements are

$$\{\delta\}^e = \begin{Bmatrix} \delta_i \\ \delta_j \\ \delta_m \\ \vdots \end{Bmatrix} \tag{37b}$$

The displacement of each point in the interior of the element e, $f(x, y)$, is defined as

$$\{f\} = [N] \begin{Bmatrix} \delta_i \\ \delta_j \\ \delta_m \\ \vdots \end{Bmatrix} \tag{38}$$

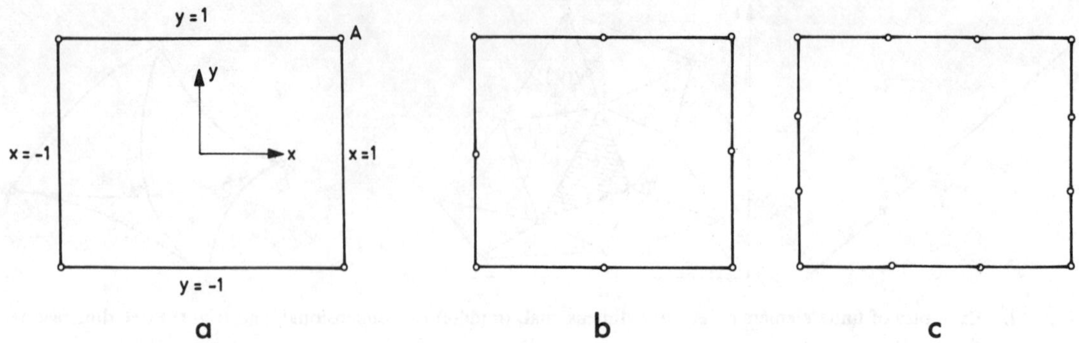

Fig. 18. Rectangular elements of the boundary node class (serendipity class) of different order: (*a*) linear, (*b*) quadratic, and (*c*) cubic.

with the so-called shape functions N. For reproduction of the nodal point displacement the property for the 'arbitrary' shape function must hold:

$$N_i(x_i, y_i) = \mathbf{I} \qquad (39a)$$

where I is a unit matrix, and

$$N_i(x_j, y_j) = N_i(x_m, y_m) = 0$$

that is,

$$N_k(x_1, y_1) = \delta_{k1} \qquad (39b)$$

where δ_{k1} is the Kronecker symbol. Equation (39*b*) is identical to the condition (7) that generally characterizes the sampling functions, practically demonstrated for the one- and two-dimensional sampling functions in sections C and D, respectively; for example,

$$W_j(\Theta_i, \lambda_i) = \delta_{ji}$$

Thus the shape functions in the finite element method conform to the method of the sampling functions as a base for the model (8).

a. Examples of two-dimensional shape functions for finite elements. The choice of arbitrary shape functions appropriate to the particular practical problem presupposes great experience on the part of the user. From the great number of possible developments, only an example of shape functions of rectangular elements of the serendipity class (named for the Prince of Serendip [*Zienkiewicz,* 1972]), is demonstrated (Figure 18).

The development of the shape function depends on the values of the nodal points. For the first element (Figure 18*a*) it starts from the product

$$\tfrac{1}{4}(x + 1)(y + 1)$$

which provides, for example, for the node A, exactly the value +1, but at all other nodal points the value is zero. Applying the variables

$$x_0 = xx_i \qquad y_0 = yy_i \qquad (40a)$$

all shape functions can be represented by

$$N_i = \tfrac{1}{4}(1 + x_0)(1 + y_0) \qquad (40b)$$

For the quadratic case (Figure 18*b*) the shape functions are defined as

Corner nodes

$$N_i = \tfrac{1}{4}(1 + x_0)(1 + y_0)(x_0 + y_0 + 1) \qquad (41a)$$

Middle side node

$$x_i = 0 \qquad N_i = \tfrac{1}{2}(1 - x^2)(1 + y_0)$$
$$y_i = 0 \qquad N_i = \tfrac{1}{2}(1 + x_0)(1 - y^2) \qquad (41b)$$

The next step in the resolution is the cubic case (Figure 18*c*), indicating the characteristic property of the finite element method that with higher order the model becomes more complicated, for example,

Corner nodes

$$N_i = \tfrac{1}{32}(1 + x_0)(1 + y_0)[9(x^2 + y^2) - 10] \qquad (42a)$$

Middle side nodes

$$x_i = \pm 1 \qquad y_i = \pm\tfrac{1}{3} \qquad N_i = \tfrac{9}{32}(1 + x_0)(1 - y^2)(1 + 9y_0)$$
$$x_i = \pm\tfrac{1}{3} \qquad y_i = \pm 1 \qquad N_i = \tfrac{9}{32}(1 + y_0)(1 - x^2)(1 + 9x_0) \qquad (42b)$$

As an additional example of a shape function development according to *Zienkiewicz* [1972], the possibility of the base of Lagrange polynomial is demonstrated (defined as a classical sampling function in (14*a*)) for example, $L_i(x)$, which results for the point of row j and column i in Figure 19, with n, m the partition along the two directions y and x. According to the results of the one-dimensional sampling function model (cf. (14*b*) and Figures 3 and 4), in this Lagrangian approach the disadvantage of unrealistic amplitudes at the boundaries also appears.

Similar to these rectangular examples of finite elements, the

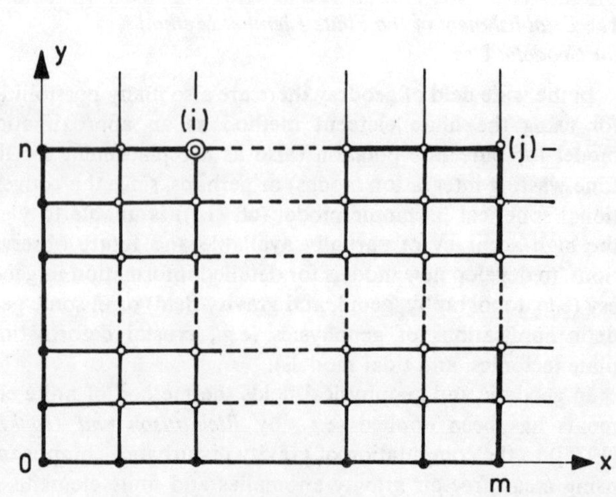

Fig. 19. Structure of a finite element for using Lagrange polynomial shape functions.

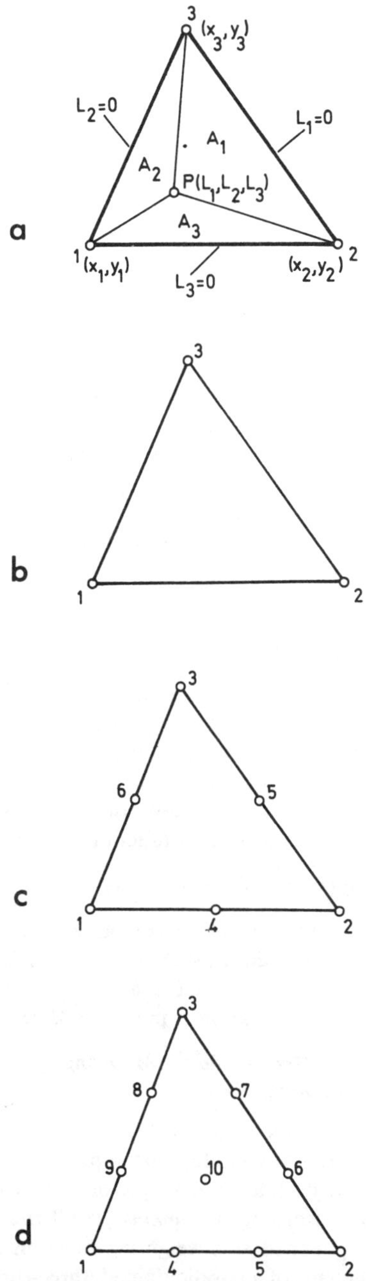

Fig. 20. (*a*) Triangular elements of Langrangian polynomial shape functions. (*b*) Linear, three nodes. (*c*) Quadratic, six nodes. (*d*) Cubic, 10 nodes.

triangular elements naturally play a major role because of the possibility of their representing more complex structures.

To avoid, for example, this disadvantage of Lagrangian interpolation or even singularity problems, the use of natural coordinates will be demonstrated: An arbitrary point P in the interior of the triangular element is fixed by the coordinates L_1, L_2, L_3 (A is the area of the triangle) [e.g., *Zienkiewicz,* 1975; *Dankert,* 1977] (Figure 20*a*)

$$L_1 = A_1/A \qquad L_2 = A_2/A \qquad L_3 = A_3/A \tag{43a}$$

with

$$2A = x_{21}y_{31} - x_{31}y_{21} \qquad x_{ij} = x_i - x_j \qquad y_{ij} = y_i - y_j$$

and the property

$$L_1 + L_2 + L_3 = 1 \tag{43b}$$

Conversely, for $P(x, y)$,

$$x = x_1L_1 + x_2L_2 + x_3(1 - L_1 - L_2)$$

$$y = y_1L_1 + y_2L_2 + y_3(1 - L_1 - L_2)$$

is valid.

Some examples of shape functions N_i for a triangular element on the basis of natural coordinates [e.g., *Dankert,* 1977; *Zienkiewicz,* 1975] are as follows (Figures 20*b*–20*d*):

Linear (three nodes)

$$N_1 = L_1 \qquad N_2 = L_2 \qquad N_3 = L_3 \tag{43c}$$

Quadratic (six nodes): corner nodes

$$N_j = (2L_j - 1)L_j \qquad j = 1, 2, 3$$

Quadratic: middle side nodes

$$N_4 = 4L_1L_3 \qquad N_5 = 4L_2L_1 \qquad N_6 = 4L_3L_2 \tag{43d}$$

Cubic (10 nodes): corner nodes

$$N_j = \tfrac{1}{2}(3L_j - 1)(3L_j - 2)L_j \qquad j = 1, 2, 3$$

Cubic: middle side nodes

$$\begin{aligned} N_4 &= \tfrac{9}{2}L_1, L_2(3L_1 - 1) \\ N_5 &= \tfrac{9}{2}L_1L_2(3L_2 - 1) \\ N_6 &= \tfrac{9}{2}L_2L_3(3L_2 - 1) \end{aligned} \tag{43e}$$

and so on

Cubic: interior node

$$N_{10} = 27L_1L_2L_3$$

b. Shape functions of three-dimensional finite elements. In further development of the two-dimensional elements of Figure 18 the three-dimensional finite elements of the serendipity class in the form of a right prism can be constructed (Figure 21). For the normalized coordinates the shape function of the node i shows, according to Figure 18*a* and the formula (40*b*) and considering (40*a*), for the three-dimensional elements (Figures 21*a*–21*c*),

Linear

$$N_i = \tfrac{1}{8}(1 - x_0)(1 + y_0)(1 + z_0) \tag{44a}$$

Quadratic: corner node

$$N_i = \tfrac{1}{8}(1 + x_0)(1 + y_0)(1 + z_0)(x_0 + y_0 + z_0 - 2) \tag{44b}$$

Quadratic: middle side node

$$N_i = \tfrac{1}{4}(1 - x^2)(1 + y_0)(1 + z_0) \tag{44c}$$

at, for example,

$$x_i = 0 \qquad y_i = \pm 1 \qquad z_i = \pm 1$$

Cubic: corner node

$$N_i = \tfrac{1}{64}(1 + x_0)(1 + y_0)(1 + z_0)[9(x^2 + y^2 + z^2) - 19] \tag{44d}$$

Cubic: middle side node

$$N_i = \tfrac{9}{64}(1 - x^2)(1 + 9x_0)(1 + y_0)(1 + z_0) \tag{44e}$$

at, for example,

$$x_i = \pm\tfrac{1}{3} \qquad y_i = \pm 1 \qquad z_i = \pm 1$$

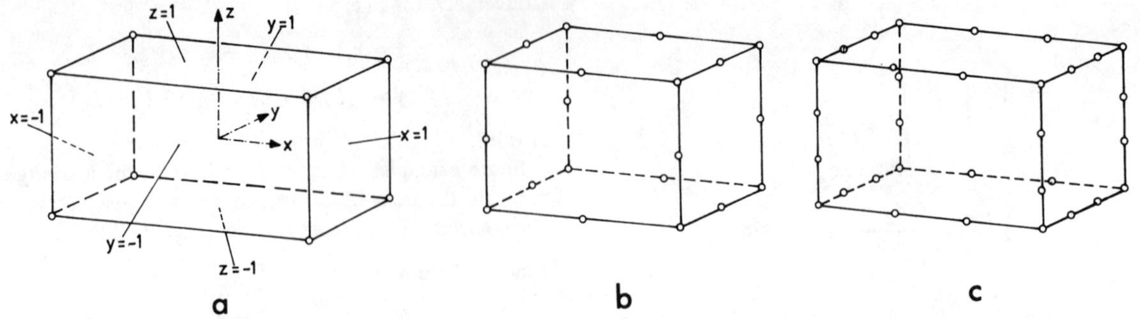

Fig. 21. Three-dimensional elements: (*a*) linear, eight nodes, (*b*) quadratic, 20 nodes, and (*c*) cubic, 32 nodes.

Similar to the two-dimensional case the complexity of the interpolating shape functions (44*a*)–(44*e*) increases with higher resolution, that is, with the growing number of nodal points. Nevertheless, by inserting the coordinates of the node $K_j(x_j, y_j, z_j)$ in (44*a*)–(44*e*) they always keep the property

$$N_i(K_j) = \delta_{ij} \tag{45}$$

identical to the condition for the one- (section C) and two-dimensional sampling functions (23). In this interpretation the shape functions in the finite element method may be seen as examples of one-, two-, or even three-dimensional locally valid sampling functions.

Just as in developing the finite element of the Lagrangian class (Figure 19) into a three-dimensional structure (additionally, for example, in the *k*th layer, the element partition in the *z* direction will be of the order of 1), it is possible to define, according to (43), the shape function

$$N_{ijk} = L_i^n(x)L_j^m(y)L_k^l(z) \tag{46}$$

In principle, similar developments are possible for the tetrahedron, the triangular prism, or the general hexahedron [*Huebner,* 1975], as well as for all the endless variety of one-, two-, and three-dimensional elements with curved boundaries, as many applications [e.g., *Szameitat,* 1979] show.

3. *Finite Element Method for Interpolation*

The shape functions $[N]$ were used in (38) for the displacement of a point within an element caused by the nodal forces. According to the displacement model (38) it is possible to approximate a function f in the interior of an element on the basis of samples of function values δ_i in the nodes and the corresponding shape functions N_i by

$$\{f\} = [N]\{\delta\} = [N_1, N_2, \cdots, N_n]\begin{Bmatrix}\delta_1\\ \delta_2\\ \vdots\\ \delta_n\end{Bmatrix} \tag{47a}$$

Some examples of the shape functions which are to be determined are given in section E2*a* (rectangular and triangular elements) and section E2*b* (three-dimensional right prism).

Likewise, the shape functions may be considered as a possibility for defining a set of base functions g_k for an approximation $\bar{f}$ [*Dankert,* 1977], for example, in the one-dimensional case based on given or computed values f_k in the form [*Moritz,* 1978]

$$\bar{f}(x) = \sum_{k=0}^{n} f_k g_k(x) \tag{47b}$$

In the sense of the finite element method the single $g_k(x)$ is a function which supports a function $\bar{f}(x)$ in a finite (compact) region and is zero outside this region. Because these $g_k(x)$ should be as simple as possible, polynomials would be best suited for this purpose [*Moritz,* 1978].

As an example of an important geodetic use, a rectangular element, that is, $n = 3$ in (37*b*), is given in Figure 22 with the simplest form of $g_k(x, y)$,

$$g_k(x, y) = a_0 + a_1x + a_2y + a_3xy \tag{48a}$$

which represents a hyperbolic paraboloid. This function is linear along all sides of the element and therefore gives a continous transition from one element to the next.

Moreover, the important fact exists that the base function (48*a*) for a rectangular element can also be represented as the product of two one-dimensional functions in x and y,

$$g_k(x, y) = (b_0 + b_2x) \cdot (c_0 + c_1y) \tag{48b}$$

as shown in the examples (40*a*)–(43) as well as in those for three-dimensional elements, (44*a*)–(44*e*) and (46).

A comparison of the finite element method with the cubic spline function approximation is made by *Moritz* [1978].

4. *Finite Element Method and the Modeling of the Gravity Field of the Earth*

The approximation techniques (47) can also be seen as a locally valid but piecewise continuous approximation $\bar{f}(x, y, z)$ of a function $f(x, y, z)$, as an example of a three-dimensional representation. According to *Junkins* [1977] the approximation $\bar{f}(x, y, z)$ is defined as a weighting function approach of linear finite elements of a parallelopiped, three-dimensional in x, y, z (section E2*b*, Figure 21*a*) with eight values $f_{ijk}(x, y, z)$, $i, j, k = 0, 1$, concerning the eight nodes and the weight function $W_{ijk}(x, y, z)$,

$$\bar{f}(x, y, z) = \sum_{i=0}^{1}\sum_{j=0}^{1}\sum_{k=0}^{1} f_{ijk}(x, y, z) \cdot W_{ijk}(x, y, z) \tag{49}$$

The $W_{ijk}(x, y, z)$ functions are constructed so that $\bar{f}$ is continuous across their mutual boundaries of validity and so that the order m of partial differentiation to which continuity is desired can be taken into consideration. As an example, *Junkins and Engels* [1979] define the $W_{111}(x, y, z)$ function with the notation $X_1 = x$, $X_2 = y$, $X_3 = z$ for the order m as

$$\prod_{i=1}^{3}\left\{\frac{(2m+1)!(-1)^m}{(m!)^2}\sum_{r=0}^{m}\frac{(-1)^r r^m}{2m-r+1}X_i^{2m-r+1}\right\} \tag{50}$$

The other $W_{ijk}(x, y, z)$ functions follow from (50), for example, $W_{001}(X_1, X_2, X_3) = W_{111}(1 + X_1, 1 - X_2, X_3)$, and so on.

It is remarkable that the maximum value of $W_{ijk}(x, y, z)$, occurring at the centroid of the corresponding element, is unity and has with that value the identical property of the sampling function definition (7), demonstrated in section C and for a two-dimensional case theoretically by (23) (Figure 14).

Furthermore, the relation

$$\sum_{i=0}^{1} \sum_{j=0}^{1} \sum_{k=0}^{1} W_{ijk}(x, y, z) = 1 \tag{51}$$

is valid similarly for the base of the trigonometric sampling function (18*c*) as well as for the spherical harmonic sampling function (24).

However, from the geodetic point of view it seems to be an important step that *Junkins* [1977] has applied the model (49) to represent both the global and the fine structure features of the geopotential and its gradients.

The application of (49) can be used as an interpolation model [e.g., *Junkins,* 1977] for the high-speed computation of the geopotential gradients necessary for satellite orbit determination as well as for modeling the geopotential in the outer part of the earth. In the later case, based on finite elements of 1° × 1° × 300 km and 5° × 5° × 300 km for gravity disturbance 'observations' derived from a given point mass model and the already defined and computed base function $W_{ijk}(x, y, z)$, the values of f_{ijk} can be determined via the standard weighted least squares techniques. Both applications show the high flexibility of the finite element method, allowing a choice of convenient base functions (also, for example, the Chebyshev polynomials, demonstrated as a possibility for a sampling function model (15*a*) and (15*b*)) or different cell sizes.

As the possibility of high-speed computation of the gravity field and its gradient is of actual interest, it seems desirable to investigate the finite element method with shape or weight functions of other types, such as trigonometric sampling functions (18*b*) or covariance functions.

Basically, it seems to be very important to use the triangular finite element (section E2*a*) not only for interpolation but also for approximation models of geodetic quantities, on the basis of the generally irregular observation distribution on the continent, on the sea, and in the outer space of the earth. Moreover, to get a smoother interpolation in the system of single elements, it would be very desirable to develop observational techniques as well as the corresponding finite element models which are able to include the first derivatives of the approximated function, existing in the nodes. Likewise, for a detailed representation of the geoid or, especially, the earth's gravity field outside its surface it would be necessary to investigate a

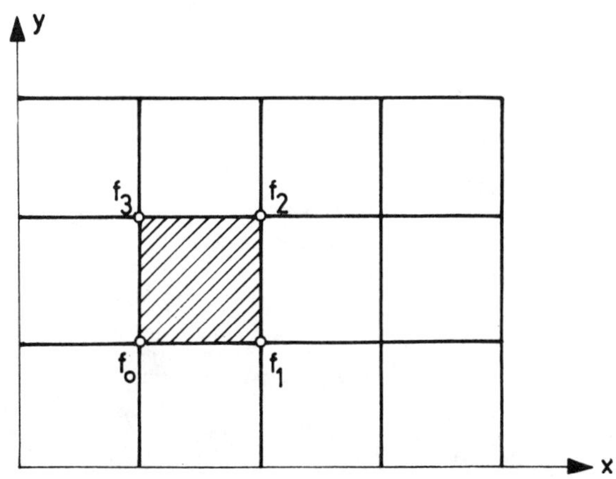

Fig. 22. Rectangular element with function values at the nodes.

possible solution for the corresponding finite element models considering the conditions of the Laplace equation, as was formulated in the finite element theory for general field problems, for example, problems in fluid mechanics [*Zienkiewicz,* 1975].

F. CONCLUSIONS

The sampling function models (sections C and D) as well as the finite element method (section E) are based on a similar fundamental idea; obtaining a realistic and effective possibility for interpolation within well-defined regions or space elements, by means of a set of convenient base functions and appropriate parameter values, respectively.

But as was demonstrated above, high-resolution global sampling function models for the gravity field, for example, on the basis of spherical harmonics (Section D2), show unsolvable problems at present because of (1) numerical difficulties, even with large computer facilities, in determining the sampling functions on a sphere from the solution of a large system of linear equations (26g) and (2) the problem of defining practically a nearly equidistant sampling point distribution on a sphere.

As a result it seems to be appropriate to represent the long-wavelength structure of the gravity field of the earth by means of existing global models, such as the standard earth models, and to model the detailed structure on the basis of locally applicable methods (e.g., mass points) or, as was indicated in section E, with the aid of the locally valid finite element method. Referring to the latter, the problem arises that the resulting parameters are not globally valid and that the Laplace equation is not fulfilled. Nevertheless, the advantage of the finite element method can be seen in its flexibility, with respect to the size and shape of the applied elements, to represent even complex structures and in the possibility of establishing a global as well as a local model. For that reason one can expect that the finite element method will offer an effective way of solving some problems in geodetic and geophysical modeling, especially the representation of the fine structure of the gravity field of the earth.

REFERENCES

Balmino, G., A new point masses model for the earth's gravity field, paper presented at the International Symposium on Earth Gravity Models and Related Problems, AGU, St. Louis, Mo., Aug., 1972.

Bjerhammar, A., A Dirac approach to physical geodesy, *Z. Vermess. Wesen, 2,* 41–44, 1976.

Brown, R. D., Geopotential modeling by binary sample functions, report, contract NAS5-11999, 22 pp., Comput. Sci. Corp., Arlington, Va., 1975.

Collatz, L., and W. Krabs, *Approximationstheorie,* 208 pp., Springer, New York, 1977.

Dankert, J., *Numerische Methoden der Mechanik,* 318 pp., Springer, New York, 1977.

De Boor, C., Bicubic spline interpolation, *J. Math. Phys. Cambridge Mass., 41,* 212–218, 1962.

Gallagher, R. H., *Finite Element Analysis Fundamentals,* Prentice-Hall, Englewood Cliffs, N. J., 1975.

Giacaglia, G. E. O., Trigonometric interpolation, *Celestial Mech., 1,* 360–367, 1970.

Giacaglia, G. E. O., and C. A. Lundquist, Sampling functions for geophysics, *Smithson. Astrophys. Observ. Spec. Rep., 344,* 94 pp., 1972.

Hahn, H. G., *Methode der finiten Elemente in der Festigkeitslehre,* 460 pp., Akademie, Frankfurt am Main, 1975.

Hardy, R. L., Geodetic applications of multiquadratic analysis, *Allg. Vermess. Nachr., 10,* 398–406, 1972.

Heiskanen, W. A., and H. Moritz, *Physical Geodesy,* 364 pp., W. H. Freeman, San Francisco, Calif., 1967.

Heitz, S., Zur interpolatorischen Darstellung von Funktionen, *Astron. Geod. Arb. 33,* pp. 109–116, Veröff. der Bayer. Komm. für die Intern. Erdmessung der Bayer. Akad. der Wiss., Munich, 1975.

Huebner, K. H., *The Finite Element Method for Engineers,* John Wiley, New York, 1975.

Junkins, J. L., Development of finite element models for the earth's gravity field, report, contract DAAG 53-76-C-0067, 27 pp., Comput. Sci. Lab., Fort Belvoir, Va., March 1977.

Junkins, J. L., and R. C. Engels, The finite element approach in gravity modelling, *Manuscr. Geod., 4,* 185–206, 1979.

Koch, K. R., and B. U. Witte, Earth's gravity field represented by a simple layer potential from Doppler tracking of satellites, *J. Geophys. Res., 76,* 8471–8479, 1971.

Lanczos, C., *Applied Analysis,* 539 pp., Pitman, London, 1957.

Laugwitz, D., *Ingenieurmathematik IV, Hochschultaschenbücher,* vol. 62/62a, 196 pp., Bibliographisches Institut, Mannheim, 1967.

Lundquist, C. A., G. E. O. Giacaglia, K. Hebb, and S. G. Mair, Possible geopotential improvement from satellite altimetry, *Smithson. Astrophys. Observ. Spec. Rep., 294,* 100 pp., 1969.

Moritz, H., Advanced least-squares methods, *Rep. 203,* 43 pp., Dep. of Geod. Sci., Ohio State Univ., Columbus, 1967*a*.

Moritz, H., Least-squares collocation as a gravitational inverse problem, *Rep. 249,* 28 pp., Dep. of Geod. Sci., Ohio State Univ., Columbus, 1976*b*.

Moritz, H., *Approximation Methods in Geodesy,* H. Wichmann, Karlsruhe, Federal Republic of Germany, 1978.

Richardson, D. A., and J. Hopkins, Gravity disturbance components computed at aircraft altitude using the finite element method, paper presented at Fall Meeting, AGU, San Francisco, Calif., Dec. 4–8, 1978.

Robinson, J., *Integrated Theory of Finite Element Methods,* 428 pp., John Wiley, New York, 1973.

Rummel, R., Determination of short-wavelength components of the gravity field from satellite-to-satellite tracking or satellite gradiometry—An attempt to an identification of a dedicated gravity satellite, *Manuscr. Geod., 4,* 107–148, 1979.

Schmidt, H. F., Untersuchungen zur Anwendung von Sampling-Funktionen in der Geodäsie, *Rep. 241,* 148 pp., Dtsch. Geod. Komm., Munich, 1978.

Schwarz, H. R. *Methode der finiten Elemente,* 320 pp., Teubner Studienbücher, Mathematik, Stuttgart, 1980.

Szameitat, H.-J., Modell einer elastischen Erde auf der Grundlage finiter Kugelschalen-Elemente, *Rep. 249,* 133 pp., Dtsch. Geod. Komm., Munich, 1979.

Tscherning, C. C., A Fortran IV program for the determination of the anomalous potential using stepwise least squares collocation, *Rep. 212,* 125 pp., Dep. of Geod. Sci., Ohio State Univ., Columbus, 1974.

Ural, O., *Finite Element Method, Basic Concept and Applications,* 272 pp., Intext Educational Publishers, New York, 1973.

Vinti, J. P., Representation of the earth's gravitational potential, *Celestial Mech., 4,* 348–367, 1971.

Whittaker, E. T., On the functions which are represented by the expansions of the interpolation-theory, *Proc. R. Soc. Edinburgh, 35,* 181–194, 1915.

Zienkiewicz, O. C. *Introductory Lectures on the Finite Element Method,* 98 pp., Springer, New York, 1972.

Zienkiewicz, O. C. *Methode der finiten Elemente,* 496 pp., Carl Hauser, Munich, 1975.

(Received September 3, 1980;
accepted February 24, 1981.)

REVIEWS OF GEOPHYSICS AND SPACE PHYSICS, VOL. 19, NO. 4, PAGES 673–686, NOVEMBER 1981

Planar Geodetic Covariance Functions

SIEGFRIED MEIER

Technische Universität Dresden, Sektion Geodäsie und Kartographie, DDR-8027 Dresden, German Democratic Republic

In the last 20 years statistical methods have been applied in geodesy with considerable success, both in physical geodesy (least squares collocation, variance-covariance propagation) and in geodetic measuring technique (error propagation and interpolation, inertial navigation, adjustment, diagnosis, and design of networks). The geodetic stochastic process is introduced per definition as the representation of a geophysical random field afflicted with observational errors, including the special cases of the (almost) error-free field (physical-geodetic process) and the pure error field (operational-geodetic process). Geodesy is an approximation science, planar approximation being of great practical importance. It can be realized by projection onto a tangential plane to the earth's surface and/or other planes of measurement and computation. The review treats the covariance functions of planar processes for different geodetic problems in the light of literature on the subject and under systematizing aspects—inter alia, gravity field, field of atmospheric refraction, error field; covariance estimation and propagation; ideal statistical structures as homogeneity, isotropy, axisymmetry, but also the physical, statistical, and geophysical significance of the prerequisites and solutions as well as problems that are still unsolved.

CONTENTS

A. INTRODUCTION

Since the first statistical analysis of the earth's gravity field made about 20 years ago [*Kaula,* 1959; *Hirvonen,* 1962], the theory of stochastic processes, especially the correlation theory of stochastic processes, has been very successfully applied in geodesy. The so-called statistical geodesy [*Nash and Jordan,* 1978] received a considerable impetus in the early seventies by the further development and generalization of least-squares estimation methods in physical geodesy [*Krarup,* 1969; *Moritz,* 1970*a, b,* 1972, 1973] and the application of the mathematical optimization to the diagnosis and design of geodetic networks (*Grafarend* [1970*c,* 1972*a, b, c*], *Baarda* [1973], *Bossler et al.* [1973], *Alberda* [1974], and others). The following reviews with extensive bibliographies are pointed out: *Grafarend* [1976], *Moritz* [1978*a*], and *Grafarend et al.* [1979]. Similar to the creation of the statistical theory of turbulence in the thirties [*Taylor,* 1935, 1938; *Kármán,* 1937; *Kármán and Howarth,* 1938] immediately multidimensional processes, especially those on the sphere, were applied according to the geodetic problems set. One-dimensional processes (time series) were introduced only later into the geodetic measuring techniques.

After an outline of the statistical fundamentals (sections B1, B2), the geodetic stochastic process is introduced per definition as the representation of a geophysical random field afflicted with observational errors including the special cases of the (almost) error-free field (physical-geodetic process) and the pure error field (operational-geodetic process) (section B3). In geodesy as an approximation science the planar approximation, realized by the projection onto a tangential plane to the earth's surface and/or different planes of measurement or computation, has a great practical importance. This review exclusively treats the covariance functions of planar geodetic processes, being subdivided into physical geodesy (sections C1–C6) and operational geodesy (sections D1–D5). Although all geodetic observations are more or less influenced by the gravity field and since an operational aspect is to be attributed also to the physical geodesy [*Grafarend,* 1978; *Moritz,* 1978*c*], the classification made within the framework of this review may be permitted. The following abbreviations will be used: CF (covariance function), ACF (autocovariance function), CCF (cross-covariance function), CT (covariance transformation), CP (covariance propagation), and MPCF (multipoint covariance function).

B. STATISTICS OF PLANAR RANDOM FIELDS

1. *Two-Point Statistics of Scalar Fields*

Random fields are described in the concept of multidimensional stochastic processes [*Obuchow,* 1954]. Some problems can be solved within the (narrower) framework of the correlation theory and/or spectral theory of stochastic processes. Let be given a random field $X(\mathbf{r})$, $\mathbf{r} = |x_1, x_2, \cdots, x_n|^T$ with the expectation value (first-order moment) $E\{X(\mathbf{r})\}$ and the ACF (second-order central moment) $C_{XX}(\mathbf{r}', \mathbf{r}'')$. $X(\mathbf{r})$ is termed homogeneous if

$$E\{X(\mathbf{r})\} = \text{const} \qquad C_{XX}(\mathbf{r}', \mathbf{r}'') = C_{XX}(\mathbf{r}'' - \mathbf{r}') \tag{1}$$

(invariance with respect to translation). Hence follows the

spectral representation

$$C_{XX}(\mathbf{r}) = \int_{-\infty}^{+\infty} \exp(j\mathbf{r}\mathbf{k})\, \tilde{C}_{XX}(\mathbf{k}) d\mathbf{k} \quad (2)$$

$$\tilde{C}_{XX}(\mathbf{k}) = \frac{1}{(2\pi)^n} \int_{-\infty}^{+\infty} \exp(-j\mathbf{k}\mathbf{r})\, C_{XX}(\mathbf{r}) d\mathbf{r} \quad (3)$$

(Wiener-Khintchine), $\mathbf{r} := \mathbf{r}'' - \mathbf{r}'$, j is an imaginary unit, $\mathbf{k}$ is the wave number vector.

$X(\mathbf{r})$ is called homogeneous and isotropic if

$$C_{XX}(\mathbf{r}) = X_{XX}(|\mathbf{r}|) = C_{XX(r)}$$
$$\tilde{C}_{XX}(\mathbf{k}) = \tilde{C}_{XX}(|\mathbf{k}|) = \tilde{C}_{XX}(k) \quad (4)$$

(invariance with respect to rotation and reflection). In this case, $n - 1$ integrations can be carried out in (2) and (3) by means of the introduction of generalized polar coordinates. Then in each case a simple integral is left that is a function of k or r only. Equation (3) is equivalent to the Hankel transform of the order $(n - 2)/2$ and (2) is equivalent to its inverse transform. In the two-dimensional case $(n = 2)$,

$$x_1'' - x_1' =: x = r\cos\varphi \qquad k_1 = k\cos\varphi$$
$$x_2'' - x_2' =: y = r\sin\varphi \qquad k_2 = k\sin\varphi$$

Equations (2) and (3) will change into

$$C_{XX}(r) = 2\pi \int_0^{\infty} J_0(rk)\, \tilde{C}_{XX}(k) k dk \quad (5)$$

$$\tilde{C}_{XX}(k) = \frac{1}{2\pi} \int_0^{\infty} J_0(kr)\, C_{XX}(r) r dr \quad (6)$$

with the Bessel function of the first kind and the zero order J_0. Equation (6) is equivalent to the Hankel transform of the zero order, (5) is equivalent to its inverse transform.

2. *Two-Point Statistics of Vector Fields*

Let $\mathbf{X}(\mathbf{r}) = [X_1(\mathbf{r}), X_2(\mathbf{r}), \cdots, X_n(\mathbf{r})]^T$, $\mathbf{r} = [x_1, x_2, \cdots, x_n]^T$ be a random vector field with the expectation value vector $E(\mathbf{r})$ and the CF $C_{ij}(\mathbf{r}', \mathbf{r}'')$, $i = j$, ACF of $X_i(\mathbf{r})$, $i \neq j$, and CCF between $X_i(\mathbf{r})$, $X_j(\mathbf{r})$. If $\mathbf{X}(\mathbf{r})$ is a first-rank tensor, the second-order moments $C_{ij}(\mathbf{r}', \mathbf{r}'')$ form a second-rank tensor. $\mathbf{X}(\mathbf{r})$ is called homogeneous if, analogous to (1),

$$E\{X_k(\mathbf{r})\} = \text{const} \qquad C_{ij}(\mathbf{r}', \mathbf{r}'') = C_{ij}(\mathbf{r}'' - \mathbf{r}') \quad (7)$$

Hence follows the spectral representation analogous to (2) and (3):

$$C_{ij}(\mathbf{r}) = \int_{-\infty}^{+\infty} \exp(j\mathbf{r}\mathbf{k})\, \tilde{C}_{ij}(\mathbf{k})\, d\mathbf{k} \quad (8)$$

$$\tilde{C}_{ij}(\mathbf{k}) = \frac{1}{(2\pi)^n} \int_{-\infty}^{+\infty} \exp(-j\mathbf{k}\mathbf{r})\, C_{ij}(\mathbf{r}) d\mathbf{r} \quad (9)$$

In the case of a homogeneous and isotropic vector $\mathbf{X}(\mathbf{r})$ holds [*Taylor,* 1935; *Kármán,* 1937]

$$C_{ij}(\mathbf{r}', \mathbf{r}'') = C_{ij}(\mathbf{r}'' - \mathbf{r}') = F(r)\,\delta_{ij} + [G(r) - F(r)]\,\Delta x_i \Delta x_j r^{-2} \quad (10)$$

with $r := |\mathbf{r}'' - \mathbf{r}'|$, $x = x'' - x'$, $\delta_{ij} = 0$ $(i \neq j)$, and $\delta_{ij} = 1$ $(i = j)$. The scalar distance functions $F(r)$, $G(r)$ are connected with each other by

$$G(r) = F(r) + rF'(r) \quad (11)$$

in the case of a nonvortical source field and by

$$F(r) = G(r) + rG'(r)/(n-1) \quad (12)$$

$$F(r) = G(r) + rG'(r) \qquad (n = 2) \quad (13)$$

in the case of a source-free rotational field. Equation (11) is a necessary and sufficient condition that $\mathbf{X}(\mathbf{r})$ possesses a potential [*Obuchow,* 1954]. The Kármán equation (12) results from the incompressibility condition div $\mathbf{X} = 0$.

Let $\mathbf{X}(\mathbf{r})$ contain nonvortical source shares $\mathbf{X}^S(\mathbf{r})$ and source-free rotational shares $\mathbf{X}^W(\mathbf{r})$, which form a homogeneous and isotropic system of fields and superimpose one upon another additively:

$$\mathbf{X}(\mathbf{r}) = \mathbf{X}^S(\mathbf{r}) + \mathbf{X}^W(\mathbf{r}) \quad (14)$$

Then the CCF between the source share and the rotational share disappears, $C_{ij}^{SW} \equiv 0$, and the decomposition [*Kármán and Howarth,* 1938; *Obuchow,* 1954]

$$C_{ij} = C_{ij}^{SS} + C_{ij}^{WW} \qquad F = F^S + F^W \qquad G = G^S + G^W \quad (15)$$

will hold. Between a vector field $\mathbf{X}(\mathbf{r})$ and a scalar field $X(\mathbf{r})$ that are contained in a homogeneous and isotropic system of fields, the CCF will disappear, $C_{XX}(\mathbf{r}', \mathbf{r}'') \equiv 0$, when $\mathbf{X}(\mathbf{r}) \equiv \mathbf{X}^W(\mathbf{r})$.

Now, we represent also the correlation functions $\Phi_{ij}(\mathbf{r}', \mathbf{r}'') = C_{ij}(\mathbf{r}', \mathbf{r}'')/C_{ij}(\mathbf{O},\mathbf{O})$ contained in the Taylor-Kármán relation (10) with $F(r)/C_{ij}(\mathbf{O}, \mathbf{O}) =: \psi(r)$, $G(r)/C_{ij}(\mathbf{O}, \mathbf{O}) =: \Omega(r)$, $|\Phi_{ij}| \leq 1$, $|\psi| \leq 1$, $|\Omega| \leq 1$, in the planar case $(n = 2)$ in polar coordinates (r, φ) with $\mathbf{X}(\mathbf{r}) = [X(\mathbf{r}), Y(\mathbf{r})]^T$, $x_1 =: x = r\cos\varphi$, and $x_2 =: y = r\sin\varphi$:

$$\Phi_{XX}(r, \varphi) = \psi(r) + [\Omega(r) - \psi(r)]\cos^2\varphi$$
$$\Phi_{XY}(r, \varphi) = \Phi_{YX}(r, \varphi) = [\Omega(r) - \psi(r)]\sin\varphi\cos\varphi \quad (16)$$
$$\Phi_{YY}(r, \varphi) = \psi(r) + [\Omega(r) - \psi(r)]\sin^2\varphi$$

where $\Omega(r)$ is the longitudinal correlation and $\psi(r)$ is the transversal correlation. The corresponding spectral representation follows (6) analogously. The invariants (against orthogonal coordinate transformations) of the symmetric second-rank tensor $\{\Phi_{ij}\}$ are

$$I_1(r) = \text{tr}\,\Phi_{ij}/2 = [\psi(r) + \Omega(r)]/2$$
$$I_2(r) = \det\Phi_{ij} = \psi(r)\Omega(r) \quad (17)$$

They uniquely define the mean correlation as well as the correlation ellipse or hyperbola [*Grafarend,* 1971*b*].

Let us assume $\mathbf{X}(\mathbf{r})$ to be homogeneous but nonisotropic and the field correlation to be characterized by a certain preferred direction λ_i. Instead of the rotation and reflection invariance of the isotropic case, now an invariance exists with respect to rotations about the direction λ_i and with respect to reflections in planes containing λ_i and a normal λ_j to this direction (axisymmetry). In general, several axes of symmetry may occur (generalized axisymmetry). The structures of the correlation functions Φ_{ij} are summarized in Table 1 for the cases of isotropy, of axisymmetry, and of bi-axisymmetry [*Robertson,* 1940; *Batchelor,* 1946; *Chandrasekhar,* 1950; *Grafarend,* 1972*a, b, c*].

3. *Geodetic Stochastic Processes*

The correlation theory of stochastic processes has been applied in geodesy mainly to the statistical analysis of the grav-

TABLE 1. Structure of the Correlation Function Φ_{ij} (r′, r″) of a Homogeneous-Isotropic, Axisymmetric, and Bi-Axisymmetric Vector Field

Property of the Vector Field	Structure of the Correlation Function $\Phi_{ij}(\mathbf{r}', \mathbf{r}'')$	Characteristic Functions a_i	Symmetry Axes λ_i, μ_i
Isotropy	$+a_1\Delta x_i\Delta x_j + a_2\delta_{ij}$	$a_i(r)$ $i = 1, 2$	$\cdots$
Axisymmetry	$+a_1\Delta x_i\Delta x_j + a_2\delta_{ij}$ $+a_3\lambda_i\lambda_j + a_4\lambda_i\Delta x_j$ $+a_5\Delta x_i\lambda_j$	$a_i(r, \alpha)$ $i = 1, 2, \cdots, 5$	λ_i $\Delta x_i\lambda_i = r\cos\alpha$
Bi-axisymmetry	$+a_1\Delta x_i\Delta x_j + a_2\delta_{ij}$ $+a_3\lambda_i\lambda_j + a_4\lambda_i\Delta x_j$ $+a_5\Delta x_i\lambda_j + a_6\mu_i\mu_j$ $+a_7\mu_i\Delta x_j + a_8\Delta x_i\mu_j$ $+a_9\lambda_i\mu_j + a_{10}\mu_i\lambda_j$	$a_i(r, \alpha_1, \alpha_2)$ $i = 1, 2, \cdots, 10$	λ_i, μ_i $\Delta x_i\lambda_i = r\cos\alpha_1$ $\Delta x_i\mu_i = r\cos\alpha_2$

ity field and in the diagnosis of geodetic networks; compare the review by *Grafarend* [1976]. In addition to these main fields of application (sections C1–C6 and D4), we will discuss some other possible applications (sections C7, D1–D3). In this section the geodetic process, its special cases and CF's are defined.

Definition 1. Let $\hat{T}(\mathbf{r})$ be a random function depending on the parameter vector **r** and representing an estimation of the geodetic disturbing potential $T(\mathbf{r})$, which is harmonic in the three-dimensional Euclidean space outside the earth. The random function $\hat{T}(\mathbf{r})$, whose moments up to the second order exist and are finite, will be called a geodetic stochastic process of the second order [*Grafarend,* 1976].

Definition 2. Let $\hat{x}_i(\mathbf{r})$ be a vector stochastic process depending on the parameter vector **r** and representing a geodetic network in Cartesian coordinates. Let $\mathbf{r}_i - \mathbf{r}_j$ be the vector connecting two points in the network. Then we call $\hat{\varepsilon}_i(\mathbf{r}) \equiv \hat{x}_i(\mathbf{r}) - E\{\hat{x}_i(\mathbf{r})\}$ the centered stochastic process of the network [*Grafarend,* 1976].

The definitions 1 and 2 can be combined and generalized. For this purpose we supplement (1) the potential field T by the fields of its derivative T_x, T_y, T_z, T_{xx}, $\cdots$, moreover by the movement and deformation field of continua and the field of atmospheric refraction, (2) the geodetic network by photographic images of parts of the celestial globe and the earth's surface. We observe, in addition, that (3) any observation error ε can be decomposed into $\varepsilon = \varepsilon^o + \varepsilon^x$, if ε^o represents the pure observation error (surveyor, instrument) and ε^x designates the error influence of a geophysical field (gravity, refraction). Thus we get to definition 3.

Definition 3. Let $\hat{\mathbf{X}}(\mathbf{r})$ be a random function depending on the parameter vector **r** and representing an estimation of a geophysical random field $\mathbf{X}(\mathbf{r})$, whose moments up to the second order exist and are finite. If also the decompositions

$$\hat{\mathbf{X}}(\mathbf{r}) = \mathbf{X}(\mathbf{r}) + \boldsymbol{\varepsilon}(\mathbf{r}) \qquad E\{\hat{\mathbf{X}}(\mathbf{r})\} = E\{\mathbf{X}(\mathbf{r})\} + E\{\boldsymbol{\varepsilon}(\mathbf{r})\}$$

$$E\{\hat{X}_i(\mathbf{r}')\hat{X}_j(\mathbf{r}'')\} = E\{X_i(\mathbf{r}')X_j(\mathbf{r}'')\} + E\{\varepsilon_i(\mathbf{r}')\varepsilon_j(\mathbf{r}'')\}$$

exist, then $\hat{\mathbf{X}}(\mathbf{r})$ will be called a geodetic stochastic process ($\hat{\mathbf{X}}(\mathbf{r}) \equiv \mathbf{X}(\mathbf{r})$ physical-geodetic process, $\hat{\mathbf{X}}(\mathbf{r}) \equiv \boldsymbol{\varepsilon}(\mathbf{r})$ operational-geodetic process) of the second order.

Each of the partial processes can be of the potential type (11) or of the solenoidal type (turbulent type) (12) [*Grafarend,* 1976]. The specialization in planar processes results in definition 4.

Definition 4. The covariance function of a geodetic stochastic process on a local plane will be called a planar geodetic covariance function. For relations to spherical processes, boundary value problem of physical geodesy, problems of existence, homogeneity, isotropy, consistency, etc., cf. *Heiskanen and Moritz* [1967], *Neiman* [1974*a*], *Rummel* [1975], *Grafarend* [1976], and *Moritz* [1976, 1980].

4. *Multipoint Statistics*

Normal random fields are of great practical importance. They are completely described by the first-order and second-order moments. All odd moments of higher orders are identical with zero, all even moments of higher orders can be calculated from the second-order moment. For the practical applications, the random fields or their realizations by suitable trend elimination—ergodicity being assumed—are occasionally normalized. It is also possible, however, to calculate higher moments.

The CF of a vector field

$$C_{ij}(\mathbf{r}', \mathbf{r}'') \equiv E\{X_i(\mathbf{r}')X_j(\mathbf{r}'')\} \tag{18}$$

depending on the parameter vectors **r**′, **r**″ of two points P', P'', is called also two-point CF. Accordingly, the n-point CF (MPCF) reads as follows:

$$C_{i_1 i_2 \cdots i_n}(\mathbf{r}', \mathbf{r}'', \cdots, \mathbf{r}^{(n)}) \equiv E\{X_{i_1}(\mathbf{r}')X_{i_2}(\mathbf{r}'') \cdots X_{i_n}(\mathbf{r}^{(n)})\} \tag{19}$$

dependent on the parameter vectors **r**′, **r**″, $\cdots$, $\mathbf{r}^{(n)}$ of the points P', P'', $\cdots$, $P^{(n)}$. If the random vector $\mathbf{X}(\mathbf{r})$ is a first-rank tensor, the central moments of the nth order of this vector field will form an n-rank tensor [*Obuchow,* 1954]. The structure of the MPCF of homogeneous and isotropic vector fields was investigated by *Grafarend* [1972*a, b, c*] for the concept of a nonlinear prediction: they contain 2 (n = 2), 4 (n = 3), 10 (n = 4), and 26 (n = 5) characteristic (distance dependent) functions. Special MPCF are obtained if certain points coincide, **r**′ = **r**″, etc.—they were given already by *Kármán and Howarth* [1938] and *Obuchow* [1954]. So far, there have not been any applications in geodesy.

Another problem for the multipoint statistics is given if the CF's of finite geodetic net elements, constituting multipoint functions, are to be calculated; e.g., the distance between two net points is a scalar-valued two-point function, the angle between the instrument setup and two sighting points is a scalar-valued three-point function, etc. Accordingly, the two-point CF of a distance (of an angle) depend on four (six) position vectors. Their derivation from the two-point CF of the point coordinates (= vector-valued one-point functions) is a problem of the linear CP (section D4).

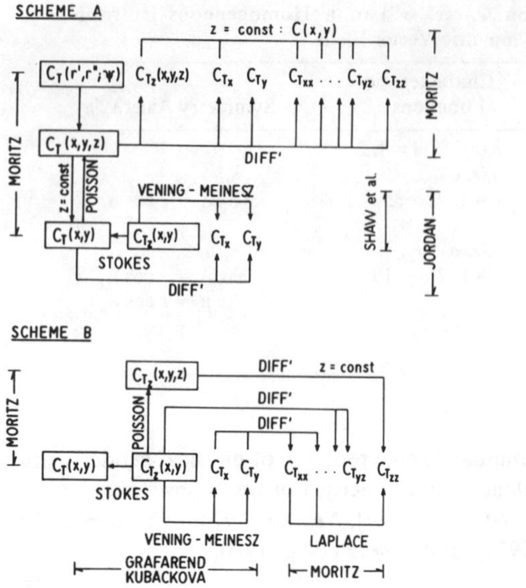

Fig. 1. Scheme of covariance transformations in the gravity field.

C. Planar Covariance Functions in Physical Geodesy

1. *Covariance Transformations in the Gravity Field (a Review)*

The disturbing potential T and its derivatives T_x, T_y, T_z, T_{xx}, T_{xy}, $\cdots$ are subjected to statistical analysis in the physical geodesy (gravimetry, gradiometry, inertial navigation). According to the particular problems (collocation, error estimation) as well as the quantity and quality of the set of data (of the geodetic realization), from which the CF's can be estimated with the required statistical significance or are already available, the scalar field that is to be subjected to a CT can be the disturbing potential T (Figure 1, scheme A) or the gravity anomaly $\Delta g = -T_z$ (Figure 1, scheme B). The transformation can be carried out in the position domain with the aid of the differentiation rules of the CP (section C2) or in the wave number domain by means of the integral formulae of the physical geodesy (Stokes, Vening-Meinesz) and the convolution theorem (section C3). Convolution operations are possible as the integrations in the case of a planar approximation cover an infinitely extended plane.

From a rotation symmetric ACF $C_T(r', r''; \psi)$ with the radius vectors r', r'' of two points in the space P', P'' and their spherical distance ψ (global ACF), one obtains the planar equivalent $C_T(x, y, z)$ in the Cartesian coordinate system x, y, z by planar approximation (local ACF; *Moritz* [1976]; Figure 1, scheme A). If the differentiation rules of the CP are applied to $C_T(x, y, z)$, one will obtain the ACF/CCF of the first and second (and higher) derivatives of T as functions of x, y, z, and with z = const their corresponding functions in a fixed plane will be obtained. The connection between $C_{T_{xx}}$, $C_{T_{yy}}$, and $C_{T_{zz}}$ is given by the Laplace equation. The advantage of this method lies in the uniform approach, which is easy to schematize [*Moritz,* 1974, 1975, 1976, 1980]; only one type of operation occurs, i.e., differentiation. The planar approximation takes place within the stochastic model. $C_T(x, y)$ can be continued by means of the Poisson integral (if it exists) in the upper half-space [*Moritz,* 1976; *Bellaire,* 1971, 1977; *Jordan and Heller,* 1978; *Keller and Meier,* 1980*a*].

CT's in the wave number domain with the aid of the Vening-Meinesz formula after planar approximation were carried out by *Shaw et al.* [1969] for the first time. *Grafarend* [1971*b–f*, 1972*a*, *b*, 1976] succeeded in making decisive generalizations with the structural analysis of the ACF/CCF of planar vector fields, especially of the anomalous vertical deflection field (T_x, T_y), and *Jordan* [1972*a*, *b*] was able to achieve them by the joint ACF/CCF analysis of the disturbing potential and the anomalous gravity vector (T; T_x, T_y, T_z); see Figure 1, schemes A, B. The planar approximation is carried out with this transformation method already in the functional model. The spectral representation of the scalar and vector signals, the convolution theorem and the planar Fourier transformation are used as mathematical aids. The existence of the solutions is not given a priori (compare section C4).

The calculation of complete ACF/CCF schemes of the disturbing potential and its first and second derivatives (and higher derivatives) is possible (1) with the differential transformation (Figure 1, scheme A), and (2) with the combination of the integral and differential transformations (Figure 1, scheme B). ACF/CCF systems of this kind were given by *Grafarend* [1971*c*] and *Jordan* [1972*b*] with regard to T; T_x, T_y, T_z, by *Moritz* [1974] and *Reilly* [1979] with regard to T; T_x, T_y, T_z; T_{xx}, T_{xy}, $\cdots$, T_{zz}, and by *Keller and Meier* [1980*a*] with regard to T_x, T_y, T_z; T_{xx}, T_{xy}, $\cdots$, T_{zz}. A simple example is given in Table 2.

2. *Covariance Transformations in the Position Domain*

To demonstrate the CT in the position domain according to Figure 1, scheme A, we assume a simple rotationally symmetric, spatial, harmonic covariance function:

$$C_T(r', r''; \psi) = \sum_{n=0}^{\infty} \sigma^{n-1} P_n(\cos \psi) \tag{20}$$

$$\sigma = R_B^2/r'r'' \qquad (R_B < R) \tag{21}$$

R_B is the radius of some sphere completely inside the terrestrial sphere of the radius R (Bjerhammar sphere), and $P_n(\cos \psi)$ are the Legendre polynomials. Equation (20) may be summed by means of the expression for the reciprocal distance [*Heiskanen and Moritz,* 1967],

$$C_T(\sigma; \psi) = \sigma(1 - 2\sigma \cos \psi + \sigma^2)^{-1/2} \tag{22}$$

defined outside the Bjerhammar sphere $r = R_B$ ($\sigma < 1$), cf. also *Krarup* [1969].

For applications in limited areas (local problems) it is useful to replace the sphere $r = R$ by its tangential plane. Then *Moritz* [1976, 1980] found the planar equivalent of (20)

$$C_T(r; z(z' + z'')) = B/z[1 + (r/z)^2]^{1/2} \tag{23}$$

with

$$r^2 = x^2 + y^2 \qquad x = x'' - x' \qquad y = y'' - y'$$
$$z = z' + z'' + z_0 \qquad z' = r' - R \qquad z'' = r'' - R \tag{24}$$
$$z_0 = 2(R - R_B) \qquad B = \text{const}$$

isotropically in the plane $z = d$ = const:

$$C_T(r) = C_T(0)/[1 + (r/d)^2]^{1/2} \qquad C_T(0) = B/d \tag{25}$$

The planar symmetry (homogeneity and isotropy) in (25) corresponds to the spherical symmetry in (20). The height dependence of the three-dimensional ACF (23) is demonstrated in

TABLE 2. Variances and Covariances $C_{ij}(0)$ of the First and Second Derivatives of the Gravity Potential in One Point, Planar Approximation, Units of Gravity Variance (Hirvonen and Gauss Models)

	Value of j								
i	x	y	z	xz	yz	zz	xx	xy	yy
x	$1/2$*	0	0	$\pi/4d$	0	0	0	0	0
y	0	$1/2$*	0	0	$\pi/4d$	0	0	0	0
z	0	0	1*	0	0	$\pi/2d$	$-\pi/4d$	0	$-\pi/4d$
xz	$\sqrt{\pi}/2d$	0	0	$2/d^2$*	0	0	0	0	0
yz	0	$\sqrt{\pi}/2d$	0	0	$2/d^2$*	0	0	0	0
zz	0	0	$\sqrt{\pi}/d$	0	0	$4/d^2$*	$-2/d^2$	0	$-2/d^2$
xx	0	0	$-\sqrt{\pi}/d$	0	0	$-2/d^2$	$3/2d^2$*	0	$1/2d^2$
xy	0	0	0	0	0	0	0	$1/2d^2$*	0
yy	0	0	$-\sqrt{\pi}/d$	0	0	$-2/d^2$	$1/2d^2$	0	$3/2d^2$*

Taken from *Keller and Meier* [1980*a*].
* Variances are equal for both models. The covariances of the Hirvonen model (above the diagonal) differ only little from those of the Gauss model (below the diagonal).

Figure 2: with growing $z' + z''$ the variance decreases and the correlation length increases. For other examples see *Schwarz* [1976], *Jordan and Heller* [1978].

The ACF (22) is designated as the 'reciprocal distance CF'; the ACF (25) is introduced in Table 3 as the Moritz model. Of the generalized Hirvonen models

$$C(r) = C(0)/[1 + (r/d)^2]^m \tag{26}$$

(Table 3), the Moritz model ($m = 1/2$) and the Poisson model ($m = 3/2$), and only these models, possess a spatial continuation of the structure $C(r; z' + z'')$ according to

$$C(r; z' + z'') = 2\pi \int_0^\infty J_0(rk)\tilde{C}(k) \exp[-(z' + z'')k]kdk \tag{27}$$

[*Moritz,* 1976, 1980]. For more details concerning the continuation in the upper half-space see *Mihelcic* [1972], *Bellaire* [1971, 1977], *Jordan and Heller* [1978], *Keller and Meier* [1980*a*].

If one has a differentiable ACF $C_T(r; z', z'')$, then the ACF of the derivatives of T and the CCF between them can be easily calculated. As an example we consider the ACF/CCF of the horizontal gravity gradient in the x-y plane, when $C_{T_zT_z} =: C_{z,z}$ are given. With the differentiation rule

$$C_{X^{(n)},Y^{(m)}} = (-1)^n \frac{d^{n+m}}{dr^{n+m}} C_{X,Y} \tag{28}$$

one will find

$$C_{xz,xz} = -\frac{\partial^2}{\partial x^2} C_{z,z} = -\left\{\frac{\partial^2}{\partial r^2} C_{z,z}\left(\frac{\partial r}{\partial x}\right)^2 + \frac{\partial}{\partial r} C_{z,z}\frac{\partial^2 r}{\partial x^2}\right\}$$

$$C_{xz,yz} = -\frac{\partial^2}{\partial x \partial y} C_{z,z} = -\left\{\frac{\partial^2}{\partial r^2} C_{z,z}\frac{\partial r}{\partial x}\frac{\partial r}{\partial y} + \frac{\partial}{\partial r} C_{z,z}\frac{\partial^2 r}{\partial x \partial y}\right\} \tag{29}$$

$$C_{yz,yz} = -\frac{\partial^2}{\partial y^2} C_{z,z} = -\left\{\frac{\partial^2}{\partial r^2} C_{z,z}\left(\frac{\partial r}{\partial y}\right)^2 + \frac{\partial}{\partial r} C_{z,z}\frac{\partial^2 r}{\partial y^2}\right\}$$

With the derivatives $\partial r/\partial x = \cos\varphi$, $\partial r/\partial y = \sin\varphi$, $\partial^2 r/\partial x^2 = r^{-1}\sin^2\varphi$, $\partial^2 r/\partial y^2 = r^{-1}\cos^2\varphi$, $\partial^2 r/\partial x\partial y = -r^{-1}\sin\varphi\cos\varphi$ from (29) follows the Taylor-Kármán structure (10) of the required ACF/CCF with

$$G(r) = -\partial^2 C_{z,z}/\partial r^2 \qquad F(r) = -r^{-1}\partial C_{z,z}/\partial r \tag{30}$$

The invariant with respect to orthogonal coordinate transformations

$$C_{xz,xz} + C_{yz,yz} = -\{\partial^2 C_{z,z}/\partial r^2 + r^{-1}\partial C_{z,z}/\partial r\} \tag{31}$$

is identical with the ACF $C_{zz,zz}$ of the vertical gravity gradient $T_{z,z}$ in the same plane, resulting from the Laplace equation $\Delta T = 0$ [*Moritz,* 1976, 1980].

As an example of a derivative of a higher order we give the ACF of the third derivative of the disturbing potential with respect to height [*Keller and Meier,* 1980*a*; *Meier,* 1980]

$$C_{zzz,zzz} = (\partial^4/\partial r^4 + 2r^{-1}\partial^3/\partial r^3 - r^{-2}\partial^2/\partial r^2 + r^{-3}\partial/\partial r)C_{z,z} \tag{32}$$

In Figure 3, the ACF of the first three derivatives of the disturbing potential with respect to height are represented for the Hirvonen model, $m = 1$ in (26). The correlation lengths of $C_{z,z}$, $C_{zz,zz}$, $C_{zzz,zzz}$ are in the ratio 1:0.42:0.27, the variances are in the ratio $1:4d^{-2}:64d^{-4}$.

3. *Covariance Transformations in the Wave Number Domain*

The CT's in the wave number domain are carried out with the aid of geodetic integral formulae (Stokes, Vening-Meinesz) after planar approximation [*Shaw et al.,* 1969; *Jordan,* 1972*a, b*; *Grafarend,* 1971*b–f,* 1975*a, b,* 1976; *Kubackova,* 1974*a, b,* 1976; *Keller and Meier,* 1980*a, b*]. As an example, we consider grad $T = [T_x, T_y, T_z]^T$, which is equivalent to the anomalous gravity vector. Let there be given a homogeneous and isotropic gravity field with $C_{z,z}(r)$ as well as T_x, T_y by the Vening-Meinesz formula in the plane (x, y)

$$\begin{Bmatrix} T_x \\ T_y \end{Bmatrix} =: \begin{Bmatrix} T_1 \\ T_2 \end{Bmatrix} = \frac{1}{2\pi}\iint_{-\infty}^{+\infty} \frac{T_z}{r^3}\begin{Bmatrix} x \\ y \end{Bmatrix} dxdy \tag{33}$$

$$f_1 := x/2\pi r^3 \qquad f_2 := y/2\pi r^3$$

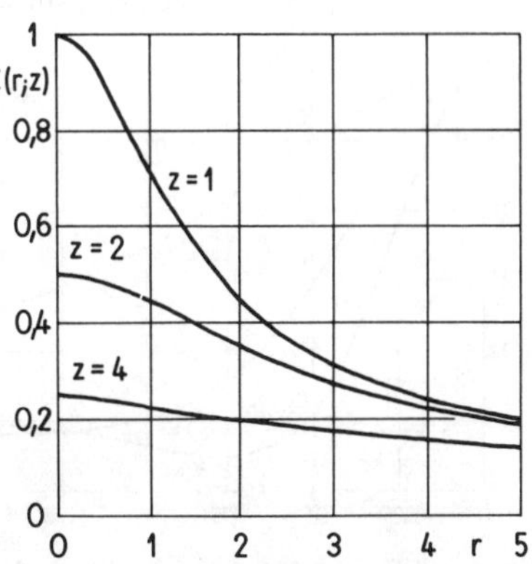

Fig. 2. Planar ACF's $C(r; z)$ for different heights (Moritz model), normalized.

TABLE 3. Some Homogeneous-Isotropic ACF Models Applied in Physical Geodesy (Planar Case)

Model	$C(r)/C(O)$	References
First-order Markov model	$(r/d)K_1(r/d)$	*Shaw et al.* [1969], *Grafarend* [1971*b*–*f*], *Grafarend and Offermanns* [1975]
Second-order Markov model	$[1 + (r/d)] \exp[-(r/d)]$	*Kasper* [1971], *Jordan* [1972*b*]
Third-order Markov model	$[1 + (r/d) + (r/d)^2/3] \times \exp[-(r/d)]$	*Bellaire* [1971, 1972], *Jordan* [1972*b*], *Kearsley* [1977], *Morrison* [1977]
Gauss model	$\exp[-(r/d)^2]$	*Heitz* [1967], *Grafarend* [1970*a*], *Meissl* [1970], *Bellaire* [1971], *Moritz* [1972], *Schwarz* [1974], *Kearsley* [1977], *Reilly* [1979], *Keller and Meier* [1980*a*]
Generalized Hirvonen models	$[1 + (r/d)^2]^{-m}$	*Moritz* [1976]
Moritz model	$m = 1/2$	*Moritz* [1976], *Schwarz* [1976], *Keller and Meier* [1980*a*]
Hirvonen model	$m = 1$	*Hirvonen* [1962], *Vyskocil* [1969], *Lauer* [1971*a*, *b*], *Moritz* [1976], *Schwarz* [1976], Kearsley [1977], *Hein and Jochemczyk* [1979], *Meier* [1980], *Keller and Meier* [1980*a*]
Poisson model	$m = 3/2$	*Bellaire* [1971, 1977], *Moritz* [1976], *Schwarz* [1976], *Jordan and Heller* [1978], *Keller and Meier* [1980*a*]
Combined models		*Heitz* [1968, 1969], *Vyskocil* [1969, 1970], *Soltau* [1970], *Kearsley* [1977], *Hein and Jochemczyk* [1979], *Stangl* [1979]

With the aid of the planar Fourier transformation and the convolution theorem one obtains

$$E\{T_iT_j\} = C_{ij} = \int_{-\infty}^{+\infty} \exp(j\mathbf{rk})\, \tilde{C}_{ij}(\mathbf{k})d\mathbf{k}$$

$$E\{T_iT\} = C_i = \int_{-\infty}^{+\infty} \exp(j\mathbf{rk})\, \tilde{C}_i(\mathbf{k})d\mathbf{k} \qquad (34)$$

$$\tilde{C}_{ij} = -\tilde{f}_i\tilde{f}_j\tilde{C}_{z,z} \qquad \tilde{C}_i = -\tilde{f}_i\tilde{C}_{z,z}$$

The CT's (34) give the ACF/CCF (C_{ij}) of the vertical deflection vector $[T_x, T_y]^T$ with a Taylor-Kármán structure (10) and

$$G(r) = C_{z,z}(O)\Omega(r)/2 \qquad F(r) = C_{z,z}(O)\psi(r)/2 \qquad (35)$$

the invariant with respect to orthogonal transformations

$$C_{x,x} + C_{y,y} = C_{z,z} \qquad (36a)$$

or

$$[\Omega(r) + \psi(r)]/2 = C_{z,z}(r)/C_{z,z}(O) \qquad (36b)$$

see Figure 4; moreover, the CCF (C_i)

$$C_{z,x} = -C_{x,z} = -C_{z,z}(O)\Theta(r)\cos\varphi$$
$$C_{z,y} = -C_{y,z} = -C_{z,z}(O)\Theta(r)\sin\varphi \qquad (37)$$

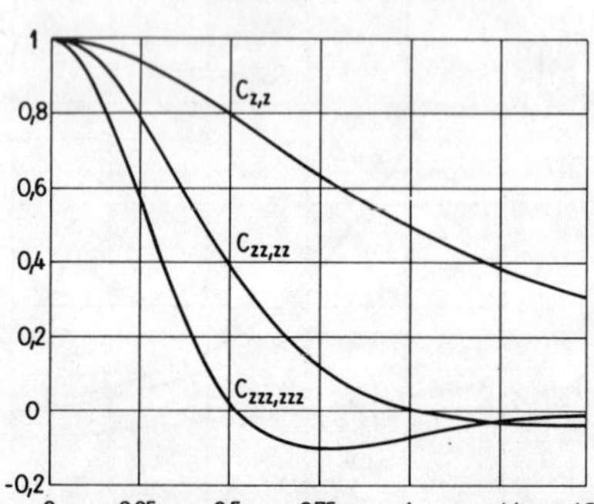

Fig. 3. Planar ACF's $C(r/d)$ of the first three derivatives of the disturbing potential with respect to height (Hirvonen model), normalized. For the numerical test of $C_{zzz,zzz}$ see *Vyskocil* [1969].

For example, the first-order Markov model (Figure 4)

$$C_{z,z}(r)/C_{z,z}(O) = (r/d)K_1(r/d)$$

$$\Omega(r) = -4(r/d)^{-2} + 2K_0(r/d) + 4(r/d)^{-1}K_1(r/d) + 2(r/d)K_1(r/d)$$

$$\psi(r) = +4(r/d)^2 - 2K_0(r/d) - 4(r/d)^{-1}K_1(r/d)$$

$$\Omega(r) + \psi(r) = 2(r/d)K_1(r/d)$$

$$\Theta(r) = \frac{\pi}{4}\left(\frac{r}{d}\right) {}_1F_2\left\{\tfrac{3}{2}, \tfrac{1}{2}, 2; \tfrac{1}{4}\left(\frac{r}{d}\right)^2\right\} + \tfrac{2}{3}\left(\frac{r}{d}\right)^2 {}_1F_2\left\{2, \tfrac{5}{2}, \tfrac{3}{2}; \tfrac{1}{4}\left(\frac{r}{d}\right)^2\right\}$$

K_0, K_1 are modified Bessel functions of the second kind, of the zero and first orders, ${}_1F_2$ are generalized hypergeometric series [*Abramowitz and Stegun*, 1965; *Gradshteyn and Ryzhik*, 1965].

The distance-dependent functions $\psi(r)$, $\Omega(r)$ have the properties of autocorrelation functions and $\Omega(r) - \psi(r)$, $\Theta(r)$ have the properties of cross-correlation functions. The integral transformation yields elementary distance functions in the case of the Moritz model and the Poisson model. With all other models according to Table 3, the solutions for ψ, Ω, and Θ contain terms of higher functions. For the symmetry of the

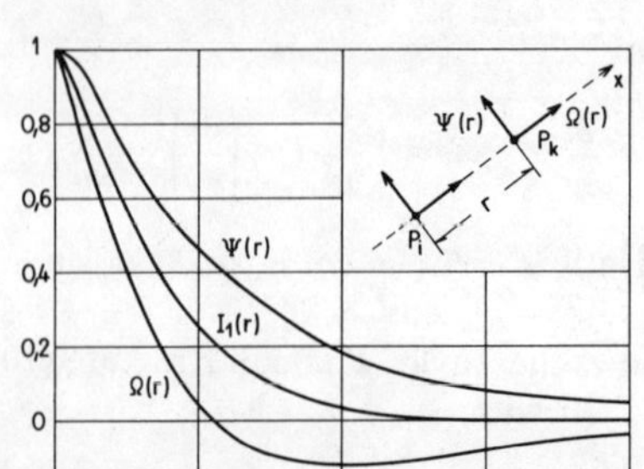

Fig. 4. Longitudinal and transversal correlations $\Omega(r)$, $\Psi(r)$ of a homogeneous and isotropic vector (vertical deflection vector). First-order Markov model. See the numerical tests with astrogeodetic data in *Grafarend* [1971*b*, *e*]. $I_1 = (\Omega + \Psi)/2$.

solutions (34)–(37), cf. also *Jordan* [1972*b*], *Keller and Meier* [1980*a*].

4. *Problems of Existence (Physical Significance)*

Of the two possible transformations (21) and (22), the first one is without any mathematical problems, if only the given field can be differentiated sufficiently frequently. The error of the planar approximation can be stated [*Moritz,* 1976]. Sometimes, however, it is appropriate to combine the two methods, e.g., if the ACF $C_{z,z}$ is given on the basis of a safe estimation from extensive observations. With regard to the integral transformation, the following special features occur.

1. The transformation $C_{z,z} \rightarrow C_{T,T}$ with the aid of the Stokes formula is possible within the position domain if and only if

$$\partial C_{x,x}/\partial y = \partial C_{x,y}/\partial x \qquad \partial C_{y,y}/\partial x = \partial C_{y,x}/\partial y \tag{38}$$

The relations (38) are satisfied in the homogeneous and isotropic vector field T_x, T_y, but the practical calculation of $C_{T,T}$ is very difficult.

2. Owing to the infinite integration limits, the transformation in the wave number domain is possible. As a rule $\tilde{C}_{T,T}$ can be calculated easily, but it is not certain whether the convolution theorem will really give $C_{T,T}$. In general, the inverse transformation $\tilde{C}_{T,T} \rightarrow C_{T,T}$ does not exist.

3. The weighting functions of the planar Vening-Meinesz formula $f_i \sim r^{-3}$ have nonintegrable algebraic singularities at zero. Therefore, the sufficient conditions for the existence of the corresponding convolution products (34) are violated. The formal solutions (35), (37) are practically useful, however. The theoretical deficiency can be overcome with the aid of generalized functions (distributions) [*Keller and Meier,* 1980*a, b*].

The random field T_z may be considered as a generalized random field, and the functions $C_{z,z}$, f_i as generalized functions. It can be proven that $f_i * T_z$ exist. (The asterisk stands for the convolution in the sense of generalized functions.) The generalized C_{ij} can be determined in an analogical way:

$$C_{ij} = f_i * f_j * C_{z,z} \tag{39}$$

If the relation (39) is subjugated a Fourier transformation, it can be seen that the distributions C_{ij} are even functions in the classical sense. The results of the integral transformations (34) are correct even in the rigorous mathematical sense, if all operations of convolutions are understood as operations in the set of distributions.

The mathematical difficulties of the CT in the wave number domain arise from the planar approximation of the geodetic integral formulae. From a physical point of view one is outside the potential theory, where the source field is not closed. The planar approximation, however, is well usable for limited areas (local CF's).

5. *Numerical Estimations and Model Functions (Statistical Significance)*

A function $C(r)$ is called ACF of a scalar homogeneous and isotropic field $X(\mathbf{r})$ if it is even, bounded, and positive definite:

$$C(r) = C(-r) \qquad C(r) \leq C(0)$$
$$\int_a^b \int_a^b f(r')f(r'')C(r'' - r')dr'dr'' \geq 0 \tag{40}$$

The properties (40) have to be taken into account in numerical estimations and model approximations (Table 3). The planar ACF $C(r)$ is positive definite if the Hankel transform according to (6) is non-negative: $\tilde{C}(k) \geq 0$. Theoretical aspects are treated by *Moritz* [1976, 1980] and *Sünkel* [1980], numerical methods for proving the positive definiteness are dealt with by *Groten et al.* [1979], the approximation of CF's by nonpositive definite functions (step functions, piecewise linear functions, spline functions) by *Sünkel* [1978].

Numerical estimations of planar CF's have been made for the gravity field (*Vyskocil* [1970, 1971], *Berljand* [1971], *Lauer* [1971*a, b*], *Hein* [1977], *Merry* [1979], *Stangl* [1979], and others), for the vertical deflection field (*Heitz* [1967, 1968], *Grafarend* [1971*b–e*, *Morrison* [1977], *Schwarz and Lachapelle* [1980], and others), for gravity gradients (*Vyskocil* [1969], *Schwarz* [1976], *Hein* [1977], *Hein and Jochemczyk* [1979], *Groten et al.* [1979], and others). In order to become able to characterize the ACF's and to compare several ACF's, *Moritz* [1976] suggested the use of three characteristic parameters: variance, correlation length, and curvature parameter. Among others, *Hein and Jochemczyk* [1979] and *Stangl* [1979] follow this concept of comparison. ACF's which are to be differentiated (section C2) must be carefully determined, especially in the neighborhood of $r = 0$ (curvature parameter). Estimation errors that are due to the limited sample and the special features of the estimation method (trend elimination, ergodicity, aliasing, etc.) are discussed by, among others, *Groten* [1973], *Hein and Jochemczyk* [1979], and *Groten et al.* [1979].

In most planar applications, statistical homogeneity and isotropy are assumed or confirmed by adequate tests. Nonisotropic effects in the gravity field and its gradients result from geological structures of the earth's crust with preferred directions (mountains, rift valleys), having been investigated by *Grafarend* [1971*b, d,* 1972*a, b, c*], *Kubackova* [1974*a, b*], *Kearsley* [1977], *Morrison* [1977], *Groten et al.* [1979], and *Stangl* [1979]. In order to filter out the anisotropies of the vertical deflection field of Western Germany from astrogeodetic observations and to assign them to the rupture directions of the earth's crust, *Grafarend* [1972*a, b, c*] applied the concept of the multi-axisymmetry (section B2).

6. *Anomalous Mass, Density, and Gravity Structure (Geophysical Significance)*

The anomalous gravity field should be regarded as the effect of an anomalous source field. The concept of representing the ACF's $C_{T,T}$, $C_{z,z}$ in terms of the sources, which are the disturbing masses, was applied by *Vasilev* [1965], *Naidu* [1967, 1968], *Grafarend* [1970*a, b*], *Jordan* [1973, 1978], *Kubackova* [1974*a*], *Schwahn* [1973, 1975*a, b, c,* 1976*a, b*], and others. Whereas for geodesy mainly the gravity field is of interest, the statistical qualities of the mass and density distribution are important for all other geosciences, too. The transformation of cause to effect, rather, is a geological and geophysical problem, so that we will discuss it only briefly.

For example, in the plane $B(x, y; z = b)$, let us assume a mass distribution with the density $\rho(\mathbf{r}; b)$ and the ACF $C_{\rho,\rho}$ $(\mathbf{r} - \mathbf{r}'; b)$; a planar homogeneous density field (Figure 5). The gravity $g(\mathbf{r}; a)$ on the plane $A(x, y; z = a)$ is given by

$$g(\mathbf{r}; a) \sim \int_{-\infty}^{+\infty} f(\mathbf{r} - \mathbf{r}'; h)\, \rho(\mathbf{r}'; b)d\mathbf{r}' \tag{41}$$

with $f(\mathbf{r}, h) = f(|\mathbf{r}|^2 + h^2)^{-3/2}$, depending on the planar param-

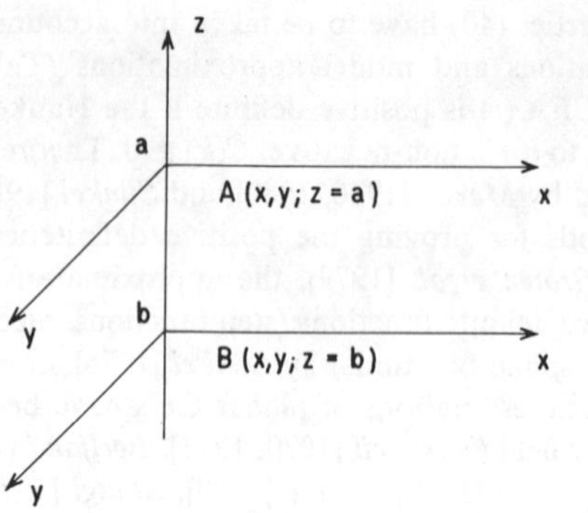

Fig. 5. Plane of sources B and gravity plane A [*Schwahn*, 1975*b*].

eter vector $\mathbf{r} = [x, y]^T$ and the depth of sources $h = a - b$. The relation (41) is of the convolution type with the corresponding gravity ACF

$$C_{g,g}(\mathbf{r}; a) \sim \int_{-\infty}^{+\infty} C_{f,f}(\mathbf{r}', h)\, C_{\rho,\rho}(\mathbf{r} - \mathbf{r}'; b) d\mathbf{r}' \tag{42}$$

being homogeneous on the plane A [*Schwahn*, 1975*a, b, c*, 1976*a, b*].

Most authors postulate $C_{\rho,\rho}$ without any precise statements concerning as to which source model or real geological situation the density ACF corresponds. *Vasilev* [1965] and *Schwahn* [1975*a, b, c*, 1976*a, b*] give an exact description of the mass distribution and the ACF of the source model. There are also assumed sources at different depths h (in different planes $B_1, B_2, \cdots$) [*Mundt*, 1969; *Schwahn*, 1975*a*, 1976*a*]. Finally, vice versa, the depth of sources h can be determined from the ACF's $C_{g,g}$, $C_{\rho,\rho}$ [*Schwahn*, 1975*b*, 1976*b*].

For example, empirical ACF's of gravity and density for a profile over Sweden (Figure 6): a repetitive pattern of the density structures at a distance of about 25 km corresponding to that of gravity. The depth of sources was estimated to be $\simeq$15–22 km.

7. *Remarks About the Anomalous Field of Movement*

A problem that is analogous to those in physical geodesy, the measurement, representation, and interpretation of anomalous fields, exists in structural mechanics and in the mechanics of continua. By means of repeated geodetic point estimations there are derived shifts, velocities, and deformations (mean values over time intervals Δt) for selected (representative) points of the continua, e.g., on structures, glaciers, and ice sheets, in the ground, and in the earth's crust. The fields of velocity and deformation may be decomposed into a mean field that can be described by the solutions of the equations of motion and an anomalous field presented in the concept of stochastic processes. In spite of the formal identity of the statistical task, there are a few special features in comparison with the gravity field.

1. Time dependence of the field observables, and thus also of the statistical parameters and functions, e.g., of the ACF/CCF (an analogon to the theory of turbulence).

2. Different material properties and geometrical structures of continua lead to different types of equations of motion and boundary conditions, and thus also to a variety of statistical structures.

3. Some field observables or their statistical signal shares are small and can be filtered out of the noise of measuring errors only with difficulty (e.g., crustal deformations); their statistical structure is still unknown.

Therefore no planar CF models exist so far, and there are a very few approaches only for the one-dimensional case: variation of quantities of motion with time (time-dependent problem), variation of quantities of motion along a profile at a fixed moment (position-dependent problem).

D. PLANAR COVARIANCE FUNCTIONS IN OPERATIONAL GEODESY

1. *Observations and Observational Equations (a Review)*

In detail, the branch of geodesy where coordinates of points on the earth's surface and in its interior or exterior space are computed from geodetic observables will be called *operational geodesy*. There are three fundaments which make geodesy operational
—the geodetic observables
—the geodetic initial-boundary value problem
—its finite element approximation by geodetic networks.
Operational geodesy needs a careful treatment of geodetic observational equations. From the set-up of geodetic observational equations, from weight estimation to their solution (inversion), *all* information has been given to operate within geodesy. [*Grafarend*, 1978]

In this section, statistical problems of operational geodesy (of surveying) are summarized that can be described and/or solved by means of planar CF's but lie outside section C or lead to mathematically and formally equal or similar solutions as in section C.—This applies to the geodetic observations afflicted with observation correlations (autocorrelations), the combination of homogeneous and/or inhomogeneous observations in the observation equations (cross-correlations) as well as the adjustment and diagnosis of geodetic networks. Differences from physical geodesy exist in two respects:

1. Field correlations (correlations of interaction) are the subject of the statistical analysis of the gravity field. In surveying, observation correlations prevail that may contain correlations of interaction (e.g., from the refraction field) and causal correlations (e.g., owing to surveyor and instrument). In the course of treatment and adjustment, mathematical correlations originate, so that, e.g., in the finite elements of geodetic networks (angles, distances, coordinates), field correlations, causal correlations, and mathematical correlations will superimpose one upon another.

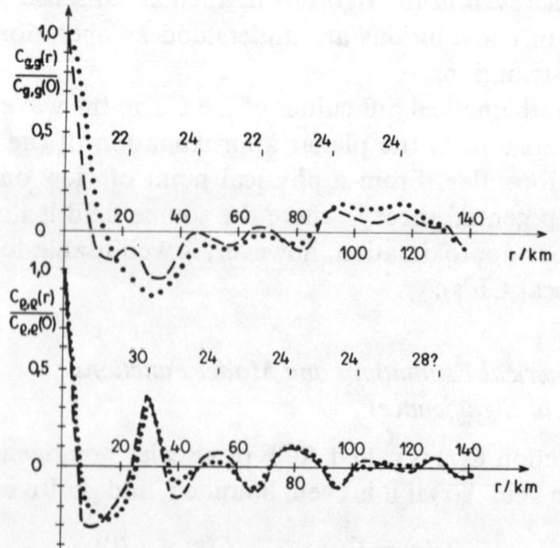

Fig. 6. Empirical ACF's of gravity (above) and density (below) for a profile in Sweden given for different high-pass filtering. The numbers above concern the distance between consecutive relative maxima [*Schwahn*, 1975*b*].

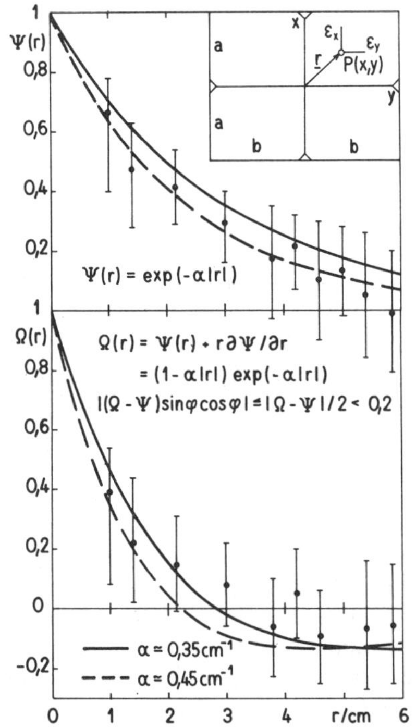

Fig. 7. Longitudinal and transversal correlations $\Omega(r)$, $\Psi(r)$ of a homogeneous and isotropic error vector on a terrestrial exposure (above). Estimated correlations with confidence intervals taken from *Schulz* [1979].

2. In the gravity field, in general, the signal variance predominates over the variance of measuring errors (low noise). On the other hand, in surveying the signal variance is often of the same order as the error variance (snr $\simeq$ 1) or the signal is masked by the measuring errors (heavy noise). Examples: terrestrial refraction within trigonometrical and geometrical leveling, deformation of aerial and terrestrial exposures, soil movements, crustal deformations, and ice movement over bedrock undulations.

A large number of statistical problems of the geodetic measuring technique can be solved within the concept of one-dimensional stochastic processes (time-dependent observations, profile measurements). Some of them cause considerable difficulties in separating signal and noise in a satisfactory way. The planar case exists in photogrammetry, astrometry, and satellite geodesy, in the field of atmospheric refraction, and in the analysis of geodetic networks.

2. *The Photographic Image Plane*

By a photographic image plane we understand a base with a film layer and a fixed Cartesian coordinate system (x, y). By means of a camera, sectors of the celestial globe or parts of the earth's surface are optically imaged on it. Owing to the process of imaging, to atmospheric influences (refraction, scintillation), to deformations of the film base and the film layer, and to plotting, the photocoordinates x, y have the errors ε_x, ε_y. The errors are partly determinate and can be presented with the aid of functional models (e.g., radial components), partly they have a statistical structure and may be described by statistical models. A random error vector $\varepsilon = [\varepsilon_x, \varepsilon_y]^T$ can be assigned to each point $P(x, y)$ of the image plane with the dimensions of $(2a, 2b)$, $-a \leq x \leq +a$, $-b \leq y \leq +b$ (Figure 7). It is reasonable to regard this planar random vector field as homogeneous and isotropic with the CF's $C_{\varepsilon_x,\varepsilon_x}$, $C_{\varepsilon_y,\varepsilon_y}$, $C_{\varepsilon_x,\varepsilon_y} = C_{\varepsilon_y,\varepsilon_x}$ of a Taylor-Kármán structure (16). In the case $\Omega \equiv \psi$, $C_{\varepsilon_x,\varepsilon_y} \equiv 0$, it will decompose into two independent scalar fields of the error components ε_x and ε_y, which can be treated separately (isotropic-chaotic field; *Steenbeck* [1963]).

In photogrammetry, planar CF's are required for (1) the interpolation of coordinate errors and (2) the two-dimensional coordinate transformation [*Kraus,* 1970, 1972*a, b; Kraus and Mikhail,* 1972; *Rampal,* 1974, 1976; *Schulz,* 1979]. According to error analysis having been made so far, the cross covariances are small, in general, so that one can work with the scalar interpolation (example in Figure 7). The ACF/CCF should be estimated not from one image but from a sample of images, if possible.

If the ACF/CCF of the coordinate errors (within the limits of the estimation errors) have a Taylor-Kármán structure (10), (16), and if (11) holds, see example Figure 7, then an error potential ε with grad $\varepsilon = [\varepsilon_x, \varepsilon_y]^T$ exists. The formal identity with physical geodesy (T, grad T) was stressed first by *Rampal* [1974, 1976].

In the photographic satellite geodesy, sectors of the celestial globe are imaged on photographic plates. Planar CF's can be used for the prediction of satellite positions (1) in a field of satellite positions (so-called trace fitting), (2) in a field of stars (so-called plate reduction). The whole process of imaging comprises atmospheric, instrumental, and photographic error components. Less important are errors of the star catalogues and of time measurement. So far, coordinate errors have been eliminated by means of suitable functional models (especially for the astronomical refraction). Random residual errors proved to be statistically homogeneous and isotropic. With a further increase of the measuring accuracy they cannot be neglected any longer and should be treated in the concept of planar stochastic processes with ACF/CCF according to (16) [*Seeber,* 1972].

3. *The Planar Refraction Field*

Geodetic angular measurements are influenced by the curvature of light beams, distance measurements are influenced by the curvature of the path of electromagnetic waves. The anomalous refraction field depends on the space coordinates x, y, z and on the time t, and it varies in the interaction of the anomalous meteorological fields of atmospheric pressure, air temperature, and air moisture, which are four-dimensional too. So far, only one-dimensional stochastic models have been developed for certain special cases, e.g., fluctuations of the local refractive index $n(t)$. We outline the position-dependent planar problem [*Meier,* 1981].

The refraction angle at the observational station is [*Fearnley,* 1883]

$$\delta = \frac{\rho}{Rs} \int_0^s (s - x)\, \kappa(x) dx \tag{43}$$

the effective refraction coefficient with approximately circular sight lines PQ (Figure 8)

$$k = \frac{2}{s^2} \int_0^s (s - x)\, \kappa(x) dx \tag{44}$$

The so-called Fearnley integral

$$I(s) = \int_0^s (s - x)\, \kappa(x) dx \tag{45}$$

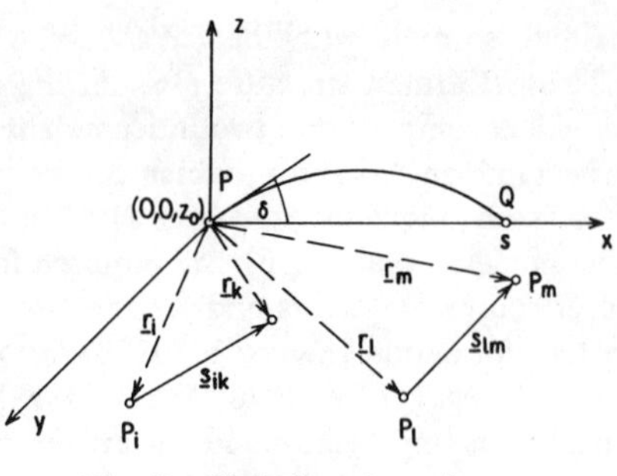

Fig. 8. Sight lines in a plane.

of the convolution type constitutes a special solution to the inhomogeneous curvature equation of the sight line PQ in the vertical plane (x, z) with the weighting function $p(x) = s - x$ $(0 \le p \le s)$ and the local refraction coefficient $\kappa(x)$ with the running position coordinate x (one-dimensional problem). R is the earth's radius, $\rho = 2 \cdot 10^5/\pi$ in mgon. To correlate two sight lines P_iP_k, P_lP_m with each other, we represent (45) in the x-y plane (Figure 8):

$$\kappa = \kappa(\mathbf{r}) \qquad \mathbf{r} = \mathbf{r}_i + t\mathbf{s}_{ik} \qquad t = x/s \qquad (0 \le t \le 1)$$

$$\mathbf{r} = [x_i + t(x_k - x_i);\ y_i + t(y_k - y_i)]^T$$

$$\mathbf{r} = [x_i + x \cos \varphi_{ik};\ y_i + x \sin \varphi_{ik}]^T \qquad (0 \le x \le s)$$

$$I(s) = s^2 \int_0^1 (1-t)\,\kappa[\mathbf{r}(t)]\,dt = \int_0^s (s-x)\kappa[\mathbf{r}(x)]\,dx \tag{46}$$

The ACF $C_{\kappa,\kappa}$ is assumed to be

$$C_{\kappa,\kappa} = C_{\kappa,\kappa}(\mathbf{r}', \mathbf{r}'') \qquad \mathbf{r}' = \mathbf{r}'(x') \qquad \mathbf{r}'' = \mathbf{r}''(x'') \tag{47}$$

The CT in the position domain gives

$$C_{\delta,\delta}(s_{ik}, s_{lm}) = \rho^2\, I(s_{ik}, s_{lm})/R^2 s_{ik}{}^2 s_{lm}{}^2 \tag{48}$$

$$C_{k,k}(s_{ik}, s_{lm}) = 4I(s_{ik}, s_{lm})/s_{ik}{}^4 s_{lm}{}^4 \tag{49}$$

$$I(s_{ik}, s_{lm}) = \int_0^{s_{ik}} \int_0^{s_{lm}} (s_{ik} - x')(s_{lm} - x'')C_{\kappa,\kappa}(\mathbf{r}', \mathbf{r}'')dx'dx'' \tag{50}$$

Analogous to δ, k, one obtains for the effective refractive index

$$\bar{n} = \int_0^1 n[\mathbf{r}(t)]dt = \frac{1}{s}\int_0^s n[\mathbf{r}(x)]\,dx \tag{51}$$

$$C_{\bar{n},\bar{n}}(s_{ik}, s_{lm}) = \frac{1}{s_{ik}s_{lm}} \int_0^{s_{ik}} \int_0^{s_{lm}} C_{n,n}(\mathbf{r}', \mathbf{r}'')dx'dx'' \tag{52}$$

Some special cases are represented in Figure 9. The ACF's $C_{k,k}$ were calculated with the following preconditions: $\kappa(\mathbf{r})$ is homogeneous and isotropic in the plane $z = z_0$ parallel to the ground, $C_{\kappa,\kappa} = C_{\kappa,\kappa}(\tau)$; $\tau =: |r'' - r'|$, Poisson model (Table 3), $s_{ik} = s_{lm} = s$.

Such CF models are suitable for modeling the refraction shares of the measuring errors in geometric and trigonometric levelings as well as in geodetic networks. The calculation of $C_{\delta,\delta}$, $C_{k,k}$, $C_{\bar{n},\bar{n}}$ is a task that is analogous to physical geodesy, if $C_{\overline{\Delta g},\overline{\Delta g}}$ is estimated from representative (mean) gravity anomalies $\overline{\Delta g}$.

4. *Diagnosis and Design of Geodetic Networks*

An optimum design and observation of a geodetic network is possible with the aid of mathematical optimization. As the objective function of the network optimization and as a criterion of quality for network diagnosis with regard to accuracy, the following can be used:

1. Scalar functions of accuracy. They combine into one number the information contained in an error covariance matrix Σ, e.g., tr(Σ), det(Σ).
2. Complete error covariance matrices, so-called criterion matrices. They contain the detailed error information in an ideal structure, e.g., error ellipses with axes of equal length, error circles.

Criterion matrices were introduced into geodesy by *Grafarend* [1970*a*, *b*, 1972*a*, *b*, *c*], *Baarda* [1973], and *Alberda* [1974] and used for diagnosing and designing geodetic networks (*Bossler et al.* [1973], *Grafarend* [1975*a*, *b*], *Grafarend et al.* [1976], *Kelm* [1976], *Schaffrin et al.* [1977], *Schmitt et al.* [1978], and others). The geometric structure of so-called multipoint tensors is used for the theory, and the theory is described in detail for two-dimensional homogeneous and isotropic geodetic networks in *Grafarend et al.* [1979], *Grafarend and Schaffrin* [1979]. Whereas it was already applied with success to the optimization of the observational process (second-order design), so far there have not been available any papers on the optimization of the network configuration (first-order design), except a few examples of the single-point determination [*Grafarend et al.*, 1979]. Criterion matrices have the advantage that the real network configuration does not enter and the drawback that the characteristic functions of a homogeneous and isotropic network remain indeterminate (e.g., the characteristic distance d in the Markov model, Figure 10).

We will give a short outline of the statistical structure of a two-dimensional homogeneous and isotropic network: CF's of the absolute point coordinates and of relative coordinates (coordinate differences) as well as of azimuths and distances that can be derived from them.

Let there be given a centered operational geodetic process (point field) $\{x_i(\mathbf{r})\}$ in the sense of definitions 2 and 3 (section B3) with the expectation values equal to zero and the CF of the absolute Cartesian coordinates x_i

$$\Sigma_{ij}(\mathbf{r}_1, \mathbf{r}_2) \equiv E\{x_i(\mathbf{r}_1)x_j(\mathbf{r}_2)\}$$
$$= \Sigma_T(r_{12})\delta_{ij} + [\Sigma_L(r_{12}) - \Sigma_T(r_{12})]\,\Delta x_i \Delta x_j r_{12}{}^{-2} \tag{53}$$

with the Taylor-Kármán structure (10), the longitudinal and transversal correlations Σ_L, Σ_T, and the planar distance $r_{12} = |\mathbf{r}_2 - \mathbf{r}_1|$. The CF of the relative Cartesian coordinates $x_i(\mathbf{r}_1) - x_i(\mathbf{r}_2)$ is calculated as follows:

$$\Phi_{ij}(\mathbf{r}_1, \mathbf{r}_2, \mathbf{r}_3, \mathbf{r}_4) \equiv E\{[x_i(\mathbf{r}_1) - x_i(\mathbf{r}_2)][x_j(\mathbf{r}_3) - x_j(\mathbf{r}_4)]\}$$
$$= \Sigma_{ij}(\mathbf{r}_1, \mathbf{r}_3) - \Sigma_{ij}(\mathbf{r}_1, \mathbf{r}_4) - \Sigma_{ij}(\mathbf{r}_2, \mathbf{r}_3) + \Sigma_{ij}(\mathbf{r}_2, \mathbf{r}_4) \tag{54}$$

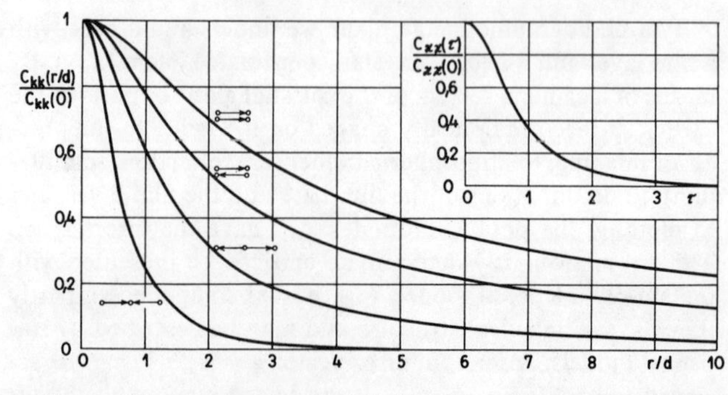

Fig. 9. Covariance functions $C(r/d)$ of the effective refraction coefficient of sight lines of equal length on a plane parallel to the earth's surface (Poisson model), normalized.

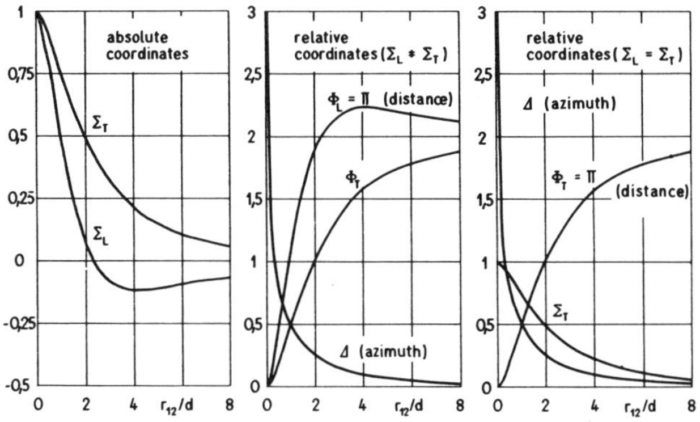

Fig. 10. Covariance functions of finite elements within a two-dimensional homogeneous and isotropic geodetic network (first-order Markov model; *Grafarend and Schaffrin* [1979], *Grafarend et al.* [1979]).

To make possible the construction of error ellipses for coordinate differences, we put $\mathbf{r}_1 = \mathbf{r}_3$, $\mathbf{r}_2 = \mathbf{r}_4$:

$$\Phi_{ij}(\mathbf{r}_1, \mathbf{r}_2) = \Sigma_{ij}(\mathbf{r}_1, \mathbf{r}_1) - \Sigma_{ij}(\mathbf{r}_1, \mathbf{r}_2) - \Sigma_{ij}(\mathbf{r}_2, \mathbf{r}_1) + \Sigma_{ij}(\mathbf{r}_2, \mathbf{r}_2)$$
$$= 2\,[\Sigma_T(0)\delta_{ij} - \Sigma_{ij}(r_{12})] \tag{55}$$

Analogous to (54) and (55), from (53) there can be derived CF's of other finite elements, summarized in Table 4 and presented in Figure 10 for a special model. With the prerequisite (53) all of them are homogeneous and isotropic.

5. *Geometrical Interpretation*

In a real geodetic network, the error ellipses are different from point to point, in a statistically homogeneous network they are equal to each other, in a homogeneous and isotropic network they are defined by the Taylor-Kármán decomposition (53). The error ellipses of local point differences are circles exactly when $\Sigma_L = \Sigma_T$ (Table 4). This ideal error situation in point fields is not aimed at in all geodetic measurements. In the construction of traffic routes (e.g., setting out of tunnels) a good directional accuracy is required rather than a good distance accuracy. Here CF's of an axisymmetric structure (sectoin B2) can be advantageous.

E. CONCLUSIONS

Planar geodetic covariance functions constitute an efficient means for the representation and geodetic realization of geophysical scalar and vector fields (physical-geodetic process), of observational errors, and of error covariance matrices of point configurations (operational geodetic process). In physical geodesy they are used for setting up the covariance matrices of signal and noise in the concept of least squares collocation within limited areas, in operational geodesy they are used for error analysis and for optimizing geodetic observational processes. In all fields of application the Taylor-Kármán decomposition of the CF of homogeneous-isotropic vectors proves to be very important. Other symmetric structures (axisymmetry) have been discussed already but not yet fully utilized with regard to their possible applications.

The planar transformation problem in physical geodesy having been solved to a considerable degree in both the position domain and the wave number domain, now the numerical problems are being increasingly investigated: CF estimations for regions with different geological structures including the anisotropy effects, approximation functions, estimation errors, positive definite functions and matrices, etc.

There are not yet available any planar models of the movement and deformation fields. Likewise, the development of planar stochastic refraction models is still in a primary stage. In the optimization of networks with criterion matrices, development work should mainly be done concerning configuration optimization (first-order design). Finally, MPCF's have been structurally investigated but not yet applied.

APPENDIX: DIVISION OF REFERENCES BY SUBJECT MATTER

Foundations. The following references fall under the heading of basic works: *Abramowitz and Stegun* [1965], *Batchelor* [1946], *Burda and Vyskocil* [1973], *Burda* [1975], *Chandrasekhar* [1950], *Fearnley* [1883], *Gradshteyn and Ryzhik* [1965], *Grafarend* [1972*a*, 1976, 1978], *Heiskanen and Moritz* [1967], *Kármán* [1937], *Kármán and Howarth* [1938], *Krarup* [1969], *Kubackova* [1975], *Moritz* [1969, 1970*a*, *b*, 1972, 1973, 1978*a*, *b*, *c*, 1980], *Mundt* [1969], *Nash and Jordan* [1978], *Obuchow* [1954], *Robertson* [1940], *Rummel* [1975], *Steenbeck* [1963], *Taylor* [1935, 1938], and *Whittle* [1954].

Physical geodesy. The following references fall under the heading of physical geodesy: *Baussus* [1960], *Bellaire* [1971, 1972, 1977], *Berljand* [1971], *Grafarend* [1970*a*, *b*, 1971*a–f*, 1972*b*, 1975*a*], *Grafarend and Offermanns* [1975], *Groten* [1973], *Groten et al.* [1979], *Hein* [1977], *Hein and Jochemczyk* [1979], *Heitz* [1967, 1968, 1969], *Heller* [1976], *Heller and Jordan* [1979], *Hirvoven* [1962], *Jordan* [1972*a*, *b*, 1973, 1978], *Jordan and Heller* [1978], *Kasper* [1971], *Kaula* [1959], *Kearsley* [1977], *Keller and Meier* [1980*a*, *b*], *Kubackova* [1974*a*, *b*, 1976], *Lauer* [1971*a*, *b*], *Meier* [1980], *Meissl* [1970], *Merry* [1979], *Mihelcic* [1972], *Moritz* [1963, 1974, 1975, 1976], *Morrison* [1977], *Naidu* [1967, 1968], *Neiman* [1974*a*, *b*], *Obenson* [1970], *Reilly* [1979], *Schwahn* [1973, 1975*a*, *b*, *c*, 1976*a*, *b*], *Schwarz* [1974, 1976], *Schwarz and Lachapelle* [1980], *Shaw et al.* [1969, 1971], *Soltau* [1970], *Stangl* [1979], *Sünkel* [1978, 1980], *Vasilev* [1965], and *Vyskocil* [1969, 1970, 1971, 1973].

TABLE 4. Covariance Functions of Finite Elements Within a Two-Dimensional Homogeneous and Isotropic Geodetic Network

Finite Elements	Symbol	Covariance Functions $\Sigma_L \neq \Sigma_T$	Covariance Functions $\Sigma_L = \Sigma_T$	
Absolute Cartesian coordinates $x_i(\mathbf{r}_1)$	$\Sigma_{ij}(\mathbf{r}_1, \mathbf{r}_2)$	$\Sigma_T(r_{12})\delta_{ij} + [\Sigma_L(r_{12}) - \Sigma_T(r_{12})]\Delta x_i \Delta x_j r_{12}^{-2}$	$\Sigma_T(r_{12})\delta_{ij}$	(covariance)
Relative Cartesian	$\Phi_{ij}(\mathbf{r}_1, \mathbf{r}_2)$	$2[\Sigma_T(O)\delta_{ij} - \Sigma_{ij}(r_{12})]$	$\Phi_T(r_{12})\delta_{ij}$	(covariance)
coordinates	$\Phi_L(r_{12})$	$2[\Sigma_T(O) - \Sigma_L(r_{12})]$	$\Phi_T(r_{12})$	(covariance)
$x_i(\mathbf{r}_1) - x_i(\mathbf{r}_2)$	$\Phi_T(r_{12})$	$2[\Sigma_T(O) - \Sigma_T(r_{12})]$	$\Phi_T(r_{12})$	(covariance)
Azimuth $\phi(\mathbf{r}_1, \mathbf{r}_2)$	$\Delta(\mathbf{r}_1, \mathbf{r}_2)$	$r_{12}^{-2}\,\Phi_T(r_{12})$	$\Phi_T(r)_{12}$	(variance)
Distance $s(\mathbf{r}_1, \mathbf{r}_2)$	$\Pi(\mathbf{r}_1, \mathbf{r}_2)$	$\Phi_L(r_{12})$	$\Phi_T(r_{12})$	(variance)

Grafarend and Schaffrin [1979]; *Grafarend et al.* [1979].

Operational geodesy. The following references fall under the heading of operational geodesy: *Alberda* [1974], *Baarda* [1973], *Borre* [1979], *Bossler et al.* [1973], *Grafarend* [1970c, 1972c, 1974, 1975b], *Grafarend and Schaffrin* [1979], *Grafarend et al.* [1976, 1979], *Kelm* [1976], *Kraus* [1970, 1972a, b], *Kraus and Mikhail* [1972], *Kubacek* [1974], *Meier* [1981], *Rampal* [1974, 1976], *Schaffrin et al.* [1977], *Schmitt* [1977], *Schmitt et al.* [1978], *Schulz* [1979], and *Seeber* [1972].

Acknowledgments. The author is grateful to E. Grafarend (Stuttgart), W. Keller (Dresden), and W. Schwahn (Potsdam) for helpful discussions, comments, and critical reviews of the manuscript and to E. Wiesend (Dresden) for the translation from German. This paper was stimulated by E. W. Grafarend, one of the pioneers in the field of geodetic application of stochastic processes, especially in development of planar models.

REFERENCES

Abramowitz, M., and I. A. Stegun (Eds.), *Handbook of Mathematical Functions,* Dover, New York, 1965.

Alberda, J. E., Planning and optimization of networks: Some general considerations, *Boll. Geod. Sci. Affini, 33,* 209–240, 1974.

Baarda, W., S-transformations and criterion matrices, *New Series 5*(1), Netherlands Geod. Comm., Delft, The Netherlands, 1973.

Batchelor, G. K., The theory of axisymmetric turbulence, *Proc. R. Soc. London Ser. A, 186,* 480, 1946.

Baussus, H., A unified isostatic and statistical theory of gravity anomalies and its significance, paper presented at 12th General Assembly, Int. Union Geod. and Geophys., Helsinki, 1960.

Bellaire, R. G., Correlation functions on the upper half-space, *Intern. Rep.,* Analytic Sci. Corp., Reading, Mass., 1971.

Bellaire, R. G., A discussion of flat-earth statistical models for the gravity disturbances and their application, *Intern. Rep.,* Analytic Sci. Corp., Reading, Mass., 1972.

Bellaire, R. G., Correlation functions on the upper half-space, *Bull. Geod., 51,* 149–161, 1977.

Berljand, N. G., O vozmoznostyach avtokorrelyacionnogo analiza pri izucenii structury gravimetacionnogo polya, *Izv. Akad. Nauk SSSR Fiz. Zemli, 1,* 68–78, 1971.

Borre, K., Covariance functions for random error propagation in large geodetic networks, *Zesz. Nauk. Akad. Górn. Hutn. (Krakow),* no. 720, 23–33, 1979.

Bossler, J. D., E. Grafarend, and R. Kelm, Optimal design of geodetic nets, *J. Geophys. Res., 78,* 5887–5897, 1973.

Burda, M., Properties of the characteristics of linear transformations of geophysical fields, *Stud. Geophys. Geod., 19,* 8–13, 1975.

Burda, M., and V. Vyskocil, Statistical characteristics of geophysical fields in Hilbert's space, *Stud. Geophys. Geod., 17,* 300–305, 1973.

Chandrasekhar, S., The theory of axisymmetric turbulence, *Proc. R. Soc. London Ser. A, 242,* 557, 1950.

Fearnley, C., Zur Theorie der terrestrischen Refraktion, in *Verhandlungen der Siebenten Allgemeinen Conferenz der Europäischen Gradmessung,* Rome, 1883.

Gradshteyn, I. S., and I. M. Ryzhik, *Table of Integrals, Series and Products,* Academic, New York, 1965.

Grafarend, E., The Gaussian structure of the gravity field, *Stud. Geophys. Geod., 14,* 159–167, 1970*a*.

Grafarend, E., Theory of the stochastic gravity field, *Phys. Lett. A, 31,* 7–8, 1970*b*.

Grafarend, E., Die Genauigkeit eines Punktes im mehrdimensionalen Euklidischen Raum, *Rep. C153,* Dtsch. Geod. Komm. bei der Bayrischen Akad. der Wiss., Munich, 1970*c*.

Grafarend, E., Discussion of paper by L. Shaw, I. Paul, and P. Henrikson, 'Statistical models for the vertical deflection from gravity-anomaly models,' *J. Geophys. Res., 76,* 627–628, 1971*a*.

Grafarend, E., Isotropietests von Lotabweichungsverteilungen in Westdeutschland, 1, *Z. Geophys., 37,* 719–733, 1971*b*.

Grafarend, E., A combined gravimetric-astrogeodetic method for telluroid and vertical deflection analysis, *Rep. B188,* pp. 23–36, Dtsch. Geod. Comm. bei der Bayrischen Akad. der Wiss., Munich, 1971*c*.

Grafarend, E., Lotabweichungsverteilungen Westdeutschlands und ihre gruppentheoretische Struktur, *Rep. 1,* Mitt. Inst. Theor. Geod., Bonn, 1971*d*.

Grafarend, E., Korrelationslängen von Schwereanomalien und Lotabweichungen, *Z. Vermessungswesen, 96,* 132–140, 1971*e*.

Grafarend, E., Statistische Modelle zur Prädiktion von Lotabweichungen, *Vermessungstechnik, 19,* 66–68, 1971*f*.

Grafarend, E., Nichtlineare Prädiktion, *Z. Vermessungswesen, 97,* 245–255, 1972*a*.

Grafarend, E., Isotropietests von Lotabweichungsverteilungen in Westdeutschland, 2, *Z. Geophys., 38,* 243–255, 1972*b*.

Grafarend, E., Genauigkeitsmasse geodätischer Netze, *Rep. A73,* Dtsch. Geod. Komm. bei der Bayrischen Akad. der Wiss., Munich, 1972*c*.

Grafarend, E., Optimization of geodetic networks, *Boll. Geod. Sci. Affini, 33,* 351–406, 1974.

Grafarend, E., Geodetic stochastic processes, in *Methoden und Verfahren der mathematischen Physik,* vol. 14, *Mathematical Geodesy,* edited by B. Brosowski and E. Martensen, Bibliographisches Institut Verlag, Mannheim, Federal Republic of Germany, 1975*a*.

Grafarend, E., Second-order design of geodetic nets, *Z. Vermessungswesen, 100,* 158–168, 1975*b*.

Grafarend, E., Geodetic applications of stochastic processes, *Phys. Earth Planet. Inter., 12,* 151–179, 1976.

Grafarend, E., Operational geodesy, in *Approximation Methods in Geodesy,* edited by H. Moritz and H. Sünkel, H. Wichmann Verlag, Karlsruhe, Federal Republic of Germany, 1978.

Grafarend, E., and G. Offermanns, Eine Lotabweichungskarte Westdeutschlands nach einem geodätisch konsistenten Kolmogorov-Wiener Modell, *Rep. A82,* Dtsch. Geod. Komm. bei der Bayrischen Akad. der Wiss., Munich, 1975.

Grafarend, E., and B. Schaffrin, Kriterion-Matrizen, 1, Zweidimensionale homogene und isotrope geodätische Netze, *Z. Vermessungswesen, 104,* 133–149, 1979.

Grafarend, E., B. Schaffrin, and G. Schmitt, Über die Optimierung lokaler geodätischer Netze, in *Seventh Int. Kurs für Ing. Vermess. hoher Präzision,* vol. 1, Technische Hochschule Darmstadt, Darmstadt, Federal Republic of Germany, 1976.

Grafarend, E., et al., *Optimierung geodätischer Messoperationen,* H. Wichmann Verlag, Karlsruhe, Federal Republic of Germany, 1979.

Groten, E., Some numerical aspects of least-squares autoregression prediction, *Boll. Geod. Sci. Affini, 4,* 331–357, 1973.

Groten, E., G. Hein, and H. Jochemczyk, On the determination of empirical covariances, *Allg. Vermessungsnachrichten, 86,* 17–32, 1979.

Hein, G., Die Berechnung von anomalen vertikalen Schweregradienten hoher Präzision, *Rep. C234,* Dtsch. Geod. Komm. bei der Bayrischen Akad. der Wiss., Munich, 1977.

Hein, G., and H. Jochemczyk, Reflections on covariances and collocation with special emphasis on gravity gradient covariances, *Manuscr. Geod., 4,* 45–86, 1979.

Heiskanen, W., and H. Moritz, *Physical Geodesy,* W. H. Freeman, San Francisco, Calif., 1967.

Heitz, S., Ein Vorschlag zur Interpolation von astronomisch-geodätisch bestimmten Lotabweichungen, *Rep. A55,* Dtsch. Geod. Komm. bei der Bayrischen Akad. der Wissenschaften, Munich, 1967.

Heitz, S., Geoidbestimmung durch Interpolation nach kleinsten Quadraten aufgrund gemessener und interpolierter Lotabweichungen, *Rep. C124* Dtsch. Geod. Komm. bei der Bayrischen Akad. der Wiss., Munich, 1968.

Heitz, S., Eine astronomisch-geodätische Geoidbestimmung für Westdeutschland, *Rep. B167,* Dtsch. Geod. Komm. bei der Bayrischen Akad. der Wiss., Munich, 1969.

Heller, W. G., A new self-consistent statistical gravity field model, paper presented at AGU Fall Meeting, San Francisco, Calif., 1976.

Heller, W. G., and S. K. Jordan, Attenuated white noise statistical gravity model, *J. Geophys. Res., 84,* 4680, 1979.

Hirvonen, R. A., On the statistical analysis of gravity anomalies, *Rep. 37,* Publ. Isost. Inst. Int. Assoc. of Geod., Helsinki, 1962.

Jordan, S. K., Discussion of paper by L. Shaw, I. Paul, and P. Henrikson, 'Statistical models for the vertical deflection from gravity-anomaly models,' *J. Geophys. Res., 77,* 971–973, 1972*a*.

Jordan, S. K., Self-consistent statistical models for the gravity anomaly, vertical deflections, and undulation of the geoid, *J. Geophys. Res., 77,* 3660–3670, 1972*b*.

Jordan, S. K., Stationary statistical models for mass anomalies and gravity disturbances, *Intern. Rep.,* Analytic Sci. Corp., Reading, Mass., 1973.

Jordan, S. K., Statistical model for gravity, topography, and density contrasts in the earth, *J. Geophys. Res., 83,* 1816–1824, 1978.
Jordan, S. K., and W. G. Heller, Upward continuation of gravity disturbance covariance functions, *J. Geophys. Res., 83,* 3382–3388, 1978.
Kármán, T., On the statistical theory of turbulence, *Proc. Nat. Acad. Sci. U. S. A., 23,* 98–105, 1937.
Kármán, T., and L. Howarth, On the statistical theory of isotropic turbulence, *Proc. R. Soc. London Ser. A, 164,* 917, 1938.
Kasper, J. F., A second-order Markov gravity anomaly model, *J. Geophys. Res., 76,* 7844–7849, 1971.
Kaula, W. M., Statistical and harmonic analysis of gravity, *J. Geophys. Res., 64,* 2041–2421, 1959.
Kearsley, W., Non-stationary estimation in gravity prediction problems, *Rep. 256,* Dep. of Geod. Sci., The Ohio State Univ., Columbus, 1977.
Keller, W., and S. Meier, Kovarianzfunktionen der 1. und 2. Ableitungen des Schwerepotentials in der Ebene, *Veröff. 60,* Zentralinst. Phys. der Erde, Potsdam, German Democratic Republic, 1980*a*.
Keller, W., and S. Meier, Geodetic integral formulas in the plane and generalized functions, paper presented at Fourth International Symposium on Geodesy and Physics of the Earth, Akad. der Wiss. der DDR, Karl-Marx-Stadt, German Democratic Republic, 1980*b*.
Kelm, R., Zur analytischen Optimierung geodätischer Netze, *Rep. C220,* Dtsch. Geod. Komm. bei der Bayrischen Akad. der Wiss., Munich, 1976.
Krarup, T., A contribution to the mathematical foundation of physical geodesy, *Publ. 44,* Dan. Geod. Inst., Copenhagen, 1969.
Kraus, K., Interpolation nach kleinsten Quadraten in der Photogrammetrie, *Z. Vermessungswesen, 95,* 387–390, 1970.
Kraus, K., Interpolation nach kleinsten Quadraten in der Photogrammetrie, *Bildmess. Luftbildw., 40,* 7–12, 1972*a*.
Kraus, K., Film deformation correction with least-squares interpolation, *Photogramm. Eng., 38,* 487–493, 1972*b*.
Kraus, K., and E. M. Mikhail, Linear least-squares interpolation, *Photogramm. Eng., 38,* 1016–1029, 1972.
Kubacek, L., On the stochastic relations between the points of a geodetic network, *Stud. Geophys. Geod., 18,* 19–32, 1974.
Kubackova, L., Disturbance of the statistical isotropy of the gravity field in the region of Czechoslovakian Carpathians, *Contrib. Geophys. Inst. Slovak. Akad. Sci., 5,* 15–20, 1974*a*.
Kubackova, L., Autocovariance function and the spectral density of the anomalous gravitational field transformed by an integral operator of convolution type, *Stud. Geophys. Geod., 18,* 329–338, 1974*b*.
Kubackova, L., Mathematical background for studying the stochastical properties of vector geophysical fields, *Stud. Geophys. Geod., 19,* 313–321, 1975.
Kubackova, L., The fundamental statistical properties of the Vening-Meinesz and Stokes integral transformations, *Stud. Geophys. Geod., 20,* 209–212, 1976.
Lauer, S., Über die stochastischen Eigenschaften lokaler Schwereanomalien und ihre Prädiktion, Ph.D. thesis, Bonn University, Bonn, 1971*a*.
Lauer, S., Kovarianzen, Power-Spektren und Prädiktionen im lokalen Schwerefeld, *Rep. B188,* 37–49, Dtsch. Geod. Komm. bei der Bayrischen Akad. der Wiss., Munich, 1971*b*.
Meier, S., Die Kovarianzfunktion der 2. Ableitung der Schwere nach der Höhe, *Stud. Geophys. Geod., 24,* 314–316, 1980.
Meier, S., Stochastisches Refraktionsmodell für Zielstrahlen in einer bodenparallelen Ebene, *Vermessungstechnik,* in press, 1981.
Meissl, P., Probabilistic error analysis of airborne gravimetry, *Rep. 138,* Dep. of Geod. Sci., The Ohio State University, Columbus, 1970.
Merry, C. L., A practical comparison of some methods of predicting point gravity anomalies, paper presented at Seventeenth General Assembly, Int. Assoc. of Geod., Canberra, 1979.
Mihelcic, M., Über eine Theorie zur Simultanauswertung von fluggravimetrischen und terrestrischgravimetrischen Messungsdaten, *Rep. C172,* Dtsch. Geod. Komm. bei der Bayrischen Akad. der Wiss., Munich, 1972.
Moritz, H., Statistische Methoden in der gravimetrischen Geodäsie, *Z. Vermessungswesen, 88,* 409–416, 1963.
Moritz, H., A general theory of gravity processing, *Rep. 122,* Dep. of Geod. Sci., The Ohio State Univ., Columbus, 1969.
Moritz, H., Eine allgemeine Theorie der Verarbeitung von Schweremessungen nach kleinsten Quadraten, *Rep. A67,* Dtsch. Geod. Komm. bei der Bayrischen Akas. der Wiss., Munich, 1970*a*.
Moritz, H., Least-squares estimation in physical geodesy, *Rep. 130,* Dep. of Geod. Sci., The Ohio State Univ., Columbus, 1970*b*.
Moritz, H., Advanced least-squares estimation, *Rep. 175,* Dep. of Geod. Sci., The Ohio State Univ., Columbus, 1972.
Moritz, H., Least-squares collocation, *Rep. A75,* Dtsch. Geod. Komm. bei der Bayrischen Akad. der Wiss., Munich, 1973.
Moritz, H., Some first accuracy estimates for applications of aerial gradiometry, *Rep. 209,* Dep. of Geod. Sci., The Ohio State Univ., Columbus, 1974.
Moritz, H., Combination of aerial gravimetry and gradiometry, *Rep. 223,* Dep. of Geod. Sci., The Ohio State Univ., Columbus, 1975.
Moritz, H., Covariance functions in least-squares collocation, *Rep. 240,* Dep. of Geod. Sci., The Ohio State Univ., Columbus, 1976.
Moritz, H., Least-squares collocation, *Rev. Geophys. Space Phys., 16,* 421–430, 1978*a*.
Moritz, H., Statistical foundations of collocation, *Rep. 272,* Dep. of Geod. Sci., The Ohio State Univ., Columbus, 1978*b*.
Moritz, H., The operational approach to physical geodesy, *Rep. 277,* Dep. of Geod. Sci., The Ohio State Univ., Columbus, 1978*c*.
Moritz, H., *Advanced Physical Geodesy,* H. Wichmann Verlag, Karlsruhe, Federal Republic of Germany, 1980.
Morrison, F., Azimuth-dependent statistics for interpolating geodetic data, *Bull. Geod., 51,* 105–118, 1977.
Mundt, W., Statistische Analyse geophysikalischer Potentialfelder hinsichtlich Aufbau und Struktur der tieferen Erdkruste, *Rep. 41,* Dtsch. Akad. Wiss., East Berlin, German Democratic Republic, 1969.
Naidu, P. S., Statistical properties of potential fields over a random medium, *Geophysics, 32,* 8898, 1967.
Naidu, P. S., Spectrum of potential field due to randomly distributed sources, *Geophysics, 33,* 337–345, 1968.
Nash, R. A., and S. K. Jordan, Statistical geodesy—An engineering perspective, *Proc. IEEE, 66,* 532–550, 1978.
Neiman, J. M., Ob usloviyach izotropnosti pri veroyatnostnom izucenii gravitacionnogo polya, *Izv. Vyssh. Uchebn. Zaved. Geod. Aerofotosemka, 1974*(1), 85–91, 1974*a*.
Neiman, J. M., Veroyatnostnaya modifikaciya formuly Stoksa pri vycislenii anomali vysot, *Izv. Vyssh. Uchebn. Zaved. Geod. Aerofotosemka, 1974*(6), 21–24, 1974*b*.
Obenson, G., The covariance matrix for deflections of the vertical and undulations based on actual gravity data, *Rep. 136,* Dep. of Geod. Sci., The Ohio State Univ., Columbus, 1970.
Obuchow, A. M., Statistische Beschreibung stetiger Felder, *Akad. Nauk SSSR, 151*(3), 1954.
Rampal, K. K., Analysis and prediction of photogrammetric residual errors, *J. Surv. Mapp. Div. Am. Soc. Civ. Eng., 100*(SU1), 33–48, 1974.
Rampal, K. K., Least-squares collocation in photogrammetry, *Photogramm. Eng. Remote Sens., 42,* 659–669, 1976.
Reilly, W. I., Mapping the local geometry of the earth's gravity field, *Rep. 143,* Geophys. Div., Dep. Sci. Ind. Res., Wellington, New Zealand, 1979.
Robertson, H. P., The invariant theory of isotropic turbulence, *Proc. Cambridge Philos. Soc., 36,* 209, 1940.
Rummel, R., Zur Behandlung von Zufallsfunktionen und -folgen in der physikalischen Geodäsie, *Rep. C208,* Dtsch. Geod. Komm. bei der Bayrischen Akad. der Wiss., Munich, 1975.
Schaffrin, B., E. Grafarend, and G. Schmitt, Kanonisches Design geodätsicher Netze, 1, *Manuscr. Geod., 2,* 263–306, 1977.
Schmitt, G., Experiences with the second order design problem in theoretical and practical networks, paper presented at International Symposium on Optimization of Design and Computation of Control Networks, Hung. Acad. of Sci., Sopron, Hungary, 1977.
Schmitt, G., E. Grafarend, and B. Schaffrin, Kanonisches Design geodätischer Netze, 2, *Manuscr. Geod., 3,* 1–22, 1978.
Schulz, H.-U., Zur Genauigkeit der Bildkoordinaten in der terrestrischen Photogrammetrie beim Einsatz von Phototheodoliten, Ph.D. thesis, Tech. Univ. Dresden, Dresden, German Democratic Republic, 1979.
Schwahn, W., Some remarks on the simulation of the empirical autocovariance functions of gravity, paper presented at Second International Symposium on Geodesy and Physics of the Earth, Akad. der Wiss. der DDR, Potsdam, German Democratic Republic, 1973.
Schwahn, W., Eine allgemeine Formulierung der Auto- und Kreuzkorrelationsfunktionen eines beliebigen statistischen Potential-

feldes in einem kartesischen Koordinatensystem, *Gerlands Beitr. Geophys., 84,* 143–154, 1975*a*.

Schwahn, W., Beziehungen zwischen den Autokovarianzfunktionen der Schwere und der Dichte und deren Anwendung bei der Tiefenbestimmung der Quellen, demonstriert anhand von zwei Profilen über Schweden, *Gerlands Beitr. Geophys., 84,* 281–296, 1975*b*.

Schwahn, W., Zwei statistische Dichtemodelle und die statistischen Momente erster und zweiter Ordnung ihrer Schwerefelder, *Gerlands Beitr. Geophys., 84,* 478–486, 1975*c*.

Schwahn, W., Die Beziehungen zwischen den Autokovarianzfunktionen der Potentialfelder und denen der Quellenverteilung unter besonderer Berücksichtigung der realen geologischen Situation, Ph.D. thesis, Akad. Wiss. DDR, Zentralinst. Phys. der Erde, Potsdam, German Democratic Republic, 1976*a*.

Schwahn, W., Stochastic source models as the base for the statistical treatment of geophysical potential fields, paper presented at Third International Symposium on Geodesy and Physics of the Earth, Weimar, German Democratic Republic, 1976*b*.

Schwarz, K. P., Combination of spatial networks using an estimated covariance matrix, *Bull Geod., 112,* 171–186, 1974.

Schwarz, K. P., Geodetic accuracies obtainable from measurements of first and second order gravitational gradients, *Rep. 242,* Dep. of Geod. Sci., The Ohio State Univ., Columbus, 1976.

Schwarz, K. P., and G. Lachapelle, Local characteristics of the gravity anomaly covariance function, *Bull. Geod., 54,* 21–35, 1980.

Seeber, G., Über das stochastische Verhalten von photographisch bestimmten Stern- und Satellitenkoordinaten, *Rep. C178,* Dtsch. Geod. Komm. bei der Bayrischen Akad. der Wiss., Munich, 1972.

Shaw, L., I. Paul, and P. Henrikson, Statistical models for the vertical deflection from gravity anomaly models, *J. Geophys. Res., 74,* 4259–4265, 1969.

Shaw, L., I. Paul, and P. Henrikson, Reply, *J. Geophys. Res., 76,* 629, 1971.

Soltau, G., Möglichkeiten zur kartographischen Darstellung des Schwerefeldes, *Rep. B179,* Dtsch. Geod. Komm. bei der Bayrischen Akad. der Wiss., Munich, 1970.

Stangl, G., Lokale Kovarianzfunktionen von Freiluftanomalien in der Bundesrepublik Deutschland, *Allg. Vermessungsnachrichten, 86,* 81–88, 1979.

Steenbeck, M., Elementare magnetohydrodynamische Behandlung chaotisch turbulenter Medien, *Monatsber. Dtsch. Akad. Wiss., 5,* 625–629, 1963.

Sünkel, H., Approximation of covariance functions by non-positive definite functions, *Rep. 271,* Dep. of Geod. Sci., The Ohio State Univ., Columbus, 1978.

Sünkel, H., On the positive definiteness of a homogeneous and isotropic covariance function, *Boll. Geod. Sci. Affini, 3,* 472–482, 1980.

Taylor, G. I., Statistical theory of turbulence, *Proc. R. Soc. London Ser. A, 151,* 421, 1935.

Taylor, G. I., Statistical theory of turbulence, 2, *Proc. R. Soc. London Ser. A, 164,* 15, 1938.

Vasilev, V. G., Optimalnaya filtraciya slucaynoy sorokupnosti plastov, neogranicennych po prostiraniyu i vertikali, *Geol. Geofiz.,* 146–148, 1965.

Vyskocil, V., Statistical characteristics of linear transformations of gravity anomalies, *Stud. Geophys. Geod., 13,* 252–258, 1969.

Vyskocil, V., On the covariance and structure functions of the anomalous gravity field, *Stud. Geophys. Geod., 14,* 174–177, 1970.

Vyskocil, V., Correlation analysis of gravity anomalies and isostasy, *Rep. 40,* U.M.P. Program in USSR, 1962–1970, Prague, 1971.

Vyskocil, V., Structure functions as characteristics of geophysical fields, *Stud. Geophys. Geod., 17,* 295, 1973.

Whittle, P., On stationary processes in the plane, *Biometrika, 41,* 434–449, 1954.

(Accepted May 26, 1981.)

REVIEWS OF GEOPHYSICS AND SPACE PHYSICS, VOL. 18, NO. 1, PAGES 27–76, FEBRUARY 1980

Gravity Fields of the Terrestrial Planets: Long-Wavelength Anomalies and Tectonics

ROGER J. PHILLIPS

Jet Propulsion Laboratory, California Institute of Technology, Pasadena, California 91103

KURT LAMBECK

Research School of Earth Sciences, Australian National University, Canberra, Australia

We present a review of the long-wavelength gravity fields of the terrestrial planets, earth, moon, Mars, and Venus with particular emphasis placed on the interrelationship between gravity anomalies and tectonic processes. After first summarizing appropriate statistical formulas, we discuss the relevant continuum mechanical solutions for elastic, viscoelastic, and convecting media in terms of the relationship to the gravity field both for predicting gravity anomalies and for the use of gravity as a constraining boundary condition. Gravity data can provide a strong constraint for the first two classes of solutions, but a significant ambiguity exists in convection interpretations because of the competing effects from flow and boundary deformation. The question of finite strength of crustal and mantle materials plays a paramount role in gravity studies because of the possibility of model discrimination based on stress states. A variety of evidence is reviewed, and it is concluded that the finite strength of the terrestrial planets is unlikely to exceed a kilobar when subject to loads over long geologic periods. Isostasy plays a dominant role in any discussion of the lithospheric contribution to planetary gravity fields. Since it tends to minimize stresses, it is a basic state to be expected, and most useful discussions of the gravity fields are centered on departures from the isostatic state, either due to dynamic forces or to finite strength. For the earth the long-wavelength anomalies arise largely from beneath the lithosphere and are related to mantle flow. Apart from the obvious correlation of gravity to the plate boundaries, the specific relationships between the gravity field and relevant flow parameters are complex, and there are at present no systematic relationships established between geophysical observables in the intraplate regions. On the moon, most of the contributions to the gravity field arise from the lithosphere and are supported by the finite strength of that layer. The mascons are due to a combination of several kilometer thick basalt layers and crustal collapse at the time of impact, forming ring structures and mantle uplift. Mars appears to be intermediate between the earth and moon and is dominated by the Tharsis anomaly, with the possibility that gravity anomalies arise from both lithospheric and sublithospheric regions. If the Tharsis Province is supported entirely by the lithosphere, then stress levels are of the order of 1 kbar, even with isostasy, which may imply partial dynamic support. Preliminary analysis of Venus gravity data suggests a power spectrum grossly similar to that of the earth, but systematic discrepanc es among some harmonics may indicate differences in the nature of interior processes for the two pla ıets. In particular, there appears to be a higher correlation of long-wavelength gravity with topograph' for Venus.

CONTENTS

1. INTRODUCTION

The purpose of this paper is to review the long-wavelength gravity anomalies of the terrestrial planets in terms of their origin and of their relationship to tectonic processes. The useful interval of the gravity spectrum for studying lateral density variations in the terrestrial planets ranges over approximately 4 orders of magnitude. In this paper we define long-wavelength anomalies, in general, as those variations with wavelengths that exceed 10–15% of a planetary radius R, or, for the earth, that exceed about 1000 km.

The interrelationship of long-wavelength gravity anomalies and tectonics may be viewed from two related points of view. In the first instance, the gravity anomaly is interpreted in terms of a lateral density variation which, along with other geophysical and geological data, is used to model the contem-

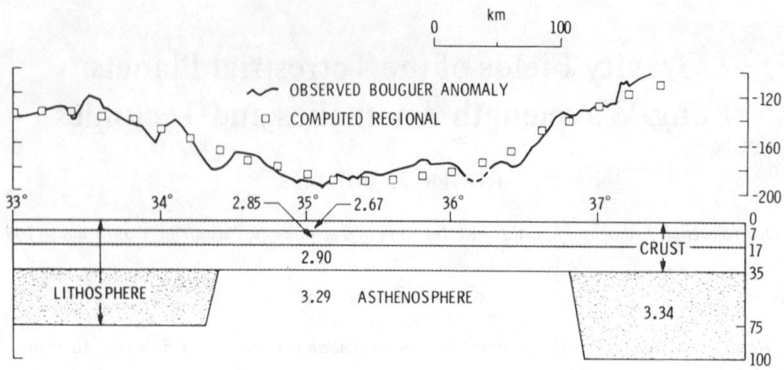

Fig. 1. Computed lithosphere model for a profile at 1.8°S across Gregory Rift of the East African Rift system. Densities are in g cm^{-3}, and computed and observed Bouguer anomalies are in mgals. From *Fairhead* [1976].

porary tectonic state and infer past tectonic events. An example of this approach is the modeling of the negative Bouguer anomaly over certain sections of the East African Rift. The resultant models have been interpreted in terms of a low-density upper mantle [e.g., *Fairhead,* 1976], which may be associated with a region of nascent plate spreading (Figure 1). On the other hand, the presence of a gravity anomaly is directly related to a state of anomalous potential energy relative to a spherically stratified planet. This implies a state of stress which is always such as to remove the lateral density variations that gave rise to the gravity anomaly in the first place. This state of stress may be intimately related to a tectonic process (e.g., as the driving force) and the gravity anomaly may be utilized to either predict or be a prediction of, via a mathematical model, another observable such as faulting pattern, gross topographic deformation, or heat flow. The simplest application of this idea is found in the elastic loading problem [e.g., *Heiskanen and Vening-Meinesz,* 1958] where the response of the elastic lithosphere to a known forcing function, the topographic load, is predicted in terms of the gravity anomaly (Figure 2).

The problems associated with the interpretation of planetary gravity fields are severe, and only by comparing them with other planetary observations can some geophysical sense be made out of the global anomalies. This is true despite the

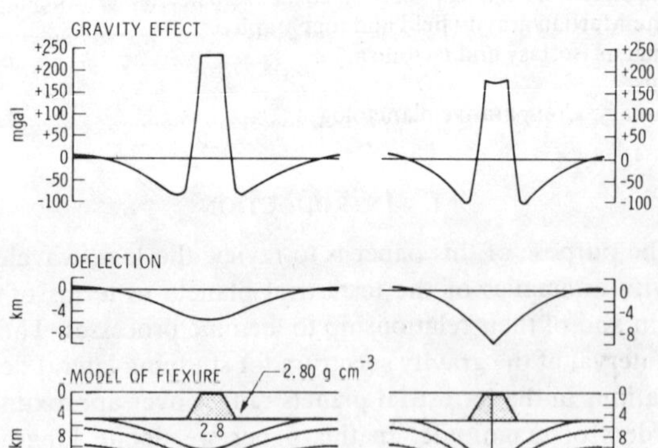

Fig. 2. Gravity effect and deflection for elastic flexure of the lithosphere due to seamount load. The effective flexural rigidity of the lithosphere is 2×10^{30} dyne-cm. Model A is for a continuous lithosphere; model B is for a fractured lithosphere. From *Watts and Cochran* [1974].

tantalizing relations often seen between terrestrial gravity anomalies and surface tectonics [e.g., *Kaula,* 1972]; on the moon, pronounced gravity anomalies are found over the ringed maria [e.g., *Muller and Sjogren,* 1968], while on Mars the Tharsis plateau dominates both gravity and topography [e.g., *Phillips et al.,* 1973]. But no other geophysical observations of the terrestrial planets, with the possible exception of topography and, on earth, the plate velocities, are known on a global basis with the same precision and spatial resolution, and the supporting data required in the gravity inversions are meager. Moreover, the interpretation of lateral density structures from analysis of gravity data suffers from an inherent nonuniqueness. The gravitational potential satisfies Laplace's equation in a source-free region, and it is well known that if a solution to this equation is specified everywhere on a boundary surface, then it is completely determined in the enclosed volume. It is clear that an infinite number of mass distributions can give rise to the same potential on the bounding surface. Parameterizing the density distribution and assigning values to one or more parameters, such as geometric shape, density contrast, or depth, often reduce the nonuniqueness to acceptable levels [e.g., *Parker,* 1975]. Ultimately, any density model that satisfies the observed gravitational potential is also required to be reasonable from geological and geophysical points of view.

In the face of nonuniqueness, one generally favors those distributions for which the density contrasts are least, as this minimizes the implications that density models may have on other geophysical observations. But, at least in the earth, the magnitude of the density anomalies required is generally much less than indicated by other observations, such as lateral seismic velocity anomalies, or by lateral temperature variations, and this suggests that there may be considerable internal compensation within the earth in which positive density anomalies are overlain by negative density anomalies and vice versa [*Lambeck,* 1976; *Kaula,* 1977*a*]. This had been suggested earlier by *Lewis and Dorman* [1970] in their regional study of experimental isostasy (but see also *Banks et al.* [1977] and section 6). *McQueen and Stacey* [1976] made a similar suggestion in their interpretation of the gravity field in terms of negatively correlated undulations in the transition zones near 400- and 650-km depth, so that the contributions to gravity from these two layers partially cancel [see also *Bott,* 1971]. Thus the observed surface gravity may well reflect second-order differences between first-order effects that are only partly observed by other geophysical techniques and which can presently only be predicted to at best a first-order accuracy.

A key problem in understanding planetary gravity fields is the separation of contributions from the crust, lithosphere, asthenosphere, and lower mantle. In any discussion, the lithosphere provides a convenient starting point for a number of reasons. Among these are (1) the density anomalies are closest to the surface and can be expected to dominate the exterior gravity field, (2) the density anomalies can be expected to correlate with topography and with geological surface observations, (3) if any part of the planet has long-term finite strength, it will be this outer portion, and the density anomalies therein can be expected to be supported elastically over geologic time.

The sublithosphere is a region where we expect the finite strength to be very low and where flow will occur by subsolidus creep processes. That part of the gravity field originating in this layer contains information on these flow processes, more so than on the material rheology, as is the case for the lithospheric layer. The problem is complicated, however, by interactions between the flow in the asthenosphere and deformations in the overlying lithosphere. For example, most of the earth's mountain ranges are a consequence of a coupling of the lithosphere to the flow below it, whether directly so, or indirectly, while the Tharsis plateau on Mars appears to be a consequence of mantle flow unless the Martian lithosphere is thick indeed (>400 km [*Lambeck,* 1979*a*]).

The separation of lithospheric and sublithospheric contributions to gravity is exacerbated by a number of interrelated processes that contribute to the overall density structure in both layers and which demonstrate some of the difficulties referred to above. Many of these processes can be attributed to lateral temperature variations. Consider a simple ocean-continent model in which the lithosphere is thicker under the continents than under the oceans. If conventional isostasy prevails, free-air gravity will be mainly positive over land. However, when both Airy compensation of the crust and thermal compensation of the lithosphere are taken into account, the difference in geoid heights between stable continents and ocean basins may be positive, negative, or zero, depending on continental lithospheric thickness [*Turcotte and McAdoo,* 1979]. The observation for the earth of essentially no difference in geoid heights for ocean basins and continents implies that these two compensation mechanisms essentially cancel. These interactions illustrate that it is futile to attempt to interpret the gravity field in terms of either compositional or thermal phenomena alone, unless it can first be demonstrated that the gravitational effect of one dominates the other.

Estimates of the contributions to the observed gravity from density anomalies in the deep mantle of a planet become tenuous because of the rapid attenuation of gravity with distance. It has often been suggested that the low-degree harmonics originate mainly in the deep mantle, while higher degree harmonics reflect the upper mantle structure. While this is approximately true, it is not very helpful. This can be seen if we consider a model for the lateral structure of a planet in which there are small density anomalies with a random spatial and magnitude distribution and of dimension less than the wavelength λ of the spherical harmonics of degree l ($= 2\pi R/\lambda$) in which the exterior gravity field is expanded. The anomalies in the lower mantle will contribute significantly only to low-degree harmonics, while such anomalies in the upper mantle contribute to all terms. (See section 7 below.) Assuming that the simple statistical model is reasonable (it does not appear to be relevant for Mars, for example, where the low-degree harmonics correlate strongly with topography and seem to be of lithospheric or upper mantle origin, section 9), this result permits only approximate limits to be established on the maximum densities anomalies that may be tolerated in the lower mantle, which is not very helpful in addressing the question as to what part of the gravity results from the lower mantle lateral structure. Other arguments, for or against a proposed low-mantle density structure, must be sought. Anomalous lateral density variations in the lower mantles of the moon and Mars do not appear to be significant and only for the earth, and possibly for Venus, need the lower mantle be considered.

Of particular importance in the discrimination among various density models is the constraint of maximum stress. Certain subsurface configurations lead to unacceptably high stress differences; i.e., they exceed what is understood to be the 'finite strength' of rock materials over the range of temperature and confining pressure assumed appropriate for the planet. Lack of evidence for recent tectonic failure then provides a bounding condition on acceptable classes of models. For example, it is a common procedure to assign certain long-wavelength terrestrial gravity anomalies to a sublithospheric origin after showing that they would lead to unacceptably high stresses if they originated in the lithosphere [e.g., *McKenzie,* 1967; *Lambeck,* 1972; *Cochran and Talwani,* 1977]. Because of this strong link to gravity modeling, we review briefly in this paper both the subjects of stress calculations and finite strength.

The association of gravity and stress is quantitatively expressed in terms of solutions of the equations of mathematical physics that relate anomalous density or temperature to strain and flow in geologic materials. In turn, gravitational potential is related to density or temperature (through the coefficient of thermal expansion) through Poisson's equation. In this paper we also review in passing the relevant relationships between gravity and the appropriate geophysical state variables and describe the response of elastic, viscoelastic, and viscous media under the action of applied stresses and buoyancy forces. In addition to the above, the first part of this paper presents a review of the statistical formulas of global planetary gravity fields which permit a simple intercomparison of those terrestrial planets for which we have statistically representative data. Also discussed are the general aspects of gravity-topography correlation and isostasy. In the last part of this review, we address ourselves specifically to the gravity fields of the earth, the moon, Mars, and Venus.

2. STATISTICAL ANALYSES OF GLOBAL PLANETARY GRAVITY FIELDS

A summary of formulas. When discussing the global aspects of the gravitational potential U of a planet, it is convenient to expand the field into spherical harmonics. The usual definition of U at latitude ϕ and longitude λ outside the planet is

$$U(r, \phi, \lambda) = \frac{GM}{r} + \Delta U$$

where the anomalous potential ΔU is expressed as

$$\Delta U(r, \phi, \lambda) = \frac{GM}{r} \sum_{l=2}^{\infty} \sum_{m=0}^{l} \left(\frac{R}{r}\right)^l \{C_{lm}{}^* \cos m\lambda + S_{lm}{}^* \sin m\lambda\} P_{lm}(\sin \phi) \qquad (1)$$

Here G is the gravitational constant, M the mass of the planet, and R the mean equatorial radius of the planet. The $P_{lm}(\sin\phi)$ are the associated Legendre polynomials and are taken here as fully normalized such that

$$\frac{1}{4\pi}\int_S [P_{lm}(\sin\phi)\cos\text{ (or sin) } m\lambda]^2\, dS = 1$$

The potential as defined by (1) is with respect to a reference frame rotating with the planet and with a center of mass origin. The Stokes coefficients C_{lm}^*, S_{lm}^* relate to the anomalous mass distribution within the planet according to

$$C_{lm}^*\text{ (or } S_{lm}^*) = \frac{1}{(2l+1)M}\int_V \left(\frac{r'}{R}\right)^l \rho(r', \phi', \lambda') \cdot P_{lm}(\sin\phi')\cos\text{ (or sin) } m\lambda'\, dV \tag{2}$$

where the integral is over all density anomalies $\rho(r', \phi', \lambda')$ within the planet of volume V. Introducing the notation

$$\begin{aligned} C_{ilm}^* Y_{ilm} &= C_{lm}^* \cos m\lambda P_{lm}(\sin\phi) \qquad i = 1 \\ C_{ilm}^* Y_{ilm} &= S_{lm}^* \sin m\lambda P_{lm}(\sin\phi) \qquad i = 2 \end{aligned} \tag{3}$$

we can express (1) compactly as

$$\Delta U = \frac{GM}{r}\sum_i \sum_l \sum_m \left(\frac{R}{r}\right)^l C_{ilm}^* Y_{ilm}$$

For rapidly rotating planets such as the earth and Mars, the Stokes coefficient C_{20}^*—related to the dynamical flattening—dominates the other coefficients, and it is usual to express the anomalous potential not relative to the central term GM/r but relative to the hydrostatic equilibrium form of the planet. The latter can be expressed in spherical harmonics as [e.g., *Heiskanen and Moritz,* 1967]

$$U^{HE} = \frac{GM}{r}\left\{1 + \sum_{l'=1}^{\infty}\left(\frac{R}{r}\right)^{2l'} C_{1(2l')0}{}^{HE} Y_{1(2l')0}\right\}$$

where, in general, only the first term in the summation is considered. The anomalous potential is now

$$\Delta U = \frac{GM}{r}\sum_i \sum_l \sum_m \left(\frac{R}{r}\right)^l C_{ilm} Y_{ilm} \tag{4}$$

with

$$C_{1lm} = C_{1lm}^*$$

if $m > 0$ or l is odd; otherwise,

$$C_{1lm} = C_{1lm}^* - C_{1lm}{}^{HE}$$
$$C_{2lm} = C_{2lm}^*$$

The relation between the anomalous potential and the gravity perturbation or gravity disturbance δg is

$$\delta g = -\frac{\partial}{\partial r}(\Delta U)$$

while the relation between ΔU and the gravity anomaly Δg is

$$\Delta g = -\frac{\partial}{\partial r}(\Delta U) - \frac{2}{r}\Delta U \tag{5}$$

which is the gravity relative to the reference ellipsoid for the planet. Hence, with (4) and $g = g(R)$ being the value of gravity at R,

$$\Delta g = g\sum_i \sum_l \sum_m (l-1) C_{ilm} Y_{ilm} \tag{6}$$

The geodetic problem is to estimate the potential coefficients, C_{ilm}, by analyzing the orbital perturbations of planetary orbiters. In the case of the earth, Δg is also observed with gravity meter measurements, and the C_{ilm} can be deduced by inverting (6), but some combination of the two methods is usually employed [*Gaposchkin and Lambeck,* 1971]. More recently, satellite radar altimeter measurements have permitted a direct geometric measurement of the equipotential surface over the oceans. The geophysical problem is to invert (2) from the known C_{ilm}, i.e., to determine the unknown $\rho(r', \phi', \lambda')$ throughout the planet subject to certain sets of the constraints to be discussed in the subsequent sections. Clearly, a unique inversion is not possible, but some order of magnitude ideas on the size of the density anomalies can be obtained if we consider density anomalies $\delta\rho(r', \phi', \lambda')$, distributed throughout a layer of radius r' and thickness δr, and examine the consequences of such a layer on gravity and on other geophysical quantities. The surface density layer can be expressed as

$$\sigma(r') \simeq \delta\rho\,\delta r = \sum_{i=1}^{2}\sum_{l=0}^{\infty}\sum_{m=0}^{l} \delta\rho_{ilm}(r') Y_{ilm}\,\delta r \tag{7}$$

where the $l = 0$ term determines the average density of the layer and contributes only to the central term GM/r of the potential. Outside of this layer the potential at r, ϕ, λ, distance L from an element σ at r', ϕ', λ', is

$$\Delta U(r, \phi, \lambda) = G\int_S \frac{\sigma(r', \phi', \lambda')}{L(r, \phi, \lambda, r', \phi', \lambda')}\, dS$$

where the integral is over the surface S of radius $r = r'$. Expanding L^{-1} into a spherical harmonic expansion, using (7) and the orthogonality relations of spherical harmonics, gives on the planet's surface $r = R$ [e.g., *Kaula,* 1968]:

$$\Delta U(R, \phi, \lambda) \simeq 4\pi G R\sum_i \sum_l \sum_m \left(\frac{r'}{R}\right)^{l+2}\frac{\delta r}{2l+1}\delta\rho_{ilm} Y_{ilm} \tag{8}$$

Equating this with (4) gives

$$\delta\rho_{ilm}\,\delta r = \frac{1}{3}\bar\rho R\,(2l+1)\left(\frac{R}{r'}\right)^{l+2} C_{ilm} \tag{9}$$

where $\bar\rho$ is the mean density of the planet. Equation (9) establishes a unique relationship between C_{ilm} and $\delta\rho_{ilm}$. If the layer δr is not infinitely thin, as is supposed in (8), but the density anomalies are distributed between two spheres of radii r_j' and r_{j+i}', then

$$\Delta U(r, \phi, \lambda) = \sum_i \sum_l \sum_m \frac{3}{\bar\rho}\frac{\delta\rho_{ilm}}{(2l+1)(l+3)} Y_{ilm} \left[\left(\frac{r_{j+1}'}{R}\right)^{l+3} - \left(\frac{r_j'}{R}\right)^{l+3}\right] \tag{10}$$

The statistical properties of the potential are conveniently summarized by the dimensionless power spectrum defined as [*Kaula,* 1967]

$$V_l^2(\Delta U) = \sum_i \sum_m C_{ilm}{}^2 \tag{11}$$

The corresponding power spectrum of surface density is

$$V_l^2(\sigma) = \sum_i \sum_m (\rho_{ilm}\, dr)^2 = \left(\frac{\bar{\rho}R}{3}\right)^2 (2l+1)^2 \left(\frac{R}{r'}\right)^{2l+4} V_l^2(\Delta U) \tag{12}$$

for a single surface density layer.

Seismic studies of the earth indicate that the actual length scale of the lateral variations in the mantle may be much less than the $2\pi R/l$ suggested by the geopotential expansions. Lateral seismic variations observed in sinking slabs appear to be more abrupt than the broad gravity anomalies seen over the subduction zones, and regional variations in v_P and v_S are observed in the upper mantle with horizontal dimensions of a few hundred kilometers. Thus while the spherical harmonic expansions are a convenient and natural choice in view of the orbital mechanics of the motions of satellites as well as for the global inversions of gravity data, they may not give a very good description of the actual mass distribution, unless the field can be expressed to a much higher degree and order than it is now. For the moon the indications are also that the mass anomalies are much smaller in size than the wavelengths of the spherical harmonic expansions. For Mars, on the other hand, this does not appear to be so, as the strong correlation between the low-degree harmonics of gravity and topography argue for long-wavelength density anomalies. For Venus we might expect a situation largely comparable to that of the earth.

The spectrum of the earth's gravitational potential (11) adheres closely to a power law first suggested by Kaula in 1963, namely, that

$$V_l^2(\Delta U) \simeq A(2l+1)10^{-10} l^{-4} \tag{13}$$

where $A \simeq 0.7$. Such a power law can be explained by a random distribution of density anomalies [*Lambeck*, 1976; *Kaula*, 1977*a*]. If we assume that the density anomalies are randomly distributed throughout n layers of radii,

$$r_j' = r_0'\left[1 + \left(\frac{n+1}{2} - j\right)\frac{dr}{r_0}\right] \qquad j = 1, 2, \cdots, n$$

where r_0' is the mean radius of the n layers and the density anomaly in each layer is

$$\delta\rho^j = \sum_i \sum_l \sum_m \delta\rho_{ilm}{}^j Y_{ilm}$$

then, with (8),

$$V_l^2(\Delta U) = \left(\frac{3}{\bar{\rho}R}\right)^2 \frac{1}{(2l+1)^2} \sum_j \left(\frac{r_j'}{R}\right)^{2l+4} V_l^2(d\rho^j)\, dr^2$$

$$= \left(\frac{3}{\bar{\rho}R}\right)^2 \frac{1}{(2l+1)^2} \left(\frac{r_0'}{R}\right)^{2l+4} \cdot \sum_j \left[1 + \left(\frac{n+1}{2} - j\right)\frac{dr}{r_0'}\right]^{2l+4} V_l^2(\delta\rho)\, \delta r^2 \tag{14}$$

This spectrum decays according to (13) if the density spectrum is the same for all layers, if the anomalies occur throughout the mantle, and if the anomalous density has a white noise spectrum for wavelengths greater than the spatial dimensions of the anomalies (Figure 15). This model is not unique, but it is the most elementary and useful for the present discussion. With such a distribution, most of the observed gravity field of the planet originates from density anomalies in the upper mantle where, in the case of the earth, seismic studies show the most important and rapid lateral variations. Density anomalies in the lower mantle are required, however, to explain the low-degree ($\lesssim 6$) spectral estimates, for without them, the decay of the spectrum would be less rapid than the above law. Hence if the random density model is valid, the slope of the potential power spectrum gives some indication of the depth range in which these anomalies occur. Much the same result was obtained by *Kaula* [1977*a*], who took a Monte Carlo approach. Kaula's investigation lends itself to further investigations of deviations of the actual spectrum from the power law (13) or of different distributions of density anomalies, but in the present context of discussions of the relatively long wavelength part of the planetary gravity fields, this refinement is probably not warranted. One conclusion reached by both Lambeck and Kaula, albeit with different arguments, is that there is a suggestion that the density anomalies may not be randomly distributed in the vertical direction and that there may be a tendency toward compensation in which positive density anomalies overlay negative density anomalies and vice versa.

Interrelationship of planetary spectra. With the above formulation, a number of convenient relations can be established between the spectrum of the gravitational potential and spectra of other geophysical quantities such as topography, seismic velocities, or temperature.

For example, the planetary topography can also be expressed in spherical harmonics according to

$$h = R \sum_i \sum_l \sum_m h_{ilm} Y_{ilm} \tag{15}$$

and the dimensionless topographic spectrum is defined as

$$V_l^2(h) = \sum_i \sum_m h_{ilm}{}^2 \tag{16a}$$

The earth's topography is measured relative to the geoid while the other planetary topographies are measured with respect to a best-fitting sphere or to an ellipsoid of revolution. Thus to make a valid comparison, the spectra of the latter must be reduced by the geoid spectrum

$$V_l^2(N) = R^2 \sum_i \sum_m C_{ilm}{}^2 \tag{16b}$$

Only in the case of the second-degree harmonics of Mars is this reduction significant. A further modification is required for the earth in that, as we are using the topography as a surface load, we must replace oceans and ice sheets by an equivalent rock layer. This is the equivalent rock topography [*Balmino et al.*, 1973].

A measure of the correlation between the topography and gravity is given by the degree correlation function as

$$V_l^2(h, \Delta U) = \sum_i \sum_m C_{ilm} h_{ilm} \Big/ [V_l^2(h) V_l^2(\Delta U)]^{1/2} \tag{17}$$

and the cumulative correlation for coefficients of degree l_1–l_2 is

$$\beta_{l_1-l_2} = \sum_{l=l_1}^{l_2} \sum_i \sum_m C_{ilm} h_{ilm} \Big/ \sum_l [V_l^2(h) V_l^2(\Delta U)]^{1/2} \tag{18}$$

The potential spectrum due to the topographic layer is, with $\delta\rho_{ilm} = \rho_c$, a constant crustal density, and $\delta r = Rh_{ilm}$,

$$V_l^2(\Delta U) = \left(\frac{3\rho_c}{\bar{\rho}}\right)^2 \frac{1}{(2l+1)^2} V_l^2(h) \tag{19}$$

If this topography is compensated by a conventional Airy mechanism

$$V_l^2(\Delta U) = \left(\frac{3\rho_c}{\bar{\rho}}\right)^2 \frac{1}{(2l+1)^2}\left[1-\left(\frac{R-D_a}{R}\right)^l\right]^2 V_l^2(h) \tag{20}$$

where D_a is the Airy compensation depth (see section 6). This expression gives essentially the same result as a Pratt model with $D_p = 2D_a \ll 2\pi R/l$.

The earth's topographic spectrum decays with increasing wave number approximately according to

$$V_l^2(h) = A/l(l+1) \tag{21}$$

as first noted by Vening-Meinesz in 1952. *Bills* [1978] further noted that such a decay also characterized the topographies of the other terrestrial planets. Substituting this into (19) gives a predicted potential spectrum that is proportional to l^{-4}, or

$$V_l^2(\Delta U) = \left(3\,\frac{\rho_c}{\bar{\rho}}\right)^2 \frac{A}{l(l+1)(2l+1)^2} \simeq \left(\frac{3}{2}\,\frac{\rho_c}{\bar{\rho}}\right)^2 \frac{A}{l^4} \tag{22a}$$

With (20) and (21) the predicted potential due to topography and its Airy compensation is

$$V_l^2(\Delta U) = \left(\frac{3\rho_c}{\bar{\rho}}\right)^2 \frac{A}{(2l+1)^2 l(l+1)}\left[1-\left(\frac{R-D_a}{R}\right)^l\right]^2$$

or, for $D \ll R$,

$$V_l^2(\Delta U) \simeq \left(\frac{3}{2}\,\frac{\rho_c}{\bar{\rho}}\,\frac{D_a}{R}\right)^2 \frac{A}{l^2} \tag{22b}$$

and decays with l according to l^{-2}.

If seismic P velocity anomalies v_P are observed in a layer and expanded as

$$v_P = \sum_i \sum_l \sum_m v_{ilm} Y_{ilm}$$

and a Birch law of the form

$$v_P = a_v\rho + b_v \tag{23}$$

is valid, then

$$V_l^2(\Delta U) = \left(\frac{3dr}{R\bar{\rho}a_v}\right)^2 \frac{1}{(2l+1)^2}\left(\frac{r'}{R}\right)^{2l+4} V_l^2(v_P) \tag{24}$$

For N layers of finite thickness, with (10),

$$V_l^2(\Delta U) = \left(\frac{3}{\bar{\rho}a_v}\right)^2 \frac{1}{(2l+1)^2(l+3)^2} \sum_j^N \left[\left(\frac{r_{j+1}'}{R}\right)^{l+3} - \left(\frac{r_j'}{R}\right)^{l+3}\right]^2 V_l^2(v_j) \tag{25}$$

The approximate pressure-temperature conditions required for a mantle phase transition (such as that from olivine to spinel) to occur are

$$P(\text{kbar}) = a_t T(°\text{C}) + b_t$$

and a change in temperature causes a change in the depth of the transition zone according to

$$dr = \frac{a_t}{\rho g}\,dT \tag{26}$$

The surface density layer is $\Delta\rho_t a_t\, dT/\rho g$, where $\Delta\rho_t$ is the density contrast across the transition zone and ρ is now the average density above this zone. Expanding the temperature fluctuations dT on a sphere of radius r as

$$dT = \sum_i \sum_l \sum_m T_{ilm} Y_{ilm}$$

the predicted potential power spectrum is

$$V_l^2(\Delta U) = \left(\frac{3a_t\rho_c}{R\bar{\rho}\rho g}\right)^2 \frac{1}{(2l+1)^2}\left(\frac{r'}{R}\right)^{2l+4} V_l^2(dT) \tag{27}$$

The above equations suffice to demonstrate how orders of magnitude can be estimated of the topography, density, seismic velocity, or temperature fluctuations that are required to explain a given gravitational potential.

The velocities (V_ϕ (ϕ, λ), V_λ (ϕ, λ)) of the terrestrial lithospheric plates have been expanded into vector spherical harmonics for spheroidal coefficients:

$$a_{ilm}{}^S = \frac{1}{l(l+1)}\int_S \left(V_\phi \frac{\partial Y_{ilm}}{\partial\phi} + \frac{V_\lambda}{\sin\phi}\,\frac{\partial Y_{ilm}}{\partial\lambda}\right) dS \tag{28a}$$

and toroidal coefficients:

$$a_{ilm}{}^T = \frac{1}{l(l+1)}\int_S \left(\frac{V_\phi}{\sin\phi}\,\frac{\partial Y_{ilm}}{\partial\lambda} - V_\lambda \frac{\partial Y_{ilm}}{\partial\phi}\right) dS \tag{28b}$$

by [*Hager and O'Connell,* 1978]. These coefficients are based on plate velocity data given by *Solomon et al.* [1975] (modified after *Minster et al.* [1974]). The power spectrum for toroidal velocities is almost as large as that for the spheroidal velocities, which is unexpected, since toroidal modes do not contain a vertical component and to first order could not be associated with convective flows. A strong coupling of toroidal and spheroidal modes is indicated.

Hager and O'Connell [1978, 1979] studied mantle convection from a kinematic point of view by using the velocity coefficients as boundary conditions for Stokes flow, neglecting buoyancy effects. The gravity and velocity spectra can be combined as observational constraints on mantle convection by including the density perturbations due to convection as the buoyancy term in the flow equation (see *Hager and O'Connell,* [1979] for the formulation). The full dynamic problem for inversion would include the energy equation for advection and conduction of temperature, which is coupled to the flow through the velocity of advection and through the Boussinesq approximation which relates perturbation density ρ and temperature T via

$$\rho = \rho_0[1 - \alpha(T - T_0)] \tag{29}$$

where T_0 and ρ_0 refer to the reference state and α is the coefficient of thermal expansion. Heat flow data are a possible additional global data set for inverting the full set of convection equations.

A further set of convenient formulas can be obtained from stress considerations, since the stress required to support the density anomalies is proportional to the surface density. A convenient criterion for characterizing the stress state is the maximum stress difference that occurs anywhere within a layer that is loaded at its surface. For a surface load $\sigma_{ilm} Y_{ilm}$ on an incompressible sphere or isostatically supported on a floating lithosphere, the maximum stress differences are of the order of

$$\tau_{\max} \simeq \beta g \sigma_{ilm} Y_{ilm} \tag{30}$$

where, typically, β ranges from 1/3 to 1 depending on the nature of the loading problem (see sections 3 and 6). The stress spectrum becomes, with (12),

$$V_l^2(\tau) = \beta^2 g^2 V_l^2(\sigma)$$
$$= \beta^2 g^2 \left(\frac{\bar{\rho} R}{3}\right)^2 (2l+1)^2 \left(\frac{R}{r}\right)^{2l+4} V_l^2(\Delta U) \tag{31}$$

The magnitude of stress occurring in the planet due to any of the above discussed processes can now be readily estimated with (31). Equation (31) also permits a scaling of the spectra from one planet to another. For example, if the ratio of the finite strength of two planets, $P^{(1)}$ and $P^{(2)}$, is denoted by

$$\alpha = \tau_{\max}{}^{(1)}/\tau_{\max}{}^{(2)}$$

and if the topography is controlled mainly by the finite strength of the crust and upper mantle, the topography scales from one planet to the next according to

$$V_l^2(h^{(1)}) = \alpha^2 \frac{[g^{(2)}\rho^{(2)}R^{(2)}]^2}{[g^{(1)}\rho^{(1)}R^{(1)}]^2} V_l^2(h^{(2)}) \tag{32}$$

where the superscripts (1) and (2) indicate the appropriate parameters for the two planets.

A similar scaling law can be obtained for the gravitational potentials. For example, if a known potential is assumed to be a consequence of an isostatically compensated topography according to (20), then the stress differences are of the order of

$$V_l^2(\tau) \simeq \beta^2 \frac{(g\bar{\rho}R)^2}{3}(2l+1)^2 \frac{V_l^2(\Delta U)}{\{1 - [(R - D_a)/R]^l\}^2} \tag{33}$$

with $\beta = 1$. It follows that the potential scales from planet $P^{(1)}$ to planet $P^{(2)}$ according to

$$V_l^2(\Delta U^{(1)}) = \alpha^2 \left(\frac{M^{(2)}}{M^{(1)}}\right)^4 \left(\frac{R^{(1)}}{R^{(2)}}\right)^8 V_l^2(\Delta U^{(2)})$$
$$= \alpha^2 (g^{(2)}/g^{(1)})^4 V_l^2(\Delta U^{(2)}) \tag{34a}$$

In a more general manner, if the anomalous gravity fields in the two planets originate in layers of radii $r^{(1)}$ and $r^{(2)}$,

$$V_l^2(\Delta U^{(1)}) = \alpha^2 \left(\frac{g^{(2)}}{g^{(1)}}\right)^4 \left(\frac{R^{(2)}}{r^{(2)}} \frac{r^{(1)}}{R^{(1)}}\right)^{2l+1} V_l^2(\Delta U^{(2)}) \tag{34b}$$

A comparison of these predicted spectra with the observed spectra provides a measure of the ratio of the effective finite strengths of the two planets. Applications of these formulas to the terrestrial planetary gravity fields are made in later parts of this paper.

3. RELATIONSHIP OF GRAVITY ANOMALIES TO ELASTIC AND VISCOELASTIC MODELS

We have alluded to relationships between gravity anomalies, stress, and mathematical models of geophysical behavior. It is through these relationships that we can use gravity data both as a constraint and as a predictive variable in improving our insight into interior processes, structure and state of the terrestrial planets. The mathematical models also provide a prediction of the interior stress state and provide the appropriate relation between gravity and stress.

Herein we review several mathematical models of geophysical behavior which are idealized but provide order of magnitude descriptions of the real behavior of planetary interiors. We will not discuss detailed solutions but, rather, point out the role of gravitational potential and describe wherein the stress distributions arise in such solutions. For the most part we will deal with spherical coordinates, in keeping with the global or long-wavelength theme of this paper.

Elastic and viscoelastic solutions. The equations describing elastic equilibrium provide a convenient start to this discussion. We expect that the outer portions of the terrestrial planets exhibit an approximately elastic behavior over long periods of geologic time. This region is described as the elastic lithosphere, the bottom boundary of which is conveniently defined as the depth at which elastic shear stresses have relaxed by a factor of $1/e$. The deeper boundary of the thermal lithosphere separates convective from conductive heat transport. The elastic lithosphere can maintain density inhomogeneities and their attendant stress distributions in a static equilibrium if the elastic limit is not exceeded. The region between the elastic and thermal boundaries of the lithosphere is subject to plastic (creep) deformation.

When applied to geophysical problems, the classical calculations of the stress state of an elastic object subject to a known load are of limited value in that the many necessary assumptions make the models only poor imitations of the real situations. Nevertheless, the manifestly oversimplified models do have some value in that they give order of magnitude estimates of the stresses, strains, and deformations involved. In particular, the stress estimates may crudely reflect those actually occurring in the planet.

Two classes of problems can be considered. One is the situation where the planet is loaded at its surface $r = R$; the other is the situation where the planet is loaded internally. The first problem is readily tractable when the loads have simple configurations [see, for example, *Jeffreys,* 1970], or when a spherical planet is harmonically loaded (see, for example, the formulations of *Longman* [1963] and *Farrell* [1972]), if it is assumed that the planet was initially in a state of hydrostatic stress. The second problem poses greater difficulties in that the internal load is usually unknown and a solution is sought for both the internal density anomalies and the stress state, subject to the boundary conditions provided by the external gravity field and topography [*Kaula,* 1963]. A unique solution is possible only if further conditions are imposed, either by constraining the nature of permissible anomalous density distributions or by minimizing the shear strain energy. This latter approach was adopted by Kaula.

Farrell [1972] has given a rather complete review of the elastic surface loading problem. A modified form of his linearized momentum equation is given by

$$\nabla \cdot \tau - \nabla(\rho_0 g_0 \mathbf{S} \cdot \mathbf{e}_r) - \rho_0 \nabla \phi + g_0 \nabla \cdot (\rho_0 \mathbf{S}) \mathbf{e}_r - \mathbf{g}_0 \delta\rho = 0 \tag{35}$$

where τ is the stress tensor, ϕ is the perturbation in gravitational potential, $\mathbf{S}$ is the displacement vector, $\mathbf{e}_r$ is a unit vector in the radial direction, and ρ_0 and $\mathbf{g}_0$ are the density and gravitational acceleration of the undeformed planet. The neglect of the gravitational coupling terms in equation (35), valid for short wavelengths where elastic forces dominate gravitational forces, would lead to the simpler Boussinesq problem. The term $\delta\rho$ was introducted by *Kaula* [1963] and is a density perturbation which can be any type of inhomogeneity, for example, undulations on a crust-mantle boundary. Because it is coupled to displacement through the momentum equation, it plays a role in the elastic deformation of the body in question. The exact nature of $\delta\rho$ depends on how the relevant equations are solved and upon the imposed boundary conditions.

The anomalous gravitational potential is related to density through Poisson's equation:

$$\nabla^2\phi = -4\pi G[\nabla \cdot (\rho_0 \mathbf{S}) - \delta\rho]$$
$$= -4\pi G(\nabla\rho_0 \cdot \mathbf{S} + \rho_0 \nabla \cdot \mathbf{S} - \delta\rho) \qquad (36)$$

The right-hand side of (36*b*) illustrates the three contributions to the anomalous gravitational potential. The first term represents the change in the anomalous potential due to the displacement in the presence of a density gradient. The second term is the contribution from the dilatation or compression of the medium. The third term is associated with the internal loading. In addition, any surface distribution of matter, σ, will contribute to the external gravitational potential through the Neumann boundary condition (continuity of the normal derivative of potential at the planet's surface). This distribution is typically taken as the product of topographic height and density of the surface load. The product of surface gravity and σ is also used as a surface normal stress in a linearized boundary condition.

The complete formulation of the elastic deformation problem requires a relation between the stress tensor and the displacement vector. For the generalized Hooke's law for an isostropic medium,

$$\tau = \lambda(\nabla \cdot \mathbf{S})\mathbf{I} + \mu[\nabla\mathbf{S} + (\nabla\mathbf{S})^T] \qquad (37)$$

where λ and μ are the Lamé constants and $\mathbf{I}$ denotes the identity matrix.

Equations (35), (36), and (37) are the most general set of equations to be solved for stress distribution subject to stress, Neumann, and Dirichlet boundary conditions. Their solution allows an answer to be given to the generally useful question regarding the application of elastic equations to geophysical problems that involve gravity data: Given an initial distribution of density in a planetary lithosphere, an assumed elastic rheology, and a prescribed surface load, what is the resulting distribution of stress, additional density distribution, and displacement that satisfies the external gravitational potential? Among the infinite number of possible solutions to this problem, it is useful to examine those in which the maximum shear stress is less than some predetermined value. Inherent in the ambiguity of the interpretation of gravity data with elastic solutions is the contribution of sublithospheric density anomalies as well as lithospheric deformation by sublithospheric forces.

The stresses, displacement, and potential described by (35), (36), and (37) are solved as a coupled set of differential equations unless treated as a Boussinesq problem. They were conveniently cast into a spherical coordinate formulation by *Alterman et al.* [1959] as a time-dependent problem to study the free oscillations of the earth. *Longman* [1962, 1963] adopted the zero frequency version of these equations to study static loading of an elastic earth, including a formulation for the liquid core. He specifically calculated load Love numbers for the Neumann boundary condition, shear stress vanishing, and normal stress equated to a unit point mass load. Other considerations of the static loading problem and appropriate treatment of the liquid core have been presented over the last 15 years [e.g., *Saito,* 1974; *Crossley and Gubbins,* 1975; *Wunsch,* 1975; *Dahlen and Fels,* 1978].

Kaula [1963] introduced the additional density term $\delta\rho$ and added the continuity of gravitational potential at the earth's surface as an additional constraint (Dirichlet boundary condition). He expanded $\delta\rho$ into spherical harmonics ρ_{ilm} (see (7) for notation) and chose to make it a continuous radial function, thus rendering the equations infinitely underdetermined. The simplest solution is, however, obtained by treating $\delta\rho_{ilm}(r)$ as a single number $\delta\rho_{ilm}$ for each spherical harmonic, yielding an even-determined set of equations which can be solved for the lateral variations in a single layer [e.g., *Cook,* 1967; *Hide and Horai,* 1968; *Arkani-Hamed,* 1973*a*]. In some problems the ensemble of $\delta\rho$ may also be taken as the undulations on some interface such as the crust-mantle boundary. Kaula chose a solution for $\delta\rho_{ilm}(r)$ that minimizes the shear strain energy in the least-squares sense, a technique also used by *Arkani-Hamed* [1970, 1973*b*]. While it is questionable whether such solutions are representative of the planet, they will certainly provide an estimate of the minimal stress differences that apply for any elastic rheology.

Examples of the solution of the spherical equations for a self-gravitating elastic Mars model loaded by topography of various harmonics $J_{lm} = (C_{1lm}{}^2 + C_{2lm}{}^2)^{1/2}$ are shown in Figure 3 for the stress difference as a function of planetary radius. All models except Figure 3*f* assume homogeneous elastic constants and the density model of *Johnston et al.* [1974]. The model in Figure 3*a* is with $\delta\rho_{ilm}$ set to zero and no Dirichlet boundary condition. All subsequent models treat $\delta\rho_{ilm}$ as a perturbation on the mean crust-mantle boundary depth (taken as 50 km) and utilize the Dirichlet condition. The effect is to partially compensate the surface load (Figure 3*b*). The model in Figure 3*c* introduces a core and the model in Figure 3*d* additionally introduces a low rigidity layer. The model in Figure 3*e* is the same as Figure 3*c* but has in-

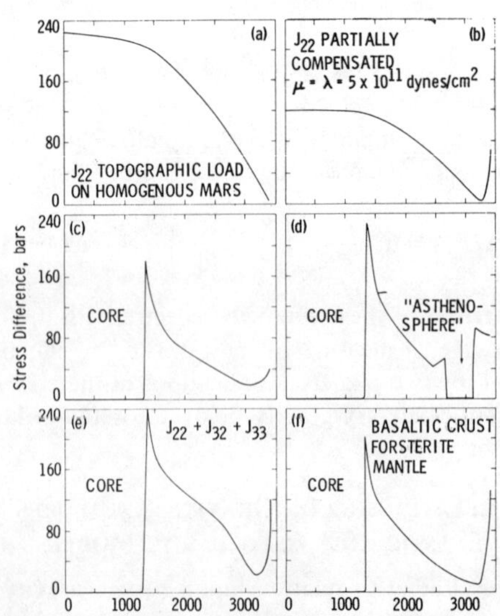

Fig. 3. Stress difference versus planetary radius for six different self-gravitating elastic Mars models. Details of the models are described in the text. All models, except Figure 3*e*, are for the J_{22} (= $(C_{122}{}^2 + C_{222}{}^2)^{1/2}$) harmonic.

troduced additional harmonics to the surface load. The model in Figure 3*f* introduces a more realistic rheology by using shear and compressional seismic wave velocities corresponding to a basaltic crust and forsterite mantle, with appropriate corrections for pressure and temperature. The stress distribution is very similar to Figure 3*c*.

We would expect that a significant volume of the mantles of the terrestrial planets would deform under the application of external loads and the models of Figure 3 could be taken as the initial stress state. A more useful rheological model is one that contains both viscous and elastic elements. The solutions of the appropriate equations are now explicitly time dependent and describe a solid state creep component of the response of a planet to a displacement or a load. A viscoelastic behavior of the terrestrial lithosphere in response to time-varying loads (glaciation, sedimentation, etc.) has been discussed by several authors [*Walcott,* 1970; *Watts,* 1976; *Sleep and Snell,* 1976; *Beaumont,* 1978], but it is not necessarily clear that the lithosphere continuously deforms or approaches some limiting value and subsequently behaves elastically. For purposes of studying rebound following Pleistocene deglaciation, the terrestrial lithosphere and asthenosphere have been modeled variously as a viscous, elastic over viscous, and viscoelastic medium [e.g. *Haskell,* 1935, 1936, 1937; *Crittenden,* 1963; *McConnell,* 1965, 1968; *Cathles,* 1975; *Peltier,* 1974; *Peltier and Andrews,* 1976]. The time scale of deglaciation is short compared to characteristic times of convection so that the two phenomena are effectively decoupled.

The addition of a viscous component introduces an additional rheological variable as well as a time element to the problem of modeling deformation, stress and gravity field, but, in essence, the viscoelastic and elastic problems are equivalent, both physically and mathematically. The utilization of the correspondence principle [*Bland,* 1960; *Fung,* 1965] allows the viscoelastic equations to be expressed explicitly as elastic equations in Laplace transform space with an appropriate redefinition of the Lamé constants as functions of the Laplace transform variable *p*. The Laplace transform solution spectra (Figure 4) are inverted to obtain the formal time-dependent solutions (Figure 5). This is not always an easy task; however, useful measures of characteristic relaxation times can be gleaned from the explicit behavior of the *p* spectra.

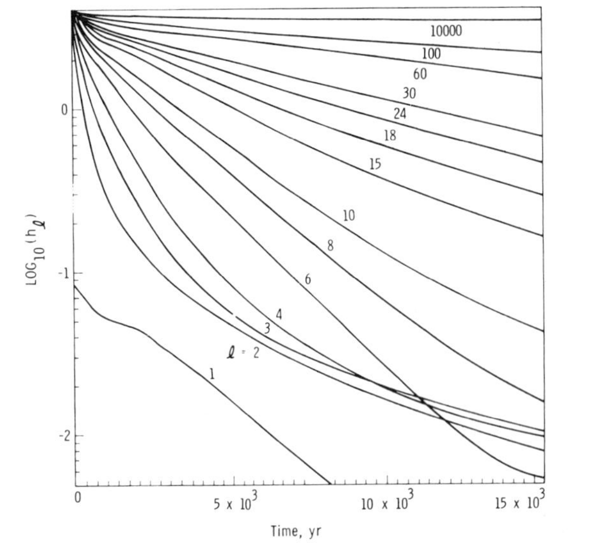

Fig. 5. Time histories corresponding to spectra of Figure 4. From *Peltier* [1974].

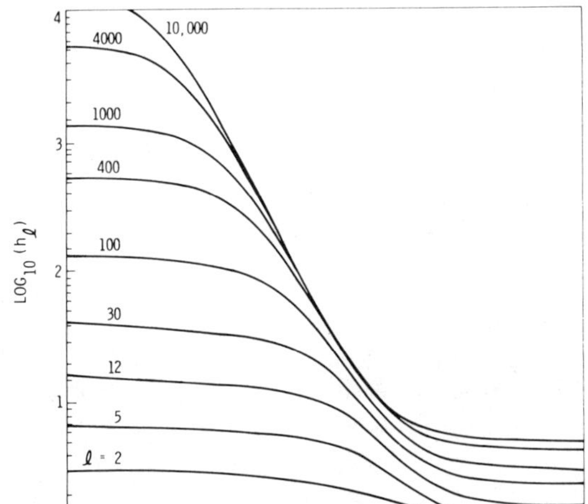

Fig. 4. Love number h_l (normalized surface displacement) for various spherical harmonics l as a function of Laplace transform variable p. The earth model has a uniform viscosity for the crust and mantle ($\eta = 10^{22}$ P) over an inviscid core. From *Peltier* [1974].

A useful linear viscoelastic description is that of a Maxwell solid, which has an instantaneous elastic response followed by a Newtonian viscous behavior. The transition in behavior occurs at the Maxwell time, η/μ, the ratio of dynamic viscosity to elastic rigidity. The p-dependent Lamé constants are given as

$$\lambda(p) = \frac{\lambda p + \mu K/\eta}{p + \mu/\eta} \qquad (38a)$$

$$\mu(p) = \frac{\mu p}{p + \mu/\eta} \qquad (38b)$$

where K, the bulk modulus, is $\lambda + \frac{2}{3}\mu$.

Much of the discussion of the previous section regarding elastic modeling applies to Maxwell viscoelastic modeling with the boundary conditions, in essence, becoming initial conditions. Fundamental relaxation times are obtained from a consideration of the response of an impulse load; i.e., a delta function load applied at $t = 0$ [*Peltier,* 1974]. The response to a load applied at $t = 0$ and left in place is obtained by convolving the impulse response with a unit step function or multiplying the Laplace transform solution by p^{-1}. The behavior of the solutions at $t = 0$ are given in the limit as $p \to \infty$ and from (38) it is seen that $\lambda(p) \to \lambda$ and $\mu(p) \to \mu$, and the solutions correspond exactly to the equivalent elastic problem. The behavior of the solutions for geologically long time periods is given for $p \to 0$, for which $\lambda(p) \to K$ and $\mu(p) \to 0$, defining an equivalent problem for a fluid material.

Viscoelastic Laplace transform spectra for the Mars model of Figure 3*f* are shown in Figure 6. The viscosity profile adopted is subdivided into four regions: (1) a 1300-km radius inviscid core, (2) a mantle zone of viscosity 10^{22} P, (3) a smoothly varying transition to the lithosphere, and (4) an elastic lithosphere with an essentially infinite Maxwell time. The characteristic time of relaxation is given by the reciprocal p values corresponding to the most rapidly changing part of the spectrum [*Peltier,* 1974]. The characteristic p value shown in

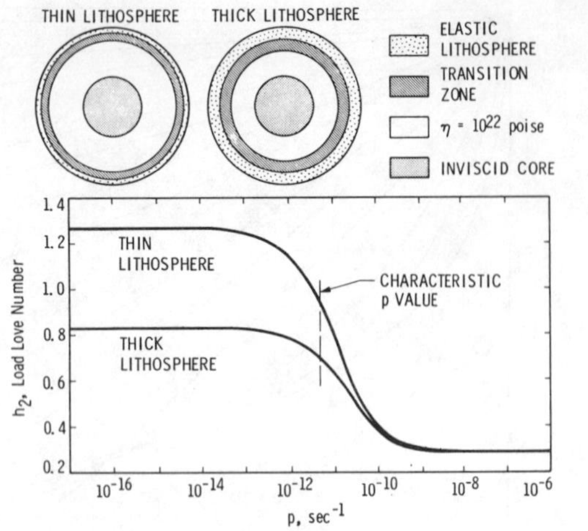

Fig. 6. Dimensionless viscoelastic Love numbers for second harmonic vertical surface displacement as a function of Laplace transform variable *p*. Mars models with elastic parameters of model of Figure 3*f* and viscous parameters as shown. Characteristic *p* value for homogeneous sphere of viscosity 10^{22} P is shown and indicates relaxation is controlled by middle and lower mantle viscosity. For a given relaxation time and flow law, the average ratio of temperature to melting temperature of the mantle can be determined. From *Phillips et al.* [1980].

Figure 6 is for the relaxation of a homogeneous sphere of viscosity 10^{22} P. For a second harmonic load, one concludes that the relaxation is strongly controlled by the deep mantle viscosity, i.e., the lithosphere is transparent to this wavelength.

The usefulness of the Maxwell description is that it describes a number of meaningful rheologies. Beneath the thermal lithosphere we expect the Maxwell time to be geologically short, and the rheology described is essentially viscous. Thus it is not surprising that the Maxwell model of *Peltier and Andrews* [1976] and the elastic over viscous model of *Cathles* [1975] are in agreement, at least in the middle and lower mantle, both suggesting a uniform mantle viscosity of the earth of the order of 10^{22} P. Within the elastic lithosphere the Maxwell time is long and the rheology is approximately elastic. In the lower thermal lithosphere, the Maxwell time is of the order of meaningful geologic time scales and the rheology is both elastic and viscous. The lithosphere as a whole is properly described as anelastic; lithospheric viscoelastic loading problems have been presented by *Walcott* [1970], *Sleep and Snell* [1976], and *Beaumont* [1978].

In practice, it is often useful to consider an elastic lithosphere with infinite Maxwell time, and the region beneath it to have undergone complete viscous relaxation so that in the limit the model describes the classical problem of the floating elastic lithosphere over a fluid interior.

In this regard, the mathematical formalism of the liquid core loading problem is adaptable, the 'core' in this case extending to the base of the elastic lithosphere, treating the sublithosphere as a hydrostatic fluid. However, when the elastic lithosphere is 'thin' (to be defined below), then certain approximations lead to a simpler analytic description and the deformation to loads is described by a single parameter, the flexural rigidity.

The problem of viscoelastic modeling during the transient or relaxation phase is fraught with difficulty for the terrestrial planets, excepting the earth where the time-history of response is observationally accessible. For the other planets, some knowledge of the time application and amplitude of the load is required. An instantaneous application of the load, such as from crater impact, simplifies the analysis but does not obviate the knowledge necessary to solve the problem. A useful approach to the transient problem is to assume that the topographic feature in question (the load) is little affected by the relaxation process since it resides in an elastic lithosphere. Rather, a viscous response takes place at those depths where both the temperature and load stresses are sufficiently high. If a significant density contrast occurs in this region and if the depth to this contrast is known, then the observed topographic spectrum is the input and the observed gravity spectrum is the output of a quantifiable process that depends on the viscosity of the interior. This approach was used for the Tharsis region of Mars to estimate a viscosity lower bound of about 10^{27} P [see *Mutch et al.,* 1976], based on the partial compensation model of *Phillips and Saunders* [1975]. As a second example, many lunar studies of size-depth relationships have advocated the long-term isostatic response of large lunar craters [*Pike,* 1967; *Baldwin,* 1970; *Kunze,* 1974]. *Dvorak and Phillips* [1978] argued, however, that the Bouguer gravity over large lunar craters indicated that, in response to load stresses, virtually no flow has taken place at the crust-mantle boundary (~60-km depth) over the last 4 billion years of lunar history (see section 8).

Flexure theory. The theory of the loading of thin shells and plates is well known [e.g., *Love,* 1944; *Nadai,* 1963; *Kraus,* 1967]; the basic postulates [*Kraus,* 1967] are that the shell is thin (thickness small relative to radius of curvature) or the plate is thin (thickness small relative to horizontal dimensions), deflections are small, the effect on bending of the normal stress transverse to the plate or shell is negligible, and normals to the reference surface (mid-plane) remain that way and undergo no change in length during deformation. From these postulates it follows that the transverse normal strain as well as the shear stresses and shear strains containing a transverse component are zero.

As an example of flexure theory, we examine the solution for a thin lithospheric shell of thickness t, Young's modulus E, and Poisson's ratio ν associated with a planet of radius R. The deflection W is described by the following differential equation [*Kraus,* 1967]:

$$\nabla^6 W + 4\nabla^4 W + \psi(1 - \nu^2)\nabla^2 W + 2\psi(1 - \nu^2) W = -\frac{R^4}{N}(\nabla^2 + 1 - \nu)q \qquad (39)$$

where ∇^2 is the Laplacian operator, $\psi = 12R^2/t^2$, N (the flexural rigidity) is equal to $Et^3/[12(1 - \nu^2)]$, and

$$q = \rho_t g_0 h + \rho_s g_0 W \qquad (40)$$

is a superimposed load at the surface (density ρ_t and topographic height h) and the buoyant response at the lithospheric base (sublithospheric density ρ_s). In terms of the spherical harmonic coefficients for topography, the displacement is given by

$$W_{ilm} = \alpha_l h_{ilm} \qquad (41)$$

where α_l is a constant depending explicitly on N, ψ, ν, ρ_s, and ρ_t. In the limit as $N \rightarrow 0$, $\alpha_l \rightarrow -\rho_t/\rho_m$, which is simple Airy isostasy (see section 6). The total displacement is given by

$$W = \sum_i \sum_l \sum_m W_{ilm} Y_{ilm} \qquad (42)$$

Appropriate differentiation of (42) to form strains and application of Hooke's law yields the three stresses $\sigma_{\phi\phi}$, $\sigma_{\lambda\lambda}$, and $\sigma_{\phi\lambda}$. In the limit as $N \rightarrow 0$, these stresses vanish, and second-order terms must be maintained to give the stresses associated with isostasy (see below).

Now, the ratio of the observed dimensionless gravitational potential power spectrum to the (topographic) load spectrum, in terms of flexural parameters of the shell, is given by

$$Z_l^2 = \frac{V_l^2(\Delta U)}{V_l^2(h)} = \left(\frac{3}{\bar{\rho}}\right)^2 \frac{1}{(2l+1)^2} (\rho_t + \alpha_l \rho_s)^2 \tag{43}$$

The ratio Z_l^2 is the power in an admittance function [cf. *McKenzie and Bowin,* 1976] and given the observed values of Z_l^2, α_l is obtained from (43), which in turn may be solved for the flexural rigidity

$$N = \gamma_l(\alpha_l, t) \tag{44}$$

where γ_l is the inverse operator. Therefore with the values of Z_l^2, or an individual harmonic basis Z_{ilm}, the flexural rigidity is obtainable by assuming a thickness t, or vice versa. For large l, γ_l becomes independent of t and N is obtainable uniquely from the observables, which is the same result as obtained by thin elastic plate theory (see *Walcott* [1976] for a derivation) and used by *Banks et al.* [1977] to obtain an expression for the isostatic response function for regional compensation. Both *McKenzie and Bowen* [1976] and *Watts* [1978] consider the relationship between sea floor topography and gravity using a finite thickness (not thin) plate, and here the elastic constants and plate thickness do not lump into a single parameter. McKenzie and Bowen were the first to use the admittance function between gravity and topography to obtain properties of the elastic lithosphere.

For thin plate theory, application of the correspondence principle yields a space-time dependent differential equation for a viscoelastic medium analogous to the space-dependent elastic equation. Solutions to the former equation are straightforward and have been exploited by *Walcott* [1970], *Watts and Cochran* [1974], *Sleep and Snell* [1976], and *Beaumont* [1978] to study the time variation of lithospheric flexure.

In spherical coordinates, the viscoelastic response of a thin shell may be studied by applying the correspondence principle to α_l. The poles of α_l in Laplace transform space yield the relaxation modes for the shell [see *Peltier,* 1976]. For small $l(<\sim 10)$, two relaxation modes result; one is always the Maxwell time. The second mode depends on the value of the membrane stress parameter [*Kuckes,* 1977] $M = \rho_s g_0 R^2/Et$. For small M, bending stresses are important, and the relaxation time is of the order of the product of a viscous sphere relaxation time and the normalized thickness (t/R) of the lithosphere. For large M, the relaxation time of the second mode is of the order of the Maxwell time.

Minimum stress distributions. *Jeffreys* [1943, 1970] argues that the elastic loading solutions may be inappropriate to many geophysical problems, and instead he seeks a solution in which the elastic stress-strain relation is abandoned and a minimum value for the maximum stress difference in a homogeneous elastic sphere or half-space is sought. This results in some reduction in the stress differences compared with the elastic solutions, or

$$\tau_{\max} \simeq (\tfrac{1}{2} \text{ to } \tfrac{1}{3})\rho g h \tag{45}$$

More appropriately, *Jeffreys* [1932, 1943] sought the mass distribution that would lead to the minimum stress in a floating crust. For both plates and spherical shells he finds the mass distribution is very nearly isostatic and the minimum value for the maximum stress is

$$\tau_{\max} \simeq \rho g h \tag{46}$$

This value also follows directly from elastic plate theory if the plate is loaded on both surfaces such that the Airy isostatic condition is maintained [e.g., *Timoshenko and Goodier,* 1951].

4. RELATIONSHIP OF GRAVITY ANOMALIES TO CONVECTION MODELS

In the previous section we illustrated some relations between gravity, stress, and displacement for a planetary interior that behaves elastically or possibly viscoelastically. The relationships are fairly direct so long as most of the density contrasts are near the surface. The lithosphere will support density anomalies as long as the resulting shear stresses are below the finite strength of the elastic layer (section 5) and, in a general way, the resulting gravity anomalies can be expected to show some correlation with topography, surface geology, and tectonics. But not all of these correlations need be due to density anomalies within the upper lithosphere; they could also be a consequence of mantle processes as well as due to lithospheric response to these processes. Conversely, the correlations may be indicative of constraints on the mantle processes by the lithospheric boundary layer. Terrestrial examples of the former situation, where both topography and gravity are, in part, a consequence of the response of the lithosphere to flow in the asthenosphere, include the outer gravity and topography highs of the oceanic lithosphere before it is subducted [*Hanks,* 1971; *Watts and Talwani,* 1974], possibly the broad positive gravity anomaly referred to as the Hawaiian swell [*Watts,* 1976; but see also *Detrick and Crough,* 1978], and possibly the residual depth anomalies over ridges [*Sclater et al.,* 1975]. A possible example of this on Mars is the Tharsis plateau [*Phillips et al.,* 1980].

Without discussing in detail the specific mechanism, there is no way of separating the contributions to gravity from density anomalies in the lithosphere from anomalies at great depth, although some general statements can be made. One is that, because of the attenuation of gravity with distance from the source, the shorter the wavelength of the gravity anomaly the more likely it is to be a consequence of near-surface density anomalies. Short-wavelength anomalies at greater depth will simply not be observed, unless the density contrasts become very large. Another is that the stress differences occurring in an elastic layer due to a density anomaly are proportional to the square of the wavelength of the anomaly. Hence the longer wavelengths, resulting in the larger stress differences, are more likely to originate at greater depths than the shorter wavelengths which are more likely to be supported elastically in the upper lithosphere. Hence if assumptions can be made about the maximum density contrasts and shear stresses that the lithosphere can tolerate, some order of magnitude separation of lithospheric and mantle contributions becomes possible.

We expect that a substantial fraction of the mantles of the terrestrial planets will not support density inhomogeneities elastically; such anomalies must flow by subsolidus creep processes. These inhomogeneities will flow, irrespective of the anomaly source, with a tendency toward stable horizontal stratification unless maintained by dynamic processes such as convection. Indeed, the curl of the momentum equation for velocity **V** for an incompressible constant viscosity flow, neglecting inertial terms, is

$$\mathbf{g}_0 \times \nabla\rho = \eta\nabla^2(\nabla \times \mathbf{V}) \tag{47}$$

showing that lateral density variations drive flow and stress fields [*McKenzie,* 1977].

Through thermal expansion, vertical temperature variations—whether maintained by appropriate temperature or heat flux boundary conditions or internal heat generation—will give rise to buoyancy forces to drive convective motions in planetary mantles. Unlike the elastic or viscoelastic problem where mantle density variations might be invoked in an ad hoc manner, the density anomalies resulting from convective flow are the results of a well-posed boundary value problem. In principle, the density distributions are exactly predictable from the mathematical model of convective flow. In practice, the problem suffers from (1) intractability in solving, with adequate resolution, fully time-dependent three-dimensional flows on the present generation of computers, (2) complications in describing the appropriate rheology (e.g., non-Newtonian, and temperature-dependence of rheology) and certain physical effects (dissipation, phase change), and (3) uncertainty in boundary conditions, initial conditions, heat source abundances, and material constants.

The two chief gravity effects of convective flow arise from the density changes due to thermal expansion and from deformation of the boundary layers (e.g., core-mantle boundary, phase change boundary, lithosphere) due to viscous stresses and pressure effects. Surface deformation due to convective flow has been summarized by *McKenzie* [1977]. The pressure field P is divided into a hydrostatic fraction P_0 [$= -\rho g_0 Z$] and a fraction associated with convective flow P_1. In practice, the pressure field is eliminated from the momentum equation with the curl operator, and perturbation density is replaced with perturbation temperature via the Boussinesq approximation. The energy equation gives a second relationship between velocity and temperature leading to a coupled set of nonlinear partial differential equations. Upon solution, P_1 is obtained from the original form of the momentum equation via

$$\nabla P_1 = \nabla \times [\eta(\nabla \times \mathbf{V})] = \nabla[2\eta\nabla \cdot \mathbf{V}] \tag{48}$$

Then on a deformed boundary, the normal stress is zero,

$$\tau_{zz} = 0 = \left[-P_1 + \rho g_0 z + 2\eta \frac{\partial V_z}{\partial z}\right]_{z=\Delta h(x,\,y)} \tag{49}$$

yielding the surface deformation

$$\Delta h(x, y) = \frac{1}{\rho g_0}\left[P_1 - 2\eta \frac{\partial V_z}{\partial z}\right]_{z=0} \tag{50}$$

Pekeris [1935] was the first to discuss distortion of the lithosphere due to mantle convection and, from the linearized solution to the convection equations, pointed out the competing gravity effects, over a rising cell, of the high-temperature fluid and the deformed boundary layer.

5. FINITE STRENGTH

As we have previously discussed, the problem of separating contributions to gravity from the different layers in a planet can be partly answered if the stresses set up by the internal and surficial loads can be calculated and if the planet response to these stresses can be predicted. To whit, the calculation of stress levels leads to typical statements such as '... gravity anomaly cannot be supported elastically by the lithosphere ...' or '... the topographic feature must be supported by flow stresses in the asthenosphere ....' In discussing the crustal and lithospheric contributions to a planetary gravity field, a convenient measure of the response of those layers is given by its finite strength which we consider here as a measure of a critical stress beyond which the response becomes distinctly nonelastic; either significant creep occurs or the material fractures. *Jeffreys* [1970] discusses some criteria for this critical stress and logically concludes that, in many situations, the maximum stress difference occurring anywhere in the material provides an adequate measure (or at least a lower bound) of the finite strength. Usually stress is not measured directly but is inferred from observations of strains, deflections, or displacements of bodies subjected to specified loads. Hence a relation between stress and strain, or between load and deformation, is required, but as has been emphasized by F. Birch, one usually settles for the tractable rather than the more realistic relation.

Evidence for the strength of the earth materials can be deduced from several sources. These include (1) laboratory measurements of the strength of rocks; (2) inferences from the physics of steady state flow; (3) in situ stress measurements in the crust; (4) stress-drop measurements across fault planes; and (5) analysis of stress fields set up by the loading of the crust by sediments, mountains or seamounts, or by the bending of the lithosphere at the oceanic trenches, referred to here as tectonic stresses. Estimates of the finite strength obtained by the above methods range over some 3 orders of magnitude, from a few bars to several kilobars.

In our restricted problem we are interested in knowing what part of a planetary gravity field can be supported elastically, for long time intervals, by the crust and lithosphere. If the finite strength of the lithosphere is of the order of several kilobars, for example, then much of the earth's gravity field can be attributed to density anomalies in the lithosphere [e.g., *Lambeck,* 1972], but if the finite strength is less than 100 bars, then these density anomalies must be relegated to the mantle and be dynamically supported [e.g., *Kaula,* 1972].

Laboratory measurements. Measurements at room temperature indicate that flow or fracture occurs at stress differences ranging from several hundred bars to several kilobars. For a number of reasons, however, these stresses must be considered as upper limits since, (1) the samples are nearly always much more homogeneous than the crust, (2) the strength measurements are for small rock samples in contrast to large rock masses—complete with faults, cracks, and other imperfections—which are more appropriate for our problem, and (3) the strain rates applied are much greater than those encountered in geological problems; at lower strain rates creep will usually occur at lower stresses. The laboratory measurements show a number of characteristics that are relevant to the present problem.

1. An increase in confining pressure increases the finite strength, presumably because porosity is reduced and because there is an improved interlocking and bonding between grains. Hence all other factors being equal, the finite strength of the crust should increase with depth.

2. The presence of water in the specimen significantly lowers the finite strength of crustal rocks because pore pressure opposes the lithostatic pressure.

3. The cycling of stress from low to high levels increases rock strength, and the strain history of the specimens may be

an important factor in assessing its strength. Old continental crustal rocks may therefore have a greater finite strength than young oceanic crust with a shorter loading history.

4. The dominant factor controlling the mechanical strength of rocks is probably the temperature T since steady state flow will generally occur when T/T_m equals or exceeds about 1/2, T_m being the melting point temperature [*Weertman and Weertman,* 1975; *Carter,* 1976]. At geological strain rates this ratio is almost certainly lower.

Inferences from steady state flow. In these studies stress-strain rate laws, based either on specific creep mechanisms or on empirical considerations, are introduced, and the stress differences are estimated for which significant solid state flow occurs. These flow stresses place upper limits on the stresses that can be supported elastically by the material. Recent reviews are by *Weertman and Weertman* [1975], *Carter* [1976], *Kirby* [1977], *Mercier et al.* [1977], *Ashby and Verrall* [1978], and *Weertman* [1978]. Most of these discussions are appropriate to the mantle rather than to the crust since the former is believed to be chemically more homogeneous and because the higher T/T_m values make tractable creep laws more applicable. Furthermore, flow in the asthenosphere and deep in the mantle is of broader geophysical significance than the crustal response. These flow laws indicate that the flow stresses in general do not exceed 100 bars or so, particularly for hydrous olivine mantle rheologies and for low strain rates, and in most instances the flow stresses do not exceed a few tens of bars. It is important to note that these flow stresses in the oceanic lithosphere will generally be inferior to continental flow stresses because of the higher geothermal gradient.

Flow stresses in crustal rock are high, ranging up to 10^4 bars for quartzite, for example, but the materials of the average upper crust will probably fail elastically before these stresses are reached.

In situ measurements of stress. Measurements from various surface, bore hole, and mine methods yield values for stress differences in the crust that are of the order of 100 bars, with the largest values appreciably lower than 1 kbar [*Sbar and Sykes,* 1973; *Haimson,* 1975; *McGarr and Gay,* 1978]. All of these measurements give only the stress state in the upper crust, and locally they may be masked by large stresses due to heterogeneities in the structure or strain history of the area. These stresses, insofar as they are representative, are the actual stress state of the crust and hence represent lower limits to the finite strength.

Stress-drop measurements. The stress drop is the difference between the initial and final shear stress on a fault and relates to the seismic moment and fault plane dimension. The observed stress drops for intraplate earthquakes are generally less than 100 bars and mostly less than 50 bars [see, for example, *Richardson and Solomon,* 1977; *Kanamori and Anderson,* 1975; *Hanks,* 1977]. Stress drops along plate margins also do not generally exceed 100 bars including intermediate- and deep-focus earthquakes in the Tonga-Kermadec Benioff zone [*Wyss and Molnar,* 1972]. These seismic stress drops may represent only lower limits to the ambient tectonic stress because we have no reason to necessarily expect total stress relief in an earthquake event and the earthquakes themselves occur in areas of weakness. The absence of a heat flow anomaly on the San Andreas fault has led to the conclusion that the frictional shear stress is on the same order as the stress drops [*Brune et al.,* 1969; *Lachenbruch and Sass,* 1973]; i.e., the seismic stress drop indeed represents the ambient stress levels along the fault. *Raleigh and Everenden* [1979] have reviewed other arguments favoring the equivalency of measured stress drops and ambient shear stress. The seismic estimates represent the response of the crust subjected to geological strain rates, and for this reason the finite strength may be less than that deduced from laboratory measurements; at the same time this is also more relevant to our problem.

Hanks [1977] argues that the actual finite strength may be an order of magnitude greater than the observed stress drops. He sees a need for this in that (1) laboratory measurements indicate kilobar finite strength, and (2) the strength of mantle materials required to support mountains is generally in the vicinity of 1 kbar [e.g., *Jeffreys,* 1970]. We return to this latter point below. Hanks suggests that the stress drops of earthquakes along plate margins may be controlled by recoverable basal stress differences of several tens of bars, and the actual stresses in the upper tens of kilometers on planes parallel to the plate margin may be on the order of 1 kbar.

Tectonic stresses. The stress differences associated with high mountains in an isostatic state will be of the order of ρgh, about 300 bars per kilometer of elevation, using the formulation discussed in section 3. Hence, for the 5-km-high Tibetan plateau, the maximum stress difference in the crust would average about 1.5 kbar. The underlying crust appears to be anomalously thick, about 70 km, yet both it and the mantle also appear to be hotter and consequently weaker than in stabler continental areas [*Bird and Toksöz,* 1977]. Therefore much of these stress differences must be supported by the finite strength of the colder upper part of the crust. The steep slopes occurring in the Himalayas also require a considerable finite strength in the upper parts of the crust, of the order of 700 bars at 25-km depth [*Lambeck,* 1979*b*].

These calculations use static solutions, and if there is one area where these solutions are inappropriate, at least for large length scales or long time scales, it is in areas of tectonic activity. Horizontal stress at the plate margins where mountains are formed is presumably of comparable magnitude to the load stresses if mountains are to form at all, and the stress differences may be less than the above calculations suggest. *Bird* [1978] and *Molnar and Tapponier* [1978] have argued that the Tibetan plateau is maintained partly by compressive forces associated with the northward motion of India relative to central Asia, and Bird estimates that the maximum stress differences associated with the plateau may only be of the order of 400–600 bars.

To be significant as a measure of finite strength, the loaded area must be essentially free from contemporary and recent tectonics. The loading of the crust at Lake Bonneville may provide one region that meets this requirement, and *Crittenden* [1967] concludes that the finite strength of the mantle immediately below the crust is at most a few bars. *Birch* [1964] gives an example of a gravity anomaly over Green Mountain, Vermont, which is associated with maximum stress differences of some 500 bars at 10-km depth. *Lambeck* [1979*b*] has discussed several large gravity anomalies in Australia, South Africa, and the United States in tectonically quiet areas which must be associated with stress differences of the order of 300–400 bars that have persisted for long periods, up to nearly 2 × 10^9 years in one case.

Probably the best estimate of the tectonic stresses in the earth is through a global analysis of the topography for, as we will discuss in section 6, the isostatic mechanism may only operate to the point where all stresses are at or below the intrinsic finite strength. For a topography that is isostatically compensated and with a power spectrum of the order of $V_l^2(h)$ (section 2), the spectrum of the stress difference is, from (46),

$$V_l^2(\tau) \simeq (\rho g R)^2 V_l^2(h) \quad (51)$$

and the maximum stress differences will be of the order of

$$\left[\sum_l V_l^2(\tau)\right]^{1/2}$$

Using the equivalent rock expansion of *Balmino et al.* [1973] gives, with l = 1–36, $\tau_{\max} \simeq 400$ bars. Insofar as topography and density anomalies are initially a consequence of dynamic processes that in many instances are still operative, the actual deviatoric stress may be somewhat less than suggested by this global analysis.

The observation that isostasy is an almost global phenomena also suggests that these estimates are upper limits of the deviatoric stresses, for if the finite strength was greatly different from the ρgh values, then one would expect greater departures from isostasy than are usually observed.

Applying (51) to the lunar topography also gives, with l – 1–36 and with the empirical spectrum discussed in section 8, $\tau_{\max} \simeq 100$ bars. If much of the upper lunar mantle can support stress differences, $\tau_{\max}$ will be about one half of this [*Lambeck and Pullan,* 1979*a*]. For Mars, the comparable value is of the order of 260 bars for l = 2–16, somewhat higher than for the earth.

A number of other tectonic problems have been analyzed for the stress state, most notable being elastic bending of the lithosphere by seamount loads and by subduction. Elastic solutions for the former generally indicate stress differences of 2 kbar per kilometer of deformation, and maximum values of as much as 10 kbar are predicted under larger loads such as the Hawaiian Islands [*Walcott,* 1970]. In these solutions the plate is generally considered to have uniform elastic properties and the maximum stresses occur at the upper surface where they are compressive and at the lower surface where they are tensile. Considerable reduction in the stress differences can be achieved, however, by models in which the finite strength decreases with depth. *Lambeck and Nakiboglu* [1979] discuss some possible solutions and conclude that the stress differences can be significantly reduced, particularly in the lower part of the plate.

The high stresses associated with the bending of a uniform elastic plate at the ocean trenches [*LePichon et al.,* 1973; *Watts and Talwani,* 1974; *Caldwell et al.,* 1976] can also be significantly reduced by introducing a partly viscous response for the lower part of the plate [*De Bremaecker,* 1977; *Melosh,* 1978]. *Forsyth* [1979] has shown that both viscous and elastic models match the observed topographic and gravity anomalies seaward of the trench, with the former model introducing bending stresses less than 1 kbar. Both types of models assume that there is no dynamic interaction between flow in the asthenosphere and the deflection of the overlying plate, but *McKenzie* [1977] has demonstrated that these gravity highs and upward deflections can also result from solid state convection in an asthenosphere of depth-dependent viscosity. From observations of the outer gravity and topographic high alone, a distinction between the various models is not possible.

Conclusion. Estimates of the average finite strength of the earth's crust range from a few hundred bars to several kilobars. Laboratory measurements, employing large strain rates and short loading cycles, will always overestimate the strength of flow stress of the samples, for any additional deformation mechanisms that operate under geological conditions but not in the laboratory will only reduce the strength. To what extent the laboratory results can be extrapolated remains problematical. The elastic plate calculations also lead to high stress differences but which can be significantly reduced by either permitting some viscous creep in the layer or by introducing depth-dependent elastic parameters. Considerations of isostasy lead to stress differences that are of the order of 500 bars, and if the finite strength of the crust greatly exceeded this, then greater departures from isostasy would be observed. We conclude that for the earth the finite strength of the crust is of the order of 500–1000 bars or possibly greater [*Kirby,* 1979] for the upper 10–20 km and for loads operating on geological time scales. Below these depths there is a rapid decrease in finite strength, approaching the 100-bar and ultimately 1-bar level as the region of subsolidus flow is reached.

The finite strength of the moon's crust can be expected to differ from that of the earth's crust in that gravity is much reduced, so that for the crust and upper mantle where confining pressure rather than temperature controls the strength of the material, the finite strength of lunar materials can be less than that for comparable terrestrial materials. The absence of water and other volatiles could, however, increase the strength. The tectonic considerations suggest a lower limit of the order of 100 bars. For Mars the tectonic considerations suggest a crustal finite strength of the order of about 300–500 bars. In view of high surface temperatures, Venus may well have a crustal finite strength that is less than that of the earth.

6. ISOSTASY

Isostasy, or the state of isostatic compensation, is usually defined as a weight balancing of surface loads by density anomalies below it, distributed in such a way that at some uniform depth the pressure is everywhere constant, and below which the planet is in a state of hydrostatic equilibrium. The manner in which the density anomalies are distributed gives rise to the specific mechanism by which this isostatic state is attained. Most models of these mechanisms are based on either the Airy or Pratt schemes. *Heiskanen and Vening-Meinesz* [1958] reviewed the classical discussions of isostasy and of the evidence for it on earth. *Woollard* [1972] has reviewed the more recent evidence for isostatic equilibrium, while *Jeffreys'* [1970] short review is also helpful. It is generally recognized that much of the earth's surface, both continental and oceanic, is close to the isostatic state, although there are some notable exceptions to this: areas of either undercompensation or overcompensation occur due to the fact that the earth's crust can support some nonhydrostatic stress for long periods and that surface deformations are also controlled by horizontal forces in the crust and upper mantle. Isostasy also appears to be an important phenomena shaping the moon [*Kaula,* 1975; *Lambeck and Pullan,* 1979*a*] and Mars [*Phillips and Saunders,* 1975].

In the traditional usage of the Airy model, a crust of uniform density but of variable thickness overlies a weak substratum which will not support nonhydrostatic stresses for long periods of time. The interface between the two layers is arranged in such a way as to reflect the topography and, in order to satisfy the isostatic condition, mountains overlie thick crustal 'roots,' and oceans overlie thin crust. This interface is usually confounded with the Moho, but there is often evidence of significant lateral structural variations below this, and the depth of compensation could in some instances be deeper. The assumption of constant density in the crust is not essential; identical depths of compensation and gravity anomalies are obtained if the crust increases in density with depth or if it is layered, provided that the crust is incompressible and that displacements of material in the crust occur only in the vertical direction. In regions where there are well-defined intermediate crustal discontinuities and where the geotherm is sufficiently high that significant depth variations in finite strength can be expected, much of the compensation may already occur above the Moho. The nature of the crust under mountains presents a particular problem in that crustal and mantle materials, at temperatures higher than the 'normal' crust and mantle, mix and differentiate such that densities and seismic velocities vary gradually with depth rather than resulting in a distinct layering. Not only will the Moho be a transition zone rather than a sharp discontinuity, but there may be low velocity layers within the crust. Both seismic evidence [e.g., *Choudhury et al.*, 1971; *Finlayson et al.*, 1979] and petrological evidence [*Ferguson et al.*, 1979] suggest that this is so. Thus, while in these orogenic regions the isostatic state should be nearly achieved since the higher temperatures result in a relatively short relaxation time, the model of compensation may be quite different from the classical Airy model. Viscous forces acting on the base of the crust or lithosphere may also contribute to the deformation, but as the asthenospheric viscosity is low, particularly in the strongly heated orogenic regions, the rates of flow required to create significant viscous stress would appear to be too high.

In the Pratt model the topography is supported by a layer of uniform thickness with respect to sea level but with a variable density such that the pressure at the base of this layer is constant. High-elevation regions are then underlain by low-density material, and low-elevation regions are underlain by denser material. The compensating depth is often taken to be the base of the lithosphere since lateral density variations cannot be supported statically below this depth. In this model, mountains are assumed to be a consequence of vertical expansion, although in most mountain formation processes compressive forces play an important if not dominant role. Also, it is not obvious that the base of the lithosphere is an appropriate measure of the compensation depth since rheological flow laws, combined with the geotherm, indicate that the lower part of the lithosphere will not be able to support significant stress differences for any length of time (section 5). Clearly, neither isostatic model represents reality, and an actual model, relevant to a particular region of the earth, will be considerably more complex and embody aspects of both: mountains underlain by roots plus lateral density variations, for example. Neither do the variants of these models, such as the Pratt-Hayford or the Airy-Heiskanen models, represent improvements insofar as representing the physics of the problem.

One reason why these classical concepts of isostasy are idealized is that vertical columns of mass are balanced independently of, and uncoupled to, neighboring masses. Such a situation can only exist in the absence of shear stresses. But this is clearly not the case for finite topography in isostatic equilibrium [*Artyushkov,* 1973], and in fact, the isostatic state is not one of exact mechanical equilibrium. However, certain solutions of rheological boundary value problems may approach isostasy. For example, a loaded elastic plate or shell floating on a fluid substratum and obeying the equations of mechanical equilibrium will approach an isostatic state when the flexural rigidity (see below) is small and when the wavelengths of the load exceed the flexural parameter (distance at which deflection of an elastic plate due to a concentrated two-dimensional load has fallen off $1/e$).

In most discussions of isostasy the stress state in the layer is ignored (see *Artyushkov* [1974] for a notable exception), it being implied not only that the stress differences are minimized but also that they are low enough for the crust or lithosphere to be able to support them. But this need not be so. Elastic plate theory predicts that for a plate loaded at its upper surface by a harmonic function and at its lower boundary by a second harmonic load such that Airy isostasy is achieved, the stress differences in the mid-plane of the plate are of the order of ρgh [*Timoshenko and Goodier,* 1951]. The maximum stress difference will occur on the upper and lower surfaces of the plate. The shearing stress is generally small but not nonzero and for a harmonic load $\rho g_0 h_{ilm}$, it will typically approach $l\rho g_0 h_{ilm} t/R$, where t is the average thickness of the plate.

Jeffreys [1970] defines the state of isostasy as the density distribution that minimizes the maximum stress difference occurring anywhere in a floating layer and below which the hydrostatic state is attained. Solving the equations of mechanical equilibrium (without reference to elasticity) for the distribution of mass that minimizes the maximum stress difference, *Jeffreys* [1932, 1943] finds a solution close to Airy isostasy. The stress differences found, as discussed in section 3, are also very close to ρgh.

Figure 7 illustrates the relaxation of a topographic load, in this case a crater of initial diameter 180 km and an initial depth of 15 km in a two-layer viscous half-space. The upper layer has a viscosity of 5×10^{24} P and a density of 2.7 g cm^{-3}, and the corresponding values for the lower layer are 5×10^{23}P and 3.2 g cm^{-3}. The evolution of the topography of the crater and of the interface between the two layers has been computed applying the correspondence principle to the elastic equations [*McConnell,* 1965] and then analytically carrying out the inverse Laplace transforms. As the stress distribution associated with the crater starts to relax, the interface rises in response. At 5×10^8 years the crater is in approximate isostatic balance, and the perturbation on the lower boundary reaches a maximum. The configuration is clearly unstable because there are buoyancy forces associated with both the crater and the lower boundary and with time the crater becomes shallower while the lower boundary deformation diminishes, yet maintaining an isostatic balance with the surface topography. As time approaches infinity, both the crater and boundary perturbation vanish. The stress history of this model follows a pattern of dominance by large normal stresses in the early stages which quickly relax as isostasy is achieved; at subsequent times, shear stresses dominate. The stable horizontal stratification at infinite time in this model would quite likely not be realized, as the stresses would drop beneath the creep strength of the material and a finite isostatic state will exist supporting stresses at about the creep threshold limit. Since, as discussed above, these stresses will be directly proportional to topographic load, then in the absence of other forces, topographic heights must be controlled by the finite strength of the lithosphere.

From these analyses and from the work done by Jeffreys we conclude that isostasy is a natural state in the evolution of a mechanical system in response to an applied stress field. It is

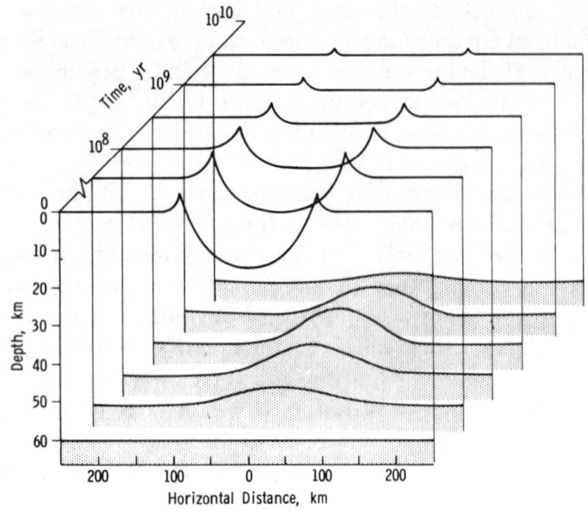

Fig. 7. Viscous relaxation of a crater as calculated from a time-dependent Stokes flow solution for a two-layer medium. Not to scale.

close to the minimum state of stress that can be achieved in a configuration of finite topography and will tend to be achieved because a mechanical system will always tend to respond in such a manner as to minimize stress. A further conclusion follows from the argument that an endogenic thermal process in a planetary interior will, at sufficiently high temperatures, not maintain shear stresses, and in the absence of all forces other than buoyancy forces, internal density configurations will form in an isostatic state; i.e., isostasy is a natural attendant of thermal-mechanical systems.

Lithospheres that are externally loaded (e.g., seamounts, sedimentary fill, plate bending at trenches) may be more properly viewed from the point of view of elastic or viscoelastic flexure. In general, the loading does not involve thermal processes in the lithosphere, and the minimum stress state configuration is not achieved. Isostatic compensation is viewed as regional in extent; i.e., a concentrated load is supported by lithospheric flexure over a wide areal distribution. For regional compensation calculations using flexure theory, the maximum stress differences may be as much as an order of magnitude higher than ρgh [but see *Lambeck and Nakiboglu,* 1979].

Despite the shortcomings of the simpler isostatic models, they serve a useful purpose for evaluating the contribution of the topography and crust to the observed planetary gravity field. Thus if ΔU^I is the isostatic potential, i.e., the potential due to topography and its compensating mass, the contribution of the crust to gravity is unimportant if $\Delta U^I \ll \Delta U_{\text{observed}}$ and the manner in which the isostatic correction is made, if made at all, is irrelevant. This is the case for the global field of the earth, where much of the observed long-wavelength gravity anomalies originate from density anomalies in the upper mantle. For the moon, ΔU^I is a comparable magnitude of the observed potential and most of the gravity field is of crustal or near-crustal origin. Mars represents an intermediate case with a significant contribution of crustal sources to the observed long-wavelength field. Venus may be more like Mars because of strong correlations observed between gravity and topography (section 10).

For regional studies a new approach to isostasy is warranted in which the mass distributions are determined either empirically or from physically plausible assumptions along the lines of geophysical prospecting [e.g., *Bullard and Cooper,* 1948]. *Dorman and Lewis* [1970] undertook such a study by assuming that the Bouguer gravity anomaly $\Delta g_B(\mathbf{r}_0)$ relates to elevation $h(\mathbf{r})$ at position $\mathbf{r}$ according to

$$\Delta g_B(\mathbf{r}_0) = \int_s q(|\mathbf{r}_0 - \mathbf{r}|)h(\mathbf{r})\, ds + g_n \tag{52}$$

where q is an unknown function of the relative position and g_n is the contribution to Δg_B that is not of isostatic origin. This relation assumes that the compensation of a point load is symmetrical beneath this point. Since the integral in (52) is a convolution, in the frequency domain,

$$\Delta G_B(\mathbf{k}) = Q(|\mathbf{k}|)H(\mathbf{k}) + g_n(\mathbf{k}) \tag{53}$$

where $\mathbf{k}$ is the wave number. Here $Q(|\mathbf{k}|)$ is the isostatic response function, which depends on the density distribution with depth. From an analysis of Δg_B and h, q or Q can be determined, and Dorman and Lewis found that, for the continental United States, the linear model (52) is satisfactory. The inversion of the response function for the density anomalies is not unique, and they assumed that the compensation is local, i.e., that the anomalous density is distributed immediately below the surface point load. Their inversion results lead to a model of an oscillatory variation of density with depth whereby elevated terrain is first underlain by a negative density anomaly and by a second, smaller, but positive anomaly at greater depth. These anomalies extend down to considerable depth, of the order of 400–500 km [*Lewis and Dorman,* 1970], and the oscillatory behavior is similar to that found by *Romanowicz* [1979] from the inversion of seismic P travel time anomalies. *Banks et al.* [1977] object to these models of Lewis and Dorman and suggest that the oscillatory behavior is a consequence of the local compensation assumption (Figures 8*a* and 8*b*). They interpret the response function in terms of a regional compensation model described by the flexure analysis theory for thin elastic plates, so that while the load is supported by the buoyancy forces in the fluid, the stresses are distributed over a larger area by the elastic plate deflection. The inversion of the response function gives a measure of the flexural rigidity of the layer. The fit of their regional model is shown in Figure 8*c*. The results of their study yield a strictly positive density gradient in the elastic lithosphere of the continental United States, but the plate thickness is far too small, 5–10 km, and the preferred explanation is that the local compensation mechanism is operative below 10-km depth.

McNutt and Parker [1978] have computed the response functions for the United States and Australia and concluded that the flexural rigidity for the Australian model is significantly less than that for the United States (Figure 9) and, in fact, a value of the flexural rigidity of zero (local compensation model) will fit the observed Q. The initial expectation for such a comparison would be that the older Australian crust and lithosphere would have a higher flexural ridigity than the younger western United States lithosphere which dominates the American response function. McNutt and Parker interpreted this paradox as a consequence of the viscoelastic response of the lithosphere, which leads to lower values of apparent flexural rigidity with increasing time [*Walcott,* 1970]. With this type of rheology, the relaxation time of the Australian upper lithosphere is greater than 10^6 years and may be as high as 10^8 years.

Banks and Swain [1978] computed the isostatic response function for East Africa (Figure 10) and showed that a local compensation model involving only negative density contrasts

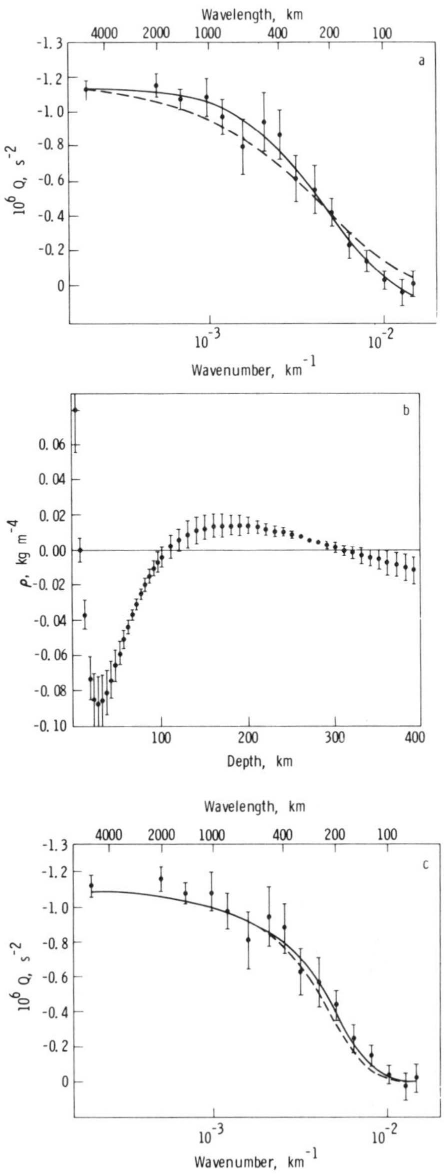

Fig. 8. (*a*) Isostatic response estimates for the United States, shown with ∓1 standard error. Solid curve is optimum local compensation model, in terms of trade-off of error and fit to data, and involves both positive and negative density contrasts in the subsurface. Dashed curve is best fitting model with only negative density contrasts. (*b*) Isostatic density model for United States corresponding to solid curve in Figure 8*a*. Ordinate is density contrast per unit topographic elevation. (*c*) Fit of regional (elastic plate) compensation models to United States isostatic response data at the 80% (dashed line) and 95% (solid line) confidence levels of acceptability. From *Banks et al.* [1977].

(compensating the positive topographic load) would fit the response function. They interpreted the density anomaly beneath the Kenya dome, a region of low seismic *P* wave velocity and high electrical conductivity, as a zone of partial melting.

The results obtained from the studies by Dorman and Lewis, Banks et al., McNutt and Parker, and Banks and Swain are encouraging in that they permit a discussion of isostasy in terms of physically plausible models. Unfortunately, the uncertainty of the response function remains significant, even for the well-surveyed areas of Australia and the United States, and this may prevent further refinements of

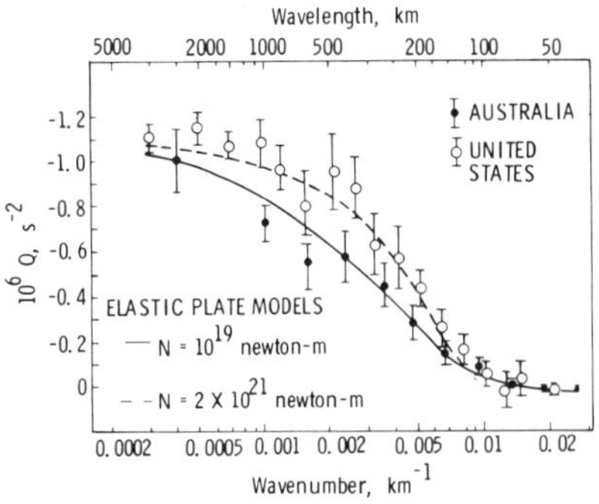

Fig. 9. Australia and United States isostatic response estimates with 1 standard deviation errors. The two curves correspond to elastic plate models with flexural rigidities (*N*) as shown [from *McNutt and Parker*, 1978].

these models. This is mainly a consequence of the noise term in (52), which contains the anomalous density variations in the crust and mantle that, on the average, do not correlate with topography. Further, lateral variations in rheology, structure, and age of loading present difficulties which must ultimately be examined; for example, the essentially horizontally averaged estimate of flexural rigidity and vertically averaged density profile may be strongly influenced by only a single physiographic province. The time sequencing of erosion as a mechanism for shifting the topographic load may play an important role in this regard.

7. EARTH

The relationship between the long-wavelength gravity anomalies and tectonics on earth can be viewed as the superpositioning of three principal phenomena: regional and plate tectonics and the closely related subjects of classical isostasy and postglacial uplift. The history of our understanding of the relation between gravity and regional tectonics goes back many years, and a recent representative discussion of the many diverse aspects of this problem is given by *De Jong and*

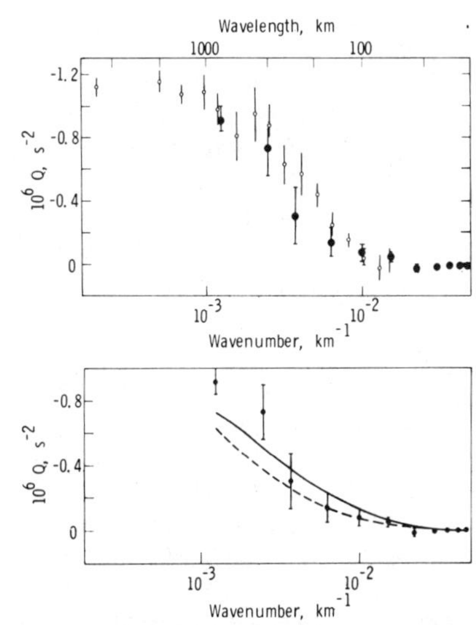

Fig. 10. East Africa (solid circles) and United States (open circles) isostatic response estimates, shown with ∓1 standard error (top). Fit of local compensation models to East African response estimates for two different measures of fit to the data (bottom). The misfit at small wave number is probably a consequence of a model constraint that maximizes the compensation below 100-km depth [from *Banks and Swain*, 1978].

Scholten [1973]. The relationship between gravity and the postglacial uplift of Fennoscandinavia and North America also has a long history of study, and the negative free-air gravity anomalies over the centers of uplift are now well understood in terms of an 'undercompensation' being removed by the response of the elastic lithosphere and viscous sublithosphere [*Heiskanen and Vening-Meinesz,* 1958; *Cathles,* 1975], although there remain diverse viewpoints on how the gravity data should be interpreted. The postglacial uplift problem on earth is unique among the terrestrial planets, and we will not discuss it further here. The general question of isostasy has been discussed in section 6, and the emphasis here will be on the relationship between gravity and the global plate tectonics hypothesis, a subject of much more recent origin than the others, originating about a decade ago with the publication of the first reliable satellite-derived gravity maps.

Observational evidence. Models for the earth's potential field are derived from analysis of orbital perturbations or from surface gravity measurements, and the most recent solutions are based on a combination of the two. Recent solutions are by *Gaposchkin* [1977], *Lerch et al.* [1979], and *Balmino et al.* [1978]. All solutions have expanded the potential to high degree and order; $l = 24$, $m = 24$ in the solution by Gaposchkin; $l = 22$, $m = 22$ in that by Lerch et al.; and $l = 30$, $m = 30$ in Balmino et al.'s work. While all of these authors have carried out extensive tests to validate their results, a comparison of the solutions indicates that the agreement between them is not entirely satisfactory. This is illustrated in Figure 11*a* in the form of the amplitude spectra and the difference spectra. In general, the power of the difference spectra exceeds the power in the potential at degrees exceeding about 16–18. Lerch et al. also give sets of coefficients based on satellite data only, and these can be compared with independent solutions based on surface gravity data. *Rapp* [1977] has expanded the latter in spherical harmonics up to high degree, and the difference spectrum between these and the satellite results exceeds the amplitude spectrum for $l \gtrsim 16$ (Figure 11*b*).

These quick and inadequate comparisons do suggest that much of the higher degree wavelength information in the published terrestrial gravity field solutions may be largely noise. This is probably a consequence of (1) the fact that, apart from resonance configurations, the satellites are insensitive to the higher degree terms and (2) the surface data are incomplete and of highly variable quality. These comparisons suggest that considerable caution should be exercised in interpreting the higher wavelength harmonics. For this reason, we limit our discussion of the global solutions to the harmonics $l \lesssim 18$. Figure 12 illustrates the gravity anomalies for the model by Lerch et al. with the spherical harmonic expansion truncated at $l = 16$, $m = 16$.

More information on the higher wavelength anomalies in the ocean geoid can be obtained from the Geos 3 and Seasat 1 radar altimeter experiments if the sea surface topography relative to a mean geoid is known. The most complete Geos 3 compilation presently available appears to be that by *Brace* [1977], which covers much of the world's oceans. A recent series of papers on geoid results with the Geos 3 altimeter is in the *Journal of Geophysical Research* (*84*(B8), 1979). Although only preliminary solutions are available, the altimeter-derived geoids readily identify geoid anomalies over major bathymetric features such as trenches and seamounts (Figures 13 and 14) and indicate that much of these anomalies are due to topography and near-surface density variations. Reduction of

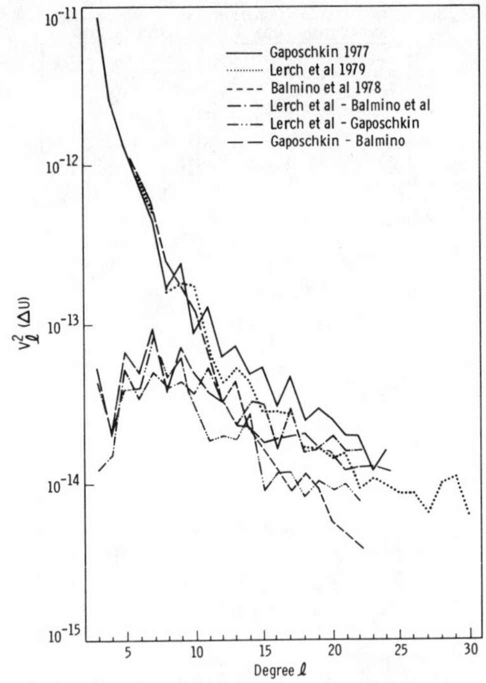

Fig. 11*a*

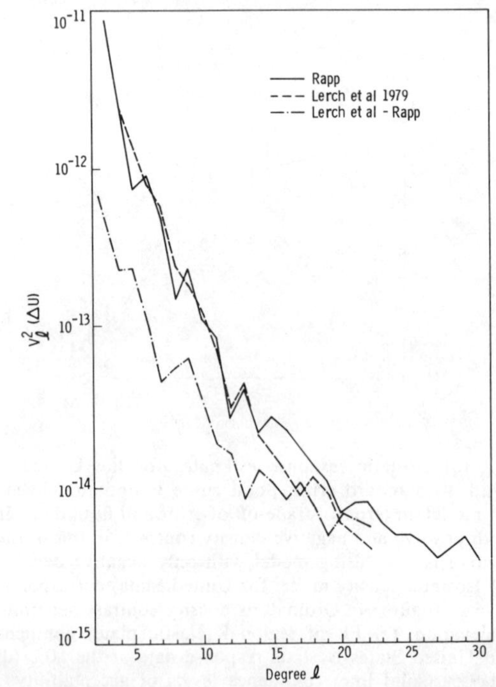

Fig. 11*b*

Fig. 11. Nondimensional normalized power spectra of the geopotential according to several authors and the corresponding difference spectra. Figure 11*a* compares solutions based on satellite and surface data; Figure 11*b* compared a purely satellite solution [*Lerch et al.*, 1979] with a surface data solution [*Rapp*, 1977].

the Geos 3 data as well as the high-quality Seasat 1 observations will result in significant improvements in our knowledge of ocean surface topography and, by assimilation or deduction, the ocean geoid. Preliminary global solutions based on a combination of conventional tracking, surface gravity data

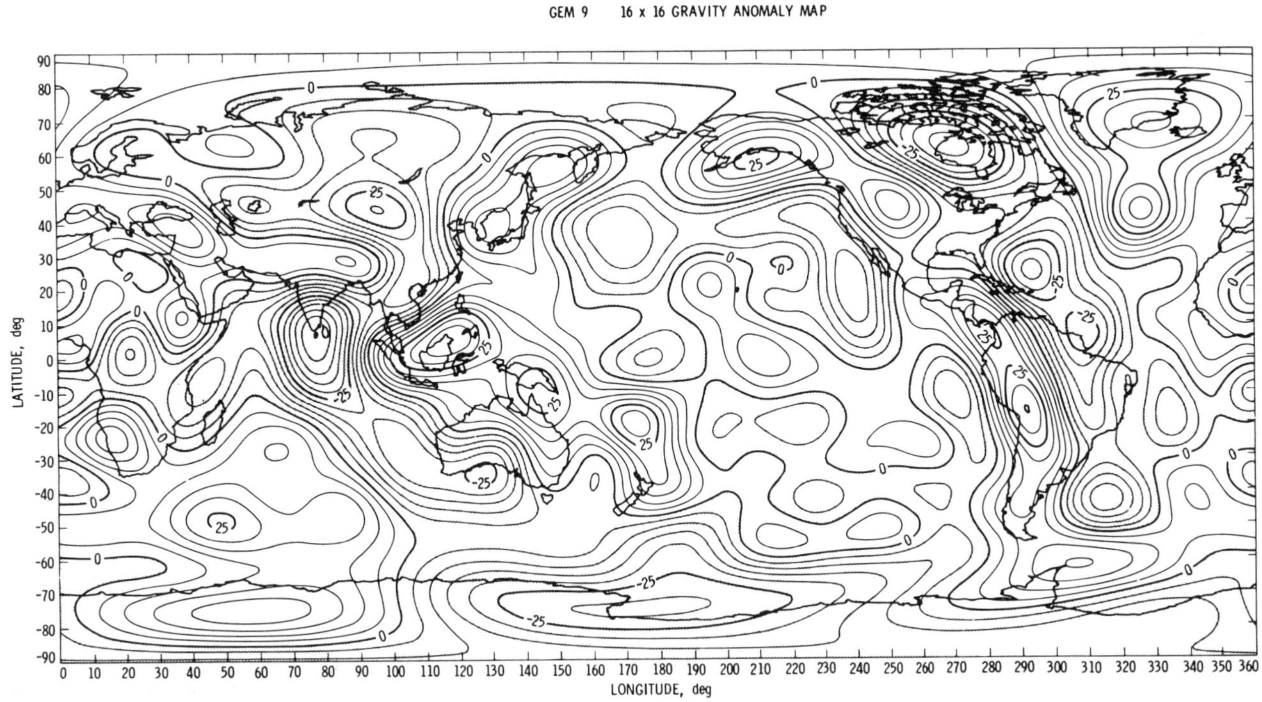

Fig. 12. Free-air gravity anomalies based on the spherical harmonic expansion of *Lerch et al.* [1979] but truncated such that $l_{max} = m_{max} = 16$.

and Geos 3 altimetry have been obtained by *Lerch et al.* [1978] and by *Gaposchkin* [1979].

Relationship with topography; isostasy. Topographic data have been expanded into spherical harmonics up to degree 36 by *Balmino et al.* [1973] using data compiled by *Lee* [1966]. Many of the ocean bathymetry data collected in the last 15 years are therefore not included in this analysis, and an updating of it is desirable. Balmino et al. also give, inter alia, expressions for the equivalent rock topography in which the oceans, large lakes, and land-ice are replaced by an equivalent rock layer.

The dependence of the equivalent rock spectrum on degree l can, with (21), be approximated by

$$V_l^2(h) \simeq 1.03 \times 10^{-7}/l(l+1) \tag{54}$$

for degrees up to at least 36. Topography is one of the most obvious contributions to the earth's gravity field. The predicted potential due to uncompensated topography follows from (22*a*) with (54) as

$$V_l^2(\Delta U) \simeq \left(\frac{3}{2}\frac{\rho_c}{\bar{\rho}}\right)^2 1.03 \times 10^{-7} l^{-4} \tag{55}$$

and this far exceeds the observed potential (Figure 15), as well as decaying more rapidly. This is indicative of the high degree of isostatic compensation of the earth's topography. If the Airy model is used, then with (22*b*) and (54),

$$V_l^2(\Delta U) \simeq \left(\frac{3}{2}\frac{\rho_c}{\bar{\rho}}\frac{D_a}{R}\right)^2 1.03 \times 10^{-7} l^{-2} \tag{56}$$

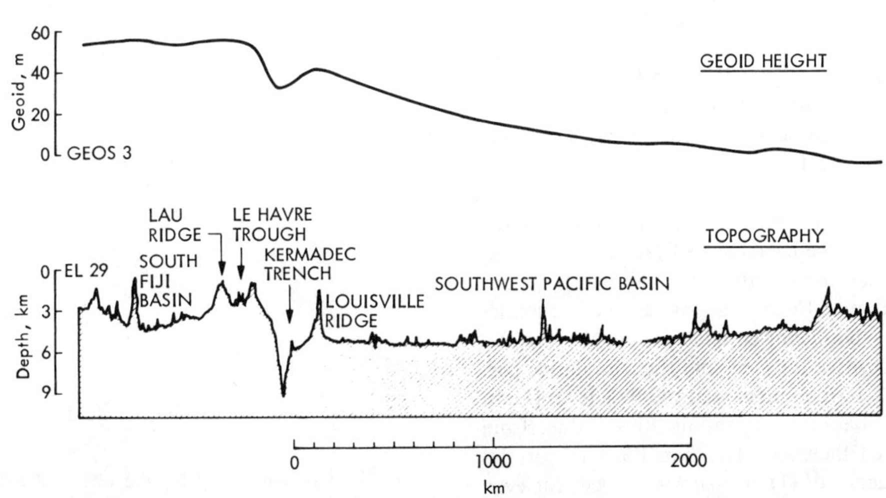

Fig. 13. Ship profile across the Tonga-Kermadec Trench (USS *Eltanin*, cruise 29) from 28°S, 172°E (left) to 28°S, 140°W (right). Geoid height is from Geos 3 satellite and referenced to an ellipsoid with 1/298.255 flattening. From *Chapman and Bodine* [1979].

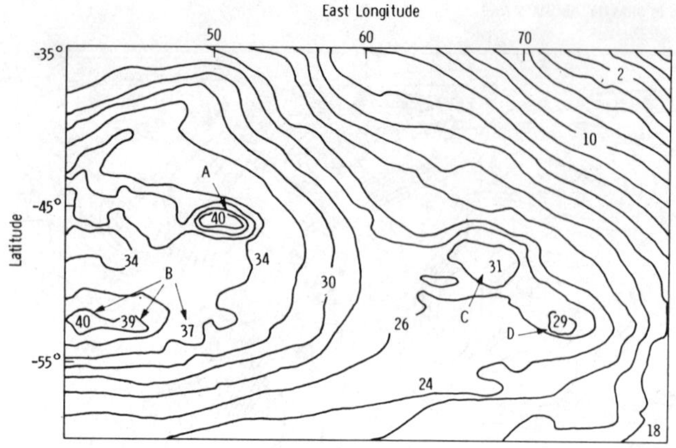

Fig. 14. An example of the ocean geoid from Geos 3 radar altimeter data for part of the southern Indian Ocean. Note small positive anomalies over the Crozet Islands (A), the Ob and Lena seamount chain (B), Kerguelen (C), and Heard (D). Contour intervals at 2 m. Adapted from *Cazenave et al.* [1978].

This potential is less than the observed potential at least up to about degree 18 (Figure 15) and also decays less rapidly than the observed spectrum (compare (56) and (13)). This may be indicative of several things: (1) compensation is incomplete, (2) the depth of compensation is considerably greater than usually suggested, (3) the chosen mechanism is inappropriate, or (4) density anomalies occur below the crust or lithosphere and have no topographic expression. Careful regional studies have indicated that, on the whole, isostasy prevails albeit that the mechanism is not everywhere the same [*Heiskanen and Vening-Meinesz*, 1958; *Woollard*, 1972]. These studies also indicate that the depths of compensation do not generally exceed 40–50 km for the Airy mechanism or about 100 km for the Pratt model, while for the predicted isostatic potential (20) to equal the observed potential, the depths of compensation must be considerably greater than these limits. Also, on the global scale considered here, the details as to how the compensating mass is distributed below the topography are unimportant. Local and regional compensation models, for example, are unlikely to yield significantly different isostatic anomalies.

The obvious reason for discrepancy between the observed and predicted potentials is the presence of density anomalies below the compensation depths which, on the whole, are not related to topography. Also Figure 15 indicates that for $l \lesssim 18$ the contribution to the potential of isostatically compensated topography is not very significant and the global anomalies such as those illustrated in Figure 12, reveal the anomalous mantle structure.

Despite these conclusions there is significant correlation between part of the low-degree harmonics of gravity and topography. The normalized cross spectrum of these two functions (17) is illustrated in Figure 16. For the low-degree harmonics there is no obviously significant correlation, but the correlation turns generally positive for $l \gtrsim 6$ and the cumulative correlation (18) for $l = 6$–18 is significant at the 99% level [*Lambeck*, 1976]; the probability of finding this value from normally distributed coefficients is less than 1%. This correlation can be a consequence of (1) near-surface effects, for even if the topography is compensated, free-air anomalies tend to be positive over elevated areas or (2) deeper lithospheric or mantle processes that have a surface expression.

A distinction between these two interpretations is possible by computing the linear regression slopes between topography and gravity and estimating reduced potential coefficients that are free of any correlation with topography [*Lambeck*, 1976]. The power spectrum of this reduced potential represents about 90% of the total power in the observed potential for $l \leq 18$, again indicating that the contribution to the potential from isostatically compensated topography is small. The residual potential coefficients (the parts that correlate with topography) may be considered as a measure of the isostatic potential and are of the same magnitude as that computed by specific isostatic models (Figure 17). This suggests that, at least globally, the correlation between gravity and topography is merely a consequence of the isostatic or near-isostatic state of the crust and that it does not necessarily reflect mantle processes. Some possible regional exceptions to this are discussed below.

As discussed earlier, in the absence of other forces, endogenic processes in the planets will tend to form isostatically, but a state of stress is associated with such configurations [*Jeffreys*, 1943, 1970; *Artyushkov*, 1973, 1974]. The state of stress is directly proportional to the height of the topography, and given some limiting finite strength of the interior, the height of topography is also limited. The isostatic support of the Himalayas and Tibetan plateau implies stress differences in excess of 1 kbar in simple isostasy, and either the upper mantle here has a finite strength on this level or these regions are in part supported by additional stresses associated with horizontal tectonics. *Bird* [1978] has presented arguments showing that the average shear stress on the intercontinental thrust fault that formed the Himalaya did not exceed 300 bars, in direct contradiction to the support of kilobar stress levels. An appropriate scenario would be that the horizontal normal stress supplied by the northward moving Indian plate approximately balances the normal 'isostatic' stress of the 5-km Tibetan plateau, imparting ~0.1-kbar level shear stresses; i.e., 'The great height of Tibet creates a hydrostatic head, which is

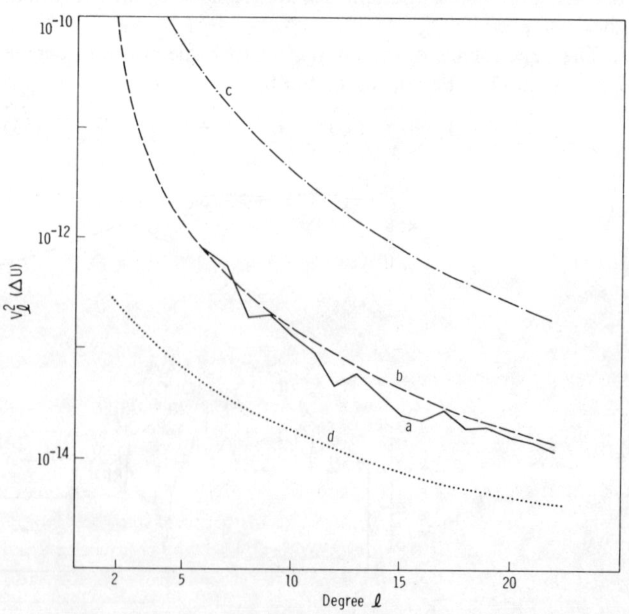

Fig. 15. Comparison of (*a*) the observed geopotential spectrum with (*b*) the approximate rule (13), (*c*) the predicted spectrum based on an uncompensated topography whose spectrum is given by (54), and (*d*) the predicted terrestrial spectrum based on a compensated topography with $D_a = 50$ km.

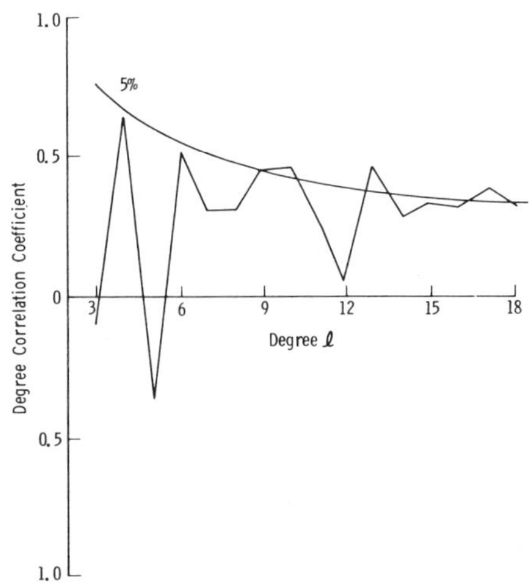

Fig. 16. Degree correlation coefficients for the earth's gravity and topography. The 5% critical limit is also illustrated.

maintained by pressure applied at its margins ...' [*Molnar and Tapponnier,* 1978]. That the Tibetan plateau is at once in isostatic equilibrium and supported by the pressure of plate motion is not necessarily a contradiction and depends on the relative timing of the thermal and mechanical events giving rise to the present structural configuration. The relatively large long-wavelength free-air gravity anomaly associated with the Himalayan/Tibetan plateau complex (Figure 12) is a consequence of a region having an unusually large horizontal dimension (~1000 × 2000 km) for such great mean height (5 km). We contend that this elevation is maintained partly by the horizontal stresses of plate motion and indeed other large continental anomalies (e.g., Andes) may be maintained in an analogous manner.

Oceanic gravity anomalies, plate tectonics, and convection. That some relation between long-wavelength gravity anomalies and plate tectonics occurs first became clearly evident in the gravity solution by *Gaposchkin and Lambeck* [1971]. This solution revealed that active plate boundaries are generally associated with positive gravity, while as a gross generalization, the interiors of plates are mainly associated with negative anomalies (see Figure 12). In particular, the subduction zones around the Pacific tend to be associated with broad positive anomalies and the complex southern margin of the Eurasian plate tends to be associated with smaller and shorter wavelength positive anomalies as do parts of the ocean ridges, most notably the North Atlantic Ridge and the East Pacific Rise. Ocean basins and old shield areas tend to be associated with negative anomalies. These observations led *Kaula* [1972] to interpret the field in terms of flow in the asthenosphere and the response of the lithosphere to this flow. *Schubert and Turcotte* [1972] showed that if counterflow to the lithospheric plate motion is confined to the asthenosphere, gravity anomalies would be more than an order of magnitude larger than actually observed. They propose that to explain both the elevation variations of the lithosphere between ridges and trenches and the gravity anomalies, that the return flow must take place at greater depth, possibly below 700 km, and that the asthenosphere actually flows in the same direction as the lithosphere itself. *Richter and McKenzie* [1978], using a simple kinematic convection description limited to the upper 700 km of the mantle, conclude that a uniform viscosity model is unsatisfactory because it predicts gravity and residual depth anomalies 2 orders of magnitude larger than those observed. The geophysical constraints can be satisfied, in terms of boundness arguments, by introducing a thin (<100 km) low-viscosity layer immediately beneath the plates.

Recent solutions on the whole confirm the existence of these gravity-plate tectonic correlations although, in many of the published anomaly maps, these relations get blurred due to the inclusion of the higher harmonics which either are less certain or do not relate directly to these global features.

While the validity of Kaula's conclusions are generally accepted, that the global gravity field reflects mantle convection (but see *Jeffreys* [1970] for a dissenting view), a more quantitative interpretation still does not exist. But gravity, with characteristic anomalies over ocean ridges, subduction zones, and ocean basins appears to be particularly telling of aspects of the convection problem. However, the relationship does not appear to be sufficiently well understood at present for the observed gravity field to provide useful constraints of global convection models. But gravity may (1) provide some useful, order of magnitude estimates on permissible density or temperature variations, (2) be indicative that further complications must be introduced into the convection models, and (3) provide useful information on particular aspects of the convection model or on rheological parameters.

Studies have been carried out relating gravity anomalies (1) quantitatively to convective flow [e.g., *Runcorn,* 1964, 1967], (2) to the planform of convection [e.g., *Marsh and Marsh,* 1976] and (3) to surface topography in a convecting system [e.g., *Mckenzie,* 1977]. By far, the majority of attempts to relate gravity anomalies to dynamic processes in the earth have been concerned with oceanic lithosphere for at least two reasons: (1) the thinner oceanic lithosphere may well respond in a more direct manner to underlying dynamic processes than the thicker continental lithosphere and (2) oceanic lithosphere is less complex than continental lithosphere and its gross features are fairly well understood in terms of contemporary theories of plate tectonics [e.g., *Parker and Oldenberg,* 1973; *Davis and Lister,* 1974]. Interpretation of gravity anomalies over the oceans is of course nonunique, and in order to establish a sublithospheric source causative to lateral density inhomogeneities, one must eliminate a lithospheric origin. The lithosphere can contribute to oceanic gravity anomalies in three ways [*Lambeck,* 1972].

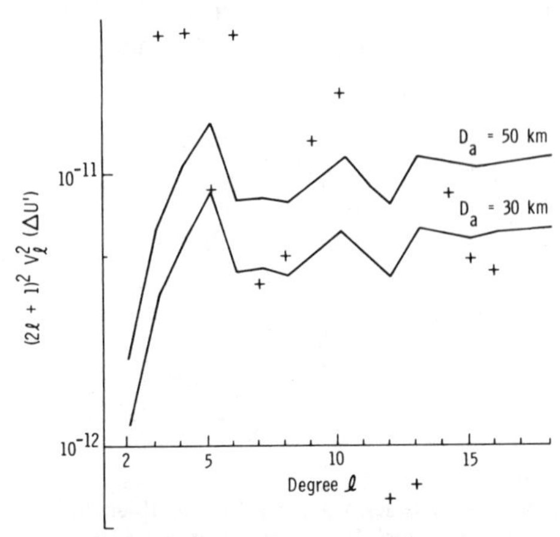

Fig. 17. Potential power spectra of the topography and its isostatic compensation from Airy compensation with $D_a = 50$ km and $D_a = 30$ km, and from linear regression between topography and gravity (crosses). From *Lambeck* [1976].

1. Mass anomalies exist in the lithosphere that are independent of the asthenosphere and are supported by the finite strength of that layer.

2. Mass anomalies exist in the lithosphere that are the consequence of asthenospheric flow, e.g., thermal contraction of the ridges, and are supported by the finite strength of that layer.

3. Mass anomalies in the lithosphere are supported by flow in the asthenosphere.

It has frequently been argued, most recently by *Cochran and Talwani* [1977], that stress levels may be used to eliminate a lithospheric origin for certain long-wavelength gravity anomalies (Figure 18). That is, if it is assumed that such anomalies as the Indian Ocean Low and the North Atlantic High arise from density anomalies in the lithosphere, then this region must be capable of supporting stresses in excess of 3 kbar. By our arguments in section 5, the lithosphere does not appear capable of supporting such large stresses. But this does not exclude at least a partial lithospheric origin for these gravity anomalies, and in particular, they could be the result of lithospheric deformation both caused by and supported by asthenospheric convection (section 4).

Heat flow and bathymetry data have lent substantial credence to the hypothesis of convection currents in the mantle underneath the oceanic plates simply due to their compatibility with an [(age)$^{1/2}$/thickness] and [(age)$^{1/2}$/heat flow] variation away from the oceanic ridges for young lithosphere [*Langseth et al.,* 1966; *Sclater and Francheteau,* 1970; *Parker and Oldenberg,* 1973]. Lithospheric parameters such as effective flexural rigidity also indicate an age dependence [*Cazenave et al.,* 1979]. *Parsons and Sclater* [1977] have demonstrated that the $t^{1/2}$ dependence is implicit in solutions of the advective heat equation, independent of the various boundary conditions imposed. More specifically, the $t^{1/2}$ dependence is a feature of both convective boundary layer theory [*Turcotte and Oxburgh,* 1967] and of moving, cooling plate models with a given upper boundary velocity. In turn, these thermal models are able to account for positive gravity anomalies, as seen in satellite solutions, over many parts of the ocean ridges, with amplitudes of a few tens of mgal and wavelengths of the order of a few thousand kilometers. *Lambeck* [1972] calculated the expected gravity anomaly over the ridges based on the thermal model of *Sclater and Francheteau* [1970] and pointed out that the predictions were satisfactory for the short-wavelength gravity anomalies measured at the surface but not for the satellite-derived long-wavelength anomalies associated with the

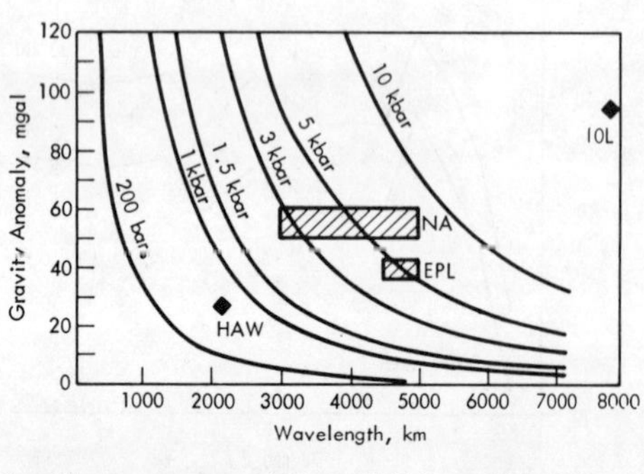

Fig. 18. Shear stresses within a 100-km thick lithosphere as a function of gravity anomaly and wavelength under the assumption that the anomalies arise from within the lithosphere. Various gravity anomalies are shown: IOL is Indian Ocean Low; EPL is East Pacific Low, off North America; NA is North Atlantic High; HAW is Hawaiian Swell. From *Cochran and Talwani* [1977].

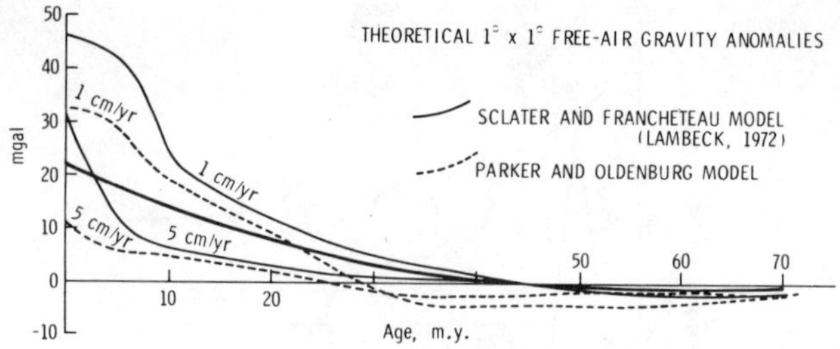

Fig. 19. Empirical 1° × 1° free-air gravity-age curve (heavy line) for mid-ocean ridge systems. Theoretical predictions are shown for two thermal models and two plate velocities. Results of Sclater-Francheteau model were calculated by *Lambeck* [1972]. From *Cochran and Talwani* [1977].

ridges. However, the *Parker and Oldenberg* [1973] model modified the earlier thermal solution by requiring finite heat flow at the ridge crest and by allowing an unlimited depth extent to the cooling. *Cochran and Talwani* [1977] recalculated the gravity anomaly based on the later model and this new result does not show the short-wavelength peak of the earlier solution and in fact shows reasonable agreement with an empirically derived gravity/age relationship out to distances at least 1000 km from the ridge axes (Figure 19). To be sure, there are certain departures from the predicted gravity behavior, the most conspicuous being the North Atlantic Gravity High.

The stress-differences associated with the oceanic ridge long-wavelength gravity anomalies are in excess of a kilobar and it is probably implausible that they are supported by the finite strength of the thin and young elastic part of the lithosphere [*McKenzie,* 1967; *Lambeck,* 1972]. More likely, these long-wavelength anomalies are supported by the asthenospheric flow beneath the ridges.

Large positive long-wavelength gravity anomalies and geoid anomalies are associated with many subduction zones and island arc structures (Figures 12, 13, and 20). The gravity data have been matched by various models of the density structure of the descending lithospheric slab plus various contributions of anomalous mantle density in both the seaward and backarc regions [*Griggs,* 1972; *Grow,* 1973; *Grow and Bowin,* 1975; *Segawa and Tomoda,* 1976]. Grow, for example, finds that a descending slab will produce too large a gravity anomaly, and he introduces a low-density zone in the crust and mantle beneath the backarc region which, along with the slab, will match the observed gravity profiles. The low topographic rise (the 'outer rise') occurring seaward of most ocean trenches contributes significantly to the overall positive gravity anomaly (Figure 20) and is quite possibly a consequence of flexure of an elastic or elastic-plastic [*Watts and Talwani,* 1974; *Caldwell et al.,* 1976; *Turcotte et al.,* 1977; *McAdoo et al.,* 1978] or viscous [*De Bramaecker,* 1977; *Forsythe,* 1979] oceanic lithosphere by vertical forces and bending moments. The outer rise could also be a response of the lithosphere to flow in the asthenosphere (Figure 21 [from *McKenzie,* 1977]). At this point, without debating the rheology associated with the maintenance of the outer rise, it is clear that the outer rise can account for all of the 'outer gravity high' (but of course not uniquely), and *Watts and Talwani* [1975] argue for little contribution to the gravity anomaly from the downgoing slab. This implies either the slab has a small density contrast with its surroundings (<0.04 g cm^{-3}) or is compensated. *Chapman* [1978] modeled both the altimeter-derived geoid anomaly and the free-air gravity anomaly over certain trenches, the first

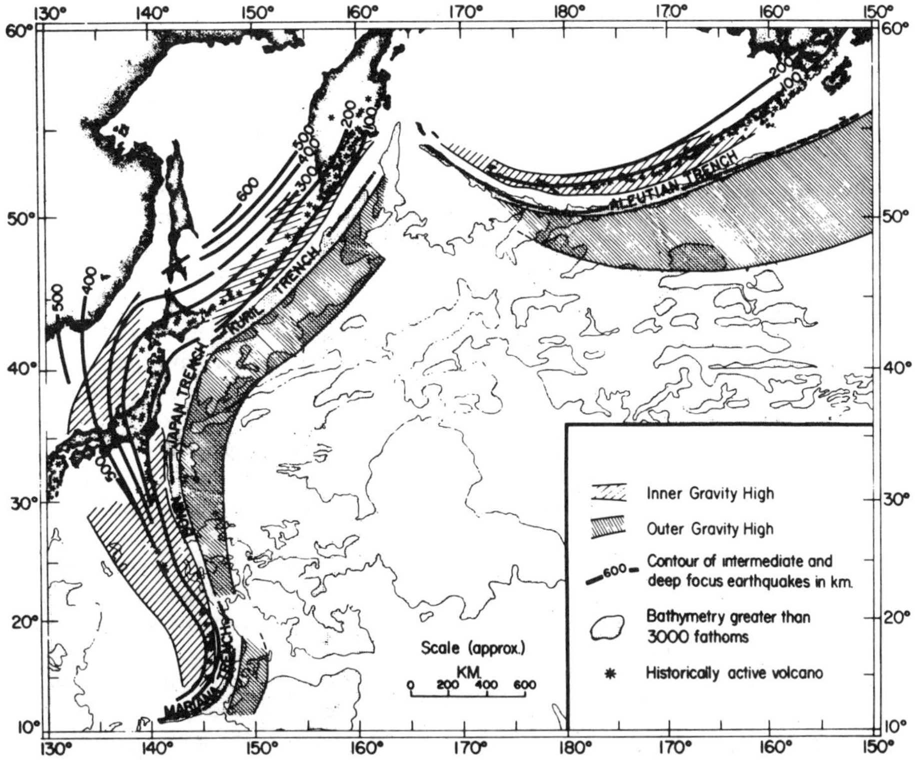

Fig. 20. The northwestern Pacific showing the relationship of gravity anomalies to trenches. From *Watts and Talwani* [1975].

data type being more sensitive to structures at depth. He finds, in general, that little information can be derived about the deep structure under the island arcs from gravity data alone. For many trench systems, the geoid anomaly is dominated by the elevated backarc region (Figure 13), and a model of thinned partly compensated backarc lithosphere will satisfy the observations. This configuration may be analogous to the mid-plate swell model of *Crough* [1978] (see below). Thus the long-wavelength gravity anomalies over trench-island arc systems may be a combination of the effect of elevated backarc lithosphere and the outer rise topographic anomaly, with little contribution from the down-going slab. The nonuniqueness of the modeling procedure must be kept in mind, however.

The simplest approach to relating gravity anomalies to asthenospheric convection is to relate organized patterns of gravity anomalies (if they exist) to organized patterns of convection as predicted by numerical and laboratory experiments. *Richter* [1973] has postulated that a second, smaller scale of convection than that implied by the overall motion of the oceanic plates is required to explain an excess heat flux (over that predicted by the $t^{1/2}$ in the deep ocean basins). The heat flow data themselves show too much scatter to determine a departure from a $t^{1/2}$ relationship. Rather, the excess heat flux is directly inferred from a departure of ocean floor depths from the $t^{1/2}$ relationship for ages greater than about 70 m.y. [*Parsons and Sclater,* 1977; *Parsons and McKenzie,* 1978]. The vertical cell dimension of the smaller scale of convection is assumed to be limited by the 650-km seismic discontinuity which also provides the horizontal cell dimension for unity aspect ratio of the flow. Richter contends, on the basis of laboratory experiments and finite-amplitude theoretical calculations [see also *Richter and Parsons,* 1975], that the preferred small-scale convective planform beneath the Pacific Ocean basin is that of rolls with axes aligned parallel to the spreading direction. Possibly only for the rapidly moving Pacific plate is the spin-up time sufficiently rapid for the rolls to form. A similar phenomenon is seen in the small-scale ocean convection induced by winds blowing from the same direction for several days. Richter's model is kinematic in the sense that plate-induced shear flow is assumed as a basic state and the buoyant finite amplitude instability is superimposed. *Phillips and Ivins* [1979] have questioned the implications of the laboratory experiments for convection in the earth, both in terms of the laboratory fluids used and the difficulty of establishing valid dynamic similarity.

Marsh and Marsh [1976] examined a preliminary geopotential solution (PGS 110) for evidence of a pattern of linear gravity anomalies that might be suggestive of the two-dimensional convection predicted by Richter. They isolate such a pattern [see also *Wagner et al.,* 1977] with harmonics 13–22, although the 2000-km wavelength they find is 50% larger than the wavelength suggested by Richter and Parsons. *Watts* [1978] carried out a preliminary study of surface ship data in the western Pacific and showed that there is little evidence that the linear belts of positive and negative anomalies seen in the global gravity models actually exist. *Marsh and Marsh* [1978] responded by arguing that single-track ship profiles are inappropriate to compare with satellite-derived gravity fields and that a reasonable comparison exists between the PGS 110 and satellite-to-satellite tracking data. It is clear, however, that the two-dimensional patterns seen in the 13–22 field as shown by *Marsh and Marsh* [1976] and also by *Wagner et al.* [1977] are artifacts of the insensitivity of the satellites to these higher wavelength anomalies and inadequate surface data coverage, as well as the least-squares estimation procedure which results in a significant correlation between certain high

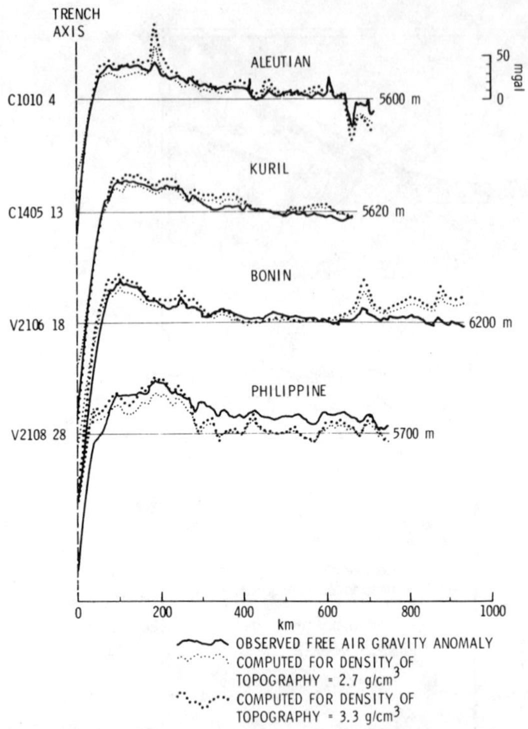

Fig. 21. Gravity anomaly profiles (solid lines) approaching trench axes from oceanic side shown with predicted curves from theoretical gravity/residual depth anomaly relationships suggested by mantle convection. From *McKenzie* [1977].

degree and order coefficients. *Lerch et al.* [1977] have used additional satellite-tracking data and a collocation technique to produce gravity field models superior to earlier solutions, and the two-dimensional patterns striking ESE in the earlier solutions are much weaker, if not absent, in the new models.

The additional constraint provided by observations of lithospheric deformation may provide a more useful approach than gravity data alone in understanding dynamic processes in the mantle. A quantitative relationship between gravity and deformation, applicable to oceanic regions, has been attempted by *McKenzie* [1977] using an admittance function (cf. section 3) defined as

$$Z(k) = \frac{\Delta g(k)}{\Delta h(k)}$$

where Δh is the residual depth anomaly; i.e., the oceanic depth relative to the predicted depth of the lithosphere cooling with distance away from the ridge axes. Both the gravity Δg and residual depth anomalies are expressed as functions of wave number k. McKenzie discusses the theoretical forms of $Z(k)$ for a number of relatively simple terrestrial interactions between convection and the boundary layers. Figure 22 illustrates some results for different boundary conditions and viscosity models. Figure 22*a* reflects the behavior of Z for the viscous convecting layer with stress-free upper and lower boundary conditions and flow driven by a horizontal temperature gradient imposed on the upper surface. The curve A is for an undeformable lower boundary, and at long wavelengths (low k) the behavior of Z is dominated by the contribution from this deeper layer. Long-wavelength gravity will be positive where the anomalous topography is negative and vice versa. The admittance changes sign at the shorter wavelengths (high k) where the deformation associated with the upper layer becomes dominant. When the lower boundary layer is deformable (curve B), compensation occurs at long wavelengths and the admittance there is zero. At high wave numbers the admittance is close to that of the previous case. If the lower boundary is deformable, then large aspect ratio flows are compensated and will lack associated gravity anomalies. Whether or not long-wavelength gravity anomalies correlate with long-wavelength depth anomalies remains uncertain. If they do not, this would argue for a deformable lower boundary. If they do, they could constrain the thickness of the convecting layer.

The temperature dependence of viscosity can change significantly the admittance function. In particular, for an intermediate temperature dependence the surface can be depressed over both rising and sinking columns while gravity is negative over the former and positive over the latter. For a strong temperature dependence both elevation and gravity can be positive, and McKenzie suggests that this may explain the highs to the seaward side of the subduction zones. Figure 22*b* illustrates the admittance for a convective layer consisting of two layers of different viscosities η_1, η_2 with the low-viscosity layer overlying the high-viscosity layer. Now $Z(k)$ may be negative for certain intermediate values of k, a behavior that is caused by a reduction in the magnitude of Δh, which in turn causes Δg to change sign. Over areas of upwelling, Δh remains positive. The low-viscosity layer significantly modifies the admittance, and a careful search of the earth's marine gravity and bathymetry data may dispel the concept that convection is constrained to the upper mantle.

In real planetary situations these simple models will be considerably complicated by the fact that the rheological properties of the boundary layer depend on the thermal structure, a structure that is itself modified by convection. Thus in the

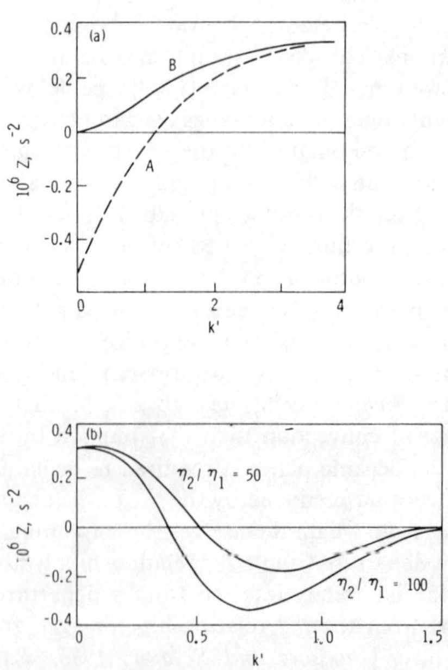

Fig. 22. Admittance function versus dimensionless wave number k' ($= 2\pi d/\lambda$) for a viscous convecting layer of thickness d with stress-free upper and lower boundaries and flow driven by a horizontal temperature gradient imposed on the upper surface. Figure 22*a* is for a layer of uniform viscosity. Curve A is for an undeformable lower boundary. Curve B is for a deformable lower boundary. Figure 22*b* illustrates the admittance for a convective layer consisting of two layers of different viscosities. From *McKenzie* [1977].

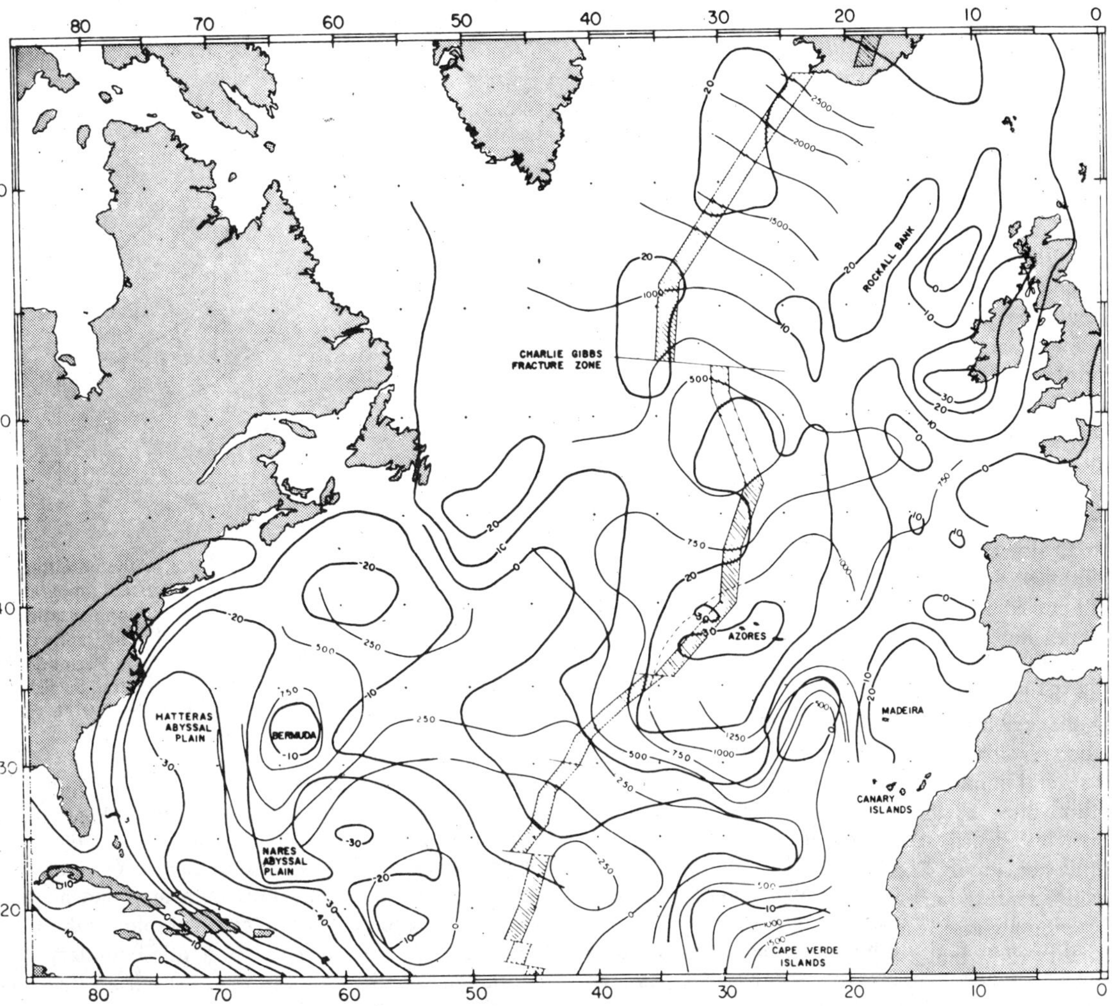

Fig. 23. Map of 5° × 5° residual gravity and depth anomalies in the North Atlantic. Gravity contour interval is 10 mgal, and depth anomaly contour interval is 250 m. From *Cochran and Talwani* [1977].

lithosphere over the rising column there will be a thinning of the elastic or stress-bearing part of the layer, while a thickening of this layer may occur over the sinking column. The density anomalies that can be passively supported over the former will be concomitantly less.

McKenzie's solutions for temperature-dependent viscosity (which should be the true state of affairs) question the notion that the spatial relationships between gravity and topography resulting from convective flow ought to be in phase. Nevertheless, a number of studies [*Anderson et al.*, 1973; *Sclater et al.*, 1975; *Watts*, 1976; *Cochran and Talwani*, 1977] have attempted to establish the correlation (or lack thereof) between the observed oceanic gravity and topography. Perhaps the most thorough review of the gravity/topography correlation is given by *Cochran and Talwani* [1977], who present residual gravity anomaly–residual depth anomaly relationships for the world's major ocean basins. The residual gravity data, analogous to the residual depth data, are obtained by removing the systematic gravity effect of the cooling lithosphere as a function of age or distance from the ridge axes. The study of Cochran and Talwani is most useful because it contains a recent compilation of gravity and is the first attempt to remove a possible masking gravity effect of the cooling lithosphere. These authors conclude that both intermediate wavelength (few 100 km $< \lambda <$ few 1000 km) and long-wavelength residual gravity anomalies are in general not correlated with residual depth anomalies. Although there are specific exceptions, there are no worldwide systematic relationships. These conclusions are perhaps best illustrated in the North Atlantic (Figure 23). On an intermediate wavelength scale,

> . . . there are locations at which there are some striking visual correlations between residual gravity and depth anomalies. Specifically, shallow depths and large positive gravity anomalies are associated with the areas around the Azores, the Cape Verde Islands and the Canary Islands. Greater depths and lower residual gravity anomalies are associated with the areas between these features and to the west of the Azores and the Cape Verde Islands. It is this general area, between 10°N and 50°N, to the east of 50°W, that the data used by *Sclater et al.* [1975] were drawn from and these local areas of unusual elevation and positive gravity anomalies appear to contribute largely to the correlation between gravity and residual depth which they noted.

The three island systems noted above are regions of recent and extensive off-ridge volcanism. Lest one jump quickly to a conclusion of hot spots and dynamic support, it is also noted that the same gravity-depth relationship exists for Bermuda, where the volcanism is 30 m.y. old.

On a broader scale in the North Atlantic there is a relatively constant positive gravity anomaly north of about 40°N, yet the depth anomalies vary from 300 to 3000 m (Figure 23). On the broadest scale, this region displays the same sign for both the residual depth and residual gravity anomalies. In contrast, the Indian Ocean Gravity Low forms a 2000 × 7500 km large-amplitude negative trough in the gravity field, yet the residual depth anomalies in this region vary from +500 to −250 m.

Thus a major conclusion to draw from *Cochran and Talwani* [1977] is that while most of the ocean basins possess negative free-air gravity anomalies, there is no systematic relation between these anomalies and ocean depth once the effects of the cooling lithosphere have been removed. But as we have discussed, simple correlations are not to be expected if these data are controlled by mantle convection and the temperature dependence of viscosity is considered [*McKenzie,* 1977].

We comment on two specific regions where there is an overall gravity-depth correlation, the Hawaiian Swell and the North Atlantic Gravity High, the latter feature showing, as noted above, an agreement on a broad scale in the sign of the two data types. Analyses of the gravity-topography relationships have suggested that both of these features are, at least in part, isostatically compensated in the asthenosphere, thus being maintained by dynamic mechanisms. *Cochran and Talwani* [1978] considered compensation models for the positive residual depth anomaly in the Atlantic north of 30°N and found that while most of the excess topography is compensated in the lithosphere, all successful models (in terms of predicting the free-air gravity) require that a portion of the compensation extend to a uniform depth, in the asthenosphere, of several hundred kilometers. About 500 m of the excess elevation must be (dynamically) supported by the asthenosphere, this density deficiency being interpreted as an upper mantle hot spot by Cochran and Talwani. *Watts* [1976] used a positive regression slope of 22 mgal/km between free-air gravity anomaly and residual depth anomaly of the Hawaiian Swell to show that a lithospheric flexure model is inappropriate and that a simple Airy model of compensation would satisfy the relationship if the depth of compensation is greater than 120 km, well within the asthenosphere. However, *Detrick and Crough* [1978] felt that Watts had overestimated the depth of compensation because of over-averaging of the data into 5° × 5° blocks. Using individual traverse profiles they conclude that the Hawaiian Swell is largely compensated within the lithosphere. *Crough* [1978] showed that lithospheric compensation is the case for four mid-plate swells, Bermuda, Cape Verde, Hawaii, and Cook-Austral, for which free-air gravity data are available (Figure 24). Crough further showed that the altimeter-derived geoid data for the Hawaiian Swell yields a depth estimate for the compensation that is quite consistent with the gravity results. He presents a convincing argument that the gravity anomalies associated with swells as well as the temporal behavior of the depth of the swells can be explained if these mid-plate features form by lithospheric reheating over hot spots. The lithosphere subsequently subsides due to cooling as it moves off the hot spot (Figure 25). The swell compensation appears to be fixed relative to the lithosphere; i.e., dynamic support due to convection is ruled out.

Upper mantle density anomalies. As we have discussed above, the low-degree harmonics of the gravitational spectrum must arise mainly from the mantle. The general lack of correlation of long-wavelength gravity anomalies with residual depth anomalies in the ocean basins further attests to the nonlithospheric origin of these terms. Given that the responsible lateral density anomalies do arise from the mantle, then there may be a first-order relationship to mantle convection, specifically temperature, via the Boussinesq relationship. Below we investigate this possibility.

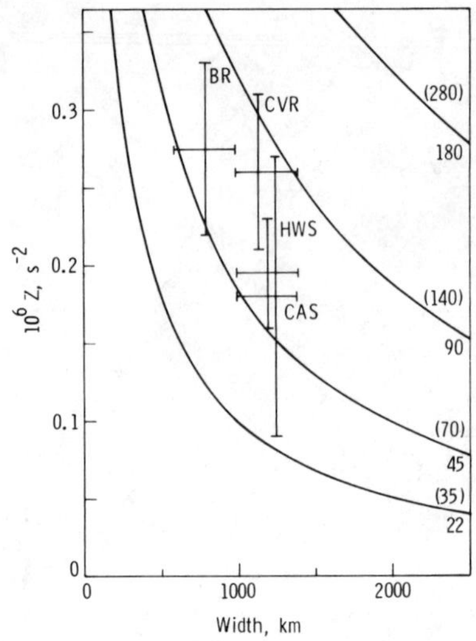

Fig. 24. Curves give the dependence of the admittance function on the 4σ width of locally compensated Gaussian topography. The solutions are for four different depths of compensation with axisymmetric topography (lower number) and topography of infinite extent along strike (parentheses). The data, from Bermuda Rise, Cape Verde Rise, Hawaiian Swell, and Cook-Austral Swell, all indicate compensation in the lithosphere. From *Crough* [1978].

For the recent gravity models the approximate potential power spectrum (13) represents quite closely the observed spectrum up to about degree 20 and possibly further and may be interpreted as being a consequence of short-wavelength density anomalies with a random spatial and magnitude distribution extending down to considerable depth in the mantle and possibly down to the core-mantle boundary. With these considerations, the (white) density spectrum that satisfies the observed gravitational spectrum is related to the average spatial dimensions (dL, dr) of the density anomalies via

$$V_l^2(\delta\rho)^{1/2} \simeq 1.7 \times 10^5 \, (dL \, dr)^{-1/2} \quad \text{g cm}^{-3} \tag{57}$$

where dL and dr are in centimeters. Much of the power in the spectrum originates from density anomalies in the upper mantle. For example, the first 400 km of upper mantle contributes about 65% to the total power in the spectrum from l = 2–18, and varies with degree from about 50% for $l = 2$ to 90% for $l = 18$. Expressed in another way, harmonics of degree higher than about 5 or 6 reflect mainly the state of the upper few hundred kilometers of mantle, while the lower harmonics reflect anomalies from near the surface down to much greater depths. Density anomalies down to 800 km contribute about 85% to the total observed power in the spectrum from l = 2–18. Below this, anomalies in density may still occur, but they do not contribute in a significant manner to the total field. Only a very approximate upper limit can be established for the depths down to which the anomalous structure may occur.

To investigate the consequences of these upper mantle anomalies, we have two options. One is to attribute any set of harmonics to surface density anomalies in a given layer (7) and the second is to use the above random small-wavelength anomaly model (14). Both lead to similar results for the orders of magnitude of the density anomalies. With (12) and (13) the relation for the power spectrum of the surface density anomalies is

$$V_l^2(\sigma) = 0.7\left(\frac{\bar{\rho}R}{3}\right)^2 \frac{(2l+1)^3}{4} 10^{-10} \left(\frac{R}{r'}\right)^{2l+4} \tag{58}$$

If these anomalies originate in the asthenosphere, then $r' \simeq$ 6200 km, $dr \simeq 200$ km, and

$$V_l^2(\delta\rho) = 3.4 \times 10^{-7}(1.03)^{2l+4}(2l + 1)^3 l^{-4} \quad (59)$$

For l = 2–20 the rms density anomaly in this layer would be of the order of

$$\left[\sum_{l=2}^{20} V_l^2(\delta\rho)\right]^{1/2} \simeq 0.005 \text{ g cm}^{-3}$$

With the model (57), taking $dr = dL = 200$ km, so that the density anomalies are of much smaller dimensions than implied by the l = 20 harmonics, the rms density anomaly is 0.009 g cm^{-3}.

If such anomalies are a consequence of lateral temperature variations—the thermal expansion coefficient of the upper mantle is of the order of 3×10^{-5} K^{-1}—a modest rms temperature fluctuation of 50 K is sufficient to explain density anomalies of 0.005 g cm^{-3}. However, as previously stressed, we cannot attribute the gravity anomalies to one source only. For example, the lateral temperatures could result in a change in the depth of a phase transition zone according to (26) or, with $a_t = 0.062$ kbar K^{-1}, $dr \sim 1.5$ km, and this will increase the gravity anomalies by about 50% over the contribution due to the thermal expansion alone, and hence permissible lateral temperatures would have to be only 35 K. Alternatively, (1) the density anomalies are of smaller extent than implied by the wavelengths $\lambda = 2\pi R/l$ of the geopotential and the random density model (57), with $dL = dr = 200$ km, results in an increase of the permissible temperature variations by a factor of about 2, (2) other contributions to the density field occur which may partly cancel out the above temperature effects, or (3) there is a complex interaction at the boundaries as previously discussed. Without further insight into these aspects, it is futile to attempt to invert the gravity field for lateral temperature variations in the upper mantle, and a significant compositional contribution to the density anomalies cannot be ruled out.

Seismic evidence for lateral variations in the structure of the upper mantle is confusing, although both surface and short-period waves show variations with horizontal dimensions of the order of a few hundred kilometers (e.g., the station anomaly maps for the United States given by *Hales and Herrin* [1972]), and regional variations of up to 5% in v_p have been suggested for the upper mantle of North America [*Herrin*, 1972]. The most detailed study of lateral variations in the upper mantle is that by *Romanowicz* [1979], who inverted station P anomalies to obtain velocity anomalies under North America. She found lateral variation of the order of 2% down to depths of about 600 km, and more significantly, she found evidence for compensation in that positive velocity anomalies near the surface tended to be underlain by negative velocity anomalies at intermediate depth and vice versa. At depths near 400–700 km the anomalies tended to be of the same sign as those near the surface. The rms velocity anomalies are of the order of 1%. If velocities can be related to density by the Birch equation (23), then this model is comparable with that obtained by *Lewis and Dorman* [1970] from gravity data. The horizontal length scale of these anomalies is of the order 2000 km, and the vertical length scale is of the order of 200 km. With (23) and

$$\frac{dv_P}{d\rho} = a_v \simeq 3.0 \text{ km s}^{-1}/\text{g cm}^{-3} \quad (60)$$

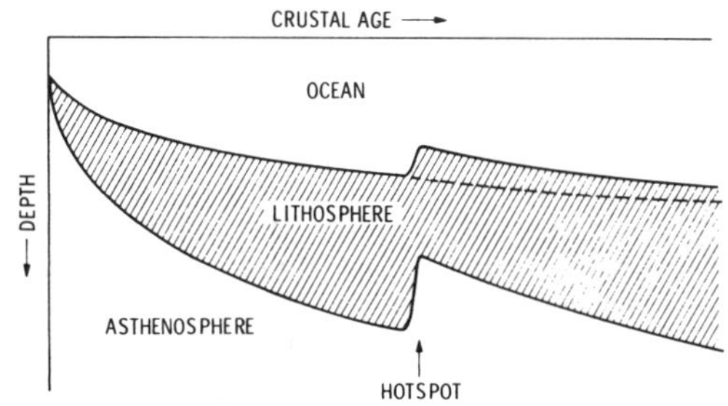

Fig. 25. Hypothesis of *Crough* [1978] for hot spot–lithosphere interaction. From the ridge to the hot spot the lithosphere thickens and subsides by cooling. At the hot spot, extra heat drives the isotherms upward, thins the lithosphere and causes uplift. Beyond the hot spot, the lithosphere cools rapidly because it is thin and thus subsides as younger lithosphere at the same depth, rather than as normal lithosphere of the same age (dashed line).

the lateral density anomalies are of the order of 0.011 g cm^{-3}. Using the 'expected' estimate from gravity (57), $d\rho \simeq 0.003$ g cm^{-3}. This discrepancy between the two estimates may be partly due to the compensation alluded to previously, to the limitations of the Birch law, or to the fact that the North American structure is not typical of the average mantle.

Lateral variations in v_S of up to 10% have been reported [e.g., *Toksöz et al.*, 1967] but these are generally less certain due to the lower reliability of S arrival times. *Romanowicz* [1979] obtained variations of the order of 4%, comparable to those found by *Cara* [1979] from the inversion of surface wave phase velocities.

The regional variability of the flow of heat through the earth's surface from its interior has been recognized for many years. Part of this is obviously a consequence of variations in the crustal content of radioactive materials and in variations in crustal thickness, while part may be a consequence of a variable flow from the mantle due to lateral temperature differences within the mantle [e.g., *Roy et al.*, 1972; *Sass and Lachenbruch*, 1979]. A recent global scale attempt at separating the crustal contributions from the surface measurements has been made by *Pollack and Chapman* [1977], and the study is suggestive of significant regional variations that follow closely the ocean-continent distribution and, within the continents, the distribution of tectonically active areas and old shields. Unfortunately, the crustal corrections are large (about 50% of the observed flow) and uncertain, and whether this correlation is a true reflection of the heat flow from the mantle, or merely a consequence of inadequate, topography-dependent corrections, remains unclear.

An inversion of regional heat flow measurements across the western part of the United States yielded lateral temperature variations of the order of about 100 K over horizontal distances of about 500–1000 km and down to depths of at least 200 km [*Roy et al.*, 1972]. This is not incompatible with the gravity and seismic anomalies for this region.

Deep mantle contributions. Because of the attenuation of potential with distance from the source, the information on the lateral density structure of the lower mantle of a planet from gravity alone is limited and, if the random density model is accepted, only the very low degree harmonics contain appreciable information on the lateral structure below about 600–800 km. In the case of the earth it has been suggested by *Hide* [1966] that the geometry of the core-mantle boundary departs significantly from a hydrostatic equilibrium spheroidal shape and, because of the large density contrast across this interface, there may be a significant surface density layer [*Hide and Horai*, 1968]. The contribution of this layer to the geopotential at the surface follows from (8), with $dr\ \delta\rho_{ilm}$ replaced by $d\rho\ \delta r_{ilm}$, where δr_{ilm} is now the topography of this surface and $d\rho$ is the density contrast across this layer. Hide and Horai suggested that the low-degree terms ($l \lesssim 5$) in the geopotential may be due to this core-mantle topography. As there is some correlation between these low-degree terms and the geomagnetic field, provided that the latter field is rotated in longitude by some 160° [*Hide and Malin*, 1970], this suggestion has considerable attraction, although it poses difficulties as well. Foremost is the fact that it is not possible to attribute any particular harmonics to density anomalies in only a single region. For example, if density anomalies elsewhere in the mantle follow a more or less random distribution, they will also contribute to these same low-degree harmonics, and *Lambeck* [1976] estimated that if the random anomalous lateral density distribution model is valid, the maximum topography is of the order of 150–200 m. One can also consider the geophysical consequence of such a topography on other mantle properties. From (31) the maximum stress difference required to support the anomalies due to the low-degree $l < 4$ terms is of the order of several kilobars, and this is excessive for a region where the temperatures may be close to the melting point. Thus lower mantle convection is required to support this nonhydrostatic state, and this results in density anomalies in the entire lower mantle which contribute to the gravity field. Now there may be some compensation between the excess mass in a 'bump' on the core-mantle boundary and the deficit of mass in the rising convection column over it and, if the convection has a substantial vertical extent, the overall anomaly at the surface could well be negative. Again, a model which at first glance is quite simple, quickly degenerates into an ambiguous situation, and all of this is not very helpful in addressing the question as to what part of gravity results from lower mantle lateral structure. Other arguments must be marshalled.

Seismic observations provide the most direct evidence of lateral heterogeneity in the deep mantle in the form of lateral seismic velocity anomalies, but at present only the earth is accessible in this way. Several studies in recent years have indicated that significant lateral heterogeneities may occur in the lower mantle, the two most detailed and global studies being by *Julian and Sengupta* [1973] and by *Dziewonski et al.* [1977]. The former analyzed $dT/d\Delta$ observations and concluded that variations in P velocities of about 1% occur at depths greater than about 2000 km, with horizontal dimensions of about 2000 km, and a vertical coherence of a few hundred kilometers. Because they measure average velocity anomalies, they view their results as indicative of lower bounds.

Dziewonski et al. inverted P and PKP travel times in order to estimate the lateral structure of the mantle. They considered P velocity anomalies in four layers, expressed in each by a spherical harmonic expansion with $l \leq 3$, $m \leq 3$. Their rms lateral velocity variations are of the order of 0.1–0.2% in the lower mantle, and while smaller, they are not incompatible with Julian and Sengupta's results, since the latter solved for only the most significant velocity anomalies and since greater averaging occurs in the low-degree spherical harmonic expansion than in the direct mapping technique.

Julian and Sengupta find no correlation between their velocity anomalies in the lower mantle and surface gravity, but this is not unexpected since the upper mantle density anomalies dominate gravity. Likewise, Dziewonski et al. do not find significant correlation (apart from the 2, 0 harmonic). Equation (25), relating the geopotential spectrum to seismic velocity anomalies in N layers, has been used to construct Table 1, which summarizes the expected spectrum for $l = 2, 3$ for Dziewonski et al.'s solution for a three-layer lower mantle model between the core-mantle boundary and a depth of 1100 km. The $[V_l^2(\Delta U)]^{1/2}$ exceed the observed values by a factor of 3 for $l = 2$ and 25 for $l = 3$, and this suggests that either (1) the proportionality coefficient a_v is too small due to a temperature or pressure dependence of the Birch relation, (2) there is a compensation between the anomalies in the upper and lower mantle, comparable to that suggested for the upper mantle alone, or (3) the seismic anomalies are overestimated. More extensive analyses of the travel time anomalies are required to permit a further discussion of these alternatives.

8. MOON

Introduction. For our purposes here, the history of the moon can be divided into three stages [e.g., *Mutch*, 1972] starting with formation 4.5–4.6 AE ago (1 AE ≡ 1 billion years) and followed by differentiation to form an anorthositic crust at 4.4 AE ago [*Tera and Wasserburg*, 1974]. This crust was subjected to intensive bombardment for a period of approximately 5×10^8 years with impacting masses ranging over a wide spectrum of magnitudes. This first stage, Pre-Imbrian time, ended approximately 3.9 AE ago with the formation of the major impact basins such as Serenitatis, Imbrium, and Orientale.

The Pre-Imbrian interval is envisaged as a time of frequent impacts of large basin-forming objects with the larger impacts fracturing and possibly foundering the young crust, at least early in this period. During this early interval, the thermal environment was such that both supersolidus and subsolidus flow took place, rapidly removing lateral density inhomogeneities, including any early surface topography. Later in Pre-Imbrian time some surface expressions of these impacts remained and are seen today as, for example, the irregular maria. The lunar interior probably responded isostatically to these impacts by both creep processes and brittle failure. The surface expressions of the large impacts which occurred at the end of this period are well preserved and are seen as both the unfilled and filled circular basins. We argue that at the end of the Pre-Imbrian period the crust and upper mantle temperature was probably low enough that any subsequent long-term viscous isostatic response to impacts and their attendant interior manifestations was small. To be sure, approximate mass balance may have been achieved due to mechanical failure (including instantaneous rebound) of the crust. The stage

TABLE 1. Spectral Estimates for $l = 2, 3$ of the Observed Geopotential and of the Predicted Potential From *Dziewonski et al.*'s [1977] Inversion of Seismic Travel Times With $a_v = 3$ (km s^{-1})/(g cm^{-3})

Harmonic	V_l^2 (Observed)	V_l^2 (Estimated): Total	Lower Mantle	Upper Mantle
$l = 2$	3.12×10^{-11}	3.69×10^{-10}	3.43×10^{-10}	4.06×10^{-11}
$l = 3$	8.64×10^{-12}	5.53×10^{-9}	5.67×10^{-9}	4.32×10^{-11}

The predicted values are based on the entire mantle, the lower mantle below 1100 km, and the mantle above 1100 km.

was set at this time for the formation of the mass concentrations associated with these ringed maria, i.e., the mascons of *Muller and Sjogren* [1968], during the next major interval of lunar history.

This thermal-mechanical model for the outer portion of the moon during Pre-Imbrian time is, of course, a continuously evolving one, and our discussion above only concerns descriptions of the dominating response to impact at specific times in this period (Figure 26). The topography toward the end of this first stage, ignoring a possible thin mare cover, can be grossly characterized as a background of middle Pre-Imbrium terrain, which on the whole is compensated. Upon this are superimposed the late Pre-Imbrian impact features, which did not induce a significant creep readjustment in the interior.

The second stage of the lunar history, Imbrian time, spanned the interval 3.9 AE ago to about 3.0 AE ago or slightly younger and is characterized as a period of major extrusive igneous activity accompanied by little tectonic activity. It is during this interval that basaltic volcanism filled both the near-side irregular and circular maria, forming mascons in the latter basins [*Phillips et al.*, 1972, 1974]. The dominant surface tectonics of this period is the brittle failure of the outer lithosphere to the loads of the mass concentrations.

The third and final stage of lunar history, the Post-Imbrian interval (formally referred to as Eratosthenian and Copernican time) extends from approximately 3 AE ago to the present, and the lunar surface shows essentially no tectonic or igneous events during this interval. Presently, seismic activity does take place in the lunar interior with moonquake zones in the upper mantle [*Nakamura et al.* 1979] as well as in a depth range from 600 to 1000 km [e.g., *Dainty et al.*, 1975]. The shallow events are thought to be tectonic in nature [*Nakamura et al.*, 1979], while there is some debate as to whether the deep events are tectonically [*Toksöz et al.*, 1977] or tidally [*Nakamura*, 1978] triggered.

The lunar gravity field. Numerous estimates of the lunar gravity field have been made since the first spacecraft orbited the moon, and *Sjogren* [1977] gives a brief review of the present status of the knowledge of the lunar gravity field. All solutions have been beset by well-known problems which have made a definitive solution of even the low-degree harmonics elusive. These problems are the absence of farside tracking data, the slow convergence of the lunar potential spectrum, the inadequate distribution of the inclinations of low-perigee satellites, and the lack of long arcs of tracking data unperturbed by spacecraft maneuvers. Different analysis techniques have been developed to offset some of these limitations, mainly by mapping directly the spacecraft accelerations [*Muller and Sjogren*, 1968] and by representing the anomalous density structure of the moon by point or disc masses at specified locations on or below the surface [*Wong et al.*, 1971]. These direct mapping methods give no information on the lunar farside gravity field.

The low-degree coefficients in the spherical harmonic expansion of the lunar gravity field have been estimated either directly, by expressing the spacecraft velocities or rates of change of the mean orbital elements as functions of the potential coefficients, or indirectly, by first estimating mass points and then relating these to the spherical harmonic coefficients. Recent solutions are by *Ananda* [1977], *Ferrari* [1977], and *Bills and Ferrari* [1980]. Some very low degree harmonic coefficients, $l = 2$ and 3, have also been determined from the lunar librations using laser ranging observations [*Williams*, 1977; *King et al.*, 1976]. The situation is unsatisfactory for the harmonics of degree 4 and higher. Even for $l = 2$ the present solutions are insufficient in that they lead to an uncertainty in the moment of inertia of about 1% and this is too large to permit conclusions to be drawn, the uniqueness question aside, about the density of the lunar core for an assumed radius. For the higher degree harmonic coefficients, the differences between the various solutions often exceed the values themselves. The earlier solutions by, for example, *Michael et al.* [1969] and *Michael and Blackshear* [1972] result in very unrepresentative models for the lunar farside and grossly overestimate the power in the potential.

Figure 27 illustrates our estimated spectrum of the lunar gravity field. For $l > 4$ it is based on the solutions by Ananda and by Ferrari, while for the lower harmonics it is based mainly on solutions by *Sjogren* [1971], *Liu and Laing* [1971], and Michael et al., and on the libration studies of Williams et al. and King et al. The spectrum based on Ferrari's solution gives power estimates that systematically exceed those of

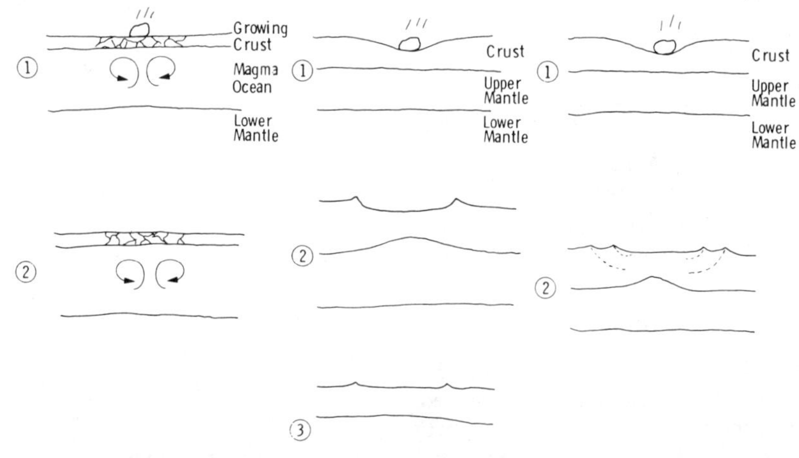

Fig. 26. Cartoons illustrating the response of the lunar lithosphere to large impacts during Pre-Imbrian time. (Left) Early Pre-Imbrian impacts fracture and founder the thin lithosphere and are assimilated by the magma ocean. (Center) Middle Pre-Imbrian lithosphere isostatically responds to impact basin, leading to irregular maria. (Right) Late Pre-Imbrian lithosphere fails mechanically at time of impact and interior density anomalies are supported by finite strength.

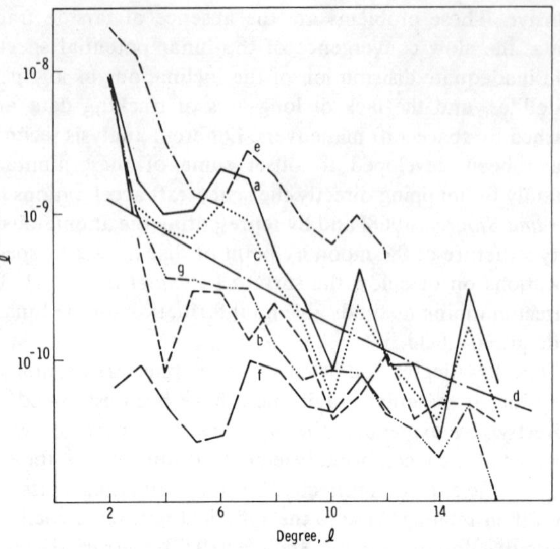

Fig. 27. Lunar potential spectra based on (*a*) *Ferrari* [1977], (*b*) *Ananda* [1977], (*c*) a composite model, (*d*) the random short-wavelength near-surface density anomaly model, (*e*) uncompensated topography, (*f*) compensated topography with $D_a = 50$ km, and (*g*) a recent solution by *Bills and Ferrari* [1980].

Ananda's solution. The latter solution results in a correlation with topography that is somewhat higher for the lower degree harmonics than Ferrari's solution (see below) and may be a more realistic one.

Perhaps the only outstanding feature of the potential power spectrum is that the second-degree spectral estimate appears to exceed that of the other degrees and that it lies above some 'regular' decay rule. Another observation is that the spectrum apparently decays more slowly than its terrestrial counterpart, suggesting that the density anomalies are relatively more superficial on the moon than on the earth. This is not to say that density anomalies do not occur in the upper mantle but, if they do, they must be proportionally smaller; otherwise the potential power spectrum would decay more rapidly than actually observed. A third observation is that gravity is less than may have been expected from an extrapolation from earth or Mars to the moon, for if the earth and moon can support the same stresses, it follows from (34*a*) that

$$V_l^2(\Delta U_M) = \left(\frac{M_E}{M_M}\right)^4 \left(\frac{R_M}{R_E}\right)^8 V_l^2(\Delta U_E)$$

$$= (g_E/g_M)^4 V_l^2(\Delta U_E) \simeq 1.1 \times 10^3 V_l^2(\Delta U_E) \qquad (61)$$

where the subscripts M and E denote quantities pertaining to the moon and earth, respectively. By comparison to Figure 11*a*, it is seen that this predicted spectrum generally exceeds the actual lunar spectrum. If the prediction is based on the Martian potential,

$$V_l^2(\Delta U_M) \simeq 22 V_l^2(\Delta U_{\text{Mars}}) \qquad (62)$$

which also exceeds the lunar spectrum (see Figure 36). The low power in the lunar gravity field, relative to that predicted by equivalent stress arguments, is indicative of a 'quiet' planet, in keeping with the lack of active tectonic processes over the last 4 billion years that would have led to large-scale density anomalies.

The lunar topogr·phy. The lunar topography, deduced from a combination of earth and Apollo mission photography, radar, and laser altimetry, has been harmonically analyzed by *Bills and Ferrari* [1977]. The spectrum of their analysis is illustrated in Figure 28, where the topography is referred to a sphere with radius equal to the mean radius of the planet.

A second analysis, based on partly independent data, has been published by *Chuikova* [1976] (Figure 28). Agreement between the two spectra is satisfactory up to degree 6, but beyond that the Bills and Ferrari spectrum gives consistently higher spectral estimates. Bills and Ferrari also carried out some extensive Fourier analyses of the lunar topography along great circles and concluded that the spectrum decayed according to (21). Specifically, they suggest that

$$V_l^2(h) \simeq 1.5 \times 10^{-6}/l(l+1) \qquad (63)$$

(Figure 28), although their observed spectrum exceeds this for $l \geq 8$. Chuikova's spectrum follows this rule more closely.

An outstanding feature of the topography spectrum is that the power in the $l = 1$ harmonics is considerably greater than that in the higher degree harmonics. Nonzero values for the $l = 1$ harmonics indicate that there is an offset between the center of mass and the center of figure which for the moon indicates that the center of mass is closer to the earth by about 2.0 km. Compared with Mars, where the offset is about 2.5 km, and with the earth, where the offset is about 1.1 km, there would be nothing very special about the lunar result were it not for the fact that, for the moon, this spectral estimate much exceeds the $l > 1$ estimates, whereas for earth and Mars all low-degree spectral estimates are of comparable magnitude (compare Figure 28 with Figure 37).

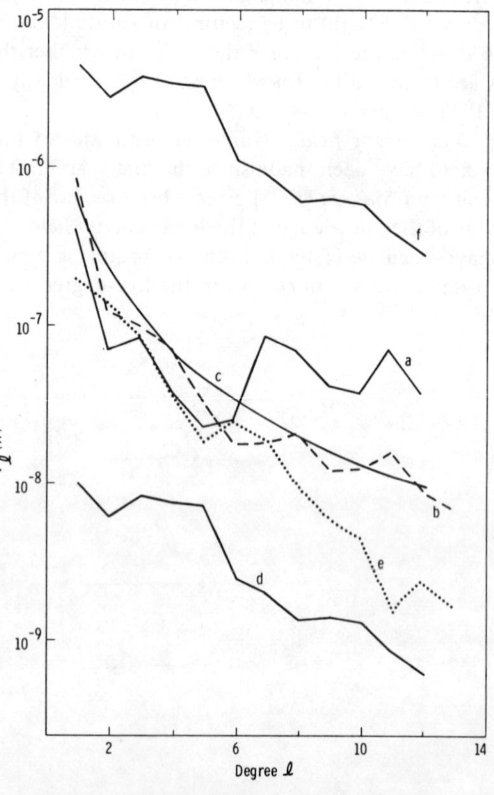

Fig. 28. Lunar topography spectra based on (*a*) *Bills and Ferrari* [1977], (*b*) *Chuikova* [1976], and (*c*) the empirical model given by (63). The spectrum (*d*) represents the terrestrial topography and the spectrum (*e*) the Martian topography. The spectrum (*f*) is the predicted topography using the earth spectrum and scaling, assuming that both planets have the same intrinsic finite strength.

Insofar as the lunar offset is mainly along the earth-moon axis, it is tempting to associate it with the dichotomy seen in the lunar surface of a nearside dominated by maria and a farside dominated by highlands. Several interpretations of this have been made. *Kaula et al.* [1974] suggested that there may be a difference in the crustal thickness between the near and far sides, the farside highlands being associated with a thicker crust than the nearside. Alternatively, adopting a Pratt-type crustal model, the crustal densities of the farside highlands would be less than those of the nearside regions. *Ransford and Sjogren* [1972] suggested that the core is offset while *Lingenfelter and Schubert* [1973] suggested that the offset may be a result of a first-degree convection cell that occurred in some early stage of the moon's history. *Kobrick* [1976] suggests that random density anomalies can explain the offset.

All these explanations consider an offset formed at an early stage in the moon's history which has been 'frozen' into the planet ever since. Thus the models have to be tested not only against the plausibility of the mechanisms causing the offset but also against the consequence of this offset on the stress state of the planet. The usual theory of the loading of a homogeneous sphere is not directly applicable for an $l = 1$ load since, in this case, one solution of the elastic equations of motion is a constant but indeterminate shift of all particles of the body which results in zero stress (see the general form of the solution for any l as given by *Takeuchi and Saito* [1972]). The solution is discussed by *Jeffreys* [1970] for the incompressible homogeneous sphere; he showed that in this case the stress is zero. A more general discussion is given by *Farrell* [1972]. *Lambeck and Pullan* [1979*a*] have computed the strain energy for different lunar models subjected to surface loads, and they find that for all reasonable models the stresses set up by the $l = 1$ harmonics remain insignificant compared with those due to the higher degree harmonics. They suggest that this harmonic may reflect a condition at the time of the final phase of the major bombardment of the lunar surface; the overall topography was much greater then than now, such that the power for low-degree harmonics approximated the power in the present $l = 1$ spectral estimate. The 'white' topographic spectrum, formed in Pre-Imbrian time, might have resulted if the characteristic wavelengths of the process responsible for the initial topography were much shorter than the wavelengths corresponding to the low-degree harmonics. This early topography, presumably in a state of near isostasy, would have caused stresses in the crust that were an order of magnitude larger than those occurring at present for $l \geqslant 2$ and would have relaxed with time to a lower energy isostatic state. The $l = 1$ harmonic, imparting much lower stresses, would have tended to remain stable. If this interpretation is accepted, then the average strain rates have been of the order of 10^{-19}–10^{-21} s^{-1} [*Lambeck and Pullan,* 1979*a*].

Another outstanding aspect of the lunar topography is that it is much less significant than may have been expected from comparisons with earth or Mars for, using the same assumption as before, namely, that the planets have the same intrinsic finite strength (i.e., support the same stresses), one obtains with (32),

$$V_l^2(h) = \frac{(g_E \rho_E{}^c R_E)^2}{(g_M \rho_M{}^c R_M)^2} V_l^2(h_E)$$

$$= 500\ V_l^2\ (h_E) \qquad (64)$$

where ρ^c is the crustal density. This predicted spectrum is illustrated in Figure 28. The smallness of the observed spectrum is indicative, inter alia, of the moon being a quiet planet whose topography is supported in a manner different from that on earth.

In view of the poor quality of the spherical harmonic expansions of the lunar gravity field, the cross correlation between gravity and topography is not a well-determined quantity at all. Figure 29 illustrates some results for the potential solutions by *Ferrari* [1977] and *Ananda* [1977]. For the former, possibly significant correlation is found for l = 2–4 only, and beyond $l = 4$ the correlation is neither significant nor consistently of one sign. The situation is somewhat better for Ananda's solution in which the correlation is consistently positive for $l < 8$. This makes Ananda's solution the more attractive since nothing has been introduced into the solution to produce such a correlation for the low-degree harmonics and since some correlation can be expected in view of the low stresses set up by the lunar topography (see below).

Isostasy and crustal stresses. If the lunar topography is entirely uncompensated, then the expected gravitational potential spectrum follows from (19), and for fully compensated topography the spectrum is given by (20), using either the observed topographic spectrum or the approximation (63). These two types of spectra are compared with the observed spectrum in Figure 30. As for the earth and Mars, the spectra for uncompensated topography (Figure 30, curves *a* and *b*) exceeds the observed spectrum (Figure 30, curve *e*) for all l, indicating that some compensation of the topography occurs, although the discrepancy is much less than for earth (see Figure 15).

Seismic data suggest that the crust in the Oceanus Procellarum region near the Apollo 12 and 14 landing sites is about 55 km thick [*Nakamura et al.,* 1974; *Toksöz et al.,* 1974]. As the highlands are on the average about 3 km higher than the maria, the expected crustal thickness under the former is of the order of 80 km if Airy isostasy prevails and less if there is a Pratt component to the compensation [*Solomon,* 1978]. Some evidence for a variation in crustal thickness is found in the Apollo 16 seismic data for the Descartes regions of the central highlands. These data have been interpreted in terms of a crust some 16 km thicker than under the Procellarum re-

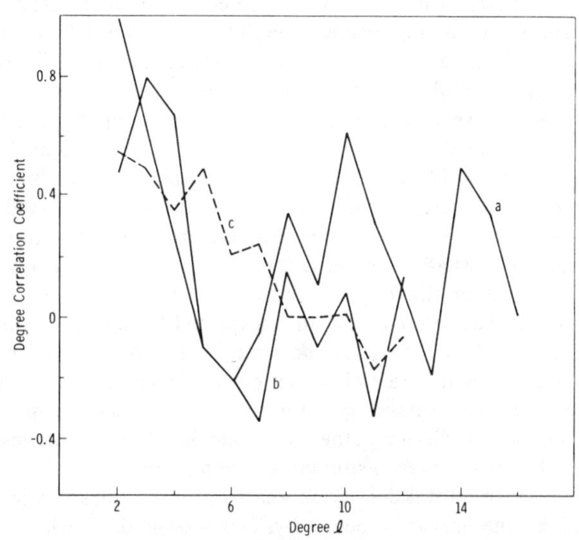

Fig. 29. Degree correlation coefficients for the lunar gravity field models of (*a*) *Ananda* [1977] and *Ferrari* [1977], (*b*) Ferrari gravity and topography, and (*c*) Ananda gravity and topography.

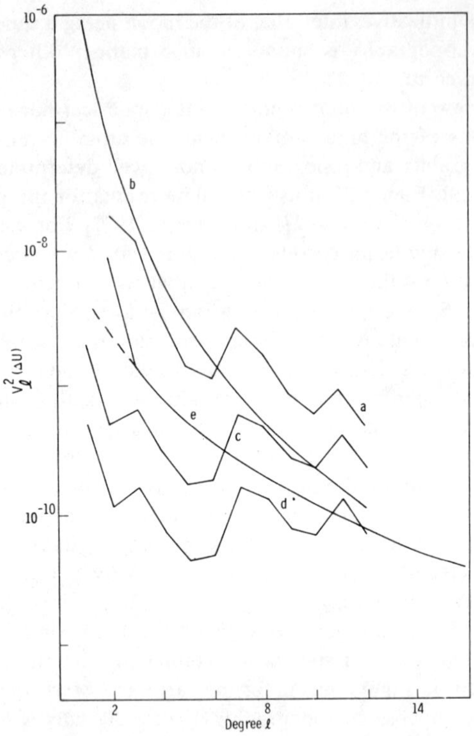

Fig. 30. Gravitational potential spectra for the moon based on (*a*) uncompensated topography of *Bills and Ferrari* [1977], (*b*) uncompensated topography given by (63), (*c*) compensated topography with depth of compensation equal to D_a = 125 km, (*d*) same as curve *c* but with D_a = 75 km, and (*e*) smoothed observed potential spectrum.

gion [*Goins et al.,* 1977], while the average elevation of the Descartes region is about 2 km above Oceanus Procellarum [*Sjogren and Wollenhaupt,* 1973].

The quantity most strongly constrained by the gravity and topography data is the integral over depth of the anomalous density variation [see, e.g., *Thurber and Solomon,* 1978]. This allows, for example, a constraint on the density distribution under an isostatic assumption. If a possible Pratt component of compensation in the crust is ignored, a maximum gravity effect results. Taking the mean depth to the level of Airy compensation, the crust-mantle boundary, as 75 km, then Figure 30, curve *d,* results and the spectral amplitudes are less than the observed spectrum, indicating that either compensation is incomplete or the depth of compensation is greater than 75 km. For D_a = 125 km, (20) results in a spectrum of similar magnitude (Figure 30, curve *c*) to that determined by Ananda for all $l > 2$. The implication is clear: either a substantial component of isostatic compensation of the lunar topography takes place in the upper mantle or compensation is incomplete. The latter is not an unexpected configuration for the lunar lithosphere since the stress differences implied by the gravitational potential do not exceed about 100 bars (see below) and, as we have argued in section 6, the isostatic process may only be driven to the point wherein all stresses are less than the intrinsic creep strength of the material.

The mean crustal density anomaly or crustal thickness implied by the gravity, topography, and seismic data, whether isostatic or near isostatic, imply an aluminum oxide content for the moon that is about 2 times the chondritic value [*Kaula,* 1977*b*]. Hence this excess in refractory lithophiles, along with other geochemical anomalies, provide a strong constraint on theories of lunar origin.

An approach to the question of stress state in the moon's interior is to invert the gravity data for both the anomalous density structure and the stress state along the lines suggested in section 3. The usual procedure is to restrict the anomalies to a single layer [*Arkani-Hamed,* 1973*a*], and this appears to be adequate for the moon, where the density anomalies are mainly confined to a near-surface layer [*Lambeck and Pullan,* 1979*a*]. In a second calculation, *Arkani-Hamed* [1973*b*] followed *Kaula* [1963] and minimized the strain energy (section 3). In the first calculation [*Arkani-Hamed,* 1973*a*] the density variations were confined to the upper 50 km of the moon and the associated stress differences were found to be of the order of 40 bars within the upper 800 km, but this calculation assumed the moon to be elastic throughout. As he used the potential solution by *Michael et al.* [1969], which we believe significantly overestimates the spectrum for $l \gtrsim 5$, these stresses are too high. In the second paper, *Arkani-Hamed* [1973*b*], again using an elastic rheology, estimated maximum stress differences to be about 100 bars for the potential model of *Michael and Blackshear* [1972], which is even less palatable for $l > 4$ than their earlier model.

A more cogent approach to the question of stress state in the moon follows from the argument that loads must be supported by a thin elastic lithosphere, as defined in section 3, below which stresses have relaxed at least $1/e$. Using the flow laws of *Durham and Goetze* [1977] and *Post* [1977] the base of the lunar lithosphere occurs at about the 0.5–0.6 melting temperature (T_m) isotherm. We argue in the following sections that this isotherm, and hence the boundary of the elastic lithosphere, occurs at a depth of 200–300 km and the attendant load stresses for both mascons and the nonhydrostatic ellipticity are of the order of 100 bars and are capable of being supported by the finite strength of the lithosphere.

The lunar ellipticities and convection. *Runcorn* [1962, 1976 and references therein] has long argued that the nonequilibrium second-degree harmonics of the lunar potential and topography present direct evidence for second-harmonic convection in the lunar interior. The lunar hydrostatic equilibrium state is governed by self attraction, spin, and the earth's attraction and with the condition that its spin equals its orbital mean motion, the hydrostatic second-degree coefficients are given by [e.g., *Kopal,* 1969]

$$5^{1/2}\, C_{20}{}^{HE} = -\frac{C - \frac{1}{2}(A + B)}{mR^2} = -\frac{5}{6}\left(\frac{R}{a}\right)^3 \frac{M}{m}(\Delta - 1) \quad (65a)$$

$$\left(\frac{5}{12}\right)^{1/2} C_{22}{}^{HE} = -\frac{A - B}{4mR^2} = \frac{1}{4}\left(\frac{R}{a}\right)^3 \frac{M}{m}(\Delta - 1) \quad (65b)$$

where A, B, C are the principal moments of inertia with $C > B > A$; m is the mass of the moon and M the mass of the earth; a is the average earth-moon distance; R is the lunar radius; and Δ is a factor that corrects for any departure from a homogeneous density distribution. For a wide range of realistic lunar models $\Delta \simeq 2.48$, compared with 2.50 for a homogeneous body. The power spectrum for this body in hydrostatic equilibrium is

$$V_2{}^2(\Delta U^{HE}) = \frac{13}{45}\left(\frac{R}{a}\right)^6 \left(\frac{M}{m}\right)^2 (\Delta - 1)^2$$

$$\simeq 4.2 \times 10^3 \left(\frac{R}{a}\right)^6 \simeq 3.2 \times 10^{-11} \quad (66)$$

compared with an observed value of about 10^{-8} and, clearly, there is a significant discrepancy between the observed and hydrostatic estimates. The observed geoid spectrum is (16*b*)

$$V_2^2(N_{\text{observed}}) \simeq 10^{-8}$$

compared with the observed power in the lunar topography of nearly 10^{-7} and, while the observed geoidal ellipticities are intrinsically large, they are smaller than those deduced from the topographic values [*Runcorn,* 1975].

Lambeck and Pullan [1979*b*] have argued that the second-degree potential harmonics of the moon are 'abnormally' large when compared with the rest of the power spectrum, for while the spectrum for $l \geq 3$ suggests a near-surface origin for the density anomalies, no simple distribution of randomly distributed density anomalies exists that can explain the second-degree harmonics and the remainder (Figure 27). They assign this anomalous part of the bulge, the difference between the observed bulge and that predicted by the random density anomaly model, as a measure of the moon's fossil bulge, acquired when the earth-moon distance was about 25 R_E.

While the geopotential bulge appears to be anomalously large, $V_2^2(h)$ is of the same magnitude as the other low-degree terms in the topography. The reason for this difference in significance between $V_2^2(\Delta U)$ and $V_2^2(h)$ is simply due to the geoid anomaly (16*b*) associated with the second-degree potential harmonics being small when compared with $V_2^2(h)$: the anomalous topography associated with the excess bulge (relative to the random density anomaly model) is less than 10^{-8}, compared with a power of about 10^{-7} in the low-degree topographic harmonics. In any discussion of the topography the $l = 2$ harmonics should not be separated from the remainder of the expansion, since all harmonics appear to have a common origin; this does not appear to be so for the potential.

Runcorn believes that the nonhydrostatic shape of the moon must be supported by dynamic processes since the stress differences required in the lunar interior to support the bulge are sufficiently large to result in creep relaxation. This conclusion does not appear to be necessary for a number of reasons.

Cassen et al. [1978] have recently presented the lunar dynamical and geometric ellipticities predicted by a mathematical model of finite amplitude convection in a spherical mantle shell, using two extreme cases of lithospheric boundary conditions. In the first case the lithosphere is allowed to conform exactly to the convective distortions, i.e., shear stresses within it are not maintained, and the dynamic ellipticity is governed by density difference due to both convection and lithospheric deformation. They find models which predict correctly the dynamic ellipticity, but they require a thermal lithosphere less than 150 km thick, which is less than that predicted from seismic and heat flow data. These models do not explain the geometric ellipticity. In the second type of boundary condition used by Cassen et al. the lithosphere is completely rigid with zero surface deformation. Models were found that predict the proper dynamic ellipticity, but the rigid lithosphere required to match the dynamic ellipticity is also capable of sustaining long-wavelength surface shapes and lithospheric density inhomogeneities, both of which would contribute to the observed value. If this model is accepted, then the anomalous part of the potential bulge could not be uniquely attributed to convection.

An alternative to convective support of the nonhydrostatic ellipticity of the moon is support by the finite strength of the interior. As mentioned above, Runcorn objects to this hypothesis on the grounds that lunar materials will creep when subjected for long time periods to stresses associated with excess ellipticity. By definition, the elastic lithosphere is not subject to creep relaxation of stresses over geologic time and will support such stresses as long as the finite strength of this layer is not exceeded. As noted above, the base of the elastic lithosphere is about at the depth of the 0.5–0.6T_m isotherm, and an estimate of this isotherm depth also yields a value for lithospheric thickness t upon which the stresses imparted by a superincumbent load depend critically. The crust and upper mantle temperature gradient has presumably been decreasing over geologic time, but the rate of change is possibly quite small over the last 2 AE. In any event, we have no understanding of the past relaxation of the excess ellipticity and can only invoke a finite strength support for the presently observed value of the ellipticity with the present selenotherm.

Observations that lead to direct estimates of the crust and upper mantle selenotherm are meager. Inversion of the lunar conductivity profile is frought with uncertainty in terms of the nonuniqueness in the estimation of the conductivity profile and also because of the extreme compositional dependence of the conductivity-temperature relationship [*Huebner et al.*, 1979; *Phillips and Ivins*, 1979]. *Keihm and Langseth* [1977] conducted a parametric study on the lunar thermal regime to 300-km depth, the basic constraints being derived from the global surface heat flow value, crustal thorium abundance, and the high seismic Q in the outer portion of the moon, the last being indicative of subsolidus temperatures. The basic assumption in their work, that the moon is near thermal steady state, has been questioned recently by *Schubert et al.* [1979] but supported by *Turcotte* [1979]. The key parameters in the analysis are the depth of differentiation of the primordial moon and the fraction of radioactive heat sources redistributed into the differentiated crust. The calculations were carried out over the range of uncertainties in heat flow, thorium concentrations in the crust, and thermal conductivity. Successful solutions yield 300-km depth temperatures from 1100 to 1600 K.

A wide range of possible solidus curves, adapted to lunar pressures, are bounded by the models of *Kennedy and Higgens* [1972] and *Kirby and Raleigh* [1973]. The solidus of *Ringwood and Essene* [1970] and of *Hodges and Kushiro* [1974] lie in this range. The intersections of the solidus range with the solution range of Keihm and Langseth yield an elastic lithosphere thickness of from 100 to 300 km. The thinner models are not preferred since they correspond to temperatures approaching solidus at 300-km depth, which is precluded by the high seismic Q of the upper mantle.

With an estimate of lithospheric thickness we return to the question of the stress level imparted by the excess ellipticity. A 500-km elastic lithosphere model of *Lambeck and Pullan* [1979*a*] creates a peak second-harmonic stress difference of 14 bars. The estimate obtained in the present paper suggests a lithosphere of about half this thickness. This thinner lithosphere will not raise the stress difference over 100 bars and by our arguments in section 6, it is perfectly reasonable and to be expected based on the calculations of *Cassen et al.* [1978] that the excess ellipticity of the moon is supported by the finite strength of an elastic lithosphere.

Mascons. The dominating positive gravity anomalies found by *Muller and Sjogren* [1968] over the ringed maria are indicative of significant mass concentrations below the lunar surface. These mascons are associated with the Serenitatis,

Imbrium, Orientale, Crisium, Smithii, Humorum, Nectaris, and Grimaldi impact basins (Figure 31). Their presence gives rise to several problems which have been summarized by *Kaula* [1969, 1971] and some of which the moon also has in common with earth or Mars. These problems include how mass was transferred to produce the excess and how this mass excess has been supported. In an attempt to resolve these problems it is useful to review the many current theories of mascon origin in terms of three dominant contributions to the gravity anomaly: the topography of the prefill basin, the basin fill, and the contribution from the mantle beneath the basin.

The mascon basins are topographic lows, with the surface of the mare fill generally being a few kilometers below the surrounding terrain. The thickness of the mare materials is probably only of the order of several hundred meters in the irregular maria [*Hörz*, 1978] but up to several kilometers in the large mascon basins [*Phillips and Maxwell*, 1978], so that the total depth of the latter is at most of the order of 5 km. This is shallow compared with the original crater of excavation which, for the large basin-forming impacts, must have been many tens of kilometers deep at the time of impact. This difference is presumably a consequence of the instantaneous rebound of the basin floor after the impact, as well as a fall-in of ejecta. If this rebound, or subsequent volcanic fill, is such as to leave

Fig. 31. Line-of-sight gravity contours of lunar nearside. Contour units are 10 mgals. From *Muller and Sjogren* [1968].

the lunar crust in an isostatic state, the free-air anomaly will be negative. The mass required for isostatic compensation can be written as

$$M^I = M_m{}^I + M_f{}^I \quad (67)$$

where $M_m{}^I$ is the mass of a displaced mantle contribution and $M_f{}^I$ is the mass of the predominantly volcanic fill in the basin. Both $M_m{}^I$ and $M_f{}^I$ refer to the situation soon after impact (Figure 32). If the isostatic state is achieved purely by basin filling [e.g., *Wood,* 1970], then $M_m{}^I$ is zero, while if isostasy is achieved by the topography of the crust-mantle boundary [e.g., *Wise and Yates,* 1970], then $M_f{}^I$ is zero. In the former model the basin-filling process would consist of early basaltic flows and a 'topping off' would have occurred as the last stage of filling which would be the cause of the excess mass that results in the positive gravity anomaly. The additional mass M^A, giving rise to the positive free-air anomalies, can be expressed generally as

$$M^A = M_m{}^A + M_f{}^A \quad (68)$$

where $M_m{}^A$ is any mass excess originating from the mantle and $M_f{}^A$ is any mass excess of the additional fill of the basin beyond isostasy. If the excess mass is due to this additional mare fill $M_f{}^A \simeq M^A$, $M_m{}^A \simeq 0$, or if the excess mass originates in the mantle $M_m{}^A \simeq M^A$, $M_f{}^A \simeq 0$. From a comparative study of gravity anomalies over the various ringed basins, some distinction among these various contributions to the mass anomaly becomes possible.

Mare Orientale, for example, is a large ringed basin with a small amount of mare fill compared to other mascon basins. *Head* [1977] suggests that for Orientale, the mantle topography resides in a superisostatic state (i.e., the mantle topography attains a height above that required for isostasy) created at the time of impact formation and which subsequently must be supported by the finite strength of the interior. In this model it is the mantle topography and not the meager amount of mare fill that contributes to the mass excess associated with the central Orientale basin ($M_m{}^A \simeq M^A$, $M_f{}^A \simeq 0$).

Gravitational evidence of a mantle component, independent of an isostatic or superisostatic contribution, has been given by *Sjogren and Smith* [1976] and *Dvorak and Phillips* [1979]. The former authors analyzed spacecraft gravity data at two altitudes over the Orientale basin and showed that the best fitting models all contained a substantial subsurface element, interpreted as mantle relief, although densification by intrusion into the crust could not be ruled out. The latter authors analyzed the Grimaldi mascon anomaly and found that here also a substantial subsurface component is required to explain the gravity data unless an inordinate amount (15 km or more) of fill resides in the central Grimaldi basin.

Phillips et al. [1972] noted that for the filled basins Serenitatis and Crisium the mass excess could be adequately described by a surface disc-like body, and they concluded that a dominant part of the positive gravity anomalies is due to the mare fill with little compensation taking place subsequently. That is, in (68) $M_f{}^A \simeq M^A$, $M_m{}^A \simeq 0$. This has been sub-stantiated by *Roth et al.* [1977], who showed that the slopes on the outer shelf of the Crisium basin could be explained in terms of a warped equipotential surface due to the presence of the mascon but only if the anomalous mass is near or at the surface. Further, concentric graben on the boundaries of the mascon basins can only be explained by loading at the surface, presumably the mare fill [*Melosh,* 1978; *Solomon and Head,* 1979].

Thus a dominant thread through most theories of mascon formation is the role of mantle topography and basin fill. Variations to this have been proposed by *Hulme* [1972] and *Kunze* [1974]. Hulme's model assumes that the crust under the mascons is in isostatic equilibrium, but this requires some 50 km of mare fill and a crustal root of some 40 km. While this model will satisfy the line-of-sight lunar Orbiter data, it is inconsistent with the lower Apollo subsatellite data. Kunze's model also presupposes an isostatic state in which a 30-km thick layer of basaltic fill is in equilibrium with a density deficiency involving some 300 km of mantle material, the deficiency resulting from a density change that accompanied the partial melting that gave rise to the mare basalts. This model has not been explicitly tested against the low-altitude Apollo data, but the discrepancies can be anticipated to exceed those for Hulme's model.

Information on the isostatic response of the lunar interior to impact events can be gained by analyzing the Bouguer gravity anomalies associated with large craters. Some of these impact features are the same size as the Grimaldi mascon, and such analyses shed light on mascon formation. A slow isostatic upwelling of the lunar mantle, in response to the topographic relief produced by impact (as in Figure 7), has been suggested as a mechanism to account for the changes in crater depths and rim heights with increasing age [e.g., *Pike,* 1967; *Baldwin,* 1970; *Kunze,* 1974]. The isostatic upwelling hypothesis has been critically tested by *Dvorak and Phillips* [1978] and *Dvorak* [1979], who analyzed the Bouguer gravity anomalies of eight Imbrian-age and older lunar craters ranging in diameter from 65 to 240 km. All of these craters have Bouguer anomalies that are essentially zero, yet they could be expected to exhibit a systematic signature for partial to complete compensation, assuming that the crust-mantle boundary is in the neighborhood of 60-km depth. In particular, the anomalies should increase systematically in magnitude from the smallest to the largest crater. Thus with complete compensation the 65-km diameter Macrobius crater would be expected to have a Bouguer anomaly of −10 mgals, the 150-km Neper crater +60 mgals, and the 200-km diameter Humboldt crater an anomaly of 140 mgals. The predicted negative anomaly for the smallest crater is due to the dominant attraction of a low-density breccia lens introduced in the model beneath the crater floor. For the larger craters the mass excess associated with the uplift of the crust-mantle boundary would dominate the contributions to the Bouguer anomaly. The observed near-zero anomalies argue against any isostatic response or for only a limited partial response.

The central crater of the mason basin Grimaldi has a diameter similar to Humboldt and Pasteur but is associated with a strongly positive Bouguer anomaly. While there is certainly more fill in Grimaldi than the other craters analyzed by Dvorak and Phillips, a substantial subsurface component is required to reproduce the gravity anomaly. The removal of the mare fill certainly does not reduce the Bouguer anomaly to the near-zero value of Humboldt. The principal morphologic difference between Grimaldi and the other craters is the presence of a concentric outer scarp or ring structure at Grimaldi, absent from the other craters. *Dvorak and Phillips* [1979] propose that the unique attributes of Grimaldi, the ring structure and excess subsurface mass, reflect an uplifting of the mantle

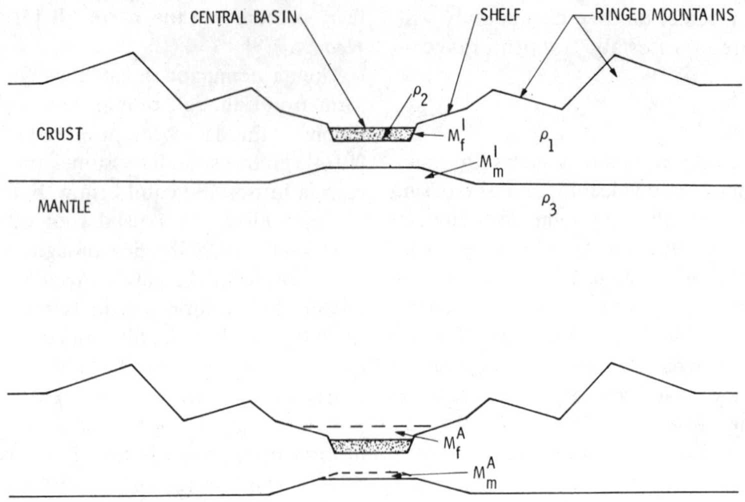

Fig. 32. Schematic illustrations of mascon basin showing contributions to (top) isostasy (M_f^I, M_m^I) and (bottom) superisostasy (M_f^A, M_m^A). Isostasy is achieved by a combination of mare fill in the central basin and relief on the crust-mantle interface. Superisostasy follows from additional contributions to either of those components.

that resulted from a collapse of the region surrounding the crater at the time of impact. This different behavior from Humboldt may reflect a different crustal thickness in the two areas, Grimaldi being on the fringe of Oceanus Procellareum and Humboldt in the southern highlands, where the crust is thicker. Alternatively, these two features were formed by impacting objects of different sizes.

From a consideration of all the analyzed mascon and nonmascon craters a mare fill is certainly an important component to produce the mass excess, but a mantle uplift is essential in bringing the crater into approximate mass balance prior to the filling stage. This uplift must have taken place as a consequence of the mechanics of impact and collapse and was not a long-term creep response. This view, i.e., that both the ring structure and the mantle uplift formed at the time of, or immediately after, impact, is strongly supported by *Head* [1977], who pointed out that the melt facies formed at the time of impact and lying within the Orientale crater show little of the deformation that would be expected if the basin floor had been subjected to a long-term rebound process.

The subsequent mare fill in the mascon basins loaded the lunar crust and led to the mild form of tectonic failure and stress readjustment observed at the surface [*Melosh*, 1978; *Solomon and Head*, 1979]. Concentric graben structures outside of, and the compressional wrinkle ridges within, the filled basins are predicted from simple models of crustal response to the mascon loads (Figure 33). Further, certain lunar seismic events have now been identified as moonquakes with upper mantle epicenters that show a modest correlation with the mascon basins [*Nakamura et al.*, 1979]. In terms of (67) and (68) we conclude that the filled ringed basins give rise to mascons with $M_m^I \simeq M^I$, $M_f^I \simeq 0$, although the mantle relief is a direct and 'instantaneous' response to impact. The excess mass arises from $M_m^A \simeq 0$, $M_f^A \simeq M^A$. Gravity anomalies over the unfilled farside basins are predicted to be essentially a consequence of $M_m^I \simeq M^I$, $M_f^I \simeq M_m^A \simeq M_f^A \simeq 0$ and the anomalies over the irregular maria are described by $M_m^I \simeq M_f^I \simeq M_m^A \simeq M_f^A \simeq 0$.

The second problem raised by Kaula involved the support of the mascons. *Kuckes* [1977] finds a maximum deviatoric stress of about 1 kbar for the Serenitatis mascon. This value follows, however, from a questionable [*Phillips and Ivins*, 1979] estimate of the elastic lithospheric thickness of 54 km. A more reasonable estimate is 200- to 300-km thickness, as discussed above, for which *Arkani-Hamed* [1973*a*] finds that the stress differences due to mascon loading reach about 100 bars at a depth of about 250 km. The actual values may be less than this, as this value is based on rather crude gravity models. This low stress value suggests that finite strength support of mascons is reasonable and to be expected. The mild tectonic features, particularly the tensile graben structures, are consistent with this stress level.

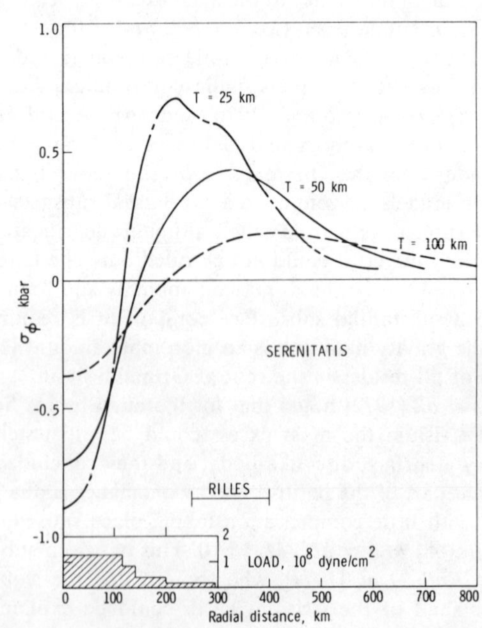

Fig. 33. Stress component σ_ϕ radial to the Serenitatis basin in response to the mascon load (insert) for several possible lithospheric thicknesses (T). Positive stress is extension, and maxima correlate with location of tensional rille structures exterior to the load. From *Solomon and Head* [1979].

9. MARS

Introduction. The early history of Mars appears to have followed that of the moon in terms of an early period of crustal differentiation and intense bombardment followed by (basaltic?) plains volcanism. Several events took place on Mars which clearly distinguishes its evolution from that of the moon. First, the primordial crust in the northern hemisphere was severely disrupted and in part removed [*Mutch et al.* 1976; *Guest et al.*, 1977]. Additionally, centers of crustal doming and shield volcanism developed, first in the Elysium region and then in the Tharsis area. The latter volcanism provides the most significant tectonic overprint on the planet. The north-south dichotomy of the crust and the Tharsis Province define the basic geological-geophysical subdivisions of Mars.

The important global north-south dichotomy of Mars is seen, inter alia, as a marked elevation difference between the northern and southern hemispheres, separated by a great circle inclined at approximately 35° to the equator [*Mutch et al.*, 1976] (see also Figure 34). In this respect, Mars is similar to earth and the moon, which also display a roughly hemispheric dichotomy in both physiography and topography. The Mars physiographic division is between an elevated, heavily cratered, generally southern region and a low, northern region covered by smooth plains. The Martian center-of-figure (CF) is offset from the center-of-mass (CM) by roughly 2.5 km in a direction 60°S, 90°E [*Bills and Ferrari*, 1978]. This vector is approximately parallel to the normal of the great circle dividing the two physiographic divisions. The significant CF-CM offsets of the earth and moon also relate intimately to major physiographic regions, the ocean and land hemispheres in the case of the earth. For all three planets the dichotomy is basically one of old and elevated terrain versus young and low terrain [*Wilhelms*, 1974].

Superficial processes alone are not sufficient to explain the Martian offset, and simple isostatic models do not suffice to explain both the topographic difference and the CM-CF offset [*Malin et al.*, 1978]. Some internal mass redistribution is required but, as for the moon, the stresses associated with the offset will be very small compared with those due to the higher degree harmonics in the topography, so that once developed, the offset can persist indefinitely with time.

Of only a slightly smaller scale than this physiographic dichotomy of the planet, the Tharsis Province represents the second major subdivision of Mars, a relatively young part of the planet that is or was volcanically and tectonically active compared with older parts that are subdued in tectonic expression and which probably have been subjected to mild plains-type volcanism (Figure 34).

The Tharsis Province, of roughly 2000-km radius centered on the equator at about 110°W longitude, lies as a plateaulike feature 8–10 km above the average height of the planet's surface. Morphologically, the topography is dominated by a domal structure, suggestive of uplift, although a significant fraction of the elevation could be due to a series of volcanic-plains units, which in turn are superimposed on the ancient planetary crust [e.g., *Mutch et al.*, 1976]. A pronounced fracture system, predominantly radial and tensional, emanates outward from the highest region of the dome. This fracture system predates most of the volcanic units superimposed on the region, including shield volcanoes, and the fracturing process appears to have terminated early in the history of the volcanic deposition [*Carr*, 1974]. The shield volcanoes reach heights of 25 km above the planet's mean surface. The vol-

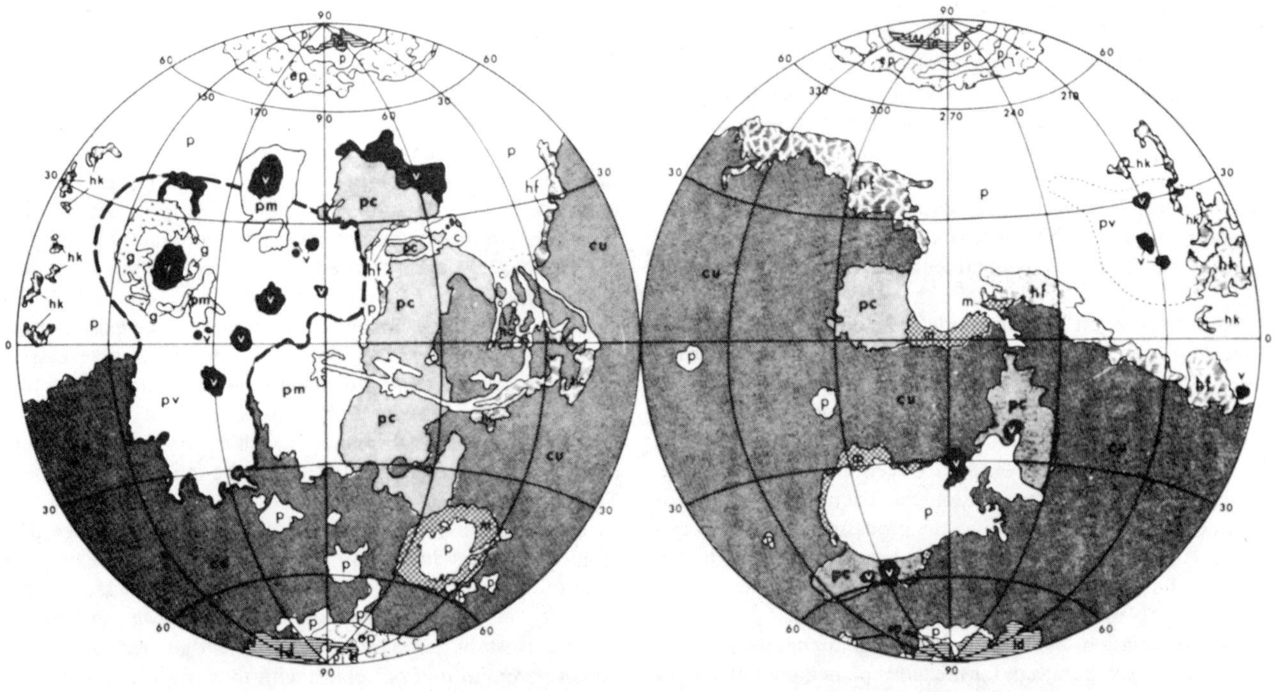

Fig. 34. Physiographic provinces of Mars. Polar units include *pi* (permanent ice), *ld* (layered deposits), and *ep* (etched plains). Volcanic units include *v* (volcanic constructs), *pv* (volcanic plains), *pm* (moderately cratered plains), and *pc* (cratered plains). Modified units include *hc* (hummocky terrain, chaotic), *hf* (hummocky terrain, fretted), *hk* (hummocky terrain, knobby), *c* (channel deposits), *p* (plains, undivided), and *g* (grooved terrain). Ancient units include *cu* (cratered terrain, undivided) and *m* (mountainous terrain). Heavy dashed line marks approximate boundary of Tharsis Province. From *Mutch et al.* [1976].

canic activity appears to have started over 3 billion years ago and has continued up to the present or the recent past [*Wise et al.*, 1979]. The heights of these volcanoes suggest that, if the lavas moved up under hydrostatic pressure, the depth of the magma source region for these shield volcanoes is of the order of 250 km.

We will first discuss the statistical aspects of the planet's gravity, topography, and stress state before returning to the specific problems posed by the Tharsis plateau.

The Martian gravity field and topography. Two recent spherical harmonic expansions of the Martian gravity field are by *Gapcynski et al.* [1977] complete to the degree and order 6 and by *Christensen and Balmino* [1979] complete to degree and order 12. Both solutions are based on Mariner 9 and Viking Orbiter 1 data. An earlier solution by *Sjogren et al.* [1975], complete to degree and order 9, was based on Mariner data only. Probably the best solution for the Mars gravity field is by *Sjogren* [1979], based on a direct mapping of line-of-sight accelerations of the Viking Orbiter 2, at times when the spacecraft was near periapsis (Figure 35). But without further Fourier analysis, this solution does not lend itself to the present statistical discussion. Figure 36 illustrates the power spectra for the solutions by *Sjogren et al.* [1975] and *Christensen and Balmino* [1979]. The agreement between the two is excellent up to degree 6, but beyond this, significant differences occur. In particular, the difference spectra for these two solutions exceeds the signal for $l > 6$.

Information on the Martian topography comes from several sources, including spacecraft occultations, spectral instruments, stereo photogrammetry, and terrestrial radar observations (see *Wu* [1978] for a review). Only the Mariner 9 data have been fully exploited so far. *Christensen* [1975] has compiled the data and expanded them into spherical harmonics. The occultation data from this spacecraft do not cover the north polar regions, and a comparison with a few published Viking results [*Fjelbo et al.*, 1977] indicates that the topography here may be in error by as much as 1 km. *Bills and Ferrari* [1978] have expanded the same data as used by Christensen to degree and order 16 using an improved estimation procedure. The corresponding power spectrum is illustrated in Figure 37 and compared with the equivalent rock spectrum of the earth. Also illustrated is the error spectrum $\epsilon_l^2(h)$ of the Martian topography based on the error estimates given by Bills and Ferrari; the error variance is obtained by dividing $\epsilon_l^2(h)$ by $(2l + 1)$.

Since the earliest Martian models, the global correlation between topography and gravity was recognized [*Lorell et al.*, 1972]. This correlation, expressed by the cospectrum (17), is illustrated in Figure 38 for the gravity solution by Christensen and Balmino. The correlation coefficient $V_l^2(\Delta U, h)$ is positive for $2 < l < 7$ but for degrees 8–12 it is equally positive or negative. The cumulative correlation coefficient for $2 < l < 6$ is significant at the 95% level, i.e., the probability that the correlation is a matter of chance only is less than 5%. For higher degrees the cumulative correlation coefficient is insignificant, but neither is the gravity spectrum very reliable.

The correlation between gravity and topography differs from that on earth in that, for the latter planet, the low-degree terms do not exhibit any significant correlation (Figure 16) and the terrestrial low-degree gravity field is dominated by processes that are not directly related to the topography. The correlation found for Mars implies one of several things, listed below.

1. Topography is fully compensated but, as the external gravity is the sum of the near-surface topography and the deeper compensating mass, correlation occurs.
2. Topography is only partially compensated.
3. The correlation is indicative of deep mantle processes that are responsible for both topography and gravity. From isostatic and stress state considerations, similar to those discussed for the earth and moon, some distinction between these alternatives is possible [*Lambeck*, 1979*a*].

The potential spectrum decays more rapidly than the $(2l + 1)/l^4$ rule (13) for the earth, and this suggests that if the random density anomaly model of section 7 is adopted, there is a proportionally larger contribution from deep mantle anomalies than in the case of the earth. In view of the low-degree correlation with the topography this is unlikely, and a more plausible interpretation is that, unlike the earth and moon, the Martian gravity and topography are dominated by a long-wavelength feature, namely, the Tharsis plateau [*Phillips and Saunders*, 1975]. Whereas on the earth and moon the observed low-degree harmonics (with the possible exception of a remnant fossil lunar bulge) are mainly a consequence of short-wavelength density anomalies, the shaping of Mars is dominated by this one feature. This not only makes Mars unique (there is no comparable topographic feature on Mercury, and the initial Pioneer Venus results do not indicate any Venus anomalies that are comparable to the Tharsis complex), but it also reduces somewhat the utility of any statistical discussion. The gravity and topography of the Tharsis plateau are described mainly by the second- and third-degree harmonics, and as a rough approximation, the corresponding spectral estimates can be considered to describe the Tharsis complex, while the higher degree spectral estimates statistically describe the remainder of the planet. In particular, for $l > 3$ the spectrum may follow the $(2l + 1)/l^4$ rule.

Scaling the topography from earth to Mars for equal stress gives, with (32),

$$V_l^2(h_{\text{Mars}}) = 25 V_l^2(h_E)$$

which equals the observed value for $l = 2$ but exceeds it for higher degree spectral estimates. Thus if the topography of Mars is isostatically compensated in a manner that is similar to the earth's topography, the stresses set up in the Martian crust and lithosphere do not exceed those required to support the isostatically compensated low-degree harmonics in the earth's topography [*Lambeck*, 1979*a*].

While an isostatic compensation model (if one exists) of the Martian topography does not appear to result in excessive stresses in the crust and lithosphere, the same may not be true if the observed gravity is the consequence of compensated density anomalies in this layer as well as compensated topography. Mars differs significantly from the moon in this respect. To obtain an order of magnitude estimate of the stress differences, we can use (33) with $\beta = 1$. For the low-degree harmonics the stress differences become excessive, of the order of 2 kbar for $D_a \simeq 100$ km (see Table 2). Rather than use the specific Airy or Pratt compensation mechanisms, we can here define the isostatic potential, $V_l^2(\Delta U^I)$, as that part of the observed potential that correlates with topography, as was done for the earth (section 7), by computing the linear regression between topography and potential. The isostatic anomaly is the difference between the observed and isostatic potential. This was done by *Lambeck* [1979*a*], who computed the linear regression separately for each degree, implying that different

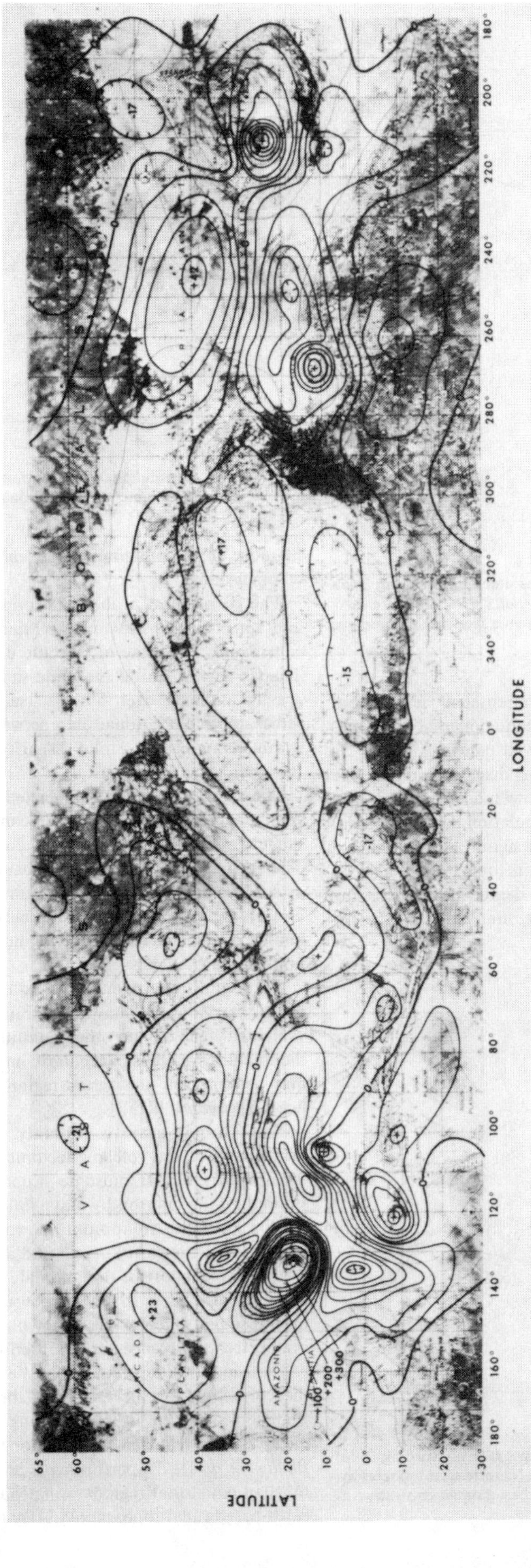

Fig. 35. Contoured line-of-sight gravity accelerations for Mars based on Viking 2 data. Contour interval is 10 mgal except at Olympus Mons (18°N, 133°W). Contours are referenced to degree and order 4 spherical harmonic field. From *Sjogren* [1979].

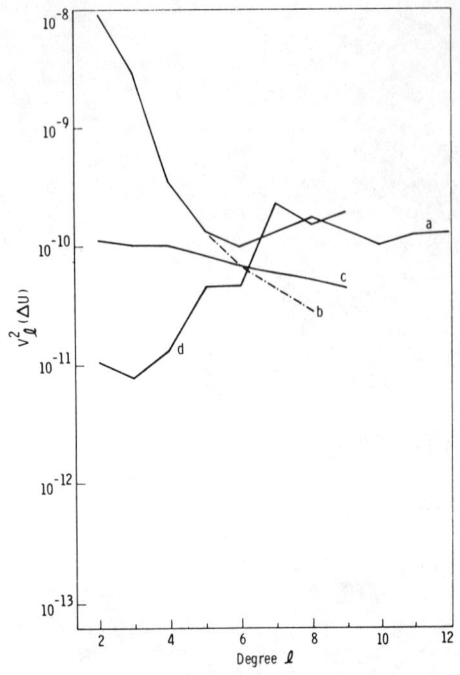

Fig. 36. Potential spectrum for Mars as determined by (*a*) *Christensen and Balmino* [1980] and (*b*) *Sjogren et al.* [1975]. (*c*) Noise spectrum of the latter solution and (*d*) the difference spectrum of these two solutions.

wavelength anomalies may be compensated at different depths. The stresses based on the isostatic potential are somewhat reduced over those of the observed potential (Table 2).

Tharsis isostasy and tectonism. As discussed above, the Tharsis region (and the adjoining lowlands of Chryse and Amazonis) shows a high degree of correlation between gravity and topography and contributes most significantly to the $l = 2, 3$ harmonics in the potential and in topography. The remainder of the better-determined low-degree harmonics show a more muted correlation, suggesting that much of the remainder of the topography is essentially in a state of isostatic compensation.

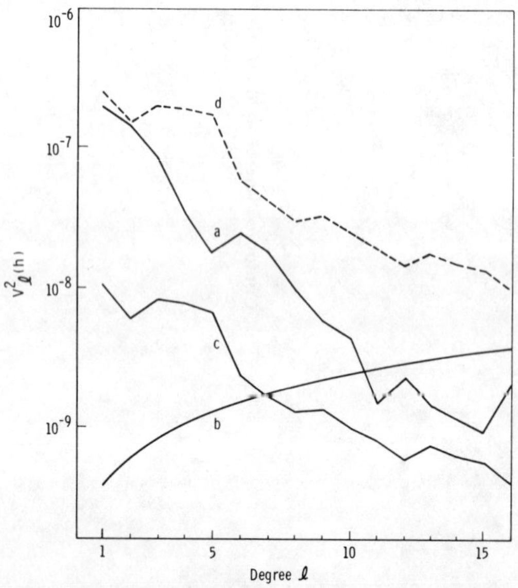

Fig. 37. Power spectrum of Martian topography according to (*a*) *Bills and Ferrari* [1978], (*b*) error spectrum, (*c*) terrestrial topography, and (*d*) the predicted Martian topography based on an equivalent finite strength assumption.

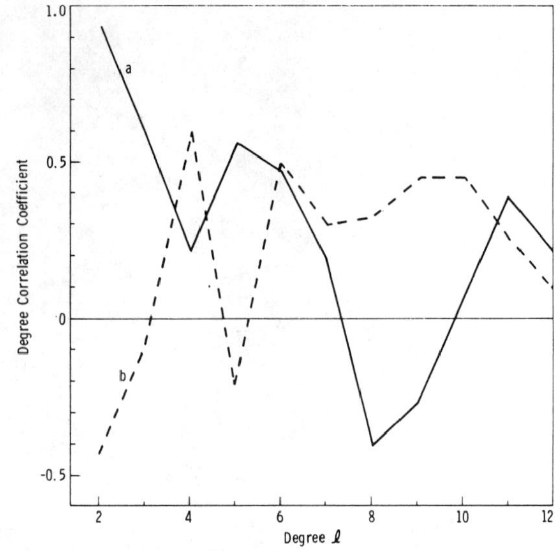

Fig. 38. (*a*) Correlation between Martian gravity and topography. (*b*) Same correlation but for the earth.

The geophysical problem in trying to understand the crust and upper mantle structure for Mars is to find models that result mainly in a state of isostatic compensation for the non-Tharsis terrain and at the same time provide a large free-air gravity anomaly over Tharsis itself. Successful models that satisfy the observational data as well as the rheological constraints may lead to a total disparity in structure between the two geological provinces.

The specific questions associated with Tharsis are in part similar to those for the lunar mascons and, for that matter, for the terrestrial anomalies: (1) What tectonic process caused the Tharsis uplift and associated volcanism? (2) How is the mass excess implied by the free-air gravity anomaly supported?, and (3) What is the relationship between the fracture pattern and origin of Tharsis in general and the gravity anomaly and uplift in particular?

It is useful to resolve the second question first since there is some observational data to constrain it, and the answers may well shed light on the other questions. An understanding of the fracture system is particularly important insofar as Tharsis may be a nascent or arrested region of lithospheric spreading [e.g., *Hartmann*, 1973].

The Tharsis gravity anomaly, if related to the commencement of the volcanic-tectonic activity, must be several billion years old. It must be supported by either the finite strength of the interior, by a dynamic mechanism such as convection, or both. An appropriate way to proceed is similar to that used for the earth, i.e., to find a plausible mechanism for the passive support of the anomaly and, failing that, turn to dynamic processes. In the first instance an isostatic mechanism is sought since, as discussed in section 6, it results in minimum stress differences in the interior.

The simplest interpretation of the gravity data results from the correction for the effects of topography (the 'complete Bouguer correction'). The resulting Bouguer anomaly is indicative of lateral density variations in the planet's interior. *Phillips et al.* [1973] performed a preliminary analysis of the Martian equatorial regions using Mariner 9 gravity data and earth-based radar topography. They found a strongly negative

TABLE 2. Maximum Stress Difference as a Function of Spherical Harmonic l for Compensation of Topography on Mars and Earth

l	Maximum Stress Difference, bars					
	(a)	(b)	(c)	(d)	(e)	(f)
2	144	132	1860	948	1420	720
3	112	154	960	500	480	280
4	70	146	320	170	16	8
5	50	142	180	96	130	70
6	60	94	120	66	74	40
7	52	84				
8	38	68				
9	30	70				
10–16	120	320				

Column (a) gives maximum stress differences occurring in a surficial layer of Mars due to the topography and its compensation. Column (b) gives the comparable stresses in the earth due to its topography and compensation. Columns (c) and (d) are the Martian stresses due to the observed gravitational potential $V_l^2(\Delta U)$, assuming that the field originates from density anomalies in a layer with Airy compensation depth D_a = 100 km (Column (c)) and D_a = 200 km (Column (d)). Column (e) gives the stresses due to the isostatic potential $V_l^2(\Delta U^I)$ with D_a = 100 km and column (f) is for D_a = 200 km.

Bouguer anomaly associated with the Tharsis plateau, indicative of a density deficiency in the interior (Figure 39). The simplest interpretation of this result is a thickened crust under Tharsis, but isostatic compensation will not explain the entire anomaly unless the depth of compensation becomes large, of the order 1000 km [*Phillips and Saunders*, 1975; *Lambeck*, 1979*a*].

Alternatively, as suggested by *Phillips et al.* [1973], Tharsis is not in isostatic balance. *Phillips and Saunders* [1975], using the spherical harmonic expansion of gravity by *Sjogren et al.* [1975] and topography by *Christensen* [1975], reemphasize this interpretation that only partially compensated Airy models are possible when using a single planet-wide depth of compensation. The mean compensation depth found is no greater than 100–150 km, and their argument for limiting the depth to this value is based on two deductions: (1) The terrain other than Tharsis appears to be largely compensated and (2) the isostatic potential exceeds the observed potential for harmonics $l > 3$ if the depth of compensation is very much greater than this. If all of the observed gravity anomalies originate from this 100- to 150-km layer, then the stress differences required to support them may become excessive [*Lambeck*, 1979*a*].

The depths required to isostatically support the Tharsis plateau (the $l = 2, 3$ harmonics) and satisfy the observed gravity data are of the order of 600 km for $l = 3$ and 1100 km for $l = 2$. If we consider only those parts of the $l = 2$ and $l = 3$ harmonics that correlate with topography, the required depth of compensation is still of the order of 400 km [*Lambeck*, 1979*a*]. This gross disparity with the result for the non-Tharsis portion of the gravity field argues against a simple Pratt or Airy compensation model for Tharsis.

Recently *Phillips et al.* [1980] evaluated the stress distribution in Mars by modeling the mantle and crust's rheological behavior as a self-gravitating Maxwell viscoelastic solid in response to a surface load (section 3). At the time of 'initiation,' Tharsis was assumed to be supported in the lower mantle because of the dominance of the second and third harmonics in the regional topography (see Figure 3). Using a low strain-rate flow law, they obtained an estimate of the upper limit on the middle and lower mantle temperatures required to prevent any significant relaxation of the anomalies (see Figure 6), and they concluded that these temperatures must be below about 0.6 of the mantle solidus temperature. (The argument is similar to that used by *Lambeck and Pullan* [1979*a*] for the moon.) Such low temperatures appear to be precluded by the extensive magmatic activity sustained throughout geologic time as evidenced by the surface volcanism. These temperatures are also inconsistent with the thermal models proposed by *Johnston and Toksöz* [1977], whose temperature profiles lead to a rapid relaxation of mantle stresses. The passive support of Tharsis can only be supplied by the outer several hundred kilometers of the planet, and models must be sought that are confined to this region.

As discussed above, an isostatic model confined to only modest depths is more complex than one-element Pratt or Airy models, and different isostatic configurations must be sought. *Sleep and Phillips* [1979] developed a heuristic model by assuming that the Tharsis gravity anomaly is a consequence of a complete upper mantle Pratt compensation of the plateau, and they then derived what the compensated topography ought to be to satisfy the gravity data. As anticipated from the

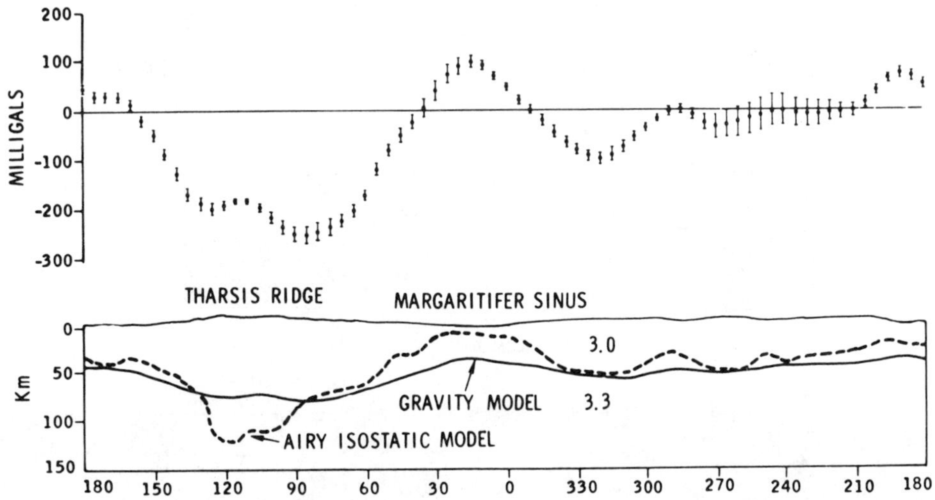

Fig. 39. East-west Bouguer gravity anomaly profile on Mars at 15°S latitude. Crustal density is 3.0 g cm^{-3}, and mantle density is 3.3 g cm^{-3}. Airy isostatic model is also shown. From *Phillips et al.* [1973].

previous discussion, these predicted heights exceed the observed heights unless a great compensation depth is assumed. Sleep and Phillips assumed that the difference between predicted and actual topography can be taken up at the crust-mantle boundary, with an appropriate correction for density contrast. The resulting thinned crustal model is one of a low-density upper mantle which compensates loads at both the surface and at the crust-mantle boundary (Figure 40). If the depth of compensation (base of the lithosphere) is taken to be of the order of 300 km or greater, then reasonable values result for both the crustal thinning and upper mantle densities. This model will accommodate a 50- to 150-km Moho depth, required for the compensation of the non-Tharsis terrain, and explains the special nature of Tharsis by lateral density variations in the upper mantle. Such a model is similar to the compensation mechanism proposed by *Garland* [1971] (also Figure 40) for the Basin and Range Province of the western United States. This region also compares favorably with the Tharsis Province insofar as both are elevated regions with volcanism and extensional tectonics, although no exact analogy in terms of origin is implied.

As discussed in section 6, the Airy model leads to a minimum lithospheric stress level and the Sleep-Phillips model will not result in a lower stress difference. Values of the order of 1 kbar still have to be entertained. In addition, the thinning of the crust is presumably a consequence of higher temperatures, implying a lower finite strength of this part of the crust and mantle. Alternatively, the elastic lithosphere is quite thick, in excess of 400 km to bring the stresses to the 1/2-kbar level. This requires a lithospheric temperature gradient that is quite low and at best matches the lunar gradient. This seems implausible for the Tharsis region, given a record of sustained volcanism.

Phillips et al. [1980] evaluated the stress state of the Martian lithosphere based on a general class of isostatic configurations whose compensation depths are limited to several hundred kilometers. The global surface pattern of the principal deviatoric stresses associated with Tharsis show a quite remarkable degree of orthogonality between tensile stresses and tensional features and between compressive stresses and compressional features (Figures 41 and 42; see also *Pines et al.* [1977]; *Arvidson et al.,* [1980]). This argues for a tectonism associated with the present-day anomalous mass associated with the Tharsis gravity anomaly. This tectonic activity appears to be a result of failure of the crust and lithosphere due to loading and not a consequence of uplift [*Carr,* 1974], as the latter would have produced the conjugate set of tectonics. It should be noted in passing that the set of stress patterns one obtains for long-wavelength anomalies loading a sphere are quite different from point loading of a half-space and that the role of membrane stress on a spherical shell is quite important.

To review, the model calculations of *Phillips et al.* [1980] are based on isostatic configurations; they lead to minimum stress states. If passive support of Tharsis is rejected on the basis of the stresses from these types of models, then presumably all other models are also rejected. The magnitude of the maximum stress differences associated with an isostatic model for the Tharsis uplift is of the order of 1 kbar (Figures 41 and 42). It is unlikely that the thermal events leading to the formation of Tharsis could have created these stress levels, and one must also consider that the cessation or waning of the formation process would isolate Tharsis in an isostatic stress state (see section 6) in the absence of other tectonic stresses. In contrast to these endogenic thermal processes, loading of the surface by volcanic extrusion need not reach a state of isostasy, and the response will be a function of the thermal state of the lith-

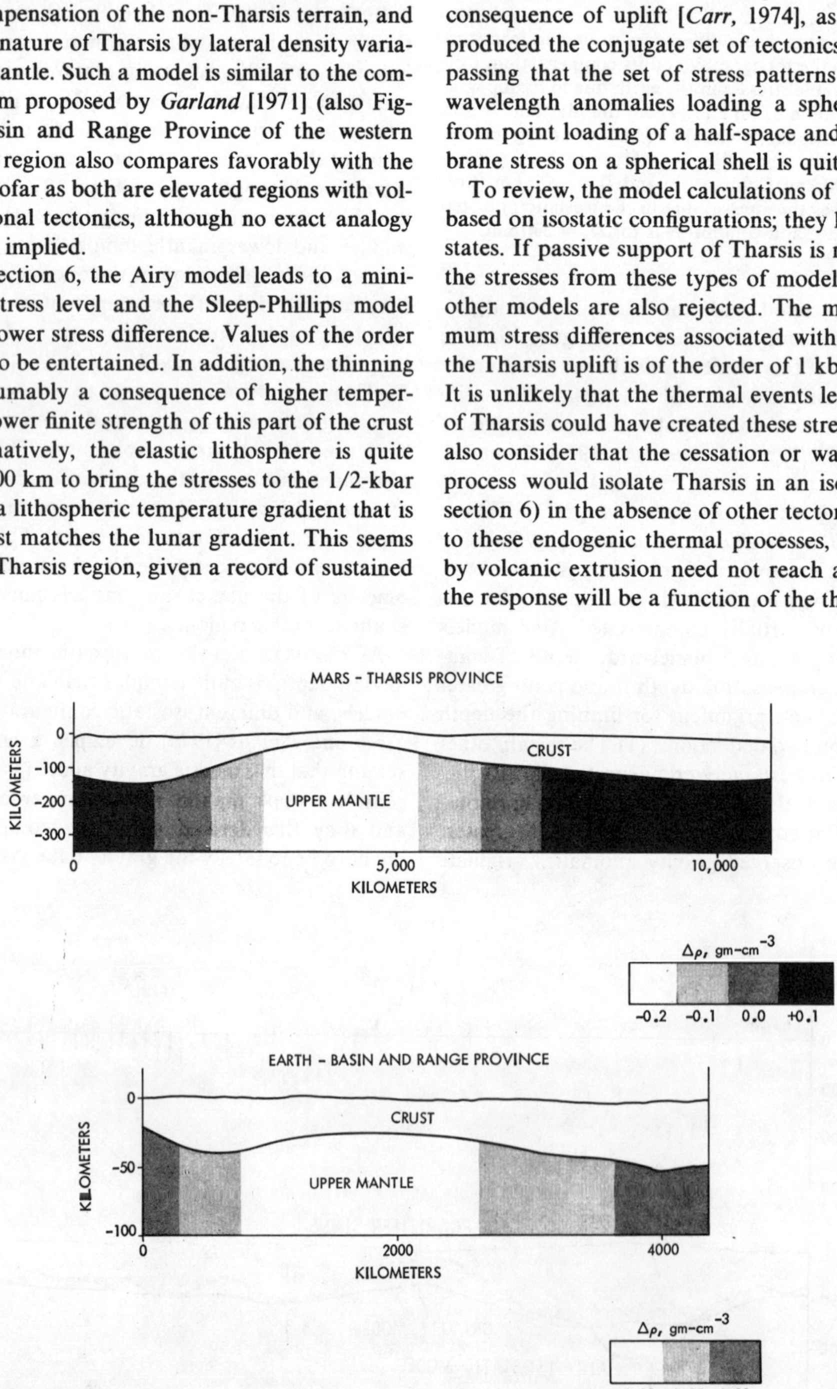

Fig. 40. Cross sections through two-element Pratt models for the Tharsis Province, Mars, and the Basin and Range Province, earth. The Tharsis solution is for a mean crustal thickness of 100 km and a mean lithospheric thickness of 350 km. Basin and Range model is derived from seismic data. From *Sleep and Phillips* [1979].

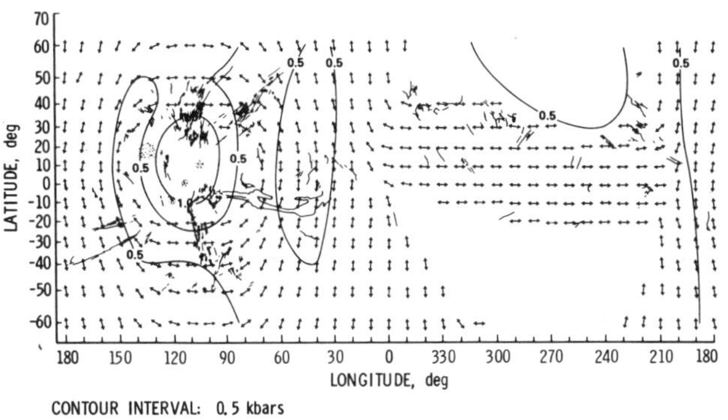

Fig. 41. Magnitude and direction of horizontal tensile (principal) deviatoric stresses at the surface of Mars. Contours are magnitude of maximum stress, in kilobars. Calculation is for a self-gravitating elastic Mars model and is typical of models with partial Airy isostasy and a 100-km depth of compensation. The topography and gravity fields, complete through the fourth spherical harmonic, are used as boundary conditions. A preliminary map of tensional geologic features shows an orthogonal relationship to stresses in the vicinity of the Tharsis uplift (left portion of map). From *Phillips et al.* [1980].

osphere. Indeed, some nonsignificant fraction of the long-wavelength free-air anomaly of Tharsis could be due to the loads of the shield volcanoes and associated volcanic plains units.

The requirement to support stress differences of the order of 1 kbar may be excessive, depending on the details of the depth distribution of these stresses in the lithosphere. As mentioned above, the stress differences can be lowered to the 1/2-kbar range at the expense of invoking an elastic lithosphere that may very well be thicker than the moon's. All in all, this might argue for at least some dynamic support of the Tharsis region [*Lambeck*, 1979*a*; *Phillips et al.*, 1980] unless significant horizontal compressive stresses occur in the lithospheric layer. The role of the Tharsis fractures in relieving the stress state is unclear, but as discussed above, they are a response to the loading, isostatic or otherwise.

It is possible that the uplift of Tharsis represents the early but arrested stage of a thermal mechanism which would have led to lithospheric breakup. The cessation of this activity would then have led to the more recent tectonism about the Tharsis region, which might on the earth be a mechanism for lithospheric rupture. That crustal fracturing has occurred and that a state of tensile stress exists in parts of the planet is clear from the photographic evidence. The most dominant tensile fracture is Valles Marineris, which *Hartmann* [1973] compares, along with the general Tharsis uplift and volcanics, to the East African Rift system tectonics. Possibly a better terrestrial analogy is the free-air gravity anomaly trending across north-central Africa [*Fairhead*, 1979]. This region, like Tharsis, is associated with crustal doming, shield volcanism, and a negative Bouguer anomaly. Fairhead has suggested that this area of Africa represents an earlier, and possibly arrested, stage of tectonic development relative to the East African Rift.

10. VENUS

The first systematic mapping of the gravity field and topography of Venus commenced in December 1978, with the orbiting of the Pioneer Venus spacecraft. Systematic evaluation of the global fields is not presently available, but certain tentative conclusions can be drawn, nevertheless. Both earth-based radar data and data acquired by the Pioneer Venus radar mapper show that Venus contains a varied terrain of cratered surfaces, rifts, volcanic constructs, and large, elevated regions [e.g., *Campbell et al.*, 1976, 1979; *Goldstein et al.*, 1976; *Malin and Saunders*, 1977; *Saunders and Malin*, 1977; *Pettengill et al.*, 1979].

A preliminary analysis of the gravity field over a 45° swath of longitude has been presented by *Phillips et al.* [1979]. They studied the isostatic compensation of a large plateau roughly 1400 km in diameter [*Campbell et al.*, 1976] of approximately 3-km average elevation [*Pettengill et al.*, 1979]. At roughly 1000-km altitude, they find no gravity signature from this feature and conclude that it is in an isostatic state with a mean depth of compensation no greater than about 100 km. The concept of isostatic compensation of the Venus topography is borne out by a 20-mgal anomaly mapped at lower spacecraft altitudes and strongly correlated with mountainous topography, which if uncompensated would produce a signal about 6 times this value.

In a second analysis, *Phillips et al.* [1979] performed Fourier analyses of the gravity data along seven orbital arcs spanning 45° of longitude and compared these results to a simulated sampling of the earth's gravity field using the Goddard Earth Model (GEM) 10 [*Lerch et al.*, 1977]. The conclusion is that the gross power in the gravity field for this limited region of Venus is about the same as for the earth, but systematic variations exist among the individual harmonics (Figure 43). In particular, the second Fourier harmonic (5000- to 6000-km wavelength) for the earth is considerably enhanced over that of Venus in this sampling due to strong anomalies associated with plate boundaries and intraplate regions of significant mantle flow. The Venus spectrum is enhanced at higher harmonics, which may indicate that important mantle flow processes act at shorter wavelengths or shallower depths on Venus or that Venus has a relatively larger lithospheric contribution to the gravity field than the earth. A strong correlation of free-air gravity with topography does exist (Figure 44) and appears to be the direct cause of the enhanced (relative to earth) power in the Venus spectrum for harmonics 3, 4, and 5. It is tempting to immediately interpret this result in terms of a relatively deeper average level of isostatic compensation for Venus, but other mechanisms, such as lithospheric flexure or

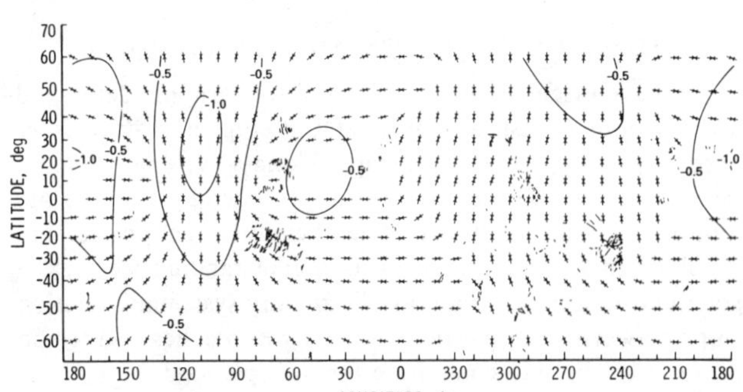

Fig. 42. Magnitude and direction of horizontal compressive (principal) deviatoric stresses at the surface of Mars. See Figure 41 for details. The stress difference is the algebraic difference of the stresses in Figure 41 and this figure; the maximum value exceeds 2 kbar. For the *Sleep and Phillips* [1979] Pratt isostatic model with a 400-km thick lithosphere, the stress difference is approximately halved, but the stress trajectories remain the same. A preliminary map of compressional geologic features shows, for those mapped circumferential to Tharsis, an orthogonal relationship to stress trajectories [see also *Ardvison et al.*, 1980]. From *Phillips et al.* [1980].

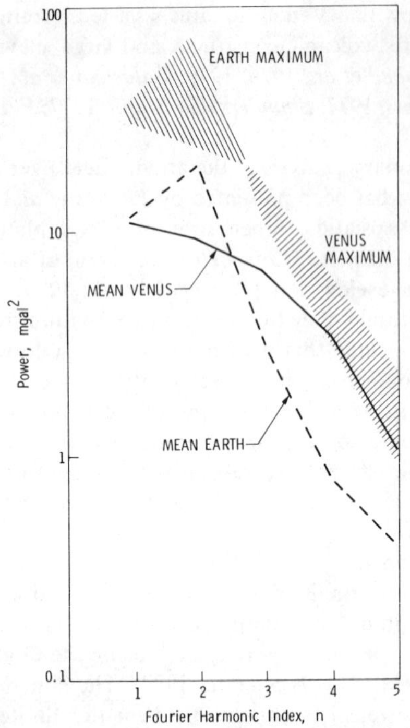

Fig. 43. Gravity spectra for Venus and earth obtained by Fourier decomposing gravity anomalies along 100° arcs of the Pioneer Venus spacecraft orbital tracks. For Venus, seven orbits covering a 45° swath of longitude were used. For the earth, the characteristics of a Pioneer Venus orbit were used to sample the GEM 10 gravity model with 16 orbits of equal node spacing. The first harmonic corresponds approximately to a 11,000-km wavelength. The mean and maximum curves for both planets are shown with the area between the two maxima hatched. The earth maximum exceeds the Venus maximum for harmonics 1 and 2, while the converse is true for harmonics 3–5. From *Phillips et al.* [1979].

dynamic support, are clearly worthy of investigation. Another concept involves the notion of smaller scale continental blocks on Venus relative to earth, shifting the power in the Venus spectrum to higher harmonics. Rigorous analysis of the Pioneer Venus data will help clarify the nature of this correlation of long-wavelength gravity to topography.

11. SUMMARY: COMPARATIVE PLANETOLOGY

We have discussed the gravity and topographic data available for four terrestrial planets and attempted to draw inferences regarding the relationship of long-wavelength anomalies to tectonics. This discourse was preceded by a review of a number of relevant topics of mathematical geophysics and rheology that are germane to the understanding we seek. A general question we might ask is, are any systematic relationships between long-wavelength gravity anomalies and tectonics revealed by intercomparing the four planets?

In Figure 45 we draw from the preceding illustrations the nondimensional normalized power spectra for the gravitational potential fields of the earth, the moon, Mars, and Venus. The Venus spectra are only approximate and representative of a limited area of the planet. They are obtained by transferring the ratio of the Fourier harmonics of Figure 43 to a spherical harmonic basis. For the earth there is minimal contribution to the spectra from the lithosphere. The increasing power from the earth to Mars to the moon is most simply explained by an increase in density anomalies from progressively thicker, colder, less volatile lithospheres and/or the lower stress levels imparted by increasingly smaller planetary gravity, g_0. The Venus spectrum is close to the earth in gross power but shows a power shift to higher harmonics. All four planetary spectra are roughly described by the 'Kaula rule' (13), derivable from random distributions of density anomalies. Departures from this rule can be explained by specific features in the planetary gravity fields, i.e., nonstochastic components superimposed on a random field. The earth shows an excess over the Kaula rule in the vicinity of the sixth harmonic, which can be specifically related to such features as the North Atlantic Gravity High, the Indian Ocean Gravity Low, and certain island arc anomalies, such as Indonesia-Java trench. Venus shows an excess of power for the higher harmonics shown, which is related to specific long-wavelength topographic features, the significance of which we do not understand at present. Mars departs significantly in the lowest harmonics, and this is directly attributable to the Tharsis complex. The lunar spectrum shows a departure in the vicinity of the fifth harmonic, relatable to the basic dichotomy between mare and highland topography, and a departure in the vicinity of the tenth harmonic, relatable to the mascon anomalies.

In general, we understand the source of the nonrandom components in the planetary gravity fields in the sense that we have a limited number of geophysical models for each feature type which satisfy the gravity field as well as other observational constraints. We wish to strongly reemphasize the nonuniqueness of interpretation using gravity data alone, and the need for further constraints.

The stochastic component of the long-wavelength gravity field is composed of topography that is for the most part compensated, plus density variations in the lithospheres and sublithospheres. For the earth we feel that much of this component must be associated with flow in the mantle, while for the moon these anomalies arise in the lithosphere and are supported by the finite strength of that layer. The source of the stochastic components of the gravity fields of Venus and Mars remains problematical.

Figure 46 summarizes the normalized topographic spectra for the earth, Mars, and the moon. The topographic information for Venus acquired by the Pioneer Venus Orbiter has not been harmonically analyzed at this point. The effects of the continents and ocean basins are clearly seen in the lower part of the earth spectrum and are absent in the corresponding potential spectrum because of isostatic compensation. We would like to argue, in analogy with the gravity spectra, that the increasing power in the spectra from the earth to Mars to the moon simply follows from a finite strength control of the topography, both in terms of decreasing volatiles and temperatures and decreasing load stresses due to lower g_0. Mars violates this simple scenario by competing with the moon in the lower harmonics. As with the gravity data, this is again related to the Tharsis Province and, ignoring Mercury and that part of Venus that has not been sampled, we can safely conclude that Tharsis is the most anomalous of the long-wavelength anomalies of the terrestrial planets. It is little comfort that a choice between definitive models for the creation and support of the Tharsis plateau may rest on a better understanding of the finite strength of rock materials.

It is clear that a major thrust in geophysical research will be the continuing attempt to understand the relationship between long-wavelength gravity anomalies and convection in the earth's mantle. To that end, the recent acquisition of long-

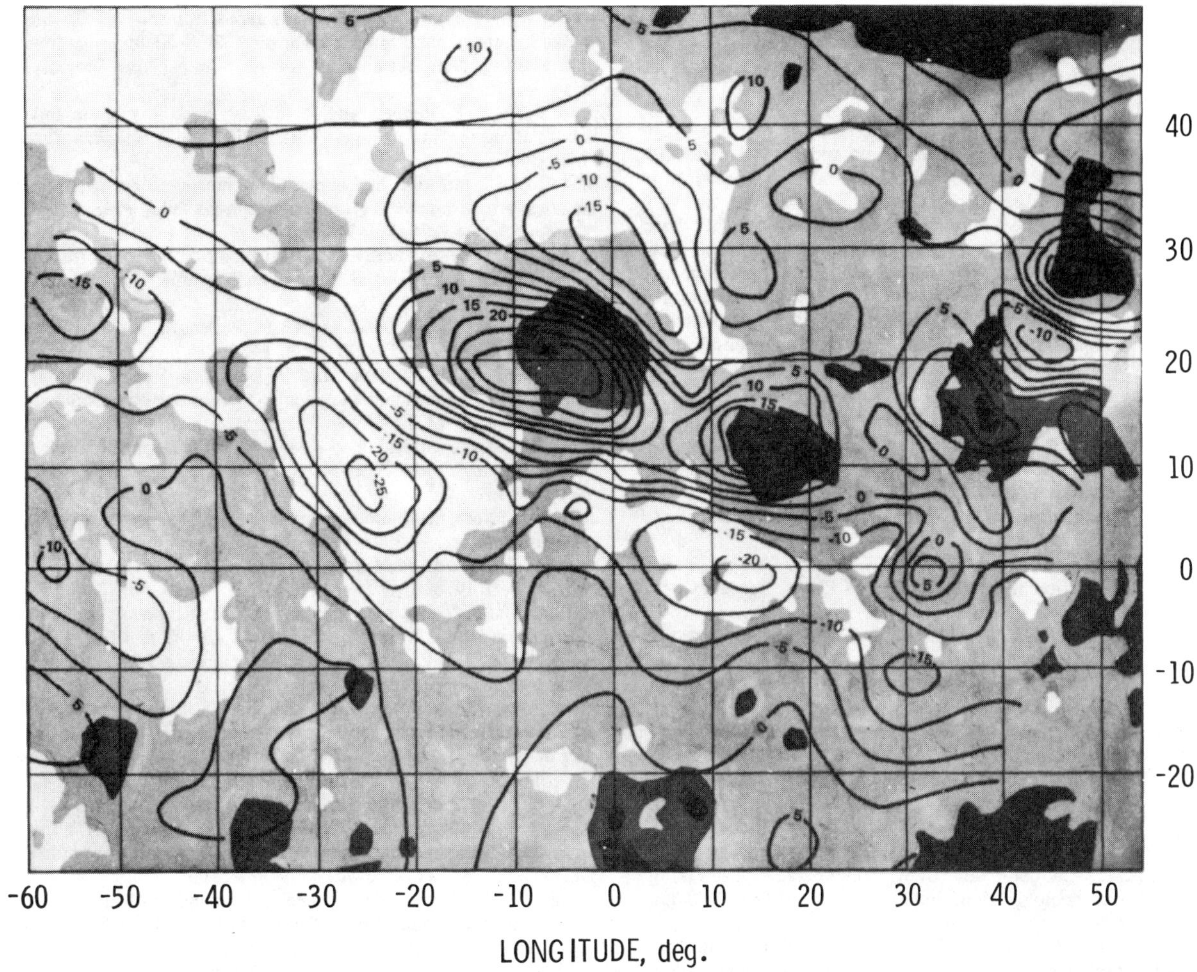

Fig. 44. Line-of-sight gravity contours (mgal) as determined by Pioneer Venus Orbiter Doppler tracking data superimposed on topography determined by Pioneer Venus Orbiter Radar Mapper. Each shading level corresponds to a vertical interval of approximately 1000 m; darker shading corresponds to higher elevation. A clear correlation of gravity to topography is seen.

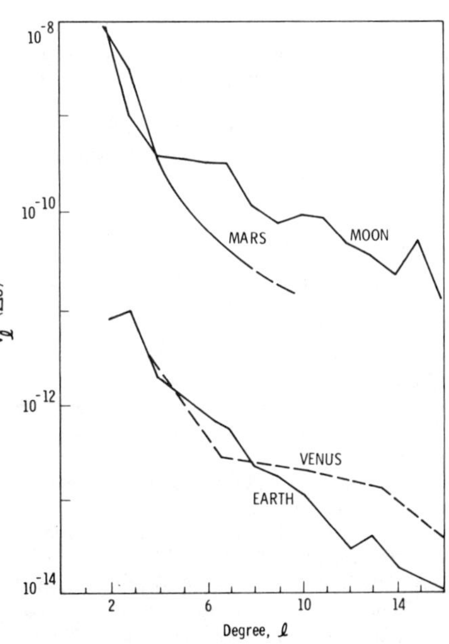

Fig. 45. Nondimensional normalized power spectra of the gravitational potential of the earth, Venus, Mars, and the moon.

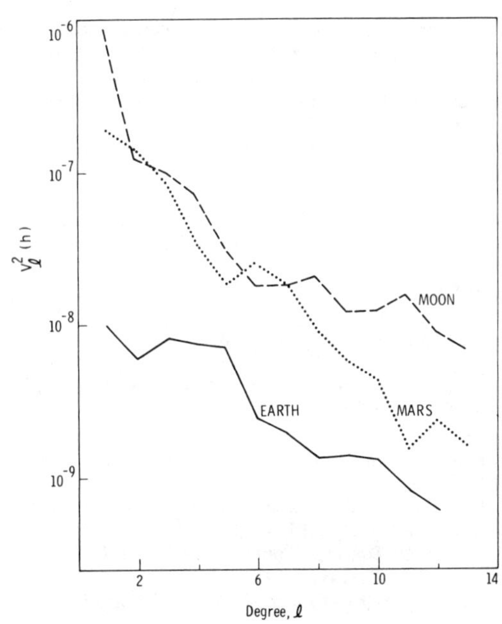

Fig. 46. Nondimensional normalized power spectra of the topography of the earth, Mars, and the moon.

wavelength data for Venus presents both a chance and a challenge to apply the concepts of comparative planetology to the understanding of dynamic processes in the interiors of the terrestrial planets.

Acknowledgments. We thank Elsa Abbott and Bette Gillette for providing their kind and competent assistance in the preparation of this manuscript. Bruce Bills, Bill Kaula, and Don Turcotte provided helpful reviews, and we look forward to friendly debates with these people over arguments of theirs we have chosen to ignore in the present paper. This paper presents the results of one phase of research carried out at the Jet Propulsion Laboratory (JPL), California Institute of Technology, under contract NAS7-100, sponsored by the National Aeronautics and Space Administration. JPL planetology publication 326-79-42 and Research School of Earth Sciences publication 1371.

REFERENCES

Alterman, Z., H. Jaroceh, and C. L. Pekeris, Oscillations of the Earth, *Proc. Roy. Soc. London, Ser. A, 252,* 80–95, 1959.

Ananda, M. P., Lunar gravity: A mass point model, *J. Geophys. Res., 82,* 3049–3064, 1977.

Anderson, R. N., D. McKenzie, and J. G. Sclater, Gravity, bathymetry and convection in the Earth, *Earth Planet, Sci. Lett., 18,* 391–407, 1973.

Arkani-Hamed, J., Lateral variations of density in the mantle, *Geophys. J. Roy. Astron. Soc., 20,* 431–455, 1970.

Arkani-Hamed, J., Stress differences in the moon as an evidence for a cold moon, *Moon, 6,* 135–163, 1973*a*.

Arkani-Hamed, J., Density and stress distribution in the Moon, *Moon, 7,* 84–126, 1973*b*.

Artyushkov, E. V., Stresses in the lithosphere caused by crustal thickness inhomogeneities, *J. Geophys. Res., 78,* 7675–7708, 1973.

Artyushkov, E. V., Can the earth's crust be in a state of isostasy?, *J. Geophys. Res., 79,* 741–752, 1974.

Arvidson, R. E., K. A. Goettel, and C. M. Hohenberg, A post-Viking view of Martian geologic evolution, submitted to *Rev. Geophys. Space Phys.*, 1980.

Ashby, M. F., and R. A. Verrall, Micromechanisms of flow and fracture and their relevance to the rheology of the upper mantle, *Phil. Trans. Roy. Soc. London, Ser. A, 288,* 59–95, 1978.

Baldwin, R. B., Absolute ages of the lunar maria and large craters, II, The viscosity of the Moon's outer layers, *Icarus, 13,* 215–225, 1970.

Balmino, G., K. Lambeck, and W. M. Kaula, A spherical harmonic analysis of the earth's topography, *J. Geophys. Res., 78,* 478–481, 1973.

Balmino, G., Ch. Reigber, and B. Moynot, Le modéle de potentiel gravitationnel terrestre GRIM 2, *Ann. Geophys. 34,* 55–78, 1978.

Banks, R. J., and C. J. Swain, The isostatic compensation of East Africa, *Proc. Roy. Soc. London, Ser. A, 364,* 331–352, 1978.

Banks, R. J., R. L. Parker, and S. P. Huestis, Isostatic compensation on a continental scale; Local versus regional mechanisms, *Geophys. J. Roy. Astron. Soc., 51,* 431–452, 1977.

Beaumont, C., The evolution of sedimentary basins on a viscoelastic lithosphere: Theory and examples, *Geophys. J. Roy. Astron. Soc., 55,* 471–497, 1978.

Bills, B., A harmonic and statistical analysis of the topography of the Earth, Moon and Mars, thesis, 263 pp., Calif. Inst. of Tech., Pasadena, 1978.

Bills, B., and A. J. Ferrari, A harmonic analysis of lunar topography, *Icarus, 31,* 244–259, 1977.

Bills, B. G., and A. J. Ferrari, Mars topography harmonics and geophysical implications, *J. Geophys. Res., 83,* 3497–3508, 1978.

Bills, B. G., and A. J. Ferrari, A harmonic analysis of lunar gravity, *J. Geophys. Res., 85,* in press, 1980.

Birch, F., Megageological considerations in rock mechanics, in *State of Stress in the Earth's Crust,* edited by W. R. Judd, pp. 55–82, Elsevier, New York, 1964.

Bird, P., Initiation of intracontinental subduction in the Himalaya, *J. Geophys. Res., 83,* 4975–4987, 1978.

Bird, P., and M. N. Toksöz, Strong attenuation of Rayleigh waves in Tibet, *Nature, 266,* 161–163, 1977.

Bland, D. R., *The Theory of Linear Viscosity,* 124 pp., Pergamon, New York, 1960.

Bott, M. H. P., The mantle transition zone as possible source of global gravity anomalies, *Earth Planet. Sci. Lett., 11,* 28–34, 1971.

Brace, K. L., Preliminary ocean-area geoid from GEOS-III satellite radar altimetry, paper presented at the GEOS-III Investigations Final Meeting, Nat. Ocean. and Atmos. Admin., New Orleans, La., Nov. 18–19, 1977.

Brune, J. N., T. L. Henyey, and R. F. Roy, Heat flow, stress and rate of slip along the San Andreas fault, California, *J. Geophys. Res., 74,* 3821–3827, 1969.

Bullard, E. C., and R. I. B. Cooper, The determination of the masses necessary to produce a given gravitational field, *Proc. Roy. Soc. London, Ser. A, 194,* 332–347, 1948.

Caldwell, J. G., W. F. Haxby, D. E. Karig, and D. L. Turcotte, On the applicability of a universal elastic trench profile, *Earth Planet. Sci. Lett., 31,* 239–246, 1976.

Campbell, D. B., R. B. Dyce, and G. H. Pettengill, New radar image of Venus, *Science, 193,* 1123–1124, 1976.

Campbell, D. B., B. A. Burns, and V. Boriakoff, Venus: Evidence of impact cratering and tectonic activity from radar observations, *Science, 204,* 1424–1427, 1979.

Cara, M., Lateral variations of *S* velocity in the upper mantle from higher Rayleigh modes, *Geophys. J. Roy. Astron. Soc., 57,* 649–670, 1979.

Carr, M. H., Tectonism and volcanism of the Tharsis region of Mars, *J. Geophys. Res., 79,* 3943–3950, 1974.

Carter, N. L., Steady state flow of rocks, *Rev. Geophys. Space Phys., 14,* 301–360, 1976.

Cassen, P., R. E. Yung, and G. Schubert, The distortion of the moon due to convection, *Geophys. Res. Lett., 5,* 294–296, 1978.

Cathles, L. M., *The Viscosity of the Earth's Mantle,* 389 pp., Princeton University Press, Princeton, N. J., 1975.

Cazenave, A., K. Lambeck, and K. Dominh, Studies of the GEOS-III altimeter derived geoid undulations over seamounts in the Indian Ocean, in *The Use of Artificial Satellites for Geodesy and Geodynamics,* vol. II, edited by G. Veis, pp. 478–499, Athens National Technical University Press, Athens, Greece, 1978.

Cazenave, A., B. Lago, K. Dominh, and K. Lambeck, On the response of the ocean lithosphere to seamount loads, submitted to *Geophys. J. Roy. Astron. Soc.*, 1979.

Chapman, M. E., Shape of the ocean surface and implications for the Earth's interior, Ph. D. thesis, Columbia Univ., New York, 1978.

Chapman, M. E., and J. H. Bodine, Considerations of the indirect effect in marine gravity modeling, *J. Geophys. Res., 84,* 3889–3892, 1979.

Choudhury, M., P. Giese, and G. de Visintini, Crustal structure of the Alps, *Boll. Geofis. Teor. Appl., 13,* 211–240, 1971.

Christensen, E. J., Martian topography derived from occultation, radar, spectral, and optical measurements, *J. Geophys. Res., 80,* 2909–2913, 1975.

Christensen, E. J., and G. Balmino, Development and analysis of a twelfth degree and order gravity model for Mars, *J. Geophys. Res., 84,* 7943–7953, 1979.

Chuikova, N. A., The geometrical figure of the Moon represented in the form of an expansion in spherical and sampling functions, *Sov. Astron., 19,* 764–771, 1976.

Cochran, J. R., and M. Talwani, Free-air gravity anomalies in the world's oceans and their relationship to residual elevation, *Geophys. J. Roy. Astron. Soc., 50,* 495–552, 1977.

Cochran, J. R., and M. Talwani, Gravity anomalies, regional elevation, and the deep structure of the North Atlantic, *J. Geophys, Res., 83,* 4907–4924, 1978.

Cook, A. H., Gravitational considerations, in *The Earth's Mantle,* edited by T. F. Gaskell, pp. 63–68, Academic, New York, 1967.

Crittenden, M. D., Effective viscosity of the earth derived from isostatic loading of Pleistocene Lake Bonneville, *J. Geophys. Res., 68,* 5517–5530, 1963.

Crittenden, M. D., Viscosity and finite strength of the mantle as determined from water and ice loads, *Geophys. J. Roy. Astron. Soc., 14,* 261–279, 1967.

Crossley, D. J., and D. Gubbins, Static deformation of the earth's core, *Geophys. Res. Lett., 2,* 1–4, 1975.

Crough, S. T., Thermal origin of mid-plate hot-spot swells, *Geophys. J. Roy. Astron. Soc., 55,* 451–469, 1978.

Dahlen, F. A., and S. B. Fels, A physical explanation of the static core paradox, *Geophys. J. Roy. Astron. Soc., 55,* 317–332, 1978.

Dainty, A. M., N. R. Goins, and M. N. Toksöz, Natural lunar seismic events and the structure of the Moon, *Proc. Lunar Sci. Conf. 6th., Geochim. Cosmochim. Acta, Suppl. 6, 3,* 2887–2897, 1975.

Davis, E. E., and C. R. B. Lister, Fundamentals of ridge crest topography, *Earth Planet. Sci. Lett., 21,* 405–413, 1974.

De Bremaecker, J. C., Convection in the Earth's mantle, *Tectonophysics, 41,* 195–208, 1977.

De Jong, K. A., and R. Scholten (Eds.), *Gravity and Tectonics,* 502 pp., John Wiley, New York, 1973.

Detrick, R. S., and S. T. Crough, Island subsidence, hot spots, and lithospheric thinning, *J. Geophys. Res., 83,* 1236–1244, 1978.

Dorman, L. M., and B. T. R. Lewis, Experimental isostasy, 1, Theory of the determination of the earth's isostatic response to a concentrated load, *J. Geophys. Res., 75,* 3357–3365, 1970.

Durham, W. B., and C. Goetze, Plastic flow of oriented single crystals of olivine, 1, Mechanical data, *J. Geophys. Res., 82,* 5737–5753, 1977.

Dvorak, J., Analysis of small scale lunar gravity anomalies: Implications for crater formation and crustal history, Ph.D. thesis, 250 pp., Calif. Inst. of Technol., Pasadena, 1979.

Dvorak, J., and R. J. Phillips, Lunar Bouguer gravity anomalies: Imbrian age craters, *Proc. Lunar Planet. Sci. Conf. 9th, Geochim. Cosmochim. Acta, Suppl. 10, 3,* 3651–3668, 1978.

Dvorak, J., and R. J. Phillips, Grimaldi: Bouguer gravity, submitted to *J. Geophys. Res.,* 1979.

Dziewonski, A. M., B. H. Hager, and R. J. O'Connell, Large scale heterogeneities in the lower mantle, *J. Geophys. Res., 82,* 239–255, 1977.

Fairhead, J. D., The structure of the lithosphere beneath the Eastern rift, East Africa, deduced from gravity studies, *Tectonophysics, 30,* 269–298, 1976.

Fairhead, J. D., A gravity link between the domally uplifted Cainozoic volcanic centres of North Africa and its similarity to the East African rift system anomaly, *Earth Planet. Sci. Lett., 42,* 109–113, 1979.

Farrell, W. E., Deformation of the earth by surface loads, *Rev. Geophys. Space Phys., 10,* 761–797, 1972.

Ferguson, J., R. J. Arculus, and J. Joyce, Kimberlite and kimberlitic intrusives of southeastern Australia: A review, *Bur. Miner. Resour. J. Aust. Geol. Geophys., 4,* 227–241, 1979.

Ferrari, A. J., Lunar gravity: A harmonic analysis, *J. Geophys. Res., 82,* 3065–3084, 1977.

Finlayson, D. M., C. D. N. Collins, and D. Denham, Crustal structure under the Lachlan fold belt, southeastern Australia, *Tectonophysics,* in press, 1979.

Fjelbo, G., D. Sweetnam, J. Brenkle, E. Christensen, D. Farless, J. Mehta, B. Seidel, W. Michael, A. Wallio, and M. Grossi, Viking radio occultation measurements of the Martian atmosphere and topography: Primary mission coverage, *J. Geophys. Res., 82,* 4317–4324, 1977.

Forsyth, D. W., Comparison of mechanical models of the oceanic lithosphere, in *Magnitude of Deviatoric Stresses in the Earth's Crust and Upper Mantle,* edited by J. Everenden, U.S. Geological Survey, Menlo Park, Calif., in press, 1979.

Fung, Y. C., *Foundation of Solid Mechanics,* 525 pp., Prentice-Hall, Englewood Cliffs, N. J., 1965.

Gapcynski, J. P., R. H. Tolson, and W. H. Michael, Mars gravity field: Combined Viking and Mariner 9 results, *J. Geophys. Res., 82,* 4325–4327, 1977.

Gaposchkin, E. M., Gravity field determination using laser observations, *Phil. Trans. Roy. Soc. London, Ser. A, 284,* 515–527, 1977.

Gaposchkin, E. M., Global gravity field to degree and order 30 from Geos 3 satellite altimetry and other data, submitted to *J. Geophys. Res.,* 1979.

Gaposchkin, E. M., and K. Lambeck, Earth's gravity field to the sixteenth degree and station coordinates from satellite and terrestrial data, *J. Geophys. Res., 76,* 4855–4883, 1971.

Garland, G. D., *Introduction to Geophysics, Mantle Core and Crust,* 420 pp., W. B. Saunders, Philadelphia, Pa., 1971.

Goins, N. R., A. M. Dainty, and M. N. Toksöz, The deep seismic structure of the moon, *Proc. Lunar Sci. Conf. 8th, Geochim. Cosmochim. Acta, Suppl. 8, 1,* 471–486, 1977.

Goldstein, R. M., R. R. Green, and H. C. Rumsey, Venus radar images, *J. Geophys. Res., 81,* 4807–4817, 1976.

Griggs, D. T., The sinking lithosphere and the focal mechanism of deep earthquakes, in *The Nature of the Solid Earth,* edited by E. C. Robertson, pp. 361–384, McGraw-Hill, New York, 1972.

Grow, J. A., Crustal and upper mantle structure of the central Aleutian arc, *Geol. Soc. Amer. Bull., 84,* 2169–2192, 1973.

Grow, J. A., and C. O. Bowin, Evidence for high-density crust and mantle beneath the Chile trench due to the descending lithosphere, *J. Geophys. Res., 80,* 1449–1458, 1975.

Guest, J. E., P. S. Butterworth, and R. Greeley, Geological observations in the Cydonia region of Mars from Viking, *J. Geophys. Res., 82,* 4111–4120, 1977.

Hager, B. H., and R. J. O'Connell, Subduction zone dip angles and flow driven by plate motion, *Tectonophysics, 50,* 111–113, 1978.

Hager, B. H., and R. J. O'Connell, Kinematic models of large-scale flow in the earth's mantle, *J. Geophys. Res., 84,* 1031–1048, 1979.

Haimson, B. C., The state of stress in the earth's crust, *Rev. Geophys. Space Phys., 13,* 350–352, 1975.

Hales, A. L., and E. Herrin, Travel times in seismic waves, in *The Nature of the Solid Earth,* edited by E. C. Robertson, pp. 172–215, McGraw-Hill, New York, 1972.

Hanks, T. C., The Kurile trench–Hokkaido rise system: Large shallow earthquakes and simple models of deformation, *Geophys. J. Roy. Astron. Soc., 23,* 173–190, 1971.

Hanks, T. C., Earthquake stress drops, ambient tectonic stresses and stresses that drive plate motions, *Pure Appl. Geophys., 115,* 441–458, 1977.

Hartmann, W. K., Martian surface and crust: Review and synthesis, *Icarus, 19,* 550–575, 1973.

Haskell, N. A., The motion of a viscous fluid under a surface load, I, *Physics, 6,* 265–271, 1935.

Haskell, N. A., The motion of a viscous fluid under a surface load, II, *Physics, 7,* 56–61, 1936.

Haskell, N. A., The viscosity of the asthenosphere, *Amer. J. Sci., 33,* 22–28, 1937.

Head, J. W., Some geologic observations concerning lunar geophysical models, The Soviet American Conference on the Cosmochemistry of the Moon and Planets, *NASA Spec. Publ., SP 370,* 407–416, 1977.

Heiskanen, W. A., and H. Moritz, *Physical Geodesy,* W. H. Freeman, San Francisco, Calif., 1967.

Heiskanen, W. A., and F. A., Vening-Meinesz, *The Earth and Its Gravity Field,* 470 pp., McGraw Hill, New York, 1958.

Herrin, E., A comparative study of upper mantle models: Canadian shield and basin and range provinces, in *The Nature of the Solid Earth,* edited by E. C. Robertson, pp. 216–231, McGraw-Hill, New York, 1972.

Hide, R., Free hydromagnetic oscillations of the Earth's core and the theory of the geomagnetic secular variation, *Phil. Trans. Roy. Soc. London, Ser. A, 259,* 615–647, 1966.

Hide, R., and R. K. Horai, On the topography of the core-mantle interface, *Phys. Earth Planet. Interiors, 1,* 305–308, 1968.

Hide, R., and S. R. C. Malin, Novel correlations between global features of the earth's gravitational and magnetic fields, *Nature, 225,* 605–609, 1970.

Hodges, F. N., and I. Kushiro, Apollo 17 petrology and experimental determination of differentiation sequences in model moon compositions, *Proc. Lunar Sci. Conf. 5th, Geochim. Cosmochim. Acta, Suppl. 5, 3,* 505–520, 1974.

Hörz, F., How thick are lunar mare basalts?, *Proc. Lunar Planet. Sci. Conf. 9th, Geochim. Cosmochim. Acta, Suppl. 10, 3,* 3311–3331, 1978.

Huebner, J. S., A. Duba, and L. B. Wiggins, Electrical conductivity of pyroxene which contains trivalent cations: Laboratory measurements and the lunar temperature profile, *J. Geophys. Res., 84,* 4652–4656, 1979.

Hulme, G., Mascons and isostasy, *Nature, 238,* 448–450, 1972.

Jeffreys, H., On the stresses in the Earth's crust required to support surface inequalities, *Mon. Notic. Roy. Astron. Soc., Geophys. Suppl., 3,* 30–41, 1932.

Jeffreys, H., The stress differences in the Earth's shell, *Mon. Notic. Roy. Astron. Soc., Geophys. Suppl., 5,* 71–89, 1943.

Jeffreys, H., *The Earth, Its Origin, History and Physical Constitution,* Cambridge University Press, New York, 1970.

Johnston, D. H., and M. N. Toksöz, Internal structure and properties of Mars, *Icarus, 32,* 73–84, 1977.

Johnston, D. H., T. R. McGetchin, and M. N. Toksöz, The thermal state and internal structure of Mars, *J. Geophys. Res., 79,* 3959–3971, 1974.

Julian, B. R., and M. K. Sengupta, Seismic travel time evidence for lateral inhomogeneity in the deep mantle, *Nature, 242,* 443–447, 1973.

Kanamori, H., and D. L. Anderson, Theoretical Basis of some empirical relations in seismology, *Bull. Seismol. Soc. Amer., 65,* 1073–1096, 1975.

Kaula, W. M., Elastic models of the mantle corresponding to variations in the external gravity field, *J. Geophys. Res., 68,* 4967–4978, 1963.

Kaula, W. M., Theory of statistical analysis of data distributed over a sphere, *Rev. Geophys., 5,* 83–107, 1967.

Kaula, W. M., *An Introduction to Planetary Physics: The Terrestrial Planets,* John Wiley, New York, 1968.

Kaula, W. M., Interpretation of lunar mass concentrations, *Phys. Earth Planet. Interiors, 2,* 123–137, 1969.

Kaula, W. M., Interpretation of the lunar gravitational field, *Phys. Earth Planet. Interiors, 4,* 185–192, 1971.

Kaula, W. M., Global gravity and tectonics, in *The Nature of the Solid Earth,* edited by E. C. Robertson, pp. 385–405, McGraw-Hill, New York, 1972.

Kaula, W. M., The gravity and shape of the moon, *Eos Trans. AGU, 56,* 309–316, 1975.

Kaula, W. M., Geophysical inferences from statistical analysis of the gravity field, Dep. Geod. Sci. Rep. 250, pp. 199–141, Ohio State Univ., Columbus, 1977*a*.

Kaula, W. M., On the origin of the moon, with emphasis on bulk composition, *Proc. Lunar Sci. Conf. 8th, Geochim. Cosmochim. Acta, Suppl. 8, 1,* 321–331, 1977*b*.

Kaula, W. M., G. Schubert, and R. E. Lingenfelter, Apollo laser altimetry and inferences as to lunar structure, *Proc. Lunar Sci. Conf. 5th, Geochim. Cosmochim. Acta, Suppl. 5, 3,* 3049–3058, 1974.

Keihm, S. J., and M. G. Langseth, Lunar thermal regime to 300 km, *Proc. Lunar Sci. Conf. 8th, Geochim. Cosmochim. Acta, Suppl. 8, 1,* 499–514, 1977.

Kennedy, G. C., and G. H. Higgins, Melting temperatures in the Earth's mantle, *Tectonophysics, 13,* 221–232, 1972.

King, R. W., C. C. Counselman, and I. I. Shapiro, Lunar dynamics and selenodesy: Results from analysis of VLBI and laser data, *J. Geophys. Res., 81,* 6251–6256, 1976.

Kirby, S. H., State of stress in the lithosphere: Inferences from the flow laws of olivine, *Pure Appl. Geophys., 115,* 245–258, 1977.

Kirby, S. H., Tectonic stresses in the lithosphere: Constraints provided by the experimental deformation of rocks, in *Magnitude of Deviatoric Stresses in the Earth's Crust and Upper Mantle,* edited by J. Everenden, U.S. Geological Survey, Menlo Park, Calif., in press, 1979.

Kirby, S. H., and C. B. Raleigh, Mechanics of high-temperature, solid-state flow in minerals and ceramics and their bearing on creep behavior of the mantle, *Tectonophysics, 19,* 165–194, 1973.

Kobrick, M., Random processes as a cause of the lunar assymetry, *Moon, 15,* 83–89, 1976.

Kopal Z., *The Moon,* 525 pp., D. Reidel, Dordrecht, Netherlands, 1969.

Kraus, H., *Thin Elastic Shells An Introduction to the Theoretical Foundations and the Analysis of Their Static and Dynamic Behavior,* John Wiley, New York, 1967.

Kuckes, A. F., Strength and rigidity of the elastic lunar lithosphere and implications for present-day mantle convection in the moon, *Phys. Earth Planet. Interiors, 14,* 1–12, 1977.

Kunze, A. W. G., Lunar mascons: Another model and its implications, *Moon, 11,* 9–17, 1974.

Lachenbruch, A. H., and J. H. Sass, Thermo-mechanical aspects of the San Andreas fault system, in *Proceedings of the Conference on Tectonic Problems of the San Andreas Fault System,* edited by R. L. Kovach and A. Nur, pp. 192–205, Stanford University Publications, Stanford, Calif., 1973.

Lambeck, K., Gravity anomalies over ocean ridges, *Geophys. J. Roy. Astron. Soc., 30,* 37–53, 1972.

Lambeck, K., Lateral density anomalies in the upper mantle, *J. Geophys. Res., 81,* 6333–6340, 1976.

Lambeck, K., Comments on the gravity and topography of Mars, *J. Geophys. Res., 84,* 6241–6247, 1979*a*.

Lambeck, K., Estimates of stress differences in the crust from isostatic considerations, in *Magnitude of Deviatoric Stresses in the Earth's Crust and Upper Mantle,* edited by J. Everenden, U.S. Geological Survey, Menlo Park, Calif., in press, 1979*b*.

Lambeck, K., and S. M. Nakiboglu, Seamount loading of the oceanic lithosphere, in *Magnitude of Deviatoric Stresses in the Earth's Crust and Upper Mantle,* edited by J. Everenden, U.S. Geological Survey, Menlo Park, Calif., in press, 1979.

Lambeck, K., and S. Pullan, Inferences on the lunar temperature from gravity, stress state and flow laws, *Phys. Earth Planet. Interiors,* in press, 1979*a*.

Lambeck, K., and S. Pullan, The lunar fossil bulge hypothesis revisited, *Phys. Earth Planet. Interiors,* in press, 1979*b*.

Langseth, M. G., X. LePichon, and M. Ewing, Crustal structure of the mid-ocean ridges, 5, Heat flow through the Atlantic Ocean floor and convention currents, *J. Geophys. Res., 71,* 5321–5355, 1966.

Lee, W. H. K., Analysis of the Earth's topography, *Final Rep. AF(601)-4171,* Inst. of Geophys. and Planet. Phys., Univ. of Calif., Los Angeles, 1966.

LePichon, X., J. Francheteau, and J. Bonnin, *Plate Tectonics,* 300 pp., Elsevier, New York, 1973.

Lerch, F. J., C. A. Wagner, S. M. Klosko, and R. P. Belott, Goddard Earth model development for oceangraphic applications, paper presented at the International Symposium on Interaction of Marine Geodesy and Ocean Dynamics, Miami Univ., Miami, Fla., Oct. 1978.

Lerch, F. J., S. M. Klosko, R. E. Laubscher, and C. A. Wagner, Gravity model improvement using Geos 3 (GEM 9 and 10), *J. Geophys. Res., 84,* 3897–3916, 1979.

Lewis, B. T. R., and L. M. Dorman, Experimental isostasy, 2, An isostatic model for the U.S.A. derived from gravity and topographic data, *J. Geophys. Res., 75,* 3367–3386, 1970.

Lingenfelter, R. E., and G. Schubert, Evidence for convection in planetary interiors from first-order topography, *Moon, 7,* 172–180, 1973.

Liu, A. S., and P. A. Laing, Lunar gravity analysis from long-term effects, *Science, 173,* 1017–1020, 1971.

Longman, I. M., A Greens' function for determining the deformation of the earth under surface mass loads, 1, Theory, *J. Geophys. Res., 67,* 845–850, 1962.

Longman, I. M., A Green's function for determining the deformation of the earth under surface mass loads, 2, Computations and numerical results, *J. Geophys. Res., 68,* 485–496, 1963.

Lorell, J., G. H. Born, E. J. Christensen, J. F. Jordan, P. A. Laing, W. L. Martin, W. L. Sjogren, I. I. Shapiro, R. D. Reasenberg, and G. L. Slater, Mariner 9 celestial mechanics experiment: Gravity field and pole direction of Mars, *Science, 175,* 317–320, 1972.

Love, A. E. H., *A Treatise on the Mathematical Theory of Elasticity,* 643 pp., Dover, New York, 1944.

Malin, M. C., and R. S. Saunders, Surface of Venus: Evidence of diverse land forms from radar observations, *Science, 196,* 987–990, 1977.

Malin, M. C., R. J. Phillips, and R. S. Saunders, The nature and origin of the Martian planetary dichotomy (is still a problem), Reports of Planetary Geology Program, 1977–1978, *NASA Tech. Mem. 79729,* 83–85, 1978.

Marsh, B. D., and J. G. Marsh, On global gravity anomalies and two-scale mantle convection, *J. Geophys. Res., 81,* 5267–5280, 1976.

Marsh, B. D., and J. G. Marsh, Reply, *J. Geophys. Res., 83,* 3555–3558, 1978.

McAdoo, D. C., J. G. Caldwell, and D. L. Turcotte, On the elastic–perfectly plastic bending of the lithosphere under generalized loading with application to the Kuril Trench, *Geophys. J. Roy Astron. Soc., 54,* 11–26, 1978.

McConnell, R. K., Isostatic adjustment in a layered earth, *J. Geophys. Res., 70,* 5171–5188, 1965.

McConnell, R. K., Viscosity of the mantle from relaxation time spectra of isostatic adjustment, *J. Geophys. Res., 73,* 7089–7105, 1968.

McGarr, A., and N. C. Gay, State of stress in the Earth's crust, *Ann. Rev. Earth Planet. Sci., 6,* 405–436, 1978.

McKenzie, D. P., Some remarks on heat flow and gravity anomalies, *J. Geophys. Res., 72,* 6261–6273, 1967.

McKenzie, D. P., Surface deformation, gravity anomalies and convention, *Geophys. J. Roy. Astron. Soc., 48,* 211–238, 1977.

McKenzie, D. P., and C. Bowin, The relationship between bathymetry and gravity in the Atlantic Ocean, *J. Geophys. Res., 81,* 1903–1915, 1976.

McNutt, M. K., R. L. Parker, Isostasy in Australia and evolution of the compensation mechanism, *Science, 199,* 773–775, 1978.

McQueen, H. W. S., and F. D. Stacey, Interpretation of low degree components of gravitational potential in terms of undulations of mantle phase boundaries, *Tectonophysics, 34,* T1–T8, 1976.

Melosh, H. J., The tectonics of mascon loading, *Proc. Lunar Planet Sci. Conf. 9th, Geochim. Cosmochim. Acta, Suppl. 10, 3,* 3513–3526, 1978.

Mercier, J. C., D. A. Anderson, and N. L. Carter, Stress in the lithosphere: Inferences from steady state flow of rocks, *Pure Appl. Geophys., 115,* 199–226, 1977.

Michael, W. H., and W. T. Blackshear, Recent results on the mass, gravitational field and moments of inertia of the Moon, *Moon, 3,* 388–402, 1972.

Michael, W. H., W. T. Blackshear, and J. P. Gapcynski, Results on the mass and the gravitational field of the Moon as determined from dynamics of lunar satellites, in *Dynamics of Satellites,* edited by B. Morando, pp. 42–56, Springer, New York, 1969.

Minster, J. B., T. H. Jordan, P. Molnar, and E. Haines, Numerical modeling of instantaneous plate tectonics, *Geophys. J. Roy. Astron. Soc., 36,* 541–576, 1974.

Molnar, P., and P. Tapponnier, Active tectonics of Tibet, *J. Geophys. Res., 83,* 5361–5376, 1978.

Muller, P. M., and W. L. Sjogren, Mascons: Lunar mascon concentrations, *Science, 161,* 680–684, 1968.

Mutch, T. A., *Geology of the Moon, A Stratigraphic View,* Princeton University Press, Princeton, N.J., 1972.

Mutch, T. A., R. E. Arvidson, J. W. Head, K. L. Jones, and R. S. Saunders, *The Geology of Mars,* 400 pp., Princeton University Press, Princeton, N. J., 1976.

Nadai, A., *Theory of Flow and Fracture of Solids,* Vol. 2, 705 pp., McGraw-Hill, New York, 1963.

Nakamura, Y., A_1 Moonquakes: Source distribution and mechanism, *Proc. Lunar Planet. Sci. Conf. 9th, Geochim. Cosmochim. Acta, Suppl. 10, 3,* 3589–3607, 1978.

Nakamura, Y., G. Latham, D. Lammlein, M. Ewing, F. Duennebier, and J. Dorman, Deep lunar interior inferred from recent seismic data, *Geophys. Res. Lett., 1,* 137–140, 1974.

Nakamura, Y., G. V. Latham, H. J. Dorman, A. K. Ibrahim, J. Koyama, and P. Horvath, Shallow moonquakes: Depth, distribution and implications as to the present state of the lunar interior, in *Lunar and Planetary Science 10th,* pp. 898–900, Lunar and Planetary Institute, Houston, Tex., 1979.

Parker, R. L., The theory of ideal bodies for gravity interpretation, *Geophys. J. Roy. Astron. Soc., 42,* 315–334, 1975.

Parker, R. L., and D. W. Oldenburg, Thermal model of ocean ridges, *Nature Phys. Sci., 242,* 137–139, 1973.

Parsons, B., and D. McKenzie, Mantle convection and the thermal structure of the plates, *J. Geophys. Res., 83,* 4485–4496, 1978.

Parsons, B., and J. G. Sclater, An analysis of the variation of ocean floor bathymetry and heat flow with age, *J. Geophys. Res., 82,* 803–827, 1977.

Pekeris, C. L., Thermal convection in the interior of the Earth, *Mon. Notic. Roy. Astron. Soc., Geophys. Suppl., 3,* 343–359, 1935.

Peltier, W. R., The impulse response of a Maxwell earth, *Rev. Geophys. Space Phys., 12,* 649–669, 1974.

Peltier, W. R., Glacial-isostatic adjustment, II, The inverse problem, *Geophys. J. Roy. Astron Soc., 46,* 669–705, 1976.

Peltier, W. R., and J. T. Andrews, Glacial-isostatic adjustment, I, The forward problem, *Geophys. J. Roy. Astron. Soc., 46,* 605–646, 1976.

Pettengill, G. H., P. G. Ford, W. E. Brown, W. M. Kaula, H. Masursky, E. Eliason, and G. E. McGill, Venus: Preliminary topographic and surface imaging results from the Pioneer Orbiter, *Science, 205,* 90–93, 1979.

Phillips, R. J., and E. R. Ivins, Geophysical observations pertaining to solid-state convection in the terrestrial planets, *Phys. Earth Planet. Interiors, 19,* 107–148, 1979.

Phillips, R. J., and T. A. Maxwell, Lunar sounder revisited: Stratigraphic correlations and structural inferences, in *Lunar and Planetary Science 9th,* pp. 890–892, Lunar and Planetary Institute, Houston, Tex., 1978.

Phillips, R. J., and R. S. Saunders, The isostatic state of Martian topography, *J. Geophys. Res., 80,* 2893–2897, 1975.

Phillips, R. J., J. E. Conel, E. A. Abbott, W. L. Sjogren, and J. B. Morton, Mascons: Progress toward a unique solution for mass distribution, *J. Geophys. Res., 77,* 7106–7114, 1972.

Phillips, R. J., R. S. Saunders, and J. E. Conel, Mars: Crustal structure inferred from Bouguer gravity anomalies, *J. Geophys. Res., 78,* 4815–4820, 1973.

Phillips, R. J., J. E. Conel, and W. L. Sjogren, The nature of circular-maria based on gravity studies, *Space Res., 14,* 635–645, 1974.

Phillips, R. J., W. L. Sjogren, E. A. Abbott, J. C. Smith, R. N. Wimberly, and C. A. Wagner, The gravity field of Venus: A preliminary analysis, *Science, 205,* 93–96, 1979.

Phillips, R. J., W. B. Banerdt, N. L. Sleep, and R. S. Saunders, Martian stress distributions: Arguments about finite strength support of Tharsis, *J. Geophys. Res., 85,* in press, 1980.

Pike, R. J., Schroter's rule and the modification of lunar crater impact morphology, *J. Geophys. Res., 72,* 2199–2106, 1967.

Pines, D., J. Shaham, and M. Cohen, Elastic energy and tectonic surface movements on Mars, *Proc. Lunar Sci. Conf. 8th, Geochim. Cosmochim. Acta, Suppl. 8, 1,* 563–571, 1977.

Pollack, H. N., and D. S. Chapman, Mantle heat flow, *Earth Planet. Sci. Lett., 34,* 174–184, 1977.

Post, R. L., High-temperature creep of Mt. Burnet dunite, *Tectonophysics, 42,* 75–110, 1977

Raleigh, B., and J. Everenden, The case for low deviatoric stress in the lithosphere, in *Magnitude of Deviatoric Stresses in the Earth's Crust and Upper Mantle,* edited by J. Everenden, U.S. Geological Survey, Menlo Park, Calif., in press, 1979.

Ransford, G., and W. Sjogren, Moon model—An offset core, *Nature, 238,* 260–262, 1972.

Rapp, R., Potential coefficient determinations from 5° terrestrial gravity data, *Dept. Geod. Sci. Rep. 215,* Ohio State Univ. Columbus, 1977.

Richardson, R. M., and S. C. Solomon, Apparent stress and stress drop for intraplate earthquakes and tectonic stress in the plates, *Pure Appl. Geophys., 115,* 317–331, 1977.

Richter, F. M., Dynamical models for sea floor spreading, *Rev. Geophys. Space Phys., 11,* 223–287, 1973.

Richter, F. M., and D. McKenzie, Simple plate models of mantle convection, *J. Geophys., 44,* 441–471, 1978.

Richter, F. M., and B. Parsons, On the interaction of two scales of convection in the mantle, *J. Geophys. Res., 80,* 2529–2541, 1975.

Ringwood, A. E., and E. Essene, Petrogenesis of Apollo 11 basalts, internal constitution and origin of the Moon, *Proc. Apollo 11 Lunar Sci. Conf., Geochim. Cosmochim. Acta, Suppl. 1, 1,* 769–797, 1970.

Romanowicz, B. A., Seismic structure of the upper mantle beneath the United States by three-dimensional inversion of body wave arrival times, *Geophys. J. Roy. Astron. Soc., 57,* 479–506, 1979.

Roth, L. E., C. Elachi, and R. J. Phillips, Equipotential doming in flooded circular basins on the Moon, *Proc. Lunar Sci. Conf. 8th, Geochim. Cosmochim. Acta, Suppl. 8, 1,* 643–654, 1977.

Roy, R. F., D. D. Blackwell, and E. R. Decker, Continental heat flow, in *The Nature of the Solid Earth,* edited by E. C. Robertson, pp. 506–543, McGraw-Hill, New York, 1972.

Runcorn, S. K., Convection in the Moon, *Nature, 195,* 1150–1151, 1962.

Runcorn, S. K., Satellite gravity measurements and a laminar viscous flow model of the earth's mantle, *J. Geophys. Res., 69,* 4389–4394, 1964.

Runcorn, S. K., Flow in the mantle inferred from the low degree harmonics of the geopotential, *Geophys. J. Roy. Astron. Soc., 14,* 375–384, 1967.

Runcorn, S. K., Solid-state convection and the mechanics of the Moon, *Proc. Lunar Sci. Conf. 6th, Geochim. Cosmochim. Acta, Suppl. 6, 3,* 2943–2953, 1975.

Runcorn, S. K., Inferences concerning the early thermal history of the Moon, *Proc. Lunar Sci. Conf. 7th, Geochim. Cosmochim. Acta, Suppl. 7, 3,* 3221–3228, 1976.

Saito, M., Some problems of static deformation of the Earth, *J. Phys. Earth, 22,* 123–140, 1974.

Sass, J. H., and A. H. Lachenbruch, Thermal regime of the Australian continental crust, in *The Earth: Its Origin, Structure and Evolution,* pp. 301–351, edited by M. W. McElhinny, Academic, New York, 1979.

Saunders, R. S., and M. C. Malin, Geologic interpretation of new observations of the surface of Venus, *Geophys. Res. Lett., 4,* 547–550, 1977.

Sbar, M. L., and L. R. Sykes, Contemporary compressive stress and seismicity in eastern North America: An example of intra-plate tectonics, *Geol. Soc. Amer. Bull., 84,* 1861–1882, 1973.

Schubert, G., and D. L. Turcotte, One-dimensional model of shallow-mantle convection, *J. Geophys. Res., 77,* 945–951, 1972.

Schubert, G., P. Cassen, and R. E. Young, Subsolidus convection cooling histories of terrestrial planets, *Icarus, 38,* 192–211, 1979.

Sclater, J. G., and J. Francheteau, The implications of terrestrial heat flow observations on current tectonic and geochemical models of the crust and the upper mantle of the Earth, *Geophys. J. Roy. Astron. Soc., 20,* 509–542, 1970.

Sclater, J. G., L. A. Lawver, and B. Parsons, Comparison of long-wavelength residual elevation and free-air gravity anomalies in the North Atlantic and possible implications for the thickness of the lithospheric plate, *J. Geophys. Res., 80,* 1031–1052, 1975.

Segawa, J., and Y. Tomoda, Gravity measurements near Japan and study of the upper mantle beneath the ocean trench–marginal sea transition zones, in *Geophysics of the Pacific Ocean Basin and Its Margin, Geophys. Monog. 19,* edited by G. P. Woollard, pp. 35–52, AGU, Washington, D. C., 1976.

Sjogren, W. L., Lunar gravity estimate: Independent confirmation, *J. Geophys. Res., 76,* 7021–7026, 1971.

Sjogren, W. L., Lunar gravity determinations and their implications, *Phil. Trans. Roy. Soc. London, Ser. A, 285,* 219–226, 1977.

Sjogren, W. L., Mars gravity: High-resolution results from Viking Orbiter 2, *Science, 203,* 1006–1010, 1979.

Sjogren, W. L., and J. C. Smith, Quantitative mass distribution models for Mare Orientale, *Proc. Lunar Sci. Conf. 7th, Geochim. Cosmochim. Acta, Suppl. 7, 3,* 2639–2648, 1976.

Sjogren, W. L., and W. R. Wollenhaupt, Lunar shape via the Apollo laser altimeter, *Science, 179,* 275, 1973.

Sjogren, W. L., J. Lorell, L. Wong, and W. Downs, Mars gravity field based on a short arc technique, *J. Geophys. Res., 80,* 2899–2908, 1975.

Sleep, N. H., and R. J. Phillips, An isostatic model for the Tharsis province, Mars, *Geophys. Res. Lett., 6,* 803–806, 1979.

Sleep, N. H., and N. S. Snell, Thermal contraction and flexure of midcontinent and Atlantic marginal basins, *Geophys. J. Roy. Astron. Soc., 45,* 125–154, 1976.

Solomon, S. C., The nature of isostasy on the Moon: How big a Pratt-fall for Airy models?, *Proc. Lunar Planet. Sci. Conf. 9th, Geochim. Cosmochim. Acta, Suppl. 10, 3,* 3499–3511, 1978.

Solomon, S. C., and J. W. Head, Vertical movement in mare basins: Relation to mare emplacement, basin tectonics, and lunar thermal history, *J. Geophys. Res., 84,* 1667–1682, 1979.

Solomon, S. C., N. H. Sleep, and R. M. Richardson, On the forces driving plate tectonics: Inferences from absolute plate velocites and intraplate stress, *Geophys. J. Roy. Astron. Soc., 42,* 769–801, 1975.

Takeuchi, H., and M. Saito, Seismic surface waves, *Methods Comput. Phys., 11,* 217–295, 1972.

Tera, F., and G. J. Wasserburg, U-Th-Pb systematics on lunar rocks and inferences about lunar evolution and the age of the moon, *Proc. Lunar Sci. Conf. 5th, Geochim. Cosmochim. Acta, Suppl. 5, 2,* 1571–1599, 1974.

Thurber, C. H., and S. C. Solomon, An assessment of crustal thickness variations on the lunar near side: Model uncertainties, and implications for crustal differentiation, *Proc. Lunar Planet. Sci. Conf. 9th, Geochim. Cosmochim. Acta, Suppl. 10, 3,* 3481–3497, 1978.

Timoshenko, S., and J. N. Goodier, *Theory of Elasticity,* 806 pp., 1951. McGraw-Hill, New York, 1951.

Toksöz, M. N., M. A. Chinnery, and D. L. Anderson, Inhomogeneities in the Earth's mantle, *Geophys. J. Roy. Astron. Soc., 13,* 31–59, 1967.

Toksöz, M. N., A. M. Dainty, S. C. Solomon, and K. R. Anderson, Structure of the moon, *Rev. Geophys. Space Phys., 12,* 539–567, 1974.

Toksöz, M. N., N. R. Goins, and C. H. Cheng, Moonquakes: Mechanisms and relation to tidal stresses, *Science, 196,* 979–981, 1977.

Turcotte, D. L., Parameterized convection within the Moon and the terrestrial planets, *Proc. Lunar Planet. Sci. Conf. 10th, Geochim. Cosmochim. Acta, Suppl. 11, 3,* 2375–2392, in press, 1979.

Turcotte, D. L., and D. C. McAdoo, Geoid anomalies and the thickness of the lithosphere, *J. Geophys. Res., 84,* 2381–2387, 1979.

Turcotte, D. L., and E. R. Oxburgh, Finite amplitude convective cells and continental drift, *J. Fluid Mech., 28,* 29, 1967.

Turcotte, D. L., J. L. Ahern, and J. M. Bird, The state of stress at continental margins, *Tectonophysics, 42,* 1–28, 1977.

Wagner, C. A., F. J. Lerch, J. E. Brown, and J. A. Richardson, Improvement in the geopotential derived from satellite and surface data (Gem 7 and 8), *J. Geophys. Res., 82,* 901–914, 1977.

Walcott, R. I., Flexural rigidity, thickness and viscosity of the lithosphere, *J. Geophys. Res., 75,* 3941–3954, 1970.

Walcott, R. I., Lithospheric flexure, analysis of gravity anomalies and the propagation of seamount chains, in *The Geophysics of the Pacific Ocean Basin and Its Margin, Geophys. Monogr. 19,* edited by G. P. Woollard, pp. 431–438, AGU, Washington, D. C., 1976.

Watts, A. B., Gravity and bathymetry in the central Pacific Ocean, *J. Geophys. Res., 81,* 1533–1553, 1976.

Watts, A. B., Comment on 'Global gravity anomalies and two-scale mantle convection' by Bruce D. Marsh and James G. Marsh, *J. Geophys. Res., 83,* 3551–3554, 1978.

Watts, A. B., and J. R. Cochran, Gravity anomalies and flexure of the lithosphere along the Hawaiian-Emperor seamount chain, *Geophys. J. Roy. Astron. Soc., 38,* 119–141, 1974.

Watts, A. B., and M. Talwani, Gravity anomalies seaward of deep-sea trenches and their tectonic implications, *Geophys. J. Roy. Astron. Soc., 36,* 57–90, 1974.

Watts, A. B., and M. Talwani, Gravity effect of downgoing lithospheric slabs beneath island arcs, *Geol. Soc. Amer. Bull., 86,* 1–4, 1975.

Weertman, J., Creep laws for the mantle of the Earth, *Phil. Trans. Roy. Soc. London, Ser. A, 288,* 9–21, 1978.

Weertman, J., and J. R. Weertman, High temperature creep of rock and mantle viscosity, *Ann. Rev. Earth Planet. Sci., 3,* 293–315, 1975.

Wilhelms, D. E., Comparison of Martian and lunar geologic provinces, *J. Geophys. Res., 79,* 3933–3942, 1974.

Williams, J. C., Present scientific achievements from lunar laser ranging, in *Scientific Applications of Lunar Laser Ranging,* edited by J. D. Mulholland, pp. 37–50, D. Reidel, Hingham, Mass., 1977.

Wise, D. U., and M. T. Yates, Mascons as structural relief on a lunar 'moho,' *J. Geophys. Res., 75,* 261–268, 1970.

Wise, D. U., M. P. Golombeck, and G. E. McGill, Tharsis province of Mars: Geologic sequence, geometry and a deformation mechanism, *Icarus, 38,* 456–472, 1979.

Wong, L., G. Buechler, W. Downs, W. L Sjogren, P. M. Muller, and P. Gottlieb, A surface representation of the lunar gravity field, *J. Geophys. Res., 76,* 6220–6236, 1971.

Wood, J. A., Petrology of the lunar soil and geophysical implications, *J. Geophys. Res., 75,* 6497–6513, 1970.

Woollard, G. P. Regional variations in gravity, in *The Nature of the Solid Earth,* edited by E. C. Robertson, pp. 463–505, McGraw-Hill, New York, 1972.

Wu, S. S. C., Mars synthetic topographic mapping, *Icarus, 33,* 417–440, 1978.

Wunsch, C., Simple models of the deformation of an Earth with a fluid core, II, Dissipation and magnetohydrodynamic effects, *Geophys. J. Roy. Astron. Soc., 41,* 165–184, 1975.

Wyss, M., and P. Molnar, Source parameters of intermediate and deep focus earthquakes in the Tonga arc, *Phys. Earth Planet Interiors, 6,* 279–292, 1972.

(Received May 22, 1979;
accepted October 15, 1979.)

REVIEWS OF GEOPHYSICS AND SPACE PHYSICS, VOL. 21, NO. 1, PAGES 27-40, FEBRUARY 1983

A REVIEW OF GEODETIC AND GEODYNAMIC SATELLITE DOPPLER POSITIONING

J. Kouba

Gravity and Geodynamics Division, Earth Physics Branch, Department of Energy, Mines and Resources
Ottawa, Ontario K1A OY3

Abstract. Since 1967, when the U.S. Navy Navigation System was released to the general public, the accuracy of geodetic and geodynamic satellite Doppler positioning using satellite Doppler has improved steadily. This improvement is due mainly to refinements in hardware, refraction, and orbital modeling. Point positioning, using a minimum of 30 passes and the U.S. Defense Mapping Agency (DMA) precise ephemeris, has reached submeter accuracy. Relative positioning among several simultaneously observing stations separated by distances up to 250 km and using a minimum of 30 passes has a typical accuracy below 0.5 m (1σ). Current hardware design and errors in modeling of delays in the wet portion of the troposphere limit the achievable accuracy in relative positioning at the 0.1- to 0.5-m (1σ) level, unless errors in receiver phase and tropospheric refraction delays are continuously measured (calibrated). Also, a crystal oscillator of lower short-term stability ($\geq 5 \times 10^{-12}$/100 s) increases the Doppler data noise and causes a positive correlation persisting over several minutes. Geodetic and geodynamic Doppler data reductions usually involve meticulous modeling for instrumental, atmospheric, and orbital errors as well as careful Doppler data editing. A short-arc orbit computation employed in such satellite Doppler reductions requires a gravitational potential expanded to at least degree and order 10 if the orbit shape accuracy is to be comparable to DMA precise ephemeris. Satellite Doppler positioning has had a significant impact on geodetic datum definition and the establishment of national and continental networks of unprecedented accuracy. Satellite Doppler methods, apart from providing one of the most accurate continuous methods of polar motion monitoring, have also been found useful in other geodynamical and ice motion studies. This review concentrates mainly on system performance, models, and computational methods.

Contents

1. Introduction

Measurements of the Doppler shift on radio signals received from the first artificial earth satellites proved useful for determination of the satellite orbital parameters. It was soon realized that Doppler shift measurements could be used to determine positions if the satellite orbits were known [McClure, 1958]. This led to the development of the U.S. Navy Navigational Satellite System (NNSS) for all-weather worldwide marine navigation with the primary objective to support Polaris submarine operations. The system, also known as TRANSIT, was introduced for military applications in 1964 and released for nonmilitary use in 1967. The navigational capabilities of the system have been described and analyzed extensively as it has evolved [Guier and Weiffenbach, 1958, 1960; Kershner and Newton, 1962; Newton, 1967; Piscane et al., 1973; Applied Physics Laboratory Staff, 1976; Black et al., 1976; Stansell, 1978; Black, 1980; Piscane, 1981; Seeber and Egge, 1981]. For users on land the TRANSIT system has become a standard for global surveying. Methods have been developed to facilitate high-precision positioning in a three dimensional terrestrial reference frame which can satisfy global as well as regional geodetic and geodynamic requirements.

The treatment of satellite Doppler measurements and the computational approach needed to obtain useful geodetic and geophysical data differ in a number of ways from those used for navigation. Geodetic solutions require careful modeling of instrumental and environmental effects on Doppler measurements and involve multipass, often multistation, data sets assembled during postprocessing predominantly for earth-fixed stations. This meticulous approach has resulted in significant improvements of our knowledge of the size and shape of the earth and its gravitational field and has led to the establishment of a truly global, nearly geocentric, reference system.

This review will concentrate on models and computational methods employed for precise positioning by means of Doppler measurements on the TRANSIT system satellite signals.

2. Satellite Doppler Measurements

Geodetic Doppler instrumentation typically integrates cycles of a beat frequency ($f_o - f_r$) which results from mixing the local reference frequency f_o and the received satellite signal f_r. The basic measurements are integrated counts D between reception times τ_1 and τ_2:

$$D = \int_{\tau_1}^{\tau_2} (f_o - f_r)d\tau \qquad (1)$$

The integration interval $\tau_2 - \tau_1$ depends on instrumentation and usually is between 4.6 and 120 s. Often it is chosen between 20 and 30 s, which is optimal with respect to geometrical strength and data-handling economy.

When f_o is considered constant in the integration of (1), then

$$D = f_o(\tau_2 - \tau_1) - \int_{\tau_1}^{\tau_2} f_r d\tau$$

The remaining integral can be evaluated by arguing that the number of cycles of f_r received between τ_1 and τ_2 has to be equal to the number of cycles of the satellite frequency f_s emitted between the corresponding satellite time marks $t_1 = \tau_1 - S_1/c$ and $t_2 = \tau_2 - S_2/c$, where S_1 and S_2 are satellite-to-receiver distances and c is the vacuum velocity of light:

$$\int_{\tau_1}^{\tau_2} f_r d\tau = \int_{t_1}^{t_2} f_s dt$$

Here we are neglecting propagation, relativistic, and equipment effects, which will be addressed later.

When f_s is also considered constant, the expression for the integrated Doppler count then becomes

$$D = (f_o - f_s)(\tau_2 - \tau_1) + (S_2 - S_1) f_s/c \qquad (2a)$$

An equivalent expression but in terms of ground frequency cycles f_o/c and emission times is obtained from (2a) when substituting $\tau_i = t_i + S_i/c$:

$$D = (f_o - f_s)(t_2 - t_1) + (S_2 - S_1)f_o/c \qquad (2b)$$

The distances

$$S_i = |\underline{r}_s(t_i) - \underline{r}_g(\tau_i)|$$

are distances between a ground station with the radius vector $\underline{r}_g$ at τ_i and a satellite position $\underline{r}_s$ at t_i.

Instrumental Effects

The Doppler ground instrumentation typically consists of an omnidirectional antenna, dual-channel phase-locked receiver, data recorder, and precision clock/oscillator assembly. Each equipment segment contributes to Doppler measurement errors. The most significant instrumental errors are: (1) - antenna phase errors, (2) - receiver phase errors, (3) - receiver time delay variations , and (4) - clock and oscillator errors.

Antenna phase errors are mainly caused by ground reflections (multipath) and can cause significant systematic errors mapping mostly into height solutions. Such height errors can reach several decimeters even for modern Doppler equipment [Kouba, 1976]. The antenna errors can be eliminated either by using a directional antenna or by a proper design of an omnidirectional antenna with zero sensitivity below the horizon.

The receiver component, apart from white phase noise, introduces a phase error (also called Doppler delay) which can change during a pass. Such slow phase variations due to changes in signal strength and temperature can reach up to 180° of the Doppler signal (beat frequency) even for modern geodetic receivers and can significantly affect the solution for position parameters. Ideally, this error source should be eliminated by improved receiver design and/or real-time receiver calibrations.

The knowledge of total receiver time delay (in addition to the Doppler phase error) is required for synchronization of the local and satellite clocks. This delay is the time required for time mark propagation and detection within the receiver/decoder assembly. It can vary from receiver to receiver and is usually between 500 and 1200 μs. Ideally, it should be constant or measured continuously. However, variations in the receiver delay occur both during a pass and from one pass to another. Even if the variations during a pass are eliminated, as is the case with some new hardware [Icenbice and Loiler, 1979], the more significant variations between passes still remain (for example, as a function of ambient temperature, signal strength, etc.). These variations can cause systematic synchronization and position biases.

If we more precisely define t_i and τ_i as the epochs at which a particular wave front has zero phase at the satellite and at the detection circuitry of the receiver, respectively, then, taking into account the above instrument sources, they can be related as follows:

$$\tau_i = t_i + S_i/c + d_i + P_i + R_i + \Delta t_a + \delta_i \qquad (3)$$

where d_i denotes instantaneous receiver time delay, P_i represents the combination of receiver and antenna phase error, R_i denotes a small relativistic correction (see section 4), Δt_a is the atmospheric (tropospheric and ionospheric) delay (see next subsection), and δ_i represents local clock errors. Δt_a, d_i, and P_i will be similar for the same receiver types simultaneously observing points with small separations (<100 km); consequently, their effect on relative positioning will be reduced.

In (3) the clock errors δ_i are small and mostly due to clock resolution because other systematic clock errors (excluding clock drift) are negligible for a typical pass of less than 1000 s. On the other hand, instability of the reference frequency standard can contribute significantly to the integrated Doppler counts. The oscillator error contribution is directly related to the two sample Allan variance $\sigma_y^2(\Delta\tau)$ [Allan et al., 1974] conventionally used to characterize frequency stability in time domain, that is,

$$\sigma_D^2(\Delta\tau)[1 - \rho(\Delta\tau)] = \sigma_y^2(\Delta\tau)c^2\Delta\tau^2 \qquad (4)$$

where $\sigma_D^2(\Delta\tau)$ and $\rho(\Delta\tau)$ are the Doppler count variance and the correlation coefficient between adjacent Doppler counts, respectively, $\Delta\tau$ is the integration interval, and c is the speed of light. The Allan variance for up to 1000 s typically has the following time dependence:

$$\sigma_y^2(\Delta\tau) = \sigma_0^2 \Delta\tau^{-q}$$

where $\sigma_0 = \sigma_y$ (1 s) and the exponent q depends on the frequency standard type. Crystal oscillators (drift removed) typically show no time dependence for $\Delta\tau$ up to 1000 s (i.e., q = 0), cesium and rubidium ($\Delta\tau$ < 100 s) standards exhibit white frequency noise (i.e., q = 1) and hydrogen maser standards contribute only white phase noise (q = 2) for $\Delta\tau$ < 100 s. Cesium and rubidium error contributions (q = 1) are statistically independent from one Doppler count to another; consequently, the correlation $\rho(\Delta\tau)$ between adjacent Doppler counts is zero and according to (4) the Doppler variance is then

$$\sigma_D^2(\Delta\tau) = \sigma_0^2 c^2 \Delta\tau \qquad (5)$$

The evaluation of variances and correlation of crystal oscillators requires error modeling [e.g., Fell, 1980]. Such modeling reveals that crystal oscillator error contributions to Doppler counting have strong positive correlation persisting over several minutes [Kouba and Orosz, 1982].

Atmospheric Delays

Both ionosphere (the ionized portion of the atmosphere extending above 50 km) and troposphere (the neutral atmosphere below 50 km) introduce significant delays in radio wave propagations.

Ionospheric delays, the larger of the two, strongly depend on solar activity, having daily variations (minimum at night and maximum at midafternoon) and annual variations (minimum in winter and maximum in summer). The ionospheric refractive index is a function of the frequency f and can be approximated by a series expansion [Clynch and Altenburg, 1979]

$$n = 1 - A_1 N_e/f^2 - A_2 N_e^2/f^4 + \ldots \qquad (6)$$

where A_1, A_2 are constants and N_e is the electron density. The ionospheric refraction causes the measured ranges S^o to be shorter than the geometrical ranges S, that is,

$$S^o = S - B_1/f^2 - B_2/f^4 + \ldots \qquad (7)$$

where the coefficient $B_q = A_q \int N_e^q dS$, q = 1, 2, ... By measuring ranges S_H^o and S_L^o (or integrated Doppler measurements) on two coherent frequencies f_H and f_L one can eliminate from (7) the most significant term with B_1:

$$S = (f_H^2 S_H^o - f_L^2 S_L^o)/(f_H^2 - f_L^2)$$

It can be easily verified that the above equation is also valid when Doppler counts D_H^o and D_L^o ((2a) or (2b)) integrated on f_H and f_L are substituted for S_H^o and S_L^o.

In the case of NNSS Doppler observations, f_H = 400 MHz, f_L = 150 MHz, and the above equation becomes

$$D_{400} = (64\, D_{400}^o - 24\, D_{150}^o)/55 \qquad (8)$$

where D_{400}^o and D_{150}^o are observed integrated Doppler counts on 400 MHz and 150 MHz, respectively. This correction is either hardwired for some receiver types or applied in preprocessing. The remaining residual ionospheric refraction error can reach up to several percent of the first-order term or up to a few meters in position. Clynch and Altenburg [1979] developed a model for estimating residual ionospheric refraction using the total electron content (TEC) evaluated from a global model; alternatively, Kouba [1981] and Hara and Sato [1982] estimated TEC from the observed two frequency corrections. An improved residual ionospheric refraction model also taking into account the earth magnetic field term is presented by Clynch and Renfro [1982], who also present some initial test results. However, the residual ionospheric refraction modeling should also be used in externally generated orbits in order to prevent small biases for point positionings utilizing the residual ionospheric model and the external orbits.

Tropospheric refraction delays are smaller than those due to the ionosphere, and in the microwave frequency region are frequency independent. The effect of tropospheric refraction is to lengthen the observed ranges to the satellite. A correction for tropospheric refraction, ΔS (to be subtracted from the measured range S^o), is given by

$$\Delta S = \int_0^S (n - 1)\, dS$$

where n is the refraction index. The refractivity N, defined as $N = (n - 1) \times 10^6$, is usually expressed as a sum of dry N_d and wet N_w components which can be evaluated from standard formulae [e.g., E.K. Smith and Weintraub, 1953]:

$$N_d = 77.6\, P/T$$
$$N_w = 3.73 \cdot 10^5\, e/T^2$$

where P is the pressure (in millibars), e is the partial vapor pressure (in millibars) and T is temperature (in degrees Kelvin). There are several tropospheric refraction models based on surface meteorological data [e.g., Hopfield, 1969; Saastamoinen, 1973; Marini, 1972]. Of these, Hopfield's model is most often used for satellite Doppler positioning.

Using the nominal temperature gradient of 6.71°/km and the perfect gas law, Hopfield [1969] derived a convenient quadratic expression for the dry refractivity profile $N_d(h)$ as a function of the height h above the earth's surface:

$$N_d(h) = N_d(0)\ [(h_d - h)/h_d]^4 \qquad h\varepsilon<0;\ h_d>$$

where $N_d(0)$ is the surface refractivity and h_d is the effective height (in meters) of the troposphere above the station given by [Hopfield, 1972]

$$h_d = 40136 + 148.72\ (T - 273.16)$$

where T is temperature in degrees Kelvin.

An analogous expression, though without any theoretical basis, was used by Hopfield [1969] for the wet refractivity $N_w(h)$:

$$N_w(h) = N_w(0)\ [(h_w-h)/h_w]^4 \qquad h\varepsilon<0;\ h_w>$$

with h_w = 11,000 m.

The tropospheric refraction correction is then evaluated from measured surface temperature, pressure, and partial vapor pressure:

$$\Delta S = N_d/h_d^4 \int_0^{h_d} (h_d-h)^4 \frac{ds}{dh}\, dh + N_w/h_w^4 \int_0^{h_w} (h_w-h)^4 \frac{ds}{dh}\, dh$$

Though the above formulae can be integrated in a closed form, neglecting only path bending, serious numerical problems are encountered for high elevation angles. Yionoulis [1970] proposed two series expansions, one for low and one for high elevation angles, which practically eliminate all the numerical problems. Although the above evaluation is not complex, the large quantity of data coupled with other model approximations permits a further simplification. ΔS can be expressed in terms of the zenith corrections ΔZ_d and ΔZ_w:

$$\Delta S = (N_d h_d I_d + N_w h_w I_w)/5 \qquad (9)$$

where

$$I_d = \Delta S_d/\Delta Z_d \text{ and } I_w = \Delta S_w/\Delta Z_w$$

Hopfield [1972] suggested the following approximations:

$$I_d = 1/\sin\ (E^2 + 6.25^o)^{1/2}$$
$$I_w = 1/\sin\ (E^2 + 2.25^o)^{1/2}$$

where E is the satellite elevation angle. Such an approximation is within 1% of the corresponding rigorous evaluation for elevations above 5° [Wells, 1974]. Black [1978] developed a similar approximation which is within 0.1% of rigorous formulae for elevations above 3° [Kouba, 1979]. All the above approximations are well below the modeling errors. The dry portion of the refraction is quite accurate, with an rms of about 2% (or 4 cm in the zenith) and an rms of 0.1% (2 mm) once the tropospheric height h_d is adjusted for a particular position and time span [Hopfield, 1972]. However, when the assumption of the perfect gas law (i.e., no horizontal gradients) is violated near weather fronts, the model errors can reach 15%. The wet portion is much more difficult to model (partly because it is difficult to measure humidity accurately). Fortunately, the wet refraction contribution is only about 10% of the total correction. The current limit of wet refraction modeling is about 10-20% or about 3-5 cm in the zenith [see Murray and Marini, 1976]. Neglected path curvature only introduces significant errors (3%) at elevation angles below 3°. To mitigate the refraction modeling errors (as well as multipath), the Doppler data below 5° or 10° elevations are usually not used.

The wet portion of the tropospheric refraction represents a fundamental limitation of Doppler (and any radio based) positioning systems unless the wet refraction delays are directly measured with sufficiently high accuracy [Resch and Claflin, 1979].

Weighting of Doppler Data

The proper assignment of weights for Doppler observations is mandatory for minimum variance solutions as well as realistic error estimates [e.g., Hamilton, 1964].

In the past it was often argued that continuously integrated Doppler (CID) data should be statistically independent, that is have a diagonal weight matrix, which also implies the serial correlation (i.e., ρ = -0.5) for intermittently integrated Doppler data (IID) [e.g., Brown, 1976; Krakiwsky et al., 1972]. We note that for instrumentation with continuous counting, the corresponding CID Doppler counts are obtained by simply summing appropriate IID Doppler counts. However, phase white noise (with σ_p) due to receiver, oscillator, and timing noise, which would be consistent with the above weighting scheme, usually constitutes only a relatively minor portion of the total IID error budget, typically about 5-10 cm for 30 s of counting [Colquitt and Anderle, 1979; Gouldman, 1982]. Most of the error sources are due to instrumentation and atmospheric refraction as discussed above and exhibit white frequency noise which is independent from IID count to count [see Kouba, 1979].

Consequently, the IID data observed with precise oscillators (1 x 10^{-12}/100 s) can be approximated by a diagonal variance-covariance matrix:

$$\sigma^2_{ii} = \sigma^2_0\ \Delta\tau \qquad \rho_{ij} = 0 \qquad i \neq j \qquad (10)$$

where σ_0^2 is the variance of unit time and $\Delta\tau$ is the integration time. The corresponding CID variance-covariance matrix is then nonstationary with a variance-covariance matrix characteristic of the random walk process:

$$\sigma^2_{ii} = \sigma^2_0\ (\tau_i - \tau_0)$$

and

$$\rho_{ij} = \left(\frac{\tau_i - \tau_o}{\tau_j - \tau_o}\right)^{1/2} \qquad \tau_j > \tau_i > \tau_o \qquad (11)$$

It can be easily confirmed that identical results can be obtained with either CID or IID data treatment, provided that corresponding variance-covariance matrices ((10) or (11)) are taken into account. When a random walk process is subjected to a straight line fit, the residual becomes nearly stationary, and the strong positive correlation diminishes more rapidly than indicated by (11) [Fell, 1980], so that a CID stationary data treatment with an appropriate correlation model as advocated by Brown and Brown [1979] is then still possible.

TABLE 1. Crystal Oscillator Error Contribution Characteristics for Intermittently Integrated Dopper Counts and Crystal Oscillator With Two Sample Allan Variance sigma $\sigma_y = 5 \times 10^{-12}$

Time Interval $\Delta\tau$, s	$\sigma_D(\Delta\tau)$, cm	Correlation Coefficients			
		$\Delta\tau$	$2\Delta\tau$	$3\Delta\tau$	$4\Delta\tau$
5	1.8	0.80	0.69	0.62	0.57
30	9.0	0.70	0.53	0.47	0.43
60	16.6	0.66	0.51	0.43	0.36
120	30.2	0.63	0.44	0.31	0.22

The σ_D for other values of σ_y is obtained by multiplying the tabulated σ_D by a factor $2\sigma_y \times 10^{11}$. The correlation coefficients remain unchanged.

However, when crystal oscillators of lower short-term stability (5×10^{-12}) are used, their error contribution dominates the total error budget, and even IID data have a significant positive correlation persisting over several periods as seen from Table 1. The CID variances then show an error growth nearly proportional to $(\tau_i - \tau_o)^2$ and correlation exceeding those of (11) [Kouba and Orosz, 1982].

The satellite instrumentation introduces a positive correlation between corresponding Doppler counts at different stations observing the same satellite pass. This correlation can be removed either by differencing the corresponding Doppler counts (if it is possible), a method analogous to the very long baseline approach [e.g., Cannon, 1978], or by introducing appropriate correlation between stations [Kouba, 1981], which can also be used for instrumentation with different Doppler gating.

3. Satellite Orbit Determination

The knowledge of satellite positions in space and time is essential for satellite navigation and point positioning. However, if the satellite is used only as a common reference during simultaneous observations from known and unknown stations, this information need not be explicitly available with great accuracy. Either the satellite positions can be treated independently (in so-called geometrical mode), or physical laws governing satellite motion are invoked (dynamical mode) to constrain and strengthen the data reduction. The complexity of such dynamical models depends on the required precision and length of the satellite trajectory. In general, the dynamical models involve either numerical or analytical integrations of the equations of motions [Anderle, 1974a].

$$\ddot{\underline{r}} = \frac{\partial V_E}{\partial \underline{r}} + \frac{\partial V_{SM}}{\partial \underline{r}} + D_g + R_p + T_{SM} \tag{12}$$

where $\ddot{\underline{r}}$ are the second time derivatives of the satellite position vector in an inertial coordinate system, $\partial V_E/\partial \underline{r}$ and $\partial V_{SM}/\partial \underline{r}$ are gravitational and lunisolar body accelerations, D_g and R_p denote air drag and radiation pressure contributions, and T_{SM} are lunisolar tidal effects. For more details, see Anderle [1974a] Kaula [1966], Mueller [1964], and O'Toole [1976].

For computations of a portion of one revolution, the so-called short-arc, all but the first term in (12) can be neglected without any significant effect on orbit shape accuracy. The first term in (12) involves the spherical harmonic expansion of the gravitational potential V_E, and this expansion can be truncated at relatively low degree and order for short-arc orbit computations.

Broadcast Ephemeris

The broadcast ephemerides, which are available (broadcast) in real time in the satellite message, are computed by the U.S. Navy Astronautics Group from Doppler data observed at the four OPNET tracking stations located in California, Minnesota, Maine, and Hawaii. Broadcast ephemerides are fitted to a 30-hour Doppler data span every 12 hours (24 hours during periods of low solar activity), and resulting orbits are then extrapolated for up to 30 hours into the future and finally uploaded into the satellite memory every 12 hours. For uploading, the extrapolated ephemerides (up to 18 hours) are factored into a precessing and osculating Keplerian ellipse (the so-called fixed parameters) with corrections to this ellipse at even minute intervals (the so-called variable parameters). Because of a limited word length the variable parameters are rounded off to the nearest 10 m or 1×10^{-5} degree, which results in positioning errors due to rounding-off errors of up to 6 m.

Even though a comprehensive modeling (equation (12)) is used for broadcast ephemeris computations, including a recent WGS72 gravitational model [Seppelin, 1974] and extrapolated Bureau International de l'Heure (BIH) pole positions, the broadcast ephemeris errors can still approach 100 m [Jenkins and Leroy, 1979], mainly due to uncertainty in orbit predictions. The largest prediction errors are in the along-track direction and are caused by unpredictable changes in the air drag. Even for balanced passes in north-south directions these errors may not average out, particularly in the case of 24-hour computation schedules [Jenkins and Leroy, 1979]. Remaining error sources are due to predicted pole positions, changes in UT1-UTC during the prediction span, initial orbit parameters, etc., all of which can cause position- and time-correlated errors which will not generally average out below 1 or 2 m. The new (NOVA) generation of NNSS satellites are air drag compensated and thus have improved along-track orbit predictions, a stronger signal, and improved clocks [Hoskins, 1979; Eisner and Yionoulis, 1981].

Accuracy of the broadcast ephemeris was investigated by Wells [1974], who also proposed a smoothing of variable parameters to reduce the rounding-off errors with the following base functions: 1, t, sin(2mt), cos(2mt) , where t is time and m is the satellite mean motion. For more recent studies of errors in broadcast orbits, see Arur [1979], Jenkins and Leroy [1979], White et al. [1975], and Ziegler [1979].

Postfitted Precise Ephemeris

Apart from the broadcast ephemeris which is available in real time from the transmitted signal from all operational NNSS satellites, the U.S. Defense Mapping Agency (DMA) regularly computes an ephemeris for at least one NNSS satellite. This so-called precise ephemeris is computed from 48 hour data spans observed by the TRANET tracking network, using different software [O'Toole, 1976] than that used for the broadcast ephemeris and with some differences in the gravity field and program constants [Anderle, 1974a; Jenkins and Leroy, 1979]. Since data are from the worldwide TRANET network [Williams, 1981] and no extrapolations are involved, the precise ephemeris is consistent at the 2-m level [Ziegler, 1979]. The precise ephemeris in the form of earth fixed Cartesian coordinates x, y, z and their time derivatives $\dot{x}$, $\dot{y}$, $\dot{z}$ at 1-min intervals is usually made available by DMA upon request from (qualified) scientific research organizations. As a by-product of the precise ephemeris generation an accurate pole position is computed for every 48-hour data span. This DMA Doppler pole monitoring has proved to be one of the most accurate continuous polar motion determinations to date [Robertson et al., 1979; King, 1981; Guinot, 1980].

Precise ephemerides also imply the coordinate system position, orientation, and scale mainly established by the set of coordinates adopted for the TRANET stations. Precise ephemeris coordinate system NWSC9Z-2 and its predecessor NWL9D were extensively compared to other independent satellite and astronomical systems [e.g., Anderle, 1975, 1980; Hothem, 1979; Kouba, 1978; Grappo, 1979; Lachapelle and Kouba, 1981]. It was shown that the NWL9D (and NWSC9Z-2) requires corrections of about -0.4×10^{-6} in scale, +0.8" in longitude east, and about +4 m in the Z coordinate in order to be consistent with international conventions for scale (the speed of light), pole, and longitude orientations. All but the last correction have been explained [Anderle, 1980]. The 4-m bias in the Z coordinate is still being investigated, and indications are that it may have been caused by a combination of the old value for GM (i.e., the product of the gravitational constant G and the earth mass M) of 398,601 km^3/s^2 instead of a more recent value of 398,600.5 km^3/s^2 and limited TRANET station distribution [Jenkins and Leroy, 1979; Anderle, 1980].

Since 1975 the coordinate system implied by the broadcast ephemeris is compatible with NWSC9Z-2 (or NWL9D) except for small scale correction of $+0.2 \times 10^{-6}$ and a -2 m Z coordinate shift [see Jenkins and Leroy, 1979].

Short-Arc Ephemeris

Short-arc orbits are usually computed within the Doppler reduction software using a rather heavily truncated expansion of the earth's gravitational potential [Brown and Trotter, 1969]. The short-arc computations only make sense when no high resolution (precise) orbits are available, since they represents a considerable additional expense, up to 10 times more than reductions using external (e.g., precise or broadcast), orbits [Nesbo, 1976; Hauck et al., 1979]. However, a short-arc computation which utilizes a least squares polynomial approximation subject to the second time derivative constraints derived from a gravitational model according to (12) does not significantly increase the computation time [Kouba, 1982]. Figure 1 indicates that for NNSS satellites and 5 cm precision in orbit shape, the earth's gravitational field should not be truncated below degree and order 10. Since most recent gravitational models agree quite well at least up to degree and order 10 [Lachapelle, 1975; Khan, 1976; Reigber, 1981; Pospisilova et al., 1982], it should not matter which model is used.

Three different broadcast orbit representations namely, short-arc, smoothed broadcast, and an orbit computed directly from the message are compared to the precise ephemeris in Figure 2 in order to examine orbit shape accuracies of each type of orbit representation. The short-arc computation [Kouba, 1982] used smoothed orbit positions and velocities at -10, 0, and +10 min with respect to the closest approach as initial conditions. The short-arc orbits (generated with a

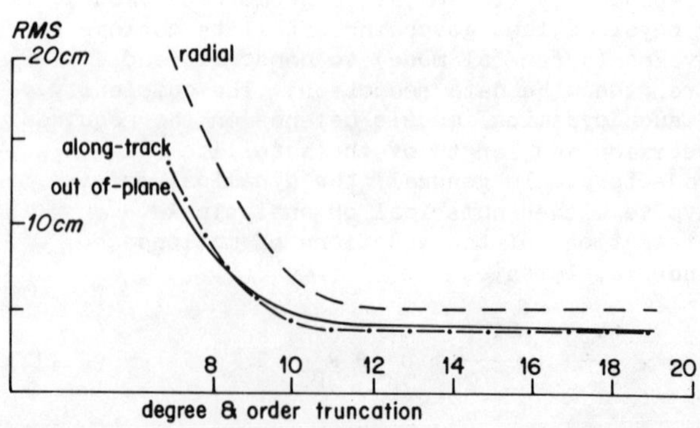

Fig. 1. Effect of earth gravitational model truncation on short-arc orbit computations. GEM9 short-arc and DMA precise ephemeris differences from [Kouba, 1982].

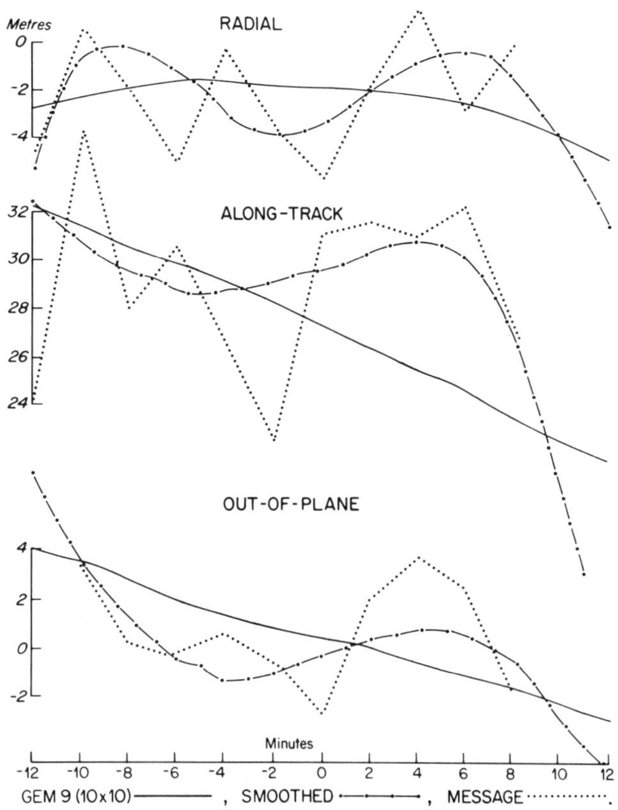

Fig. 2. Comparisons of three types of broadcast orbit representation with DMA precise ephemeris [from Kouba, 1982].

gravitational model truncated at degree and order 10) approximate the precise ephemeris well, except for position biases and small slope, both of which can, to a large extent, be accounted for by changes in the initial orbital elements. The orbit shape errors, in particular the long-wavelength features, can for a small number of passes correlate with some position parameters.

4. Satellite Doppler Positioning

Precise Doppler positioning for a stationary observer (called point positioning) is based on many satellite passes and the assumption that orbital, propagation, and instrumental errors will average out over many passes. This may not be the case for broadcast ephemeris or unfavorable atmospheric and solar conditions if repeatability levels below 1 or 2 m are required. Most of the orbital error and part of the propagation error can be eliminated from relative positioning for short (<250 km) separations when data from several, simultaneously observing stations are utilized. In the simplest case, often called translocation, the simultaneous point positioning solutions are subtracted without any regard to commonality of passes or Doppler counts. Some translocation positioning may also involve trimming to simultaneous Doppler counts and/or passes [Wells et al., 1979; Wells, 1976], which may be beneficial for short distances but may have some harmful effects over longer distances by distorting data balance (east-west and north-south). Logical extensions of translocation are short-arc and semi-short-arc positioning which utilize either locally computed (short-arc) or externally generated orbits (semi-short-arc) in a rigorous least squares adjustment involving data from several stations and many satellite passes.

Doppler Reduction Characteristics

The measured IID counts, corrected for atmospheric propagation delays (section 2), are related to the unknown position (or orbit) parameters through the following observation equations (see (2a),(2b), and (3)):

$$D_i+\varepsilon_i = (S_k - S_j)f_o/c + (t_k - t_j)(f_o - f_s) + d.(\dot{S}_k - \dot{S}_j)f_o/c + R_i + \Delta_i + s_i \qquad (13)$$

where S_k, S_j are distances from a ground station with reference frequency f_o and radius vector $\underline{r_g}$ to satellites at $\underline{r_s}$ transmitting frequency f_s; s_i represents instrumental and atmospheric biases which can be modeled either as random biases (uncorrelated between passes) or by time processes reflecting correlation between passes and stations [e.g., Gelb, 1975; Kouba, 1981]; ε_i denotes observational errors, c is the speed of light; d is the mean pass receiver delay; and $\dot{S}_k$, $\dot{S}_j$ are time derivatives of S_k, S_j.

The relativistic correction R_i is given by Boucher [1978] as

$$R_i = -[GM(1/r_g - 1/r_s) - (|\dot{\underline{r}}_s|^2 - |\dot{\underline{r}}_g|^2)/2] .(t_k - t_i).f_o/c^2$$

and the earth's rotation correction [e.g., Kouba and Boal, 1976a] is

$$\Delta_i = (\underline{\omega} \times \underline{r_g}).[\underline{r_s}(t_k) - \underline{r_s}(t_j)]f_o/c^2$$

where ω is the vector of earth rotation. The CID data treatment is analogous to range observations subject only to an initial range unknown S_o and frequency offsets for each station and satellite pass:

$$D_i + \varepsilon_i = (S_i - S_o)f_o/c + (t_i - t_o)(f_o - f_s) + d.(\dot{S}_i - \dot{S}_o)f_o/c + R_i + \Delta_i + s_i \qquad (14)$$

The simplest reduction models include at least three position unknowns and for each pass a frequency offset unknown. Geodetic and geodynamic reductions typically utilize many passes (usually a minimum of 30) from a number of simultaneously observing stations, systematic error modeling which may include several bias parameters for each pass (e.g. six orbital unknowns) and for each pass and station (e.g., refraction scaling, time synchronization, frequency, and frequency drift) [e.g., Kouba and Wells, 1976; Brown, 1976; Kouba, 1981]. The solution for position and bias parameters is obtained by a standard least squares analysis [e.g., Hamilton, 1964].

The set of observation equations (13) or (14) may be expressed in a vector form

$$\underline{L} + \underline{E} = \underline{F}(\underline{X})$$

where $\underline{L}$ and $\underline{E}$ denote the vectors of observation and measurement errors, respectively, $\underline{F}(\underline{X})$ is the functional relationship of the right-hand side of the observation equations ((13) and (14)), and the vector $\underline{X}$ consists of unknown position and bias parameters. The above observation equations are linearized around the approximate values $\underline{X}^o$ of parameters $\underline{X}$ as follows:

$$A \quad + \underline{W} = + \underline{V}$$

where the design matrix $A = \partial \underline{F}/\partial \underline{X}$ and $\underline{W}$ is the misclosure vector (i.e., $\underline{W} = \underline{F}(\underline{X}^o) - \underline{L}$). The vector of residuals $\underline{V}$ and corrections $\underline{\delta}$ are obtained from minimizing the quadratic sum:

$$\underline{V}^T P \underline{V} + \underline{\delta}^T P_X \underline{\delta} \qquad (15)$$

The weight matrices P and P_X correspond to observations $\underline{L}$ and parameters $\underline{X}$ and are proportional to the inverses of their respective variance-covariance matrices Σ_L^{-1} and Σ_X^{-1}. The solution for $\underline{\delta}$ is then [e.g,. Mikhail, 1976]

$$\underline{\delta} = -(A^T PA + P_X)^{-1} A^T P \underline{W}$$

and finally the adjusted parameters are $\underline{X} = \underline{X}^o + \underline{\delta}$. Such solutions usually require several iterations, and each iteration can be done in a batch mode involving all stations and passes [e.g., R.W. Smith et al., 1976] or partitioned in a sequential manner, usually pass by pass, to economize computations and statistical data editing [e.g., Kouba and Boal, 1976a]. Statistical data editing may include normal distribution tests on residuals $\underline{V}$ and the quadratic sum equation (15), usually at or below the 1% significance level. Since gross errors arising from decoding, recording, propagation delays, etc., can hardly be expected to conform to the normal distribution, the statistical data rejections are usually higher than the expected significance level. Usually, the rejections are between 1% and 10%.

For the optimal least squares solution, proper assignments of weight matrices P_X and P are important. The assignment of the matrix P_X is not so crucial as the assignment of P, since the position and frequency unknown parameters do not require any a priori weighting and the remaining bias parameters tend to become deterministic for a number of simultaneously observing stations. In such cases the solution is relatively insensitive to different assignments of P_X.

Proper assignment of the weight matrix P, on the other hand, is mandatory for minimum variance solution [Hamilton, 1964] as well as for realistic error estimates. As discussed in the previous section, the matrix P tends to be diagonal and stationary for IID measurements with precise oscillators when all error sources are considered. On the other hand, the CID treatment may require a fully populated weight matrix P [Kouba and Orosz, 1982; Brown and Brown, 1979].

Software Standardization

Data editing and various error and orbit modeling can potentially influence the final results [e.g., White, 1976; Hothem, 1975]. It is important that different Doppler software give the same results under similar conditions, particularly for applications of a global character [Hothem, 1979; Langley et al., 1979; Kouba, 1978]. Kouba and Hothem [1978] studied compatibility of the U.S. and Canadian Doppler software used by the two national geodetic surveys [R.W. Smith et al., 1976; Kouba and Boal, 1976a] and concluded that both programs agreed at the 20 cm level when comparable editing and other options were used. Subsequent studies identified additional minor differences, and when these were eliminated, the results of point positioning agreed within 10 cm. For this purpose the Special Study Group on Doppler Positioning of the International Association of Geodesy assembled three test data sets which can be used to study Doppler software compatibility and verification [Strange, 1980].

There is no firm agreement on what the best reduction options are, although the practical consensus which has emerged over the years includes (1) - solution for a frequency offset unknown for each pass and station, (2) - solution for the receiver delay, (3) - 3 σ statistical limit for Doppler residual testing, (4) - Hopfield tropospheric refraction model with surface meteorological data input, usually subject to a scaling bias solution with a priori σ of 5-10%, (5) - Doppler data deleted below 5°-10° elevation angles, and (6)- short-arc or semi-short-arc reduction methods for relative positioning of several, simultaneously observing stations.

5. Satellite Doppler Applications

Brown [1970] was one of the first to recognize the potential of Doppler positioning for submeter accuracy, advocating the short-arc method, the reduction method involving short-arc orbit computations and data from several simultaneously observing stations. Anderle [1972] and subsequently the Department of Defense [1972] demonstrated convincingly, using precise ephemeris and independent point positioning, that a 1-m accuracy level was achievable with 50-100 observed passes. From these two approaches, several applications and refinements evolved over the years. In Canada, three organizations (the Bedford Institute of Oceanography, Shell Canada, and the University of New Brunswick) pooled their resources to conduct initial tests in Doppler point positioning and translocation [Hittel et al., 1971; Hittel and Kouba, 1971; Wells et al., 1976].

The first geodetic applications provided a badly needed zero-order geodetic control in Canada at a spacing of 200-500 km (see Figure 3) and supplemented the transcontinental geodimeter traverse in the United States [Bossler, 1978], for which the U.S. National Geodetic Survey of NOAA adhered to precise point positioning using the DOPPLR software [R.W. Smith et al., 1976]. The Geodetic Survey of Canada attempted to use precise ephemeris satellites for absolute positioning as well as data from satellites for which only the broadcast ephemeris was available, in order to strengthen relative

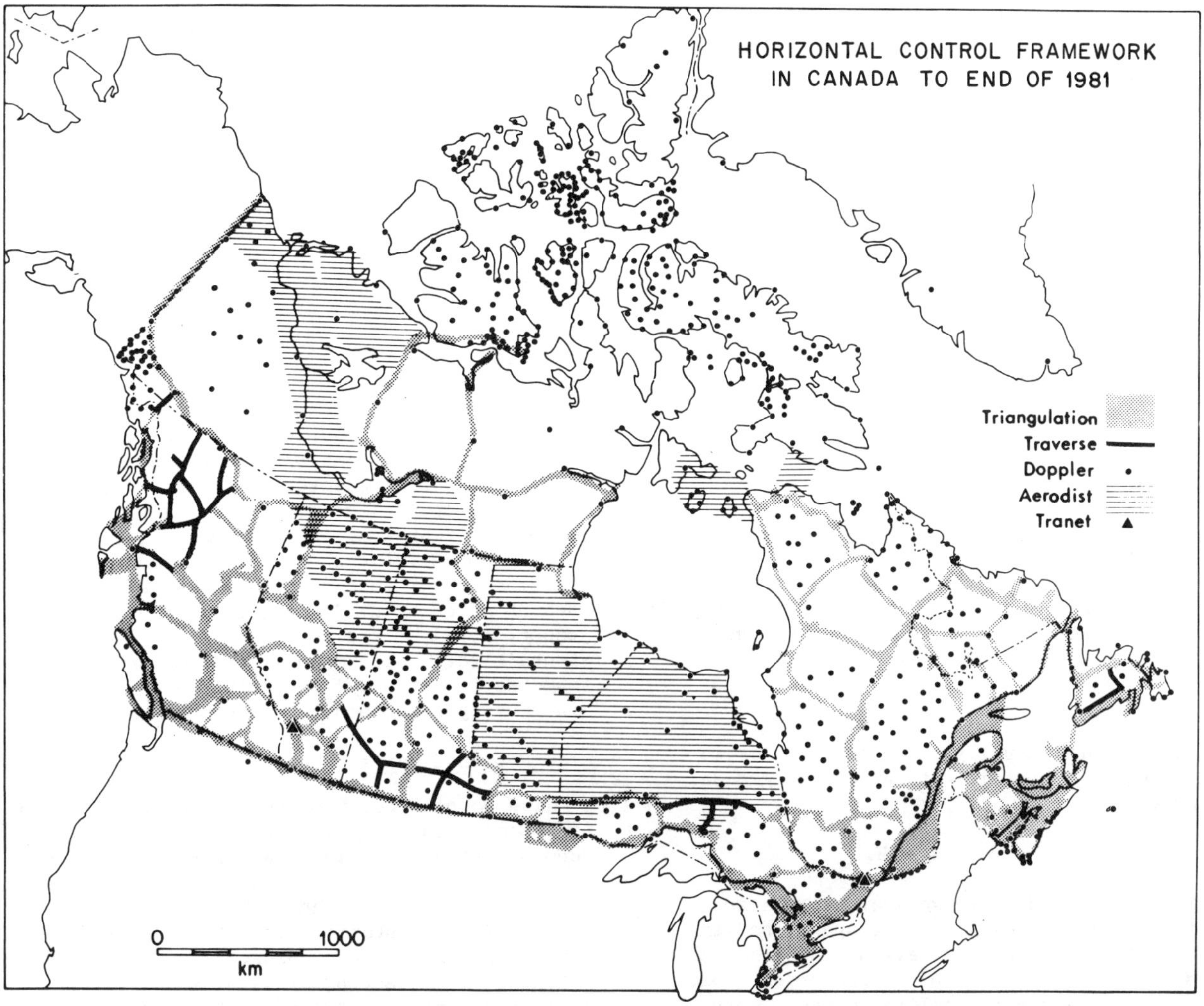

Fig. 3. Horizontal control framework in Canada to the end of 1981, including zero order Doppler network at a spacing of 200-500 km and Doppler densification network at 50- to 100-km spacing.

positioning by using semi-short-arc software [Kouba and Wells, 1976; Kouba and Boal, 1976a]. Similar work was carried out in Europe and Africa [Boucher et al., 1979; Seeger et al., 1979; Rinner and Pesec, 1982; Walker, 1979].

Geodetic Datum

Doppler point positioning utilizing the precise ephemeris (precise point positioning) implies a position and orientation of a geodetic network to much higher accuracy than was possible using classical methods that is, by astronomical and gravity observations. Anderle [1974b] recognized this important aspect of precise Doppler positioning for a geodetic datum realization, that is, positioning and orientation of a reference ellipsoid. Subsequently, precise Doppler positions, scaled and oriented to international conventions, were proposed for a datum realization [Kouba, 1978, 1980; Ashkenazi, 1981]. This new concept of datum realization was tested in several trial adjustments of Canadian networks [Beattie et al., 1978] and is likely to be adopted for the planned readjustment of the North American Datum scheduled for 1983-1984.

Doppler Geodetic Control Densification

With improved Doppler hardware [Blaha, 1979; Icenbice and Loiler, 1979; Stansell, 1979; Humble, 1981] and precise, short-arc and semi-short-arc reduction methods, relative positioning with accuracies below 0.5 m become possible with relatively small data samples of 30-50 passes and 2- to 3-day observing periods (see Figure 4) [Brown, 1976; Kouba and Boal, 1976b; Seeger et al., 1979; Hatch et al., 1979; Usandivaras et al., 1976]. At that accuracy level, Doppler positioning can also be used for densification of geodetic networks at spacings of less than 200 km. In Canada, densification at 50- to 100-km spacing was found to be useful in providing the necessary control for detail surveys by inertial surveying methods (Figure 3) [Babbage, 1977; Carrière et al., 1977].

Doppler Leveling

Doppler positioning is three dimensional, so it also provides a direct determination of the geometrical height h above the reference ellipsoid. When the orthometric heights H (i.e., above the mean sea level) are also known

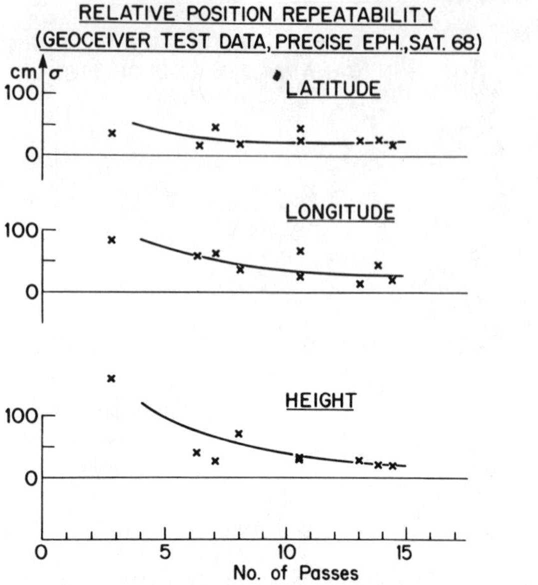

Fig. 4. Typical relative position repeatability for short separations (<250 km) based on short-arc and semi-short-arc methods [after Kouba, 1979].

at some points, usually obtained by geodetic leveling, it is possible to determine geoidal heights:

$$N_o = h - H.$$

Such Doppler-derived geoidal heights are useful for controlling error propagations in continental geoid determinations and also can be used for determinations of a zero order geoid undulation, mean earth ellipsoid and Doppler coordinate shift [e.g., Lachapelle and Kouba, 1981; Grappo, 1979]. Comparisons of Doppler derived geoidal heights with the most accurate geoids are in agreement to well below the 0.5-m level [Seeger et al., 1979]. Conversely, when the geoid is known to a sufficient accuracy, the Doppler ellipsoid heights can be used for orthometric height determinations (H). This method is particularly viable for remote and inaccessible regions.

Geodynamic Applications

As a by-product of the precise ephemeris, since 1972 DMA has solved continuously for the pole positions for every 48 hour span. The Doppler pole positions were forced to coincide (by appropriate orientation of the TRANET station network) with BIH pole positions at 1966.7 [Anderle, 1975]. DMA 2-day pole positions are estimated to be consistent to about 0.5 m and are regularly being included in the final BIH pole coordinates [Oesterwinter, 1979]. The high accuracy of Doppler pole positions was confirmed by a number of independent studies [e.g., King, 1981; Robertson et al., 1979; Guinot, 1980].

Although in principle, given errorless orbits, the solution for variations in the rotation of the earth (UT1 - UTC) is also possible, for polar satellites the UT1-UTC (DUT1) time difference couples almost exactly with the correction $d\Omega$ to the right ascension of the ascending node Ω. In the case of the precise ephemeris the departures from the extrapolated DUT1 are absorbed by $d\Omega$ for every 48-hour span. Anderle and Oesterwinter [1980] analyzed $d\Omega$ corrections in view of the possibility of DUT1 determination with some degree of success, in particular for the high-frequency variations, as low frequencies are corrupted by orbital errors. Research into Doppler DUT1 determination continues [Williams, 1981].

Another potentially useful by-product of precise ephemeris generation may be crustal movement determinations obtained from analyses of long series of Doppler positions of TRANET tracking stations [Malyevac and Anderle, 1979]. Anderle [1968] predicted the possibility of detecting crustal movements at the 5-cm/yr level from a 20-year-long, continuous series of Doppler observations. With some success the feasibility of such studies was already demonstrated by R. J. Anderle (unpublished paper, 1978), Malyevac and Anderle [1979] and Kouba [1981]. The main limitation of such studies is again due to orbital errors, that is, in gravity field and air drag modeling.

Ice Motion Studies

The Doppler positioning is also well suited for continuous ice dynamics monitoring as demonstrated by Wells et al. [1979] and Popelar and Kouba [1982]. Figure 5, from Popelar and Kouba [1982], shows the high degree of resolution for station separations of the three camps in the Lomonosov Ridge Polar Expedition (LOREX) in the spring of 1979 [Weber, 1979]. Here the station motions, up to 20 km/d, between and during passes were modeled as the first-order Gauss Markov processes [see Kouba, 1981].

6. Summary

The original satellite Doppler positioning design accuracy of 100 m (real-time navigation) has improved to the 1-m level for a stationary observer using postfitted precise ephemeris and observation over a period of several days with 30 or more satellite passes. This improvement results from improvements in gravity modeling, hardware, and software. Accuracy in relative positioning below the 0.5-m (1σ) level is now possible with 1- or 2-day observations and at least 30 passes, using either externally generated (e.g., broadcast or precise) ephemeris or short-arc orbits and several simultaneously observing stations.

Potential accuracy of Doppler positioning is estimated to be between 5 and 20 cm (1σ level in relative positioning). The major limitation is the water vapor influence on the radio wave propagation and, for larger separations, residual ionospheric refraction. Doppler hardware design currently limits the achievable accuracies to between 10 and 50 cm. Oscillators with improved stability, improved hardware design, and hardware calibrations would mitigate variations in receiver delay and antenna phase

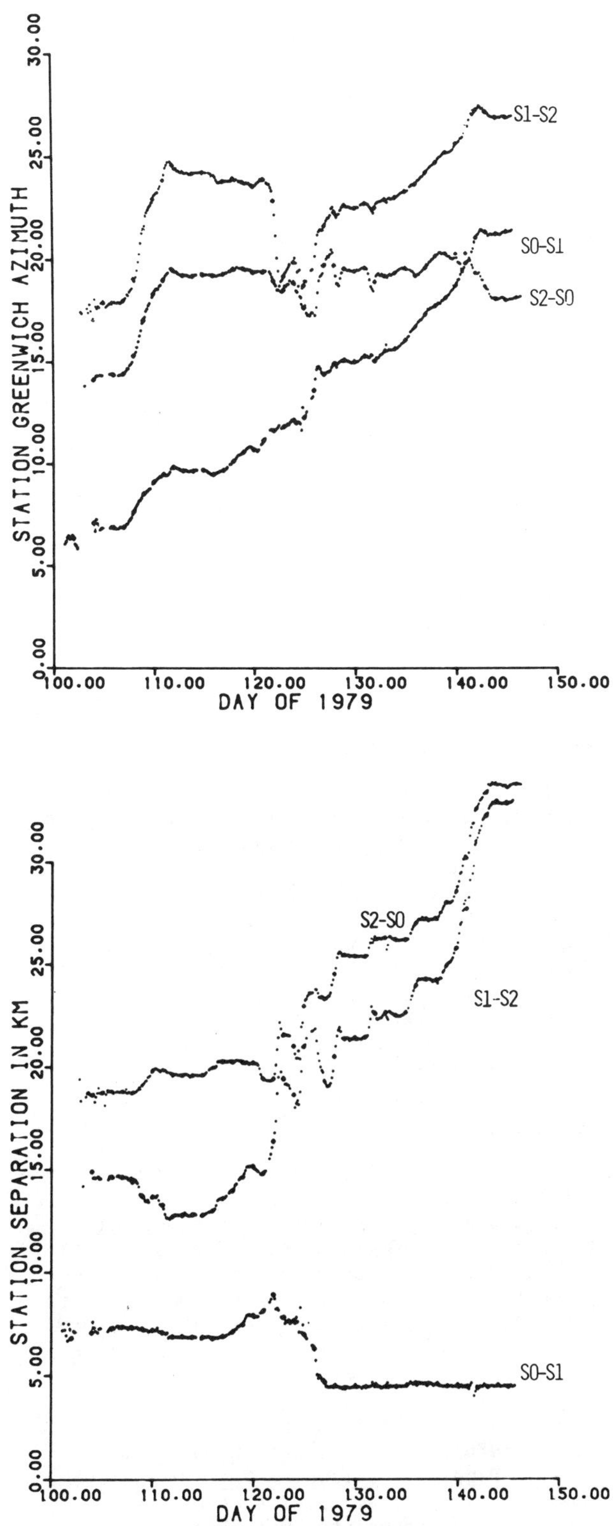

Fig. 5. Relative changes of the LOREX triangle side lengths and orientations (in degrees) [from Popelar and Kouba, 1982], SO is Main Camp; S1, Snowsnake; and S2, Iceman. (Reprinted by permission of Crane, Russak & Co.)

center and should permit Doppler relative positioning at the 5 cm (1σ) level. This would lead to additional geophysical applications, in particular monitoring of local crustal movements.

Acknowledgments. D. E. Wells of the University of New Brunswick and his students Ken Alexander, Harold Janes, and Furaha Lugoe reviewed the manuscript and supplied the author with many useful suggestions and comments. Contribution 1027 of the Earth Physics Branch.

References

Allan, D. W., J. H. Shoaf, and D. Haford, Statistics of time and frequency data analysis, in Time and Frequency: Theory and Fundamentals, NBS Monogr. 140, edited by B. E. Blair, 151-204, National Bureau of Standards, Washington, D.C., 1974.

Anderle, R. J., Accuracy of determination of continental drift from Doppler observations, paper presented at Annual Spring Meeting, AGU, Washington, D. C., April 18-11, 1968.

Anderle, R. J., Accuracy of Doppler determinations of station positions, in Rotation of the Earth, edited by P. Melchior and S. Yumi, 101-103, International Astronomical Union, Marioka, Japan, 1972.

Anderle, R. J., Transformation of terrestrial survey data to Doppler satellite datum, J. Geophys. Res., 79 (35), 5319-5331, 1974a.

Anderle, R. J., Role of artificial earth satellites in redefinition of the North American Datum, Can. Surv., 28(5), 590-597, 1974b.

Anderle, R.J., Long term consistency in positions of sites determined from Doppler satellite observations, Rep. NSWC/DL TR-3433, Nav. Surface Weapons Center, Dahlgren, Vir., April 1975.

Anderle, R. J., Accuracy of mean earth ellipsoid based on Doppler, laser and altimeter observations, Bull. Geod., 54(4), 521-527, 1980.

Anderle, R. J., and C. Oesterwinter, Determination of high frequency variation in earth rotation from Doppler satellite observations, Bull. Geod., 54(4), 544-552, 1980.

Applied Physics Laboratory Staff, Planned improvements in the TRANSIT system, 1975, Navigation Washington, 22, 352-360, 1976.

Arur, M. G., Experiments for improved positioning by means of integrated Doppler satellite observations and NNSS broadcast ephemeris, Proc. Int. Geod. Symp. Satell. Doppler Positioning 2nd, 83-112, 1979.

Ashkenazi, V., Models for controlling national and continental networks, Bull. Geod., 55(1) 49-58, 1981.

Babbage, G., Operation of the inertial survey system: The manager's view, Proc. Int. Symp. Inertial Technol. Surv. Geod. 1st, 312-321, 1977.

Beattie, D. B., J. A. R. Blais, and M. C. Pinch, Test adjustments of Canadian primary horizontal network, Proc. Int. Symp. Probl. Related Redefinition N. Am. Geod. Networks 2nd, 41-54, 1978.

Black, H. D., An easily implemented algorithm for tropospheric range correction, J. Geophys. Res., 83(B4), 1825-1828, 1978.

Black, H. D., The TRANSIT system 1977: Performance, plans and potential, Philos.

Trans. R. Soc. London, Ser. A, 294, 217-236, 1980.
Black, H. D., R. E. Jenkins, and L. L. Pryor, The TRANSIT system, 1975, Tech. Memo. TG1305, Appl. Phys. Lab., Johns Hopkins Univ., Laurel, Md., Dec. 1976.
Blaha, F., Features and capabilities of a new generation CMA-751 satellite Doppler field survey system, Proc. Int. Geod. Symp. Satell. Doppler Positioning 2nd, 535-556, 1979.
Bossler, J. D., Status of the new adjustment in the United States, Proc. Int. Symp. Probl. Related Redefinition N. Am. Geod. Networks 2nd, 31-403, 1978.
Boucher, C., Relativistic correction to satellite Doppler observation, paper presented at Symposium on the Mathematical Structure of Gravity Field, Int. Assoc. of Geod., Lagonissi, Greece, 1978.
Boucher, C., P. Pâquet, and P. Wilson, The second European Doppler observation campaign (EDOC-2): Results and conclusions obtained by EDOC-2 computing centres, Proc. Int. Geod. Symp. Satell. Doppler Positioning 2nd, 819-850, 1979.
Brown, D. C., Near term prospects for positional accuracies of 0.1 to 1.0 m from satellite geodesy, Rep. AFCRL-70-0501, U.S. Air Force Cambridge Res. Lab., Bedford, Mass., 1970.
Brown, D. C., Doppler positioning by the short-arc method, Proc. Int. Geod. Symp. Satell. Doppler Positioning 1st, 97-140, 1976.
Brown, D. C., and J. D. Brown, Short-arc adjustment of 36 Doppler stations, Tech. Rep. 79-015A, Indiaatlantic, Florida, Aug. 1979.
Brown, D. C., and J. E. Trotter, SAGA, a computer program for short-arc geodetic adjustment of satellite observations, Rep. AFCL-69-0080, U.S. Air Force Cambridge Res. Lab., Bedford, Mass., 1969.
Cannon, W. H., The classical analysis of the response of long baseline radio interferometer, Geophys. J. R. Astron. Soc., 53, 503-530, 1978.
Carrière, R. J., J. Kouba and R. C. Penney, Experience with inertial survey system at the Geodetic Survey of Canada, Proc. Int. Symp. Inertial Technol. Surv. Geod. 1st, 146-197, 1977.
Clynch, J. R., and R. A. Altenburg, Ionospheric residual range error model, Proc. Int. Geod. Symp. Satell. Doppler Positioning 2nd, 113-134, 1979.
Clynch, J. R., and B. A. Renfro, Evaluation of ionospheric residual range error model, Proc. Int. Geod. Symp. Satell. Doppler Positioning 3rd, in press, 1982.
Colquitt, E. S., and R. J. Anderle, Effect of oscillator performance on Doppler Geodesy, Proc. Int. Geod. Symp. Satell. Doppler Positioning 2nd, 409-422, 1979.
Department of Defense, Report of the DOD GEOCEIVER test program, DMA Rep. 0001, Def. Mapp. Agency, Washington, D. C., 1972.
Eisner, A., and S. M. Yionoulis, Neutral density variations in 900-1200 km region of upper atmosphere, Ann. Geophys., 37(1), 235-240, 1981.
Fell, P. J., Geodetic positioning using a global positioning system of satellites, Dep. of Geod. Sci. Rep. 299, Ohio State Univ., Columbus, June 1980.
Gelb, A., (Ed.), Applied Optimal Estimation, MIT Press, Cambridge, Mass., 1975.
Gouldman, M. W. S, Oscillator studies, Proc. Int. Geod. Symp. Satell. Doppler Positioning 3rd, in press, 1982.
Grappo, G. A, Determination of the earth's mean equatorial radius and center of gravity from Doppler derived and gravimetric geoid heights, paper presented at 17th General Assembly, Int. Union of Geod. and Gephys., Canberra, Australia, 1979.
Guier, W. H., and G.C. Weiffenbach, Theoretical analysis of Doppler radio signals from earth satellites, Rep. JHU/APL BB-176, Appl. Phys. Lab., Johns Hopkins Univ., Laurel, Md., 1958.
Guier, W. H., and G.C. Weiffenbach, A satellite Doppler navigation system, Proc. Inst. Radio Electron. Eng. Aust., 48, 407-516, 1960.
Guinot, B., Determination of the motion of the pole, and comparison with astrometry, Philos. Trans. R. Soc. London, Ser. A, 294, 119-124, 1980.
Hamilton, W. C., Statistics in Physical Science, Ronald, New York, 1964.
Hara, T., and K. Sato, Electron content distribution and variation around Mizusawa, Japan, Proc. Int. Geod. Symp. Satell. Doppler Positioning 3rd, in press, 1982.
Hatch, R., S. Chamberlain, and J. Moore, MX 1502 Doppler survey software, Proc. Int. Geod. Symp. Satell. Doppler Positioning 2nd, 187-208, 1979.
Hauck, H., K. Herzberger, D. Lelgemann, and W. Schluter, Comparisons of single-point, semi-short arc and short-arc solutions using Doppler data, paper presented at 17th General Assembly, Int. Union of Geod. and Geophys., Canberra, Australia, 1979.
Hittel, A., and J. Kouba, Accuracy tests of doppler satellite positioning in orbital, translocation and simultaneous modes, Proc. Int. Congr. Surv. 17th, 503.1-503.13, 1971.
Hittel, A., J. Kouba, E. J. Krakiwsky, D. E. Wells, and R.M. Eaton, Report on a doppler satellite positioning in Canada, paper presented at the 64th Annual Meeting, Can. Inst. of Surv., Ottawa, Ont., Feb. 3-5, 1971.
Hopfield, H. S., Two-quadratic tropospheric refractivity profile for correcting Satellite Data, J. Geophys. Res., 74 4487-4499, 1969.
Hopfield, H. S, Tropospheric refraction effects on satellite range Measurements, APL Tech. Dig., 11(4), 11-19, 1972.
Hoskins, G. W, The future Navy Navigation Satellite System, Proc. Int. Geod. Symp. Satell. Doppler Positioning 2nd, 21-38, 1979.
Hothem, L. D., Evaluation of precision and error sources associated with Doppler Positioning, paper presented at 16th General Assembly, Int. Union of Geod. and Geophys., Grenoble, France, Aug. 1975.
Hothem, L. D., Determination of accuracy, orientation and scale of satellite Doppler point-positioning coordinates, Proc. Int. Geod. Symp. on Satell. Doppler Positioning 2nd, 609-630, 1979.
Humble, A., The Mini-Ranger satellite survey system Motorola position determining systems, IAG/COSPAR CSTG Bull. 3, 98-105, Int. Assoc.

of Geod., Ohio State Univ., Columbus, July 1981.

Icenbice, P. J., and R. D. Loiler, Major improvements in satellite receiver time recovery for (cm) accuracy, Proc. Int. Geod. Symp. Satell. Doppler Positioning 2nd, 557-573, 1979.

Jenkins, R. E., and C. F. Leroy, "Broadcast" versus "precise" ephemeris - Apples and oranges, Proc. Int. Geod. Symp. Satell. Doppler Positioning 2nd, 39-62, 1979.

Kaula, W. M., Theory of Satellite Geodesy, Blaisdell, Waltham, Mass., 1966.

Kershner, R. B., and R. R. Newton, The TRANSIT system, J. Inst. Navig., 15, 129-144, 1962.

King, R. W., Lunar ranging validation of LAGEOS and Doppler polar motion results, IAG/COSPAR CSTG Bull. 3, 95-96, Int. Assoc. of Geod., Ohio State Univ., Columbus, July 1981.

Khan, M. A., Comparative evaluations of recent global representations of earth's gravity field, Geophys. J. R. Astron. Soc., 46, 435-554, 1976.

Kouba, J., Doppler levelling, Can. Surv., 30(1), 21-31, 1976.

Kouba, J., Datum considerations for test adjustments of Canadian primary horizontal network, Proc. Int. Symp. Probl. Related Redefinition N. Am. Geod. Networks 2nd, 55-62, 1978.

Kouba, J., Improvements in Canadian geodetic Doppler programs, Proc. Int. Geod. Symp. Satell. Doppler Positioning 2nd, 63-82, 1979.

Kouba, J., Geodetic satellite Doppler positioning and application to Canadian test adjustments, Philos. Trans. R. Soc. London, Ser. A, 294, 271-276, 1980.

Kouba, J., Satellite Doppler positioning for geodynamics, Ann. Geophys., 37(1), 205-213, 1981.

Kouba, J., An efficient short-arc orbit computation, submitted to Bull. Geod., 1982.

Kouba, J., and J. D. Boal, Program GEODOP, Publ. 1213E, Surv. and Mapp. Br., Ottawa, Ont., 1976a.

Kouba, J., and J. D. Boal, The Canadian Doppler satellite network, Proc. Int. Geod. Symp. on Satell. Doppler Positioning 1st, 187-206, 1976b.

Kouba, J., and L. D. Hothem, Compatiblity of Canadian and United States Doppler station networks, Proc. Int. Symp. Probl. Related Redefinition N. Am. Geod. Networks 2nd, 195-208, 1978.

Kouba, J., and J. A. Orosz, Evaluation of ELECTRAC receiver and oscillator effects on Doppler data quality at TRANET station 128 Ottawa, Proc. Int. Geod. Symp. on Satell. Doppler Positioning 3rd, in press, 1982.

Kouba, J., and D.E. Wells, Semi-dynamical Doppler satellite positioning, Bull. Geod., 50(1), 27-42, 1976.

Krakiwsky, E. J., D. E. Wells, and B. P. Kirkham, Geodetic control from Doppler satellite observations, Can. Surv., 26(2), 146-162, 1972.

Lachapelle, G., Determination of the geoid using heterogeneous data, Mitt. Geod. Inst. Tech. Univ. Graz, 19, 1975.

Lachapelle, G., and J. Kouba, Relationship between terrestrial and satellite Doppler systems, in Reference Coordinate Systems for Earth Dynamics, edited by E.M. Gaposchkin and B. Kolaczek, 195-203, D. Reidel, Hingham, Mass., 1981.

Langley, R. B., W. H. Cannon, W. T. Petrachenko, and J. Kouba, LBI and satellite Doppler: Baseline comparisons, Proc. Int. Geod. Symp. Satell. Doppler Positioning 2nd, 589-608, 1979.

Malyevac, C., and R. J. Anderle, Determination of plate tectonic motion from Doppler observations of Navy navigation satellites, Proc. Int. Geod. Symp. on Satell. Doppler Positioning 2nd, 695-742, 1979.

Marini, J. W., Correction of satellite tracking data for an arbitrary tropospheric profile, Radio Sci., 7(2), 223-231, 1972.

McClure, T. F., Method of Navigation, U.S. Patent, no 3, 172, 108, filed May 12, 1958. (Issued March 2, 1965).

Mikhail, E. M., Observation and Least Squares, Dun-Donnelley, New York, 1976.

Mueller, I. I., Introduction to Satellite Geodesy, Frederick Ungar, New York, 1964.

Murray, C. W., and J. W. Marini, A comparison of methods using surface meteorological measurements to predict altimeter bias caused by atmospheric water vapor, report Goddard Space Flight Center, Greenbelt, Md., June 1976.

Nesbo, I., Comparison of mathematical models for Doppler satellite positioning,M.Sc. thesis, Dep. of Surv. Eng., Univ. of New Brunswick, Fredericton, Aug. 1976.

Newton, R. R., The United States Doppler geodetic system and its observational accuracy, Philos. Trans. R. Soc. London, Ser. A, 292, 50-66, 1967.

Oesterwinter, C., Polar motion from astrometric, laser and Doppler, Proc. Int. Geod. Symp. Satell. Doppler Positioning 2nd, 751-766, 1979.

O'Toole, J. W., CELEST Computer Program for computing satellite orbits, Rep. NSWC/DL TR 365, Nav. Surface Weapons Center, Dahlgren, Vir., 1976.

Piscane, V. L., Navy Navigation Satellite System status, IAG/COSPAR CSTG Bull. 3, 106-107, Int. Assoc. of Geod., Ohio State Univ., Columbus, July 1981.

Piscane, V. L., B. B. Holland, and H. D. Black, Recent (1973) improvements in Navy Navigation Satellite System, Navigation Washington, 20, 224-229, 1973.

Popelar, J., and J. Kouba, Satellite Doppler determination of differential sea ice motion in the vicinity of the north pole, submitted to Mar. Geod., 1982.

Pospisilova, L., J. Klokocnik, and J. Kostelecky, A comparison of earth models by degree of their harmonic coefficients, Bull. Astron. Inst. Czech., 33, 33-36, 1982.

Reigber. C., Determination of the earth potential and possible improvements, Ann. Geophys., 37(1), 15-23, 1981.

Resch, G. M., and E.S. Claflin, Microwave radiometry as a tool to calibrate tropospheric water-vapor Delay, NASA Conf. Publ.2115, 377-384, 1979.

Rinner, K., and P. Pesec, West-east European Doppler observation campaign, Manuscr. Geod., 7(2), 117-132, 1982.

Robertson, D. S., et al., Polar motion and UT1: Comparison of VLBI, lunar laser, satellite laser, satellite doppler, and conventional astrometric determination, Radio Interferometry Techniques for Geodesy, NASA Conf. Publ. 2115, 33-44, 1979.

Saastamoinen, J., Contribution to the theory of atmospheric refraction, Bull. Geod., (107), 13-34, 1973.

Seeber, G., D., and Egge, Positionsbestimmung mit Satelliten - Doppler verfuren in bewegter Situation, Z. Vermess., 106(3), 123-131, 1981.

Seeger, H., W. Schluter, K. Herzberger, F. Kramer, and G. Soltau. The German-Austrian (German part) and EROS Doppler observation campaigns (DODOC and EROS-DOC), Proc. Int. Geod. Symp. Doppler Satell. Positioning 2nd, 889-924, 1979.

Seppelin, T.O., The Department of Defence World Geodetic Datum 1972, Can. Surv., 28(5), 496-506, 1974.

Smith E.K., Jr., and S. Weintraub, The constants in the equation for atmospheric refraction index of radio frequencies, Proc. Inst. Radio Electron. Eng. Aust., 41, 1035-1037, 1953.

Smith, R. W., G. R. Schwarz, and W. D. Googe, DOPPLR - A point positioning program using integrated doppler satellite observations, Proc. Int. Geod. Symp. Satell. Doppler Positoning 1st, 839-890, 1976.

Stansell, T. A., The many faces of TRANSIT, Navigation Washington, 25, 55-70, 1978.

Stansell, T. A., The 1502 satellite surveyor, Proc. Inter. Geod. Symp. Satell. Doppler Positioning 2nd, 497-534, 1979.

Strange, W. E., International Association of Geodesy, Special Study Group 2.54 -Satellite Radio Techniques (IAG SSG 2.54), IAG/COSPAR CSTG Bull. 1, 63-64, Int. Assoc. of Geod., Ohio State Univ.,Columbus, May 1980.

Usandivaras, J., P. Pâquet, and R. Verbeiren, An ORB Doppler program analysis and its application to European data. Proc. Int. Geod. Symp. Satell. Doppler Positioning 1st, 707-726, 1976.

Walker, J. W., Transformation of Doppler positions to local or national datum, Proc. Int. Geod. Symp. Satell. Doppler Positioning 2nd, 925-946, 1979.

Weber, J. R., The Lomonosov Ridge Experiment: LOREX 79, Eos Trans. AGU, 60(42), 715-721, 1979.

Wells, D. E., Doppler satellite control, Tech. Rep. 29, Dep. of Surv. Eng., Univ. of New Brunswick, Fredericton, Sept. 1974.

Wells, D. E., Concept of satellite Doppler positioning using translocation techniques, Proc. Int. Geod. Symp. Satell. Doppler Positioning 1st, 77-96, 1976.

Wells, D. E., E. J. Krakiwsky, D. B. Thomson, and J. Kouba, Review of Doppler satellite positioning in Canada, Bull. Geod., 50(4), 307-321, 1976.

Wells, D. E., C. R. Gillespie, P. Martin, and A. Thorndike, Error analysis of satellite doppler positioning of AIDJEX manned camps, Proc. Int. Geod. Symp. Satell. Doppler Positioning 2nd, 379-346, 1979.

Williams, O. W., DMA - U.S. Defense Mapping Agency, IAG/COSPAR CSTG Bull. 3, 13-22, Int. Assoc. of Geod., Ohio State Univ., Columbus, July 1981.

White, H. L., D. Huber, and J. Taylor, Comparison between Naval Surface Weapons Center and Navy Astronautic Group ephemeris for GEOCEIVER positioning, Rep. TR-75-001, Def. Mapp. Agency Aerosp. Center, St. Louis, Mo., 1975.

White, H. L., Variation in Doppler positions resulting from differences in computer programs and tropospheric refraction computations, Proc. Int. Geod. Symp. Satell. Doppler Positioning 1st, 551-576, 1976.

Yionoulis, S. M., Algorithm to compute tropospheric refraction effects on range measurements, J. Geophys. Res., 75, 7636-7637, 1970.

Ziegler, R. E., Ephemeris comparisons, Proc. Int. Geod. Symp. Satell. Doppler Positioning 2nd, 135-156, 1979.

(Received July 20, 1982;
accepted September 22, 1982.)

REVIEWS OF GEOPHYSICS AND SPACE PHYSICS, VOL. 18, NO. 2, PAGES 505–524, MAY 1980

Geodetic Leveling and Its Applications

PETR VANICEK

Department of Surveying Engineering, University of New Brunswick, Fredericton, New Brunswick, Canada E3B 5A3

ROBERT O. CASTLE

U.S. Geological Survey, Menlo Park, California 94025

EMERY I. BALAZS

National Geodetic Survey, Rockville, Maryland 20852

The inherent precision of spirit leveling has preserved its utility as a geodetic measurement system for over a century. While various instrumental and procedural modifications designed to enhance this precision have been introduced over the years, the basic measurement system has remained virtually unchanged since the mid-nineteenth century. Possible systematic error has dictated the majority of the procedural and instrumental requirements associated with geodetic leveling; the physical source(s) of several of these errors remain poorly understood. Statistically independent random errors, which accumulate according to the square root of the survey distance, are generally controlled through redundancy and procedural randomization; they range from 0.5 mm $L^{1/2}$ for the highest-order modern leveling to about 6 mm $L^{1/2}$ for the lowest-order nineteenth-century geodetic surveys, where L is the survey distance in kilometers. Height differences are conceptually distinct from observed or measured elevation differences in the sense that the former are uniquely defined, whereas the latter are path dependent, a distinction that arises from the nonparallelism of the equipotential surfaces of the earth's gravity field. The number of possible height systems is virtually limitless. They include the systems of geopotential numbers and dynamic heights; although neither of these systems is geometrically informative, each provides perfectly valid height characterizations that may be especially useful in the solution of certain physical problems. The most generally used system of heights is the orthometric height system; the resulting heights are true geometric heights above the geoid. Normal height systems are referred to the quasi-geoid rather than the geoid. Each of the various height systems meets the requirement of uniqueness, and none can be viewed as being conceptually superior. Conversion of the observed elevation differences obtained from leveling into uniquely defined height differences requires the application of a gravity-dependent correction. Because gravity coverage in North America was generally sparse until recently, an approximation for this correction, which provides for the effects of the poleward covergence of the equipotential surfaces, has usually been used on this continent. Heights have been traditionally referred to mean sea level as a datum, a usage that implies coincidence between mean sea level and the geoid (or quasi-geoid). Because the determination of mean sea level is dependent on the length of the observation period, because its definition makes no allowance for vertical crustal displacements or changes in eustatic sea level, and because its definition disregards the demonstrable existence of sea surface topography, local mean sea level generally departs from the geoid. This introduces errors in computed heights that probably equal or exceed those due to leveling. Repeated levelings continue to provide the best basis for determining terrestrial vertical displacements. These displacements are necessarily measured with respect to the time-dependent (instantaneous) geoid. Analysis of changes in the orthometric correction and changes in the geoid as a function of changes in gravity indicates that height determinations are almost insensitive to temporal variations in gravity. The simplest and probably the most accurate procedure for determining vertical displacements is based on direct comparisons between observed elevations derived from repeated levelings along the same line. The estimated errors associated with these displacements are a function of the precision of the utilized levelings and have been shown to be much smaller than those associated with heights. A smoothed representation of the vertical displacement field can be obtained by the fitting of a mathematical surface to the results of segmented relevelings. Surface-fitting techniques usually tend to subdue short-wavelength features, but they are especially useful in depicting artificially induced subsidence and broadly defined tectonic uplift or collapse.

INTRODUCTION

Spirit leveling is a remarkably simple yet inherently precise measurement system that has remained virtually unchanged for over a century. During this period the vertical control needs of the geodetic, cartographic, and engineering communities have been well served by this system. Nonetheless, only within the past several decades has its value as a geophysical tool begun to be appreciated by most earth scientists. This growing appreciation can be closely correlated with two phenomena: (1) the widespread recognition of surface subsidence associated with the extraction of underground fluids [*Poland and Davis,* 1969; *International Association of Scientific Hydrology,* 1969; *Castle and Yerkes,* 1976; *International Association of Hydrological Sciences,* 1977] and (2) the expanding bulk of the increasingly detailed and increasingly accurate geodetic record that has permitted the detection of relatively small, clearly tectonic vertical displacements generated over relatively short increments of historic time [*National Geodetic Survey,* 1973; *Pavoni and Green,* 1975]. Moreover, while these displacements are generally small in human terms, they commonly have proved much larger than could have been anticipated from geologic or geophysical considerations. Although the disparity between contemporary vertical displacements deduced from

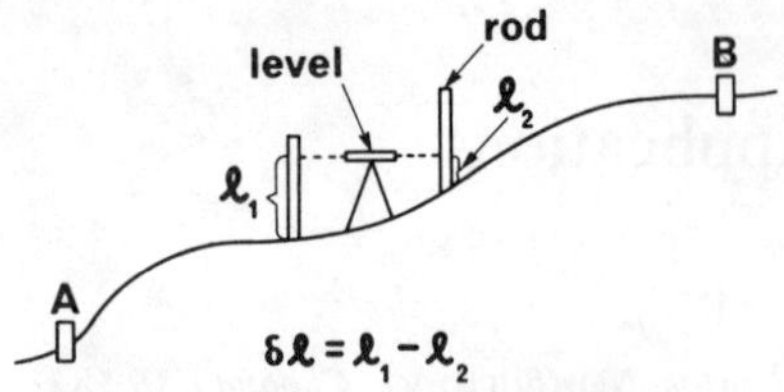

Fig. 1. Schematic diagram illustrating one setup in the leveling procedure designed to measure the elevation difference between two established bench marks, A and B.

repeated geodetic measurements and those that might otherwise have been inferred seemingly challenges the accuracy of these measurements, the growing number of geodetically determined, very subtly defined signals that closely agree in both sense and spatial distribution with the predicted displacements has greatly enhanced the credibility of the reported precision of geodetic leveling.

This review focuses on four related aspects of leveling: (1) measurement procedures and precision, (2) height concepts, (3) the use of mean sea level as a reference datum, and (4) the utilization of the results of leveling as a vehicle for detecting and describing the vertical displacement field that characterizes the present terrestrial surface.

MEASUREMENT SYSTEM

We begin by distinguishing between measured (observed) elevation differences and height differences. This is not a pedantic distinction, for as we show in the following section, the determination of height differences must consider factors other than the results of the leveling process. Nevertheless, because height determinations are overwhelmingly dependent on the measured elevation differences obtained from leveling, it is especially appropriate that we consider first the principles and procedures involved in geodetic leveling.

The basic instruments used in leveling are a level and a graduated rod. The fundamental measurement is the observed elevation difference δl between two nearby points occupied by a pair of rods or a single rod successively positioned on either side of the level (Figure 1). In this simplified representation, it is assumed that the rods are equidistant, in order to minimize refraction error from the instrument, and held precisely vertically (that is, perpendicular to the equipotential surface passing through the instrument) and that the level is perfectly collimated and precisely leveled (that is, aligned with the equipotential surface passing through the instrument). Thus provided that the rods are both invariant in length and calibrated against an accepted standard and that atmospheric conditions are both uniform and unchanging during each setup, the observed elevation difference between A and B, Δl_{AB}, consists simply of the summation of the separately determined elevation increments δl obtained during each individual setup:

$$\Delta l_{AB} = \sum_{i=1}^{n} \delta l_i \qquad (1)$$

The accuracy of the computed elevation difference Δl_{AB} obviously is a function of the accuracy of δl at each successive setup (Figure 1). The instrumental and procedural requirements stipulated for geodetic leveling have been introduced in order to minimize errors associated with each of the incremental measurements and to randomize the distribution of those errors that inevitably persist. Accordingly, we are concerned here chiefly with those instrumental and procedural specifications that impinge on the precision of the measurements, whether at a setup (δl), over a section (Δl_{AB}), or around a loop.

Geodetic leveling consists of several orders and classes, each of which has been developed to serve a particular need or group of needs. The higher the order and, in general, the more recent the leveling, the greater the precision. The newly formulated specifications of the U.S. *Federal Geodetic Control Committee* [1974] for each of the orders and classes within these orders of geodetic leveling comprise a state-of-the-art set of standards designed to achieve maximum precision in new leveling at reasonable cost. These specifications, which we summarize in Table 1, provide a convenient vehicle for enumerating those procedural and instrumental requirements controlling the precision estimates identified with modern geodetic leveling. Nevertheless, while certain general aspects of these specifications are germane to this review, a detailed consideration of instrumental design features and field practices, including bench mark installation, is clearly beyond the scope of this review; comprehensive discussions, together with explanations supporting the adoption of particular instrumental features and field procedures, can be found, for example, in the writings of *Birdseye* [1928], *Rappleye* [1948], *Bomford* [1971, pp. 232–243], and the *Federal Geodetic Control Committee* [1975].

Because leveling precision is a function of the nature and magnitude of the errors associated with the individual measurements, it is considered here in the context of error sources and their control. Recognized errors identified with geodetic leveling range from those that are trivial and can be easily controlled, through those that can be virtually eliminated through meticulous adherence to field and laboratory procedures, to those whose nature is so complex that they continue to cause concern [*Entin*, 1959; *Karren*, 1964]. Leveling errors are usually categorized as blunders, systematic errors, and random errors (Figure 2). Our experience to date indicated that blunders are generally easily detected. Moreover, double-run and double-rodded leveling (comprising forward and backward runnings made with a pair of leveling rods used alternately for foresight and backsight) has virtually eliminated blunders during uninterrupted leveling.

Systematic errors tend to accumulate in proportion to (1) distance and (2) height or height gradient and may become very large over relatively short distances. It is the recognition or suspicion of these potentially large errors that has dictated many of the procedural requirements stipulated for geodetic leveling (Table 1 and Figure 2). Certain types of systematic error, such as those associated with imperfect collimation of the instrument, can be virtually eliminated through rigid adherence to prescribed field procedures (Figure 2). Still others, such as errors in rod length, of whatever origin, can be tightly controlled through a combination of field and laboratory procedures, including frequent calibration against an invariant standard. A third type of systematic error, however, is less easily controlled owing to the physical complexity of the source or even, as in the case of a postulated directional bias in spirit leveling, uncertainty as to its existence. Probably the most troublesome of the recognized systematic errors is the error attributed to residual refraction, in which the refraction in the foresight is unmatched by that in the backsight. Although systematic refraction error probably is generally very small, *Bomford* [1971, pp. 240–241] suggests that under certain ex-

TABLE 1. Newly Adopted Specifications for Vertical Control Surveys in the United States and Estimated Maximum Closures and Standard Deviations Associated With Surveys That Meet These Specifications (modified after *Federal Geodetic Control Committee* [1974, pp. 3 and 9])

Classification	First Order, Class 1 and Class 2	Second Order, Class 1	Second Order, Class 2	Third Order
		Specifications		
Instrument standards	automatic or tilting levels with parallel plate micrometers; invar scale rods	automatic or tilting levels with optical micrometers or three-wire levels; invar scale rods	geodetic levels and invar scale rods	geodetic levels and rods
Field procedures	double run; forward and backward, each section	double run; forward and backward, each section	double or single run	double or single run
Section length	1–2 km	1–2 km	1–3 km for double run	1–3 km for double run
Maximum length of sight	50 m class 1; 60 m class 2	60 m	70 m	90 m
Field procedures				
Maximum difference in lengths forward and backward sights per setup	2 m class 1; 5 m class 2	5 m	10 m	10 m
Per section (cumulative)	4 m class 1; 10 m class 2	10 m	10 m	10 m
		*Maximum Closures**		
Section; forward and backward	3 mm $K^{1/2}$ class 1; 4 mm $K^{1/2}$ class 2	6 mm $K^{1/2}$	8 mm $K^{1/2}$	12 mm $K^{1/2}$
Loop or line	4 mm $K^{1/2}$ class 1; 5 mm $K^{1/2}$ class 2	6 mm $K^{1/2}$	8 mm $K^{1/2}$	12 mm $K^{1/2}$
		Standard Deviations		
Accuracy of height difference, between directly connected bench marks	0.5 mm$K^{1/2}$ class 1; 0.7 mm $K^{1/2}$ class 2	1.0 mm $K^{1/2}$	1.3 mm $K^{1/2}$	2.0 mm $K^{1/2}$

*Check between forward and backward runnings where K is the distance in kilometers.

treme conditions it may be as large as 2×10^{-4} times the height difference. It is, in fact, the potentially large magnitude of this error that has produced many of the standard leveling specifications (especially sight length restrictions and minimum readings on the higher rod) and various attempts to compute corrections based on the technique of *Kukkamäki* [1938].

Random error stems from inaccurate observations, nonsystematic imperfections in the equipment, and ever present vagaries in atmospheric conditions (Figure 2). Moreover, short runs of systematic error that tend to randomize over a few kilometers may be treated as random error. Random error generally can be diminished through redundance, such as that achieved through double running and the use of double-scale rods.

The errors accumulated during leveling may be assessed through statistical examination of the section results (that is, the discrepancies between forward and backward runs) or loop misclosures. If the random errors in the measured elevation difference Δl between two bench marks a distance L apart are truly statistically independent, then their standard deviation is proportional to $L^{1/2}$. While the results of single-run leveling are less amenable to statistical analysis than are those of double-run surveys, random error is reasonably assumed to accumulate in about the same way in both cases if only because the procedures stipulated for both single-run and double-run surveys are otherwise virtually identical. This assumption is supported, moreover, by the experience of the U.S. Geological Survey in assessing the results of literally thousands of relatively small loop or line closures (M. R. Elliott, personal communication, 1978).

Random error estimates associated with geodetic leveling are functions of both the order and the vintage of the survey, and because this review is concerned with utilization of the full geodetic record, it is especially pertinent that we show here how these estimates have changed through time. First-order leveling procedures in the United States currently are such that where L is given in kilometers, the estimated standard deviation in the measured elevation difference between any two bench marks, σ, for class 1 and class 2 surveys is about 0.5 mm $L^{1/2}$ and 0.7 mm $L^{1/2}$, respectively (Table 1). However, the cumulative experience of the U.S. National Geodetic Survey with first-order leveling indicates that σ was about 2.5 mm $L^{1/2}$ prior to 1901, about 2.0 mm $L^{1/2}$ during the period 1901–1916, about 1.5 mm $L^{1/2}$ during the period 1917–1955, and about 1.0 mm $L^{1/2}$ during the period 1956–1974 (S. R. Holdahl, personal communication, 1974). The increase in precision that occurred after 1900 and again after 1916 can be correlated with significant improvements in equipment (namely, successive introduction of the Fisher level and the invar rod), whereas the increased precision that occurred after 1955 can be traced to major procedural changes, especially reductions in both maximum sight lengths and forward and backward sight length differences. Hence one standard deviation in the measured elevation difference between two bench marks at either end of a 500-km line would be about 56 mm had this leveling been carried out prior to 1901, about 48 mm had it been carried out between 1901 and 1916, about 34 mm had it been carried out between 1917 and 1955, about 22 mm had it been carried out between 1956 and 1974, and about 11 mm had it been carried out since 1974 in accordance with class 1 procedures. Second-order leveling specifications are such that σ is today given as 1.0 mm $L^{1/2}$ and 1.3 mm $L^{1/2}$ for class 1 and class 2 surveys, respectively (Table 1). Again, however, the ex-

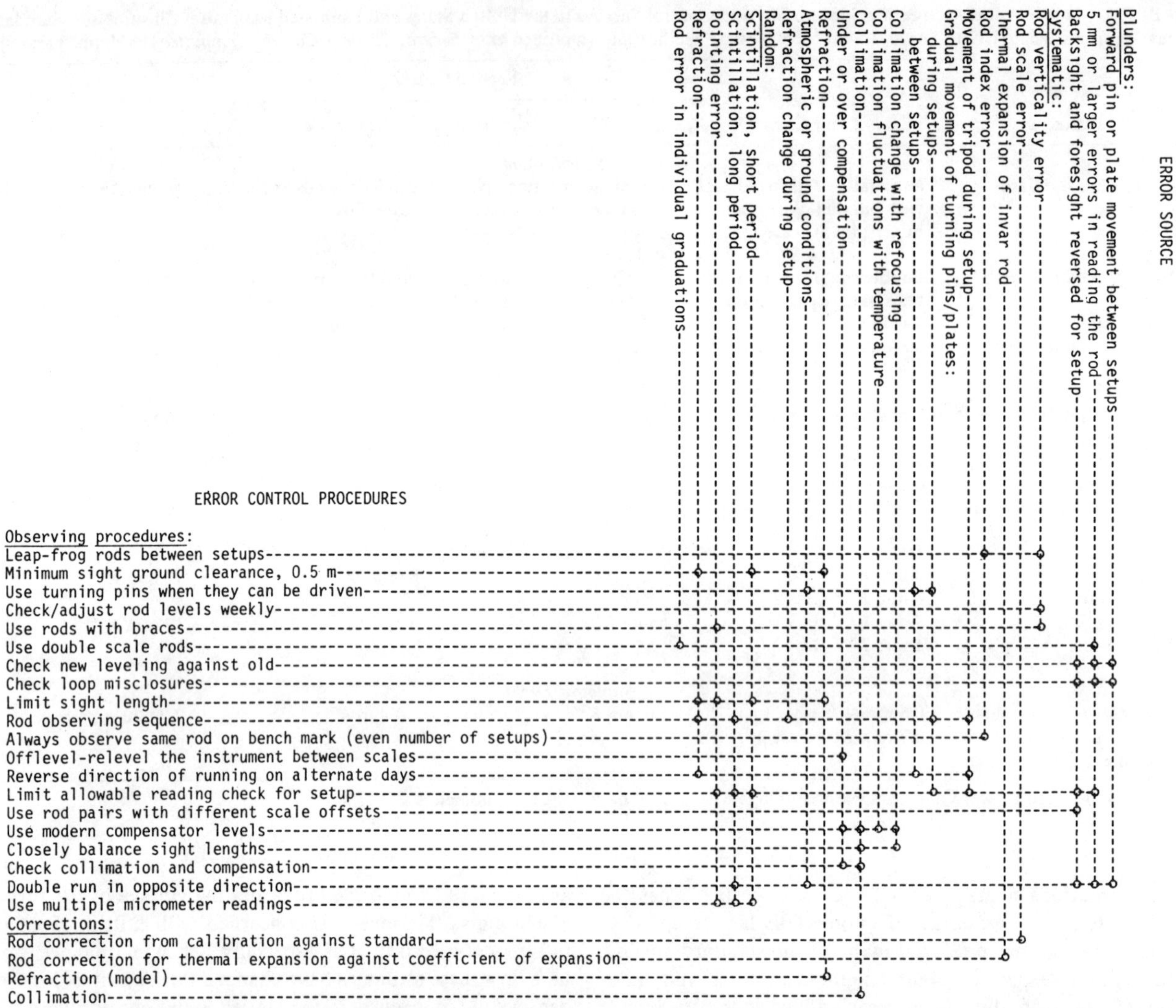

Fig. 2. Matrix showing error sources associated with spirit leveling and procedures and corrections adopted by the National Geodetic Survey to detect and minimize or eliminate these errors (modified after *Whalen and Balazs* [1977]).

perience of the National Geodetic Survey with all second-order leveling indicates that σ has dropped progressively from about 5.0 mm $L^{1/2}$ prior to 1900, to about 4.0 mm $L^{1/2}$ during the interval 1901–1916, to about 3.0 mm $L^{1/2}$ during the interval 1917–1955, and to about 2.0 mm $L^{1/2}$ during the period 1956–1974 (S. R. Holdahl, personal communication, 1974). These successive decreases in the random error estimates associated with second-order leveling are generally attributable to the same changes in equipment and procedures responsible for the increasing precision in first-order leveling; that the first- and second-order values differ at all is only because second-order procedures are the more permissive. Contemporary specifications established for third-order leveling, the least precise category of geodetic leveling, are such that σ is given as 2.0 mm $L^{1/2}$ (Table 1). The experience of the U.S. Geological Survey, based chiefly on the results of countless third-order loop and line misclosures, indicates that prior to 1956, σ was generally close to 6 mm $L^{1/2}$, whereas between 1956 and 1974 it was about 4 mm $L^{1/2}$. The specifications for third-or-

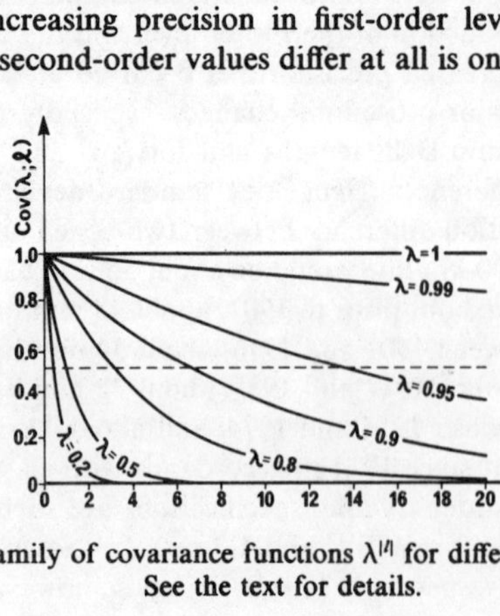

Fig. 3. Family of covariance functions $\lambda^{|l|}$ for different values of λ. See the text for details.

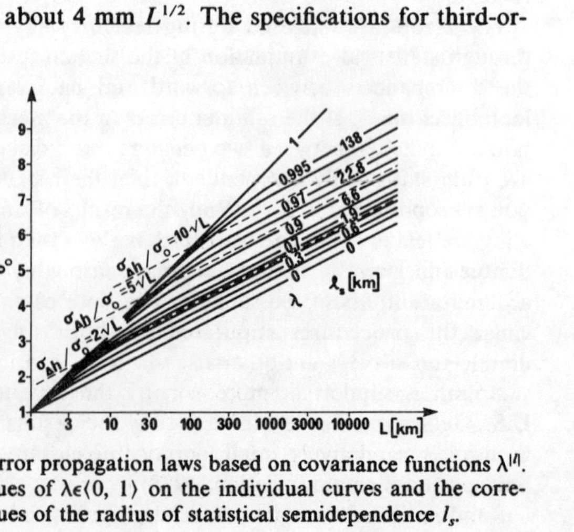

Fig. 4. Error propagation laws based on covariance functions $\lambda^{|l|}$. Note the values of $\lambda \epsilon \langle 0, 1 \rangle$ on the individual curves and the corresponding values of the radius of statistical semidependence l_s.

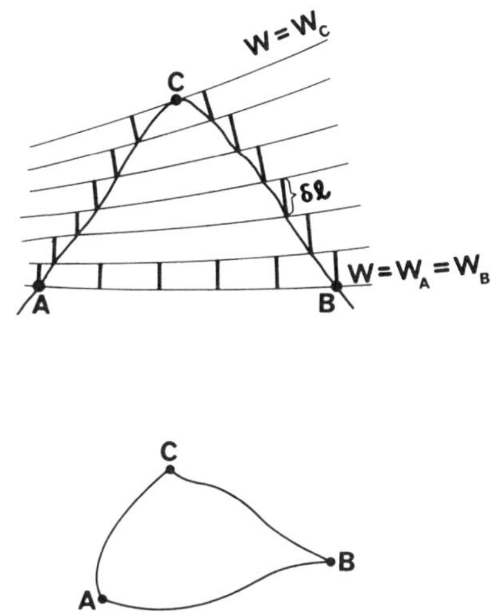

Fig. 5. Observed elevation differences (A) around level circuit ABCA(B), where A and B both lie on the same gravity equipotential surface.

der (or what has in the past been called 'primary') leveling are such, however, that these estimates almost certainly err on the conservative side.

Calculation of the statistically independent random error estimate associated with a measured elevation difference between the endpoints of a single line composed of several segments of differing class or order requires, in effect, that the standard deviation be calculated for each segment. Thus in the most general case, $\sigma_{\Delta l}$, or one standard deviation in the measured elevation difference between any two marks based on n combinations of orders or classes of leveling, is given as

$$\sigma_{\Delta l} = \left(\sum_{i=1}^{n} L_i \sigma_{\Delta \delta l_i}{}^2 \right)^{1/2} \tag{2}$$

where $\sigma_{\Delta \delta l_i}$ is the estimated standard deviation applicable to the leveled height difference δl_i of each successive segment of length L_i. For example, one standard deviation in the measured elevation difference between the endpoints of a 1970 survey consisting of a 100-km segment of first-order leveling joined to a second 100-km segment of second-order leveling would be about 22 mm.

There is an explicit assumption behind the acceptance of (2) and thus behind the acceptance of the error estimates described in the preceding paragraph: the errors in the measured elevation differences are truly statistically independent. That is, it is only when these errors are indeed statistically independent that the square root law,

$$\sigma_{\Delta l} = \sigma_0 L^{1/2} \tag{3}$$

on which (2) is based, actually applies. In fact, it is very unlikely that these errors are statistically independent [*Müller and Schneider,* 1968]. Because the closures around a leveled circuit are generally larger than those predicted on the basis of the repeatability of δl_i (that is, on the basis of $\sigma_{\delta l_i}$), a varying degree of statistical dependence among the errors in the δl_i is implied.

The theoretical aspects of this problem were investigated by *Vaníček and Grafarend* [1979], who discovered that the form of the propagation of statistically dependent errors stems from the form of the covariance function (autocorrelation) governing the statistical dependence of those errors. For example, if the covariance function, as postulated by *Lucht* [1972], is given as

$$\mathrm{Cov}\,(\lambda;\, l) = \lambda^{|l|} \tag{4}$$

where $\lambda \epsilon \langle 0,\ 1\rangle$ is a parameter (Figure 3), then the corresponding error propagation laws take on the forms shown in Figure 4. We note that the two limiting cases, $\lambda = 0$ and $\lambda = 1$, correspond to total statistical independence and total statistical dependence. In between are the cases of varying degrees of statistical dependence. Moreover, in investigating the related problem of the correlation between the backward (B) and forward (F) runnings in double-run leveling, *Remmer* [1975] found that the level of this correlation in the Danish first-order geodetic network is equal to 0.73. This correlation ρ is presumed to be the consequence of common influences identified with the locale of the observations which adversely affect the leveled elevation differences, $\delta l = \frac{1}{2}(F + B)$. Use of the covariance law [*Mikhail,* 1976] produces

$$\sigma_{\delta l}{}^2 = \tfrac{1}{4}(\sigma_F{}^2 + 2\rho\sigma_F\sigma_B + \sigma_B{}^2) \tag{5}$$

which is clearly larger than would be the case were the two runnings statistically independent. Assuming, without any detriment to the generality, that $\sigma_F = \sigma_B$, we obtain

$$\sigma_{\delta l}{}^2 = \tfrac{1}{2}\sigma_F{}^2(1 + \rho) \tag{6}$$

Thus the results of these several investigations show that the precision (in effect, the repeatability of the results) of leveling over the same survey route is generally and significantly higher than its accuracy (as implied by circuit misclosures).

The assessment of systematic errors associated with spirit leveling persists as a major geodetic problem. Although a variety of systematic errors may be revealed through misclosures, many others seem to have little effect on network closures. Moreover, because significant misclosures may be due to blunders or the occurrence of crustal deformation during the completion of a loop, misclosures alone are generally inadequate indices of systematic error. Cumulative divergence between forward and backward runs in double-run leveling affords still another index of systematic error; it usually does not, however, provide any direct insight into the source of the error. Because rod and refraction errors are height and slope dependent, they are most easily detected by searching for a correlation between topography and measured elevation discrepancies. If the results of one or more surveys are contaminated by rod error, the correlation between elevation and elevation discrepancies should be nearly perfect. Refraction errors are somewhat less strongly correlated with terrain, because refraction is a function of atmospheric density along the line of sight and thus dependent on the vertical temperature gradient, humidity, and so forth. Short-term meteorologic variations superimposed on normal diurnal and seasonal variations should tend to obscure the correlation between slope and elevation discrepancies produced by repeated surveys.

Consideration of all the known sources of survey error, together with the techniques that have been adopted to control and detect these errors, indicates that geodetic leveling is inherently a very precise and in a certain sense, as will be seen later, very accurate measurement system. While height determinations may be seriously contaminated locally by systematic errors, examination of a large volume of data indicates that overall errors rarely rise above the expected levels indicated in Table 1.

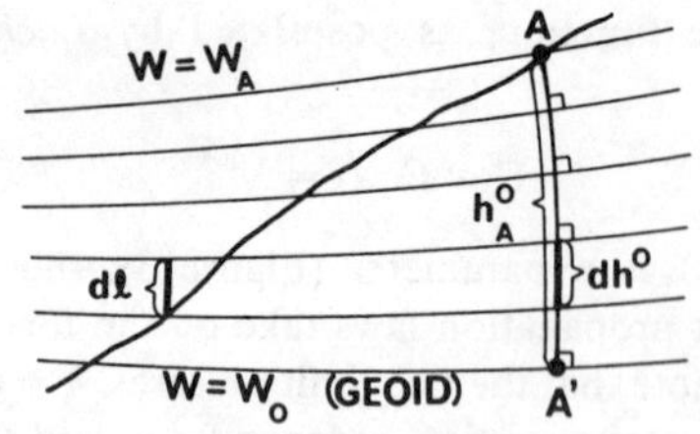

Fig. 6. Definition of the orthometric height $h_A{}^O$ for a point A at the surface of the earth.

HEIGHT CONCEPTS

With the rudimentary background developed in the preceding section on how observed elevation differences may be obtained through spirit leveling and the precision with which these measurements can be made, we turn now to a consideration of how these field measurements can be converted into heights (or, more specifically, height differences). Because observed elevation differences are conceptually distinct from height differences, the conversion process is less straightforward than is generally imagined.

Let us begin by considering a closed leveling loop ABCA (Figure 5), where the subparallel lines shown in Figure 5 represent the intersections of surfaces of equal gravity potential with the terrain. If the individual observed elevation differences or leveling increments, represented by the symbol δl (see Figure 1), can be treated as being infinitesimally small, direct examination of Figure 5 permits the formulation of the following inequality:

$$\int_A^B dl + \int_B^C dl + \int_C^A dl \neq 0 \tag{7}$$

Alternatively, (7) can be written as

$$\oint dl \neq 0 \tag{8}$$

an expression that simply states that the sum of the observed elevation differences around a closed loop is generally not equal to zero. This integral property of observed elevation differences is evidently a consequence of the nonparallelism of the gravity equipotential surfaces, where an equipotential surface is defined simply as one along which the gravity potential W equals a constant.

Implicit in (7) and (8) is that the use of observed elevation differences, even starting from the same initial point, will generally produce different elevations or 'heights' for the endpoint of a level line, depending on the survey route. Thus it follows that observed elevations are nonunique—they are path dependent. Yet if heights are to serve as vertical coordinates, and this is, after all, their chief role, they must be (1) defined uniquely and (2) referred to a well-defined terrestrial framework (reference surface, datum). To consider the second point first, the usual approach has been to refer heights to the earth's gravity field, which in this context is assumed to be invariant with respect to time. Specifically, the generally adopted reference datum is the geoid, which may be defined as the gravity equipotential surface that most closely approximates mean sea level.

Heights may be defined in a variety of useful ways. Probably the simplest system is the one that bypasses the geometric problem and describes the height of any point in terms of its gravity potential. The potential may be related to the observed height differential dl by means of

$$dW/dl = -g \tag{9}$$

where g is the gravity at the surface of the earth. This equation can be rewritten in a more useful form as

$$dW = -g\, dl \tag{10}$$

If we take the potential on the geoid (W_0) as our reference, then the negative potential difference between the point of interest (A) and the geoid is known as the geopotential number of this point; it is usually denoted by the symbol C and defined by the equation

$$C_A = \int_{\text{geoid}}^A g\, dl \tag{11}$$

Accordingly, the difference in geopotential numbers between any two points A and B on the surface of the earth is given as

$$\Delta C_{AB} = \int_A^B g\, dl \doteq \sum_i g_i \delta l_i \tag{12}$$

where g is measured at the earth's surface. If we choose to convert geopotential numbers into units of length, we must divide them by a value of gravity, an operation that produces what is known as dynamic heights. The value of gravity, G, used in this definition is usually chosen such that it represents the mean value of gravity for the region of interest. The defining equation for the difference in dynamic height between points A and B thus becomes

$$\Delta h_{AB}{}^D = \frac{1}{G}\int_A^B g\, dl = \frac{\Delta C_{AB}}{G} \tag{13}$$

The alleged disadvantage associated with the utilization of both geopotential numbers and dynamic heights is that they have no obvious geometric significance. Neither geopotential numbers nor dynamic heights can be 'plotted' above the reference surface by using a ruler of fixed graduation. On the other hand, they are certainly physically sensible quantities, since they describe points on the same equipotential surface as being equally elevated (that is, as having the same height). Thus, disregarding any dynamic effect, the surface of a lake represented in terms of either of these two systems is flat. Therefore if solutions to certain physical problems are anticipated, either of these height systems may be preferred.

The third and most generally utilized system of heights is based on so-called orthometric heights. These are true geometric heights in the sense that the orthometric height $h_A{}^O$ of a point A on the surface of the earth is measured from the geoid along the actual plumb line by using a scale of fixed graduation (Figure 6). Denoting gravity along the plumb line of A as $g_A{}'$, we may write

$$dW = -g\, dl = -g_A{}'\, dh^O \tag{14}$$

from which we obtain

$$dh^O = \frac{g}{g_A{}'}\, dl \tag{15}$$

Integrating between the appropriate limits, we have

$$h_A{}^O = \frac{1}{\bar{g}_A{}'}\int_{\text{geoid}}^A g\, dl = \frac{C_A}{\bar{g}_A{}'} \tag{16}$$

where $\bar{g}_A{}'$ denotes the mean gravity (in the integral sense) along the plumb line A'A (Figure 6).

The general problem implicit in the use of orthometric heights is that $\bar{g}_A{}'$ cannot be obtained directly and must be based on a hypothesized or postulated gravity gradient. Thus,

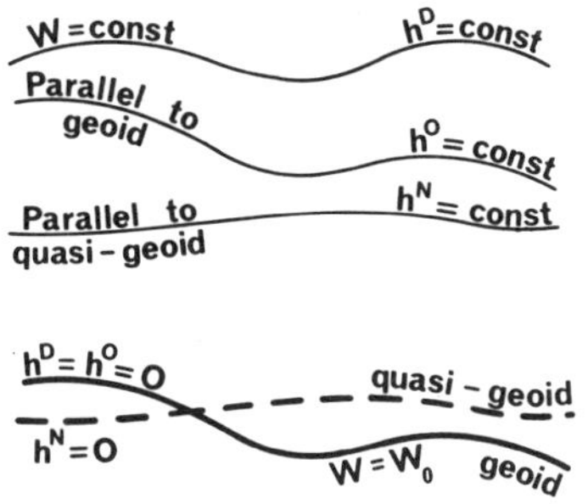

Fig. 7. Schematic comparisons between the three main height systems.

as is clearly evident from (16), we may obtain any number of orthometric height systems, depending on the value selected for $\bar{g}_A'$ [*Bomford,* 1971]. Among the systems of orthometric heights developed to date, the most generally used are Helmert's heights, which are based on the acceptance of the Poincaré-Pray gradient,

$$\partial g/\partial h = -0.0848 \text{ mGal/m} \tag{17}$$

as being representative of the value for any point between the geoid and the surface of the earth [*Heiskanen and Moritz,* 1967, pp. 163–165]. It assumes a uniform density of 2.67 g cm^{-3} for subsurface material to the depth of geoid. Because the Poincaré-Pray gradient is constant, the mean gravity $\bar{g}'$ may be related to the surface gravity g by means of the linear equation

$$\bar{g}' = g + \frac{\partial g}{\partial h}\frac{h}{2} \tag{18}$$

A fourth and distinctly different family is that of normal heights. They are distinguishable chiefly in that they use the quasi-geoid [*Molodenskij et al.,* 1962] rather than the geoid as a reference surface. The quasi-geoid is a surface (not an equipotential surface) that coincides with the geoid on the open seas and departs from it elsewhere, typically by a few decimeters and up to a few meters under a high mountain. Moreover, although the significance is no more than academic, normal heights are measured along the plumb line of the normal gravity (see (20)) field rather than the actual gravity field. By adopting arguments that parallel those used in developing the orthometric height system the defining equation for normal heights is given as

$$h_A{}^N = \frac{1}{\bar{\gamma}_A{}'}\int_{\text{quasi-geoid}}^{A} g\,dl = \frac{C_A}{\bar{\gamma}_A{}'} \tag{19}$$

where $\bar{\gamma}_A'$ is the mean normal gravity between the mean-earth reference ellipsoid and the telluroid, which is the locus of normal heights with respect to the mean-earth ellipsoid. Mean normal gravity can be evaluated from the normal gravity on the ellipsoid, γ_0, and the vertical gradient of normal gravity. The pertinent equations are

$$\gamma_0(\phi) = \gamma_a(1 + \alpha \sin^2\phi + \beta \sin^2 2\phi) \tag{20}$$

$$\frac{\partial \gamma}{\partial h} \doteq -\frac{2\gamma_0}{a}(1 + m + f\cos 2\phi) \tag{21}$$

which are accurate to about 0.3%, and

$$\bar{\gamma}' = \gamma_0 - \frac{\partial \gamma}{\partial h}\frac{h^N}{2} \tag{22}$$

[*Heiskanen and Moritz,* 1967; *International Association of Geodesy,* 1971]. In (20)–(22), γ_a, α, and β are the coefficients of the normal gravity formula, a is the semimajor axis of the reference ellipsoid, f is the flattening of the reference ellipsoid (~0.003), $m = a^3b\omega^2/(\kappa M)$ is the so-called geodetic parameter, ϕ is the latitude, κ is the gravitational constant, and M is the mass of the earth. Heights defined in this way are known as Molodenskij's heights and are used in the Soviet sphere of influence in lieu of orthometric heights.

The (Vignal) normal height system, used in western Europe, differs from Molodenskij's in that it is based on a less rigorously determined normal gravity gradient, namely,

$$\frac{\partial \gamma}{\partial h} = -\frac{\kappa M}{(a^2b)^{1/3}} - 2\omega^2 = 0.3086 \text{ mGal/m} \tag{23}$$

which is simply the free-air gradient. Thus Vignal's heights can be regarded as approximations of Molodenskij's.

Finally, it is instructive to compare, in at least schematic fashion, the several systems of described heights (Figure 7). Thus the 'isoheight' surface identified with the dynamic height is the local equipotential surface, whereas the isoheight surfaces of the two geometric heights are congruent with their respective datums. Moreover, it is worth emphasizing at this point that normal heights do not depend on postulated gravity gradients and hence are as rigorously defined as the accuracy of the formula for normal gravity (including its gradient) permits. And indeed, normal gravity can be evaluated as precisely as one wishes. This is obviously not the case in dealing with orthometric height systems. Hence, strictly speaking, none of the orthometric heights can be explicitly referred to the geoid and should be viewed instead as heights referred to a surface implied by the selected vertical gradient of gravity. Hence orthometric heights offer no real advantage over normal heights and vice versa. In this same connection it should be noted that any one of the described height systems fulfills the requirement of uniqueness. Thus each of these systems satisfies the condition

$$\oint dh = 0 \tag{24}$$

and the sum of the errorless height differences along a closed loop theoretically goes to zero (see (40)–(42)), a point that is commonly misunderstood even by the professionally trained.

Parenthetically, in any practical mapping or surveying application all that are actually needed are height or potential differences on the earth's surface; the position of the datum is of no real concern, and the actual height differences, whether orthometric or normal, have no physical meaning anyway. The choice of datum and thus the choice between orthometric and normal heights becomes significant only where the heights together with their horizontal positions are converted into three-dimensional coordinates such that either geoidal heights or height anomalies emerge as a result.

Given even the limited number of height systems described in the preceding paragraphs, we need now contend directly with the implicit question with which we introduced this section: how can the observed elevation differences be related to height differences of whatever system? If gravity measurements along the leveling route are readily available, the geopotential difference can be evaluated by means of (12). The geopotential difference, in turn, can be converted directly into any height difference that we may be interested in, provided only that the height of one of the endpoints is also known. A generally more useful approach, however, is to develop theoretical expressions that represent the differences between the observed elevation difference and the corresponding height difference. These differences, or 'corrections,' can then be used to compute height differences from observed elevation differences.

Turning first to dynamic heights, (13) can be rewritten as

$$\Delta h_{AB}{}^{D} = \frac{1}{G}\int_A^B g\,dl = \int_A^B \frac{g - G + G}{G}\,dl = \int_A^B dl + \int_A^B \frac{g - G}{G}\,dl \qquad (25)$$

Because the first term is evidently nothing more than the observed height difference Δl_{AB}, the second term must be the dynamic correction DC_{AB}. Thus

$$DC_{AB} = \int_A^B \frac{g - G}{G}\,dl \doteq \sum_i \frac{g_i - G}{G}\,\delta l_i \qquad (26)$$

Because

$$|(g - G)/G| \ll g/G \qquad (27)$$

the correction is normally much smaller than the height difference and hence relatively easy to evaluate precisely.

The orthometric correction OC_{AB} can be derived in a comparable way and is given by *Heiskanen and Moritz* [1967, pp. 166–169] as

$$OC_{AB} = DC_{AB} + \frac{\bar{g}_A{}' - G}{G}\,h_A - \frac{\bar{g}_B{}' - G}{G}\,h_B \qquad (28)$$

where h_A and h_B are the heights of A and B, respectively, and need be no better than close approximations. Similarly, the normal correction NC_{AB} is given as

$$NC_{AB} = DC_{AB} + \frac{\bar{\gamma}_A{}' - G}{G}\,h_A - \frac{\bar{\gamma}_B{}' - G}{G}\,h_B \qquad (29)$$

Because these corrections are generally larger than the errors in measured first-order elevation differences between the endpoints of the survey, A and B, it is mandatory that they be utilized in transforming the results of first-order leveling into height differences. However, until very recently, appropriate surface gravity measurements were not generally available on this continent, such that the various height corrections could not be precisely evaluated. A solution to this dilemma was sought in the modeling of at least the bulk of the corrections through the use of normal rather than actual or observed gravity values. The expressions for approximate height corrections based on normal rather than observed gravity were developed by *Bowie and Avers* [1914] and remain in use in North America to this day.

The approximate orthometric correction based on normal gravity, which we refer to here as the U.S. Coast and Geodetic Survey (USCGS) orthometric correction, is given as

$$\widetilde{OC} = -C(\bar{\phi})\bar{h}\Delta\phi \qquad (30)$$

where $\bar{h}$ is the mean elevation of the two endpoints, $\Delta\phi$ is the difference in latitude, and

$$C(\bar{\phi}) = 2\alpha \sin 2\bar{\phi}\left[1 + \left(\alpha - \frac{2\beta}{\alpha}\right)\cos 2\bar{\phi}\right] \qquad (31)$$

In (31), $\bar{\phi}$ is the mean latitude of the two endpoints, and α and β, which equal 0.002644 and 0.000007, respectively, are the coefficients in the USCGS normal gravity formula

$$\gamma_0(\phi) = \gamma_{45°}(1 - \alpha \cos 2\phi + \beta \cos^2 2\phi) \qquad (32)$$

[*Bowie and Avers,* 1914].

The equation for the corresponding USCGS dynamic correction is given as

$$\widetilde{DC} = -(D(\bar{\phi}) + 3.147 \times 10^{-7}\,\bar{h})\,\Delta h \qquad (33)$$

where

$$D(\bar{\phi}) = \alpha \cos 2\bar{\phi} - \beta \cos^2 2\bar{\phi} \qquad (34)$$

It contains an error of the order of $\widetilde{OC}$ [*Nassar,* 1977] and is thus not used in practice. Instead, USCGS dynamic heights $\tilde{h}^D$ are computed from the USCGS orthometric height $\tilde{h}^O$ by means of

$$\tilde{h}^D = \tilde{h}^O - D(\phi)\tilde{h}^O - D'(\phi, h)(\tilde{h}^O)^2 \qquad (35)$$

where

$$D'(\phi, h) = 1.5735 \times 10^{-7}(1 + 7.1 \times 10^{-4} \cos 2\phi - 1.53 \times 10^{-7}h) \qquad (36)$$

for each individual point.

If normal gravity is used instead of actual gravity, the distinction between orthometric and normal heights disappears; if $\bar{\gamma}_A{}'$ is substituted for $\bar{g}_A{}'$ in (16), it becomes indistinguishable from (19). Similarly, the distinctions between the various members of the system of orthometric heights also lose their significance.

During the past few decades the gravity coverage in North America has increased to the point that en route gravity values can be obtained either directly or through sufficiently precise interpolation from adjacent observations [*Vaníček et al.,* 1972], so that the next adjustment of all first-order leveling networks probably will be based on observed gravity values. Let us at this point consider the magnitude of the error introduced into height determinations through the use of normal rather than observed gravity, specifically as it applies to the three most important height systems: dynamic, Helmert's orthometric, and Vignal's normal. *Nassar and Vaníček* [1975] have shown that the relation between the dynamic height difference based on actual gravity and the USCGS dynamic height difference based on normal gravity for two nearby points A and B is given as

$$\Delta h_{AB}{}^{D} - \Delta\tilde{h}_{AB}{}^{D} = (\Delta h_{AB}/G)\overline{\Delta g}_{AB} \qquad (37)$$

where ∇g_{AB} is the mean free-air anomaly between A and B with respect to USCGS normal gravity (equation (32)) and G is the reference gravity for dynamic heights, taken by the United States and Canada as normal gravity on the ellipsoid at latitude 45°. Similarly, the discrepancy between the Helmert height difference based on observed gravity and the corresponding USCGS orthometric height difference is

$$\Delta h_{AB}{}^{O} - \Delta\tilde{h}_{AB}{}^{O} = -\frac{\bar{h}_{AB}}{G}(\Delta g_B - \Delta g_A - 0.2238\Delta h_{AB}) \qquad (38)$$

where 0.2238 is in milligals per meter. Finally, the discrepancy between the Vignal normal height difference and the USCGS orthometric height difference is

$$\Delta h_{AB}{}^{N} - \Delta\tilde{h}_{AB}{}^{N} = \Delta h_{AB}{}^{N} - \Delta\tilde{h}_{AB}{}^{O} = (1/G)(\Delta h_{AB}\overline{\Delta g}_{AB} + h_B\delta\gamma_B - h_A\delta\gamma_A) \qquad (39)$$

where $\delta\gamma$ represents the difference between normal gravity used in the definition of normal height (equation (20)) and USCGS normal gravity.

We are now in a position to prove the validity of (24), which we will do for orthometric heights (based on actual gravity). For normal heights the argument is much the same. Let us consider, without any detriment in generality, a closed leveling loop composed of two segments AB and BA. Thus we have

$$\oint dh^O = \Delta h_{AB}{}^O + \Delta h_{BA}{}^O \tag{40}$$

and substituting for the orthometric height differences, we get (cf. (28))

$$\oint dh^O = \Delta l_{AB} + DC_{AB} + \Delta l_{BA} + DC_{BA} \tag{41}$$

since the remaining four terms cancel out. But (41) is clearly (cf. (25))

$$\oint dh^O = \oint dh^D \tag{42}$$

and because the dynamic heights (based on actual gravity) close around a loop, so do the orthometric heights.

It can be shown (from (37) and (39)) that an evaluation of the corrections for dynamic and Vignal heights to such accuracies that their standard deviations are smaller than the absolute value of the corrections themselves requires that the standard deviations of the free-air anomalies satisfy the following inequality:

$$\sigma_{\nabla g} < |\nabla g| \tag{43}$$

In the case of Helmert heights we arrive at the more complicated formula

$$\sigma_{\Delta g} < \frac{|\Delta g_B - \Delta g_A - 0.2238\Delta h_{AB}|}{[2(1 - \mathrm{Cov}\,(\Delta g_A, \Delta g_B)/(\sigma_{\Delta g_A}\sigma_{\Delta g_B}))]^{1/2}} \tag{44}$$

Here the critical quantity is the horizontal gradient of gravity, $\nabla_s g$, in the direction of south. Using this gradient, we can rewrite (44) as

$$\sigma_{\nabla_s g} < |\nabla_s g - 0.2238\beta| \tag{45}$$

where 0.2238 is in milligals per meter and β is the slope of the terrain between A and B.

Numerical evaluation of (37)–(39) indicates that the USCGS dynamic and orthometric height differences closely approximate the corresponding dynamic and Vignal height differences (based on observed gravity), whereas significant discrepancies, of the order of decimeters, commonly exist between USCGS orthometric heights and true Helmert heights. Comparisons between heights based on normal gravity and those based on observed gravity are meaningless: since USCGS heights (based on normal gravity) are path dependent, no analytical solution can be devised to obtain meaningful comparative values.

The magnitude of the discrepancy between precisely defined height differences and their counterparts based on normal gravity remains a source of concern. Although these discrepancies may randomize over distances of thousands or even many hundreds of kilometers, this is certainly not the case over distances of tens or several hundreds of kilometers. Over these shorter distances the gravity-related discrepancies successfully compete for predominance with cumulative first-order leveling errors, a point that has been confirmed by several investigators [*Vignal and Kukkamäki*, 1954; *Krakiwsky*, 1966; *Vaníček et al.*, 1972].

VERTICAL REFERENCE DATUM

Mean sea level is a familiar surface that traditionally has served as a zero reference datum for orthometric heights. Its use as a height datum implies coincidence with the geoid (or quasi-geoid), a generally valid approximation but one that breaks down in detail, partly because of the way in which mean sea level is defined. Thus the use of mean sea level as a geodetic reference surface is not without problems, particularly where it provides the primary constraint in the adjustment of higher-order leveling networks.

Until several decades ago it was believed that mean sea level coincided with the geoid or at least that the departure between the two was negligible. Hence the task of locating the position of the geoid with respect to a reference bench mark at the shoreline reduced simply to determining the position of local mean sea level [*U.S. Coast and Geodetic Survey,* 1965]. Thus the variations of local sea level H_{ISL} (with respect to the arbitrarily defined zero mark on the recording tide gauge) were recorded, the local mean sea level H_{MSL} was subsequently calculated, and the height of the reference bench mark above sea level, $H_{MSL} + \Delta H_{BM-TG}$, was established (Figure 8). The heights of all the other points in the network were then obtained from the heights of the reference bench marks by accumulating the height differences along the interconnected level lines (see, for example, *Cannon* [1929]). In fact, the resultant heights above local mean sea level are only approximately equal to the true heights above the geoid (or quasi-geoid). The difference between these two values is the sea surface topography (Figure 8) and may easily amount to many decimeters. By forcing the height of local mean sea level to zero (that is, neglecting the sea surface topography) we evidently warp the network in such a way that the heights of all the points are distorted.

A second problem arises from the basic determination of local mean sea level. Simple arithmetic averaging of the hourly readings may result in a bias due to aliasing, a consideration that requires that we look more closely at the actual causes of sea level variations. The short-period variations due to tsunamis, waves, and semidiurnal and diurnal tides can be effectively filtered out by using a low-pass filter. Hence we may concentrate on the secular and long-period monthly or annual effects normally involved in mean sea level determinations.

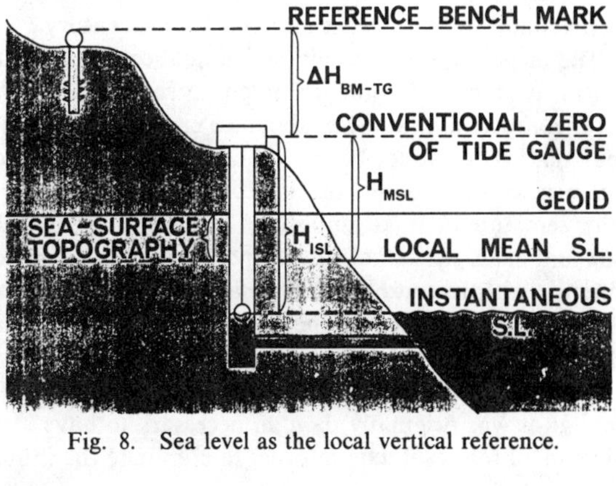

Fig. 8. Sea level as the local vertical reference.

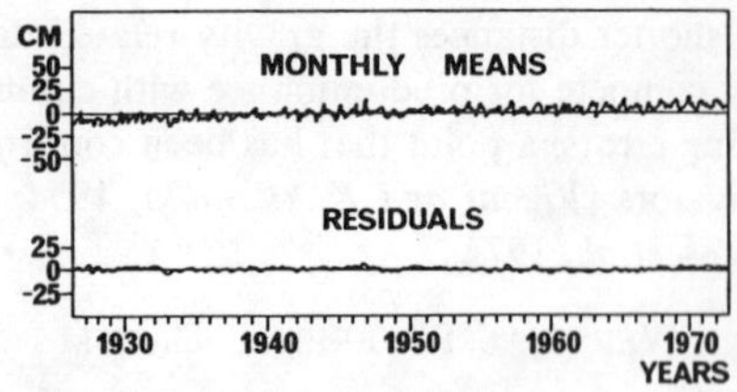

Fig. 9. Sea level variations and residuals for the port of Halifax, Nova Scotia, Canada, based on unpublished data of the Canadian Marine Data Service [after *Anderson,* 1979]. See the text for details.

Montgomery [1937] considered the following long-period phenomena as the chief contributors to long-period variations in sea level: (1) atmospheric pressure variations, (2) dynamic effects of sea current changes, (3) wind variations, (4) thermohaline changes, (5) river discharge fluctuations, (6) changes in bathymetric configuration, (7) glacial melting, and (8) long-period tides. Changes due to variations in atmospheric pressure may reach several decimeters under certain conditions [*Roden,* 1966]. The greater the pressure, the greater the depression of the sea surface; the coefficient of proportionality averages about 1 cm/mbar. The dynamic effects of variations in sea currents are poorly known. On the other hand, the wind stress effect has been studied extensively and may amount to several meters [*Miller,* 1958]. Long-period wind effects are, of course, much smaller. For example, the effect of the wind component normal to the shore pileup amounted to a maximum of a few decimeters in the monthly averages for the port of Halifax, Nova Scotia [*Anderson,* 1979]. Changes in the thermohaline structure of the oceans are among the most important causes of long-period changes in sea level. Fortunately, however, while significant seasonal variations of as much as several decimeters in sea level due to thermohaline structure are known to occur, the structure is generally stable. *Roden* [1966] has computed the temperature effect to be between 1 and 3 cm/°C along the west coast of the United States. Little is known of temporal variations in salinity and their effect on sea level. Fluctuations in river discharge can contribute significantly to changes in sea sevel and may produce sea level changes in the decimeter range [*Vaníček,* 1978]. *Meade and Emery* [1971] concluded that variations in discharge along the east coast of the United States accounted for 7–21% of the variations in the annual means. The effects of changes in bathymetry are generally thought to be small, but they are extremely difficult to evaluate without numerical simulation. Glacial melting and crustal response to the induced changes in water load are presumed to be the main constituents of continuing changes in eustatic sea level. The actual value of the eustatic rise is still a matter of considerable uncertainty; estimates range between 5 and 15 cm/century.

The magnitudes of the long-period tidal constituents are so small as to be of little practical importance. The annual tide has a theoretical amplitude of about 0.5 cm [*Rossiter,* 1966]; its actual amplitude cannot be determined precisely owing to interference with the temperature effect. The equilibrium value of the semiannual tidal amplitude is about 3 cm [*Rossiter,* 1966], and again, its actual value is poorly known owing to interference with meteorologically related effects such as river discharge. The motions of the lunar perigee and the lunar node, identified with periods of 8.85 and 18.6 years, respectively, produce variations of the order of 1 cm [*Vaníček,* 1978]. Although it was originally thought necessary to have at least 18.6 years of sea level data in order to eliminate the influence of the nodal cycle, it now appears that this precaution is unnecessary.

We are now in a position to consider a possible linear filter for obtaining meaningful mean sea level values that may be based on observational periods shorter than 18.6 years. Denoting by $P_i(\tau)$, $i = 1, n$ the available time series of meteorologic and other data (water temperature, barometric pressure, etc.) recorded at or near the tide gauge, we can write the following linear filter:

$$H_{MSL}(\tau_0) = l(\tau) - c_E(\tau - \tau_0) - \sum_{i=1}^{n} c_i[P_i(\tau) - \bar{P}_i] - \sum_{i=1}^{5} (a_i \cos \omega_i \tau + b_i \sin \omega_i \tau) \qquad (46)$$

Here $l(\tau)$ is the discretized sea level record (a time series consisting of, say, monthly averages), τ_0 is the epoch for which the value of local mean sea level is sought, c_E is the rate of the linear part of the eustatic water rise combined with the linear part of the local crustal movements, and the ω_i $(i = 1, 5)$ are the frequencies of the long-period tidal constituents and the Chandler effect. The parameters c_E, c_i $(i = 1, n)$, a_i, b_i $(i = 1.5)$ and H_{MSL} (τ_0) can be determined from least squares regression. The problem of the separation of c_E into the linear eustatic water rise and the (linear part of) crustal movement will be discussed in the last section of the paper. Given the problem at hand it is H_{MSL} (τ_0) that is defined as the signal; the rest of the terms are noise and need not be evaluated here. It should be clear, however, that the value of H_{MSL} necessarily depends on the selected epoch τ_0. To illustrate this point, we show the sea level record from the port of Halifax, together with the residuals created by filtering out the noise due to linear trend, seawater temperature, barometric pressure, river discharge, wind pileup, and the five chief tidal constituents (Figure 9). The mean square error in *Anderson*'s [1979] determination of H_{MSL} was as low as 0.3 cm.

There remains an additional problem related to eustatic changes in sea level that impinges on the definition of the geoid. Evidently, if mean sea level is understood to be varying with time, then so does the geoid and, presumably, the value of the actual potential defining it. This would be an awkward concept to live with, since contemporary changes in eustatic sea level would reduce the heights of points around the world by 0.5–1.5 mm/yr. Alternatively, the epochal definition of mean sea level may be taken as coinciding with the date of the network adjustment, recognizing that this convention fixes the geoid (at least conceptually) in both space and time.

Among the problems relating to the use of mean sea level as a reference datum for heights the existence of sea surface topography is by far the most serious; it cannot be fully resolved on the basis of our present knowledge of the actual behavior of the seas. Three conceptually distinct techniques have been developed to measure sea surface topography: steric leveling, global circulation analysis, and satellite altimetry. Steric leveling is a technique that makes use of variations in seawater density with depth at a number of points. Specifically, on the basis of the questionable assumption that horizontal ocean currents cease at depths of more than a few kilometers the sea surface topography can be evaluated directly by integration over the sampled vertical density profile at each point [*Bjerkness and Sandstrom,* 1910]. Global circulation patterns can be obtained by measuring the three-dimensional distribution of water velocity. On the basis of these data the differential

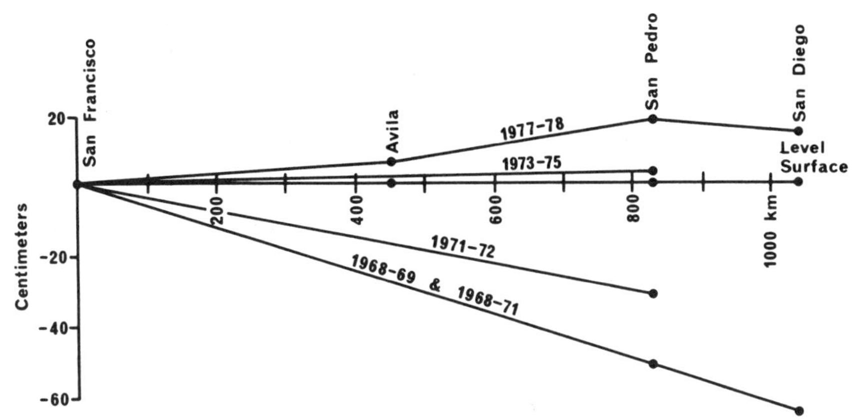

Fig. 10. Relative heights of 1941–1959 local mean sea level as indicated by four successive geodetic levelings, referred to the actual equipotential surface passing through the San Francisco tide station. Orthometric corrections are based on observed gravity [after *Balazs and Douglas*, 1979].

equations of motion of the oceans can be formulated and solved, and the sea surface topography is obtained as a by-product (see, for example, *Hela and Lisitzin* [1967]). Satellite altimetry is the most modern of the three techniques, but its accuracy is as yet too low to yield useful data on sea surface topography except, perhaps, for wavelengths longer than 1500 km [*Mather et al.,* 1976].

All of the techniques described in the preceding paragraph suffer from a common failing: they measure at best the departures of 'instantaneous' sea level from the geoid, or what might be called the 'dynamic sea surface topography.' The 'permanent' sea surface topography can be obtained only after the observations and calculations have been repeated many times over a relatively long period. Moreover, the accuracy of the oceanographic techniques decreases toward the coast; steric leveling cannot even be used in shelf areas, and results from the deep-sea measurements must be transferred toward the coast by using other techniques and information. Since the coastal regime is of particular interest in the present context, it is easy to see why a correction for sea surface topography has not generally been introduced into the definition of the vertical datum.

Owing to the inherent problems involved in oceanographic determinations of sea level topography and the inevitable inaccuracies in geodetically measured height differences between tide stations, it is hardly surprising that there exist occasionally significant differences between the results of geodetic and steric leveling. Probably the most widely publicized discrepancies between geodetically and sterically determined sea slopes occur along the west coast of the United States, particularly along the coast of California [*Balazs,* 1973; *Sturges,* 1974; *Douglas,* 1978; *Balazs and Douglas,* 1979; W. Sturges and R. B. Montgomery, unpublished manuscript, 1974]. The magnitudes of several of these discrepancies are illustrated in the results of a recent study by *Balazs and Douglas* [1979], who show both the geodetically determined sea slopes between San Francisco and San Diego (Figure 10) and how these slopes have been seemingly changing with time. Not only is the sense of the down-to-the-south slopes shown in Figure 10 at variance with that based on steric leveling (W. Sturges and R. B. Montgomery, unpublished manuscript, 1974), but the apparent changes in slope are undisclosed in the sea level records obtained from the identified primary (continually recording) tide stations. Because the bases for these discrepancies are still a matter of considerable controversy, a detailed discussion is beyond the scope of this report. Directionally dependent systematic errors in leveling, questionable assumptions involved in steric leveling, intrasurvey crustal movement, and perhaps a host of other possible bases for the reported sea slope discrepancies are currently under investigation. Nevertheless, while it is certain that intrasurvey movement cannot everywhere account for the separately determined differences in sea slope, vertical displacements during the time-consumptive level surveys and resultant errors in the accumulated height differences between tide stations probably have produced inaccurate characterizations of the sea surface topography along several especially active coastlines.

We have labored the problems surrounding the precise characterization of sea surface topography because it is fundamental to the network adjustment procedure and, ultimately, to the utilization of adjusted heights. An alternative to the standard approach might be based on the selection of a single point somewhere in the middle of the network as the origin of the network such that the heights of local mean sea level at the connected tide gauges would emerge as the product of the leveling and resultant adjustment. The heights of all of the points, including the origin, could, of course, be corrected by some constant value in order to bring the geodetically determined values of mean sea level into approximate agreement with those based on local observations. There exist still other possibilities, moreover, that minimize or bypass the problems of crustal instability and changes in eustatic sea level.

We turn now to a consideration of the adjustment itself. The procedure described here is the one used in the latest adjustment of the North American network in 1928 [*Cannon,* 1929]. Its application depends on the acceptance of three assumptions, all of which are now known to be incorrect:

1. Observed elevation differences Δl between bench mark pairs are contaminated only by random errors characterized by a symmetrical distribution of probability.

2. Mean sea level heights h_k, $k = 1, m$, at the utilized m tide stations are equal to zero; that is, mean sea level is coincident with the geoid at each of the incorporated tide stations.

3. There is crustal stability during the development of the network.

The adjustment of the network should be preceded by the conversion of the nonunique observed elevation differences to height differences based on actual gravity. This was not done in 1928, however, and USCGS approximate orthometric height differences were used instead. Also, no notice was taken of possible crustal movements, and the network was treated as being static. As a result the adjusted heights must be used with caution, since they are clearly affected by a variety of systematic errors. What the magnitude of these errors is will not be known until after the vertical networks are redefined and readjusted in the next decade.

Each leveled segment between bench marks provides one linear observation equation of the type

$$h_j - h_i = \Delta h_{ij} + r_{ij} \quad i, j = 1, \cdots, n$$

or

$$\mathbf{Ah} = \mathbf{\Delta h} + \mathbf{r} \tag{47}$$

where the $\mathbf{r}$ are corrections to observations. Each geodetically connected tide gauge provides a second linear observation equation of the type

$$h_k = 0 \qquad k = 1, \cdots, m$$

or

$$\mathbf{Bh} = \mathbf{0} \tag{48}$$

where $\mathbf{B}$ is a diagonal matrix of ones (those corresponding to the tide gauges) and zeros. Imposition of the minimum condition min $\mathbf{r}^T\mathbf{C}_{\Delta h}^{-1}\mathbf{r}$, where $\mathbf{C}_{\Delta h}$ is the covariance matrix of observed height differences—in 1928 $\mathbf{C}_{\Delta h}$ was a diagonal matrix, implying the acceptance of the hypothesis that the observation errors were statistically independent (with weights inversely proportional to the length of the observed leveled segments)—leads to the following system of normal equations:

$$(\mathbf{A}^T\mathbf{C}_{\Delta h}^{-1}\mathbf{A})\mathbf{h} = \mathbf{A}^T\mathbf{C}_{\Delta h}^{-1}\mathbf{\Delta h} \tag{49}$$

which combined with the constraints implicit in the use of (48) produce

$$(\mathbf{A}^T\mathbf{C}_{\Delta h}^{-1}\mathbf{A} + \mathbf{B})\mathbf{h} = \mathbf{A}^T\mathbf{C}_{\Delta h}^{-1}\mathbf{\Delta h}$$

or simply

$$\mathbf{Nh} = \mathbf{S} \tag{50}$$

Thus

$$\mathbf{h} = \mathbf{N}^{-1}\mathbf{S} \tag{51}$$

The covariance matrix of adjusted heights $\mathbf{h}$ is given as

$$\mathbf{C}_h = \hat{\sigma}_0^2\mathbf{N}^{-1} \tag{52}$$

where

$$\hat{\sigma}_0^2 = \frac{\mathbf{r}^T\mathbf{C}_{\Delta h}^{-1}\mathbf{r}}{n + m - p} \tag{53}$$

and n is the number of leveled segments; p is the number of bench marks identified with adjusted heights. Note that if the second assumption is not enforced, then $\mathbf{N}$ in (50) simply equals $\mathbf{A}^T\mathbf{C}_{\Delta h}^{-1}\mathbf{A}$.

Vertical Crustal Movements Disclosed Through Repeated Levelings

The results of repeated level surveys are being used with increasing frequency as a basis for assessing continuing deformation of the earth's crust. While the use of repeated levelings offers significant advantages over other presently available techniques as a vehicle for the detection of contemporary vertical displacements (owing largely to the inherently high precision of geodetic leveling), it carries with it several problems.

Vertical displacements derived from the results of repeated levelings, whether based on comparisons between observed elevation differences, computed Helmert height differences, adjusted heights, or whatever, are necessarily measured with respect to the geoid (or the quasi-geoid), and because the geoid is a gravity-dependent, time-varying surface [*Whitcomb,* 1976; *Walsh and Rice,* 1979], it is appropriate that we begin this discussion with an analysis of changes in height and changes in the geoid as a function of changes in gravity. We shall attempt to show the extent to which changes in gravity impinge on changes in height in two steps. We begin by comparing temporal variations in orthometric height as a function of changes in gravity; this comparison shows that for all practical purposes, changes in observed elevation and orthometric height differences are equal (that is, the orthometric correction is insensitive to temporal variations in gravity). We then turn to an assessment of variations in the geoid in time (with respect to a fixed geocentric reference ellipsoid) as a function of both changes in elevation and changes in gravity; this analysis shows that over limited areas, changes in the position of the geoid with respect to the reference ellipsoid generally are no more than tiny fractions of the corresponding changes in height with respect to the geoid.

The orthometric height difference between two points A and B is given by

$$\Delta h_{AB}{}^O = \Delta l_{AB} + OC_{AB} \tag{54}$$

where the orthometric correction OC is given in (27). Hence the temporal variation in Δh_{AB} can be written as

$$\delta\Delta h_{AB}{}^O = \delta\Delta l_{AB} + \delta OC_{AB} \tag{55}$$

Since $\delta\Delta l_{AB}$ is the temporal variation in the measured elevation difference, the necessary and sufficient condition for the statement that the two variations are approximately equivalent is that $\delta OC_{AB} \rightarrow 0$. Differentiation of OC yields

$$\delta OC_{AB} = \sum_i \frac{\delta g_i}{G}\Delta l_i + \sum_i \frac{g_i - G}{G}\delta\Delta l_i + \frac{\delta\bar{g}_A{}'}{G}h_A + \frac{\bar{g}_A{}' - G}{G}\delta h_A - \frac{\delta\bar{g}_B{}'}{G}h_B - \frac{\bar{g}_B{}' - G}{G}\delta h_B \tag{56}$$

Because the ratios $(g - G)/G$ are all smaller than 10^{-3} (and usually much smaller), the second, fourth, and sixth terms contribute at most 0.1% to the observed temporal variation in height and can thus be neglected. Let us denote by δg^* the average surface gravity change along the level line between A and B. The mean gravity $\bar{g}'$ along the plumb lines is given by (18). Denoting $\partial g/\partial h$, the Poincaré-Pray gradient (equation (17)), by ∇g,

$$\delta\bar{g}' = \delta g + \delta\nabla g\,\frac{h}{2} + \nabla g\,\frac{\delta h}{2} \tag{57}$$

Equation (56) can then be rewritten as

$$\delta OC_{AB} \doteq \delta g^*\frac{\Delta l_{AB}}{G} + \left(\delta g_A + \delta\nabla g_A\,\frac{h_A}{2} + \nabla g\,\frac{\delta h_A}{2}\right)\frac{h_A}{G} - \left(\delta g_B + \delta\nabla g_B\,\frac{h_B}{2} + \nabla g\,\frac{\delta h_B}{2}\right)\frac{h_B}{G} \tag{58}$$

Since δg_A, δg_B, and all of the other changes will tend to have the same sign, the extreme change should occur where either

h_A or h_B is equal to 0. Assume, for example, that $h_B = 0$ and $h_A = 10^3$ m; assume in addition an extreme observed change in surface gravity of 0.1 mGal ($\delta g^* = \delta g_A = 0.1$ mGal). Substituting in (55), we obtain

$$\delta OC_{AB} \doteq 10^{-4}\text{ m} + 10^{-4}\text{ m} + 0.5\ 10^6\text{ m}^2\ \frac{\delta\nabla g_A}{G} + 0.5\ 10^{-4}\ \delta h_A \tag{59}$$

It is inconceivable that $\delta\nabla g$ could be larger than a few percent of the gravity gradient ∇g itself, so that the third term should contribute at most something of the order of 10^{-3} m. Thus the maximum change in the orthometric correction, even under these extreme conditions, should be of the order of 1 mm; hence this change can be safely neglected in considering height changes based on repeated levelings.

We turn now to a consideration of variations in the height of the geoid above the geocentric ellipsoid in response to temporal variations in the gravity field; our analysis is based on the acceptance of Stokes' formula (see, for example, *Heiskanen and Moritz* [1967])

$$N \doteq \frac{R}{4\pi\bar{g}}\int_0^{2\pi}\int_0^{\pi} \Delta g(\psi, \alpha)S(\psi)\sin\psi\, d\psi\, d\alpha \tag{60}$$

Stokes' formula shows the relation between the separation N between the geoid and the fixed geocentric reference ellipsoid that generates the normal gravity γ used in evaluating the gravity anomaly Δg and the actual gravity g at the earth's surface (through the gravity anomaly Δg) at an angular distance ψ and azimuth α; $\bar{g}$ is the global mean gravity, R is the mean radius of the earth, and S is the symbol for Stokes' function. We identify the free-air gravity anomaly as Δg; as long as the indirect effect of this anomaly is small, we need not be concerned about which particular anomaly is used.

It is expedient at this point to introduce the mean gravity anomaly within the spherical distance ψ, i.e.,

$$\overline{\Delta g}(\psi) = \frac{1}{2\pi}\int_0^{2\pi} \Delta g(\psi, \alpha)\, d\alpha \tag{61}$$

where $\frac{1}{2}S(\psi)\sin\psi$ is identified as $F(\psi)$ (see, for example, *Lambert and Darling* [1936]). Substituting in (60), we can write the temporal variation in N, δN, as

$$\delta N \doteq \frac{R}{\bar{g}}\int_0^{\pi} \delta\overline{\Delta g}(\psi)F(\psi)\, d\psi \tag{62}$$

where $\delta\overline{\Delta g}(\psi)$ is the mean temporal variation of the free-air gravity anomalies over the distance ψ.

Let us now limit this discussion to those cases where changes in both gravity and elevation occur entirely within a radius ψ_{max}. Equation (62) can then be evaluated through integration by parts; denoting $\int F(\psi)\, d\psi$ by $\Phi(\psi)$ and realizing that for $\psi < 10°$ (corresponding to about 1100 km), $\Phi(\psi)$ is very nearly equal to $0.02\psi°$ [*Lambert and Darling,* 1936], the equation for temporal variations in the configuration of the geoid can be written as

$$\delta N \doteq -0.128\text{ m mGal}^{-1}\int_0^{\psi_{max}} \psi°\left(\frac{\partial\delta\bar{g}(\psi)}{\partial\psi} + \nabla g\,\frac{\partial\delta\bar{h}(\psi)}{\partial\psi}\right) d\psi \tag{63}$$

The first term in parentheses is the horizontal gradient of the mean variation in surface gravity. The second term represents the horizontal gradient of the mean variation in height multiplied by the free-air vertical gradient of gravity.

To obtain numerical estimates of this effect, we must now postulate a mathematical model for changes in $\delta\bar{g}$ and $\delta\bar{h}$. Let us assume a linear decrease from maximum values at $\psi = 0$ (the point for which we seek δN) to zero, where $\psi = \psi_{max}$. This 'conical' model is described by

$$\delta\bar{g}(\psi) = (\delta g_{max}/\psi_{max})(\psi_{max} - \psi) \tag{64}$$

$$\delta\bar{h}(\psi) = (\delta h_{max}/\psi_{max})(\psi_{max} - \psi) \tag{65}$$

Substituting these expressions in (58), we obtain

$$\delta N \doteq 0.064\text{ m mGal}^{-1}\ (\delta g_{max} + 0.309\text{ m mGal}^{-1}\ \delta h_{max})\psi_{max}° \tag{66}$$

For comparison, let us obtain a similar estimate for the same conical model of $\delta\bar{g}(\psi)$ combined with a 'tapered cylindrical' model (see Figure 11) for $\delta\bar{h}(\psi)$; $\delta\bar{h}$ must be written as

$$\begin{aligned}\delta\bar{h}(\psi) &= \delta h_{max} && \psi < \psi_1\\ \delta\bar{h}(\psi) &= \frac{\delta h_{max}}{\psi_1 - \psi_2}(\psi - \psi_2) && \psi_1 \le \psi \le \psi_2\\ \delta\bar{h}(\psi) &= 0 && \psi_2 < \psi\end{aligned} \tag{67}$$

This model yields

$$\delta N \doteq 0.064\text{ m mGal}^{-1}\left(\delta g_{max} + 0.309\text{m mGal}^{-1}\ \delta h_{max}\ \frac{\psi_2 - \psi_1}{\psi_{max}}\right)\psi_{max}° \tag{68}$$

evidently a value smaller than that given in (62).

Finally, we turn to the calculation of numerical estimates of changes in the height of the geoid as a function of temporal variations in gravity. Again assuming that $\delta g_{max} = 0.1$ mGal and using the conical (or worst) model of radius 110 km ($\psi_{max}° = 1°$), we obtain

$$\delta N \doteq 0.0064\text{ m} + 0.020\ \delta h_{max} \tag{69}$$

Hence it is clear that even in this worst case, where the change in gravity and the change in height have the same sign, the changes in the separation between the geoid and the reference ellipsoid are but small fractions of the corresponding changes in height with respect to the geoid, provided that the affected area does not extend significantly beyond 1000 km.

Changes in height (which for practical purposes can be equated precisely with vertical displacements) preferably are measured with respect to some control point, such as the reference bench mark associated with a primary tide station, whose vertical stability can be independently assessed. Such assessments, of course, are again relative ones in the sense that

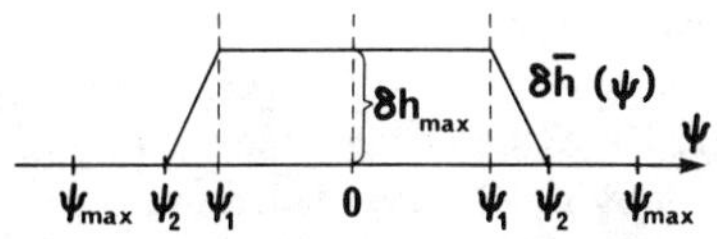

Fig. 11. Schematic illustration showing 'conical-cylindrical' deformational model used in calculating temporal variations in the configuration of the geoid.

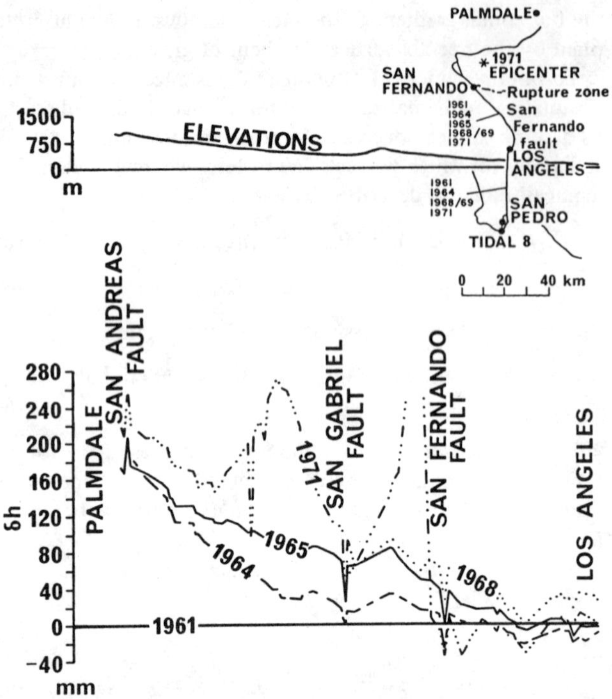

Fig. 12. Profiles showing topography and changes in observed elevation δh since 1961 along a level line extending northward from Los Angeles to Palmdale, California, based on comparisons with bench mark Tidal 8, San Pedro, as invariant in height. Sharp peaks in the 1971 profile are coseismic effects associated with the 1971 San Fernando earthquake (modified after *Castle et al.* [1974]).

the stability of the selected tide station can be evaluated only with respect to a second tide station outside the area of interest. Nevertheless, evaluations of this sort generally provide adequate control for most regionally limited studies in crustal dynamics. While the signal that we seek (i.e., the linear part of the crustal movements (part of c_E) plus possibly episodic movements) may be obscured in the sea level records, notably by eustatic water rise, differencing the records against those of a second tide station within the same oceanographic regime usually will reduce this noise to geodetically tolerable levels. For example, differencing of the annual means at San Pedro, California, against those at San Diego, California, produces maximum excursions of about 30 mm off the regression line through the differenced data [*Wood and Elliott,* 1979, p. 258]. Thus the relative change in sea level between these two tide stations (0.9 ± 0.2 mm/yr) probably closely approximates the inverse of the relative rate of crustal movement between San Pedro and San Diego. Moreover, because the differenced curve is smooth and straight, it is reasonably certain that neither station has sustained significant episodic movement.

In the absence of a nearby tide station we can resort to the conceptually similar artifice of referring any vertical displacements to a control point whose history has been established through repeated levelings connecting this point to some distant set of bench marks of known history. The choice of the control point and its required stability obviously will be governed in part by the magnitude and character of the expected displacements referred to this point. Where the displacements are large, the choice of the reference mark may be dictated largely by geologic considerations. For example, the commonly large displacements associated with the extraction of underground fluids [*Poland and Davis,* 1969] can be quite usefully measured with respect to a bedrock mark adjacent to the disturbed area, provided that this mark does not occur within a landslide or a zone of possibly active faults. Thus even if the region including the selected mark were to sustain tectonic uplift or collapse, the effects of this movement generally could be dismissed as being trivial, since the whole of the investigated system would be more or less uniformly affected.

Probably the simplest and least ambiguous procedure used to detect vertical displacements based on repeated levelings consists of direct comparisons between successively generated sets of observed elevations along the same survey route. Because it is simple and unambiguous, at least in the sense that the observed elevation changes are at once almost precisely equivalent to both height changes and vertical displacements measured with respect to some geometrically invariant surface and independent of spatial variations in the gravity field, profiling of changes in observed elevations along the same line probably is the most widely used of the several available techniques (see, for example, *Pavoni and Green* [1975]). A representative product, which shows changes in observed elevations derived from a series of first-order levelings between Los Angeles and Palmdale, California, is illustrated in Figure 12. The observed elevations used in calculating the vertical displacements shown in Figure 12 have been reconstructed with respect to a fixed height for bench mark Tidal 8, San Pedro. However, because Tidal 8 is adjacent to a primary tide station that has been characterized by a history of relatively positive movement, comparisons between the results of successive sets of observed elevations emanating from Tidal 8 are biased against the recognition of uplift measured with respect to an absolutely invariant datum [*Castle et al.,* 1974].

If it is assumed that leveling errors are random and statistically independent, one standard deviation in the derived vertical displacement $\delta\Delta h$ between any two adjacent bench marks is given simply as

$$\sigma_{\delta\Delta h} = (\sigma_1{}^2 + \sigma_2{}^2)^{1/2} \tag{70}$$

where σ_1 and σ_2 are the standard deviations in $\Delta h(\tau_1)$ and $\Delta h(\tau_2)$, the elevation differences obtained at times τ_1 and τ_2, respectively. For example, if we consider the results shown in Figure 12, where all of the reconstructed elevations were based on first-order levelings carried out between 1961 and 1971, σ_1

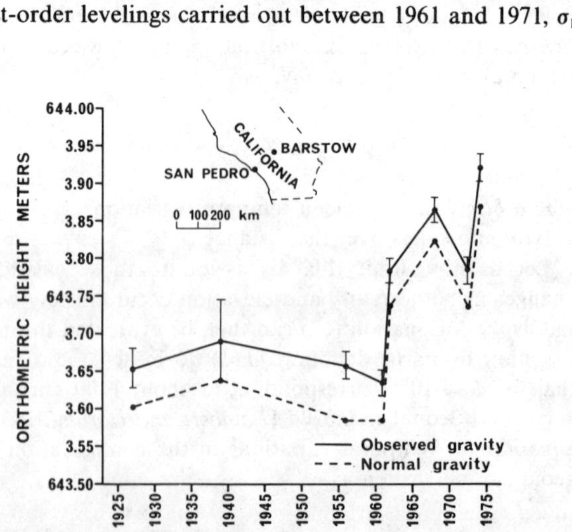

Fig. 13. Changes in orthometric height at bench mark E43, Barstow, California, based on levelings over four separate routes and reconstructed with respect to bench mark Tidal 8, San Pedro, as invariant in height. One standard deviation error bars are shown for conventionally estimated random error only.

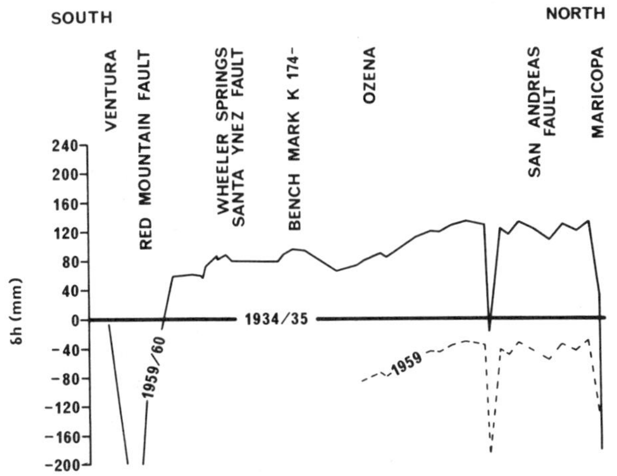

Fig. 14. Profile showing vertical displacements δh during the interval 1934/1935–1959/1960 along a level line extending northward from Ventura to Maricopa, California, based on comparisons with bench mark Tidal 8, San Pedro, as invariant in height. Subsidence toward the southern end of the line is centered over the Ventura oil field; that at the northern end of the line lies within the Midway-Sunset oil field. See the text for details.

$= \sigma_2 = 1.0$ mm $L^{1/2}$ (see section on the measurement system). Accordingly, one standard deviation in the measured displacements at Palmdale (with respect to San Pedro) should be about 16 mm. Had either of the comparative levelings been to only third-order standards, the corresponding standard deviation would have jumped to about 46 mm, a change that shows the extent to which accuracy of the measured displacement is controlled by the results of the lower-order survey.

The detection of elevation- or slope-dependent systematic errors in height changes commonly is sought by searching for a correlation between the apparent vertical displacements and the configuration of the terrain (see section on the measurement system). While this technique is certainly a useful one, a rough correlation between height changes and topography is reasonably expected in areas of continuing tectonic activity. To categorically dismiss measured vertical displacements in areas such as southern California as the artifactual products of systematic error only because of the above correlation would be unbelievably myopic. In fact, exhaustive comparisons between vertical displacement and terrain profiles from southern California, together with examination of numerous network misclosures [*Castle et al.,* 1980] suggest that neither rod nor refraction errors have significantly affected measured elevation differences in this topographically and climatically diverse regime. It is conceivable, of course, that refraction error is so overwhelmingly slope dependent that it tends to persist regardless of weather, length of day, and so forth. If this is in fact the case, residual refraction may produce significant errors in measured elevation differences yet contribute very little to errors in measured vertical displacements based on repeated surveys. Such a residual refraction tends to be larger in the case of older measurements, when the sights tended to be longer.

This subjective conclusion is strongly supported by the following theoretical argument. As we have already shown (see section on the measurement system), there is probably a strong positive statistical dependence between the results of backward and forward runs [*Remmer,* 1975]. Thus we should expect to see a strong positive statistical dependence between the results of any two levelings (from which the vertical displacements may be derived) over the same route. If we identify the corresponding covariance as σ_{12}, one standard deviation in the computed relative displacement becomes

$$\sigma_{\delta\Delta h} = (\sigma_1^2 - 2\sigma_{12} + \sigma_2^2)^{1/2} < (\sigma_1^2 + \sigma_2^2)^{1/2} \qquad (71)$$

Denoting the standardized covariance $\sigma_{12}/(\sigma_1 \cdot \sigma_2)$ by ρ_{12} and assuming, without any concession to the generality, that $\sigma_1 = \sigma_2$, we obtain

$$\sigma_{\delta\Delta h} = [2\sigma_1(1 - \rho_{12})]^{1/2} \qquad (72)$$

which is evidently $(1 - \rho_{12})^{1/2}$ times smaller than if there were no statistical dependence. Clearly, those errors dependent on height, height gradient, azimuth of sight, latitude, or simply location tend to cancel through this statistical dependence, and *Remmer*'s [1975] figure for ρ_{12} of 0.73 seems very reasonable. Hence the vertical displacements computed from repeated levelings along the same route are much less likely to be contaminated by systematic errors than the heights themselves.

Where the computed vertical displacements are based on levelings over different routes, significantly larger errors in the resultant displacements may be expected. Moreover, in order to minimize these errors it is especially important that observed rather than normal gravity be used in computing successive heights. By way of comparison, calculation of successively determined heights based on both normal and observed gravity at Barstow, California (Figure 13), suggests that these gravity-dependent errors can be of the order of several centimeters. While the actual heights based on observed gravity differ substantially from those based on normal gravity, the maximum discrepancy between height changes based on observed gravity and those based on normal gravity is about 20 mm. Although this is thought to be a fairly representative value, these discrepancies may range up to 40 mm or more in areas characterized by steep and irregular gravity gradients.

Intrasurvey movement rather than survey measurement error may be the most significant source of error encountered in the detection of vertical displacements based on repeated levelings. The potential magnitude of this error can be shown by

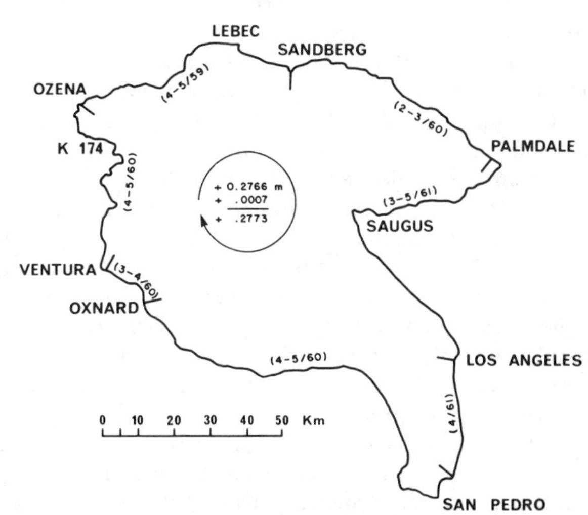

Fig. 15. Misclosure around the circuit San Pedro–Oxnard–Ventura–Ozena–Lebec–Sandberg–Palmdale–Saugus–Los Angeles–San Pedro, California, based on first-order levelings completed during the period April 1959 to May 1961. Summation consists of the observed misclosure plus an orthometric correction based on normal gravity; this figure enlarges to 0.2923 m where the orthometric correction is based on observed gravity.

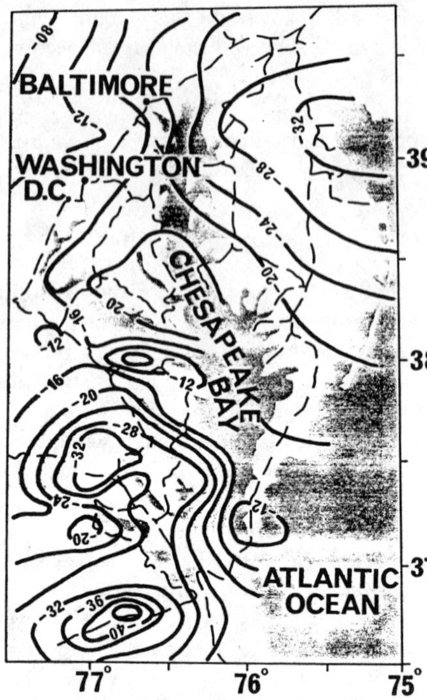

Fig. 16. Spatially smoothed interpolation of vertical movements (in centimeters per century) in the Chesapeake Bay area based on a dynamic velocity adjustment (modified after *Holdahl and Morrison* [1974]).

means of a relatively simple example, again drawn from southern California. In this particular example we consider displacements generated during the 26-year interval 1934–1960 along a line extending northward from Ventura to Maricopa (Figure 14). Both sets of observed elevations from which these displacements were derived have been reconstructed from discontinuous levelings. The 1934/1935 elevations between Ventura and bench mark K 174 were developed from the results of leveling completed during the period April–May 1934; those between bench mark K 174 and Maricopa are based on surveys completed during the period December 1934 to February 1935. The 1959/1960 observed elevations between Maricopa and Ozena are based on levelings completed during the period March–April 1959; those between bench marks Ozena and Ventura were developed from surveys completed during the period April–May 1960. It has been assumed in reconstructing the elevations from which these height changes were derived that heights in the areas of bench mark K 174 and Ozena remained invariant during the intervals April 1934 to February 1935 and April 1959 to May 1960, respectively. Neither area has been associated with the extraction of underground fluids, and neither was associated with significant earthquake activity during these intervals; hence to assume that the junction bench marks remained invariant during these two short intervals is both reasonable and conservative. It is also wrong.

While it can be shown with considerable confidence that the 1934/1935 reconstructed elevations are free of significant error of any sort, the reconstructed 1959/1960 elevations are seriously flawed. A level circuit based on surveys completed during the period April 1959 to May 1961, which included both Tidal 8 and the 1960 leveling between Ventura and Ozena, produced an unusually large misclosure (Figure 15). Because this nearly 0.3-m misclosure is more than 15 times the one standard deviation figure and more than 3 times the rejection limit for first-order leveling of this vintage, it is virtually certain that the relevant leveling was contaminated by one or more blunders or that the area was affected by significant intrasurvey movement. Comparisons against the results of earlier and later levelings indicate that the misclosure cannot be attributed to survey blunders; thus it is more likely that the area embraced by this circuit (Figure 15) sustained major aseismic movement during the completion of the leveling. Because we can show that the established 1953 elevation of a control point north of Lebec, from which the 1959 leveling was propagated southward, remained virtually invariant with respect to Tidal 8 between 1953 and at least the middle of 1959, the 1959 observed elevations can be reconstructed as if the 1959 leveling had emanated directly from Tidal 8 [*Castle et al.,* 1980]. This reconstruction indicates that the 1959 elevations of Ozena (with respect to Tidal 8) were about 0.16–0.17 m below the 1960 elevations. Moreover, we can also show that the 1960 height of Boundary Monument 2, Lebec (Figure 15), was at least 0.11 m below that which obtained in 1959. In other words, between the spring of 1959 and the following spring the region between Ozena and Lebec sustained an up-to-the-west tilt of 0.27–0.28 m, a tilt that accounts almost precisely for the indicated misclosure. Accordingly, the actual height changes along the northern end of the Ventura-Maricopa line during the period 1934/1935–1959 are thought to be closely approximated by the dashed line representation shown in Figure 14. This representation, unlike that based on our 'conservative' approach (Figure 14), shows that the region between Ozena and Maricopa apparently collapsed between 1934/1935 and 1959, a collapse that we interpret as a coseismic effect associated with the 1952 Kern County earthquake [*Castle et al.,* 1980].

The preceding discussion suggests that assumptions of stability during the course of a specified leveling may be meaningless where these assumptions cannot be verified independently. Useful verification may be based on either geodetic or geologic considerations. Significant geodetic tests include those based on misclosures, comparative heights derived from levelings introduced during the intrasurvey period and emanating from areas of otherwise demonstrable stability, repeated levelings through the area of concern, and so forth. Geologic indices of potential instability are less objective but may be no less valuable. For example, an explicit knowledge of the surficial deposits and changes in the groundwater regime along the survey route may permit the detection of subsidence prone junctions even if, as is commonly the case, there is no previous history of geodetically defined subsidence. Similarly, a perception of the geologically recent tectonic history could contribute to the recognition or at least the suspicion of various sorts of aseismic disturbances during the intrasurvey period. For example, were the leveling to originate or to be interrupted for extended periods within an area of Holocene or even late Quaternary faulting, the occurrence of aseismic slip (fault creep) at depth might be reasonably anticipated such that the possibility of height changes during the intrasurvey period should be at least entertained. Again, where geologic criteria suggest recent tectonic uplift or subsidence, especially where the vertical displacement rates can be shown to have changed through relatively short distances or over relatively short periods [*McCrory and Lajoie,* 1979; *Wehmiller et al.,* 1979], otherwise unverified assumptions of stability are clearly suspect.

Because interrupted or discontinuous levelings may provide

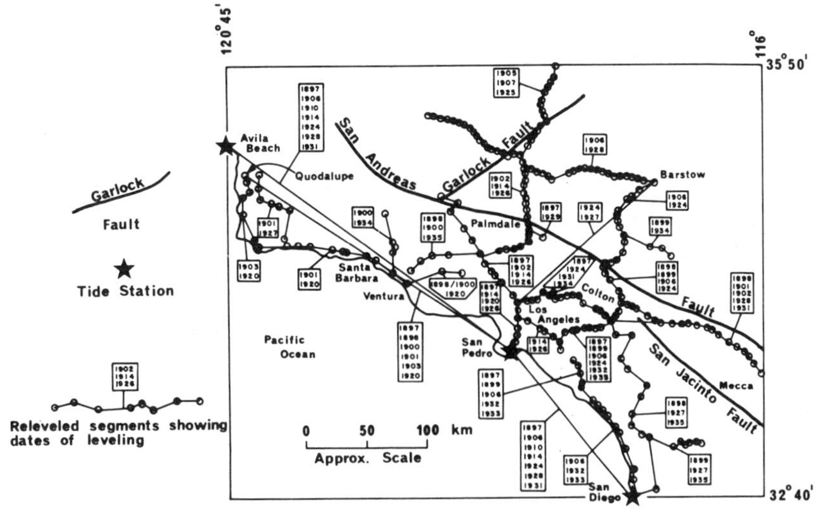

Fig. 17*a*

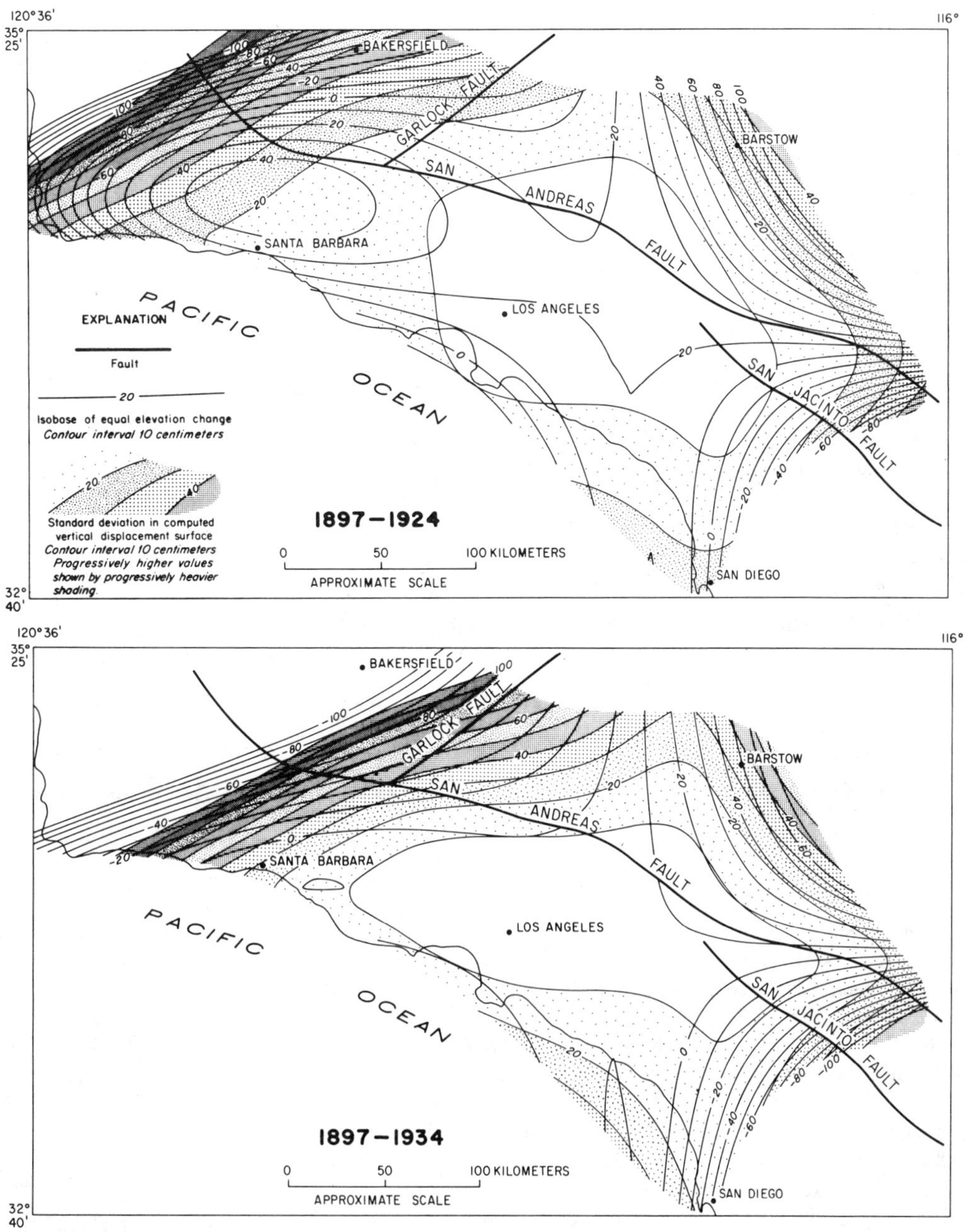

Fig. 17*b*

Fig. 17. Modeled vertical displacements in the area of the southern California uplift during selected periods within the interval 1897–1935. (*a*) Map showing distribution of releveled segments incorporated in the modeled solution. (*b*) Computed vertical displacement surfaces incorporating a power of 2 in latitude, a power of 3 in longitude, a power of 3 in time, and no episodes for the periods 1897–1924 and 1897–1934 [after *Vaníček et al.*, 1979].

the only bases for comparisons, vertical crustal movements commonly have been characterized as vertical velocities [*Holdahl and Morrison,* 1974; *Reilinger and Oliver,* 1976; *Reilinger et al.,* 1977]. Although this approach rather neatly bypasses the problem of intrasurvey movement, it is built on an assumption of constant vertical velocity, where the velocity at any point is assumed to be linear with respect to time. The least ambiguous indication of the validity of this assumption can be obtained through examining velocity misclosures. That is, provided that the area is indeed characterized by constant vertical velocities and that the survey data are parts of error-free network relevelings, the velocity closures around any loop should sum to zero regardless of the temporally disconnected nature of the levelings. Because these circuit closures generally do not sum to zero, various 'dynamic' adjustment procedures have sometimes been used to obtain the most representative, if not the most realistic, portrayal of the vertical velocity field. For example, this technique has been used with seeming success in the Chesapeake Bay area (Figure 16), where both the sea level data [*Holdahl and Morrison,* 1974, p. 386] and various geologic considerations support the notion of constant vertical velocity. Nonetheless, where there is a reasonable likelihood that the vertical displacements have been episodic or spasmodic, adjustments of this sort are generally of little value or even meaningless.

In the absence of continuous line or network relevelings, a smoothed reconstruction of the vertical displacement field can be obtained through the fitting of a mathematically defined surface to the successive results of segmented relevelings where each segment is treated as a tilt element (that is, as a horizontal gradient of velocity between the two endpoints). The simplest and most generally utilized procedure is the surface-fitting technique, which is based on least squares regression [*Korn and Korn,* 1968]. Specifically, let us assume that we seek a vertical velocity surface in the form

$$v(\phi, \lambda) = \sum_{i=1}^{m} \phi_i(\phi, \lambda)c_i = \mathbf{\Phi}^T(\phi, \lambda)\mathbf{c} \tag{73}$$

where the ϕ_i are some selected base functions of positions ϕ, λ. Then for each releveled segment A, B the following observation equations can be written as

$$\delta v_{AB} = v(\phi_B, \lambda_B) - v(\phi_A, \lambda_A) = [\mathbf{\Phi}^T(\phi_B, \lambda_B) - \mathbf{\Phi}^T(\phi_A, \lambda_A)]\mathbf{c} \tag{74}$$

or, in a more compact fashion,

$$\delta v_{AB} = \mathbf{\Delta\Phi}^T(\phi_A, \lambda_A, \phi_B, \lambda_B)\mathbf{c} \tag{75}$$

[*Vaníček and Christodulides,* 1974].

It will be observed that this formulation is not dissimilar to the formulation of a partial differential equation. While *Vaníček and Christodulides* [1974] utilized two-dimensional algebraic functions of up to degree 4 for the base function, other base functions, such as quadrics, also may be used [*Holdahl and Hardy,* 1979].

Given a sufficiently high density of releveled segments, so distributed that the selected functions $\mathbf{\Phi}$ are linearly independent, the coefficients $\mathbf{c}$ can be estimated by means of a least squares technique. The weight matrix $\mathbf{P}_t$ (diagonal) is assembled from the inverse values of variances computed for each releveled segment (compare (70)–(72)). The normal equations thus become

$$\mathbf{\Delta\Phi P}_t\mathbf{\Delta\Phi}^T\mathbf{c} = \mathbf{\Delta\Phi P}_t\,\mathbf{\delta v} \tag{76}$$

Tilt data obtained from the releveled segments can be combined with tide gauge data, where each station supplies one point velocity equation. The two systems of observation equations (tilt and point velocity) merge as

$$[\mathbf{\Delta\Phi}^T \vdots \mathbf{\Phi}^T]^T\mathbf{c} = [\mathbf{\delta v} \vdots \mathbf{v}]^T \tag{77}$$

The appropriate weight matrix $\mathbf{P}$ for the corresponding system of normal equations is evidently

$$\mathbf{P} = \begin{bmatrix} \mathbf{P}_t & \vdots & \mathbf{O} \\ \cdots & \vdots & \cdots \\ \mathbf{O} & \vdots & \mathbf{P}_v \end{bmatrix} \tag{78}$$

where $\mathbf{P}_v$ is the diagonal weight matrix for the point velocities, with weight inversely proportionate to their variances. The normal equations then can be written as

$$[\mathbf{\Delta\Phi}^T \vdots \mathbf{\Phi}^T]\mathbf{P}[\mathbf{\Delta\Phi}^T \vdots \mathbf{\Phi}^T]^T\mathbf{c} = [\mathbf{\Delta\Phi}^T \vdots \mathbf{\Phi}^T]\mathbf{P}[\mathbf{\delta v} \vdots \mathbf{v}]^T \tag{79}$$

The covariance matrix $\mathbf{C}_c$ of the resulting coefficients $\mathbf{c}$ is obtained as a properly scaled inverse of the matrix of normal equations. It can then be used in the evaluation of the standard deviation of the predicted velocity at any point (ϕ, λ). The covariance law yields

$$\sigma_v(\phi, \lambda) = \{[\mathbf{\Phi}^T(\phi, \lambda)\mathbf{C}_c\mathbf{\Phi}(\phi, \lambda)]\}^{1/2} \tag{80}$$

Where the movements cannot be assumed to have been linear with respect to time, the approach outlined in the preceding paragraphs provides 'average' velocities only. Nonetheless, this technique can be generalized to model accelerated movements as well and is especially useful where the data are randomly scattered in time as well as space. In developing a four-dimensional model it is expedient that we seek vertical displacements rather than velocities. Denoting the vertical displacement of the point (ϕ, λ) during the period τ_0, τ by (ϕ, λ, τ) and assuming

$$u(\phi, \lambda, \tau_0) = 0 \tag{81}$$

everywhere, this four-dimensional model may be written as

$$u(\phi, \lambda, \tau) = \mathbf{u}^T\mathbf{\Phi}^T(\phi, \lambda)\mathbf{T}^T(\tau)\tilde{\mathbf{c}}\mathbf{u} \tag{82}$$

where $\mathbf{T}$ is the vector of base functions of time such that

$$\mathbf{T}(\tau_0) = \mathbf{0} \tag{83}$$

$\mathbf{u}$ is a vector of ones, and $\tilde{\mathbf{c}}$ is a matrix of coefficients. Repeated levelings provide the spatial difference of the displacements of the two end points for each releveled segment; that is,

$$\begin{aligned} \delta u_{AB}(\tau_1, \tau_2) &= u(\phi_B, \lambda_B, \tau_1, \tau_2) - u(\phi_A, \lambda_A, \tau_1, \tau_2) \\ &= \Delta H_{AB}(\tau_2) - \Delta H_{AB}(\tau_1) \end{aligned} \tag{84}$$

during the period τ_1, τ_2. Thus the observation equation becomes

$$\begin{aligned} \delta u_{AB}(\tau_1, \tau_2) &= \mathbf{u}^T[\mathbf{\Phi}^T(\phi_B, \lambda_B) - \mathbf{\Phi}^T(\phi_A, \lambda_A)][\mathbf{T}^T(\tau_2) - \mathbf{T}^T(\tau_1)]\tilde{\mathbf{c}}\mathbf{u} \\ &= \mathbf{u}^T\mathbf{\Delta\Phi}^T(\phi_A, \lambda_A, \phi_B, \lambda_B)\mathbf{\Delta T}^T(\tau_1, \tau_2)\,\tilde{\mathbf{c}}\mathbf{u} = \mathbf{B}^T\mathbf{c} \end{aligned} \tag{85}$$

Merging these equations with the corresponding observation equations describing the accelerated variations of sea level,

$$u_A(\tau_1, \tau_2) = \mathbf{u}^T\mathbf{\Phi}^T(\phi_A, \lambda_A)[\mathbf{T}^T(\tau_2) - \mathbf{T}^T(\tau_1)]\mathbf{c}\mathbf{u} \tag{86}$$

we obtain normal equations for **c**. These equations are solved for **c**, which can be used in turn to predict the uplift of any point (ϕ, λ, τ) in both space and time. The standard deviations of the predicted uplifts are calculated from formulas similar to those used in calculating the standard deviations associated with the presumed linear movements. Where the occurrence of episodic movement is precluded, and the power in time is specified as being equal to one, the general four-dimensional model automatically reduces to one characterized by constant velocity.

The surface-fitting approach offers several significant advantages over other techniques: (1) it permits the use of otherwise unusable, disconnected releveled segments, and (2) it provides for an objective characterization of variations in the vertical velocity field and hence does not depend on assumptions of intrasurvey stability or constant vertical velocity. The chief disadvantage of any surface-fitting technique is that the smoothing process tends to discard or subdue conceivably significant short-wavelength features; conversely, it is not recommended that the number of the base functions be increased solely for the purpose of finding these local features. The experience of the authors has been that an increase does not usually result in an improvement in statistical confidence in the solution.

The selection of the base functions, including the selection of the power of the polynomial, may be either subjective, in the sense that it is based on some preknowledge of the deformational history, objective, in the sense that it depends on various statistical tests for judging the goodness of fit, or some combination of both. In modeling an early twentieth-century deformational feature in southern California (Figure 17) the initial choice of the power of the algebraic polynomial was based on an earlier perception of the evolution of this feature developed from a classical reconstruction. The selection of this power can be further refined through a statistically controlled trial and error process. For example, while it seems likely that the evolution of the modeled feature was accompanied by at least one sharply defined displacement episode, because it failed to improve on the statistical confidence, the inclusion of an episodic event was rejected in the reconstruction shown here (Figure 17). The statistical index that provides the best insight into the goodness of the fit is the standard deviation field calculated for each stipulated displacement surface (Figure 17). Although the standard deviations do not reflect the effect of systematic errors, they comprise an especially useful vehicle for establishing the reality of the represented displacement field and obviously closely correlate with the distribution (in both space and time) and quality of the releveled segments.

While existing surface-fitting techniques are ill suited to the description of sharply or irregularly defined vertical displacement fields, such as those associated with coseismic rupturing athwart active faults, they are especially useful in describing smoothly varying displacements. Specifically, subsidence associated with the extraction of underground fluids can be both adequately and relatively inexpensively described by means of four-dimensional modeling. Because this technique can be used on any scale and because extraction-induced subsidence characteristically proceeds smoothly in both space and time (see, for example, *Poland and Davis* [1969]), the fitting of a time-varying surface to the results of segmented relevelings should produce a much better representation of the changing surface configuration within a recognized subsidence domain than can be obtained through conventional network relevelings. Similarly, changes in broadly defined tectonic displacement fields of the sort epitomized by the southern California uplift [*Castle,* 1978] can be monitored at minimal expense through randomly distributed (again, in both space and time) segmented relevelings and four-dimensional modeling of the results. Moreover, surface-fitting techniques probably can be adapted to the monitoring of height changes even on a continental scale, particularly if the analyses are based on the results of modeling within successively overlapping study areas of much smaller dimensions.

Acknowledgments. We thank C. T. Whalen, S. R. Holdahl, and M. Kumar of the U.S. National Geodetic Survey; R. C. Jachens, J. P. Minta, and M. R. Elliott of the U.S. Geological Survey; and E. G. Anderson of the University of New Brunswick (Canada) for numerous suggestions and comments that led to significant improvements in an earlier version of this paper.

REFERENCES

Anderson, E. G., Modelling of physical influences in sea level records for vertical crustal movement detection, in *Proceedings of GEOP-9 Conference, Columbus, Ohio, October, 1978,* pp. 145–152, Department of Geodetic Science, Ohio State University, Columbus, 1979.

Balazs, E. I., Local mean sea level in relation to geodetic leveling along the United States coastlines, report, 9 pp., *Nat. Geod. Surv.,* Rockville, Md., 1973.

Balazs, E. I., and B. C. Douglas, Status report on elevation differences between mean sea levels in California, submitted to *J. Geophys. Res.,* 1979.

Birdseye, C. H., Topographic instructions of the United States Geological Survey, *U.S. Geol. Surv. Bull., 788,* 432 pp., 1928.

Bjerkness, V., and J. W. Şandstrom, *Dynamic Meteorology and Hydrography,* vol. 1, Carnegie Institution, Washington, D. C., 1910.

Bomford, G., *Geodesy,* 731 pp., Oxford University Press, New York, 1971.

Bowie, W., and H. G. Avers, Fourth general adjustment of the precise level net in the U.S. and the resulting standard deviations, *U.S. Coast Geod. Surv. Spec. Publ., 18,* 1914.

Cannon, J. B., Adjustment of the precise level net of Canada 1928, *Geod. Surv. Can. Spec. Publ., 28,* 41 pp., 1929.

Castle, R. O., Leveling surveys and the southern California uplift, *U.S. Geol. Surv. Earthquake Inform. Bull., 10,* 88–92, 1978.

Castle, R. O., and R. F. Yerkes, Recent surface movements in the Baldwin Hills, Los Angeles County, California, *U.S. Geol. Surv. Prof. Pap., 882,* 125 pp., 1976.

Castle, R. O., J. N. Alt, J. C. Savage, and E. I. Balazs, Elevation changes preceding the San Fernando earthquake of February 9, 1971, *Geology, 2,* 61–66, 1974.

Castle, R. O., J. P. Church, M. R. Elliott, S. H. Wood, and E. I. Balazs, The evolution of the southern California uplift, 1955–76, submitted to *U.S. Geol. Surv. Prof. Pap.,* 1980.

Douglas, B. C., Geodetic and steric leveling: Together at last? (abstract), *Eos Trans. AGU, 59,* 1052–1053, 1978.

Entin, I. I., Main systematic errors in precise leveling, *Bull. Geod., 52,* 37–45, 1959.

Federal Geodetic Control Committee, Classification, standards of accuracy, and general specifications of geodetic control surveys, report, 12 pp., Nat. Oceanic and Atmos. Admin., Washington, D. C., 1974.

Federal Geodetic Control Committee, Specifications to support classification, standards of accuracy, and general specifications of geodetic control surveys, report, 30 pp., Nat. Oceanic and Atmos. Admin., Washington, D. C., 1975.

Heiskanen, W. A., and H. Moritz, *Physical Geodesy,* 364 pp., W. H. Freeman, San Francisco, Calif., 1967.

Hela, I., and E. Lisitzin, A world mean sea-level and marine geodesy, in *Proceedings of 1st Marine Geodesy Symposium U.S.,* pp. 71–73, Environmental Science Service Administration, Washington, D. C., 1967.

Holdahl, S. R., and R. Hardy, Solvability and multi-quadric analysis as applied to investigations of vertical crustal movements, *Tectonophysics, 52,* 139–156, 1979.

Holdahl, S. R., and N. L. Morrison, Regional investigations of vertical crustal movements in the U.S., using precise relevelings and mareograph data, *Tectonophysics, 23,* 373–390, 1974.

International Association of Geodesy, Geodetic reference system 1967, *Spec. Publ. 3,* 116 pp., Int. Ass. Geod., Paris, 1971.

International Association of Hydrological Sciences, *Land Subsidence Symposium,* 669 pp., Paris, 1977.

International Association of Scientific Hydrology, *Land Subsidence,* vols. 88 and 89, 661 pp., International Association of Scientific Hydrology, Unesco, Paris, 1969.

Karren, R. J., Recent studies of leveling instrumentation and procedures *Surv. Mapp., 24*(3), 383–397, 1964.

Korn, G. A., and T. M. Korn, *Mathematical Handbook for Scientists and Engineers,* 2nd ed., McGraw-Hill, New York, 1968.

Krakiwsky, E. J., Proposed establishment of a first order height system in the U.S.A., *Geod. Sci. Rep. 72,* 157 pp., Ohio State Univ., Columbus, 1966.

Kukkamäki, T. J., Über die nivellitische refraction, *Finn. Geod. Inst. Spec. Publ., 25,* 48 pp., 1938.

Lambert, W. D., and F. W. Darling, Tables for determining the form of the geoid and its indirect effect on gravity, *U.S. Coast Geod. Surv. Spec. Publ., 199,* 130, 1936.

Lucht, H., Korrelation in Präzisionsmivellement, *Sci. Ref. 48,* Dep. of Geod., Photogramm. and Cartogr., Tech. Univ. of Hannover, West Germany, 1972.

Mather, R. S., R. Coleman, and O. L. Colombo, On the recovery of long-wave features of the sea surface topography from satellite altimetry, *Unisurv G 24,* pp. 21–46, Sch. of Surv., Univ. of N.S.W., Sydney, Australia, 1976.

McCrory, P., and K. R. Lajoie, Marine terrace deformation San Diego County, California, *Tectonophysics, 52,* 407–408, 1979.

Meade, R. H., and K. O. Emery, Sea-level as affected by river runoff, eastern U.S., *Science, 173,* 425–428, 1971.

Mikhail, E. M., *Observations and Least Squares,* 497 pp., IEP, New York, 1976.

Miller, A. R., The effects of winds on water levels on the New England coast, *Limnol. Oceanogr., 3*(1), 1–14, 1958.

Molodenskij, M. S., V. F. Eremeev, and M. I. Yurkina, *Methods for Study of the External Gravitational Field and the Figure of the Earth,* 248 pp., translated from Russian, Israel Program for Scientific Translations, Jerusalem, 1962.

Montgomery, R. B., Fluctuations in monthly sea-level on eastern U.S. coast as related to dynamics of western North Atlantic, *J. Mar. Res., 1,* 165–185, 1937.

Müller, K., and E. Schneider, Nivellementswidersprüche in statistischer Sicht, *Vermessungstechnik, 16,* 8–15, 1968.

Nassar, M. M., Gravity field and levelled heights in Canada, Ph.D. dissertation, 318 pp., *Dep. of Surv. Eng.,* Univ. of N. B., Fredericton, 1977.

Nassar, M. M., and P. Vaníček, Levelling and gravity, *Tech. Rep. 33,* 133 pp., Dep. of Surv. Eng., Univ. of N. B., Fredericton, 1975.

National Geodetic Survey, Reports on geodetic measurements of crustal movements, Rockville, Md., 1973.

Pavoni, N., and R. Green (Eds.), Recent crustal movements, *Tectonophysics, 29,* 552 pp., 1975.

Poland, J. F., and G. H. Davis, Land subsidence due to withdrawal of fluids, *Rev. Eng. Geo., 2,* 187–269, 1969.

Rappleye, H. S., Manual of geodetic leveling, *U.S. Coast Geod. Surv. Spec. Publ., 239,* 94 pp., 1948.

Reilinger, R. E., and J. E. Oliver, Modern uplift associated with a proposed magma body in the vicinity of Socorro, New Mexico, *Geology, 4,* 583–586, 1976.

Reilinger, R. E., G. P. Citron, and L. D. Brown, Recent vertical crustal movements from precise leveling data in southwestern Montana, western Yellowstone National Park, and the Snake River plain, *J. Geophys. Res., 82,* 5349–5359, 1977.

Remmer, O., Leveling errors in Statu Nascendi, *Geod. Inst. Rep. Den., 51,* 43 pp., 1975.

Roden, G. I., Low-frequency sea level oscillations along the Pacific coast of North America, *J. Geophys. Res., 71,* 4755–4776, 1966.

Rossiter, J. R., Long-term variations in sea-level, in *The Sea,* vol. 1, edited by M. N. Hill, pp. 590–610, Interscience, New York, 1966.

Sturges, W., Sea level slope along continental boundaries, *J. Geophys. Res., 79,* 825–830, 1974.

U.S. Coast and Geodetic Survey, Manual of tide observations, *U.S. Coast Geod. Surv. Spec. Publ., 30-1,* 72 pp., 1965.

Vaníček, P., To the problem of noise reduction in sea-level records used in vertical crustal movement detection, *Phys. Earth and Planet. Interiors, 17,* 265–280, 1978.

Vaníček, P., and D. Christodulides, A method for the evaluation of vertical crustal movements from scattered geodetic relevellings, *Can. J. Earth Sci., 11*(5), 605–610, 1974.

Vaníček, P., and F. W. Grafarend, On the weight estimation in levelling, technical report, Nat. Oceanic and Atmos. Admin., Washington, D. C., in press, 1979.

Vaníček, P., J. D. Boal, and T. A. Porter, Proposals for a more modern system of heights for Canada, *Tech. Rep. 72-3,* 22 pp. plus appendices, Surv. and Mapp. Br., Ottawa, Ont., 1972.

Vaníček, P., M. R. Elliott, and R. O. Castle, Four-dimensional modeling of recent vertical movements in the area of the southern California uplift, *Tectonophysics, 52,* 287–300, 1979.

Vignal, J., and T. J. Kukkamäki, Comptes rendus des travaux de la section des nivellements de précision, *Bull. Geod. 34,* suppl., 299–493, 1954.

Walsh, J. B., and J. R. Rice, Local changes in gravity resulting from deformation, *J. Geophys. Res., 84,* 165–170, 1979.

Wehmiller, J. F., A. Sarna-Wojcicki, R. F. Yerkes, and K. R. Lajoie, Anomalously high uplift rates along the Ventura–Santa Barbara coast, California—tectonic implications, *Tectonophysics, 52,* 380, 1979.

Whalen, C. T., and E. I. Balazs, Test results of first-order class III leveling, *Tech. Rep. 68NGS4,* 30 pp., Nat. Oceanic and Atmos. Admin., Washington, D. C., 1977.

Whitcomb, J. H., New vertical geodesy, *J. Geophys. Res., 81,* 4937–4944, 1976.

Wood, S. H., and M. R. Elliott, Early 20th century uplift of the northern Peninsular Ranges Province of southern California, *Tectonophysics, 52,* 249–265, 1979.

(Received May 2, 1979;
accepted June 28, 1979.)

REVIEWS OF GEOPHYSICS AND SPACE PHYSICS, VOL. 21, NO. 1, PAGES 41–50, FEBRUARY 1983

Applications of Deformation Analysis in Geodesy and Geodynamics

ATHANASIOS DERMANIS AND EVANGELOS LIVIERATOS

Higher Geodesy and Cartography, Faculty of Engineering, University of Thessaloniki
Thessaloniki, Greece

The role of deformation analysis is discussed with respect to its existing or possible future applications in geodesy and geodynamics. Expressions for strain tensors are given in the more general case of Riemannian spaces and specialized for Euclidean spaces and the case of infinitesimal deformation. Among the various applications, special emphasis is given to the study of crustal deformations of the earth, deformations of the gravity field, and gravity field related deformations. Other applications are also considered.

INTRODUCTION

Deformation, loosely speaking, is the alteration of form and shape. Deformation is historically linked to the study of deformable material bodies, originally within the theory of elasticity and later on within continuum mechanics [*Love*, 1927; *Malvern*, 1969; *Truesdell*, 1977; *Eringen*, 1980]. The abstraction of relevant concepts and mathematical tools developed in mechanics leads to their broader use in a variety of problems which are not related to material bodies.

Since form and shape are connected to the metric characteristics of bodies, deformation, in a broad sense, is also concerned with the alteration of such characteristics of general physical or even abstract entities. It is only necessary that the elements of such entities can be brought into a one-to-one correspondence, in the same way that through the identification of material points a correspondence is established between two different states of a deformable body.

Applying deformation analysis in the geosciences, for example, in geodesy, the previous discussion meets a clear example: From its definition, geodesy studies among other topics the material shape and the gravity field of the earth as well as their alteration in space and time. The earth body deformation corresponds to the classical treatment in mechanics of deformable bodies, while the deformation of the gravity field of the earth calls for an abstract application of the same tools. Another, even more obvious, example of abstract deformation is illustrated by mapping the geographic grid of parallels and meridians from the globe onto a plane grid, a standard technique in cartography; the spherical and plane grids under comparison are obviously abstract entities of a nonmaterial nature.

In mechanics, deformation (strain) is not studied by itself but mainly in connection with the underlying forces (stress-strain relation). Although the forces acting to cause earth deformation are of interest in both geodesy and geophysics, geodesists are primarily concerned with the development of methods and techniques for the determination of such deformation.

The object of geodesy is the study of the size, shape, and gravity field of the earth, position determination, time variation of all the above, and their representation. Consequently, deformation methods find manifold applications in geodetic work.

MATHEMATICAL TOOLS

Since form and shape are described mathematically by the concept of metric, deformation can be represented by mathematical objects which depend on metric alterations.

Usually, in classical mechanics the relevant metric spaces are Euclidean. This is due to the model accepted for the description of the physical three-dimensional space in the neighborhood of the earth. The most convenient way of describing Euclidean space is by means of Cartesian orthogonal coordinate systems. However, curvilinear coordinates within Euclidean spaces are also useful in some applications.

In order to apply the above mathematical techniques outside the realm of mechanics, a generalization to Riemmanian metric spaces is sometimes necessary, for example, in cartographic applications.

Therefore the development of our formulation will start from curvilinear Riemmanian spaces. Euclidean descriptions with curvilinear or Cartesian coordinate systems will then be derived as special cases.

The Riemmanian Case

Let there be two Riemmanian spaces with corresponding metrics dS^2 and ds^2 which describe their metrical properties in the infinitesimal neighborhood of points of the two spaces. Curvilinear coordinate systems $(U^1\ U^2\ U^3)$ and $(u^1\ u^2\ u^3)$ are assigned to the two spaces. The metrics with respect to the chosen coordinate system are given by

$$dS^2 = \sum_{ij} G_{ij}\, dU^i\, dU^j = d\mathbf{U}^T\mathbf{G}\, d\mathbf{U} \tag{1}$$

$$ds^2 = \sum_{ij} g_{ij}\, du^i\, du^j = d\mathbf{u}^T\mathbf{g}\, d\mathbf{u} \tag{2}$$

where G_{ij}, g_{ij} are the corresponding metric tensors and $\mathbf{G}$, $\mathbf{g}$ their matrix representations.

It is furthermore assumed that a one-to-one correspondence exists between the two spaces described by

$$\mathbf{u} = \mathbf{u}(\mathbf{U}) \tag{3}$$

This mapping function must also be continuous and have a continuous inverse, that is, a homeomorphism in mathematical terms. In classical mechanics the sets $\mathbf{U}$ and $\mathbf{u}(\mathbf{U})$ correspond to the same point in two different states of a material body.

For the description of deformations we look at the differences $(ds^2 - dS^2)$ between corresponding elements of the two spaces under comparison. These differences can be expressed either in terms of the first coordinate system $\mathbf{U}$ (Lagrangian approach) or in terms of the second coordinate system $\mathbf{u}$ (Eulerian approach).

When the Lagrangian approach is followed, the Lagrangian strain tensor (strain matrix) **E** is introduced by means of

$$ds^2 - dS^2 \overset{\text{def}}{=} 2d\mathbf{U}^T\mathbf{E}\,d\mathbf{U} \tag{4}$$

It follows that

$$\mathbf{E} = \tfrac{1}{2}[(\partial\mathbf{u}/\partial\mathbf{U})^T\mathbf{g}(\partial\mathbf{u}/\partial\mathbf{U}) - \mathbf{G}] \tag{5}$$

When the Eulerian approach is followed, the Eulerian strain tensor (strain matrix) $\mathbf{E}^*$ is introduced by means of

$$ds^2 - dS^2 \overset{\text{def}}{=} 2d\mathbf{u}^T\mathbf{E}^*\,d\mathbf{u} \tag{6}$$

so that

$$\mathbf{E}^* = \tfrac{1}{2}[\mathbf{g} - (\partial\mathbf{U}/\partial\mathbf{u})^T\mathbf{G}(\partial\mathbf{U}/\partial\mathbf{u})] \tag{7}$$

We have called **E** and $\mathbf{E}^*$ in (5) and (7) tensors. However, some clarifications are needed. Tensors obey well-known transformation rules under changes of coordinate systems. These rules apply to **E** and $\mathbf{E}^*$ when the first coordinate system **U** undergoes a transformation, while the second coordinate system **u** undergoes the corresponding transformation induced by (3). On the one hand, function (3) depends on an underlying correspondence between the elements of the two compared Riemmanian spaces; on the other hand, its specific mathematical representation is a consequence of the more or less arbitrary selection of the coordinate systems **U** and **u**. A general discussion of this aspect is given by *Boucher* [1980]. The point will be clarified later when specific applications are analyzed.

Note that all components of the strain tensor must have the same physical dimensions. The reason for this restriction is that scalar invariants of the strain tensor are sometimes used as deformation parameters. Since these invariants (i.e., the trace and determinant) are combinations of the tensor components, they are physically meaningful when the particular components have the same physical dimensions. Recalling (4), the above restriction necessitates the use of curvilinear coordinates of the same physical units in contrast to some usual choices (e.g., spherical coordinates with both angular and linear units). In order to avoid this inconvenience, all coordinates may be transformed to coordinates of the same units through multiplication with suitable scalar factors. For example, when angular coordinates are used in the same set with linear coordinates, they can be transformed to linear ones through multiplication with the radii of curvature of the corresponding coordinate lines.

Equation (4) can be written as

$$ds^2/dS^2 \overset{\text{def}}{=} 1 + 2(d\mathbf{U}^T/dS)\mathbf{E}(d\mathbf{U}/dS) \tag{8}$$

or

$$m_L{}^2 \overset{\text{def}}{=} 1 + 2\mathbf{L}^T\mathbf{E}\mathbf{L} \tag{9}$$

where m_L is the linear modulus of deformation along the direction of the unit vector $\mathbf{L} = d\mathbf{U}/dS$. Replacing $\mathbf{L}^T\mathbf{L} = 1$ in (9), we obtain

$$m_L{}^2 = \mathbf{L}^T(\mathbf{I} + 2\mathbf{E})\mathbf{L} \tag{10}$$

which represents a quadratic form. The associated quadric surface, being the geometric locus of all points with $m_L{}^2 = 1$, is the well-known Cauchy strain ellipsoid in three dimensions or the Tissot strain ellipse in two dimensions used in cartography.

In the usual description of this quadratic form, the restriction that all curvilinear coordinates in **U** be of the same physical dimension is relaxed, and (9) is now written as

$$m_L{}^2 = 1 + 2\mathbf{L}^T\mathbf{K}\mathbf{E}\mathbf{K}\mathbf{L} \tag{11}$$

where **K** is the diagonal matrix with elements the corresponding curvatures of the coordinate lines. Similarly, (10) becomes

$$m_L{}^2 = \mathbf{L}^T(\mathbf{K}^2 + 2\mathbf{K}\mathbf{E}\mathbf{K})\mathbf{L} \tag{12}$$

The Euclidean Case

In the special but common case where the spaces under comparison are Euclidean, it is always possible to assign global orthogonal Cartesian coordinate systems, e.g., $(X^1\ X^2\ X^3)$ and $(x^1\ x^2\ x^3)$. Then the corresponding metrics become

$$dS^2 = d\mathbf{X}^T\,d\mathbf{X} \tag{13}$$

$$ds^2 = d\mathbf{x}^T\,d\mathbf{x} \tag{14}$$

that is, the metric tensors become identity ($\mathbf{G} = \mathbf{g} = \mathbf{I}$).

For a given one-to-one correspondence

$$\mathbf{x} = \mathbf{x}(\mathbf{X}) \tag{15}$$

the Langrangian and Eulerian strain tensors become

$$\mathbf{E} = \tfrac{1}{2}[(\partial\mathbf{x}/\partial\mathbf{X})^T(\partial\mathbf{x}/\partial\mathbf{X}) - \mathbf{I}] \tag{16}$$

$$\mathbf{E}^* = \tfrac{1}{2}[\mathbf{I} - (\partial\mathbf{X}/\partial\mathbf{x})^T(\partial\mathbf{X}/\partial\mathbf{x})] \tag{17}$$

It is also possible to describe Euclidean spaces by curvilinear orthogonal coordinate systems, e.g., $(Q^1\ Q^2\ Q^3)$ and $(q^1\ q^2\ q^3)$. In this case the metrics are written as

$$dS^2 = d\mathbf{Q}^T\mathbf{C}\,d\mathbf{Q} \tag{18}$$

$$ds^2 = d\mathbf{q}^T\mathbf{c}\,d\mathbf{q} \tag{19}$$

Comparison with (13) and (14) gives

$$\mathbf{C} = (\partial\mathbf{X}/\partial\mathbf{Q})^T(\partial\mathbf{X}/\partial\mathbf{Q}) \tag{20}$$

$$\mathbf{c} = (\partial\mathbf{x}/\partial\mathbf{q})^T(\partial\mathbf{x}/\partial\mathbf{q}) \tag{21}$$

Similar expressions in tensor notation are given by *Hotine* [1969].

For a given one-to-one correspondence

$$\mathbf{q} = \mathbf{q}(\mathbf{Q}) \tag{22}$$

the Lagrangian and Eulerian strain tensors are

$$\mathbf{E} = \tfrac{1}{2}[\mathbf{C} - (\partial\mathbf{q}/\partial\mathbf{Q})^T\mathbf{c}(\partial\mathbf{q}/\partial\mathbf{Q})] \tag{23}$$

$$\mathbf{E}^* = \tfrac{1}{2}[(\partial\mathbf{Q}/\partial\mathbf{q})^T\mathbf{C}(\partial\mathbf{Q}/\partial\mathbf{q}) - \mathbf{c}] \tag{24}$$

All the above derivations hold equally well for any dimensions, but our applications are limited to two and three dimensions.

Strain tensors are powerful tools in studying deformations, for they allow a pointwise illustration of alteration of the metric properties. In particular, the tensor field **E(U)** describes deformation in the infinitesimal vicinity of each element of the first space with coordinates $(U^1\ U^2\ U^3)$. $\mathbf{E}^*(\mathbf{u})$ does the same for elements of the second space.

Infinitesimal Deformation

When deformations are small, the transformation $\mathbf{u} = \mathbf{u}(\mathbf{U})$ is a transformation close to the identity. In this case the displacements $(\xi^1\ \xi^2\ \xi^3)$

$$\xi = \mathbf{u} - \mathbf{U} = \mathbf{u}(\mathbf{U}) - \mathbf{U} = \xi(\mathbf{U}) \quad (25)$$

are quantities of the first order. Differentiating (25), we obtain

$$\frac{\partial \mathbf{u}}{\partial \mathbf{U}} = \frac{\partial \xi}{\partial \mathbf{U}} + \mathbf{I} \quad (26)$$

and replacing into (5), the Lagrangian strain tensor becomes

$$\mathbf{E} = \frac{1}{2}\left[\left(\mathbf{g}\frac{\partial \xi}{\partial \mathbf{U}}\right)^T + \left(\mathbf{g}\frac{\partial \xi}{\partial \mathbf{U}}\right) + \mathbf{g} - \mathbf{G} + \left(\frac{\partial \xi}{\partial \mathbf{U}}\right)^T \mathbf{g}\left(\frac{\partial \xi}{\partial \mathbf{U}}\right)\right] \quad (27)$$

The previous equation is decomposed into a first-order part, linear in $\partial\xi/\partial\mathbf{U}$,

$$\boldsymbol{\varepsilon} \stackrel{\text{def}}{=} \frac{1}{2}\left[\left(\mathbf{g}\frac{\partial \xi}{\partial \mathbf{U}}\right)^T + \left(\mathbf{g}\frac{\partial \xi}{\partial \mathbf{U}}\right) + \mathbf{g} - \mathbf{G}\right] \quad (28)$$

and into a second-order part

$$\delta\boldsymbol{\varepsilon} = \frac{1}{2}\left[\left(\frac{\partial \xi}{\partial \mathbf{U}}\right)^T \mathbf{g}\left(\frac{\partial \xi}{\partial \mathbf{U}}\right)\right] \quad (29)$$

so that

$$\mathbf{E} = \boldsymbol{\varepsilon} + \delta\boldsymbol{\varepsilon} \quad (30)$$

Similarly,

$$\frac{\partial \mathbf{U}}{\partial \mathbf{u}} = \mathbf{I} - \frac{\partial \xi}{\partial \mathbf{u}} \quad (31)$$

and replacing into (7), the Eulerian strain tensor becomes

$$\mathbf{E}^* = \frac{1}{2}\left[\mathbf{g} - \mathbf{G} + \left(\mathbf{G}\frac{\partial \xi}{\partial \mathbf{u}}\right)^T + \left(\mathbf{G}\frac{\partial \xi}{\partial \mathbf{u}}\right) - \left(\frac{\partial \xi}{\partial \mathbf{u}}\right)^T \mathbf{G}\left(\frac{\partial \xi}{\partial \mathbf{u}}\right)\right] \quad (32)$$

with first-order part

$$\boldsymbol{\varepsilon}^* \stackrel{\text{def}}{=} \frac{1}{2}\left[\mathbf{g} - \mathbf{G} + \left(\mathbf{G}\frac{\partial \xi}{\partial \mathbf{u}}\right)^T + \left(\mathbf{G}\frac{\partial \xi}{\partial \mathbf{u}}\right)\right] \quad (33)$$

and second-order part

$$\delta\boldsymbol{\varepsilon}^* = -\frac{1}{2}\left[\left(\frac{\partial \xi}{\partial \mathbf{u}}\right)^T \mathbf{G}\left(\frac{\partial \xi}{\partial \mathbf{u}}\right)\right] \quad (34)$$

being

$$\mathbf{E}^* = \boldsymbol{\varepsilon}^* + \delta\boldsymbol{\varepsilon}^* \quad (35)$$

Here $\boldsymbol{\varepsilon}$ and $\boldsymbol{\varepsilon}^*$ are called the infinitesimal strain tensors. Not only are they first-order approximations of $\mathbf{E}$ and $\mathbf{E}^*$, but they also stand by themselves as independently defined entities.

In the particular case of Euclidean spaces with associated orthogonal Cartesian coordinate systems $\mathbf{X}$ and $\mathbf{x}$ the strain tensors referred to the displacement $\xi = \mathbf{x} - \mathbf{X}$ become

$$\mathbf{E} = \frac{1}{2}\left[\left(\frac{\partial \xi}{\partial \mathbf{X}}\right)^T + \left(\frac{\partial \xi}{\partial \mathbf{X}}\right) + \left(\frac{\partial \xi}{\partial \mathbf{X}}\right)^T\left(\frac{\partial \xi}{\partial \mathbf{X}}\right)\right] \quad (36)$$

$$\mathbf{E}^* = \frac{1}{2}\left[\left(\frac{\partial \xi}{\partial \mathbf{x}}\right)^T + \left(\frac{\partial \xi}{\partial \mathbf{x}}\right) - \left(\frac{\partial \xi}{\partial \mathbf{x}}\right)^T\left(\frac{\partial \xi}{\partial \mathbf{x}}\right)\right] \quad (37)$$

The infinitesimal strain tensors are

$$\boldsymbol{\varepsilon} = \frac{1}{2}\left[\left(\frac{\partial \xi}{\partial \mathbf{X}}\right) + \left(\frac{\partial \xi}{\partial \mathbf{X}}\right)^T\right] \quad (38)$$

$$\boldsymbol{\varepsilon}^* = \frac{1}{2}\left[\left(\frac{\partial \xi}{\partial \mathbf{x}}\right) + \left(\frac{\partial \xi}{\partial \mathbf{x}}\right)^T\right] \quad (39)$$

From the above developments one can observe that the infinitesimal strain tensors are the symmetric part of the Jacobians of the displacements. The antisymmetric parts are

$$\boldsymbol{\Omega} \stackrel{\text{def}}{=} \frac{1}{2}\left[\left(\frac{\partial \xi}{\partial \mathbf{X}}\right) - \left(\frac{\partial \xi}{\partial \mathbf{X}}\right)^T\right] \quad (40)$$

$$\boldsymbol{\Omega}^* \stackrel{\text{def}}{=} \frac{1}{2}\left[\left(\frac{\partial \xi}{\partial \mathbf{x}}\right) - \left(\frac{\partial \xi}{\partial \mathbf{x}}\right)^T\right] \quad (41)$$

Thus the corresponding Jacobians, also called deformation gradients, are decomposed as usual into symmetric and antisymmetric parts

$$\partial\xi/\partial\mathbf{X} = \boldsymbol{\varepsilon} + \boldsymbol{\Omega} \quad (42)$$

$$\partial\xi/\partial\mathbf{x} = \boldsymbol{\varepsilon}^* + \boldsymbol{\Omega}^* \quad (43)$$

Note that since the transformation $\mathbf{x} = \mathbf{x}(\mathbf{X})$ is close to the identity, in the first-order approximation

$$\partial\mathbf{x}/\partial\mathbf{X} \simeq \mathbf{I} \quad (44)$$

and

$$\partial\mathbf{X}/\partial\mathbf{x} = (\partial\mathbf{x}/\partial\mathbf{X})^{-1} \simeq \mathbf{I} \quad (45)$$

hold. Consequently,

$$\partial\xi/\partial\mathbf{X} \simeq \partial\xi/\partial\mathbf{x} \quad (46)$$

and

$$\boldsymbol{\varepsilon} \simeq \boldsymbol{\varepsilon}^* \quad (47)$$

$$\boldsymbol{\Omega} \simeq \boldsymbol{\Omega}^* \quad (48)$$

which means that the distinction between the Langrangian and Eulerian approaches practically disappears. For more details on the infinitesimal case, see, for example *Eringen* [1980].

Strain Rates

When metric alterations are associated with a scalar parameter, t being the independent variable, deformation may be studied by comparing the metrics at two infinitesimally closed values t and $t + dt$. In this case the difference of metrics $ds^2 - dS^2$ is replaced by the square of the derivative, $(dS/dt)^2$, in which case

$$(dS/dt)^2 \stackrel{\text{def}}{=} 2d\mathbf{U}^T(d/dt)\dot{\mathbf{E}}\, d\mathbf{U} \quad (49)$$

Usually, the parameter t represents time, and

$$\dot{\mathbf{E}} \stackrel{\text{def}}{=} d/dt\ \mathbf{E} \quad (50)$$

is called the strain rate tensor, in terms of the used curvilinear coordinates $\mathbf{U}$.

Stretch-Rotation Decomposition

An advantage of the analysis concerning infinitesimal deformation is the determination of the antisymmetric part of the Jacobian of the displacements $\partial\xi/\partial\mathbf{U}$ (see (40) and (41)), which represents the infinitesimal rotation of a small neighborhood of the point in question with respect to its original orientation.

In the case of noninfinitesimal deformation a similar approach is possible through the polar decomposition theorem of Cauchy [*Truesdell*, 1977]. Restricting ourselves to the Euclidean case, the Jacobian of the displacements can be uniquely decomposed in two ways,

$$\partial\xi/\partial\mathbf{X} = \mathbf{R}\mathbf{V}_R = \mathbf{V}_L\mathbf{R} \tag{51}$$

where $\mathbf{R}$ is an orthogonal matrix called the rotation matrix and $\mathbf{V}_R$, $\mathbf{V}_L$ are positive symmetric matrices called the right and the left stretch matrices, respectively. The rotation matrix represents the change of orientation of a small neighborhood of the point in question.

The independent elements of $\mathbf{R}$ can be computed from the displacement Jacobian matrix by solving one of the following rather complicated matrix equations:

$$\mathbf{R} = \frac{\partial\xi}{\partial\mathbf{X}}\left[\left(\frac{\partial\xi}{\partial\mathbf{X}}\right)^T \frac{\partial\xi}{\partial\mathbf{X}}\right]^{-1/2} = \left[\frac{\partial\xi}{\partial\mathbf{X}}\left(\frac{\partial\xi}{\partial\mathbf{X}}\right)^T\right]^{-1/2} \frac{\partial\xi}{\partial\mathbf{X}} \tag{52}$$

DEFORMATION OF THE EARTH

General Discussion

As is well known, geodetic observations are of a geometric and physical nature. They are also limited to points on and/or outside the surface of the earth. It is beyond the scope of this paper to discuss deformation problems related to the interior of the earth [*Bullen*, 1975], since such deformations do not significantly affect geodetic observations. On the contrary, deformations of the crust are directly connected with geodetic observables. Geometric observations are related to the alteration of the two-dimensional surface of the crust as embedded into three dimensions.

Deformations of the crust, beyond the surface, cause changes in the gravity field of the earth and can be sensed through gravity-dependent geodetic observations.

It must be noted that changes in the gravity vector and gravity potential at discrete points of the surface can be ascribed either to the surface motion or to changes in the gravity field per se. These two effects can be separated when the whole surface is covered by observations, in which case geodynamic boundary value problems are formulated for the determination of variation in the geometry of the boundary surface and variations of the gravity field outside the surface [see *Dermanis and Sansò*, 1982; *Heck*, 1981].

The study of global deformations of the crust surface can be primarily done by space geodetic techniques (laser ranging, very long baseline interferometry, etc.), while traditional geodetic techniques (triangulation, trilateration) are valuable for the study of deformations on a local scale. Global crust deformation and plate tectonic motion determination are the subject of extensive research both theoretical and applied, but there are still problems to be solved (for a discussion see, for example, *Baarda* [1975] and *Livieratos* [1979]).

Local crustal deformation studies are referred to two dimensions, following the classical separation of geodetic practice, that is, the duality planimetry-elevation. In this case, leveling is tied through gravity with internal changes. Since height is anholonomic, that is, path-dependent, geopotential numbers must rather be used through the combination of leveling with gravity. Geopotential differences may be caused by changes of the gravity field due to density redistributions within the crust without any actual change of the geometric shape of the surface. A discussion of the discrepancies between actual and apparent height differences as related to crustal movement is given by *Biró* [1975].

In order to compute strains in a three-dimensional medium (e.g., crust embedded in a Euclidean space) it is necessary to know the continuous field of displacements at every point of the medium or, at least, in the neighborhood of specific points where strain is to be calculated (see (36) and (37)).

Geodetic observations are usually discrete and limited to points on the two-dimensional surface. Even if the three components of the displacement vectors in a network of surface points could be evaluated by analyzing geodetic observations, strains could not be directly computed. In order to evaluate the space derivatives of displacements appearing in (36) and (37) a continuous field of displacements is needed. When only discrete values of displacements are available, a continuous field $\xi(\mathbf{X})$ can be obtained by interpolation. The interpolation concerns the determination of displacements not only on the surface but also within and outside the crust. For points within the crust the strains are computed by analytically differentiating the interpolated function $\xi(\mathbf{X})$, and consequently, the strains depend not only on the discrete geodetic observations but also on the chosen interpolation technique. The more the interpolation technique reflects physical reality, the closer the computed strains are to the real ones.

The same holds for strain computation at points on the crust surface, but an additional problem exists: In physical reality, displacements outside the medium are not defined, and stress-strain relation problems can be faced, introducing boundary conditions on the surface.

Mathematically speaking, interpolation techniques predict displacements even outside the medium, and computed surface strains depend on such predictions and consequently do not reflect physical reality. They have only a descriptive value and should not be used, for example, for conclusions on stress-strain relations.

In practice, the problems connected with deformations in three dimensions are bypassed by computing separately two-dimensional plane strains and vertical motions. Since it is natural to be interested in three-dimensional deformations, one may think of combining two-dimensional plane displacements with vertical motions in order to derive displacements in three dimensions and the relevant strains. However, as already mentioned, vertical motions are severely dependent on changes of the gravity field even when these changes are not accompanied by motion of the terrestrial surface.

As is well known to geodesists, the really estimable quantities are differences of the geopotential, which can be transformed into orthometric height differences and into their time variations only when the gravity field and its time variation are independently known.

On the other hand, the effect of the gravity field variations on the plane horizontal coordinates of triangulation points is negligible. When analyzing a local triangulation network, a common horizontal plane of reference is assumed for the perpendicular projection of network points. In this case, though observations (angular horizontal measurements) at each point depend on the local direction of the plumb line, they are not sensitive to the expected small changes in this direction. This allows the determination of horizontal displacements which essentially reflect actual horizontal motions with negligible effects from gravity field variations.

Works and discussions related to the above can be found in several collective volumes [e.g., *Pavoni and Green*, 1975; *Rikitake*, 1976, 1981; *Mueller*, 1978; *Hieber and Guyenne*, 1978; *Vogel*, 1979; *NASA*, 1979, 1980, 1981; *Isikara and Vogel*, 1981; *U.S. National Research Council*, 1981].

For vertical movements and associated gravity variations see, for example, *Faytel'son and Yurkina* [1974], *Yurkina* [1977], *Hein* [1978], *Drewes* [1980], and *Vanicek et al.* [1980].

Further elaborations with emphasis on analysis are given by *Brunner* [1979], *Kumar* [1979], *Wassef* [1979], *Brunner et al.* [1981], *Dermanis* [1981], *Hein and Kistemann* [1981], *Kelm* [1981], *Koch and Fritsch* [1981], *Livieratos and Vlachos* [1981], and *Niemeier* [1981].

Two-Dimensional Plane Strain

For the study of crustal deformations in two dimensions, usually followed in practice, intrinsic two-dimensional strains are defined for the description of the geometric alterations in the positions of the projections of the surface points onto the local reference plane.

Again, as in the three-dimensional case, strains, being local differential quantities, require continuous displacement information. Geodetic techniques provide only discrete displacements, and strains can be computed only after an explicit or implicit interpolation of displacements.

The definitions of the Lagrangian strain tensor **E**, the infinitesimal strain tensor **ε**, the infinitestimal rotation matrix **Ω**, and their Eulerian counterparts apply equally well in the two-dimensional case. Since the analysis is carried out on the plane (Euclidean space), the equations (36)–(41) are appropriate.

The numerical values of strain tensor elements depend on the reference frame used. Thus one should look for scalar functions of the strain tensor elements which are invariant with respect to the chosen reference frame and moreover have an evident physical interpretation. It is well known that there exists a particular reference frame with respect to which the strain tensor, being symmetric, takes a diagonal form:

$$\mathbf{\Lambda}_E = \begin{bmatrix} E_{\max} & 0 \\ 0 & E_{\min} \end{bmatrix} \tag{53}$$

The quantities $E_{\max}$, $E_{\min}$ are called the (Lagrangian) principal strains. In order to compute the principal strains from the components of the strain tensor

$$\mathbf{E} = \begin{bmatrix} E_{xx} & E_{xy} \\ E_{xy} & E_{yy} \end{bmatrix} \tag{54}$$

in any two-dimensional Cartesian coordinate system, well-known spectral techniques must be applied. Indeed, the principal strains $E_{\min}$, $E_{\max}$ are the eigenvalues of **E** and can be computed as the roots of the characteristic equation

$$\det(\mathbf{E} - \lambda \mathbf{I}) = \lambda^2 - \mathrm{tr}\,(\mathbf{E})\lambda + \det(\mathbf{E}) = 0 \tag{55}$$

which are

$$\begin{aligned} \lambda_1 = E_{\max} &= \tfrac{1}{2}\{\mathrm{tr}\,(\mathbf{E}) + [\mathrm{tr}\,(\mathbf{E})^2 - 4\det(\mathbf{E})]^{1/2}\} \\ \lambda_2 = E_{\min} &= \tfrac{1}{2}\{\mathrm{tr}\,(\mathbf{E}) - [\mathrm{tr}\,(\mathbf{E})^2 - 4\det(\mathbf{E})]^{1/2}\} \end{aligned} \tag{56}$$

The scalar quantities

$$\begin{aligned} \Delta &= E_{\max} + E_{\min} \\ \gamma &= E_{\max} - E_{\min} \end{aligned} \tag{57}$$

are called the dilatation and the maximum shear strain, respectively. Δ represents the isotropic part and γ the anisotropic part of deformation in the infinitesimal vicinity of the point of interest. They are both point functions with the following physical interpretation: Δ is the areal change per unit area (positive for an increase in area), and γ is the shear across the direction of its maximum value (always positive).

The scalar invariants Δ and γ can be expressed, according to *Truesdell and Noll* [1965], in terms of the so-called principal invariants $I_1(\mathbf{E})$ and $I_2(\mathbf{E})$, which in the two-dimensional case are

$$I_1(\mathbf{E}) = \mathrm{tr}\,(\mathbf{E}) \qquad I_2(\mathbf{E}) = \det(\mathbf{E}) \tag{58}$$

Combination of (56) with (57) gives

$$\begin{aligned} \Delta &= I_1(\mathbf{E}) = \mathrm{tr}\,(\mathbf{E}) = E_{xx} + E_{yy} \\ \gamma^2 &= I_1{}^2(\mathbf{E}) - 4I_2(\mathbf{E}) = [\mathrm{tr}\,(\mathbf{E})]^2 - 4\det(\mathbf{E}) \\ &= (E_{xx} - E_{yy})^2 + (2E_{xy})^2 \end{aligned} \tag{59}$$

Introducing the shear components

$$\gamma_1 = E_{xx} - E_{yy} \tag{60}$$

$$\gamma_2 = 2E_{xy} \tag{61}$$

the maximum shear strain becomes

$$\gamma = (\gamma_1{}^2 + \gamma_2{}^2)^{1/2} \tag{62}$$

The above shear components have the following interpretation [*Prescott et al.*, 1979]: γ_1 is the shear across any line parallel to the y direction (north), and γ_2 is the shear across any line parallel to the x direction (east).

The azimuth of the maximum extension $E_{\max}$ is computed from the eigenvector problem

$$\mathbf{E}\mathbf{n} = \lambda \mathbf{n} \tag{63}$$

where

$$\mathbf{n} = \begin{bmatrix} \sin\varphi \\ \cos\varphi \end{bmatrix} \tag{64}$$

Eliminating λ from (63), it follows that

$$\varphi = \tfrac{1}{2}\arctan(-\gamma_2/\gamma_1) \tag{65}$$

When the infinitesimal strain **ε** only is considered, corresponding infinitesimal quantities Δ, γ, etc. are defined.

From the Jacobian of the displacements (46), the infinitesimal antisymmetric rotation matrix

$$\mathbf{\Omega} = \begin{bmatrix} 0 & \omega \\ -\omega & 0 \end{bmatrix} \tag{66}$$

is also defined, where ω is the angle (in radians) of infinitesimal rotation. Similar quantities are analogously defined for the Eulerian mode of description.

There is an extensive literature concerning applications of the above two-dimensional plane strain analysis from geodetic results. Two approaches are basically followed. In the first approach, angular or distance observations at two different epochs are used (observation methods). In the second approach, the coordinates from independently adjusted triangulations at two epochs are used (coordinate method). This is essentially a finite element method where strains are considered constant within the triangular elements.

Concerning horizontal deformations see, for example, *Frank* [1966], *Sato* [1973], *Bibby* [1975, 1981], *Savage* [1976], *Prescott* [1976], *Harada* [1977], *Gerasimenko et al.* [1977], *Livieratos* [1978, 1980], *Baldi et al.* [1979], *Cohen and Cook* [1979], *Harada and Shimura* [1979], *Nyland et al.* [1979], *Prescott et al.* [1979], *Reilly* [1979], *Savage et al.* [1979], *Snay and Cline* [1980], *Vlachos* [1980], *Baldi and Marson* [1981], *Cacon* [1981], *Dermanis et al.* [1981], *Lisowski and Thatcher* [1981], and *Bencini et al.* [1982].

DEFORMATION OF THE GRAVITY FIELD OF THE EARTH AND GRAVITY FIELD RELATED DEFORMATIONS

As already discussed in the introduction, while the analysis in the previous chapter concerned the classical treatment of material deformation, the same tools can be abstracted for the study of deformations related to the gravity field of the earth.

In this context, three alternative cases are studied:

1. The most straightforward way of defining deformation of the gravity field is to compare two different gravity vector fields in the vicinity of a fixed point in geometry space.
2. Since traditionally in geodesy we deal with positioning by means of observations related to the gravity field, an alternative approach can be formulated by studying the change of geometric positions identified by the same value of the gravity vector, or any three other appropriate gravity field related parameters, with respect to two different gravity fields under comparison.
3. The third alternative is to study the alterations of the geometric and gravity characteristics, as well as their interrelation, in the vicinity of material points of the earth's surface.

For the study of these cases the following Jacobians are used:

$$\begin{aligned} \partial\boldsymbol{\Gamma}/\partial\mathbf{X} &= \mathbf{B} & \partial\boldsymbol{\Gamma}/\partial\mathbf{x} &= \mathcal{B} \\ \partial\boldsymbol{\gamma}/\partial\mathbf{X} &= \mathbf{b} & \partial\boldsymbol{\gamma}/\partial\mathbf{x} &= \boldsymbol{\beta} \end{aligned} \tag{67}$$

where $\mathbf{x}$ and $\mathbf{X}$ are the 'old' and 'new' position vectors, respectively, and $\boldsymbol{\gamma}$ and $\boldsymbol{\Gamma}$ the corresponding gravity vectors.

Gravity Deformation at Fixed Space Position

Let us consider a point $\mathbf{X}$ fixed in geometry space and two different potential fields W and w assigning corresponding gravity vectors $\boldsymbol{\Gamma} = \text{grad } W(\mathbf{X})$ and $\boldsymbol{\gamma} = \text{grad } w(\mathbf{X})$. Since $\boldsymbol{\Gamma}$ and $\boldsymbol{\gamma}$ are Cartesian coordinates in gravity space, they can be used for the definition of gravity strain in the same way that the space Cartesian coordinates are used in (16) and (17) for the definition of geometry strains. The Lagrangian gravity strain is

$$\mathbf{E}(\boldsymbol{\Gamma}) = \tfrac{1}{2}[(\partial\boldsymbol{\gamma}/\partial\boldsymbol{\Gamma})^T(\partial\boldsymbol{\gamma}/\partial\boldsymbol{\Gamma}) - \mathbf{I}] \tag{68}$$

and the Eulerian gravity strain is

$$\mathbf{E}^*(\boldsymbol{\gamma}) = \tfrac{1}{2}[\mathbf{I} - (\partial\boldsymbol{\Gamma}/\partial\boldsymbol{\gamma})^T(\partial\boldsymbol{\Gamma}/\partial\boldsymbol{\gamma})] \tag{69}$$

In the more general case when curvilinear coordinates $\boldsymbol{\Phi}(\boldsymbol{\Gamma})$ and $\boldsymbol{\varphi}(\boldsymbol{\gamma})$ are used for the description of gravity space, the above relations are written as in (5) and (7),

$$\mathbf{X} \equiv \mathbf{x} \xrightarrow{W} \boldsymbol{\Gamma}$$

$$\mathbf{X} \equiv \mathbf{x} \xrightarrow{w} \boldsymbol{\gamma}$$

$$\mathbf{E}(\boldsymbol{\Phi}) = \tfrac{1}{2}[(\partial\boldsymbol{\varphi}/\partial\boldsymbol{\Phi})^T\mathbf{m}(\partial\boldsymbol{\varphi}/\partial\boldsymbol{\Phi}) - \mathbf{M}] \tag{70}$$

$$\mathbf{E}(\boldsymbol{\varphi}) = \tfrac{1}{2}[\mathbf{m} - (\partial\boldsymbol{\Phi}/\partial\boldsymbol{\varphi})^T\mathbf{M}(\partial\boldsymbol{\Phi}/\partial\boldsymbol{\varphi})] \tag{71}$$

$\mathbf{m}$, $\mathbf{M}$ being the metric tensors associated with the curvilinear frames used. Since gravity space is Euclidean,

$$\mathbf{m} = \mathbf{d}^T\mathbf{d} \qquad \mathbf{M} = \mathbf{D}^T\mathbf{D} \tag{72}$$

where

$$\mathbf{d} = \partial\boldsymbol{\gamma}/\partial\boldsymbol{\varphi} \qquad \mathbf{D} = \partial\boldsymbol{\Gamma}/\partial\boldsymbol{\Phi} \tag{73}$$

Introducing the matrices of second-order space derivatives of the potential (gravity gradients),

$$\frac{\partial}{\partial\mathbf{X}}\left(\frac{\partial W}{\partial\mathbf{X}}\right)^T = \frac{\partial\boldsymbol{\Gamma}}{\partial\mathbf{X}} \overset{\text{def}}{=} \mathbf{B} = \frac{\partial\boldsymbol{\Gamma}}{\partial\mathbf{x}} \overset{\text{def}}{=} \mathcal{B} \tag{74}$$

$$\frac{\partial}{\partial\mathbf{X}}\left(\frac{\partial w}{\partial\mathbf{X}}\right)^T = \frac{\partial\boldsymbol{\gamma}}{\partial\mathbf{X}} \overset{\text{def}}{=} \mathbf{b} = \frac{\partial\boldsymbol{\gamma}}{\partial\mathbf{x}} \overset{\text{def}}{=} \boldsymbol{\beta} \tag{75}$$

the gravity strain tensors referred to Cartesian coordinates become

$$\mathbf{E}(\boldsymbol{\Gamma}) = \tfrac{1}{2}(\mathbf{B}^{-1}\mathbf{b}^2\mathbf{B}^{-1} - \mathbf{I}) \tag{76}$$

$$\mathbf{E}^*(\boldsymbol{\gamma}) = \tfrac{1}{2}(\mathbf{I} - \mathbf{b}^{-1}\mathbf{B}^2\mathbf{b}^{-1}) \tag{77}$$

and the gravity strain tensors referred to curvilinear coordinates are

$$\begin{aligned} \mathbf{E}(\boldsymbol{\Phi}) &= \tfrac{1}{2}[\mathbf{D}^T\mathbf{B}^{-1}\mathbf{b}(\mathbf{d}^{-1})^T\mathbf{m}\mathbf{d}^{-1}\mathbf{b}\mathbf{B}^{-1}\mathbf{D} - \mathbf{M}] \\ &= \tfrac{1}{2}[\mathbf{D}^T\mathbf{B}^{-1}\mathbf{b}^2\mathbf{B}^{-1}\mathbf{D} - \mathbf{D}^T\mathbf{D}] \end{aligned} \tag{78}$$

and

$$\mathbf{E}^*(\boldsymbol{\varphi}) = \tfrac{1}{2}[\mathbf{d}^T\mathbf{d} - \mathbf{d}^T\mathbf{b}^{-1}\mathbf{B}^2\mathbf{b}^{-1}\mathbf{d}] \tag{79}$$

Comparing with (76) and (77),

$$\mathbf{E}(\boldsymbol{\Phi}) = \mathbf{D}^T\mathbf{E}(\boldsymbol{\Gamma})\mathbf{D} \tag{80}$$

$$\mathbf{E}^*(\boldsymbol{\varphi}) = \mathbf{d}^T\mathbf{E}(\boldsymbol{\gamma})\mathbf{d} \tag{81}$$

in agreement with the well-known tensor transformation rules.

The two potential fields under comparison could be either the actual instantaneous fields of the earth at two distinct epochs or the actual field and a model field (normal potential).

Deformation of Geometry Space at Fixed Gravity

This case is dual to the previous one, since it follows by simply interchanging geometry with gravity space:

$$\mathbf{X} \xrightarrow{W} \boldsymbol{\Gamma} \equiv \boldsymbol{\gamma}$$

$$\mathbf{x} \xrightarrow{w} \boldsymbol{\Gamma} \equiv \boldsymbol{\gamma}$$

The situation now is as follows: Consider a position $\mathbf{X}$ and a potential field W assigning a gravity vector $\boldsymbol{\Gamma} = \text{grad } W(\mathbf{X})$. For a second potential field w there exists a position $\mathbf{x}$ such that the same gravity vector $\boldsymbol{\Gamma} = \text{grad } w(\mathbf{x})$ is assigned. In this abstract way, positions $\mathbf{X}$ and $\mathbf{x}$ are brought to one-to-one correspondence, in the same way that in continuum mechanics, positions are paired as positions of the same material point at different epochs.

The strain tensors are thus defined as in (16) and (17):

$$\mathbf{E}(\mathbf{X}) = \tfrac{1}{2}[(\partial\mathbf{x}/\partial\mathbf{X})^T(\partial\mathbf{x}/\partial\mathbf{X}) - \mathbf{I}] \tag{82}$$

$$\mathbf{E}^*(\mathbf{x}) = \tfrac{1}{2}[\mathbf{I} - (\partial\mathbf{X}/\partial\mathbf{x})^T(\partial\mathbf{X}/\partial\mathbf{x})] \tag{83}$$

Introducing the matrices of second-order derivatives of the potential,

$$\begin{aligned} \frac{\partial}{\partial\mathbf{X}}\left(\frac{\partial W}{\partial\mathbf{X}}\right)^T &= \left(\frac{\partial\boldsymbol{\gamma}}{\partial\mathbf{X}}\right) \overset{\text{def}}{=} \mathbf{b} = \frac{\partial\boldsymbol{\Gamma}}{\partial\mathbf{X}} \overset{\text{def}}{=} \mathbf{B} \\ \frac{\partial}{\partial\mathbf{x}}\left(\frac{\partial w}{\partial\mathbf{x}}\right)^T &= \left(\frac{\partial\boldsymbol{\gamma}}{\partial\mathbf{x}}\right) \overset{\text{def}}{=} \boldsymbol{\beta} = \frac{\partial\boldsymbol{\Gamma}}{\partial\mathbf{x}} \overset{\text{def}}{=} \mathcal{B} \end{aligned} \tag{84}$$

the strain tensors become

$$\mathbf{E}(\mathbf{X}) = \tfrac{1}{2}(\mathbf{B}\boldsymbol{\beta}^{-2}\mathbf{B} - \mathbf{I}) \tag{85}$$

$$\mathbf{E}^*(\mathbf{x}) = \tfrac{1}{2}(\mathbf{I} - \boldsymbol{\beta}\mathbf{B}^{-2}\boldsymbol{\beta}) \tag{86}$$

Generalizations can be easily derived for the case when curvilinear coordinates are used in geometry space.

An application of the above treatment is found in the case where W is the actual potential of the earth and w the normal potential. In this case, $\mathbf{x}$ is the image of $\mathbf{X}$ under the telluroid mapping $\mathbf{X} \mapsto \mathbf{x}$ defined by

$$\text{grad } w(\mathbf{x}) = \text{grad } W(\mathbf{X}) \tag{87}$$

The relevant strain tensors (85) and (86) describe deformation induced by, for example, mapping the surface of the earth into the telluroid [*Bocchio*, 1976, 1979*a*, *b*; *Osada*, 1978, 1980; *Marussi*, 1973, 1974; *Grafarend*, 1978*a*, 1978*b*; *Teisseyre*, 1969].

A similar analysis can be carried out by identifying positions through their natural coordinates $\boldsymbol{\Phi} = (\Phi\ \Lambda\ W/w_0)^T$ or normal coordinates $\boldsymbol{\varphi} = (\varphi\ \lambda\ w/w_0)^T$ instead of the gravity vectors $\boldsymbol{\Gamma}$ or $\boldsymbol{\gamma}$, respectively, where w_0 is a reference constant with potential dimension, that is, a reference normal potential value. The choice of pure numbers W/w_0 and w/w_0 instead of the corresponding potentials W and w is suggested in order to obtain strain tensor components of the same units.

$$\mathbf{X} \xrightarrow{W} \boldsymbol{\Phi} \equiv \boldsymbol{\varphi}$$

$$\mathbf{x} \xrightarrow{w} \boldsymbol{\Phi} \equiv \boldsymbol{\varphi}$$

The strain tensors are defined exactly as in (16) and (17) or (82) and (83). Introducing the matrices

$$\mathbf{F} = \partial\boldsymbol{\Phi}/\partial\mathbf{X} \qquad \mathbf{f} = \partial\boldsymbol{\Phi}/\partial\mathbf{x} \tag{88}$$

the strain tensors become

$$\mathbf{E}(\mathbf{X}) = \tfrac{1}{2}[\mathbf{F}^T(\mathbf{f}\mathbf{f}^T)^{-1}\mathbf{F} - \mathbf{I}] \tag{89}$$

$$\mathbf{E}^*(\mathbf{x}) = \tfrac{1}{2}[\mathbf{I} - \mathbf{f}^T(\mathbf{F}\mathbf{F}^T)^{-1}\mathbf{f}] \tag{90}$$

This case finds application in studying deformations induced by the telluroid mapping $\mathbf{X} \mapsto \mathbf{x}$ defined by

$$w(\mathbf{x}) = W(\mathbf{X}) \qquad \frac{\text{grad } w(\mathbf{x})}{|\text{grad } w(\mathbf{x})|} = \frac{\text{grad } W(\mathbf{X})}{|\text{grad } W(\mathbf{X})|} \tag{91}$$

The Apparent Gravity Deformations

The case most closely connected to observational reality concerns the change of the gravity field at material points which is due to both change of the gravity field itself and change of position due to motion of the surface of the earth. The situation can be shown as

$$P \to \mathbf{X} \xrightarrow{W} \boldsymbol{\Gamma}$$

$$P \to \mathbf{x} \xrightarrow{w} \boldsymbol{\gamma}$$

where $\mathbf{X}$ and $\mathbf{x}$ are two different positions of the same material point P and $\boldsymbol{\Gamma}$ and $\boldsymbol{\gamma}$ the corresponding apparent (observable) gravity vectors induced by two different potential fields W and w.

The geometry strain tensors $\mathbf{E}(\mathbf{X})$ and $\mathbf{E}(\mathbf{x})$ are defined as in (16) and (17) or (82) and (83). Introducing the matrix of second-order derivatives of the potential

$$\mathbf{B} = \partial\boldsymbol{\Gamma}/\partial\mathbf{X} \tag{92}$$

and recalling definition (75), we obtain

$$\frac{\partial\mathbf{x}}{\partial\mathbf{X}} = \frac{\partial\mathbf{x}}{\partial\boldsymbol{\gamma}}\frac{\partial\boldsymbol{\gamma}}{\partial\boldsymbol{\Gamma}}\frac{\partial\boldsymbol{\Gamma}}{\partial\mathbf{X}} = \boldsymbol{\beta}\frac{\partial\boldsymbol{\gamma}}{\partial\boldsymbol{\Gamma}}\mathbf{B} \tag{93}$$

Thus the strain tensors become

$$\mathbf{E}(\mathbf{X}) = \frac{1}{2}\left[\mathbf{B}^T\left(\frac{\partial\boldsymbol{\gamma}}{\partial\boldsymbol{\Gamma}}\right)^T\boldsymbol{\beta}^2\frac{\partial\boldsymbol{\gamma}}{\partial\boldsymbol{\Gamma}}\mathbf{B} - \mathbf{I}\right] \tag{94}$$

$$\mathbf{E}^*(\mathbf{x}) = \frac{1}{2}\left[\mathbf{I} - \boldsymbol{\beta}^{-1}\left(\frac{\partial\boldsymbol{\Gamma}}{\partial\boldsymbol{\gamma}}\right)^T\mathbf{B}^{-2}\frac{\partial\boldsymbol{\Gamma}}{\partial\boldsymbol{\gamma}}\boldsymbol{\beta}^{-1}\right] \tag{95}$$

Similar expressions can be derived for the case where instead of $\boldsymbol{\Gamma}$ and $\boldsymbol{\gamma}$ the natural coordinates $\boldsymbol{\Phi}$ and $\boldsymbol{\varphi}$, respectively, are available.

Earth Tidal Deformation

A particular case of the previous formulation concerns the deformation of the earth's crust associated with the lunisolar tides. A material point in a relaxed state has a position $\mathbf{X}$, and the potential W is due only to earth attraction and rotation. An additional potential reflecting lunisolar attraction leads to a new disturbed state where the material point moves to a new position $\mathbf{x}$ and the final potential w depends on the original potential, the potential of the lunisolar attractions, and the variations caused by mass redistribution after elastic yielding. The final position $\mathbf{x}$ of any material point is a function of its original position $\mathbf{X}$, the original and final potential W and w, and a set of parameters (Love and Shida numbers). In consequence, strains depend on the above function. For derivations of tidal strains, see *Melchior* [1978]. For an interesting application see *Marussi* [1979].

OTHER APPLICATIONS

A case completely analogous to the temporal variation of network coordinates caused by crustal deformation is the one where two coordinate sets of the same time-invariant network are available from different sources. Typical examples are the inconsistencies in coordinates as derived from classical adjustment and space techniques (see, for example, *Mueller* [1974]), differences in coordinates as adjusted on a reference ellipsoid and on a projective plane, coordinate differences between the results of compatible adjustments based on data from different observational campaigns, coordinate differences in common parts of networks, alterations in the form of a lower-order network when superfluous, fixed, higher-order control points are used, etc. The physical meaning of deformation in such cases varies with each particular application.

When three-dimensional coordinates from space techniques are compared with classical ellipsoidal geodetic coordinates, deformation parameters may be related to systematic errors caused, for example, when normal parameters of the gravity field are used instead of their actual values in the classical procedures of reduction of the actual observations to the reference surface [*Meissl*, 1973; *Thomson et al.*, 1974].

Another example is the comparison of two forms of a network (or parts of a network) resulting from the utilization of different subsets of the total originally available observations, in which case, deformation parameters may be used for the detection of gross observation inconsistencies (blunders) [*Thapa*, 1980; *Vanicek et al.*, 1981].

A very interesting aspect of the study of deformation appears in engineering surveying [*Pelzer*, 1971, 1974; *Platek*, 1974; *Van Mierlo*, 1977; *Hallerman*, 1981], which concerns the temporal alterations in the shape of man-made structures. Beyond the need for generalization to three dimensions we notice the possibility of relating observed strains with stresses, since the elastic behavior of such structures are more or less known.

A simple two-dimensional Euclidean case is the study of film deformation in photogrammetric applications.

A well-known two-dimensional case is the deformation involved in cartographic projections, when original figures on a Riemmanian space (sphere or ellipsoid) are deformed into corresponding figures on a Euclidean space (plane map) (see, for example, *Marussi* [1951*a*, *b*, 1959], *Bernasconi* [1953, 1956], *Caputo* [1956], *Bocchio* [1969], *Chovitz* [1979], *Hojovec and Jokl* [1981], and *Dermanis and Livieratos* [1982]).

Other geodetic applications involving comparison of figures in two-dimensional Riemmanian spaces are the study of deformations occurring in mapping the geoid onto the reference ellipsoid and similar comparisons between alternative choices of reference ellipsoids (see, for example, *Marussi* [1951*a*, *b*, 1957], *O'Keefe* [1953], *Caputo* [1959], and *Zadro and Carminelli* [1966]).

Since deformation parameters are connected to comparison of metrics, one can study deformation between different metrizations of one and the same physical space, without the need of establishing point-to-point correspondences. An example can be found in the study of atmospheric refraction, where the three-dimensional Euclidean earth space is compared with itself when endowed with a metric related to the index of refraction (see, for example, *Marussi* [1953] and *Grafarend* [1975]).

A related approach is based on the exploitation of certain analogies between stress-strain relations of elasticity theory and error propagation in geodetic networks [*Borre and Krarup*, 1974; *Grafarend*, 1977; *Borre*, 1979].

REFERENCES

Baarda, W., Difficulties in establishing a model for testing crustal movements, in *Progress in Geodynamics*, pp. 45–51, North-Holland, Amsterdam, 1975.

Baldi, P., and I. Marson, Gravity and geodetic networks for the study of crustal deformations in seismic area (Ancona), *Boll. Geod. Sci. Affini*, *40*(3), 189–196, 1981.

Baldi, P., D. Postpischl, M. Unguendoli, and E. Boschi, Strain measurements in the Ancona area, in *Optimization of Design and Computation of Control Networks*, edited by F. Halmos and J. Somogyi, pp. 17–22, Akademiai Kiado, Budapest, 1979.

Bencini, P., A. Dermanis, E. Livieratos, and D. Rossikopoulos, Crustal deformation at the Friuli area from discrete and continuous geodetic prediction techniques, *Boll. Geod. Sci. Affini*, *41*(2), 137–148, 1982.

Bernasconi, C., Carte geografiche secondo il modello della membrana elastica, *Geofis. Pura Appl.*, *26*, 16–19, 1953.

Bernasconi, C., Rappresentazioni parziali e totali dell' ellissoide di rotazione sulla sfera, *Geofis. Pura Appl.*, *33*, 1–8, 1956.

Bibby, H. M., Crustal strain from triangulation in Marlborough, New Zealand, *Tectonophysics*, *29*, 529–540, 1975.

Bibby, H. M., Geodetically determined strain across the southern end of the Tonga-Kermadec-Hikurangi subduction zone, *Geophys. J. R. Astron. Soc.*, *66*, 513–533, 1981.

Biró, P., Vertical crustal movements and time changes of the gravity field, paper presented at Symposium on Recent Crustal Movement, XVIth General Assembly, Int. Union of Geod. and Geophys., Grenoble, France, 1975.

Bocchio, F., Sui campi di curve superficiali a transformata geodetica nelle rappresentazioni cartografiche di tipo generale, *Rend. Accad. Naz. Lincei, Ser. VIII*, *46*(2), 92–96, 1969.

Bocchio, F., Linear and angular deformations in three-dimensional potential invariant representations of the actual gravity field of the earth and applications, *Geophys. J. R. Astron. Soc.*, *46*, 155–164, 1976.

Bocchio, F., Three-dimensional representations of the geopotential field and related topics, *Boll. Geod. Sci. Affini*, *38*(1), 25–43, 1979*a*.

Bocchio, F., Isoparametric and isozenithal type representations of the actual gravity field of the earth, *Geophys. J. R. Astron. Soc.*, *57*, 343–351, 1979*b*.

Borre, K., Elasticity and stress-strain relations in geodetic networks, *Boll. Geod. Sci. Affini*, *38*(3), 417–456, 1979.

Borre, K., and T. Krarup, Foundation of a theory of elasticity for geodetic networks, paper presented at 7th Nordiske Geodaetmøde, Copenhagen, May 6–10, 1974.

Boucher, C., The general theory of deformation and its applications in geodesy, *Boll. Geod. Sci. Affini*, *39*(1), 13–35, 1980.

Brunner, F. K., On the analysis of geodetic networks for the determination of the incremental strain tensor, *Surv. Rev.*, *25*, 192, 1979.

Brunner, F. K., R. Coleman, and B. Hirsch, A comparison of computation methods for crustal strains from geodetic measurements, *Tectonophysics*, *71*, 281–298, 1981.

Bullen, K. E., *The Earth's Density*, Chapman and Hall, London, 1975.

Cacon, S., Local spatial network as a fragment of the system of geodetic networks for complex observation of crustal movement, *Sci. Bull. Stanislaw Staszic Univ. Min. Metall.*, no. 842,67, 65–80, 1981.

Caputo, M., Su alcuni problemi al contorno della teoria delle rapresentazioni conformi, *Atti Ist. Veneto Sci. Lett. Arti*, *115*, 9–21, 1956.

Caputo, M., Conformal projection of an ellipsoid of revolution when the scale factor and its normal derivative are assigned on a geodesic line of the ellipsoid, *J. Geophys. Res.*, *64*(11), 1867–1873, 1959.

Chovitz, B. H., A general theory of map projections, *Boll. Geod. Sci. Affini*, *38*(3), 457–479, 1979.

Cohen, S. C., and G. R. Cook, Determining crustal strain rates with spaceborne geodynamics ranging system data, *Manuscr. Geod.*, *4*, 245–260, 1979.

Dermanis, A., Geodetic estimability of crustal deformation parameters, *Quat. Geod.*, *2*(2), 159–169, 1981.

Dermanis, A., and E. Livieratos, Dilatation, shear, rotation and energy analysis of map projections, *Boll. Geod. Sci. Affini*, in press, 1982.

Dermanis, A., and F. Sansò, The geodynamic boundary value problem, *Boll. Geod. Sci. Affini*, *41*(1), 65–87, 1982.

Dermanis, A. E., Livieratos, D. Rossikopoulos, and E. Vlachos, Geodetic prediction of crustal deformations at the seismic area of Volvi, in *Proceedings of the International Symposium on Geodetic Networks and Computations, Munich, 31 Aug.–5 Sept. 1981*, vol. 5, pp. 234–248, Deutsche Geodaetische Kommission, Munich, 1981.

Drewes, H., Problems related to earthquake premonitory gravity observations, *Boll. Geofis. Teor. Appl.*, *22*(88), 261–268, 1980.

Eringen, A. C., *Mechanics of Continua*, R. E. Krieger, New York, 1980.

Faytel'son, A. Sh., and M. I. Yurkina, On secular variation of gravity and recent vertical crustal movements, *Geod. Mapp. Photogramm*, Engl. Transl., *16*(3), 168–171, 1974.

Frank, F. C., Deduction of earth strains from survey data, *Bull. Seismol. Soc. Am.*, *56*(1), 35–42, 1966.

Gerasimenko, M. G., V. V. Zlotin, and Ya. V. Naumov, A study of recent movements of the earth's crust by geodetic methods (from conference materials), *Geod. Mapp. Photogramm.*, Engl. Transl., *19*(3), 152–155, 1977.

Grafarend, E., Three-dimensional geodesy, III, Refraction, *Proc. Hotline Symp. Math. Geod., 6th*, 105–112, 1975.

Grafarend, E., Stress-strain relations in geodetic networks, *Commun. Geod. Inst. Uppsala Univ.*, no. 16, 1977.

Grafarend, E., Dreidimensionale geodätische Abbildungsgleichungen und die Näherungsfigur der Erde, *Z. Vermess.*, *103*, 132–140, 1978*a*.

Grafarend, E., The definition of the telluroid, *Bull. Geod.*, *52*, 25–37, 1978*b*.

Hallermann, L. (Ed.), *Proceedings of the II. International Symposium of Deformation Measurements by Geodetic Methods, Bonn, Sept. 25–28, 1978*, K. Wittwer, Stuttgart, 1981.

Harada, T., Quiet and violence in horizontal movement of the crust, *J. Phys. Earth*, *25*, suppl., S79–S83, 1977.

Harada, T., and M. Shimura, Horizontal deformation of the crust in western Japan revealed from first-order triangulation carried out three times, *Tectonophysics*, *52*, 469–478, 1979.

Heck, B., Combination of leveling and gravity data for detecting real crustal movements, in *Proceedings of the International Symposium on Geodetic Networks and Computations, Munich, 31 Aug.–5 Sept., 1981*, vol. 7, pp. 20–30, Deutsche Geodaetische Kommission, Munich, 1981.

Hein, G., Requirements and statistics of modern levelling networks with respect to vertical crustal movements, in *Proceedings of the 2nd International Symposium on Problems Related to the Redefinition of North American Geodetic Networks*, pp. 629–636, U.S. Department of Commerce, Washington, D. C., 1978.

Hein, G., and R. Kistemann, Mathematical foundation of non-tectonic effects in geodetic recent crustal movement models, *Tectonophysics*, *71*, 315–334, 1981.

Hieber, S., and T. D. Guyenne (Eds.), *Space Oceanography, Navigation and Geodynamics: Proceedings of ESA European Workshop, Elmau*, European Space Agency, Paris, 1978.

Hojovec, V., and L. Jokl, Relation between the extreme angular and areal distortion in cartographic representation, *Stud. Geophys. Geod.*, *25*, 132–151, 1981.

Hotine, M., *Mathematical Geodesy, ESSA Monogr.*, U.S. Department of Commerce, Washington, D. C., 1969.

Isikara, A. M., and A. Vogel (Eds.), *Multidisciplinary Approach to Earthquake Prediction*, Vieweg, Wiesbaden, Federal Republic of Germany, 1981.

Kelm, R., A geometric-gravimetric estimation procedure for most accurate determination of relative point displacements in local areas—Proposal and first study of sensitivity, in *Multidisciplinary Approach to Earthquake Prediction*, edited by A. M. Isikara and A. Vogel, Vieweg, Wiesbaden, Federal Republic of Germany, 1981.

Koch, K.-R., and D. Fritsch, Multivariate hypothesis tests for detecting recent crustal movements, *Tectonophysics*, *71*, 301–313, 1981.

Kumar, M., Significance of β error in the assessment of crustal movements, *Surv. Mapp.*, *39*(2), 133–136, 1979.

Lisowski, M., and W. Thatcher, Geodetic determination of horizontal deformation associated with the Guatemala earthquake of 4th February 1976, *Bull. Seismol. Soc. Am.*, *71*(3), 845–856, 1981.

Livieratos, E., Techniques and problems in geodetic monitoring of crustal movements at tectonically unstable regions, paper presented at International Workshop on Monitoring Crustal Dynamics in Earthquake Zones, Eur. Seismol. Comm. and Eur. Geophys. Soc., Strasbourg, Aug. 29 to Sept. 5, 1978. (Published in *Terrestrial and Space Techniques in Earthquake Prediction Research*, edited by A. Vogel, pp. 515–530, Vieweg, Wiesbaden, Federal Republic of Germany, 1979.)

Livieratos, E., On the error analysis of geodetically derived strains in seismic zones, *Eur. Space Agency Spec. Publ.*, *ESA SP 149*, 27–32, 1978.

Livieratos, E., Crustal deformation from geodetic measurements, *Boll. Geofis. Teor. Appl.*, *22*(88), 255–260, 1980.

Livieratos, E., and D. Vlachos, The influence of correlations in computing crustal strains from trigonometric network results, *Tectonophysics*, *77*, 323–332, 1981.

Love, A. E. H., *The Mathematical Theory of Elasticity*, Cambridge University Press, New York, 1927.

Malvern, L. E., *Introduction to the Mechanics of a Continuum Medium*, Prentice-Hall, Englewood Cliffs, N. J., 1969.

Marussi, A., Sulla rappresentazione del geoide sull' ellisoide, *Boll. Geod. Soc. Affini*, *10*(3), 1–7, 1951*a*.

Marussi, A., Determinazione a priori del modulo di deformazione lineare nella rappresentazione conforme di Gauss, *Rend. Accad. Naz. Lincei, Ser. VIII*, *11*, 3–4, 1951*b*.

Marussi, A., Un' analogia fra le leggi della propagazione della luce in mezzi rifrangenti continui isotropi, e le rappresentazioni conformi, *Atti Fond. Giorgio Ronchi Contrib. Ist. Naz. Ottica*, *8*(1), 1953.

Marussi, A., Sulle rappresentazioni fra superfici definite mediante la forma quadratica che ne determina il modulo di deformazione, in *Festschrift C.F. Baeschlin*, pp. 201–210, Zürich, 1957. (Also available as Publ. 33, Ist. di Topogr. e Geod., Univ. di Trieste, Trieste, Italy.)

Marussi, A., Une nouvelle position dans la theorie des cartes geographiques, *Commun. Observ. R. Belg.*, no. 152, Ser. Geophys., no 49, 1959.

Marussi, A., On the three-dimensional computation of geodetic networks and anomalies, *Rend. Accad. Naz. Lincei, Ser. VIII*, *55*, 5, 1973.

Marussi, A., On the representation of the actual gravity field of the earth on the normal ellipsoidal field, *Geophys. J. R. Astron. Soc.*, *37*, 347–352, 1974.

Marussi, A., The tidal field of a planet and the related intrinsic reference systems, *Geophys. J. R. Astron. Soc.*, *56*, 409–417, 1979.

Meissl, P., Distortion of terrestrial networks caused by geoid errors, *Proc. Symp. Math. Geod. 5th*, 41–52, 1973.

Melchior, P., *The Tides of the Planet Earth*, Pergamon, New York, 1978.

Mueller, I. I., Review of problems associated with conventional geodetic datums, *Can. Surv.*, *28*(5), 514–523, 1974.

Mueller, I. I., (Ed.), Applications of Geodesy to Geodynamics, Proceedings of the 9th GEOP Conference, *Rep. 280*, Dep. of Geod. Sci., Ohio State Univ., Columbus, 1978.

NASA, Application of space technology to crustal dynamics and earthquake research, *NASA Tech. Pap.*, *1464*, 1979.

NASA, NASA geodynamics program: Annual report for 1979, *NASA Tech. Memo.*, *81978*, 1980.

NASA, NASA geodynamics program: Annual report for 1980, *NASA Tech. Memo.*, *84010*, 1981.

Niemeier, W., Statistical tests for detecting movements in repeatedly measured geodetic networks, *Tectonophysics*, *71*, 335–351, 1981.

Nyland, E., et al., Measurement and analysis of ground movement using microgeodetic networks on active faults, *Geofis. Int.*, *18*(1), 53–71, 1979.

O'Keefe, J. A., The isoparametric method of mapping one ellipsoid on another, *Eos Trans. AGU*, *34*(6), 869–875, 1953.

Osada, E., Non-conformal transformations of three-dimensional geodetic networks, *Boll. Geod. Sci. Affini*, *37*(4), 635-649, 1978.

Osada, E., Deformation of geodetic networks in gravity invariant and potential invariant representations, *Bull. Geod.*, *54*, 510–520, 1980.

Pavoni, N., and R. Green, *Recent Crustal Movements*, Elsevier, New York, 1975.

Pelzer, H., Zur Analyse geodaetisher Deformationsmessungen, *Rep. C164*, Dtsch. Geod. Komm., Munich, 1971.

Pelzer, H., Neuere Ergebnisse bei der statistischen Analyse von Deformationsmessungen, in *Proceedings of the 14th International Congress of Surveyors*, pap. 608.3, Washington, D. C., 1974.

Platek, A., On the application of geodetic methods in the measurements of spatial deformations of structures, in *Proceedings of the 14th International Congress of Surveyors*, pap. 604.1, Washington, D. C., 1974.

Prescott, W. H., An extension of Frank's method for obtaining crustal shear strains from survey data, *Bull. Seismol. Soc. Am.*, *66*(6), 1847–1853, 1976.

Prescott, W. H., J. C. Savage, and T. Kinoshita, Strain accumulation rates in the western United States between 1970 and 1978, *J. Geophys. Res.*, *84*(B10), 5423–5435, 1979.

Reilly, W. I., Determination of geometric and gravimetric earth deformation parameters from geodetic observations, *Rep. 140*, Geophys. Div., Dep. of Sci. and Ind. Res., Wellington, New Zealand, 1979.

Rikitake, T., *Earthquake Prediction*, Elsevier, New York, 1976.

Rikitake, T. (Ed.), *Current Research in Earthquake Prediction*, vol. I, D. Reidel, Hingham, Mass., 1981.

Sato, H., A study of horizontal movement of the earth crust associated with destructive earthquakes in Japan, *Bull. Geogr. Surv. Inst.*, *19*(1), 89–130, 1973.

Savage, J. C., Geodetic measurements of crustal motions, The Changing World of Geodetic Science, *Rep. 250*, pp. 99–118, Dep. of Geod. Sci., Ohio State Univ., Columbus, 1976.

Savage, J. C., W. H. Prescott, M. Lisowski, and N. King, Geodetic measurements of deformation near Hollister, California, 1971–1978, *J. Geophys. Res.*, *84*(B13), 7599–7615, 1979.

Snay, R., and M. W. Cline, Geodetically derived strain at Shelter Cove, California, *Bull. Seismol. Soc. Am.*, *70*(3), 893–901, 1980.

Teisseyre, R., Note on the approximation for gravity and magnetic field determined by deformational anomalies, *Acta Geophys. Pol.*, *17*(3), 221–231, 1969.

Thapa, K., Strain as a diagnostic tool to identify inconsistent observations and constraints in horizontal geodetic networks, *Tech. Rep. 68*, Dep. of Surv. Eng., Univ. of New Brunswick, Fredericton, 1980.

Thomson, D. B., M. M. Nassar, and C. L. Merry, Distortions of Canadian geodetic networks due to the neglect of deflections of the vertical and geoidal heights, *Can. Surv.*, *28*(5), 598–605, 1974.

Truesdell, C., *A First Course in Rational Continuum Mechanics*, vol. 1, Acadamic, New York, 1977.

Truesdell, C., and W. Noll, The non-linear field theories of mechanics, in *Encyclopedia of Physics*, edited by S. Fluegge, vol. III/3, Springer, New York, 1965.

U.S. National Research Council, *Geodetic Monitoring of Tectonic Deformation—Toward a Strategy*, National Academy of Sciences, Washington, D. C., 1981.

Vanicek, P., R. O. Castle, and E. I. Balazs, Geodetic leveling and its applications, *Rev. Geophys. Space Phys.*, *18*(2), 505–524, 1980.

Vanicek, P., K. Thapa, and D. Schneider, The use of strain to identify incompatible observations and constraints in horizontal geodetic networks, *Manuscr. Geod.*, *6*(3), 257–281, 1981.

Van Mierlo, J., Systematic investigations on the stability of control points, in *Proceedings of the 15th International Congress of Surveyors*, pap. 606.2, Stockholm, 1977.

Vlachos, D., Land deformation control network in the epicentral region of the 1978 earthquake in northern Greece, paper presented at International Symposium on Earthquake Prediction in the North Anatolian Fault Zone, Eur. Geophys. Soc., Istanbul, March 31, to April 5, 1980. (Published in *Multidisciplinary Approach to Earthquake Prediction*, edited by M. Işikara and A. Vogel, pp. 315–324, Vieweg, Wiesbaden, Federal Republic of Germany, 1981.)

Vogel, A. (Ed.), *Terrestrial and Space Techniques in Earthquake Prediction Research*, Vieweg, Wiesbaden, Federal Republic of Germany, 1979.

Wassef, A. M., The concept of goodness of geodetic observations restated in view of the requirements of studies of recent crustal movements, *Bull. Geod.*, *53*, 53–60, 1979.

Yurkina, M. I., Combined determination of changes in the earth's gravitational field and of vertical displacements of the earth's crust, *Geod. Mapp. Photogramm.*, Engl. Transl., *19*(2), 59–62, 1977.

Zadro, M., and A. Carminelli, Rappresentazione conforme del geoide sull elissoide internazionale, *Boll. Geod. Sci. Affini*, *25*(1), 25–36, 1966.

(Received June 15, 1982;
accepted July 6, 1982.)

REVIEWS OF GEOPHYSICS AND SPACE PHYSICS, VOL. 21, NO. 4, PAGES 878–890, MAY 1983

Inertial Surveying and Geodesy

K. P. SCHWARZ

Division of Surveying Engineering, University of Calgary, Calgary, Alberta, Canada T2N 1N4

Inertial survey systems offer a novel and, in many ways, unique approach to the geodetic problems of positioning and gravity field determination. This paper outlines the basic theory of using the output of an inertial measuring unit to obtain geodetically meaningful results. Emphasis is on the development of appropriate mathematical models and estimation procedures rather than on hardware aspects. A discussion of results achieved to date and an outlook on future applications concludes the presentation.

CONTENTS

1. INTRODUCTION

Surveying systems based on the principle of inertial navigation are relatively new in geodetic practice. Developed in the early '70's for military agencies, they became available to nonmilitary users in 1975. After a short period of testing, they were almost immediately employed in geodetic production work, where they proved quite successful [see, e.g., *Babbage*, 1977].

Initially, inertial survey systems were mainly used as black boxes, delivering results in a somewhat mysterious but reliable fashion. The process of demystifying the black box started when geodesists realized that these systems were unique in that they provided information on both major problems of classical geodesy. The positioning problem, traditionally falling into the realm of geometric geodesy, could be solved by extracting vehicle acceleration from the system output. Gravity field determination, the main problem of physical geodesy, could similarly be solved by extracting gravitational acceleration. What made inertial survey systems unique was the fact that, in principle, both sets of data could be obtained from the same sensor output, thus providing a prime example for the necessity of an integrated approach to geodesy. In addition, they exhibited a versatility of applications not available from any other geodetic measuring device. For a discussion of this aspect, see, e.g., *Mueller* [1981].

To model the sensor output of these systems, methods of integrated geodesy are most appropriate and lead to rigorous nonlinear formulas. If the linearization of these formulas is done by using the normal equipotential ellipsoid as reference model, the resulting operational formulas are those of classical geodesy. Since the actual mechanization of the system is usually based on an ellipsoidal reference model, this operational approach provides an easy interpretation of the actual sensor output. It has therefore been used in the following. A more general treatment can be found in *Grafarend* [1981].

2. PRINCIPLE OF OPERATION

The following discussion is based on classical mechanics. In such a setting an inertial coordinate system is one in which the Newtonian equations of motion hold. Systems of this type are characterized by Galilean transformations and can be visualized as nonaccelerating and nonrotating with respect to the distant galaxies. The use of relativistic mechanics is an unnecessary complication at this point in time because the relativistic effects are orders of magnitude smaller than the measuring accuracy of the system. For a discussion of these effects, see, e.g., *Moritz* [1981].

In essence, inertial survey systems consist of three orthogonally mounted accelerometers whose orientation in space is known at any instant of time. If the basic measuring frame is inertial, the measurements are of the simple form

$$\mathbf{f} = \mathbf{a} - \mathbf{G} \tag{1}$$

where $\mathbf{f}$ is a three-dimensional specific force vector, $\mathbf{a}$ is a three-dimensional vector of vehicle accelerations, and $\mathbf{G}$ is a three-dimensional vector of gravitational accelerations. Equation (1) is the fundamental equation of inertial geodesy. It implies that positioning and gravity field determination cannot be separated. To get vehicle accelerations, which twice integrated give position differences, $\mathbf{G}$ must be known; to obtain gravitational acceleration, the vector $\mathbf{a}$ must be given. The usual procedure is to linearize $\mathbf{a}$ and $\mathbf{G}$ around known values and to iterate if the linear approximation is not sufficient.

The mechanical implementation of this concept requires that an orthogonal measuring frame be established that will maintain its orientation in space when transported. This can be achieved by a system of gyroscopes. Since a gyroscope will preserve directions in inertial space unless acted upon by a torque, an array of them can be used to either measure or control orientation changes of the measurement frame. The initial orientation of this frame is determined during the system alignment. It relates the arbitrary orientation of the frame to the local astronomic frame shown in Figure 1. The alignment involves accelerometer leveling and gyrocompassing. The leveling procedure aligns the Z axis of the system along the local plumb line. Gyrocompassing orients one of the leveled axes toward astronomic north. If the deflections of the vertical are known at the initial point, the ellipsoidal system of coordinates used in geodesy can be obtained by a simple transformation. The alignment is not necessarily done by torquing the measuring frame into the local astronomic frame. It is sufficient to determine the rotation angles between the two frames and to include them in the subsequent computations. During transport, the orientation in

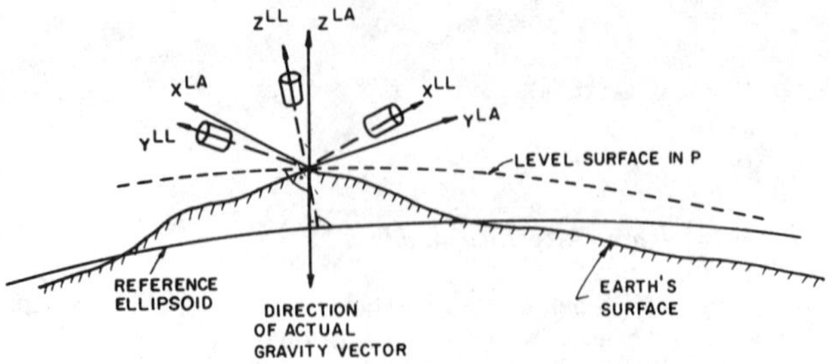

Fig. 1. Definition of the local astronomic (LA) and the local level frame (LL).

space can be maintained in two ways. Either all orientation changes of the instrument unit with respect to inertial space are measured (strapdown systems), or they are controlled in such a way that they can be described by a well-defined mathematical function (gimballed systems). This second possibility is used in all inertial survey systems presently available. It requires the mechanical implementation of a so-called stable element—a stabilized platform that isolates the measuring unit from the vehicle motions. In the space stable system the platform frame is fixed with respect to inertial space; in the local level system it rotates against inertial space in such a way that the *Z* axis of the measuring system remains orthogonal to an initially defined ellipsoid. This second system, which can be related in a simple way to the ellipsoidal coordinate systems used in geodesy, is also shown in Figure 1.

Three problems arise when implementing (1) for systems moving on the surface of the earth or its vicinity. First, since the measurement frame rotates with the earth, the sensor output contains terms generated by the rotation of the earth-fixed frame with respect to an inertial frame. Examples are the centripetal, the tangential, and the Coriolis acceleration. Thus (1) has to be replaced by

$$\mathbf{f} = \mathbf{a} + \sum_j \mathbf{a}_j - \mathbf{G} \tag{2}$$

where the sum denotes the so-called inertial accelerations. The vector $\mathbf{f}$ now describes the sensor output of a rotating system.

Second, the linearization of (2) leads to

$$\mathbf{G} = \mathbf{\Gamma} + \boldsymbol{\delta}\mathbf{g} \tag{3}$$

for the gravitational term and to

$$\mathbf{a} = \mathbf{a}_0 + \boldsymbol{\delta}\mathbf{a} \tag{4}$$

for the acceleration term, where $\mathbf{\Gamma}$ is the vector of normal gravitational acceleration, $\delta\mathbf{g}$ is the gravity disturbance vector, $\mathbf{a}_0$ is the acceleration vector giving the trajectory, and $\boldsymbol{\delta}\mathbf{a}$ contains perturbations of the trajectory. Thus (1) now reads

$$\mathbf{f} = \mathbf{a}_0 + \sum_j \mathbf{a}_{j-1} - \boldsymbol{\gamma} + \boldsymbol{\delta}\mathbf{a} - \boldsymbol{\delta}\mathbf{g} \tag{5}$$

where $\boldsymbol{\gamma} = \mathbf{\Gamma} + \mathbf{a}_c$ is the normal gravity vector and $\mathbf{a}_c$ is the vector of centripetal acceleration. In this equation term $\boldsymbol{\delta}\mathbf{a}$ models the known system errors, while $\boldsymbol{\delta}\mathbf{g}$ models the deviations of the actual gravity vector from the normal reference field. The actual trajectory of the measuring unit in space is thus perturbed by time-dependent alignment errors and system noise as well as by position-dependent gravity field errors. The integration of the sensor output will therefore result in a perturbed trajectory. Figure 2 shows in a schematic way the typical differences between the actual and the perturbed trajectory. In order to separate alignment errors, system noise, and gravity field effects, additional measurements are necessary to determine the model parameters in $\boldsymbol{\delta}\mathbf{a}$ and $\boldsymbol{\delta}\mathbf{g}$. Alignment errors and system noise are usually modeled by a system of first-order differential equations of the form

$$\dot{\mathbf{x}} = \mathbf{F}\mathbf{x} + \mathbf{w} \tag{6}$$

while the modeling of the anomalous gravity field is still very much a problem of current research, see, e.g., *Schwarz* [1980*a*].

Once adequate models for $\boldsymbol{\delta}\mathbf{a}$ and $\boldsymbol{\delta}\mathbf{g}$ are available, suitable estimation procedures have to be found to determine the various effects in an optimal way. This is by no means a trivial problem because of the rather large perturbations accumulating over the length of a survey. Some type of real-time procedure is necessary to secure the linearity of the error model. On the other hand, post-mission procedures, like optimal and empirical smoothing and adjustment, are needed to make best use of the inherent accuracy of the data. The optimal combination of the different procedures and the rigorous error propagation from one step to the next are presently under investigation [see, e.g., *Hannah and Mueller,* 1981; *Schwarz and Gonthier,* 1981*b*].

3. The Mathematical Model

The sensor output of an inertial survey system refers to a measurement frame that, in general, is not inertial. The derivation of an appropriate mathematical model thus involves the transformation of (1) into the measurement frame. Major steps in this derivation are the definition of an operational inertial frame, the formulation of the transformation equations for the radius vector and its time derivatives, and the modeling of the gravitational vector. We will use the local level frame as measurement frame because it is the most common mechanization for inertial survey system. The derivation of the model equations for the space stable and the strapdown mechanization follows along similar lines.

To simplify derivations, both the inertial and the local level frame will be related to an intermediate system, the average terrestrial (AT) system, which is defined in the following way: Its origin is at the centre of mass of the earth; its *Z* axis coincides with the mean spin axis of the earth, which in turn defines the mean equatorial plane orthogonal to it. In this plane the *X* axis points towards the mean

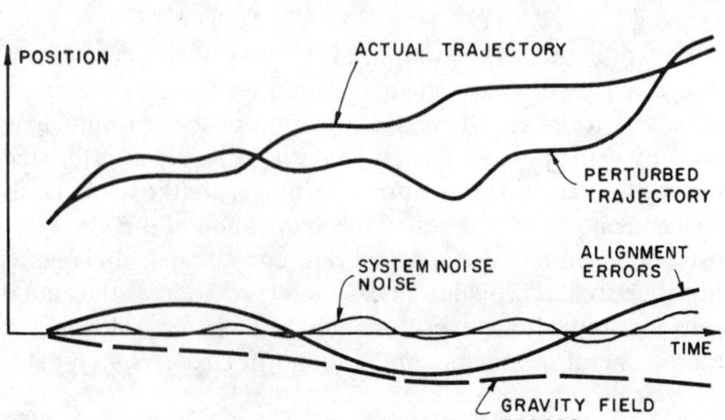

Fig. 2. Perturbations of the actual trajectory by alignment errors, gravity field errors, and system noise.

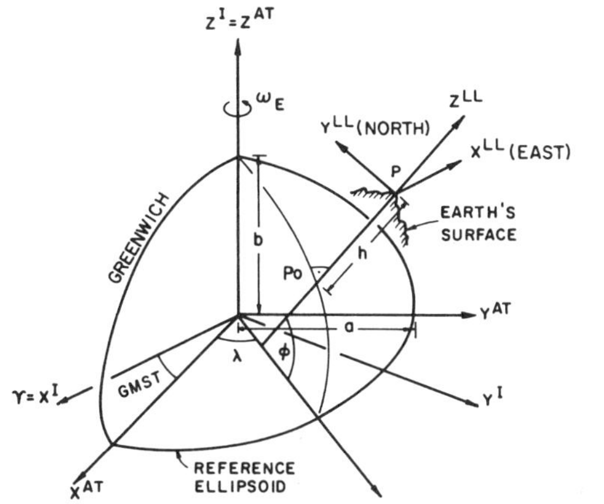

Fig. 3. Relations between the inertial (I), the average terrestrial (AT), and the local level system (LL).

meridian of Greenwich, and the Y axis is given by the right-handedness of the system. This system rotates with respect to an inertial frame. The rotation can be defined with respect to an astronomical star catalog system which, at present, is the best realization of an inertial system in the Newtonian sense. The transformation equations without the proper motion terms are

$$\mathbf{r}^{\mathrm{I}} = \mathbf{P}^{-1}\mathbf{N}^{-1}\mathbf{R}_3(-\mathrm{GAST})\mathbf{R}_1(y_p)\mathbf{R}_2(x_p)\mathbf{r}^{\mathrm{AT}} \tag{7}$$

where $\mathbf{R}_i$ are rotation matrices as defined, for example, in *Vaníček and Krakiwsky* [1982, p. 38], and $\mathbf{r}^{\mathrm{I}}$ is the radius vector in the inertial frame; $\mathbf{P}$ is a rotation matrix containing the elements of general precession; $\mathbf{N}$ is a rotation matrix containing the elements of astronomic nutation; GAST is Greenwich Apparent Siderial Time; x_p, y_p are the components of polar motion; and $\mathbf{r}^{\mathrm{AT}}$ is the radius vector in the AT system. A definition of the astronomical terms can be found in *Mueller* [1969, chap. 4]. The accuracy of this definition of an inertial frame is about $5 \cdot 10^{-7}$ deg/h in angular motion and about $2 \cdot 10^{-7}$ g in linear acceleration; see *Kayton* [1963] for a detailed discussion. The stability of gyros currently used in inertial survey systems is of the order of 10^{-3} deg/h and the accuracy of the corresponding accelerometers about $5 \cdot 10^{-5}$ g. Thus, in light of present-day measuring accuracy, a less accurate definition would be acceptable if it results in major simplifications of the equations. This can be achieved by approximating the terms for precession, nutation, and polar motion through a mean time definition. Equation (7) can then be written as

$$\mathbf{r}^{\mathrm{I}} = \mathbf{R}_3(-\mathrm{GMST})\mathbf{r}^{\mathrm{AT}} \tag{8}$$

where GMST is Greenwish Mean Siderial Time. This operational definition of an inertial frame, which is accurate to about $5 \cdot 10^{-5}$ deg/h in angular motion, will be used subsequently. It obviously includes a shift of origin from the barycenter of the sun-moon system to the mass center of the earth. However, in terms of linear acceleration this error does not exceed $2 \cdot 10^{-7}$ g and can be easily taken care of if a higher accuracy is required.

After defining an operational inertial frame and relating it to the AT system, the same will be done for the local level system, which is defined as a Cartesian frame whose Z axis is orthogonal to the reference ellipsoid pointing outward; its Y axis is pointing toward geodetic north, i.e., in the north direction of the ellipsoidal meridian; its X axis is defined by the right-handedness of the frame, i.e., it points east. Its origin is topocentric and is defined by the position of the inertial measuring unit. The advantage of using such a system is that it refers to a smooth mathematical surface—the ellipsoid—and at the same time is very close to the natural system of coordinates which can be physically realized by the inertial sensors (see Figure 1). The deviations between the mathematical and the natural frame—the so-called deflections of the vertical—are, on the average, not larger than 10 arc sec and rarely exceed 30 arc sec for a suitably defined reference ellipsoid. A graphical representation of the inertial, the average terrestrial, and the local level system is given in Figure 3. This figure can be used to derive the transformation between the Average Terrestrial and the Local Level System as

$$\mathbf{r}^{\mathrm{AT}} = \mathbf{R}_3(-90° - \lambda)\mathbf{R}_1(\phi - 90°)\mathbf{r}^{\mathrm{LL}} \tag{9}$$

where ϕ is the geodetic latitude, λ is the geodetic longitude, $\mathbf{r}^{\mathrm{LL}}$ is the radius vector in the local level (LL) system.
The origin of the LL system can be expressed in AT coordinates by

$$\mathbf{r}_0^{\mathrm{AT}} = \begin{pmatrix} (N+h)\cos\phi\cos\lambda \\ (N+h)\cos\phi\sin\lambda \\ \{N(1-e^2)+h\}\sin\phi \end{pmatrix} \tag{10}$$

where $N = a\{1 - e^2\sin^2\phi\}^{-1/2}$, $e^2 = (a^2 - b^2)/a^2$, and N is the prime vertical radius of curvature; a, b are the semimajor and semiminor axes of the ellipsoid; e is the first eccentricity of the ellipsoid, and h is the ellipsoidal height. The vector (10) can be transformed to the LL frame by inverting the rotations in (9), i.e.,

$$\mathbf{r}^{\mathrm{LL}} = \mathbf{R}_1(90° - \phi)\mathbf{R}_3(90° + \lambda)\mathbf{r}^{\mathrm{AT}} \tag{11}$$

Substituting (10) into (11) results in

$$\mathbf{r}_0^{\mathrm{LL}} = \begin{pmatrix} 0 \\ -e^2N\sin\phi\cos\phi \\ N(1-e^2\sin^2\phi)+h \end{pmatrix} \tag{12}$$

Using (8) and (9) we thus get the transformation between the local level and the inertial frame as

$$\mathbf{r}^{\mathrm{I}} = \mathbf{R}_3(-\mathrm{GMST})\mathbf{R}_3(-90° - \lambda)\mathbf{R}_1(\phi - 90°)\mathbf{r}^{\mathrm{LL}} \tag{13}$$

Using the substitution

$$\mu = \mathrm{GMST} + \lambda$$

we get

$$\mathbf{r}^{\mathrm{I}} = \mathbf{R}_3(-\mu - 90°)\mathbf{R}_1(\phi - 90°)\mathbf{r}^{\mathrm{LL}} \tag{14}$$

which is equivalent to

$$\mathbf{r}^{\mathrm{I}} = \mathbf{R}_3(-\mu)\mathbf{R}_2(\phi - 90°)\mathbf{R}_3(-90°)\mathbf{r}^{\mathrm{LL}} \tag{15}$$

The matrix

$$\mathbf{C} = \mathbf{R}_3(-\mu)\mathbf{R}_2(\phi - 90°)\mathbf{R}_3(-90°) \tag{16}$$

has the explicit form

$$\mathbf{C} = \begin{pmatrix} -\sin\mu & -\sin\phi\cos\mu & \cos\phi\ \psi o\sigma\ \mu \\ \cos\mu & -\sin\phi\sin\mu & \cos\phi\sin\mu \\ 0 & \cos\phi & \sin\phi \end{pmatrix} \tag{17}$$

Using the transformation (13) in the form

$$\mathbf{r}^{\mathrm{I}} = \mathbf{C}\,\mathbf{r}^{\mathrm{LL}} \tag{18}$$

we can now express Newton's second law in the local level system. Denoting the time derivatives by $\dot{\mathbf{r}} = (d\mathbf{r}/dt)$ and $\ddot{\mathbf{r}} = d^2\mathbf{r}/dt^2$, we get

$$\dot{\mathbf{r}}^{\mathrm{I}} = \dot{\mathbf{C}}\,\mathbf{r}^{\mathrm{LL}} + \mathbf{C}\,\dot{\mathbf{r}}^{\mathrm{LL}}$$

and

$$\ddot{\mathbf{r}}^{\mathrm{I}} = \ddot{\mathbf{C}}\,\mathbf{r}^{\mathrm{LL}} + 2\dot{\mathbf{C}}\,\dot{\mathbf{r}}^{\mathrm{LL}} + \mathbf{C}\,\ddot{\mathbf{r}}^{\mathrm{LL}} \tag{19}$$

The time derivative of an orthogonal matrix can be expressed by

$$\dot{\mathbf{C}} = \mathbf{C}\,\mathbf{\Omega} \tag{20}$$

where $\mathbf{\Omega}$ is the skew-symmetric matrix

$$\mathbf{\Omega} = \begin{pmatrix} 0 & -\omega_3 & \omega_2 \\ \omega_3 & 0 & -\omega_1 \\ -\omega_2 & \omega_1 & 0 \end{pmatrix} \tag{21}$$

containing the elements of the angular rotation vector

$$\boldsymbol{\omega} = \begin{pmatrix} \omega_1 \\ \omega_2 \\ \omega_3 \end{pmatrix} \tag{22}$$

see, e.g., *Britting* [1971]. Using (20) twice in (19), we get

$$\ddot{\mathbf{r}}^{\mathrm{I}} = \mathbf{C}\{(\mathbf{\Omega}\mathbf{\Omega} + \dot{\mathbf{\Omega}})\mathbf{r}^{\mathrm{LL}} + 2\mathbf{\Omega}\dot{\mathbf{r}}^{\mathrm{LL}} + \ddot{\mathbf{r}}^{\mathrm{LL}}\} \tag{23}$$

The first two terms on the right-hand side contain the centripetal, the tangential, and the Coriolis acceleration, which had been denoted by Σa_j in (2). Using (17) and (20) the angular velocities of the LL system with respect to the inertial system can be determined as

$$\mathbf{\Omega} = \begin{pmatrix} 0 & -\sin\phi\,\dot{\mu} & \cos\phi\,\dot{\mu} \\ \sin\phi\,\dot{\mu} & 0 & \dot{\phi} \\ -\cos\phi\,\dot{\mu} & -\dot{\phi} & 0 \end{pmatrix} \tag{24}$$

or

$$\boldsymbol{\omega} = \begin{pmatrix} -\dot{\phi} \\ \cos\phi\,\dot{\mu} \\ \sin\phi\,\dot{\mu} \end{pmatrix} \tag{25}$$

where

$$\dot{\phi} = \frac{v_{\mathrm{N}}}{(M+h)}$$

$$\dot{\mu} = \omega_{\mathrm{E}} + \dot{\lambda}$$

$$\dot{\lambda} = \frac{v_{\mathrm{E}}}{(N+h)\cos\phi}$$

$$M = \frac{a(1-e^2)}{(1-e^2\sin^2\phi)^{3/2}}$$

and v_{E}, v_{N} are the velocities in east and north direction; ω_{E} is the rotation rate of the earth, and M is the meridian radius of curvature. Within the accuracy assumptions made, ω_{E} can be considered constant.

Premultiplying (1) by $\mathbf{C}^{-1}$ yields the sensor output in the LL system

$$\mathbf{f}^{\mathrm{LL}} = \mathbf{C}^{-1}\ddot{\mathbf{r}}^{\mathrm{I}} - \mathbf{G}^{\mathrm{LL}} \tag{26}$$

Substituting (23) into this equation results in

$$\mathbf{f}^{\mathrm{LL}} = (\mathbf{\Omega}\mathbf{\Omega} + \dot{\mathbf{\Omega}})\mathbf{r}^{\mathrm{LL}} + 2\mathbf{\Omega}\dot{\mathbf{r}}^{\mathrm{LL}} + \ddot{\mathbf{r}}^{\mathrm{LL}} - \mathbf{G}^{\mathrm{LL}} \tag{27}$$

Because of the equivalence

$$\mathbf{\Omega}\mathbf{r} = \boldsymbol{\omega} \times \mathbf{r} \tag{28}$$

this formula is often written as

$$\mathbf{f}^{\mathrm{LL}} = \boldsymbol{\omega} \times (\boldsymbol{\omega} \times \mathbf{r}^{\mathrm{LL}}) + \dot{\boldsymbol{\omega}} \times \mathbf{r}^{\mathrm{LL}} + 2\boldsymbol{\omega} \times \dot{\mathbf{r}}^{\mathrm{LL}} + \ddot{\mathbf{r}}^{\mathrm{LL}} - \mathbf{G}^{\mathrm{LL}} \tag{29}$$

Substituting $\mathbf{r}_0^{\mathrm{LL}}$ from (12) and $\boldsymbol{\omega}$ from (25), and performing the differentiations and cross products, results in

$$\mathbf{f}^{\mathrm{LL}} = \dot{\mathbf{v}}^{\mathrm{LL}} + \boldsymbol{\rho}^{\mathrm{LL}} \times \mathbf{v}^{\mathrm{LL}} - \mathbf{g}^{\mathrm{LL}} \tag{30}$$

where

$$\mathbf{v}^{\mathrm{LL}} = \begin{pmatrix} v_{\mathrm{E}} \\ v_{\mathrm{N}} \\ v_{\mathrm{u}} \end{pmatrix}$$

$$\boldsymbol{\rho}^{\mathrm{LL}} = \begin{pmatrix} -\dot{\phi} \\ l\cos\phi \\ l\sin\phi \end{pmatrix}$$

$$\mathbf{g}^{\mathrm{LL}} = \begin{pmatrix} \eta g \\ \xi g \\ g \end{pmatrix}$$

$$l = 2\omega_{\mathrm{E}} + \dot{\lambda}$$

and ξ, η are the deflections of the vertical in N-S and E-W direction; v_{u} is the velocity orthogonal to the ellipsoid outward; and $\mathbf{g}$ is the gravity vector. For a detailed derivation, see *Farrell* [1976] or *Adams* [1979]. All other quantities are defined in the notation following (25). It should be noted that the gravitation vector $\mathbf{G}^{\mathrm{LL}}$ has been replaced by the gravity vector $\mathbf{g}^{\mathrm{LL}}$. This can be achieved by adding the AT-referenced centripetal acceleration, which is part of $\boldsymbol{\omega} \times (\boldsymbol{\omega} \times \mathbf{r}^{\mathrm{LL}})$ to $\mathbf{G}^{\mathrm{LL}}$. The gravity vector can be expressed as

$$\mathbf{g}^{\mathrm{LL}} = \boldsymbol{\gamma}^{\mathrm{LL}} + \delta\mathbf{g}^{\mathrm{LL}} \tag{31}$$

where

$$\boldsymbol{\gamma}^{\mathrm{LL}} = \begin{pmatrix} 0 \\ 0 \\ \gamma \end{pmatrix}$$

is the normal part and

$$\delta\mathbf{g}^{\mathrm{LL}} = \begin{pmatrix} \eta g \\ \xi g \\ \delta g \end{pmatrix}$$

is the anomalous part. Normal gravity is defined by

$$\boldsymbol{\gamma}^{\mathrm{LL}} = \operatorname{grad} U \tag{32}$$

where U is the potential of the equipotential ellipsoid having the mass and rotation rate of the earth. Since the geometric dimensions of this ellipsoid agree with those of the reference ellipsoid used for the LL system, the normal gravity vector points in the direction opposite to z^{LL}. Normal gravity can in good approximation be expressed by

$$\gamma = c_1 - c_4 h + (c_2 - c_5 h)\sin^2\phi + c_3 \sin^4\phi + c_6 h^2 \tag{33}$$

where the constants c_1 to c_6 depend on the ellipsoid chosen. For a derivation of (33), reference is made to *Heiskanen and Moritz* [1967].

Substituting (31) into (30) gives, after rearrangement, the so-called mechanization equations for the local level system

$$\dot{\mathbf{v}}^{\mathrm{LL}} = \mathbf{f}^{\mathrm{LL}} - \boldsymbol{\rho}^{\mathrm{LL}} \times \mathbf{v}^{\mathrm{LL}} + \boldsymbol{\gamma}^{\mathrm{LL}} + \delta\mathbf{g}^{\mathrm{LL}} \tag{34}$$

or explicitly

$$\begin{pmatrix} \dot{v}_{\mathrm{E}} \\ \dot{v}_{\mathrm{N}} \\ \dot{v}_u \end{pmatrix} = \begin{pmatrix} f_{\mathrm{E}}^{\mathrm{LL}} \\ f_{\mathrm{N}}^{\mathrm{LL}} \\ f_u^{\mathrm{LL}} \end{pmatrix} - \begin{pmatrix} 0 & -l\sin\phi & l\cos\phi \\ l\sin\phi & 0 & \dot{\phi} \\ -l\cos\phi & -\dot{\phi} & 0 \end{pmatrix} \cdot \begin{pmatrix} v_{\mathrm{E}} \\ v_{\mathrm{N}} \\ v_u \end{pmatrix} - \begin{pmatrix} 0 \\ 0 \\ \gamma \end{pmatrix} - \begin{pmatrix} \eta g \\ \xi g \\ \delta g \end{pmatrix} \tag{35}$$

The gravity disturbance vector, which is usually unknown, is often not implemented in the mechanization. The equations give vehicle acceleration in the LL system as functions of the sensor output, velocity, and gravity and are used to obtain coordinate differences along the LL axes by integration. They also form the basis for the derivation of the error model in the next section.

4. THE ERROR MODEL

Error growth in a dynamic system is usually characterized by a quasisystematic part stemming from errors in the initial values and a stochastic part generated by system noise during the operation. In our case the initial errors are due to the alignment procedure and to imperfections in the physical definition of the instrument frame. All these errors are time dependent and are therefore best described by a system of differential equations. The linear, first-order system

$$\dot{\mathbf{x}}(t) = \mathbf{F}(t)\mathbf{x}(t) + \mathbf{w}(t) \tag{36}$$

with initial values

$$\mathbf{x}(0) = \mathbf{x}_0 \tag{37}$$

is used as the standard model in many engineering applications. It is often called the space state model, and its components are named $\mathbf{x}(t)$, the state vector; $\mathbf{F}(t)$, the dynamics matrix; and $\mathbf{w}(t)$, the driving noise. The derivation of such a model starts with the linearization of the basic mathematical model. In our case this is (34), which we write symbolically

$$\dot{\mathbf{v}} = F\{\mathbf{f}(t), \mathbf{v}(t), \mathbf{r}(t), \mathbf{g}(\mathbf{r})\} \tag{38}$$

The linearization is performed about the approximation

$$\dot{\mathbf{v}}_0 = F\{\mathbf{f}_0(t), \mathbf{v}_0(t), \mathbf{r}_0(t), \boldsymbol{\gamma}(\mathbf{r}_0)\} \tag{39}$$

which describes the trajectory, as obtained from the integration of the system output, using normal gravity. It results in

$$\mathbf{d}\dot{\mathbf{v}}^{\mathrm{LL}} = \mathbf{df}^{\mathrm{LL}} + \mathbf{v} \times \mathbf{d}\boldsymbol{\rho}^{\mathrm{LL}} - \boldsymbol{\rho}^{\mathrm{LL}} \times \mathbf{dv}^{\mathrm{LL}} + \mathbf{dg}^{\mathrm{LL}} \tag{40}$$

where the error terms on the right-hand side are usually dependent on a number of individual error sources. They are discussed in detail in *Britting* [1971] and can be summarized by writing

$$\mathbf{df} = \mathbf{f} \times \boldsymbol{\varepsilon} - \mathbf{f} \times \boldsymbol{\theta} + \mathbf{b}_0 + \mathbf{sI} \cdot \mathbf{f} + \mathbf{n}_{\mathrm{v}} \tag{41}$$

and

$$\mathbf{d}\boldsymbol{\rho}^{\mathrm{LL}} = \mathbf{D}\,\mathbf{dr} \tag{42}$$

where $\boldsymbol{\varepsilon}$ is a vector of platform-dependent attitude errors, $\boldsymbol{\theta}$ is a vector of accelerometer misalignments, $\mathbf{b}_0$ is a vector of accelerometer biases, $\mathbf{s}$ is a vector of accelerometer scale factors, $\mathbf{n}_{\mathrm{v}}$ is a vector of random noise, $\mathbf{dr}$ is a vector of position errors.

Combining (41) and (42) with (40), we can write symbolically

$$\mathbf{d}\dot{\mathbf{v}}^{\mathrm{LL}} = -\boldsymbol{\rho}^{\mathrm{LL}} \times \mathbf{dv}^{\mathrm{LL}} + \mathbf{w}\{\boldsymbol{\varepsilon}, \boldsymbol{\theta}, \mathbf{b}_0, \mathbf{s}, \mathbf{dr}, \mathbf{dg}, \mathbf{n}_v\} \tag{43}$$

which gives one set of equations in the standard form (36). Similar equations can be derived for $\dot{\boldsymbol{\varepsilon}}$ and $\mathbf{d}\dot{\mathbf{r}}$, which after reordering (43) result in

$$\begin{pmatrix} \dot{\boldsymbol{\varepsilon}} \\ \mathbf{d}\dot{\mathbf{v}} \\ \mathbf{d}\dot{\mathbf{r}} \end{pmatrix} = \mathbf{F} \begin{pmatrix} \boldsymbol{\varepsilon} \\ \mathbf{dv} \\ \mathbf{dr} \end{pmatrix} + \mathbf{w}\{\boldsymbol{\theta}, \mathbf{b}_0, \mathbf{s}, \mathbf{dg}, \mathbf{n}_v\} \tag{44}$$

This is the simplest form of the state space model describing the error behavior of a three-dimensional inertial survey system. The dynamics matrix for this specific formulation is given below.

$$\mathbf{F} = \begin{bmatrix} 0 & l\sin\phi & -l\cos\phi & 0 & 0 & 0 & -1 & 0 & 0 \\ -l\sin\phi & 0 & -\dot{\phi} & -l\sin\phi & 0 & 0 & 0 & \cos\phi & 0 \\ l\cos\phi & \dot{\phi} & 0 & l\cos\phi & 0 & 0 & 0 & \sin\phi & 0 \\ 0 & 0 & 0 & 0 & 0 & 0 & 1 & 0 & 0 \\ 0 & 0 & 0 & 0 & 0 & 0 & 0 & 1 & 0 \\ 0 & 0 & 0 & 0 & 0 & 0 & 0 & 0 & 1 \\ f_u/r & 0 & -f_{\mathrm{E}}/r & 0 & 0 & 0 & 0 & -l\sin 2\phi & -2\dot{\phi}/r \\ 0 & -f_u \sec\phi/r & f_{\mathrm{N}}\sec\phi/r & 0 & 0 & 0 & (\dot{\lambda} + 2\omega_e)\tan\phi & 0 & -(\dot{\lambda} + 2\omega_e)/r \\ -f_{\mathrm{N}} & f_{\mathrm{E}} & 0 & 0 & 0 & 2\gamma/r - k2 & 2r\dot{\phi} & 2rl\cos^2\phi & 0 \end{bmatrix} \tag{45}$$

where $k2 = 2\alpha\gamma/r$ is a damping constant and the state vector x is defined by $\mathbf{x}^T = (\varepsilon_i, \varepsilon_N, \varepsilon_{uP}, d\phi, d\lambda, dh, d\dot{\phi}, d\dot{\lambda}, d\dot{h})$. For a derivation of the matrix and a more detailed discussion of individual terms, see *Wong* [1982].

It should be noted that the choice of the vector $\mathbf{x}$ is not unique. The error models used in the actual systems, therefore, often add accelerometer biases and misalignments as well as platform drifts to the state vector. A major consideration for including a certain state into $\mathbf{x}$ is that numerical values of the error described by it are needed. This can be important for real-time corrections of the system or for the post-mission determination of specific error sources.

The usefulness of the standard error model (36) becomes apparent when we ask how errors of this type can be controlled in real time. Equation (36) is the starting point for most procedures in optimal estimation and control of dynamic systems. Once the errors have been cast into this form, standard algorithms can be used to get optimal estimates. Before these algorithms are presented in the next section, the solvability of (36) must be discussed. This will be done in the second part of this section, using again an implementation-oriented approach.

It is well-known that a general solution for the equation system (36) does not exist [see, e.g., *Hochstadt*, 1975]. However, solutions are possible for matrices $\mathbf{F}$ with constant or periodic coefficients. We will only consider the constant coefficient case, which is more important for the applications discussed in this paper. The solution to the homogeneous set of equations

$$\dot{\mathbf{x}} = \mathbf{F}\mathbf{x} \tag{45}$$

is in general form

$$\mathbf{x}(t) = \mathbf{\Phi}(t, t_0)\mathbf{x}_0 \tag{46}$$

where $\mathbf{\Phi}(t, t_0)$ describes the transition from state $\mathbf{x}(t_0)$ to $\mathbf{x}(t)$ and is therefore called the transition matrix. Another term frequently used in mathematical references for $\mathbf{\Phi}\mathbf{x}_0$ is complementary function. The significant feature of (46) is that the shape of the error curves is completely determined by $\mathbf{\Phi}(t, t_0)$, while their amplitude is given by $\mathbf{x}_0$. Since $\mathbf{\Phi}(t, t_0)$ is known, provided $\mathbf{F}$ is a constant coefficient matrix, the problem reduces to obtaining the initial state vector $\mathbf{x}_0$. In principle this can be done by measuring the elements of $\mathbf{x}(t)$ at one instant t_i. Once $\mathbf{x}_0$ has been obtained, $\mathbf{x}(t)$ is predictable for any future time t. This considerable simplification of the problem became possible because the term $\mathbf{w}(t)$ in (36) has been neglected. The solution of the inhomogeneous equation (36) is

$$\mathbf{x}(t) = \mathbf{\Phi}(t, t_0)\mathbf{x}_0 + \int_{t_0}^{t} \mathbf{\Phi}(t, \tau)\mathbf{w}(\tau)\, d\tau \tag{47}$$

For the type of problems discussed here, $\mathbf{w}(t)$ is a set of stochastic variables. Thus the predictable initial state errors are disturbed by a stochastic component that grows rapidly with time. This means that the estimation of $\mathbf{x}_0$ is now done from data that are disturbed by system noise. To reduce the effect of system noise, frequent measurements of the error states are required. The use of these measurements in the general model (47) leads to optimal filtering and smoothing.

Equations (46) and (47) show clearly the central role of the transition matrix in the solution of (36). Characteristic features of the system errors can be deduced by analyzing its structure. A brief discussion on how to compute it either numerically or analytically will therefore conclude this section.

A general expression for $\mathbf{\Phi}$, in case of a constant coefficient matrix $\mathbf{F}$, is given by the matrix exponential

$$\mathbf{\Phi}(t, t_0) = e^{\mathbf{F}(t-t_0)} \tag{48}$$

which alternatively can be written as an infinite series in $\mathbf{F}$

$$\mathbf{\Phi}(\Delta t) = \sum_{n=0}^{\infty} \frac{1}{n!} \mathbf{F}^n \Delta t^n \tag{49}$$

where

$$\Delta t = t - t_0$$

A truncated form of this series is often used to determine the transition matrix numerically [see, e.g., *Gelb*, 1974]. If Δt is chosen small enough, the simple approximation

$$\mathbf{\Phi}(\Delta t) = \mathbf{I} + \mathbf{F}\Delta t$$

will often be sufficient. This is the case in the inertial survey applications where the internal integration loop is as small as 16 ms.

Although a numerical computation of the transition matrix is usually preferred in production-type problems, an analytical expression is usually superior in error analysis. In simple cases it can be obtained by an analytical development of (49) for a number of terms. Closed expressions can then be deduced by inspection of the first few terms in the series. However, for larger matrices $\mathbf{F}$ this approach becomes extremely awkward. It is often replaced by the inverse Laplace transform technique, which gives a solution by forming

$$\mathbf{\Phi}(t, t_0) = L^{-1}\{s\mathbf{I} - \mathbf{F})^{-1}\} \tag{50}$$

for individual elements of the inverse matrix on the right-hand side. Extensive tables of inverse Laplace transforms L^{-1} exist, and usually it is possible to transform the elements of the inverse matrix in such a way that they correspond to a table entry. However, the analytical inversion of $(s\mathbf{I} - \mathbf{F})$ is a problem, especially for larger matrices $\mathbf{F}$. A third way to arrive at an analytical expression for $\mathbf{\Phi}$ is by way of the eigenvalue problem. If $\mathbf{F}$ is a matrix with distinct eigenvalues λ_i, its canonical matrix $\mathbf{\Lambda}$ can be obtained by the transformation

$$\mathbf{T}^{-1}\mathbf{F}\mathbf{T} = \mathbf{\Lambda} \tag{51}$$

where the columns of $\mathbf{T}$ are the eigenvectors of $\mathbf{F}$ and

$$\mathbf{\Lambda} = \lambda_i \mathbf{I} \tag{52}$$

The transition matrix is then of the simple form

$$\mathbf{\Phi}(t, t_0) = \mathbf{T}e^{\mathbf{\Lambda}t}\mathbf{T}^{-1} \tag{53}$$

In case of multiple eigenvalues the situation is more complicated, and reference is made to textbooks like *Gantmacher* [1959] and *Hochstadt* [1975]. Although this approach is the most elegant, the practical difficulties of deriving $\mathbf{T}$ and $\mathbf{\Lambda}$ in analytical form are again considerable. Expressions corresponding to a simplified form of the dynamics matrix shown on page 882 have been published in *Wong and Schwarz* [1979].

5. ESTIMATION PROCEDURES

Rewriting (47) in discrete form for a constant coefficient matrix results in

$$\mathbf{x}_k = \mathbf{\Phi}_{k,k-1}\, \mathbf{x}_{k-1} + \mathbf{w}_{k,k-1} \tag{54}$$

where $\mathbf{x}_k = \mathbf{x}(t_k)$, $\mathbf{\Phi}_{k,k-1} = \mathbf{\Phi}(t_k, t_{k-1})$, and $\mathbf{w}_{k,k-1} = \int_{t_{k-1}}^{t_k} \mathbf{\Phi}(t_k, \tau)\, \mathbf{w}(\tau)\, d\tau$. This equation is the model used to describe the error behavior of an inertial system. Its state vector $\mathbf{x}_k$ contains individual error sources stemming from the initial alignment (see, e.g., (44)), while the driving noise $\mathbf{w}_{k,k-1}$ is a conglomerate of stochastic errors. In order to separate the errors in the state vector from the stochastic part, measurements are needed that are accurate enough to eliminate a large portion of $\mathbf{w}_{k,k-1}$. Let us assume that these measurements are of the linear form

$$\mathbf{l}_k = \mathbf{A}_k\, \mathbf{x}_k + \mathbf{e}_k \tag{55}$$

where $\mathbf{l}_k$ is the vector of measurements at time t_k, $\mathbf{A}_k$ is the design matrix at time t_k, $\mathbf{e}_k$ is the measurement noise, uncorrelated in time, with fully populated covariance matrix $\mathbf{C}_k^e$.

Since the measurements $\mathbf{l}_k$ are a linear combination of the elements of $\mathbf{x}_k$, (54) and (55) can be combined to give an optimal estimate of $\mathbf{x}_k$. If optimality is defined such that the resulting estimate is unbiased and the trace of the corresponding covariance matrix is minimum, the Kalman filtering equations result. They can be subdivided into the prediction equations and the update equations. The first set is used when no measurements of type (55) are available. The second set is used in the actual filtering procedure when an estimate of type (54) is updated by measurements of type (55). To distinguish between the estimate before and after update, we will use the convention $\hat{\mathbf{x}}_k(-)$ for estimate before update and $\hat{\mathbf{x}}_k(+)$ for estimate after update.

The prediction equations can then be written as

$$\hat{\mathbf{x}}_k(-) = \mathbf{\Phi}_{k,k-1}\, \hat{\mathbf{x}}_{k-1} \tag{56}$$

$$\mathbf{C}_k^x(-) = \mathbf{\Phi}_{k,k-1}\, C_{k-1}^x\, \mathbf{\Phi}_{k,k-1}^T + C_{k,k-1}^w \tag{57}$$

where $\mathbf{C}_k^x$ is the covariance matrix of $\hat{\mathbf{x}}_k$ and $\mathbf{C}_{k,k-1}^w$ is the covariance matrix of the driving noise $\mathbf{w}_{k,k-1}$.

The state vector $\hat{\mathbf{x}}_{k-1}$ and the covariance matrix $\mathbf{C}_{k-1}^x$ on the right-hand side of the equations may come either from an update or from a prediction. The symbols (+) or (−) have therefore been omitted. The derivation of these equations is straightforward. Equation (56) comes directly from equation (54), and (57) is obtained by standard covariance propagation.

The update equations are of the form

$$\hat{\mathbf{x}}(+) = \hat{\mathbf{x}}(-) + \mathbf{K}\{\mathbf{l} - \mathbf{A}\, \hat{\mathbf{x}}(-)\} \tag{58}$$

$$\mathbf{C}^x(+) = \{\mathbf{I} - \mathbf{K}\, \mathbf{A}\}\, \mathbf{C}^x(-) \tag{59}$$

$$\mathbf{K} = \mathbf{C}^x(-)\, \mathbf{A}^T\{\mathbf{A}\, \mathbf{C}^x(-)\, \mathbf{A}^T + \mathbf{C}^e\}^{-1} \tag{60}$$

They express the new estimate of the state vector $\hat{\mathbf{x}}(+)$ in terms of the old estimate $\hat{\mathbf{x}}(-)$ and the measurement $\mathbf{l}$. The new estimate, the measurement, and the old estimate refer to the same instant t_j. The subscript has therefore been omitted. The second term on the right of equation (58) shows how the measurement $\mathbf{l}$ is used for the update. It is compared to its predicted value $\mathbf{A}\, \hat{\mathbf{x}}(-)$, and the difference between the two is multiplied by a weight matrix $\mathbf{K}$. Thus the effect of the measurement on the estimate $\hat{\mathbf{x}}(+)$ is determined by $\mathbf{K}$. This matrix is called the gain matrix and is given by (60). Its specific form is due to the optimality criterion mentioned earlier. For a concise derivation of these equations, see *Gelb* [1974].

A major advantage of Kalman's formulation of the dynamic estimation problem is that (58) to (60) are recursive. They give the optimal estimate of the state vector on the basis of all measurements up to the time t_j, without the necessity of storing these data. The information on previous measurements is contained in the state vector and its covariance matrix. Thus, besides the update measurement and its covariance matrix, no other information is needed. Kalman filtering is often used to obtain real-time estimates in situations where a small onboard computer is the rule rather than the exception. With these computers a recursive algorithm is ideal in minimizing the requirement for fast-access storage without impairing the accuracy of the estimate.

Kalman filtering as a typical real-time procedure gives optimal estimates based on all measurements available at time t_j. If a new measurement becomes available at time t_{j+1}, all previous estimates can be improved by the new information. The process of improving previous estimates by a new measurement is called smoothing. If again an optimality criterion is employed, we speak of optimal smoothing. Since smoothing goes backward in time it is a typical post-mission process.

There are different forms of the optimal smoothing algorithm. The one used in this paper goes back to *Rauch et al.* [1965] and relates the Kalman estimates $\hat{\mathbf{x}}_j(+)$ directly to the smoothed estimates $\hat{\mathbf{x}}_j^s$. The formulas are

$$\hat{\mathbf{x}}_j^s = \hat{\mathbf{x}}_j(+) + \mathbf{D}_{j+1,j}\{\hat{\mathbf{x}}_{j+l} - \hat{\mathbf{x}}_{j+1}(-)\} \tag{61}$$

$$\mathbf{C}_j^s = \mathbf{C}_j^x(+) + \mathbf{D}_{j+1,j}\{\mathbf{C}_{j+1}^s - \mathbf{C}_{j+1}^x(-)\}\mathbf{D}_{j+1,j}^T \tag{62}$$

$$\mathbf{D}_{j+1,j} = \mathbf{C}_j^x(+)\, \mathbf{\Phi}_{j+1,j}(\mathbf{C}_{j+1}^x(-))^{-1} \tag{63}$$

If the last update measurement has been made at time t_N, then for this point we have

$$\hat{\mathbf{x}}_{j+1}^s = \hat{\mathbf{x}}_N^s = \hat{\mathbf{x}}_N(+) \tag{64}$$

$$\mathbf{C}_{j+1}^s = \mathbf{C}_N^s = \mathbf{C}_N^x(+) \tag{65}$$

because the smoothed estimate is equal to the filtered estimate. Equations (64) and (65) can be used to initiate the smoothing algorithm. The weight matrix is again derived by using optimality criteria. It differs from the Kalman gain matrix because it does not refer to the same point in time but relates the estimates at time t_{j+1} and t_j. The quantities required for optimal smoothing are the state vectors and covariance matrices before and after filtering and the transition matrices between update points. In applications it often happens that adequate buffers and tape drives to store the large amount of data at the updates are not available. In such cases, optimal smoothing cannot be applied, and empirical smoothing methods are used instead.

Empirical smoothing uses a number of simplifying assumptions to reduce the real-time storage requirements and the complexity of the model. It is only used in the context of inertial positioning. The main assumption is that zero velocity updates at vehicle stops correct a different set of error states than coordinate updates. Thus the two types of update information are treated separately. Zero velocity updates are used to update the Kalman filter. Coordinate updates at the end of the traverse are used to determine parameters not contained in the Kalman state vector. Let us assume that the parameters $\mathbf{c}$ are random in character and can be described by their first two moments

$$E\{\mathbf{c}\} = 0 \qquad E\{\mathbf{c}\,\mathbf{c}^T\} = \mathbf{C}_c \tag{66}$$

where $E\{\cdot\}$ is the statistical expectation operator. Let us assume further that the corrections to the filtered coordinates are given by the linear model

$$\mathbf{y}_k = \mathbf{B}_k\mathbf{c} \tag{67}$$

where $\mathbf{y}_k$ is the coordinate correction at point k, $\mathbf{c}$ are the random parameters, and $\mathbf{B}_k$ is the error design matrix. Then the covariance matrix of the coordinate corrections is

$$\mathbf{C}_{kj} = E\{\mathbf{y}_k\,\mathbf{y}_j^T\} = \mathbf{B}_k\,\mathbf{C}_c\,\mathbf{B}_j^T \tag{68}$$

and by applying the Gauss-Markov theorem to this case, we obtain

$$\hat{\mathbf{c}} = \mathbf{C}_c\,\mathbf{B}_j^T(\mathbf{B}_j\,\mathbf{C}_c\,\mathbf{B}_j^T)^{-1}\,\mathbf{z}_j \tag{69}$$

as an estimate for the parameters and, using (67),

$$\hat{\mathbf{y}}_k = \mathbf{B}_k\,\mathbf{C}_c\,\mathbf{B}_j^T\,(\mathbf{B}_j\,\mathbf{C}_c\,\mathbf{B}_j^T)^{-1}\,\mathbf{z}_j \tag{70}$$

as an estimate for the coordinate corrections at an arbitrary point k. The vector $\mathbf{z}_j$ contains the differences between the filtered and the control coordinates at the given station j. For a detailed discussion of (66) to (70) and their relation to optimal smoothing, see *Schwarz and Gonthier* [1981*a*]. This approach gives satisfactory results when the matrix $\mathbf{B}$ is chosen judiciously and a good estimate of $\mathbf{C}_c$ is available.

Kalman filtering and optimal smoothing solve the linear dynamic estimation problem along the time line, i.e., along a traverse where each instant of time corresponds to a unique point in space. If the same station is occupied twice, two distinct estimates will be obtained, and it is not possible to reformulate the above methods in such a way that only one estimate will result. Thus for a network of intersecting traverses, and even for a double traverse run, least squares adjustment must be used in addition to the above methods. The formulation of the mathematical model for either the filtered or smoothed data is not a major problem. However, to get a rigorous solution, the complete covariance matrix of the input coordinates is needed. It is not available from the usual filtering and smoothing algorithms that compute the autocovariances at the points but not the cross-covariances between points. The solution of this problem for the filtered coordinates is given in *Schwarz* [1980*b*]. A solution for the smoothed coordinates is not available at the moment because of undefined initial values in the backward filter [see, e.g., *Gelb*, 1974]. The solution for the empirically smoothed data is again simple and can be obtained with the least computational effort. A more detailed discussion of these problems is given in *Schwarz and Gonthier* [1981*b*].

6. INERTIAL POSITIONING

At present, inertial survey systems are mainly used for positioning purposes, covering a wide range of applications [see, e.g., *Kurtz*, 1981]. Major advantages are versatility in the field, independence of lines of sight and refraction, and speed of survey, even under adverse climatic conditions.

A typical survey mission starts with the alignment procedure described in section 2, which usually takes about 1 to $1\frac{1}{2}$ hours. This is followed by an accurate coordinate update on a known station. The known coordinates are used as initial values for the integration process. The system then moves along the line to be surveyed, stopping at intervals of about 4 min to make zero-velocity updates (ZUPT) for error control. Each update takes from 1 to 2 min and is required at each station to be surveyed. When a monument with known coordinates is reached, another coordinate update takes place, and the vehicle starts on the next leg. Progress is fairly rapid, typical values are 80 km to 100 km of traverse per hour for the helicopter mode and 30 km to 40 km/h for the van.

The basic mathematical model used in inertial positioning is

$$\mathbf{r}_2 - \mathbf{r}_1 = \mathbf{v}_1\Delta t + \iint_{t_1}^{t_2} \dot{\mathbf{v}}\, ds\, dt \tag{71}$$

where $\dot{\mathbf{v}}$ is given by (34), ds and dt denote integrations with respect to time, and $\Delta t = t_2 - t_1$. The integration thus provides coordinate differences with respect to the initial point $\mathbf{r}_1$. The model is implemented through a hierarchy of loops which update the parameters in $\dot{\mathbf{v}}$ according to their expected time rate of change. A typical implementation has a fast loop of 16–20 ms, a slow loop of 100 ms, and a turtle loop of 1 s. If the integrations in (71) are done with $\dot{\mathbf{v}}$ as defined in (34), large velocity and position errors will accumulate over a relatively short period of time. Regular zero-velocity updates are therefore used to control these errors in real time and to provide a feedback to the system that secures the linearity of the error model. At present, two approches to real-time error control are used: one is Kalman filtering, the other the method of simple ZUPT. Both of them correct the terms $\mathbf{v}_1$ and $\dot{\mathbf{v}}$ in (71) by small amounts $\boldsymbol{\delta}\mathbf{v}$ and $\boldsymbol{\delta}\dot{\mathbf{v}}$:

$$\mathbf{r}_2 - \mathbf{r}_1 = (\mathbf{v}_1 + \boldsymbol{\delta}\mathbf{v})\Delta t + \iint_{t_1}^{t_2} (\dot{\mathbf{v}} + \boldsymbol{\delta}\dot{\mathbf{v}})\, ds\, dt \tag{72}$$

However, the estimation of the corrections is fundamentally different and has important consequences on the system output. The two approaches will therefore be treated separately.

Kalman filtering uses the available update measurements to give optimal real-time estimates for all elements in the state vector. We thus have

$$\begin{aligned}\boldsymbol{\delta}\mathbf{v} &= \mathbf{x}_v\\ \boldsymbol{\delta}\dot{\mathbf{v}} &= \delta\{\mathbf{x}\}\end{aligned} \tag{73}$$

where $\mathbf{x}_v$ is the velocity state portion of the state vector and $\delta\{\mathbf{x}\}$ indicates that $\boldsymbol{\delta}\dot{\mathbf{v}}$ is dependent on all elements of the state vector. Both quantities are computed from (58) during updates and from (56) between updates. A comparison of Figures 4 and 5 shows the dramatic increase in accuracy brought about by Kalman filtering. Both show the coordinate errors as functions of time obtained along a test course with known coordinate control. Figure 4 shows the errors for (71), i.e., without any corrections, while Figure 5 shows the errors for (72), i.e., after Kalman filtering. A more detailed discussion of these results can be found in *Wong* [1981].

The state vector used in inertial positioning applications typically contains the elements

$$\mathbf{x} = (\boldsymbol{\epsilon}, \mathbf{dv}, \mathbf{dr}, \mathbf{b}_0, \boldsymbol{\theta}, \boldsymbol{\alpha}) \tag{74}$$

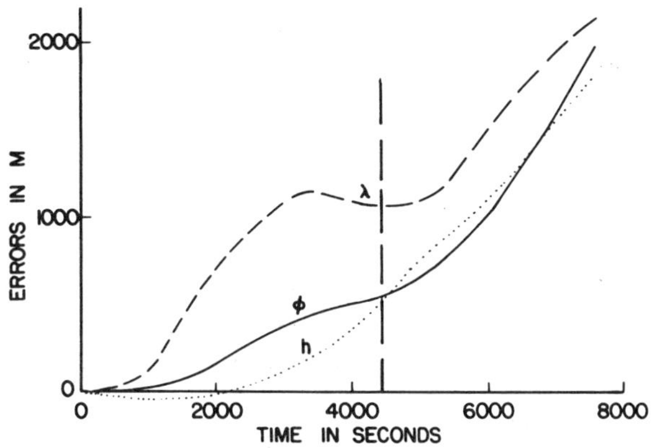

Fig. 4. Position errors before filtering (equation (71)).

where $\boldsymbol{\alpha}$ is a vector of gyro drifts, and all other terms are explained in connection with (41). The design matrix $\mathbf{A}$ for zero-velocity updates therefore has the simple form

$$\mathbf{A} = \left[\begin{array}{c|ccc|c} \mathbf{0} & 1 & 0 & 0 & \mathbf{0} \\ (3, 3) & 0 & 1 & 0 & (3, 12) \\ & 0 & 0 & 1 & \end{array} \right] \tag{75}$$

where $\mathbf{0}(3, 12)$ denotes a zero matrix of three rows and twelve columns. Similarly, the design matrix for coordinate updates will be a combination of unit matrices and zero matrices. A more complicated form of the matrix $\mathbf{A}$ results when the update measurement is a linear combination of a number of states, as, e.g., in the case of satellite ranges. A derivation of the design matrix for this case is given in *Wong and Schwarz* [1982].

Kalman filtering not only provides good real-time control but is also a method of data compression, because storage between updates is not required. Optimal smoothing uses only the prediction and filtering information at updates. Thus, given a correct error model, optimal estimates are obtained without storage of additional data. At present, none of the existing systems stores all the information necessary for optimal smoothing. Empirical smoothing methods are therefore widely used.

The second method, simple ZUPT, models the velocity error by a set of quadratics

$$\delta\mathbf{v}(t) = \mathbf{a} + \mathbf{b}t + \mathbf{c}t^2 \tag{76}$$

where the coefficients $\mathbf{a}$, $\mathbf{b}$, and $\mathbf{c}$ are determined by a fit to three successive zero-velocity measurements. The corrections in this case are

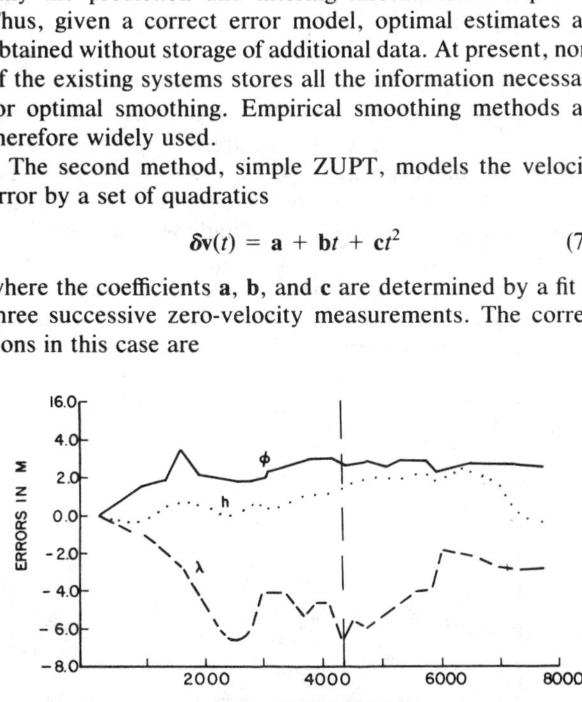

Fig. 5. Position errors after Kalman filtering with zero-velocity updates only.

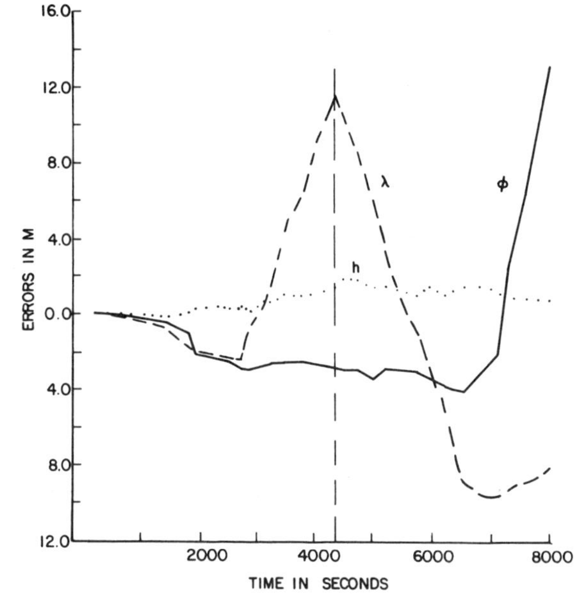

Fig. 6. Position errors after use of the 'simple ZUPT' method.

$$\begin{aligned} \delta\mathbf{v} &= \mathbf{a} \\ \delta\dot{\mathbf{v}} &= \mathbf{b} + 2\mathbf{c}t \end{aligned} \tag{77}$$

where $\mathbf{v}_1$ has been assumed at the first zero-velocity measurement. To avoid discontinuities between groups of three updates, cubic spline fitting has been proposed [*Cross and Webb,* 1981], but results have so far not been published. Although the error model contains considerable simplifications, results for positioning applications are usually good because of the strong correlation between velocity and position. Figure 6 shows the error of the simple ZUPT method for the same case treated in Figures 4 and 5. Therefore, comparison of Figures 5 and 6 shows the difference in the real-time estimate obtained from Kalman filtering and from simple ZUPT.

The fitting of zero-velocity measurements to quadratic curves is used for real-time error control. Post-mission treatment usually goes back to the raw velocity and coordinate data stored at 1-s to 10-s intervals between updates and at about 0.5-s intervals during updates. This data storage is necessary because no data compression is provided by the simple ZUPT method. It offers the advantage that extensive data analysis can be performed post mission. A variety of methods has been used for this, including simple curve fitting plus subsequent adjustment, Kalman filtering and optimal smoothing, and rigorous simultaneous adjustment of all zero velocity and coordinate updates.

A comparison of the two real-time approaches shows the characteristic difference between a general and a specialized method. Kalman filtering can accept any type of update measurement, while simple ZUPT is restricted to zero velocities. Kalman filtering provides estimates for the individual error sources, while simple ZUPT lumps all error sources together. Kalman filtering uses a unified error model with correlations between the three sensor axes, while simple ZUPT uses three separate curve fits with little consideration of the underlying physical model. Kalman filtering gives the estimate and its standard error, while no accuracy estimate is obtained from simple ZUPT. Despite these deficiencies in the error model and estimation procedure, the simple ZUPT method provides reliable results

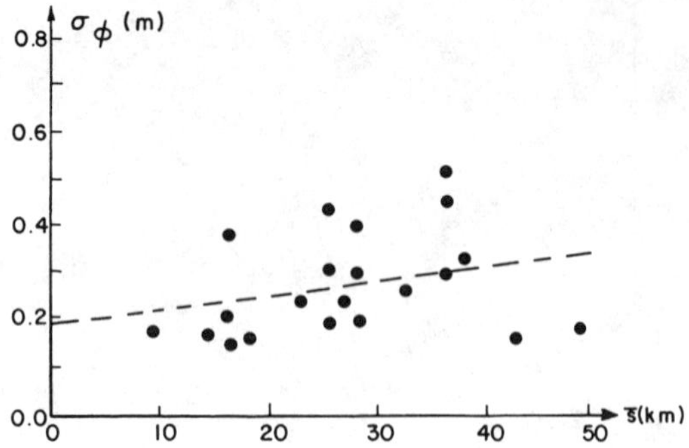

Fig. 7. Regression line for standard error of latitude after empirical smoothing.

when applied to the specific case it was designed for—land-based inertial positioning with frequent zero-velocity updates.

For many applications the treatment of individual traverses is still the standard procedure. Figures 7 and 8 show the accuracy of such traverses as derived from about 30 runs over a test course of 100-km length. Each circle gives the standard error obtained from seven to eight sets of forward and backward runs after filtering and empirical smoothing. Least squares regression lines, fitted to these data, give the results

$$\sigma_\phi = 0.19 + 2.9 \cdot 10^{-6}\,\bar{s}$$
$$\sigma_\lambda = 0.16 + 6.5 \cdot 10^{-6}\,\bar{s} \quad (78)$$

where $\bar{s}$ is the distance in m from the closest end point of the traverse. Thus the standard errors in the middle of a traverse of 100-km length are $\sigma_\phi = 0.34$ m and $\sigma_\lambda = 0.48$ m.

These results can be improved when a network of traverses with crossing lines is available. Although the best adjustment procedures to be used in such a case are a matter of current research [see, e.g., *Schwarz and Gonthier*, 1981*b*], preliminary results show that for a control point spacing of 100 km the standard error in each coordinate may be as small as 0.20 cm $\leq \sigma \leq$ 0.30 cm. Actual results can be found in *Hannah* [1982*a*] and *Todd and Tindall* [1981], using filtered coordinates as adjustment input, and in *Zimmerman and Harris* [1981], using empirically smoothed data as input.

7. INERTIAL GRAVIMETRY

As mentioned in the introduction, inertial survey systems are unique in that their output can be used to determine the three-dimensional position of a point as well as the gravity vector at that point. Since in most systems the normal gravity vector $\boldsymbol{\gamma}$ is used as the reference model, the task is to determine the gravity disturbance vector $\boldsymbol{\delta g}^{LL}$, which is defined by

$$\boldsymbol{\delta g}^{LL} = \begin{pmatrix} \dfrac{1}{r}\dfrac{\partial T}{\partial \phi} \\ \dfrac{1}{r\cos\phi}\dfrac{\partial T}{\partial \lambda} \\ \dfrac{\partial T}{\partial h} \end{pmatrix} \quad (79)$$

where T is the anomalous gravity potential, and $\{\phi, \lambda, h\}$ are the ellipsoidal coordinates shown in Figure 2. In spherical approximation, $\boldsymbol{\delta g}^{LL}$ can be written as

$$\boldsymbol{\delta g}^{LL} = \begin{pmatrix} \gamma \cdot \eta \\ \gamma \cdot \xi \\ \delta g \end{pmatrix} \quad (80)$$

where $\eta = (\Lambda - \lambda)\cos\phi$, $\xi = \Phi - \phi$, $\delta g = g - \gamma$, and Φ, Λ are astronomic latitude and longitude, and γ is given by (33).

It should be noted that in geodetic references the gravity disturbance vector is usually defined without the factor γ for ξ and η. The determination of $\boldsymbol{\delta g}^{LL}$ from the output of an inertial survey system will subsequently be called inertial gravimetry.

The principle of obtaining $\boldsymbol{\delta g}$ is again obvious from the fundamental equation (1):

$$\mathbf{f} = \mathbf{a} - \mathbf{G}$$

If the vehicle acceleration is known at any point in time, $\mathbf{G}$ can directly be determined from the sensor output. To derive $\boldsymbol{\delta g}^{LL}$, position and velocity must also be known. While the vehicle is moving, usually no outside information on $\mathbf{a}$ is available. At vehicle stops, however, we have $\mathbf{v}^{LL} = 0$ and $\dot{\mathbf{v}}^{LL} = 0$, and (34) simplifies to

$$\boldsymbol{\delta g}^{LL} = -(\mathbf{f}_{STOP}{}^{LL} + \boldsymbol{\gamma}^{LL}) \quad (81)$$

Thus the gravity disturbance vector can be determined whenever the vehicle is at rest on the earth's surface. This statement hinges on two major assumptions: first, that the orientation of the local-level frame inside the system is perfect; second, that the position of the vehicle is accurately known. The first assumption is necessary to obtain the correct output for $\mathbf{f}_{STOP}{}^{LL}$, the second to determine $\boldsymbol{\gamma}^{LL}$. Both assumptions are only approximately satisfied.

The correct measurement of $\mathbf{f}_{STOP}{}^{LL}$ in (81) would require that the gravity disturbance vector be known at the initial point, so that the system can be aligned to the ellipsoidal reference figure. Furthermore, perfect torquing to the ellipsoid is required while the system is moving. Because of alignment errors, system noise, and torquing errors, these conditions can never be perfectly met. In addition, the gravity disturbance vector is often not known at the initial point. In that case, the local-level frame refers to a slightly tilted ellipsoid. Fortunately, the errors caused by this approximation are very small for the difference of the gravity disturbance vector between two stations if the survey is restricted to a radius of about 100 km. The determination of γ is especially sensitive to height errors because the term c_4 in (33) is about 0.3 mGal m^{-1}. Since the vehicle position in real time has to be derived from the system output itself,

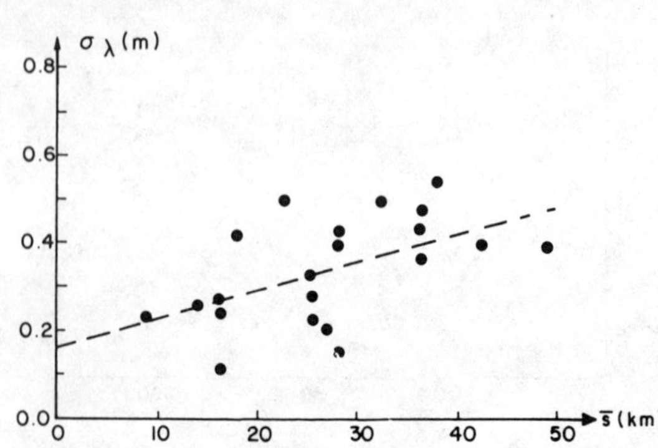

Fig. 8. Regression line for standard error of longitude after empirical smoothing.

considerable errors are possible. In this case an iterative procedure can be used, however, to get better post-mission results.

Therefore, inertial gravimetry has to solve two major problems: (1) to formulate a model for the change of the gravity disturbance vector between vehicle stops; (2) to develop an estimation model that separates gravity field effects from system errors. The solution to the first problem is given in *Moritz* [1975], where the change between stations 1 and 2 is expressed by

$$\gamma(\xi_2 - \xi_1) = \frac{1}{r}\int_{t_1}^{t_2} (T_{\phi\phi}v_\phi + T_{\phi\lambda}v_\lambda + T_{\phi r}v_r)\,dt$$

$$\gamma(\eta_2 - \eta_1) = \frac{1}{r\cos\phi}\int_{t_1}^{t_2} (T_{\lambda\phi}v_\phi + T_{\lambda\lambda}v_\lambda + T_{\lambda r}v_r)\,dt \qquad (82)$$

$$(\delta g_2 - \delta g_1) = -\int_{t_1}^{t_2} (T_{r\phi}v_\phi + T_{r\lambda}v_\lambda + T_{rr}v_r)\,dt$$

where $T_{\phi\phi}, T_{\phi\lambda}, \cdots, T_{rr} = T_{ij}$ are second-order gravitational gradients of the anomalous potential T. The nine gradients satisfy the three symmetry conditions

$$T_{ij} = T_{ji} \qquad i \neq j \qquad (83)$$

and the Laplace condition

$$\sum_{i=\phi,\lambda,r} T_{ii} = 0 \qquad (84)$$

Thus, (82) can be written in terms of five independent quantities. It should be noted, however, that each element in the gravity disturbance vector is dependent on three gradients. This means that each component in $\boldsymbol{\delta}\mathbf{g}^{\mathrm{LL}}$ changes whenever the system moves. A model that assumes changes in ξ, η, and δg to be only dependent on changes in the respective coordinate directions is clearly inadequate, albeit frequently used.

Equations (82) require the knowledge of the second-order gravitational gradients along the vehicle path. Although considerable effort has gone into the development of dynamic gravity gradiometers (see, e.g., *Heller* [1981] for an overview and *Metzger and Jincitano* [1981] and *Trageser* [1981] for a discussion of the state of the art), the enormous costs of these systems will most likely be prohibitive for applications in standard survey operations. Assuming therefore that the T_{ij} are unknown, they have to be estimated from the system output at vehicle stops. For this it is necessary to introduce a model for the behavior of the T_{ij} between vehicle stops. The simplest assumption is that they are constant between zero-velocity updates. Such an assumption is not overly restrictive because it can be achieved by a judicious spacing of the update points. We will therefore use this model in the following. Equations (82) can be rewritten as

$$\boldsymbol{\Delta\delta}\mathbf{g}^{\mathrm{LL}} = \begin{pmatrix} \gamma\Delta\eta \\ \gamma\Delta\xi \\ \Delta\delta g \end{pmatrix}$$

$$\frac{1}{r\cos\phi}\{a_{\lambda\lambda}\Delta\lambda + a_{\lambda\phi}\Delta\phi + a_{\lambda r}\Delta h\}$$

$$= \frac{1}{r}\{a_{\lambda\phi}\Delta\lambda + a_{\phi\phi}\Delta\phi + a_{\phi r}\Delta h\} \qquad (85)$$

$$-\{a_{\lambda r}\Delta\lambda + a_{\phi r}\Delta\phi + a_{rr}\Delta h\}$$

with obvious substitutions.

The use of (85) for the estimation problem is rather straightforward when the coefficients a_{ij} can be considered as stochastic quantities with a known covariance function. Functions of this type, derived from actual data, have been published in *Schwarz* [1980*a*] and *Hein* [1981]. The coefficients a_{ij} are added to the state vector as

$$\dot{\mathbf{a}} = \mathbf{0} \qquad (86)$$

and their spatial correlation is described by the given covariance function. Real-time corrections are made to the measurements by replacing $\boldsymbol{\delta}\mathbf{g}^{\mathrm{LL}}$ in (34) by the right-hand side of (85). The addition of (86) to the original state vector results in an augmented system that is nonlinear in nature. Thus, in principle, iterative procedures are always required. Since the nonlinearities are very small in this case, the first iterative step will often be sufficient. For a brief discussion of the nonlinear problem and possible solution approaches, see *Schwarz* [1980*a*].

Equations (82) give the general structure of the gravity disturbance vector. They are therefore not only useful as error model for a real-time filter but can also be used for a direct post-mission approach. Such a procedure is advantageous for all systems that store velocity information in regular time intervals. In this case the simplifying assumptions made for (85) are not necessary, and the second-order gradients can be represented by a set of suitable base functions. It is also possible to express the elements of the gravity disturbance vector directly as three-dimensional functions if care is taken that the functional relationships between them are satisfied.

Results available at the moment are all based on a simplified real-time filter and various types of post-mission smoothing. The filter models the anomalous gravity field in each axis by one initial attitude error and one term that changes with the motion along the specific axis. Such a filter will work well along straight traverses but is problematic for a more general traverse design. Repeatability of results is at the level of 0".5, but considerable systematic differences exist between the forward and the backward run [see *Schwarz,* 1978]. Simple traverse smoothing will therefore result in a higher standard error, typically about 0".8 for the deflections and 2 mGal for the gravity disturbance when using the helicopter mode. Results for the vehicle mode are slightly worse, see *Todd* [1981]. An area adjustment with known deflection points about every 30 km will eliminate most of the systematic effects at intermediate stations and will result in standard errors of 0".4–0".5 and 1 mGal, respectively [see *Todd*, 1981; *Hannah,* 1982*b*]. The models used in the adjustment are one-dimensional in character and do not use the structural information contained in (82). A first attempt to model $\{\xi, \eta, \delta g\}$ as two-dimensional functions using the known functional relationships between them has been made by *Bose and Huddle* [1981]. They use an optimal smoothing algorithm based on a Karhunen-Loeve expansion, the coefficients of which are determined from measurements in the specific area. Standard errors of about 0".2 for the deflections of the vertical and 0".7 mGal for the gravity disturbance are reported for the same data used by Todd. It is not quite clear, however, whether or not the stations used to compute the coefficients of the expansion have also been used for the computation of the standard errors. In that case they could not be considered as independent determinations characterizing the accuracy of the method. The standard error of the known deflections, which is $\sigma_{\xi,\eta} = 0''.25$ in this case, indicates that the estimated standard errors may be slightly optimistic. The results show, however, the potential of refined post-mission methods and the difficulty of finding areas with deflection information accurate enough to test this potential.

Although inertial gravimetry can definitely be used as a stand-alone method, its greatest potential is perhaps in the refinement of existing local gravity field models. *Forsberg and Tscherning* [1981] have pointed out that deflections of the vertical and gravity anomalies can be obtained with standard errors of about $\sigma_{\xi,\eta} = 0.8$ arc sec and $\sigma_{\Delta g} = 4$ mGal from a combination of gravity and topographical data. Since this type of data is available in many areas, a detailed local gravity approximation can be computed before starting the inertial survey. Real-time use of such a model is possible if simple representations for the onboard processor can be found. Zero-velocity measurements based on such a model will give valuable geophysical information and will result in refinements of local, three-dimensional gravity field approximations which cannot easily be achieved by any other method.

8. OUTLOOK

Inertial survey systems have been accepted in geodetic practice after only a brief period of adaption. The two major reasons for this (versatility and economy) are summarized in two quotes from *Babbage* [1981]. On versatility:

> Inertial systems have been mounted in cars, trucks, tracked vehicles, helicopters and fixed-wing aircraft. They have operated well in climatic extremes ranging from the heat of the Arabian desert to the cold slopes of northern Canada. These systems have positioned land parcels in the Alaskan wilderness as readily as bench marks in the suburbs of Chicago. They have provided data for mapping and for the route selection and building of pipelines, hydro-electric transmission lines, dams, roads and railroads. In addition, they have facilitated petroleum exploration on land and in the off-shore and the positioning of gravity and seismic data.

On economy:

> Given a fair opportunity to compete, inertial systems have proven to be at least as accurate as, and much quicker and cheaper than, the best conventional methods for many important surveying tasks of large scope. In other cases they have provided the only practical solution to a unique problem.

While the use of these systems is thus well established, a number of applications will profit from future developments in the hadware sector and from refinements of the modeling techniques. According to *Draper* [1977], the present accuracy of hardware components can be improved by at least 2 orders of magnitude within the next decade. This would bring positioning accuracies to the centimeter level and open new possibilities for inertial gravimetry in geophysical prospecting. Such improvements of the sensors would require refinements in the definition of the inertial frame, in the gravity field model, and in the real-time and post-mission error models. As has been pointed out in sections 6 and 7, developments in the areas of post-mission models and gravity field models are most urgent at present.

In positioning, such improvements would lead to the standard use of the system in first-order control point densification. In conjunction with highly accurate satellite positioning methods, especially GPS interferometry, the conventional classification into first-, second-, and third-order control could be abandoned, and a base network of uniform accuracy could be created. Thus the criterion would be the required density of stations in a certain area rather than the accuracy of the stations. As *Mueller* [1981] has pointed out, satellite and inertial techniques will complement each other well in these applications. Accurate positioning for large engineering projects and monitoring of such projects and of local geodynamic phenomena are other possible applications. In the area of monitoring the speed of the survey and the capability of determining changes in positions and in the gravity disturbance vector will be an asset. A number of other positioning applications will profit from the fact that the ZUPT periods can be considerably extended if a lower accuracy is sufficient and that the requirements for control point density can be relaxed.

In inertial gravimetry, improvements of the hardware components would make it worthwhile to investigate the use of the total gravity disturbance vector for geophysical prospecting. Present gravimetric exploration methods that use the magnitude of the disturbance vector only could be enhanced by the vital information on the direction of this vector. The capability of inertial survey systems of giving dense area coverage in a relatively short time is an attractive feature for these applications.

Integration of inertial navigation systems and satellite positioning systems like NAVSTAR-GPS will be used for accurate dynamic positioning. One application that is presently investigated is offshore positioning for seismic surveys [see *Wong and Schwarz,* 1982]. Another is the elimination of ground control for photogrammetric block adjustments by providing inflight positions or attitude angles for the exposure stations [see, e.g., *Gibson and Masry,* 1981]. Aerial gravimetry may finally become a reality by combining an improved inertial system with accurate range updates from either satellites or ground stations.

Finally, as with any new measuring device, new applications will be envisaged, once its capabilities are understood and accepted by a larger group of users. A typical example, presently under investigation, is the regular monitoring of pieplines in the Arctic to detect critical bending of pipes. In this case the use of the system's attitude information is of prime importance. The significance of this application in the prevention of major oilspills is obvious. Other possible applications are discussed by *Kurtz* [1981]. Thus it appears that at present the potential of these systems to solve geodetic problems in an efficient and unconventional manner is far from being exploited and that interesting applications can be expected in the future.

Acknowledgments. This research was supported by a contract between the Geodetic Survey of Canada and the University of Calgary as well as by grants from the National Sciences and Engineering Research Council of Canada. Software development necessary to obtain the results in Figures 4–8 was done by R. V. C. Wong and D. Arden. All this support is gratefully acknowledged.

REFERENCES

Adams, J. R., Description of the Local Level Inertial Survey System and Its Simulation, *Tech. Rep. 59,* Dep. Surv. Eng., Univ. New Brunswick, Fredericton, 1979.

Babbage, G., Operation of the inertial survey system: The manager's view, paper presented at First International Symposium on Inertial Technology for Surveying and Geodesy, Can. Inst. Surv., Ottawa, October 12–14, 1977.

Babbage, G., Inertial surveying—A study in accuracy and versatility, paper presented at Second International Symposium on Inertial Technology for Surveying and Geodesy, Can. Inst. Surv., Banff, June 1–5, 1981.

Bose, S. C., and J. R. Huddle, Regional adjustment of inertial gravity disturbance vector measurements by optimal two-dimensional smoothing, paper presented at Second International Symposium on Inertial Technology for Surveying and Geodesy, Can. Inst. Surv., Banff, June 1–5, 1981.

Britting, K. R., *Inertial Navigation Systems Analysis,* Interscience, New York, 1971.

Cross, P. A., and J. P. Webb, Post-processing experiments with FILS2 test data, paper presented at Second International Symposium on Inertial Technology for Surveying and Geodesy, Can. Inst. Surv., Banff, June 1–5, 1981.

Draper, C. S., Inertial technology for surveying and geodesy, paper presented at First International Symposium on Inertial Technology for Surveying and Geodesy, Can. Inst. Surv., Ottawa, October 12–14, 1977.

Farrell, J. L., *Integrated Aircraft Navigation,* Academic, New York, 1976.

Forsberg, R., and C. C. Tscherning, Deflection and gravity anomaly prediction for inertial positioning using collocation, paper presented at Second International Symposium on Inertial Technology for Surveying and Geodesy, Can. Inst. Surv., Banff, June 1–5, 1981.

Gantmacher, F. R., *The Theory of Matrices,* vol. 2, Chelsea Publishing Company, New York, 1959.

Gelb, A., *Applied Optimal Estimation,* MIT Press, Cambridge, Mass., 1974.

Gibson, J. R., and S. E. Masry, Application of inertial systems in aerotriangulation, paper presented at Second International Symposium on Inertial Technology for Surveying and Geodesy, Can. Inst. Surv., Banff, June 1–5, 1981.

Grafarend, E. W., From kinematical geodesy to inertial positioning, paper presented at Second International Symposium on Inertial Technology for Surveying and Geodesy, Can. Inst. Surv., Banff, June 1–5, 1981.

Hannah, J., The Development of Comprehensive Error Models and Network Adjustment Techniques for Inertial Surveys, *Rep. 330,* Dep. Geod. Sci. Surv., Ohio State Univ., Columbus, 1982*a*.

Hannah, J., Inertial rapid geodetic survey system (RGSS) error models and network adjustment, *Rep. 332,* Dep. Geod. Sci. Surv., Ohio State Univ., Columbus, 1982*b*.

Hannah, J., and I. I. Mueller, Improvement of inertial surveys through post-mission network adjustment and self-calibration, paper presented at Second International Symposium on Inertial Technology for Surveying and Geodesy, Can. Inst. Surv., Banff, June 1–5, 1981.

Hein, G. W., Untersuchungen zur terrestrischen Schweregradiometrie, *Heft 264,* Deut. Geod. Komm., Reihe C, München, 1981.

Heiskanen, W. A., and H. Moritz, *Physical Geodesy,* W. H. Freeman, San Francisco, Calif., 1967.

Heller, W. G., Prospects for gradiometric aiding of inertial survey systems, paper presented at Second International Symposium on Inertial Technology for Surveying and Geodesy, Can. Inst. Surv., Banff, June 1–5, 1981.

Hochstadt, H., *Differential Equations,* Dover, New York, 1975.

Kayton, M., Fundamental limitations on inertial measurements, in *Guidance and Control of Aerospace Vehicles,* edited by C. T. Leondes, McGraw-Hill, New York, 1963.

Kurtz, M. K., New applications for inertial surveying and positioning systems—A Challenge, paper presented at Second International Symposium on Inertial Technology for Surveying and Geodesy, Can. Inst. Surv., Banff, June 1–5, 1981.

Metzger, E. H., and A. Jincitano, Application of Bell rotating accelerometer gravity gradiometers and gravity meters to airborne or land vehicle gravity surveys, paper presented at Second International Symposium on Inertial Technology for Surveying and Geodesy, Can. Inst. Surv., Banff, June 1–5, 1981.

Moritz, H., Combination of aerial gravimetry and gradiometry, *Rep. 223,* Dep. Geod. Sci., Ohio State Univ., Columbus, 1975.

Moritz, H., Relativistic effects in reference frames, in *Reference Coordinate Systems for Earth Dynamics,* edited by E. M. Caposchkin and B. Kolaczek, D. Reidel, Hingham, Mass., 1981.

Mueller, I. I., *Spherical and Practical Astronomy as Applied to Geodesy,* Frederick Ungar Publishing Co., New York, 1969.

Mueller, I. I., Inertial survey systems in the geodetic arsenal, paper presented at Second International Symposium on Inertial Technology for Surveying and Geodesy, Can. Inst. Surv., Banff, June 1–5, 1981.

Rauch, H. E., F. Tung, and C. T. Striebel, Maximum likelihood estimates of linear dynamic systems, *AIAA J., 3*(8), 1445–1450, 1965.

Schwarz, K. P., Accuracy of vertical deflection determination by present-day inertial instrumentation, Applications of Geodesy to Geodynamics, *Rep. 280,* pp. 273–281, Dep. Geod. Sci., Ohio State Univ., Columbus, 1978.

Schwarz, K. P., Gravity field approximation using inertial survey systems, *Can. Surv., 34*(4), 383–395, 1980*a*.

Schwarz, K. P., Error propagation in inertial positioning, *Can. Surv., 34,* 265–276, 1980*b*.

Schwarz, K. P., and M. Gonthier, Analysis of the heading sensitivity in the Litton inertial system, *Bull. Géod., 55*(4), 315–328, 1981*a*.

Schwarz, K. P., and M. Gonthier, Adjustment problems in inertial positioning, Symposium on Geodetic Networks and Computations, *Heft 2581 IV,* pp. 97–110, Deut. Geod. Komm., Reihe B, Munich, 1981*b*.

Todd, M. S., Rapid geodetic survey system (RGSS) White Sands test for position, height, and the anomalous gravity vector components, paper presented at Second International Symposium on Inertial Technology for Surveying and Geodesy, Can. Inst. Surv., Banff, June 1–5, 1981.

Todd, M. S., and T. O. Tindall, Inertial survey adjustment, paper presented at Second International Symposium on Inertial Technology for Surveying and Geodesy, Can. Inst. Surv., Banff, June 1–5, 1981.

Trageser, M. B., Floated gravity gradiometer status report, paper presented at Second International Symposium on Inertial Technology for Surveying and Geodesy, Can. Inst. Surv., Banff, June 1–5, 1981.

Vanicek, P., and E. J. Krakiwsky, *Geodesy: The Concepts,* North-Holland, Amsterdam, 1982.

Wong, R. V. C., A Kalman filter/smoother for the Ferranti inertial land surveyor system, paper presented at Second International Symposium on Inertial Technology for Surveying and Geodesy, Can. Inst. Surv., Banff, June 1–5, 1981.

Wong, R. V. C., A Kalman filter-smoother for an inertial survey system of local level type, *Publ. 20001,* Div. Surv. Eng., Univ. of Calgary, 1982.

Wong, V. C., and K. P. Schwarz, Investigations on the analytical form of the transition matrix in inertial geodesy, *Tech. Rep. 58,* Dep. Surv. Eng., Univ. New Brunswick, Fredericton, 1979.

Wong, R. V. C., and K. P. Schwarz, Offshore positioning with an integrated GPS/inertial navigation system, paper presented at the Third International Symposium on Satellite Doppler Positioning, Can. Inst. Surv., Las Cruces, N.M., February 7–12, 1982.

Zimmerman, D. S., and H. C. Harris, General purpose least-squares adjustment of IPS data, paper presented at Second International Symposium on Inertial Technology for Surveying and Geodesy, Can. Inst. Surv., Banff, June 1–5, 1981.

(Received December 7, 1982;
revised January 26, 1983;
accepted February 17, 1983.)

REVIEWS OF GEOPHYSICS AND SPACE PHYSICS, VOL. 18, NO. 4, PAGES 725–745, NOVEMBER 1980

On Using Satellite Altimetry to Determine the General Circulation of the Oceans With Application to Geoid Improvement

CARL WUNSCH

Department of Earth and Planetary Sciences, Massachusetts Institute of Technology, Cambridge, Massachusetts 02139

E. M. GAPOSCHKIN

Smithsonian Astrophysical Observatory, Cambridge, Massachusetts 02138

We describe the problem of combining hydrography with marine geodesy and satellite altimetry for the purpose of determining the general circulation of the oceans, defining the eddy field, and improving the marine geoid. The critical problem is to understand the error budgets of four fields: orbit, height measurement, geoid, and ocean water density. Corrections must be made for atmospheric load, tides, tropospheric water vapor, wave height, and other parameters. A general formalism for deducing the geoid and ocean circulation is obtained in terms of inverse theory and applied to some limited examples.

INTRODUCTION

With the launching of the Geos 3 and Seasat satellites the possibility of global measurements of the absolute elevation of the sea surface relative to the center of the earth has become a reality. For many years it has been appreciated that if this measurement could be done sufficiently accurately, then with a sufficiently accurate measurement of the earth's gravity field, one could begin to determine the general circulation of the ocean. What has not been widely appreciated are (1) that the precise meaning of the words 'sufficiently accurate' is far from obvious and (2) that to actually determine the mean circulation, a third field in addition to surface height and gravity is necessary—the three-dimensional density distribution of the ocean.

Despite the early demise of Seasat (see the mission overview by *Born et al.* [1979*a*]) there does exist at the present time a global data set consisting of the elevation of sea level relative to the satellite, acquired over the approximately 3 months that the satellite functioned. In addition, the nearly 3 years of data from the less accurate Geos 3 altimeter system is also available. A number of discussions of satellite altimeter measurements are available (e.g., *Huang et al.* [1978], *Leitao et al.* [1979], *Gordon and Baker* [1980], *Cheney and Marsh* [1980], *Douglas and Gaborski* [1979], and much of the July 30, 1979, issue of the *Journal of Geophysical Research*). Many of these results are intriguing, seemingly showing real oceanographic features; it is clear that there will now be efforts to take these measurements beyond the mainly qualitative use to which they have hitherto been put and to use them for a variety of quantitative purposes. It is our purpose here to set out, in a somewhat pedagogical fashion, a scheme for discussing the role of satellite altimetry in studying the general circulation of the oceans and determining the geoid (these are only two of a number of possible applications). Despite the long discussions extant of applications of satellite altimetry to the ocean circulation there seems to be no adequate description of how actually to use the data and of its required accuracy. The procedure requires a working knowledge of three fields: physical oceanography, geodesy, and satellite dynamics. At the risk of boring many readers we have tried here to review some of the very basic features of each of these fields, because exploitation of altimeters requires at least a working knowledge of all of them. But this paper is not intended as an exhaustive review of the large, complex literature on the details of the satellite systems and hardware. We have tried to provide a few key references which will allow the reader to begin to penetrate the technical as opposed to the scientific questions. Much of this literature is 'grey,' and a comprehensive review would be very valuable, but is beyond the scope of the present paper.

THE PROBLEM

Ocean Water Movement

For simplicity we will work largely in local Cartesian coordinates and discuss a basically two-dimensional problem. The generalization to the sphere and three dimensions is reasonably obvious. We will accept the observed conclusion (which is also supported by theory) that on the large scale the ocean is nearly in geostrophic equilibrium and hydrostatic balance. In the coordinate system of Figure 1, with x horizontal and z positive upward, this implies

$$fv = -\frac{1}{\rho}\frac{\partial p}{\partial x} \tag{1}$$

$$0 = -\frac{\partial p}{\partial x} - g\rho \tag{2}$$

where $f = 2\Omega \sin$ (latitude) is the Coriolis parameter, with Ω the rotation rate of the earth, p is the pressure, v is the velocity normal to the x, z plane, g is the local vertical component of gravity, and ρ is the water density. We can combine (1) and (2) into

$$f\frac{\partial}{\partial z}(\rho v) = g\frac{\partial \rho}{\partial x} \tag{3}$$

the 'thermal wind' equation. Integrating once, we have

$$\rho v = \frac{g}{f}\int_{z_0}^{z}\frac{\partial \rho}{\partial x}\,dz + b \tag{4}$$

where z_0 is an arbitrary starting point ('reference level') for the integration and b is an integration constant ('reference velocity') which is not normally known. The term involving the density derivative is the 'relative velocity.'

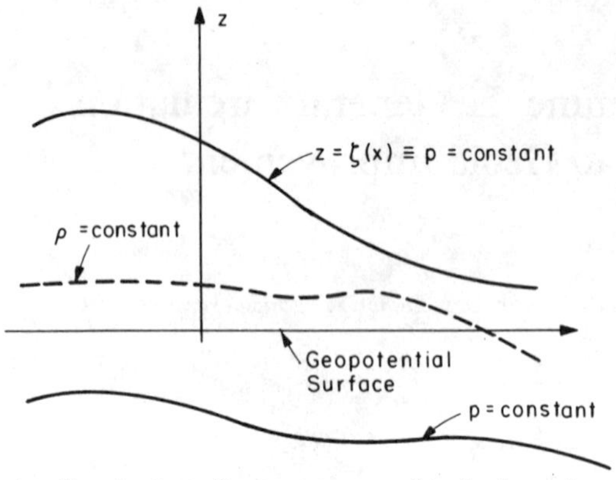

Fig. 1. Sketch of coordinate systems used in the dynamic method. The verticle coordinate may be either p or z, their relationship coming through the hydrostatic equation (2).

Oceanographers have found it more convenient to use pressure rather than z as a vertical coordinate (the conversion between them coming from (2)) and to introduce an auxiliary variable D, called the dynamic height, defined as

$$D(x, p) = \int_{p_R(x)}^{p} \frac{dp}{\rho(x, p)} \tag{5}$$

where p_R is an arbitrary reference pressure [see *Sverdrup et al.,* 1942]. The unit of D is the dynamic meter (dyn-m), such that the magnitude of 1 dyn-m is about 1 m of physical displacement in the vertical. Equation (4) then becomes

$$v(x, p) = \frac{10}{f}\frac{\partial D}{\partial x} + b \tag{6}$$

(The factor of 10 derives from the definition of the units of D in dynamic meters.) The dynamic height is the geopotential distance between two pressure levels in the sea. Because the density of seawater is never far from 1.03 g/cm^3, it is conventional to use the anomaly of dynamic height

$$D = D_0 + \Delta D(x, p)$$

relative to an arbitrary constant reference value D_0. We thus have finally

$$v(x, p) = \frac{10}{f}\frac{\partial(\Delta D)}{\partial x} + b \tag{7a}$$

To use relationship (7*a*), oceanographers with ships measure the density as a function of depth (computed via an equation of state from temperature and salinity). This is normally done at positions called 'stations' separated by a distance Δx. A line of such stations is called a 'section' (see Figure 2). The subject is called 'hydrography.' We label each adjacent pair of stations with an index i and compute the derivative in (7*a*) as a simple difference and the integral (5) as an approximating sum. We then have for station pair i,

$$v(x_i, p) \equiv v_i(p) = v_{Ri} + b_i \tag{7b}$$

where b_i is the reference velocity and v_{Ri} the relative velocity (because it is computed in relation to the initial reference level p_R). Note that the value of b_i depends upon the value of p_R.

A fundamental problem of physical oceanography is that the reference velocity is usually unknown. If we could conclude that some pressure level p_R coincided with a gravitational equipotential surface, then we could begin the integration in (5) from that level, and we would have

$$v_i(p) = \frac{10}{f}\frac{\partial}{\partial x}\Delta D \tag{8}$$

the integration constant being zero because $v = 0$ at $p = p_R$ (i.e., the isobaric surface and the 'level' surface coincide). This is the hypothesis of a 'level of no motion' at $p = p_R$. Given a true level of no motion, we can integrate (5) to the surface (p = 0—atmospheric pressure is taken as constant):

$$D_s(x) = \int_{p_R}^{0} \frac{dp}{\rho} = D_0 + \Delta D_s \tag{9}$$

This function would represent the elevation of the surface of the water in dynamic meters relative to the geoid (or, more precisely, with respect to an arbitrary geopotential surface). Notice that p_R may be a function of x. Because of the specific choice of units a change of $D_s(x)$ by 1 dyn-m is very close to 1 m of elevation relative to the geoid.

With the assumption of a level of no motion, one can map D_s or ΔD_s as a function of position in the world ocean whereever there are measurements of ρ. A good example is *Wyrtki's* [1975] Pacific Ocean charts of ΔD_s, one of which is displayed in Figure 3, where he assumed that p_R = 1000 dbar (1 decibar = 1 dbar = 10^{-1} bar ~ 1 m of water). Wyrtki's chart suggests that the elevation change of the sea surface relative to the geoid is of the order of 1.5 m over the entire width of the Pacific Ocean. This order of magnitude is undoubtedly correct, but we wish to emphasize two important points about Figure 3.

1. There is no reason to believe the hypothesis that the velocity vanishes at p_R = 1000 dbar (*Wyrtki* [1979] disclaims any implication that his reference levels are levels of no motion). It is a convenient fiction to allow a display of the spatial variation of ΔD_s. No theory suggests that the hypothesis might be correct, and it is possible that the correct reference pressure varies from its value at the surface to that at the bottom of the ocean. Observations [e.g., *Wunsch,* 1978; *Roemmich,* 1979; *Schmitz,* 1978, 1980; *Fuglister,* 1963] suggest that isotachs of

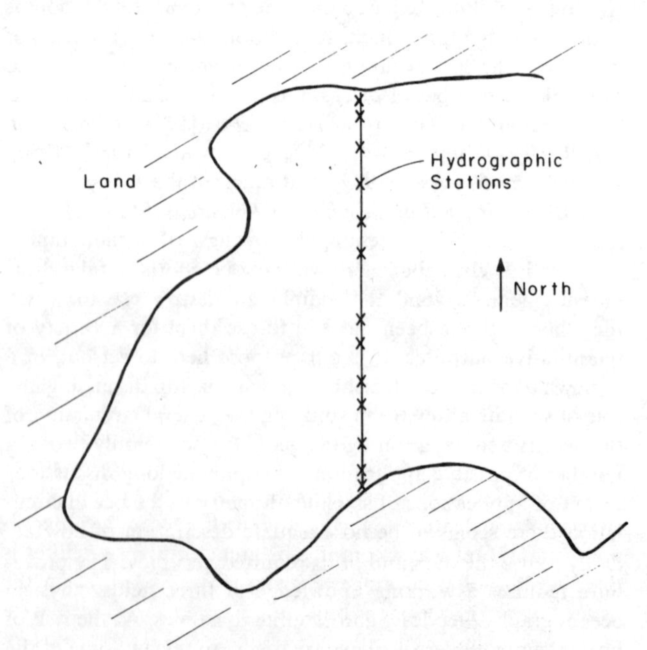

Fig. 2. Schematic diagram of a line of hydrographic stations (a section) bounding a volume of ocean to which certain conservation statements must apply [*Wunsch,* 1978].

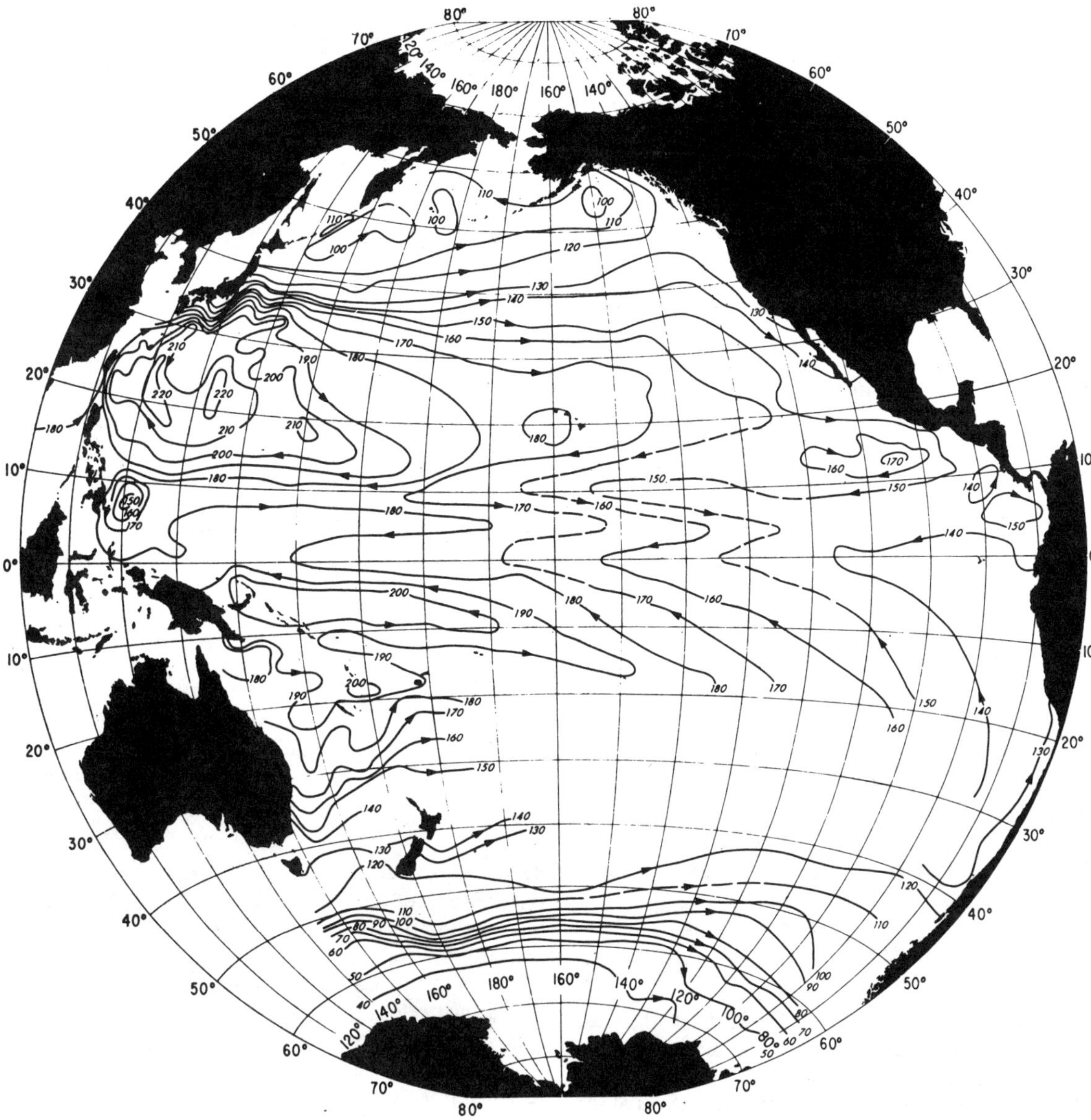

Fig. 3. Mean annual dynamic topography of the Pacific sea surface, in dynamic centimeters, relative to the 1000-dbar pressure surface [from *Wyrtki,* 1975]. If the 1000-dbar surface corresponds to a gravitational equipotential surface, then this chart would be an estimate of the absolute elevation of the sea surface relative to the geoid, interpretable in centimeters (with a constant offset). However, 1000 dbar is almost certainly not an equipotential surface.

zero horizontal velocity intersect the sea surface and sea bottom in a very complex spatial structure. An example [*Roemmich,* 1979] in the Gulf Stream system is shown in Figure 4.

2. Figure 3 represents an inhomogeneous time average of data required over many years in different seasons from a large number of different ships. It is known [e.g., *Mode Group,* 1978; *Mode Atlas Group,* 1977] that the ocean changes through time with a complex and not well described frequency/wave number spectrum. Even if the hypothesis of a time mean reference level at $p_R = 1000$ dbar were correct, an instantaneous picture of the dynamic height field relative to p_R would, while sharing some features in common with Figure 3, differ in many significant ways from it. The great virtue of satellites is their rapid global coverage; charts such as Wyrtki's, on the other hand, have been built up over many years with enormous painstaking effort. But there are large areas of ocean, especially in the South Pacific, where there are almost no measurements of water density to this day.

At this stage it is convenient, however, to accept the fiction that the dynamic height field is globally known and is not time varying. We will relax this drastic hypothesis later. Referring back to (7), we can evaluate it at $p = 0$ and obtain

$$v_{si} \equiv v_i(p = 0) = \frac{10}{f} \frac{\partial}{\partial x} \Delta D(x_i, p = 0) + b_i \tag{10a}$$

From (1), v_s is proportional to the surface pressure gradient, and from (2) this must be

$$v_{si} = \frac{g}{f} \frac{\partial \zeta}{\partial x}(x_i) \tag{10b}$$

where $\zeta = S - N$ is the difference between the height of the ocean surface, S, relative to a reference surface, and of the geoid, N, referred to the same surface. If we could measure ζ (notice that it is only the spatial derivative $\partial\zeta/\partial x$ that matters), then the surface velocity would be known. The surface velocity fixes the values of the integration constant b_i and thus fixes

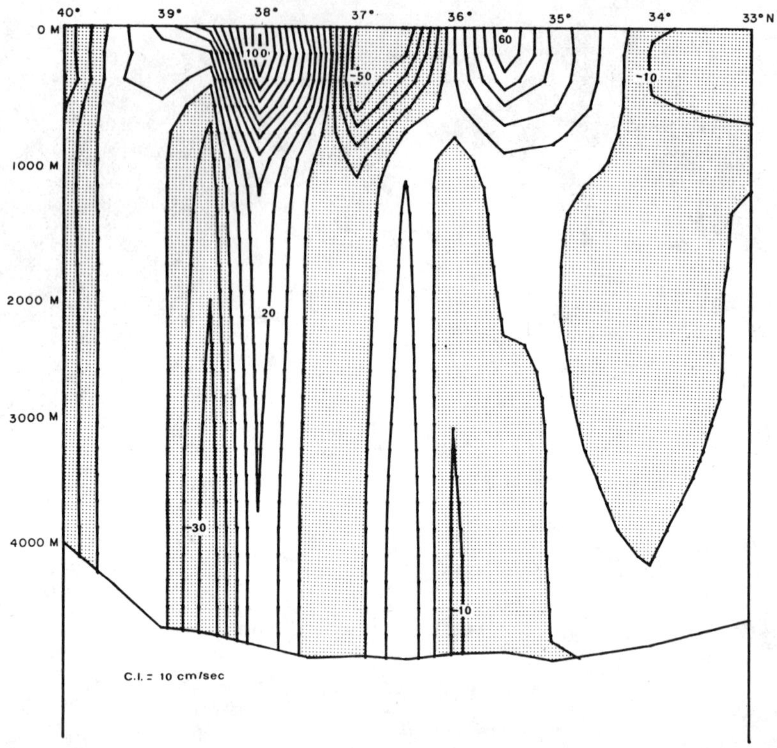

Fig. 4. Velocity section across the Gulf Stream at 57°W [from *Roemmich,* 1979]. The hydrography was from *Fuglister* [1963], and absolute velocity was inferred from a modified beta spiral vorticity balance. Stippled areas show westward flow. The structure is much more one of vertically oriented cells than one of a constant level of no motion. The contour interval is 10 cm/s.

the water velocity everywhere in the water column. It would be simplest then to take $p_R = 0$, and we have

$$b_i = \frac{g}{f}\frac{\partial \zeta}{\partial x}(x_i) \tag{11}$$

Thus if ζ can be measured directly, we can eliminate the need to make ad hoc assumptions like levels of no motion and determine the general circulation absolutely.

The relationship between velocity and the derivatives of the pressure field implies that under some circumstances there is an especially simple (somewhat degenerate) relationship between the pressure field and the total water transport across the section. Let $h(x_i)$ be the water depth (in pressure units) at station pair i. Then the total geostrophic mass transport of water across the section is (from (7) and (11))

$$V = \int_0^{x_L} dx \int_0^{h(x)} \rho v \, dp$$

$$= \int_0^{x_L}\int_0^{h(x)} \left(\frac{10\rho}{f}\frac{\partial}{\partial x}\Delta D + g\rho\frac{\partial \zeta}{\partial x}\right) dp\, dx$$

(the integral would be performed as a sum). Here $x = 0$ and x_L are the ends of the line. If the water depth is constant and if f does not vary significantly (it is constant along latitude lines), this integral reduces to

$$V = \int_0^h \left\{\frac{10\rho}{f}[\Delta D(x = x_L) - \Delta D(x = 0)] + g\rho[\zeta(x = x_L) - \zeta(x = 0)]\right\} dp$$

which depends upon the values of ΔD and ζ only at the two endpoints. In general, of course, the ocean is not of constant depth. Also, even in the constant depth case a calculation of the transport of the different water types in the section (rather than the total) will depend upon the intermediate values of ζ and ΔD.

We should pause here to clarify an important point about the surface velocities. The surface velocity which appears in (10) is only a part of the total velocity with which the water is moving in the upper ocean—it is the geostrophic part. A direct measurement of fluid velocity near the surface would yield

$$v_s' = v_s + \delta \tag{12}$$

In the upper layers of the ocean the term δ represents in part currents driven directly and locally by the wind and other meteorological forces and balanced by turbulent stresses, friction, buoyancy, and other fluid forces. The magnitude of δ may exceed that of v_s. Generally speaking, the nongeostrophic stress-dominated field is confined, to a surface boundary layer of thickness of the order of $E^{1/2} = (\nu/f)^{1/2}$, where ν is a crude measure of the turbulent dissipation of momentum (eddy or Austausch coefficient). Reasonable numbers suggest that the depth $E^{1/2}$ of the boundary layer (Ekman depth) is of the order of 100 m at mid-latitudes. It may be shown [e.g., *Greenspan,* 1968] that the surface pressure gradients associated with the surface boundary layer will be of the order of $E^{1/2}/D$ times smaller than those associated with the geostrophic flows, where D is the total water depth. Thus to a good first approximation we can ascribe the surface slope, and consequent pressure gradient, wholly to the geostrophic component of (12), as is done in (10) and (11).

This conclusion must be modified at the boundaries of the ocean. Surface pressure contours will generally intersect the oceanic boundaries (roughly definable as the edges of the con-

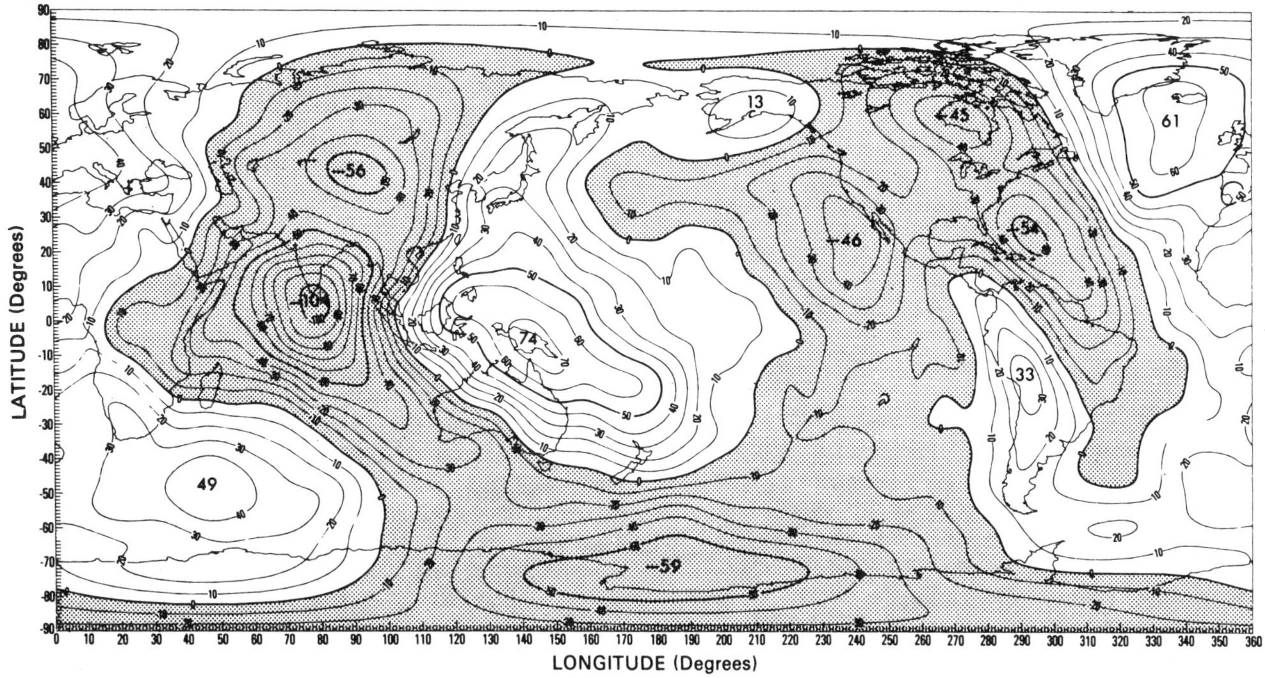

Fig. 5*a*. Geoid surface (GEM 9) in meters above the mean ellipsoid, computed by *Lerch et al.* [1979], using satellite tracking data.

tinental shelves), and if the flow were strictly geostrophic, there would be an implied massive flow into the continents. There are longshore pressure gradients on the coasts (see for example, *Csanady* [1979]) maintained by the deep-sea circulation. The existence of these gradients implies that the flow must become significantly nongeostrophic in the immediate vicinity of the boundaries. Generally speaking, in nondimensional units this deviation from geostrophy is of the order of the Rossby number, defined as $Ro = U/fL$, where U is a characteristic velocity, L is a characteristic length scale, which is $O(100\text{ km})$ at the coast, and f is still the Coriolis parameter. Over the vast bulk of the oceans the Rossby number is very small (typically $\lesssim 10^{-2}$). In the Gulf Stream itself it may reach as high as a few tenths. What this means [see *Stommel,* 1965] is that the downstream component of pressure in the flow cannot immediately be interpreted as being balanced by a cross-stream velocity, as geostrophy would imply; other forces, in particular the nonlinear ones, play a major role. Physically, there is a nonlinear boundary layer (the Gulf Stream) that is flowing slightly downhill; the resulting pressure head can be deduced from the conventional Bernoulli relation [see *Charney,* 1955]. On the other hand, the cross-

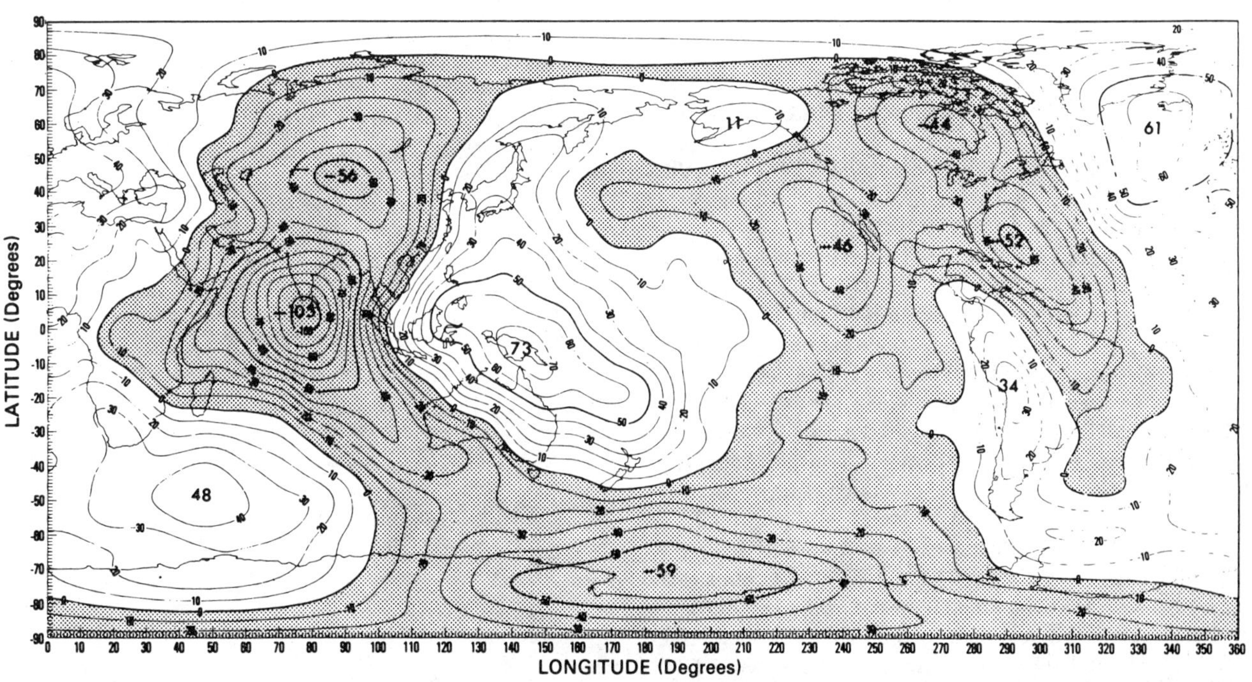

Fig. 5*b*. GEM 10 geoid computed by *Lerch et al.* [1979] by combining GEM 9 (Figure 5*a*) with surface gravity data. The geoid is relative to the reference ellipsoid.

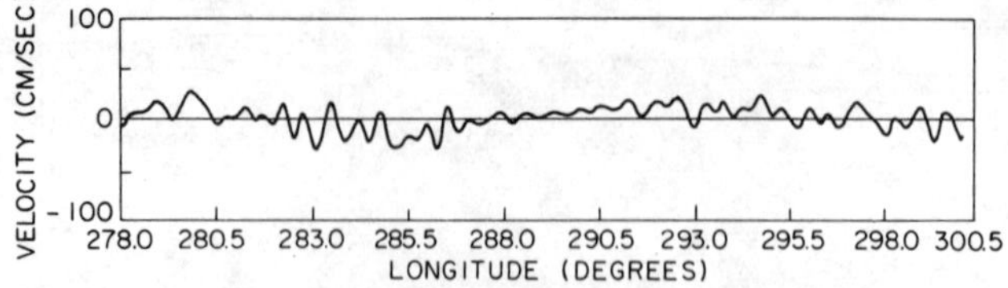

Fig. 6*a*. Derivative of geoid difference [*Marsh and Chang,* 1978; *Mader,* 1979] expressed in terms of a geostrophic velocity error (equation (10*b*)) along latitude 31.75°N.

stream pressure gradient and resulting downstream component of velocity (which is by far the major component) remain in geostrophic balance to a very high order of approximation.

Another region where one anticipates modification of the geostrophic relationship is in the immediate vicinity of the equator [*Moore and Philander,* 1977], because the Coriolis parameter goes to zero there. The region of breakdown of geostrophy is surprisingly narrow (little more than 1° or 2° of latitude on either side of the equator), although the measurement accuracy required to use the geostrophic balance this close to 0° latitude is much greater than it is at mid-latitudes. On the equator and in the vicinity of boundary currents one may still use the surface pressure gradients, if known, to deduce the surface velocity fields; one simply must use a more complex relationship between the velocity and the pressure force than the geostrophic one. But these regions occupy a very small fraction of the oceanic volume, and for our present purposes we will not pursue the details.

In addition to the stress-dominated nongeostrophic flows, there are surface velocities due to ordinary wind waves and swell, internal waves [*Phillips,* 1977], and tides. Surface waves and tides will be discussed below. It is characteristic of internal waves [e.g., *Phillips,* 1977] that they tend not to move the sea surface vertically. The very longest internal waves (wavelengths of the order of 100 km) can generate surface elevations which might reach 1 or 2 cm; for the foreseeable future these signals will remain well down in the noise and are not significant. Generally speaking then, the surface pressure gradients manifesting themselves at the sea surface can be attributed wholly to the geostrophic component of flow (setting aside for the moment the contributions from tides and atmospheric pressure loads).

Because it is the geostrophic component of flow beneath the surface boundary layer which we are discussing here, it is thus desirable that an altimeter not yield the total surface velocity but only the geostrophic fraction. For other purposes one may wish to find the total surface flow, but that requires different instrumentation, and the results could not be used as we are doing here.

(The large-scale geostrophic flow is, of course, largely wind driven too, but much more indirectly than the surface boundary layer. Essentially, it depends upon the small convergences and divergences of fluid vertically out of the surface boundary layer. A full discussion would take us too far afield; the reader should consult *Stommel* [1965].)

Given that with an altimeter we can discuss the surface geostrophic flow, there are some further clarifying points that must be made. The ocean circulation is a complex fluid system which varies spatially and temporally on a very large range of space and time scales (the subject is reviewed by *Wunsch* [1980]). Along any satellite pass, the altimeter will measure a geostrophic surface velocity which is the sum of a great variety of space and time scales. For many purposes it is convenient to distinguish especially two particular scales: (1) mesoscale eddies with diameters of the order of 200 km and time scales of the order of 100 days [see *Mode Group,* 1978] and (2) the time mean circulation. There are, however, some problems with this distinction. The time mean circulation probably includes space scales in the same range as that found in mesoscale eddies; the time mean is not well defined; and there is probably a complete range of time variability lying between the mesoscale and the time average. Some aspects of the ocean circulation have a known annual periodicity [e.g., *Niiler and Richardson,* 1973], which is often simply ignored in discussions of the mean flows. It is a potentially significant alias. The problem of isolating the time average from the variability is one that has never been solved adequately by oceanographers because of immense sampling problems. We do know that some aspects of the larger-scale oceanic structure have remained unchanged over the 100 years of historical record, but whether there are nonetheless significant interannual fluctuations on scales between the mesoscale and the basin dimensions is unknown. There is normally, however, at least an implicit assumption that the larger scales in the density field change more slowly than the 100-km scales. The fuzziness of the distinction between 'eddies' and 'general circulation' thus should be kept in mind in what follows.

To proceed, we must examine in turn the determination of N and S.

The Geoid N

There are a number of ways to obtain an estimate of the particular gravitational equipotential surface defined as the geoid. We intend to discuss the two methods that are of immediate concern.

Given a description of the geopotential, it is possible to calculate the orbit of a close earth satellite (the direct problem). The three-dimensional position and velocity (the state vector) can be calculated as a function of time (an ephemeris). Conversely, one can take observations of the position and velocity and obtain a description of the geopotential (the inverse problem). This is now a well-known procedure which has been discussed extensively [*Kaula,* 1966; *Gaposchkin,* 1973]. For these methods of satellite geodesy it has been conventional, though

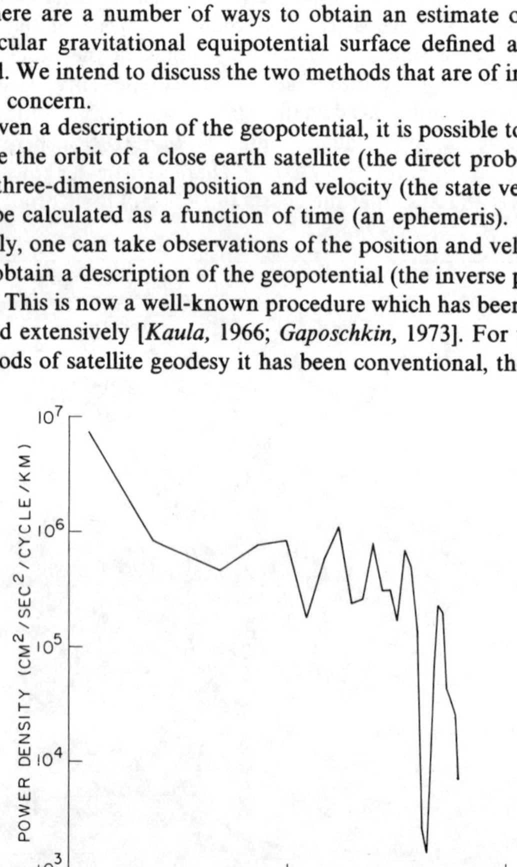

Fig. 6*b*. Spectrum of data in Figure 6*a*. Notice the nearly white spectrum of the velocity error due to geoid discrepancies.

not necessary, to represent the geopotential in terms of the associated Legendre functions (spherical harmonics). It is easy to show that the satellite position is only sensitive to the low-degree and low-order harmonics (the long-wavelength components) of the geopotential. The satellite acts as a low-pass filter; observation of several satellites has led to many determinations of the geoid. An example is the Goddard Earth Model 9 (GEM 9) (see Figure 5*a*, taken from *Lerch et al.* [1979]). The expansion of the potential in spherical harmonics has been carried to degree and order 20 (approximately 440 coefficients), which represents features in the geoid with a wavelength of $2\pi R/20 = 2000$ km (R is the radius of the earth). To obtain reasonable least squares geopotential solutions, the normal equations are generally scaled by an arbitrary amount, and the formal error estimates are not easy to interpret. From comparisons with other data, recent global spherical harmonic geoids probably have an rms accuracy of about 2 m for wavelengths greater than 2000 km [*Lerch et al.,* 1979; *Gaposchkin,* 1979*a*], decreasing with wave number. The uncertainty of the quadrupole moment (J_2) is about 2 mm.

To obtain the geoid at shorter wavelengths, one must seek other methods. Direct gravity measurements are available on the earth's surface. Like hydrography, the distribution is very uneven, and the accuracy variable. On some continents there are abundant measurements [*Tscherning and Rapp,* 1974], but there are large land areas, including most of Asia, Africa, and South America, where there are almost no data. (Some of the 'data' sometimes shown in these regions are actually based upon models of the surface topography with assumptions about compensation depth.) At sea, measuring gravity is much more difficult [*Talwani et al.,* 1972]. The principal error source is the Eötvös correction, which depends on ship navigation. Only recently has satellite navigation (and gyro-stabilized platforms) allowed this correction to be made with an accuracy approaching 1 mGal (for a 50-km block, one needs global gravity measurements of 1.67-mGal rms accuracy to obtain a geoid with a formal error of 10 cm). The ocean is large, and most of it has never been covered by ship-

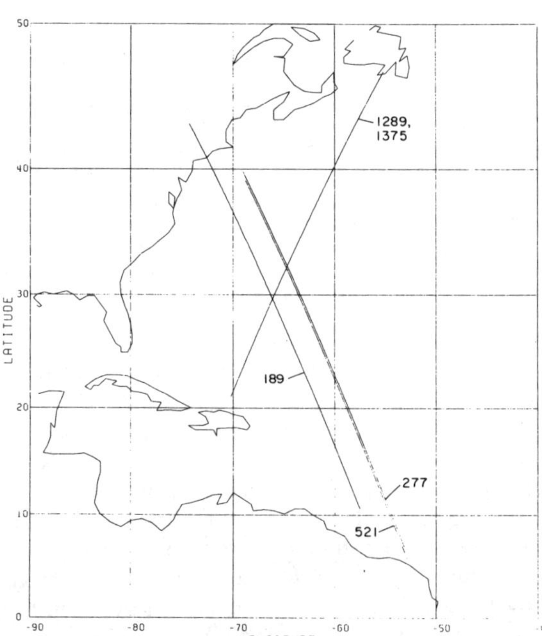

Fig. 7*a*. Track chart showing locations of some altimetric passes (revolutions) in the western North Atlantic.

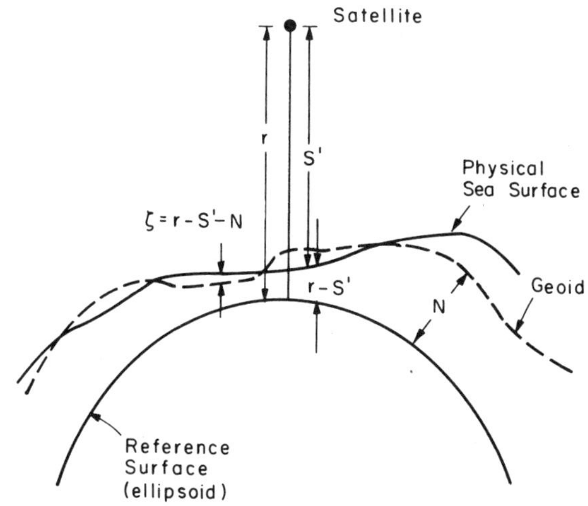

Fig. 7*b*. Definition sketch for the physical variables involved in the discussion of altimetric measurement of sea surface topography.

borne gravity measurements. Most of the early work suffers from navigation errors so large that they make the data relatively useless.

To obtain the geoid from surface gravity, one can employ an approximate solution of the boundary value problem of physical geodesy [*Heiskanen and Moritz,* 1967], called Stokes' formula. One uses measurements of gravity anomalies (Δg) to obtain the geoid (N) with

$$N(\theta, \lambda) = \frac{R}{4\pi g}\int_{\sigma} S(\Psi)\Delta g(\theta', \lambda')\, d\sigma' \qquad (13)$$

where the Stokes kernel is

$$S(\psi) = \frac{1}{\sin(\psi/2)} - 6\sin(\psi/2) + 1 - 5\cos\psi - 3\cos\psi \ln[\sin(\psi/2) + \sin^2(\psi/2)] + \cdots$$

θ and λ are the colatitude and longitude of the point of interest, θ', λ' is the integration point, ψ is the angular distance between them, and g is normal gravity.

There exist both local and global gravimetric geoids based on (13). While it is relatively easy to estimate the random errors from a straightforward application of Stokes' formula, the systematic errors have never been treated adequately. These errors are of a variety of types. First are the errors in (13) itself. It is a spherical approximation, which immediately introduces errors of the order of $1/f$, where f is the flattening of the best fitting ellipsoid [*Lelgemann,* 1970]. It assumes that the solid earth topography above the geoid can be neglected and the attraction of the atmosphere can be disregarded. Then, the discretization method necessary for performing the numerical quadrature is critical [*Uotila,* 1960]. More fundamentally, Stokes' equation is obtained from an improperly posed mathematical problem, which really should lead to recasting the boundary value problem of physical geodesy [*Molodenskii et al.,* 1962]. Second come the problems with the data, which are not uniform and are of uneven quality. Assumptions must be made because of the finite sampling of gravity. Documentation of the systematic errors in actual gravity data, as well as the instruments used and reduction procedures, is not available. To treat these issues in a systematic way [*Moritz,* 1974], approaches that combine statistical

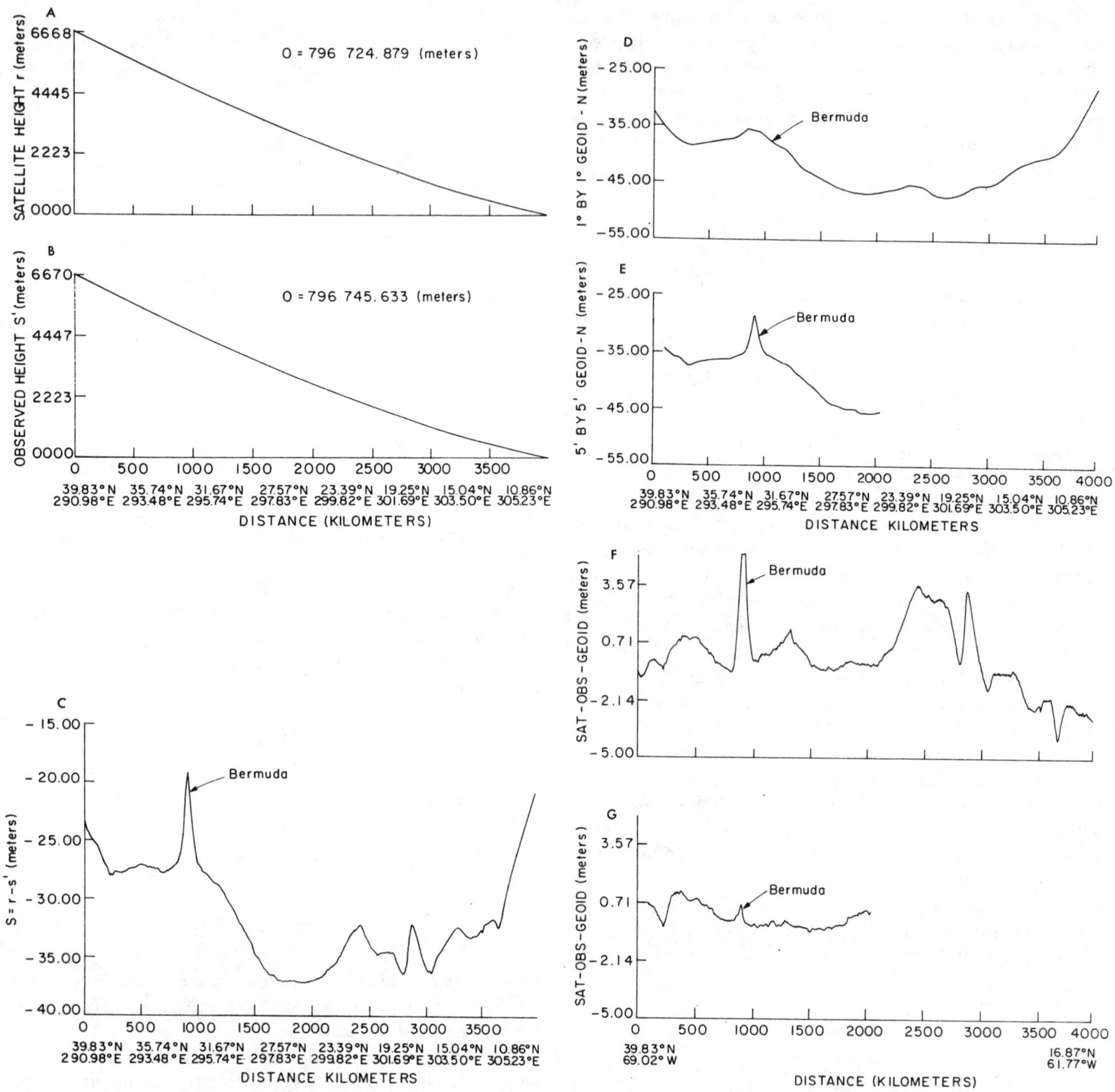

Fig. 8. (*a*) Satellite height r above the reference ellipsoid, determined by laser tracking and orbit dynamics along rev 277. The measurement is relative to a zero at 796,725 m. (*b*) Altimetric height S' of satellite above the sea surface. The height is referred to a zero of 796,746 m. (*c*) Sea surface elevation $S = r - S'$ relative to the reference ellipsoid along rev 277. The major features are the Bermuda Rise and the Puerto Rico trench system at the southern end of the line. (*d*) GEM 10 geoid (see Figure 5*b*) along rev. 277. (*e*) *Marsh and Chang* [1978] geoid along rev. 277. (*f*) Nominal value of $\zeta = r - S' - N$ of the sea surface relative to the GEM 10 geoid along rev 277. At Bermuda the difference is off scale, suggesting that GEM 10 does not resolve Bermuda, as one anticipates from the large-scale averaging involved in that geoid. (*g*) Sea surface elevation ζ relative to the Marsh/Chang geoid (Figure 8*e*).

estimation theory with physical geodesy have been developed, such as least squares collocation [*Moritz,* 1978]. An alternative approach would be to use geophysical inverse theory, and such an effort is now underway but is not reported here.

A global geoid giving 1° × 1° (110 km) average geoid heights based upon (13) is available [*Marsh and Chang,* 1979]. It uses a spherical harmonic representation for areas that are not observed, resulting in a horizontal resolution varying between about 1° and 20°. There are two local gravimetric geoids available over the ocean [*Marsh and Chang,* 1978; *Mader,* 1979]. These are limited to the western North Atlantic and are given in 5′ × 5′ (~10 km) averages. Neither gives a complete error analysis, though Mader concludes that the very long wavelength error is between 2 and 3 m.

Both geoids were computed supposedly using basically the same surface gravity data and computational methods. In order to obtain a rough idea of their accuracy we have compared the two geoids along the latitude circle 31.75°N. Values of the two geoids were subtracted, and the horizontal gradient was then computed from (10*b*) to give a measure of the velocity error introduced by geoid error. The result, taken from a running average over 12.8 km, is displayed in Figure 6*a*. Discrepancies between these two geoids are thus equivalent to velocity errors of tens of centimeters per second. The spectrum of the result is shown in Figure 6*b* and is almost 'white' at wavelengths longer than about 100 km. The published information on the two geoids is inadequate to determine whether one of them is substantially more accurate than this comparison suggests. It is also true, to the extent that the data and procedures are common, that common errors are subtracted out in our comparison. It is thus difficult to be entirely sure of the error fields in these best of the available geoids. Global coverage even approaching this problematic accuracy must await a dedicated gravity mission [*National Research Council,* 1979].

These local and global accuracy estimates lead to an important conclusion. Taking charts like Wyrtki's (Figure 3) as an order of magnitude estimate, it appears that the departure of the sea surface from the geoid (accounting for tides separately) is at most 1–2 m and generally less than 0.5 m. Therefore a much improved global geoid in ocean areas could be obtained in wavelengths from 20 to 5000 km from an accurate altimetric measurement of the sea surface, completely ignoring the oceanographic effects. That is, the errors in present global ocean geoids greatly exceed departures of the sea surface from the geoid owing to water motion. This procedure is followed by *Kahle et al.* [1978], who compare over the Indian Ocean the Geos 3 altimeter geoid and the GEM 6 geoid [*Lerch et al.,* 1974], and by *Gaposchkin* [1980], who combines Geos 3 altimeter data, surface gravity data, and satellite dynamics to obtain a global spherical harmonic representation.

The Combined Problem

The combined problem of determining the ocean circulation and geoid is best illustrated by a specific example. Before going on we wish to reemphasize that our purpose here is to set out a general procedure and not to obtain a definitive determination of the geoid or of the ocean circulation (although we will obtain estimates in limited regions). The data sets we will use are definitely suboptimum in ways we will describe, and our procedures have been chosen to be transparent rather than necessarily the best.

Figure 7*a* displays a few track lines in the western North Atlantic along which Seasat obtained altimeter records in the summer and autumn of 1978. We will concentrate initially on the line denoted revolution (rev) 277, which passed almost overhead of Bermuda. The height S' measured by the altimeter from the satellite to the sea surface is shown in Figure 8*b*. The only visible feature is a mean trend indicating a change of about 5 km from one end of the track to the other. This overwhelming feature is the result of the flattening of the earth—the polar radius is shorter than the mean equatorial radius by about 21 km [*Kaula,* 1966]. From tracking of the satellite and analysis of its orbit we can obtain the height r of the satellite above the reference ellipsoid. This estimate along the track is displayed in Figure 8*a* (the physical relationship of the various variables is shown in Figure 7*b*). The difference $S = r - S'$ between the tracking height and the altimetric height is the elevation of the sea surface (Figure 8*c*). In the absence of ocean currents or tides, eddies, storm surges, etc., $r - S'$ would be the geoid relative to the reference ellipsoid (we will simply call this the 'geoid'). S is negative here, because it may be seen from Figure 5 that in the western North Atlantic the geoid, and hence S, is below the reference ellipsoid. Note that the satellite was at a nominal altitude of about 800 km. The radar return on the sea surface was from an area, 'footprint,' of

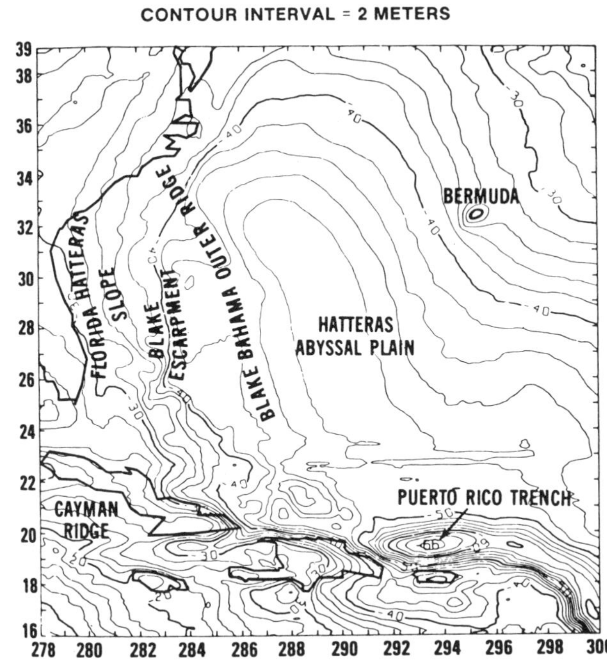

Fig. 9. *Marsh and Chang* [1978] gravimetric geoid with nominal resolution of 5′. The geoid exists only over this limited region.

about 8 km, although as we will see below, the resolution of the system is not this fine.

As was noted above, at the 1-m level of error, S is the geoid. But for studying the ocean circulation we are interested in the differences between S and the true geoid N. Before proceeding, let us note the large amplitude fluctuations in S at around latitude 15°N in Figure 8*c*. This is a manifestation of the Puerto Rico trench system. Evidently, in the region of the trench the geoid undergoes fluctuations of the order of 10 m, far greater than any reasonable oceanographic deflection in S.

To proceed beyond this point to calculate geostrophic surface velocities, we need a geoid determined independently of the altimeter. In Figure 8*d* we display the GEM 10 geoid (this geoid, constructed by *Lerch et al.* [1979], is a combination of GEM 9 and surface gravity data and is displayed on a global scale in Figure 5*b*) along revolution 277, and in Figure 8*f* the difference $\tilde{\zeta} = S - N$ ($\tilde{\zeta}$ is an estimate of the true value of ζ appearing in (10*b*)). Because GEM 10 cuts off at degree and order 22, it effectively represents an average geoid over 16° squares. The difference between the altimetric sea surface and this geoid displays elevation changes of some tens of meters, far too large to be attributed to oceanographic phenomena. The major differences between the altimetric surface and the GEM 10 geoid occur at the southern end of the line in the Puerto Rico trench region, where one would anticipate that large-scale averaging could not resolve the geoid over the trench. At the northern end of the line the variations are more plausible, but clearly, there is a major discrepancy between the spatial resolution of the altimeter and the large-scale geoid. One might attempt to proceed by averaging the altimeter over the same distance as the geoid. Alternatively, we could much improve GEM 10 along rev 277 simply by defining $N = S$ there.

For the present we will use instead the 5′ resolution gravimetric geoid produced by *Marsh and Chang* [1978] discussed above. The price we will pay for doing this is that we are limited to the comparatively small region of the western North

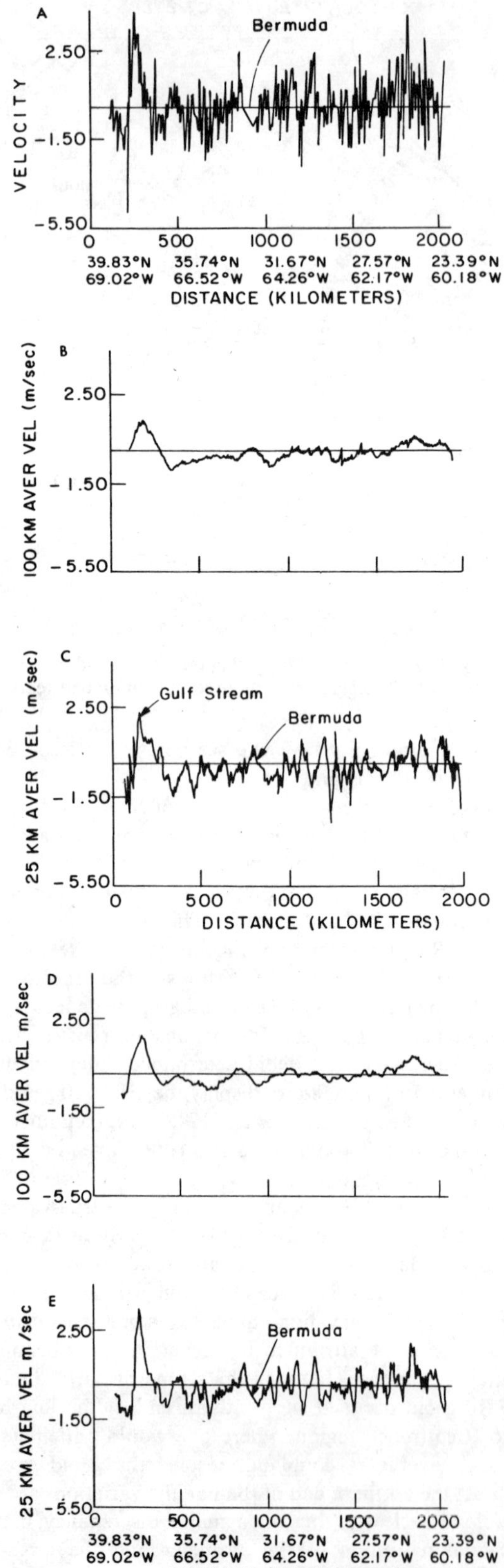

Fig. 10. (*a*) Raw geostrophic velocity v_s derived from Figure 8*g*. Much high-wave number noise is apparent in the results. The Bermuda region has been interpolated by a straight line. (*b*) Geostrophic surface velocity averaged over 100 km. (*c*) Geostrophic surface velocity averaged over 25 km. (*d*) Same as Figure 10*b*, a running 100-km average velocity, except along rev 521, which nearly coincides with rev 277 (shown in Figures 9*a*–9*e*), except 17 days later and about 30 km displaced (see Figure 7*a*). There are distinct similarities and differences between this figure and Figure 10*b*, consistent with the time interval. (*e*) Same as Figure 10*c*, except along rev 521.

Atlantic where this geoid exists (Figure 9). In addition, there remains the problem of geoid errors discussed above.

The values of the Marsh/Chang geoid along rev 277 are displayed in Figure 8*e*, and the difference ξ in Figure 8*g*. We are now using a small segment of rev 277. A conspicuous feature both in the geoid and in ξ is the high which corresponds to the near transit of Bermuda. The large feature remaining in ξ at Bermuda suggests that the geoid does not adequately resolve this feature. In what follows we will simply interpolate across the Bermuda region, begging the question for the moment of whether any of the other, smaller features in N are also artifacts of a lack of a correct geoid.

Assuming then that the geoid is otherwise without error, we can compute (Figures 10*a*–10*c*) an estimate $\tilde{v}_s$ of the geostrophic velocity v_s along the line from expression (10*b*). The different figures represent different spatial averaging of the result.

Figures 10*a*–10*c* are intriguing. There is an identifiable Gulf Stream with about the anticipated width and intensity; we see on the flanks the westward 'recirculation' now widely expected [*Worthington*, 1976; *Wunsch*, 1978; *Schmitz*, 1978; *Luyten*, 1977]; the oscillations one sees are on a scale of $O(100)$ km and have about the right sizes and intensities for known mesoscale eddy fields [*Mode Group*, 1978]. The *Gulf Stream Bulletin* for July 1978 suggests that the axis of the Gulf Stream was about 37°N during this period at this longitude, in good agreement with what we see in Figure 10. (*Cheney and Marsh* [1980] have made more detailed comparisons.) The results have a pleasing oceanographic verisimilitude. Evidently, the errors in the geoid do not swamp the oceanographic features in Figure 10.

We display in Figure 11 the spectrum of the altimetric heights (Figure 11*a*), the geoid N (Figure 11*b*), the relative topography ξ (Figure 11*c*), and the velocity v_s (Figure 11*d*). The rising power that occurs in the spectrum of velocity at wavelengths shorter than about 25 km comes from the scheme used for differentiating the altimeter data. Notice that the altimetric spectrum is white at wavelengths shorter than 25 km; hence its derivative has an approximate k^2 spectrum beyond this point. We suggest that the white region of the altimeter spectrum represents the effective system noise and consequently there is no geophysical information at shorter wavelengths, at least in this particular orbit. We should therefore average out wavelengths shorter than 25 km, which thus represents an effective system spatial resolution and leads to the averaging distances applied in Figures 10*a*–10*c*.

Simple theory (see, for example, *Blumen* [1972]) implies that geostrophic currents should not normally occur on spatial scales much shorter than the so-called Rossby radius of deformation

$$R_D \propto (gh\Delta\rho/\rho f^2)^{1/2}$$

where $\Delta\rho/\rho$ is the gross fractional vertical density difference in the ocean. $R_D \sim$ 30–50 cm in the western North Atlantic, and we thus anticipate a spectral cutoff roughly like what we see in Figures 11*c* and 11*d*.

QUANTIFICATION

In examining Figure 10 and others like it from other spacecraft revolutions we have only qualitative results that we tend to accept because they confirm our a priori prejudices about what the ocean should look like near the Gulf Stream. But the major attraction of the altimetric system is in its potential

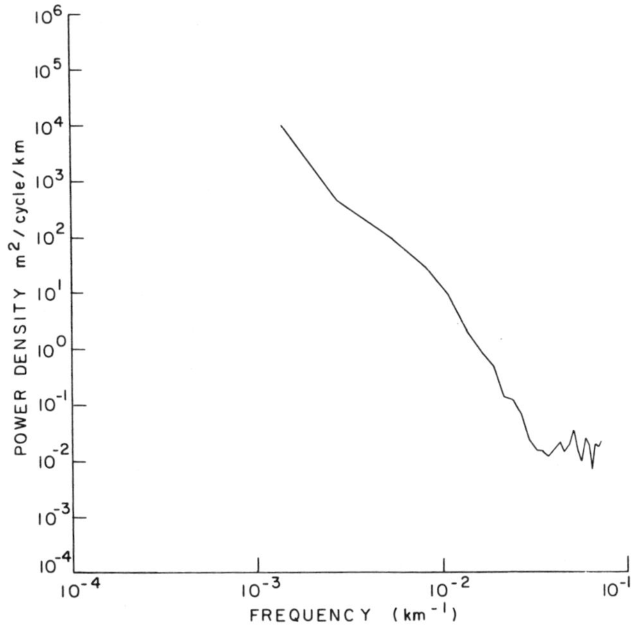

Fig. 11*a*. Spectrum of surface elevation S' along rev 277.

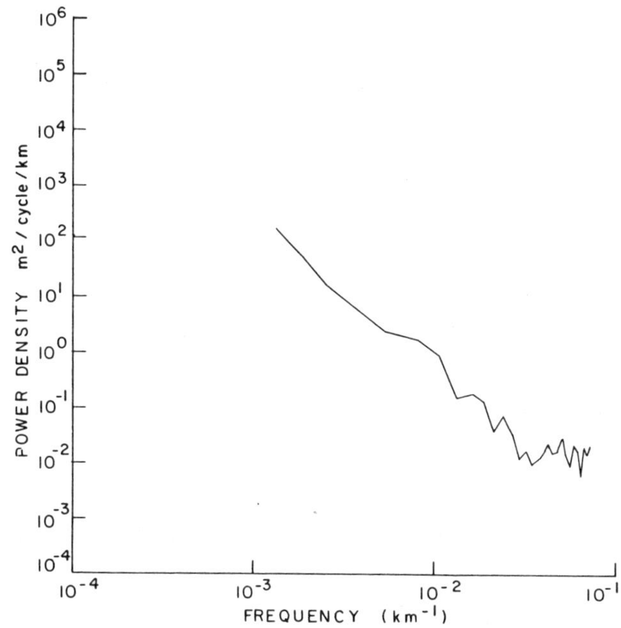

Fig. 11*c*. Spectrum of elevation ζ along rev 277.

ability to yield useful new, quantitative information about the ocean and its circulation. That is, we are able to proceed beyond the stage where much previous satellite data stops—the qualitative pictorial stage. But to exploit altimeters quantitatively, we must pursue a number of issues, not all of which can be disposed of here.

Errors

There are a large number of potential errors in ζ. Fortunately, none of them appears to make any substantial change in these particular results. Nonetheless, in general, we must completely understand the sources of error and how they will ultimately affect the estimates that we will make about the ocean. We anticipate wide geographical variability in the error sources.

The tide. Tides are a time-dependent fluctuation in the geoid to which the oceans respond dynamically and which introduce several errors in results. A gravimetric geoid determined at sea will include tidal errors (0.15-mGal amplitude) owing to the redistribution of mass, which changes the local gravity and which then appears as an alias. The surface along which the ship is making its measurements is also oscillating vertically with the tide with an amplitude of 0.6 μGal. These are negligible.

A more serious error is that in an altimetric pass the elevation of the sea surface will be distorted from that controlled by the geostrophic relation (equation (10)) by the presence of the tidal fluctuation. There are two ways to proceed. In principle, with adequate data one could [*Zetler and Maul,* 1971] determine the tidal elevation from the altimeter as a time periodic term. The Seasat data set is far too short to permit this to be done.

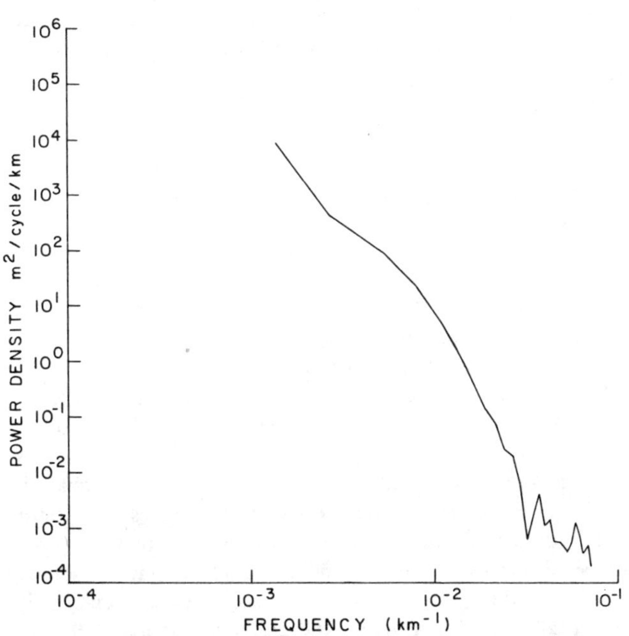

Fig. 11*b*. Spectrum of the Marsh/Chang geoid along rev 277.

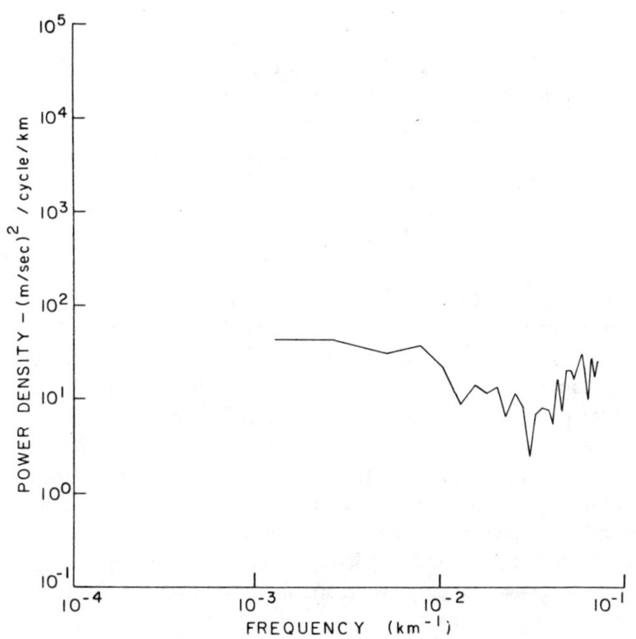

Fig. 11*d*. Spectrum of v_s along rev 277. Note the nearly white spectrum and drop in power at wavelengths shorter than about 100 km.

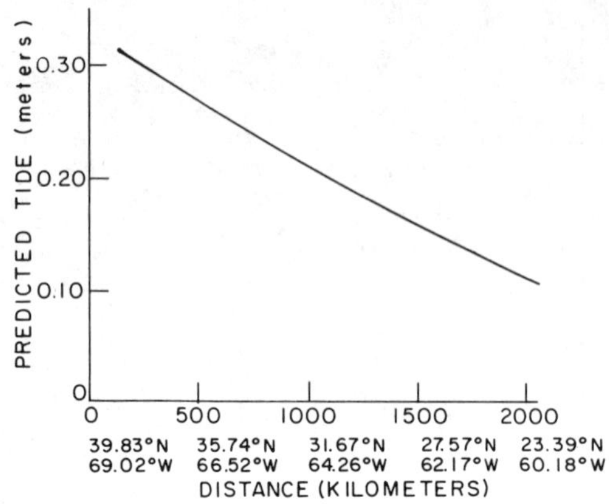

Fig. 12. *Mofjeld* [1975] model tide along rev 277.

We are thus forced back upon models of the tide in order to subtract it from the altimetric height. Several models [*Schwiderski,* 1980*a, b; Estes,* 1977; *Hendershott,* 1972; *Accad and Pekeris,* 1978; *Parke and Hendershott,* 1980] of global ocean tides exist and have varying errors which are position dependent. Some are limited to M_2, and others compute a number of tidal lines. Several models also include body tides and tidal loading. Tidal loading and body tides are important, because the satellite altimeter measures the distance from the satellite to the sea surface, which is referred to the earth's center of mass from the satellite ephemeris. Recent reviews of the tidal models are by *Parke and Hendershott* [1980] and *Schwiderski* [1980*c*]. For our purposes here, we have chosen to use the *Mofjeld* [1975] empirical model of the western North Atlantic, which he adjusted to reproduce with eight tidal lines the measured tidal heights in this region. This model does a good job over a limited region. In Figure 12 we display for rev 277 the Mofjeld model tidal amplitude along the track. As one might anticipate, the tide (in the open ocean) is a long-wavelength phenomenon (relative to the scale of Figure 10). In subtracting the tide from ζ we find a small correction to the geostrophic velocity at the long wavelengths (the velocities displayed in Figure 10 were corrected for the tide, although the correction is insignificant to the eye). For future analysis we expect to use the *Schwiderski* [1980*a, b*] model, which is global and has a claimed accuracy better than 10 cm.

Atmospheric load. Studies of the relationship between atmospheric pressure fluctuations and sea level at mid-ocean islands [e.g., *Wunsch,* 1972] show that at periods longer than 1 day there is an 'inverted barometer' effect. That is, the ocean responds isostatically to changes in pressure load; each millibar of increasing (decreasing) atmospheric pressure decreases (increases) sea level by 1.01 cm.

Short-period (hours) changes such as storm surges are shallow water phenomena and have little effect on open ocean sea level. Studies of open ocean records [e.g., *Wunsch,* 1972] show that there tends to be little response to meteorological forcing, either pressure or winds, at periods of a few days and shorter (the spectrum of sea level variation drops very rapidly with increasing frequency). At periods of many months the inverted barometer becomes of less significance, and wind effects tend to predominate. Wind effects on large spatial scales at low frequencies are presumably related to forced fluctuations in the large-scale general circulation and are part of the sea level signal that we are seeking.

For the single satellite pass rev 277, and over the comparatively short distance of 2000 km that we are using, the barometric correction will be nearly constant and will yield, through the high-pass filter operation of the derivative, a nearly negligible correction to the velocity field. As we proceed in the future to use longer lines and to compare long wavelengths at different times, we will need to determine atmospheric pressure loads from surface analyses of weather or develop satellite systems for obtaining the field. At long periods one will want to study the relationship between satellite determinations of the wind field [*Mognard and Lago,* 1979] and fluctuations of the surface elevations, but in the present data we do not need to make any correction for atmospheric load.

Wave height. If the sea surface were smooth, the radar return would be a specular reflection from a true mean sea surface. But the surface is a rough, random, moving surface. The effect of ocean waves is to stretch the leading edge of the returned radar pulse. The slope of the leading edge responds inversely to the height of the waves and is used to determine the so-called significant wave height $H_{1/3}$, a slightly peculiar quantity defined as the mean of the highest one third of the waves [*Kinsman,* 1965]. The effect of surface winds (and hence waves) is to change the backscatter cross section σ_0; higher winds result in a rougher surface, reducing σ_0. Seasat altimeter data provided measurements of $H_{1/3}$ and σ_0.

There are two effects of concern here. The radar altimeter used an on-board software tracker to sample the return wave form. The midpoint of the leading edge (the tracking point) is taken as the return time. Lengthening the pulse changes the tracking point. This dependence of observed altitude on $H_{1/3}$ (and satellite attitude) is known from preflight calibration. This purely instrumental correction, supplied with the data, is known with sufficient accuracy, assuming the values of $H_{1/3}$ are valid. There is also a height bias, because the mean reflecting surface is not the mean wave height; that is, more energy is reflected from wave troughs than wave crests, thus always giving too great a value of apparent satellite height. There is some uncertainty about how to determine the correction. *Yaplee et al.* [1974] and *Jackson* [1979] estimate that it is approximately

$$\text{bias}_{\text{SWH}} = 0.055 H_{1/3} \simeq 0.2 \text{ rms wave height}$$

Born et al. [1979*b*] report additional data on this bias. The subject needs further study, but for the moment, using 10% of $H_{1/3}$ seems a workable correction.

Finally, we should point out a small correction that needs to be made. The tracker is only a second-order tracker. Therefore there are effects depending on d^2h/dt^2. For a very rapidly changing sea surface a correction should be applied. It is easily obtained as

$$\delta\zeta \text{ (m)} = 6.51 \, d^2h/dt^2 \text{ m/s}^2$$

A measurement of $H_{1/3}$ along rev 277 is shown in Figure 13*a*. We see there, in these summertime conditions, a comparatively low sea state with an intensification under what one takes to be the trade wind field of the subtropical Atlantic (we are not prepared, at this time, to vouch for the accuracy of Figure 13*a*).

In the present case, rather than proceed directly to make the wave height correction, we will content ourselves with testing the magnitude of the bias. In Figure 13*b* we display the power density spectrum of the wave field; Figure 13*c* is the coherence (amplitude and phase) between the geostrophic velocity (Figure 10*a*) and $H_{1/3}$. Overall, the coherence is about 0.2–0.3, suggesting that between 4% and 10% of the apparent geostrophic current energy is coherent with wave height and may be erroneous. The phase tends to hover around 180°, which is puzzling, because one anticipates a 90° phase if one field is the derivative of the other.

We could proceed by removing from the current field that part which is coherent with wave height. This procedure is fraught with potential difficulties. Both wave height and current are at least weakly related to local winds and are thus legitimately coherent. It is probably better to attempt to use the actual correction algorithms, even though there remain some questions about their universal applicability (in the presence of strong shear flows, for example). For the present case it can be seen that the correction is small and again on long wavelengths; for now, we have ignored it.

Tropospheric correction. The transmission time of the altimeter pulse is lengthened by passing through the atmosphere. Total refractive bias at zenith is approximately 2.3 m [*Dickinson et al.,* 1970]. The dry troposphere correction depends only on the weight of the column of air above the subsatellite point and can be calculated with adequate accuracy (±3 cm) from hindcast surface pressure maps provided by the Fleet Numerical Weather Center (FNWC). The wet troposphere correction is generally about 10% of the dry troposphere correction and much harder to determine. The wet troposphere also could be estimated roughly from the FNWC hindcast. Fortunately, another method is available. Seasat had a scanning microwave multichannel radiometer (SMMR) which yields an independent water vapor estimate. Figure 14 shows a comparison of the wet troposphere correction [*Born et al.,* 1976*b*] derived from FNWC data, and two candidate algorithms for calculating the correction. Also shown is the calculation from radiosonde measurements. The two SMMR algorithms agree to about 3 cm. However, the difference between the SMMR value and the FNWC value can be more than 10 cm; the SMMR wet troposphere correction must ultimately be used. Even so, precipitation is probably not properly accounted for. Unfortunately, the Seasat SMMR data have not yet been distributed, and for the moment we cannot make the wet troposphere correction.

Ionospheric correction. The bias in range introduced by propagation through the ionosphere is well within the uncertainty of the basic altimeter measurement. The Seasat radar altimeter operated at 13.5 GHz. At this frequency the total refractive bias at zenith due to the ionosphere is approximately 5.5 cm [*Dickinson et al.,* 1970]. Currently, no correction

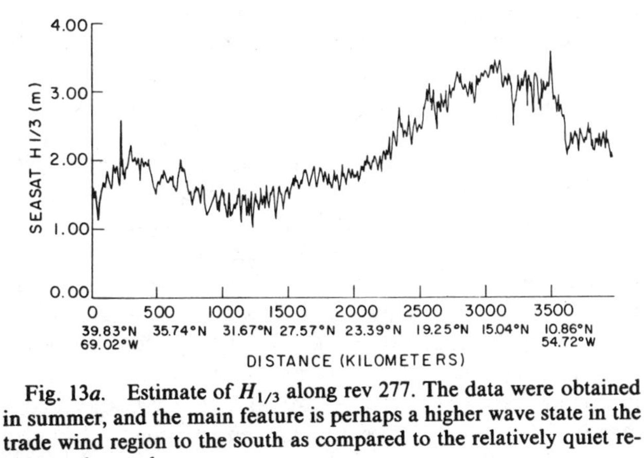

Fig. 13*a*. Estimate of $H_{1/3}$ along rev 277. The data were obtained in summer, and the main feature is perhaps a higher wave state in the trade wind region to the south as compared to the relatively quiet region to the north.

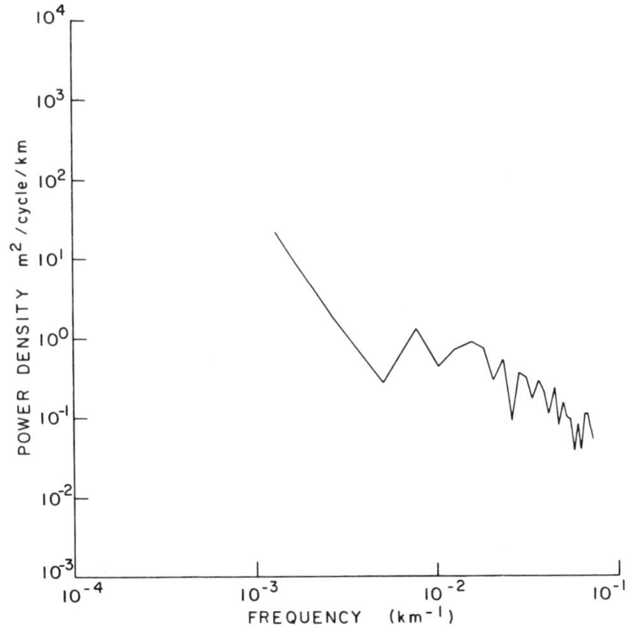

Fig. 13*b*. Spectrum of $H_{1/3}$.

is applied to the Seasat data. Ultimately, an ionosphere refraction correction will be computed by using simplified Appleton-Hartree equations. The model electron content is adjusted daily by using values of the electron content observed at Goldstone, California, and Armidale, Australia, based on Faraday rotation data.

Orbital errors. Satellites respond to a variety of forces: gravitational forces due to the geopotential and the sun and moon, the solid earth and ocean tides, and nongravitational forces due to atmospheric drag, solar radiation pressure, and reflected albedo pressure. Each force presents unique problems to the orbit analyst, and none can yet be considered to be

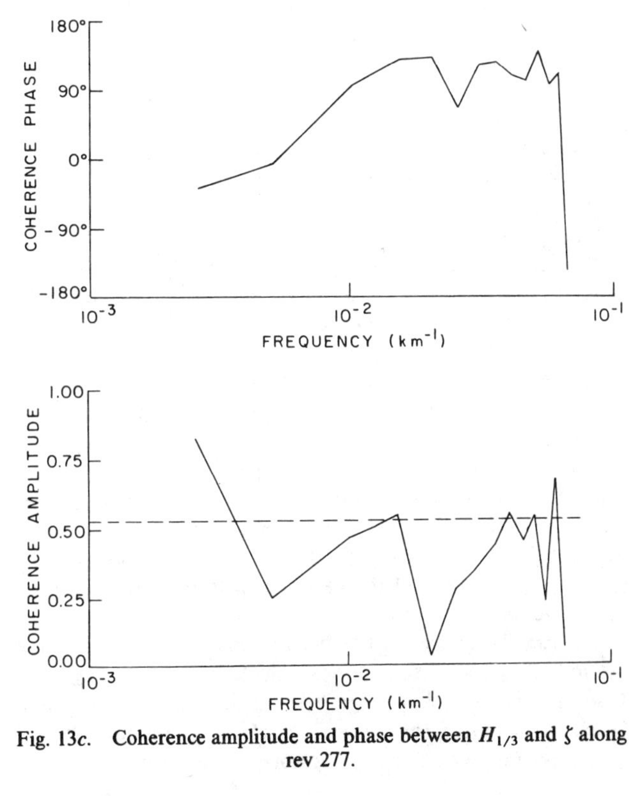

Fig. 13*c*. Coherence amplitude and phase between $H_{1/3}$ and ζ along rev 277.

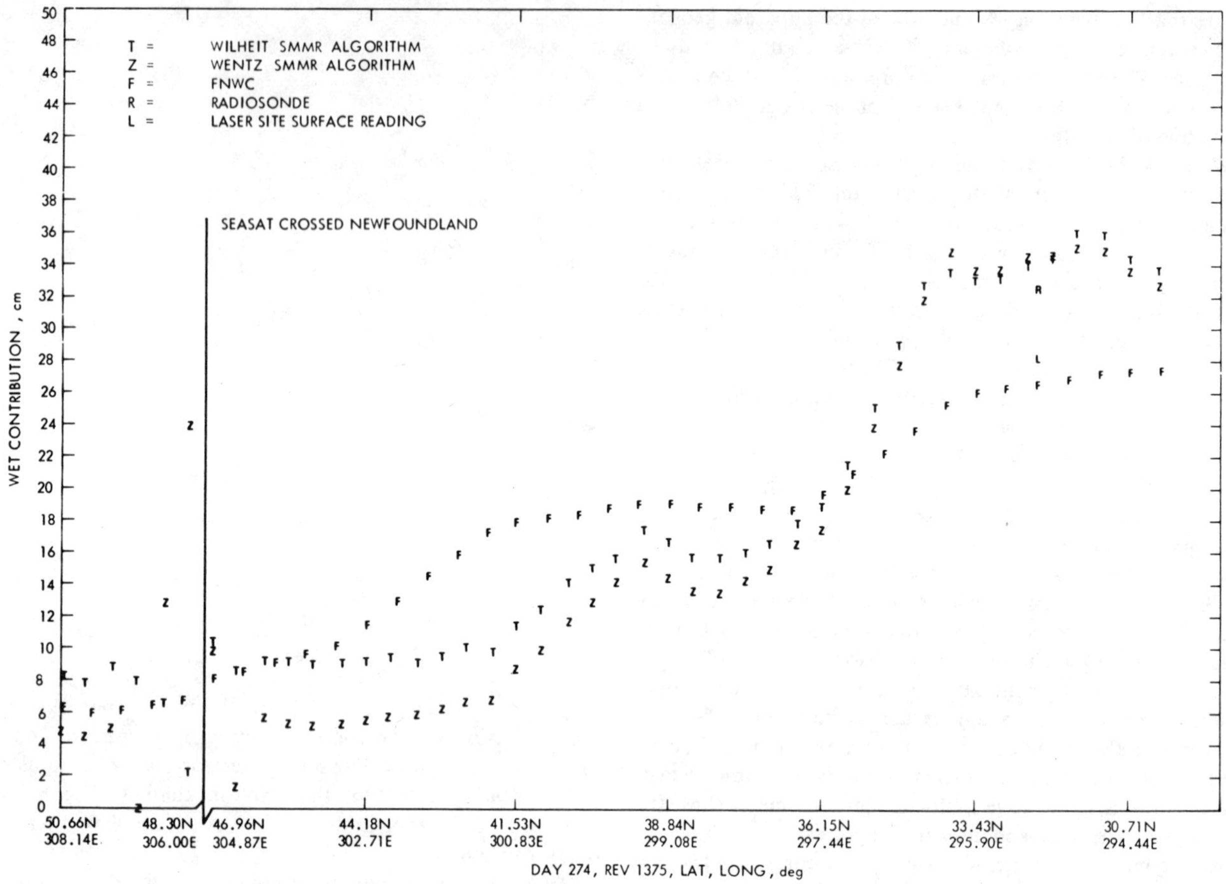

Fig. 14. Comparison [from *Born et al.*, 1979*b*] of two algorithms for making a wet troposphere correction from SMMR data. The trajectory is from Newfoundland to Bermuda.

completely treated [*Gaposchkin,* 1979*b*]. For analysis of satellite to sea surface altimetry data we are concerned with short-period effects with horizontal scales of up to 10,000 km. For this purpose the major difficulties in modeling the orbit come from the earth's geopotential, the atmospheric drag, and possibly the solar radiation pressure. It should be clear that for analysis of satellite altimeter data the critical quantity is the geocentric radius, which one would like to know more accurately than 5 cm. Such a requirement exceeds the present capability of orbital dynamics. Currently, for satellites such as Geos 3 and Seasat the orbital accuracy is 100–200 cm in radius.

The error sources are threefold. First the effects of radiation pressure and drag can be modeled only approximately. The Seasat spacecraft had a very large and nonisotropic cross section for calculating nongravitational perturbations. In addition, the most comprehensive atmospheric density models [*Jacchia,* 1977] can provide the density only to 20% accuracy. The change in radius due to drag is 35 cm/rev, giving an integrated unmodelable random change of 7 cm/rev. Current practice is to find drag and radiation pressure scale factors as additional parameters when determining the orbit from tracking data. These scale factors can absorb errors in the drag, the radiation pressure, and the earth's albedo pressure, resulting in an overall improvement in the accuracy of the trajectory.

The gravity field error has been estimated in two ways. An intercomparison of trajectories calculated with two gravity field models is shown in Figure 15. Each trajectory was calculated, using a given set of tracking data, and subtracted by the Naval Surface Weapons Laboratory (NSWL), using numerical integration methods [*Anderle,* 1974]. The gravity field models compared are the GEM 10 [*Lerch et al.,* 1979] and the NSWL 10E models. These are two comprehensive geopotential models, used for precision orbit determination and based mainly on analysis of satellite tracking data. The results of this comparison indicate differences of 2 m rms, the largest component of this difference having a wavelength of almost 20,000 km. The size of the difference with 7000-km wavelength is estimated to be about 20 cm rms.

Such a comparison, however, does not estimate the error common to both geopotential models. Such a common error could be due to, among other things, the geopotential coefficients ignored in both models. Any errors in the gravity field model (either of omission or commission) will produce errors in satellite position correlated with geographic position. *Anderle and Hoskin* [1977] found systematic errors in satellite position with a correlation distance of 600 km and an amplitude of as much as 50 cm. This error will not be removed by averaging many tracks of data or using a crossing arc calculation to remove systematic track to track biases. Improved gravity field models are essential for reducing this systematic error.

The third error source is the accuracy of tracking data and knowledge of the geocentric tracking station coordinates. Seasat was tracked by a ground-based laser network which was incompletely deployed at the time of spacecraft failure. In addition, a set of Doppler tracking stations obtained data. Orbits have been computed by using the laser data and a gravity field model determined at Goddard Space Flight Center, and a second set of orbits have been computed by using the Doppler data and a gravity field determined at the Naval Surface

Weapons Laboratory. These two ephemerides have been compared, and the radial agreement is better than 2 m rms [*Tapley et al.*, 1979].

These comparisons display the effects of gravity field errors and tracking data distribution and errors. Attempts have been made to use the altimeter observations as tracking data [*Piscane et al.*, 1979]. One can imagine a general solution where the geopotential model, the ocean circulation model, and the trajectory are all simultaneously improved as described below.

In summary, the major orbit error occurs on scales of the orbit, 20,000 km, and decreases with increasing wave numbers. For the moment, with a very short track, we ignore this error. In the future, as longer tracks are used (e.g., trans-Pacific), the orbit errors will dominate and will limit the usefulness of the system. A quantified error spectrum must eventually be constructed.

Determining the Ocean Circulation

Assuming that we have optimally corrected the altimetric measurement to yield S and that we have a perfect geoid, we can then postulate the following hypothetical situation. An investigator at sea makes a line of hydrographic stations and computes the relative geostrophic velocity from (8), using the surface as the reference level. If the line of stations coincides with an altimetric track, the surface velocity is obtained from (10), and the full velocity field is known. This velocity field, of course, includes both the large-scale means plus any eddies or fluctuations that are also present.

Current reality differs from this situaton in a variety of ways. There is no altimetric satellite now in orbit; during the lifetime of Seasat no significant number of stations was taken with the altimeter in mind (some work was done in the Gulf of Alaska [see *Born et al.*, 1979*b*] and in the Kuroshio (H. M. Byrne et al., unpublished manuscript, 1979)). The altimeter measurement is not without error, as we have seen, and the geoid contains major errors too. Hydrography is imperfect owing to navigation problems, internal waves, etc.

In order to make quantitative use of existing and future altimetry we must deal with these problems. We will divide the problem into two parts. We first deal with (at the moment hypothetical) simultaneous hydrography and altimetry along a common track, which is useful as a formalism and as a framework for discussion of potential future systems. Finally, we will deal with the problem of using the existing altimetry to learn something from existing hydrography.

Simultaneous data. Consider a satellite orbit under which we have the altimetric field S, the dynamic height field ΔD_s, and the geoid N. We suppose that each of these fields has a known covariance matrix

$$\mathbf{C}_\zeta = \langle \zeta_i \zeta_j \rangle \qquad \mathbf{C}_D = \langle \Delta D_{si} \Delta D_{sj} \rangle \qquad \mathbf{C}_N = \langle N_i N_j \rangle \quad (14)$$

that the observations have known error covariances $\mathbf{B}_\zeta$, $\mathbf{B}_D$, $\mathbf{B}_N$, $\cdots$, and that the covariance of the observation error and signal is zero.

(As recognition that our procedures will all deal with sampled quantities, N_i, ζ_i, etc., refer to the value of $N(x_i)$, etc., at discrete position i. We are explicitly going to use the discrete version of inverse theory in what follows rather than the Hilbert space version of *Backus and Gilbert* [1967]. This choice is a matter of taste, and the procedures that we will use can be reformulated to apply to continuous functions.) Note that the quantities of physical interest are really the spatial derivatives of ζ, N, ΔD_s; for example, we use the deflection of the vertical, rather than N itself, in all our equations. We will assume further that given the correlation function of either the function or its derivative, we can obtain the correlation function of the other. If C_f is the correlation function of a quantity f, then $-d^2/dx^2[C_f(x)] = C_{f'}$ is the correlation function of the derivative of f.

Our problem is twofold: to obtain the best estimate of the full velocity field along the section and to make the best estimate of the corrected geoid (for simplicity, we will ignore the possibility of correcting the altimetry S', that is, the orbit; doing so is an obvious generalization of the procedure to be described and will ultimately be necessary).

Our procedure is based upon the assumption that knowledge of the ocean circulation can be used to improve the geoid and that a knowledge of the geoid leads to improved ocean circulation models. To take a crude qualitative example, if we had a geoid that led to a Gulf Stream flowing the wrong way, we would be inclined to use what knowledge we do have of the Gulf Stream to correct the geoid. Similarly, in a region where the geoid was believed known with high accuracy, one would use the expression (10) directly to fix the water velocity.

But relative knowledge of geoid, of hydrography, and of the water circulation is a complex function of position. We need a formalism to become quantitative. The one we will use here is based upon the inverse method calculations described by *Wunsch* [1978]. The reader is referred to that paper for details and for elaboration; in essence, one uses known ocean kinematics and dynamics to make inferences about the overall circulation.

Perfect geoid: determining the ocean circulation: Consider a simple example. Suppose that we have a line of hydrographic stations extending from land to land, as shown schematically in Figure 2. We choose a reference level at the surface, and we suppose, following *Wunsch* [1978] and *Roemmich* [1979], that conservation of certain water properties flowing across the line is plausible. For example, we expect that water of a certain character flowing into the region west of the line must flow out again simply because the property (e.g., density or heat content) is conserved to a good first approximation. Following *Wunsch* [1978], let $\mathbf{b} = \{b_i\}$ be the unknown surface velocity; then the conservation requirements are a set of linear constraints on the unknown $\mathbf{b}$, which we may write as

$$\mathbf{Ab} = -\mathbf{\Gamma} \qquad (15)$$

where Γ_i is the amount of property i that is not conserved under the initial assumption that the surface is a level of no motion and the elements of the matrix $\mathsf{A} = \{A_{lm}\}$ are the areas occupied by property l in station pair m. In general, equations (15) will be highly underdetermined and, owing to errors, possibly inconsistent.

With hydrography alone, one proceeds as described by *Wunsch* [1978] and *Roemmich* [1979] to find the determined and nondetermined parts of $\mathbf{b}$, bringing to bear the general machinery of inverse theory. Although we will not reproduce the details here, it is a key point that the inverse machinery yields a solution $\mathbf{b}$ with several properties: (1) an explicit error variance at each location, (2) a statement of the resolution of the solution at each location, and (3) a determination of the parts of $\mathbf{b}$ which are uniquely and fully determined. (In one interpretation the nonunique part of the field is defined as the eddies.)

Suppose that in addition to the hydrography we have an es-

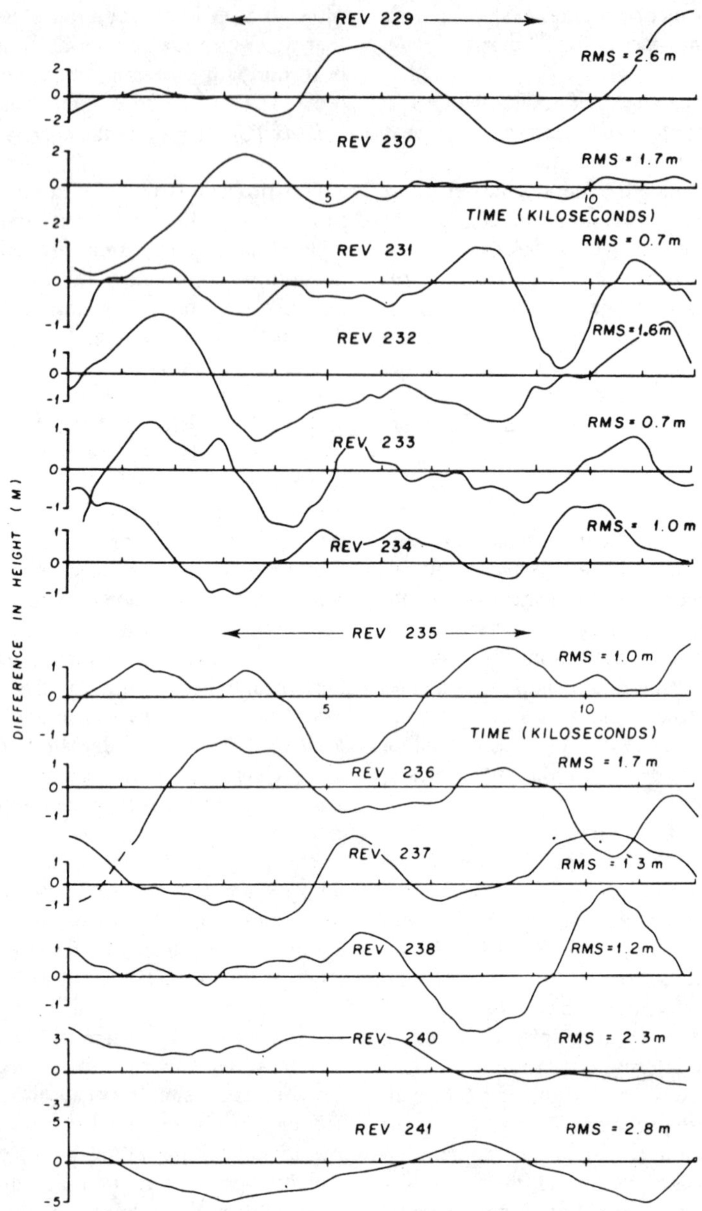

Fig. 15. Satellite height differences for several revolutions using two different gravity fields [from *National Research Council*, 1979].

timate from the altimeter of the surface geostrophic velocity $\mathbf{v}_s$ given by the vector

$$\Delta\zeta = \left\{ \frac{g}{f} \frac{\Delta(S-N)_i}{\Delta x_i} \right\}$$

We append to (15) the additional set of equations

$$\mathbf{b} = \mathbf{v}_s = \Delta\zeta \qquad (16)$$

The combined set of equations, which we can write as

$$\mathsf{A}'\mathbf{b} = -\Gamma' \qquad \Gamma' = \begin{Bmatrix} \Gamma \\ -\Delta\zeta \end{Bmatrix} \qquad \mathsf{A}' = \begin{Bmatrix} \mathsf{A} \\ \mathsf{I} \end{Bmatrix} \qquad (17)$$

where I is the identity matrix, is now formally overdetermined. In this noise-free situation one requires that the equations be compatible [see *Lanczos*, 1961]. If all measurements were perfect, then subject to the validity of the conservation assumptions underlying (15) the altimetrically defined values of **b** automatically would satisfy the conservation conditions, and we are finished, having a fully determined ocean circulation.

Perfect hydrography: correcting the geoid: But the real situation is not this simple: there are errors, as we have seen in both Γ and in $\Delta\zeta$. Suppose that we wish to correct N by an amount δN, so that the new and more accurate geoid is $N' = N + \delta N$. Assume that the hydrography (and the altimetry S) is perfect. That is, we postulate that the only noise is in the geoid. Because **b** would be wholly determined from (15), we have from (16) simply

$$\frac{\Delta \delta N_i}{\Delta x_i} = \frac{f}{g} \Delta\zeta_i - \frac{f}{g} b_i \qquad (18)$$

As noted, **b** is here derived from an underdetermined system. Thus the correction N_i will reflect the resolution and (more generally) the variance of **b**. We can, if we wish, treat (18) as a general linear relationship in which the b_i have errors and solve the system subject to the presumed known covariance matrix for N by the usual methods of inverse theory and least squares.

Given δN_i, we obtain by integration

$$N'(x_j) = N(x_j) + \sum_{i=1}^{j} \Delta\delta N_i \tag{19}$$

N' is a new estimate of the geoid obtained from the previous estimate N and from knowledge of the ocean circulation. (There is a constant of integration missing in (19) which cannot be determined by the present method and must be found, if required, by other means.)

Other investigators have noted that where the geoid is poorly known, the hydrography can be used to make a correction. H. M. Byrne et al. (unpublished manuscript, 1979), using a segment of stations across the Kuroshio, constructed a geoid estimate deduced from the dynamic height relative to a fixed reference level. This estimate is, however, only as good as the estimate of a fixed reference level; data from the corresponding region of the Gulf Stream [e.g., *Fuglister*, 1963; *Roemmich*, 1979] do not support the hypothesis of a fixed reference level (see Figure 4). But with only a segment of hydrography we can apply no simple constraints for a more realistic solution.

In the opposite extreme, where the altimetric surface velocity is known perfectly (no geoid error) but the $\mathbf{\Gamma}$ in (15) have errors, one can construct a system in which one infers corrections to the dynamic height field by similar procedures.

General case: correcting the geoid and finding the ocean circulation: In the more general case there will be errors in both the $\mathbf{\Gamma}$ and the $\mathbf{\Delta\zeta}$, and we want to give some weight to both (15) and (16). Strictly speaking, equations (15) are of the form

$$\mathsf{A}\mathbf{b} = -\mathbf{\Gamma}' + \mathbf{n} \tag{20}$$

where $\mathbf{\Gamma}'$ is the correct value of $\mathbf{\Gamma}$ and $\mathbf{n}$ is the noise. We adjoin the equation sets

$$\mathsf{A}\mathbf{b} = -\mathbf{\Gamma} \tag{21a}$$

$$\mathbf{b} + \mathbf{\Delta\nu} = \mathbf{\Delta\zeta} \tag{21b}$$

where

$$\mathbf{\Delta\nu} = \left\{\frac{g}{f}\,\frac{\Delta\delta N_i}{\Delta x_i}\right\}$$

and write one system

$$\mathsf{A}''\mathbf{c} = -\mathbf{\Sigma} \tag{22}$$

in which $\mathbf{c}$ represents both $\mathbf{b}$ and $\mathbf{\Delta\nu}$, $\mathbf{\Sigma}$ subsumes $\mathbf{\Gamma}$ and $\mathbf{\Delta\zeta}$ in the obvious way, and A'' contains both A and the elements appropriate to (21*b*).

Using the covariance matrices (equation (14)), we can determine the optimum solution for b_i and $\Delta\delta N_i/\Delta x_i$ as a function of position depending upon the relative error in the different fields at each location.

We have not dealt with many of the necessary details which are discussed in the references. We will content ourselves here with pointing out that the strategy is to use as many constraints as possible to improve both geoid and ocean circulation. Such constraints could come from many sources, for example, current meters, measurements of chemical tracers, or surface gravity measurements.

But the critical point is that the entire process rests upon our ability to define the errors in the fields that we are working with. Making those estimates is a major unfinished piece of business.

Perhaps we should make the somewhat obvious point in this hypothetical case of simultaneous altimetry and hydrography that studies of the purely time-dependent ocean circulation can be conducted with very poor, or no, geoids at all. In addition, if attention is focused solely upon the time variability of the surface geostrophic currents, then even the ocean density field becomes irrelevant. We are paying little attention to these possibilities here, because they are almost easy in comparison to the problems of dealing with the mean or very slowly changing parts of the circulation and because ultimately, understanding of the ocean will come only through determination of the full three-dimensional fields. Studies of the time-dependent components will, of course, have to take into account the time- and wave number–dependent error fields of orbital height, tides, waves, moisture, etc., much as discussed here.

To give some idea of both the repeatability of the altimetry and the temporal changes that one can study with it, we display in Figures 10*d* and 10*e*, the counterparts to Figures 10*b* and 10*c*, the surface velocity deduced from the altimetry and the Marsh and Chang geoid along rev 521. This pass nearly coincides with rev 277 (see Figure 7*a*) but occurred 17 days later and is displaced slightly to the southwest. Much of the structure is repeatable in the two passes, but much has changed as well. With many such repeated passes it becomes possible, in an obvious way, to study the space-time structure of the oceanic variability (see, for example, *Gordon and Baker* [1980] and *Cheney and Marsh* [1980]). The study of the differences of repeating passes allows one to eliminate the geoid problem, but at the sacrifice of the mean field. Many Seasat passes repeated exactly.

Nonsimultaneous data. As was pointed out above, the preceding discussion supposes that the altimetric velocity was determined simultaneously with the hydrography. We have almost no such data sets. What we do have is the Seasat 3-month altimeter record and the 100-year history of hydrographic measurements. The question that we pose here is whether there is any useful way to combine them.

If the ocean circulation were truly steady, there would be no problem. But the ocean fluctuates on all known time scales with a wide range of space scales [*Wunsch*, 1980]. A hydrographic section across the Gulf Stream (say) in 1958 will encounter the Gulf Stream at a different position and perhaps with a different strength than would the altimeter in 1978. Obviously, we cannot impose upon the constraints (15), from 1958 data, the specific surface velocities (16) measured in 1978, even if the lines were geographically coincident.

In order to proceed we make the following critical assumption: the statistics of the surface velocities as measured in 1978 were the same as those in 1958 (or any other year). This assumption is a plausible one; it merely asserts that the ocean has been a stationary system, and it would be surprising if the assumption were not correct.

Perfect geoid: determining the ocean circulation: Let us suppose that we deduce from Seasat the spatial correlation function $\mathsf{C}_v = \{\langle v_{si}v_{sj}\rangle\}$ of the surface geostrophic velocity in some region of the ocean (the spectrum in Figure 11*d* suggests that C_v is nearly diagonal). The surface velocity is the sum of the mean circulation and the eddy field. Let us suppose that our hydrography is in the same region. Then we wish a solution to (15) which will yield a surface velocity with the same correlation function. One way to write the appropriate solution is [*Franklin*, 1970; *Wiggins*, 1972; *Wunsch*, 1978]

$$\mathbf{b} = \mathsf{C}_v\mathsf{A}^T(\mathsf{A}\mathsf{C}_v\mathsf{A}^T + \mathsf{B}_\Gamma)^{-1}(-\mathbf{\Gamma}) \tag{23}$$

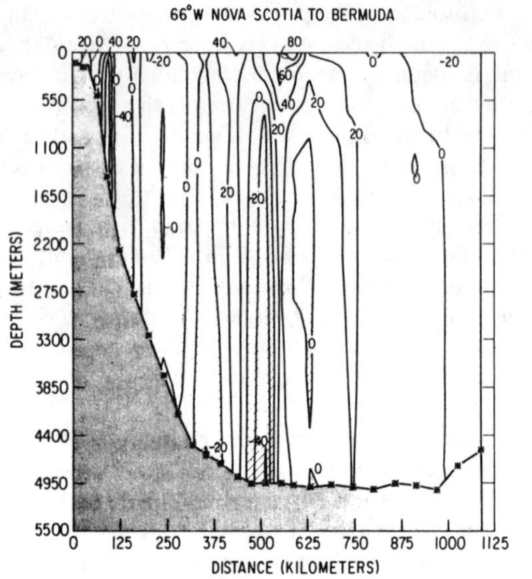

Fig. 16*a*

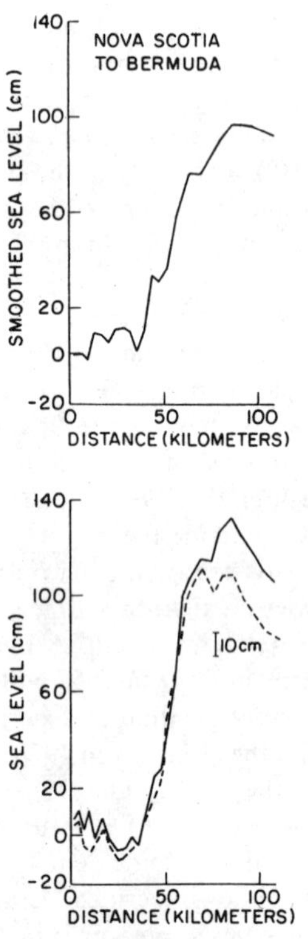

Fig. 16*b*

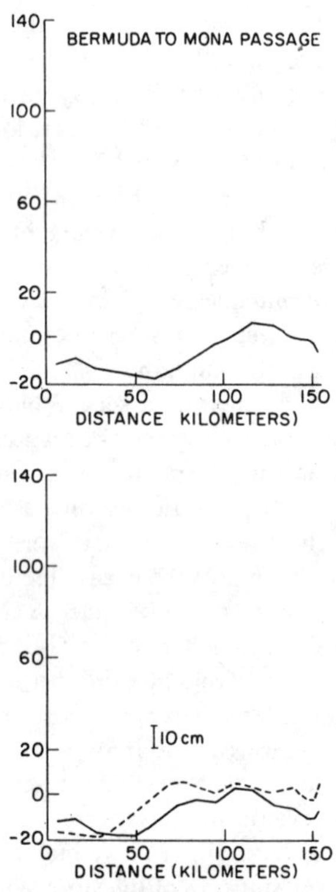

Fig. 16*c*

Fig. 16. (*a*) Velocity field along 66°W computed by inverse methods, where surface velocity was required to have the same covariance as altimetrically derived surface velocities obtained in the same region (not along this same line). (*b*) The solid curve in the lower panel is sea surface elevation from Nova Scotia to Bermuda determined from the inverse solution displayed in Figure 16*a*. The dashed curve is the elevation if one assumes arbitrary level of no motion at the bottom. The upper panel shows the elevation from the unique component of the inverse solution (the part independent of the initial reference level) and may correspond more closely to the long-term time average flow. (*c*) Same as Figure 16*b*, but along 66°W meridian from Bermuda to Mona Passage into the Caribbean Sea.

where $\mathbf{B}_\Gamma$ is the covariance matrix of the noise in $\mathbf{\Gamma}$.

In Figure 16*a* we show the velocity field for a hydrographic section along 66° in the western North Atlantic where the altimetric statistics from Seasat were imposed upon the inverse solution.

The variance of the solution is actually somewhat less than that implied by the altimeter, presumably because the ocean contains energy in the null space of $\mathbf{A}$ and/or because the geoid is in error [see *Wunsch,* 1978]. The altimeter sees both the determined and the undetermined fields and should always find a greater variance in the surface velocity. (Figure 16*a* is meant as an example and not as a final version of the absolute flow across this section (C. Wunsch, unpublished manuscript, 1980).

Perfect hydrography: improving the geoid: The above procedure is a good one where the geoid is believed to have small errors. It needs to be changed where the geoid is believed to be badly in error and the hydrography well determined. Suppose that we solve (15) with no altimetric input at all. We obtain a best estimate of $\mathbf{b}$ and hence of $\zeta = S - N$ from the hydrography alone. To the extent that the hydrographic inversion determines time-independent components of ζ we can determine an improved geoid from time-averaged altimetry. Let ζ be known from the hydrography, and let S be a time average value. Then

$$N = S - \zeta \tag{24}$$

We do not have any simple constraints that we can apply to N that would be analogous to (15) for $\mathbf{b}$, and thus we can only accept this solution for N subject to a priori correlation information. An example of the procedure is shown in Figure 17. We have chosen a line of hydrographic stations in the central Caribbean Sea, running along 79°W from Cuba to South America. The hydrography along this line was used by *Roemmich* [1979] as part of an inversion by the method of *Wunsch* [1978] to obtain the absolute geostrophic flow field. Given Roemmich's solution, one obtains from (10*b*) the sea surface elevation ζ shown in the top panel of Figure 17. (The elevation displayed is the 'smoothed' solution, which is the uniquely defined part of the elevation (see *Roemmich* [1979] or *Stengel* [1979]). The altimetric elevation along this line was taken not from Seasat data, but from the Geos 3 time mean sea surface [*Marsh et al.,* 1980], which only extends over the northern portion of the Caribbean. The result is shown in the middle panel of Figure 17. (We associate here the unique part of the hydrographic inversion with the large-scale time average; the identification is somewhat controversial but will nonetheless serve to demonstrate our general method.) We can solve easily for N, which is shown in the bottom panel of Figure 17. This then is an altimetrically determined geoid corrected systematically for oceanic hydrography. The difference between S and N is obviously small in this case (the range of ζ is only about 50 cm in the presence of a geoid fluctuation of nearly 16 m). Nonetheless, N is a more accurate geoid than is S. From the inverse apparatus used to deduce ζ, one can deduce both the resolution and the variance of these results, but we will not discuss them further here.

General case: correcting the geoid and determining the ocean circulation: The general case is simply a concatenation of equations (15) with (24) in which the statistics of both $\mathbf{b}$ and N and their error budgets are prescribed. The final solution thus simultaneously makes a best estimate of $\mathbf{b}$ and corrects the geoid, the allocation of the solution between them being determined by the a priori statistics. As both the Geos 3 and the Seasat altimetric data sets become optimum under the current editing procedures, we anticipate being able to carry out this

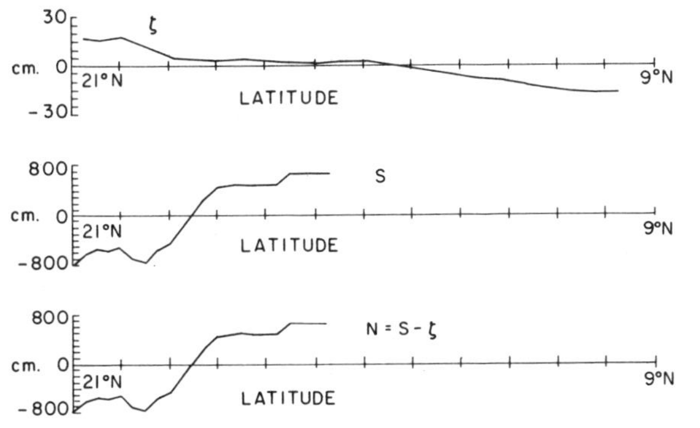

Fig. 17. (Top) Smoothed (uniquely determined portion) surface elevation relative to the geoid in the Caribbean Sea obtained by *Roemmich* [1979] from hydrography alone. (Middle) Surface elevation S obtained along the same arc from the Geos 3 mean surface [*Marsh et al.*, 1980]. (Bottom) Best estimate of the geoid obtained by combining ζ and S. The difference between the middle and the bottom curves would be greater if we displayed derivatives instead.

recipe globally in the not-too-distant future. Because of the enormous variability in the spatial coverage of the ocean density field, there will be a corresponding variation in the reliability of the solutions.

Final Remarks

In practice, one can proceed to improve globally both the geoid and the circulation models. The details will depend upon the relative confidence in each field in each area. We need a long-duration altimetric satellite, independent improvement of the geoid both regionally by ship and globally by gravimetric satellites, and regional and global improvements of measurements of the oceanic mass field.

A rough idea of the requisite minimal accuracy required can be obtained from Figure 16*b*. For the hydrographic data used to construct Figure 16*a* we display in Figure 16*b* (dashed curve) the sea surface elevation, relative to the geoid, under the initial assumption that the level of no motion is arbitrarily placed at the bottom of the ocean (a purely ad hoc assumption). The solid curve is the sea surface elevation determined from the flow field in Figure 16*a* after the conservation constraints had been imposed. (Both curves were integrated from the northern end of the line.) Clearly, any combined altimetric/geoidal system with an accuracy exceeding 10 cm on the spatial scales evident in Figure 16 would strongly constrain the possible ocean circulation. The upper part of Figure 16*b* represents the uniquely determined part of the sea surface elevation in Figure 16*b* and is effectively a spatial low-pass filter version of it. If this is indeed the time-independent part as well, it would be determinable with a long-duration altimetric mission. Figure 16*c* shows the same variables as Figure 16*b*, but continued from Bermuda to the Caribbean Sea.

Both altimetric fields and gravity fields are accessible, at least in principle, with space-borne instrumentation. From the discussion here it should be clear that to exploit these two fields fully, one needs a third field—the three-dimensional density distribution of the ocean. There are no known practical methods for deducing this latter field from space, and ultimate resort to ships may always be necessary. But there are some possible approaches for making inferences about the density from space-borne measurements alone.

Suppose that the altimetric and gravity fields are known sufficiently well that we may regard the surface geostrophic pressure $P_S(x, y, t) = g\zeta(x, y, t)$ as a given quantity. To demonstrate what might be done, let us make the over-simplification that the time changes in the ocean are described by linear dynamics. Let $\mathbf{P}_s = \mathbf{P}_0(x, y) + \mathbf{P}_s'(x, y, t)$; that is, we separate the time average surface pressure from its fluctuations. Then for the time-dependent portion we can write

$$\mathbf{P}_s'(x, y, t) = \sum_{k=0}^{\infty} A_n \frac{dF_n}{dz} F_n(z) \exp\,[ik_n x + il_n y - i\sigma_n t]|_{z=0}$$

that is, a sum over the Rossby wave modes of the sea. The unknown coefficients A_n can then be determined by a three-dimensional inverse Fourier transform of P_s' or by a least squares fit. Observations [e.g., *Mode Group,* 1978] suggest that only A_0 and A_1 are likely to be significant for temporal fluctuations of periods from weeks to many months. Given A_n, the density field in the sea is then

$$\rho(x, y, z, t) = \rho_0(x, y, z) + \rho'(x, y, z, t)$$

where

$$\rho' \propto \sum_n A_n F_n(z) \exp\,[ik_n x + il_n y - i\sigma_n t]$$

and ρ_0 is the time average density (see, for example, *Moore and Philander* [1977]).

More complex models than the linear one may be used to describe the temporal evolution of the density field subject to the surface boundary condition that $P = P_s'$. The mathematically more interesting question is whether a similar process can be used for the time-independent mass field. A differential operator of some complexity can be written down for the time average mass or pressure field, for example, the P equation of *Needler* [1967]. The solution to this equation can be found, numerically if necessary, subject to an upper boundary condition that $P = P_s$ and to lower boundary conditions involving zero normal velocity and mass flux. The degree to which the solution is actually determinable from these conditions and auxiliary knowledge is not immediately clear. But it is certainly possible that useful quantitative statements can be made about the density field, both time varying and time independent, if an adequate knowledge of the dynamical behavior of the ocean can be brought to bear. There will undoubtedly be efforts made in this direction in the future.

Thus with the global fields one can, from the methods described above, envision a determination of the long–time average ocean circulation on large and small scales and the time-dependent ocean circulation on large and small scales, including interannual, annual, and mesoscale variability. The impact of such a system in oceanography and geodesy is potentially very great.

Acknowledgments. Supported in part by contract MO-A01-78-00-4327 with the National Oceanographic and Atmospheric Administration. We thank Barbara Grant for the computing and our many colleagues who were responsible for producing the remarkable Seasat data sets. Some computation was supplied by the Smithsonian Astrophysical Observatory.

References

Accad, Y., and C. L. Pekeris, Solution of the tidal equations for the *M*2 and *S*2 tides in the world oceans from a knowledge of the tidal potential alone, *Phil. Trans. Roy. Soc. London, Ser. A, 290,* 235–266, 1978.

Anderle, R. J., Geodetic analyses through numerical integration, in *The Use of Artificial Satellites for Geodesy and Geodynamics,* edited by G. Veis, pp. 253–265, National Technical University of Athens, Athens, 1974.

Anderle, R. J., and R. L. Hoskin, Correlated errors in satellite altimetry geoids, *Geophys. Res. Lett., 4,* 421–423, 1977.

Backus, G. E., and J. F. Gilbert, Numerical applications of a formalism for geophysical inverse problems, *Geophys. J. Roy. Astron. Soc., 13,* 247–276, 1967.

Blumen, W., Geostrophic adjustment, *Rev. Geophys. Space Phys., 10,* 485–528, 1972.

Born, G. H., J. A. Dunne, and D. B. Lame, Seasat mission overview, *Science, 204,* 1405–1406, 1979a.

Born, G. H., J. Wilkerson, and D. Lame, Seasat Gulf of Alaska workshop report, vol. 1, Panel Reports, *Rep., 622-701,* Jet Propul. Lab., Pasadena, Calif., 1979b.

Charney, J. G., The Gulf Stream as an inertial boundary layer, *Proc. Nat. Acad. Sci. U.S., 41,* 731–740, 1955.

Cheney, R. E., and J. G. Marsh, Seasat altimeter observations of dynamic topography in the Gulf Stream region, *J. Geophys. Res., 85,* in press, 1980.

Csanady, G. T., The pressure field along the western margin of the North Atlantic, *J. Geophys. Res., 84,* 4905–4915, 1979.

Dickinson, D. F., M. D. Grossi, and M. R. Pearlman, Refractive corrections in high-accuracy radio interferometry, *J. Geophys. Res., 75,* 1619–1621, 1970.

Douglas, B. C., and P. D. Gaborski, Observations of sea surface topography with Geos 3 altimeter data, *J. Geophys. Res., 84,* 3893–3896, 1979.

Estes, R. H., A computer software system for the generation of global ocean tides including self-gravitation and crustal loading effects, *NASA/GSFC X Doc., 920-77-82,* 1977.

Franklin, J. N., Well-posed stochastic extensions of ill-posed linear problems, *J. Math. Anal. Appl., 31,* 682–716, 1970.

Fuglister, F. C., Gulf Stream '60, *Progr. Oceanogr., 1,* 265–373, 1963.

Gaposchkin, E. M., Satellite dynamics, 1973 Smithsonian Standard Earth (III), *Spec. Rep. 353,* pp. 85–192, Smithson. Astrophys. Observ., Cambridge, Mass., 1973.

Gaposchkin, E. M., Estimation of 550-km × 550-km mean gravity anomalies and mean geoid heights, *Spec. Rep. 384,* 66 pp., Smithson. Astrophys. Observ., Cambridge, Mass., 1979*a*.

Gaposchkin, E. M., Recent advances in analytical satellite theory, Applications of Geodesy to Geodynamics, II, *Rep. 280,* pp. 197–206, Dep. of Geod. Sci., Ohio State Univ., Columbus, 1979*b*.

Gaposchkin, E. M., Global gravity field to degree and order 30 from Geos 3 satellite altimetry and other data, *J. Geophys. Res., 85,* in press, 1980.

Gordon, A. L., and T. N. Baker, Ocean transients as observed by Geos 3 coincident orbits, *J. Geophys. Res., 85,* 502–506, 1980.

Greenspan, H. P., *The Theory of Rotating Fluids,* 321 pp., Cambridge University Press, New York, 1968.

Heiskanen, W. A., and H. Moritz, *Physical Geodesy,* 364 pp., W. H. Freeman, San Francisco, Calif., 1967.

Hendershott, M. C., The effects of solid earth deformation on global ocean tides, *Geophys. J. Roy. Astron. Soc., 29,* 389–403, 1972.

Huang, N. E., C. D. Leitao, and C. G. Parra, Large-scale Gulf Stream frontal study using Geos 3 radar altimeter data, *J. Geophys. Res., 83,* 4673–4682, 1978.

Jacchia, L. G., Thermospheric temperature, density, and composition, New Models, *Spec. Rep. 375,* 106 pp., Smithson. Astrophys. Observ., Cambridge, Mass., 1977.

Jackson, F., The reflection of impulses from a nonlinear random sea, *J. Geophys. Res., 84,* 4939–4943, 1979.

Kahle, H.-G., M. Chipman, and M. Talwani, Detailed 1° × 1° gravimetric Indian Ocean geoid and comparison with Geos-3 altimeter geoid profiles, *Geophys. J. Roy. Astron. Soc., 55,* 703–720, 1978.

Kaula, W. M., *Theory of Satellite Geodesy,* 124 pp., Blaisdell, Waltham, Mass., 1966.

Kinsman, B., *Wind Waves: Their Generation and Propagation on the Ocean Surface,* 676 pp., Prentice-Hall, Englewood Cliffs, N. J., 1965.

Lanczos, C., *Linear Differential Operators,* 564 pp., Van Nostrand, New York, 1961.

Leitao, C. D., N. E. Huang, and C. G. Parra, A note on the comparison of radar altimetry with IR and in situ data for the detection of the Gulf Stream surface boundaries, *J. Geophys. Res., 84,* 3969–3973, 1979.

Lelgemann, D., Untersuchungen zu einer genaueren Lösung des Problems von Stokes, *Publ. C155,* Ger. Geod. Comm., Munich, 1970.

Lerch, F. J., C. A. Wagner, J. A. Richardson, and J. E. Brownd, Goddard earth models (5 and 6), *NASA/GSFC X Doc., 921-74-145,* 310 pp., 1974.

Lerch, F. J., S. M. Klosko, R. E. Laubscher, and C. A. Wagner, Gravity model improvement using Geos 3 (GEM 9 and 10), *J. Geophys. Res., 84,* 3897–3916, 1979.

Luyten, J. R., Scales of motion in the deep Gulf Stream and across the continental rise, *J. Mar. Res., 35,* 49–74, 1977.

Mader, G. L., A revised 5′ gravimetric geoid and associated errors for the North Atlantic calibration area, *NASA Contract. Rep., 156851,* 32 pp., 1979.

Marsh, J. G., and E. S. Chang, 5′ Detailed gravimetric geoid in the northwestern Atlantic Ocean, *Mar. Geod., 1,* 253–261, 1978.

Marsh, J. G., and E. S. Chang, Global detailed gravimetric geoid, *Mar. Geod., 2,* 145–159, 1979.

Marsh, J. G., T. V. Martin, J. J. McCarthy, and P. S. Chovitz, Mean sea surface computation using Geos-3 altimeter data, *Mar. Geod., 3,* 359–378, 1980.

Mode Atlas Group, *Atlas of the Mid-Ocean Dynamics Experiment,* edited by V. Lee and C. Wunsch, Polymode, Massachusetts Institute of Technology, Cambridge, 1977.

Mode Group, The Mid-Ocean Dynamics Experiment, *Deep Sea Res., 25,* 859–910, 1978.

Mofjeld, H. O., Empirical model for tides in the western North Atlantic Ocean, *Tech. Rep. ERL 340-AOML 19,* 24 pp., Nat. Oceanic and Atmos. Admin., Washington, D. C., 1975.

Mognard, N., and B. Lago, The computation of wind speed and wave heights from Geos 3 data, *J. Geophys. Res., 84,* 3979–3986, 1979.

Molodenskii, M. S., V. F. Eremeev, and M. I. Yurkina, *Methods for the Study of the External Gravitational Field and Figure of the Earth,* translated from Russian, 248 pp., Israel Program for Scientific Translations, Jerusalem, 1962.

Moore, D. W., and S. G. H. Philander, Modelling of the tropical oceanic circulation, in *The Sea,* vol. 6, edited by E. D. Goldberg et al., pp. 319–361, John Wiley, New York, 1977.

Moritz, H., Precise gravimetric geodesy, *Rep. 219,* 75 pp., Dep. of Geod. Sci., Ohio State Univ., Columbus, 1974.

Moritz, H., Least squares collocation, *Rev. Geophys. Space Phys., 16,* 421–430, 1978.

National Research Council, Applications of a dedicated gravitational satellite mission, report, Panel on Gravity and Sea Level, Comm. on Geod., Nat. Acad. of Sci., Washington, D. C., 1979.

Needler, G. T., A model of the thermohaline circulation in an ocean of finite depth, *J. Mar. Res., 25,* 329–342, 1967.

Niiler, P. P., and W. S. Richardson, Seasonal variability of the Florida Current, *J. Mar. Res., 31,* 144–167, 1973.

Parke, M. E., and M. C. Hendershott, M_2, S_2, K_1 models of the global ocean tide on an elastic earth, *Mar. Geod., 3,* 379–408, 1980.

Phillips, O. M., *Dynamics of the Upper Ocean,* 336 pp., Cambridge University Press, New York, 1977.

Piscane, V. L., A. Eisner, S. M. Yionoulis, R. J. McConahy, H. D. Black, and L. L. Pryor, A Geos 3 orbit determination experiment, *J. Geophys. Res., 84,* 3926–3932, 1979.

Roemmich, D. H., The application of inverse methods to problems in ocean circulation, Ph.D. thesis, 193 pp., Mass. Inst. of Technol./ Woods Hole Oceanogr. Inst., Cambridge, 1979.

Schmitz, W. J., Jr., Observations of the vertical distribution of low frequency kinetic energy in the western North Atlantic, *J. Mar. Res., 36,* 295–310, 1978.

Schmitz, W. J., Jr., Weakly depth-dependent segments of the North Atlantic circulation, *J. Mar. Res., 38,* 111–133, 1980.

Schwiderski, E. W., Ocean tides, I, Global ocean tidal equations, *Mar. Geod., 3,* 161–218, 1980*a*.

Schwiderski, E. W., Ocean tides, II, A hydrodynamical interpolation model, *Mar. Geod., 3,* 219–255, 1980*b*.

Schwiderski, E. W., On charting global ocean tides, *Rev. Geophys. Space Phys., 18,* 243–268, 1980*c*.

Stengel, K. C., An inverse technique for obtaining geostrophic ocean currents independent of reference level, *J. Phys. Oceanogr., 9,* 856–857, 1979.

Stommel, H., *The Gulf Stream, A Physical and Dynamical Description,* 248 pp., University of California Press, Berkeley, 1965.

Sverdrup, H. V., M. W. Johnson, and R. H. Fleming, *The Oceans,* 1060 pp., Prentice-Hall, Englewood Cliffs, N. J., 1942.

Talwani, M., H. R. Poppe, and P. D. Rabinowitz, Gravimetrically determined geoid in the western North Atlantic, Sea Surface Topography from Space, vol. II, *Tech. Rep. ERL 228-AOML 7-2,* edited by J. Apel, pp. 23-1–23-33, Nat. Oceanic and Atmos. Admin., Washington, D. C., 1972.

Tapley, B. D., B. E. Schutz, J. G. Marsh, W. F. Townsend, and G. H. Born, Accuracy assessment of the Seasat orbit and altimeter height measurement, *Rep. TR79-5,* 120 pp., Inst. for Advan. Study of Orbit Mech., Univ. of Tex., Austin, 1979.

Tscherning, C. C., and R. H. Rapp, Closed covariance expressions for gravity anomalies, geoid undulations and deflections of the vertical implied by anomaly degree variance models, *Rep. 208,* 89 pp., Dep. of Geod. Sci., Ohio State Univ., Columbus, 1974.

Uotila, U. A., Investigations on the gravity field and shape of the earth, *Publ. 10,* 92 pp., Inst. Geod. Photogramm. and Cartogr., Ohio State Univ., 1960.

Wiggins, R. A., The general linear inverse problem: Implication of surface waves and free oscillations for earth structure, *Rev. Geophys. Space Phys., 10,* 251–285, 1972.

Worthington, L. V., *On the North Atlantic Circulation,* 110 pp., Johns Hopkins University Press, Baltimore, Md., 1976.

Wunsch, C., Bermuda sea level in relation to tides, weather, and baroclinic fluctuations, *Rev. Geophys. Space Phys., 10,* 1–49, 1972.

Wunsch, C., The North Atlantic general circulation west of 50°W determined by inverse methods, *Rev. Geophys. Space Phys., 16,* 583–620, 1978.

Wunsch, C., Low frequency variability of the sea, in *Evolution of Physical Oceanography, Scientific Surveys in Honor of Henry Stommel,* edited by B. A. Warren and C. Wunsch, pp. 342–374, MIT Press, Cambridge, Mass., 1980.

Wyrtki, K., Fluctuations of the dynamic topography in the Pacific Ocean, *J. Phys. Oceanogr., 5,* 450–459, 1975.

Wyrtki, K., Sea level variations: Monitoring of the breath of the Pacific, *Eos Trans. AGU, 60,* 25–27, 1979.

Yaplee, B. S., A. Shapiro, D. L. Hammond, B. D. Au, and E. A. Uliana, Nanosecond radar observations of the ocean surface from a stable platform, *IEEE Trans. Geosci. Electron., GE-9,* 170–174, 1974.

Zetler, B. D., and G. A. Maul, Precision requirements for a spacecraft tide program, *J. Geophys. Res., 76,* 6601–6605, 1971.

(Received November 28, 1979;
accepted April 29, 1980.)

REVIEWS OF GEOPHYSICS AND SPACE PHYSICS, VOL. 21, NO. 8, PAGES 1657–1666, NOVEMBER 1983

Recent Progress in the Application of Satellite Altimetry to Observing the Mesoscale Variability and General Circulation of the Oceans

LEE-LUENG FU

Jet Propulsion Laboratory, California Institute of Technology

Recent progress in the application of satellite altimetry from Seasat and GEOS 3 to the observation of the oceanic mesoscale variability and general circulation is reviewed. The lack of accurate geoid models has been the major obstacle in the study of the general ocean circulation from altimetry. The use of geoid-independent methods that utilize the temporal differences in altimetric measurements taken at fixed locations, however, has made significant contributions to our knowledge of the mesoscale variability of the ocean. The mesoscale energies of the sea surface height and geostrophic current have been mapped on a global basis. Their distributions in wave number space have also been analyzed. Because of many of the deficiencies of existing altimeter data (short duration, inadequate orbit, poor accuracy, etc.) most of these results describe only a small portion of the frequency–wave number spectrum of the variability, but they have nonetheless demonstrated the great value of an optimally designed altimetric mission in advancing our knowledge of the global mesoscale variability. The current technology allows satellite altimetry to detect oceanic variability at periods from a few days to 3–5 years, and wavelengths from 50 to 10,000 km. Determining the time-averaged general ocean circulation from altimetry is more problematic because an accurate geoid is indispensable. The currently available global geoid models have useful accuracies only at wavelengths greater than about 7000 km. There have been several attempts at mapping the global ocean circulation at those scales using existing altimeter data and geoids. When these results are compared with hydrographic surveys, some qualitative agreement can be observed, but the quantitative differences are mostly inconclusive because of the geoid and orbit errors. It has been suggested, however, that an altimetric mission that is optimally designed with the current technology, when complemented by a state-of-the-art gravimetric mission to map the earth's gravity field, is able to determine the ocean circulation quantitatively at scales from the ocean basin to about 200 km.

1. INTRODUCTION

Through the enormous observational effort made in the past two decades, it has been established that the water movement of the ocean varies over a wide range of spatial and temporal scales: from tens to thousands of kilometers and from days to years (see *Wunsch* [1981] for a recent review of the subject). To obtain an accurate description of the general (time-averaged) circulation of the global ocean and eventually understand its dynamics, one thus needs to properly sample all of the energetic variabilities of the ocean, resulting in a sampling requirement of taking synoptic measurements roughly every 50 km horizontally and 500 m vertically for at least several years. It is obvious that observations made by conventional shipboard measurements will never meet this requirement. Because of its fast global coverage the newly developed satellite altimetry (see *Stewart* [1983] for a general introduction), however, holds the promise of providing the required surface observations. When satellite altimetry is complemented by information on the density structure at depths, which can be provided by the currently developing acoustic tomography [*Ocean Tomography Group*, 1982], a global three-dimensional description of the ocean circulation and its variability will begin to emerge [see *Munk and Wunsch*, 1982].

Although it has been only 10 years since the launch of the first spaceborne altimeter on Skylab in 1973, there has been a large and growing body of literature on the applications of satellite altimetry in various branches of geophysics. The focus of this review is recent progress in the application of satellite altimetry to the observation of the mesoscale (wavelengths from a few tens to 1000 km) variability and general circulation of the oceans, with discussions on the limitations of existing data and on the potential of better-designed altimetric missions in the future.

A satellite altimeter measures, among other variables, the altitude of the satellite above the sea surface through radio pulse ranging. With the height of the satellite above a reference surface (the orbit height) being determined independently by tracking and modeling, the height of the sea surface above the same reference surface is obtained by subtracting the satellite altitude from the orbit height. The horizontal gradient of the difference between the height of the sea surface and of the geoid (the equigeopotential surface), both referred to the same reference surface, is thus a measure of the surface geostrophic current. Such a measurement is inevitably contaminated by a large number of errors: those of orbit, geoid, and altitude [e.g., *Wunsch and Gaposchkin*, 1980; *Stewart*, 1983]. Of these the altitude error is due to a combination of instrument errors, atmospheric and sea state effects, and the ocean and earth tides. To obtain useful measurements of ocean currents, these errors must be reduced to acceptable levels.

There have been three satellites carrying an altimeter for earth observations: Skylab (1973), GEOS 3 (1975), and Seasat (1978). The noise level of the altimeter on board Skylab was so high (60 cm for 1-s average) that even the Gulf Stream boundary could not be unambiguously detected in the data [*Leitao et al.*, 1974]. With much improved precision (25 cm for 1-s average), GEOS 3 altimeter data began to reveal surprisingly detailed information on the Gulf Stream meanders and rings [e.g., *Huang et al.*, 1978], among other features. A disadvantage of GEOS 3 data, however, is the lack of corrections of errors due to effects of the ionospheric free electrons, the tropospheric water vapor, and the atmospheric pressure loading (the inverse barometer effect), resulting in unknown errors at all scales. Nevertheless, the 3.5 years of GEOS 3 data have generated many interesting and useful results. With further improved precision (5 cm for 1-s average) and correction algorithms the accuracy of Seasat altimeter data (altitude) has reached the 10-cm level required for ocean circulation studies [*Tapley et al.*, 1982]. Despite the short 3.5-month duration of Seasat, numerous studies of its high-quality data have demonstrated the potential value of a well-designed altimeter in improving our knowledge of the global mesoscale variability and general circulation of the oceans.

In addition to its application to ocean circulation problems, satellite altimetry can also be applied to various other oceanographic problems [see *Greenwood et al.*, 1969]. For instance, the shape and intensity of the return radio pulse can be used to measure ocean wave height and surface wind speed, respectively [e.g., *Chelton et al.*, 1981]. The altitude measurement can also be used to study ocean tides [e.g., *Parke*, 1980; *Cartwright and Alcock*, 1981]. However, these subjects are beyond the scope of this review.

In the following I will discuss the progress to date in the application of GEOS 3 and Seasat data to ocean circulation problems. I will first discuss the problem of the mesoscale variability and then the more difficult problem of the general circulation.

2. Mesoscale Variability

The major difficulty in applying satellite altimetry to ocean circulation problems lies in the fact that the spatial variability of the sea surface height is dominated by that of the geoid. The amplitude of the geoid variability is of the order of 100 m, whereas that of the oceanic variability is of the order of only 1 m. Since existing global geoid models are not sufficiently accurate for studying the general circulation (discussed in section 3), satellite altimetry has primarily been applied to studying the mesoscale variability using differences between altimetric measurements taken at the same earth location at different times, whereby the time-invariant geoid is removed. There are three methods frequently used in this kind of study. One of them is the "repeat track method," where the differences between measurements taken along nearly repeating tracks are used. Another is the "mean sea surface method," which utilizes the differences between individual measurements and a mean sea surface constructed from altimeter data. The third is the "crossover difference method," which utilizes the differences between measurements taken at the intersections of altimeter tracks. Most of the studies discussed below are based on one of these methods. In these methods a tilt and a bias are usually removed by one way or another over a distance of several thousand kilometers along each track to minimize the long-wavelength errors of orbit and tides; however, these procedures minimize also the long-wavelength oceanic variability, making the results applicable only for studying the mesoscale variability.

To make the following discussion more orderly, results from GEOS 3 and Seasat are discussed in separate subsections.

GEOS 3 Results

Because of the lack of recording capability on board GEOS 3, continuous data coverage is limited to the western North Atlantic Ocean, and so have been most studies using GEOS 3 data. For a mission overview the reader is referred to *Stanley* [1979]. The first attempt at using GEOS 3 data to study mesoscale variability was made by *Huang et al.* [1978] in the Gulf Stream area. After minimizing the radial orbit error by a linear adjustment (through a bias and a tilt) of each pass such that the crossover differences in the study area were minimized, they produced monthly maps of the sea surface topography using data at crossover points and studied the monthly variations of the Gulf Stream and eddies by differencing the monthly map from a 6-month mean map. Their results showed good agreement with concurrent infrared images. The same method was applied to a longer data set in the same area by *Mather et al.* [1979, 1980] with similar results. *Robinson et al.* [1983] modified the method to include all the data (not exclusive to crossover points) in the computation of a mean sea surface and applied the procedure to 3 years of data, resulting in a 3-year time series of sea surface topography maps. Their results also showed qualitative agreement with in situ observations and infrared images. A disadvantage of these methods is that residual geoid errors have been introduced into the monthly maps as a result of their much sparser data coverage than the mean map from which the differencing is made.

Spatial profiles of mesoscale variability can be obtained by studying individual tracks of data. Strong features like the Gulf Stream boundaries can be detected even in raw altimeter profiles [*Leitao et al.*, 1979]. The papers by *Douglas and Gaborski* [1979] and by *Mather et al.* [1980] are among the earliest demonstrations of the utility of repeat track data in detecting individual mesoscale features. *Douglas and Gaborski* [1979] also demonstrated that the Gulf Stream boundaries and rings could be detected by subtracting from altimeter data the high-resolution geoid of *Marsh and Chang* [1978].

In addition to individual features, the geographic distribution of mesoscale variability can also be obtained from a dense net of repeat tracks. Using data in the western North Atlantic, *Douglas and Cheney* [1981] made the first of such attempts. To maximize the number of tracks in each repeat group, they included all the tracks lying within 20 km of each other in the computation of the sea surface height variability and used the geoid model of *Mader* [1979] to compensate the geoid variability resulting from the track spacings. Their result showed qualitative agreement with the eddy potential energy map of *Dantzler* [1977] based on shipboard measurements.

Using only pairs of tracks lying within a few kilometers of each other, *Douglas et al.* [1983*a*] recomputed the mesoscale variability in the western North Atlantic and extended the computation to cover the Gulf of Mexico. Their results should be relatively free from geoid contamination. Two figures summarizing their results are reproduced here in Figures 1*a* and 1*b*, showing the sea surface height variability and the eddy kinetic energy, respectively. The kinetic energy was computed from the slope of the sea surface height using the geostrophic relation. Note that the maxima of both the sea surface height variability and the kinetic energy correspond to the meandering of the Gulf Stream and the Loop Current. They made detailed comparisons of Figures 1*a* and 1*b* with measurements made by expendable bathythermographs (XBT's) and surface drifters, respectively, and found close qualitative agreement in most of the main features; however, the shipboard results showed systematically higher values than did the altimetric results in the Gulf Stream area south of Cape Hatteras along the coast. Their explanation was that the eddy energy based on shipboard measurements was computed as the standard deviation in a finite gridded area and hence was seriously contaminated by the high spatial variability near the Gulf

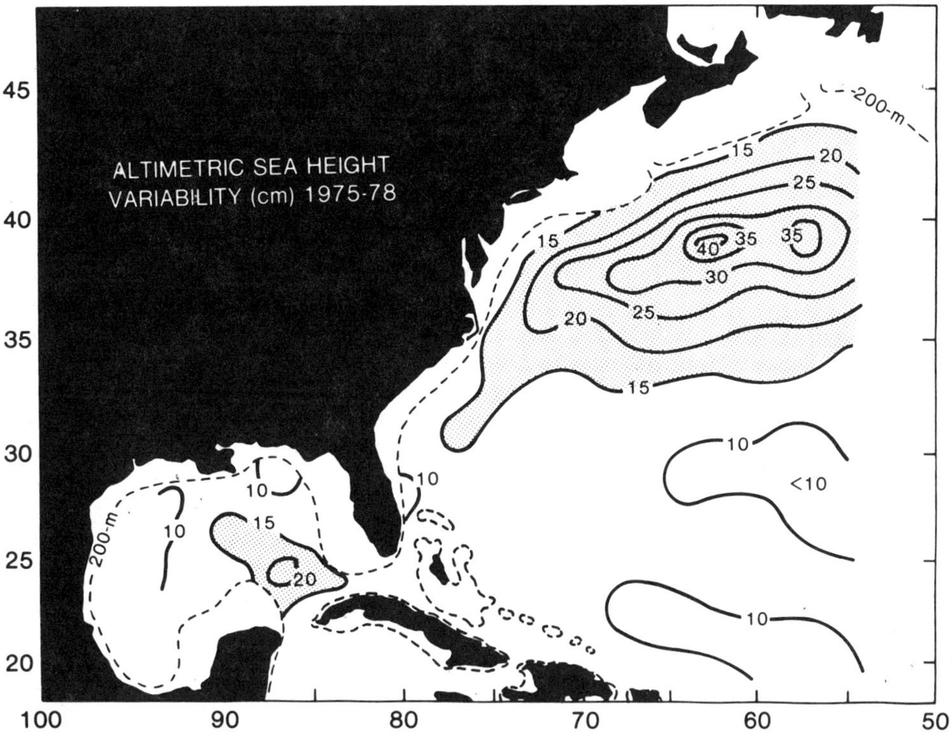

Fig. 1*a*. Sea surface height variability in centimeters from GEOS 3 altimeter repeat pairs [from *Douglas et al.*, 1983*a*].

Stream area, whereas the altimetric measurement sampled only the temporal variability. Therefore the maps shown in Figures 1*a* and 1*b* bear a clearer interpretation than most existing eddy energy maps based on shipboard measurements.

Maps of mesoscale variability can be obtained from the crossover difference method as well. After a linear adjustment which minimizes the crossover differences in a particular area, the magnitude of the residual crossover differences, when divided by the square root of 2, is a measure of the mesoscale variability. Using 1.5 years of data, *Cheney and Marsh* [1981*a*] produced the first variability map thus obtained of the western North Atlantic. The result agreed with Figure 1*a* within about 10 cm, the crossover values being systematically higher than the repeat track results. It is not clear if the dis-

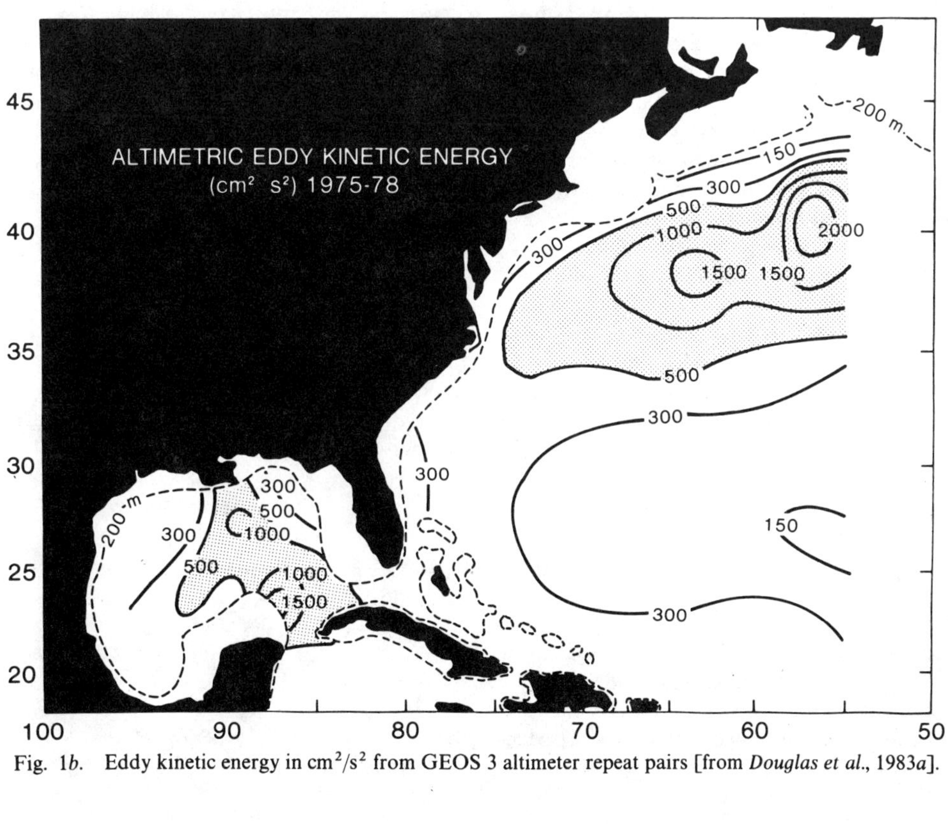

Fig. 1*b*. Eddy kinetic energy in cm^2/s^2 from GEOS 3 altimeter repeat pairs [from *Douglas et al.*, 1983*a*].

crepancy can be totally ascribed to the difference in sampling scheme between the two methods, because the data samples used in the two studies are also different. *Gordon et al.* [1983] applied the same method to the data in the southern hemisphere and produced maps showing the variability of the major currents there.

The fast along-track sampling of an altimeter (a few thousand kilometers in 10 min) provides nearly synoptic measurements of the sea surface height for estimating the wave number spectrum of mesoscale variability, a quantity that has been difficult to estimate because of the lack of synoptic measurements with sufficient spatial resolution. Using a number of repeat tracks in the western North Atlantic, *Gordon and Baker* [1980] demonstrated the utility of altimetry in providing the along-track component of the wave number spectrum of mesoscale variability. At wavelengths shorter than 200 km their spectrum was dominated by instrument noise; at longer wavelengths the spectral energy increased with decreasing wave number following a -2 power dependence until the spectrum leveled off at a wavelength of about 500 km. Because their results were based on a small set of data, the geographic variability of the spectrum was not examined.

Seasat Results

The advantages of the Seasat altimeter over the GEOS 3 altimeter are its high precision (5 cm versus 25 cm) and global repeat track coverage (see *Born et al.* [1979] for a mission overview), whereas these advantages are frequently overshadowed by the unfortunate fact that Seasat lasted only 3.5 months, in contrast to 3.5 years for GEOS 3. Therefore one cannot obtain from Seasat data long-term statistics of the oceans; most studies using Seasat data are of a demonstrative nature, showing the kind of information one could obtain from a well-designed altimetric mission.

Subtracting the $5' \times 5'$ gravimetric geoid of *Marsh and Chang* [1978] from Seasat altimeter data, *Cheney and Marsh* [1981*b*] demonstrated how well individual oceanic features (Gulf Stream boundaries and eddies) were detected by the altimeter. They showed excellent agreement between the altimetric measurements and concurrent observations from infrared imagery, XBT's, and surface drifters. They also demonstrated the utility of the repeat track data which lasted for about 24 days during the last month of the mission. In addition, *Cheney* [1982] assembled a comprehensive data set taken in the western North Atlantic during the Seasat mission, providing a valuable data set for comparisons with the altimetric measurements. Other examples of similar studies demonstrating the utility of Seasat altimeter data include those of *Wunsch and Gaposchkin* [1980] and *Byrne and Pullen* [1981].

Bernstein et al. [1982] conducted an experiment in the Kuroshio area east of Japan to verify the sea surface height measurements by the Seasat altimeter. They dropped airdropped expendable bathythermographs along the satellite's ground tracks and used the data to compute the dynamic height of the sea surface relative to 1000 dbar. They found that the dynamic height agreed with the altimeter-measured sea surface height within 10 cm. Part of the discrepancy was due to the fact that only a portion of the baroclinic component of the sea surface height variability was reflected in the 0/1000-dbar dynamic height measurement, whereas the altimeter-measured height contained both the baroclinic and barotropic components. *Byrne and Pullen* [1983] also conducted a verification experiment in the Kuroshio area. They compared the altimeter data with coincident conductivity-temperature-depth (CTD) and deep XBT (1800 m) observations and reached similar conclusions.

Kao and Cheney [1982] compared altimetric observations of the variation of sea surface height across the Gulf Stream with the predictions of several theoretical models. They found that the altimeter data were able to resolve the differences between the various models. *Thompson et al.* [1983] demonstrated that the temporal variability of the Loop Current in the Gulf of Mexico measured by the altimeter agreed closely with the result of a numerical model. These results have demonstrated the potential value of altimetry in validating numerical and analytical models of the ocean.

Using the global 24-day repeat track data, *Cheney et al.* [1983] constructed the first map of the global mesoscale variability based on altimetry. Their map is reproduced here in Figure 2. The repeat tracks form a uniform global coverage with a cross-track spacing of 900 km at the equator. Each repeat track group contains 8 to 10 closely spaced (within a band of 2.5 km) tracks with a repeat period of 3 days. The analysis techniques were similar to those used by *Douglas and Cheney* [1981]. For each group of repeat tracks, a mean sea surface height profile was first formed and then subtracted from each individual profile. After the removal of a linear trend from the residual profiles, the root-mean-squares of the residuals were computed, interpolated, and gridded in $2° \times 2°$ to produce the map shown in Figure 2. Coastal areas of high tides were excluded from the map.

As one would expect, a local maximum of variability is associated with each of the major western boundary currents. The maximum value of 19 cm occurs in the southwestern Atlantic. A series of local maxima are found in the Antarctic Circumpolar Current; most of them are located downstream of prominent topographic features. *Colton and Chase* [1983] studied in some detail the interaction of current with topography in these regions from Seasat altimetry. Another notable feature on the map is the secondary maxima in the North Equatorial currents of the Pacific and the Atlantic oceans. No prominent variability is found in the South Equatorial currents. Also notable on the map is the absence of significant variability along the equator, whereas a narrow belt of maximum variability is found right along the equator on the eddy kinetic energy map of *Wyrtki et al.* [1976]. Perhaps this is due to the fact that the Coriolis parameter diminishes on the equator so that smaller sea surface height variations are produced by equatorial motions than by mid-latitude eddies of the same kinetic energy.

Because there are only 24 days of repeat track data, only the energies at periods shorter than 24 days have been well represented by Figure 2. The energy of the sea surface height variability (square the values on Figure 2) can be expressed by the following integral:

$$E = \int_0^\infty F(f)R(f)\,df \tag{1}$$

where $F(f)$ is the frequency spectrum of the sea surface height variability, f is frequency in cycles per day, and

$$R(f) = \left(1 - \frac{\sin 24\pi f}{24\pi f}\right)^2 \tag{2}$$

which is the transfer function of the 24-day high-pass filter. This transfer function is plotted in Figure 3. Note that the

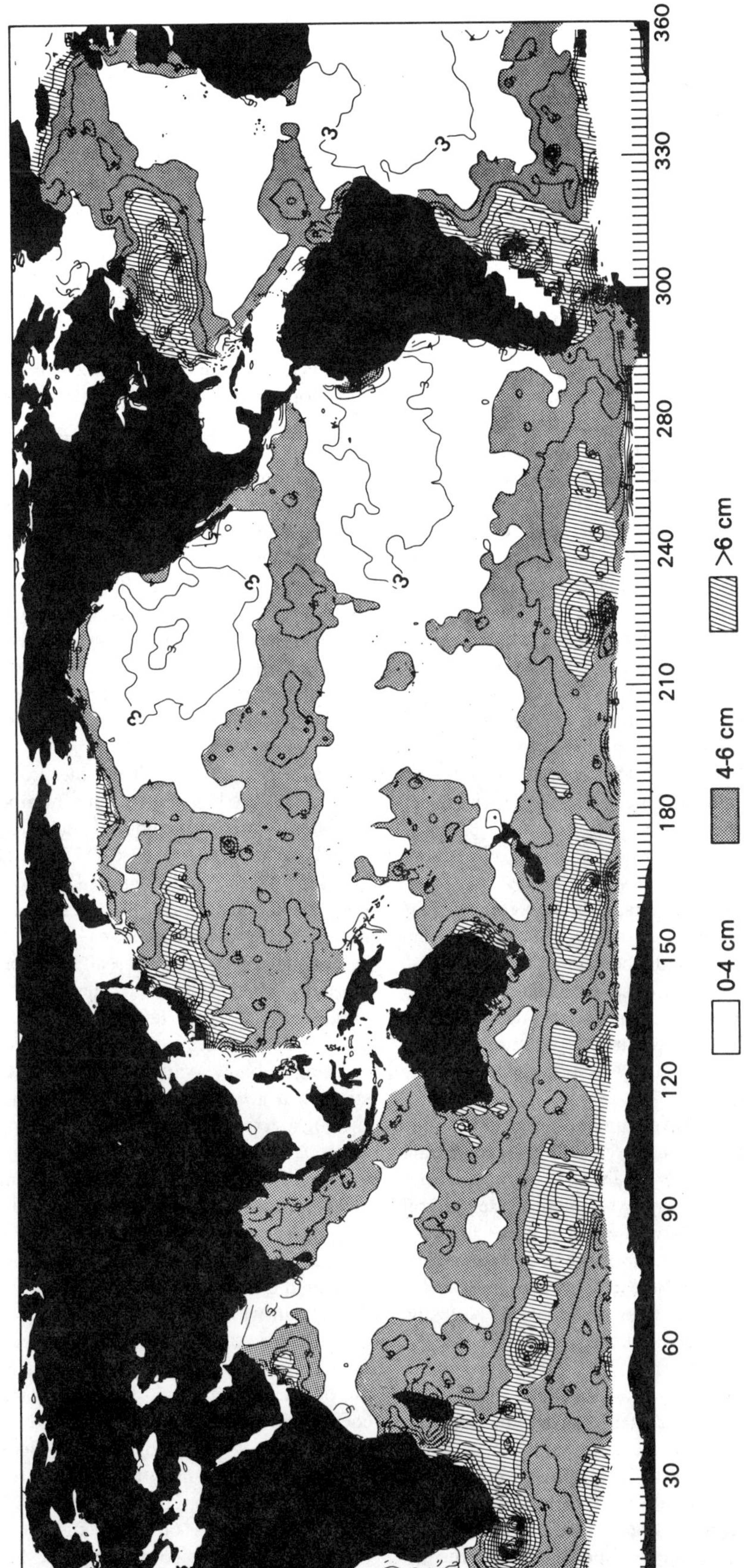

Fig. 2. Sea surface height variability from the repeat tracks of Seasat altimeter data. Contour interval is 1 cm [from *Cheney et al.*, 1983].

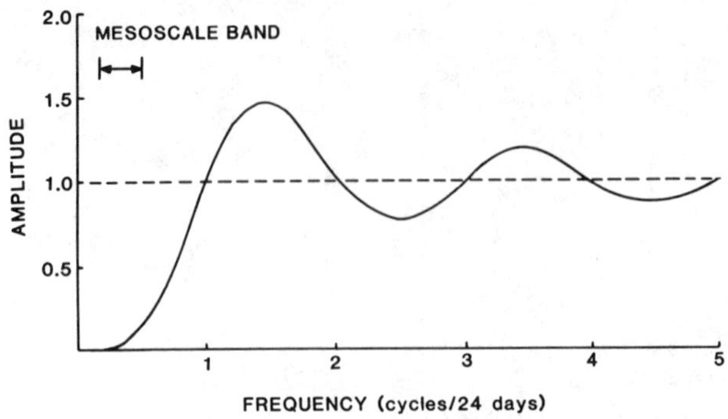

Fig. 3. The transfer function of the 24-day high-pass filter. The frequency band of the mesoscale variability is indicated.

mesoscale variability whose periods are from 50 to 150 days is severely suppressed by the transfer function. However, the low-frequency tail of the transfer function does pick up some of the mesoscale energy. The maximum variability of 12 cm in the Gulf Stream area shown in Figure 2 is about one third of the value shown in Figure 1*a*, which is based on 3.5 years of data. This implies that the energy sampled by the 24-day data is about one tenth of the total energy. This fraction is consistent with what one would obtain from (1) and (2) using existing frequency spectra such as those discussed by *Wunsch* [1981].

Using the crossover difference technique, *Parke and Stavert* [1983] also produced a map of the global mesoscale variability from Seasat altimeter data. Their result agreed with Figure 2 in most of the main features; however, their values were slightly higher in general because they used only the crossover differences resulting from tracks separated by more than 30 days.

Menard [1983] used the repeat track data to compute the eddy kinetic energy in the regions of the Gulf Stream and Kuroshio. His results agreed with the map of *Wyrtki et al.* [1976] within a factor of 2. This close agreement is puzzling because the energy resulting from the 24-day data used by *Menard* [1983] presumably represents only a small fraction (about one tenth) of the total energy as represented by the map of *Wyrtki et al.* [1976].

Fu [1983] computed the wave number spectrum of the variability shown in Figure 2, using the same repeat tract data. He divided the area of the oceans into two categories according to their mesoscale energy levels for separate spectral computation: the "high-energy areas" close to major currents and the "low-energy areas" remote from major currents. The resulting spectra are reproduced here in Figures 4*a* and 4*b* for the high-energy and low-energy areas, respectively. Note the difference in spectral characteristics between the two categories.

In the high-energy areas the spectra basically follow a -5 power law at wavelengths shorter than 250 km and begin to level off at longer wavelengths, indicating that the energy-containing eddies have wavelengths longer than 250 km. The -5 power law regime was not detected by *Gordon and Baker* [1980] because of the high noise level of the GEOS 3 altimeter. As a result of geostrophy, the -5 power law for the sea surface height spectrum implies a -3 power law for the kinetic energy spectrum, which is predicted by the theory [*Charney*, 1971] and numerical models [e.g., *McWilliams and Chow*, 1981] of geostrophic turbulence. This -3 power law has also been observed on the kinetic energy spectra computed directly from Seasat altimater data [*Menard*, 1983].

In the low-energy areas the spectra follow a -1 power law from 100 to 1000 km, implying a "blue" spectrum, with the spectral level increasing with k, for the kinetic energy. This blue spectrum is striking; however, one has to keep in mind that the frequency content of the spectra shown in Figure 4 is limited by the transfer function expressed by (2). Neither theoretical results nor in situ observations are available yet for interpreting wave number spectra of this frequency content in the low-energy areas.

In summary, as revealed by the studies discussed above, the precision and coverage of Seasat data are good enough for advancing our knowledge of the global mesoscale variability; however, the utility of the data is limited severely by its short duration, as stated in the beginning of the section.

3. General Circulation

To determine the general circulation of the ocean from satellite altimetry, an accurate, independently determined geoid is required. As demonstrated by *Roemmich and Wunsch* [1982], the combined error of altimetry and geoid must be less than 10 cm in order to make altimetry a useful constraint for determining the ocean circulation quantitatively. Unfortunately, existing geoids have errors much larger than 10 cm; therefore most studies of the problem to date are of a qualitative nature. Optimal strategies for determining the ocean circulation in the presence of geoid error have been discussed by *Wunsch and Gaposchkin* [1980].

The dense coverage of marine gravity data in the western North Atlantic has resulted in a number of high-resolution (5′ × 5′) gravimetric geoids of that region [e.g., *Marsh and Chang*, 1978; *Mader*, 1979]. Using GEOS 3 data and the geoid of Marsh and Chang, *Huang et al.* [1978] produced maps of the deviation of the mean sea surface height from the geoid, that is, the dynamic sea surface topography (referred to as dynamic topography hereafter). Their results showed strong gradients at locations consistent with the mean Gulf Stream path. Because of the lack of a rigorous error analysis of the geoid model the accuracy of the results cannot be assessed.

Using publicly available gravity data, *Zlotnicki* [1983] constructed a geoid model for the North Atlantic Ocean with a rigorous error estimate. He then applied the strategies of *Wunsch and Gaposchkin* [1980] to Seasat altimeter data with his geoid to estimate the general circulation of the North Atlantic. Because of the large errors in the geoid model, only a very crude gyre was detected in the Sargasso Sea. However, the main thrust of his study was the error analysis of the estimated dynamic topography.

At basin-wide scales, gravimetric geoids are no longer available, but geoids based on satellite tracking data have barely useful accuracies. These geoids are usually expressed in terms of spherical harmonics. The error of the spherical harmonic coefficients increases with degree and order. One such example is the geoid of Goddard Earth Model (GEM) 9 [*Lerch et al.*, 1979], which is complete up to degree and order 20 (about 2000 km in wavelength). The error of GEM 9 is about 1.9 m (root sum squares of all coefficient errors); the neglect of higher-degree terms contributes another error of about 2.5 m, making a total error of about 3.1 m. Nevertheless, the error is only 16 cm for terms up to degree and order 4 (10,000 km in wavelength) and 30 cm for terms up to degree and order 6 (6700 km in wavelength).

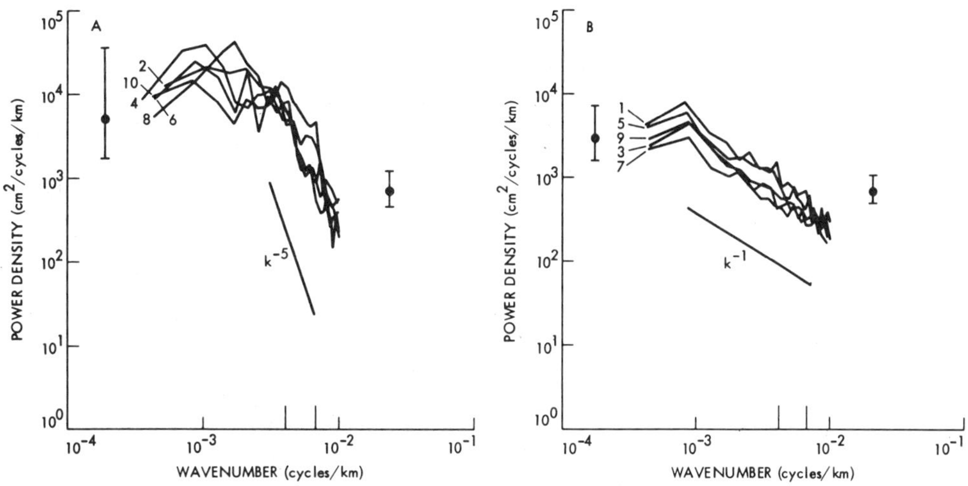

Fig. 4. Spectra of the sea surface height variability in (*a*) the high-energy areas and (*b*) the low-energy areas. The 10 labeled curves are from different areas in the world's oceans. The error bars are 95% confidence limits; the left one in each figure is for wavelengths greater than 250 km, and the right one is for wavelengths shorter than 250 km. The two large tick marks on the wave number axis indicate the wavelengths of 150 km and 250 km, respectively [from *Fu*, 1983].

GEM 9 was first applied to GEOS 3 data by *Mather et al.* [1979] to study the global general circulation. They found, however, that the error of GEM 9 dominated the variability of the ocean circulation (based on hydrography) at spherical harmonics of degree and order greater than 4. Realizing this fact, they constructed a mixed representation of the global dynamic topography in spherical harmonics complete to degree and order 16, the terms up to degree and order 4 being derived from altimetry and the remaining terms from hydrography. Because the altimeter data used in the study were not subject to crossover adjustments, the altimetric part of the topography contained a substantial amount of orbit error in addition to geoid error.

Using GEM 9 and the crossover adjusted Seasat data of *Rapp* [1983], *Tai and Wunsch* [1983*a*] constructed an altimetric dynamic topography of the Pacific Ocean at a resolution of 20°. *Tai* [1983] made detailed comparisons of the altimetric result with the hydrography-derived dynamic topography of *Wyrtki* [1975] at each degree of spherical harmonics up to degree 6. Beyond degree 6 he found that the geoid error dominated the variability of the ocean circulation. Reproduced in Figures 5*a* and 5*b* are the spherical harmonic expansions up to degree and order 6 of the dynamic topography from the altimetric and the hydrographic data, respectively [from *Tai and Wunsch*, 1983*a*]. To my knowledge, Figure 5*a* is the first published dynamic topography based solely on altimetric data. Note that the subtropical gyre of the North Pacific has been detected on the altimetric map and that both maps suggest the absence of a simple gyre in the South Pacific, where the flow is predominantly zonal. Despite the overall resemblance between the two maps, some anomalous features are readily apparent on the altimetric map: the northward displacement of the center of the North Pacific subtropical gyre, the abrupt southward deflection of the gyre east of its center, the low in the northeast Pacific, and the high in the southeast Pacific. In addition, the altimetric map shows a circulation twice as strong as that of the hydrographic map. These differences can easily be accounted for by the geoid and orbit errors, of which the latter contains a systematic error that has not been removed by the crossover adjustments. On the other hand, they might just indicate that the large-scale circulation of the Pacific Ocean during the Seasat mission was somewhat different from that shown on Wyrtki's map based on some 70 years of hydrographic data and an arbitrarily chosen "level of no motion" at 1000 dbar.

Goad et al. [1980] developed a procedure to correct satellite radial orbit error on a global basis. *Douglas et al.* [1983*b*] applied this procedure to 3 days of Seasat data and used the corrected data and a geoid model called GEM L2 [*Lerch et al.*, 1982] to construct a global dynamic topography at a resolution of 20°. *Cheney and Marsh* [1982] used a combination of the complete Seasat data and 1.5 years of GEOS 3 data (without crossover correction to orbit error, however) and GEM L2 to construct a global dynamic topography at similar resolution. Despite the great difference in the period of averaging between the two studies, the results revealed some common large-scale features of the global general circulation. The GEM L2 geoid has less error in low-degree terms than the GEM 9 geoid; for instance, complete to degree and order 4, the rms error is 8 cm for GEM L2 and 16 cm for GEM 9. However, because spherical harmonic analysis was not performed in these two studies, the geoid contamination in the results could not be rigorously assessed at various scales as in the work by *Tai* [1983].

Using Rapp's adjusted Seasat data and GEM L1 geoid, which was an earlier version of GEM L2 with a similar accuracy, *Engelis* [1983] constructed a global dynamic topography in terms of spherical harmonics complete to degree and order 6. Because the rms error of GEM L1 was about 20 cm for terms complete to degree and order 6, the rms error of his global map was estimated to be 20–25 cm. His map showed a high correlation with the global hydrographic result of *Levitus* [1982] with correlation coefficients ranging from 0.91 at degree 1 to 0.54 at degree 6.

Recognizing the fact that the dynamic topography constructed from altimetry and geoid models like GEM 9, GEM L1, and GEM L2 was contaminated by the large, high-degree errors of the geoid models, *Tai and Wunsch* [1983*b*] thus constructed a composite geoid that used GEM 9 for terms up to degree and order 6 and a high-resolution altimatric geoid for terms of degree and order higher than 6. Using this geoid and Rapp's mean sea surface, they constructed a global dy-

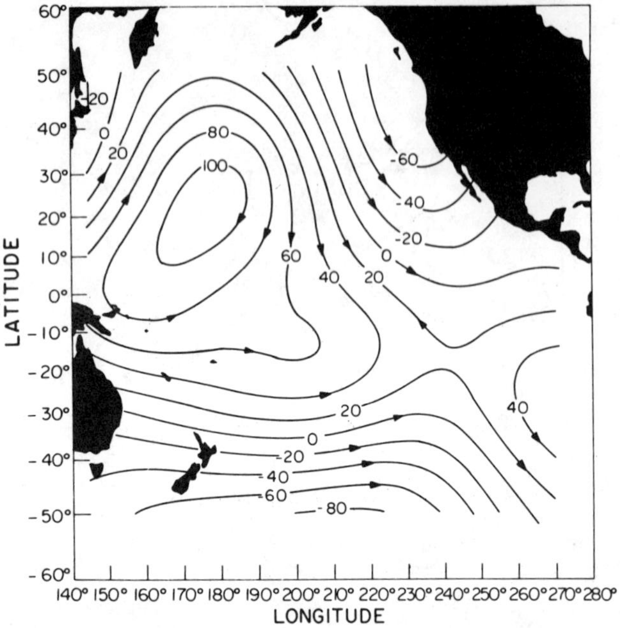

Fig. 5*a*. Spherical harmonic expansion complete to degree and order 6 of the dynamic topography of the Pacific Ocean determined from Seasat altimeter data. Contours are in centimeters. The arrows represent the directions of the surface geostrophic velocity [from *Tai and Wunsch*, 1983*a*]. (Reprinted by permission from *Nature*. Copyright © 1983 Macmillan Journals Limited.)

namic topography in spherical harmonics complete to degree and order 6 and found some notable improvement over the result based on GEM 9 alone.

In summary, the accuracies of the presently available geoid models allow one to obtain only a rather crude picture of the general circulation of the ocean. This picture is only marginally correct at the scales of the ocean basin and is dominated by geoid error at smaller scales. As has been pointed out by Wunsch (unpublished manuscript, 1983), however, knowing only the largest components of the dynamic topography provides useful constraints for ocean circulation models and allows one to make inferences through models about the three-dimensional structure of the large-scale ocean circulation.

4. Concluding Remarks

Because of many of the deficiencies of existing altimeter data (short duration, inadequate orbit, poor accuracy, etc.), only a small portion of the frequency–wave number spectrum of the oceanic variability has been observed by satellite altimetry. The results of the various studies discussed above, however, have demonstrated that an optimally designed altimetric mission with a lifetime of several years will improve our knowledge of the global oceanic variability to an extent unattainable by any other practical means. Such a mission, called TOPEX (Ocean Topography Experiment [see *TOPEX Science Working Group*, 1981]), has been studied by the National Aeronautics and Space Administration as a prospective flight project in the near future. The proposed altimeter has a precision level of 2 cm, extending the lower limit in wavelength of detectable oceanic variability from 100 km for Seasat to about 50 km. With a reduction of orbit error to the 10-cm level by improved tracking technology, even the variability of basin-wide scales can be measured in addition to the mesoscale variability. The nominal TOPEX orbit has a 10-day repeat cycle with equatorial separation of 300 km, a compromise between sampling requirements in space and time. The mission is planned to last 3 years with an option for a 2-year extension. With 5 years of such data, one can then measure the global oceanic variability at periods from 20 days to 5 years and wavelengths from 50 to 10,000 km.

Determining the general ocean circulation from altimetry is more problematic because a single altimetric mission cannot solve the problem and a highly accurate geoid must be determined independently. As noted earlier, the accuracies of the presently available geoid models are not adequate for determining the general ocean circulation; however, as indicated in the recent studies of *Kahn* [1983] and *Lerch and Putney* [1983], the proposed satellite gravimetric mission called GRM (Geopotential Research Mission; formerly known as Gravsat) will provide, through the mutual tracking of two low-flying satellites, a geoid with an accuracy of about 8 cm for $1° \times 1°$ averages. Such an accuracy is good for determining the ocean circulation down to scales as small as 200 km.

At large scales the satellite radial orbit error becomes another important concern for determining the general ocean circulation. In addition to tracking and modeling effort, the use of altimeter data itself as a constraint for orbit improvement has proven effective [e.g., *Goad et al.*, 1980; *Rapp*, 1983; *Parke and Stavert*, 1983]. However, some theoretical work is needed here to determine an optimal method for correcting orbit from altimetry and for estimating its residual error. This problem is currently being investigated by C. Wunsch and V. Zlotnicki (personal communication, 1983) using objective mapping techniques. Ultimately, one has to treat the three coupled problems together, that is, the determinations of the ocean circulation, the geoid error, and the orbit error, by optimal procedures such as the inverse method outlined by *Wunsch and Gaposchkin* [1980]. Research in this direction needs to be advanced before the promise of TOPEX/GRM becomes a reality.

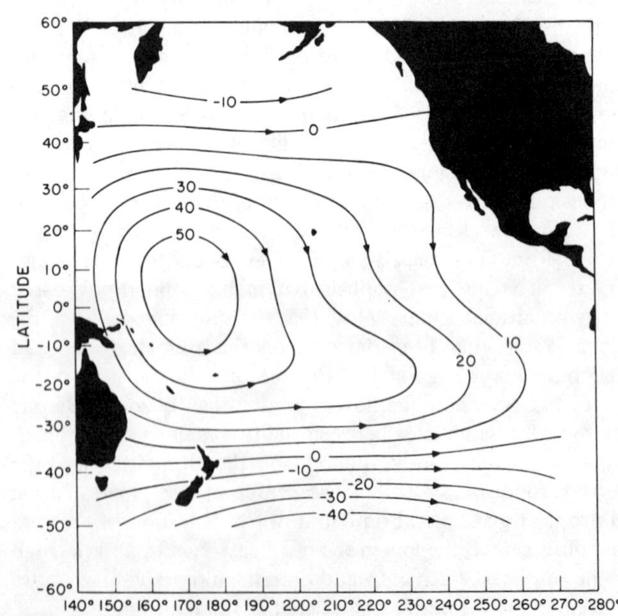

Fig. 5*b*. As Figure 5*a* except that the dynamic topography is determined from Wyrtki's hydrographic data [from *Tai and Wunsch*, 1983*a*]. (Reprinted by permission from *Nature*. Copyright © 1983 Macmillan Journals Limited.)

Acknowledgments. I am grateful to Chang-Kou Tai and Carl Wunsch of Massachusetts Institute of Technology and to Bruce Douglas and Robert Cheney of the National Ocean Service/NOAA for providing me with their figures for reproduction in this paper. Valuable comments from Robert Cheney and Carl Wunsch are also gratefully acknowledged. The research described in this paper was performed at the Jet Propulsion Laboratory, California Institute of Technology, under contract with the National Aeronautics and Space Administration.

REFERENCES

Bernstein, R. L., G. H. Born, and R. H. Whritner, Seasat altimeter determination of ocean current variability, *J. Geophys. Res., 87*, 3261–3268, 1982.

Born, G. H., J. A. Dunne, and D. B. Lame, Seasat mission overview, *Science, 204*, 1405–1406, 1979.

Byrne, H. M., and P. E. Pullen, Western boundary current variability derived from Seasat altimetry data, in *Oceanography From Space*, edited by J. F. R. Gower, pp. 877–884, Plenum, New York, 1981.

Byrne, H. M., and P. E. Pullen, Seasat-derived ocean surface topography: Comparison with coincident Kuroshio hydrographic data, *J. Geophys. Res., 88*, 2621–2625, 1983.

Cartwright, D. E., and G. A. Alcock, On the precision of sea surface elevations and slopes from Seasat altimetry of the northeast Atlantic Ocean, in *Oceanography From Space*, edited by J. F. R. Gower, pp. 885–895, Plenum, New York, 1981.

Charney, J. G., Geostrophic turbulence, *J. Atmos. Sci., 28*, 1087–1095, 1971.

Chelton, D. B., K. J. Hussey, and M. E. Parke, Global satellite measurements of water vapour, wind speed, and wave height, *Nature, 294*, 529–532, 1981.

Cheney, R. E., Comparison data for Seasat altimetry in the western North Atlantic, *J. Geophys. Res., 87*, 3247–3253, 1982.

Cheney, R. E., and J. G. Marsh, Oceanic eddy variability measured by GEOS 3 altimeter crossover differences, *Eos Trans. AGU, 62*(45), 743–752, 1981*a*.

Cheney, R. E., and J. G. Marsh, Seasat altimeter observations of dynamic topography in the western North Atlantic, *J. Geophys. Res., 86*, 473–483, 1981*b*.

Cheney, R. E., and J. G. Marsh, Global ocean circulation from satellite altimetry (abstract), *Eos Trans. AGU, 63*(45), 997, 1982.

Cheney, R. E., J. G. Marsh, and B. D. Beckley, Global mesoscale variability from collinear tracks of Seasat altimeter data, *J. Geophys. Res., 88*(C7), 4343–4354, 1983.

Colton, M. T., and R. R. P. Chase, The interaction of the Antarctic Circumpolar Current with bottom topography: An investigation using satellite altimetry, *J. Geophys. Res., 88*, 1825–1843, 1983.

Dantzler, H. L., Jr., Potential energy maxima in the tropical and subtropical North Atlantic, *J. Phys. Oceanogr., 7*, 512–519, 1977.

Douglas, B. C., and R. E. Cheney, Ocean mesoscale variability from repeat tracks of GEOS 3 altimeter data, *J. Geophys. Res., 86*, 10,931–10,937, 1981.

Douglas, B. C., and P. D. Gaborski, Observation of sea surface topography with GEOS 3 altimeter data, *J. Geophys. Res., 84*, 3893–3896, 1979.

Douglas, B. C., R. E. Cheney, and R. W. Agreen, Eddy energy of the northwest Atlantic and Gulf of Mexico determined from GEOS 3 altimetry, *J. Geophys. Res.*, in press, 1983*a*.

Douglas, B. C., R. W. Agreen, and D. T. Sandwell, Observing global ocean circulation with Seasat altimeter data, *Mar. Geod.*, in press, 1983*b*.

Engelis, T., Analysis of sea surface topography using Seasat altimeter data, *Rep. 343*, 97 pp., Dep. of Geod. Sci. and Surv., Ohio State Univ., Columbus, 1983.

Fu, L.-L., On the wave number spectrum of oceanic mesoscale variability observed by the Seasat altimeter, *J. Geophys. Res., 88*(C7), 4331–4342, 1983.

Goad, C. C., B. C. Douglas, and R. W. Agreen, On the use of satellite altimater data for radial ephemeris improvement, *J. Astronaut. Sci., 28*(4), 419–428, 1980.

Gordon, A. L., and T. N. Baker, Ocean transients as observed by GEOS 3 coincident orbits, *J. Geophys. Res., 85*, 502–506, 1980.

Gordon, A. L., K. Horai, and M. Donn, Southern hemisphere western boundary current variability revealed by GEOS 3 altimeter, *J. Geophys. Res., 88*, 755–762, 1983.

Greenwood, J. A., et al., Oceanographic applications of radar altimetry from a spacecraft, *Remote Sens. Environ., 1*, 71–80, 1969.

Huang, N. E., C. D. Leitao, and C. G. Parra, Large-scale Gulf Stream frontal study using GEOS 3 radar altimeter data, *J. Geophys. Res., 83*, 4673–4682, 1978.

Kahn, W. D., Error analysis studies for a geopotential research mission (abstract), *Eos Trans. AGU, 64*, 207, 1983.

Kao, T. W., and R. E. Cheney, The Gulf Stream front: A comparison between SEASAT altimeter observations and theory, *J. Geophys. Res., 87*, 539–545, 1982.

Leitao, C. D., L. S. Miller, and N. E. Huang, Detecting the Gulf Stream using satellite altimetry (abstract), *Eos Trans. AGU, 55*, 1107, 1974.

Leitao, C. D., N. E. Huang, and C. G. Para, A note on the comparison of radar altimetry with IR and in situ data for the detection of the Gulf Stream surface boundaries, *J. Geophys. Res., 84*, 3969–3973, 1979.

Lerch, F. J., and B. H. Putney, Gravsat simulations for analysis of geopotential (abstract), *Eos Trans. AGU, 64*, 465, 1983.

Lerch, F., S. M. Klosko, R. E. Laubscher, and C. A. Wagner, Gravity model improvement using GEOS 3 (GEM 9 and 10), *J. Geophys. Res., 84*, 3897–3916, 1979.

Lerch, F. J., S. M. Klosko, and G. B. Patel, A refined gravity model from LAGEOS (GEM-L2), *Geophys. Res. Lett., 9*, 1263–1266, 1982.

Levitus, S., Climatological atlas of the world ocean, *NOAA Prof. Pap. 13*, 173 pp., U.S. Dep. of Commerce, Rockville, Md., 1982.

Mader, G. L., A revised 5′ gravimetric geoid and associated errors for the North Atlantic calibration area, *NASA Contract Rep. 156851*, 32 pp., 1979.

Marsh, J. G., and E. S. Chang, 5′ detailed gravimetric geoid in the northwestern Atlantic Ocean, *Mar. Geod., 1*, 253–261, 1978.

Mather, R. S., C. Rizos, and R. Coleman, Remote sensing of surface ocean circulation with satellite altimetry, *Science, 205*(4401), 11–17, 1979.

Mather, R. S., R. Coleman, and B. Hirsh, Temporal variations in regional models of the Sargasso Sea from GEOS 3 altimetry, *J. Phys. Oceanogr., 10*, 171–185, 1980.

McWilliams, J. C., and J. H. S. Chow, Equilibrium geostrophic turbulence, I, A reference solution in a beta-plane channel, *J. Phys. Oceanogr., 11*, 921–949, 1981.

Menard, Y., Observation of the eddy field in the northwest Atlantic and northwest Pacific by Seasat altimeter data analysis, *J. Geophys. Res., 88*, 1853–1866, 1983.

Munk, W., and C. Wunsch, Observing the ocean in the 1990's, *Philos. Trans. R. Soc. London, Ser. A, 307*, 439–464, 1982.

Ocean Tomography Group, A demonstration of ocean acoustic tomography, *Nature, 299*, 121–125, 1982.

Parke, M. E., Detection of tides on the Patagonian Shelf by the Seasat satellite radar altimeter: An initial comparison, *Deep Sea Res., 27*, 297–300, 1980.

Parke, M. E., and L. R. Stavert, Mean sea surface and variability from Seasat radar altimeter, *J. Geophys. Res.*, in press, 1983.

Rapp, R. H., The determination of geoid undulations and gravity anomalies from Seasat altimeter data, *J. Geophys. Res., 88*, 1552–1562, 1983.

Robinson, A., N. E. Huang, C. D. Leitao, and C. G. Parra, A study of the variability of ocean currents in the northwestern Atlantic using satellite altimetry, *J. Phys. Oceanogr., 13*, 565–585, 1983.

Roemmich, D., and C. Wunsch, On combining satellite altimetry with hydrographic data, *J. Mar. Res., 40*(suppl.), 605–619, 1982.

Stanley, H. R., The GEOS 3 project, *J. Geophys. Res., 84*, 3779–3783, 1979.

Stewart, R. H., *Methods of Satellite Oceanography*, University of California Press, Berkeley, in press, 1983.

Tai, C. K., On determining the large-scale ocean circulation from satellite altimetry, *J. Geophys. Res.*, in press, 1983.

Tai, C. K., and C. Wunsch, Absolute measurement by satellite altimetry of dynamic topography of the Pacific Ocean, *Nature, 301*(5899), 408–410, 1983*a*.

Tai, C. K., and C. Wunsch, An estimate of global absolute dynamic topography, *J. Phys. Oceanogr.*, in press, 1983*b*.

Tapley, B. D., G. H. Born, and M. E. Parke, The SEASAT altimeter data and its accuracy assessment, *J. Geophys. Res., 87*, 3179–3188, 1982.

Thompson, J. D., G. H. Born, and G. A. Maul, Collinear-track altimetry in the Gulf of Mexico from Seasat: Measurements, models, and surface truth, *J. Geophys. Res., 88*, 1625–1636, 1983.

TOPEX Science Working Group, Satellite altimetric measurements of the ocean, *Doc. 400-111*, Jet Propul. Lab., Pasadena, Calif., 1981.

Wunsch, C., Low-frequency variability of the sea, in *Evolution of Physical Oceanography: Scientific Surveys in Honor of Henry Stommel*, edited by B. A. Warren and C. Wunsch, pp. 342–374, MIT Press, Cambridge, Mass., 1981.

Wunsch, C., and E. M. Gaposchkin, on using satellite altimetry to determine the general circulation of the oceans with application to geoid improvement, *Rev. Geophys. Space Phys.*, *18*, 725–745, 1980.

Wyrtki, K., Fluctuations of the dynamic topography in the Pacific Ocean, *J. Phys. Oceanogr.*, *5*, 450–459, 1975.

Wyrtki, K., L. Magaard, and J. Hagar, Eddy energy in the oceans, *J. Geophys. Res.*, *81*, 2641–2646, 1976.

Zlotnicki, V., The oceanographic and geoidal components of sea surface topography, Ph.D. thesis, 194 pp., Woods Hole Oceanogr. Inst., Mass. Inst. of Technol., Woods Hole, Mass., February 1983.

Lee-Lueng Fu, Jet Propulsion Laboratory, California Institute of Technology, 4800 Oak Grove Drive, Pasadena, CA 91109.

(Received June 2, 1983;
accepted August 22, 1983.)

REVIEWS OF GEOPHYSICS AND SPACE PHYSICS, VOL. 18, NO. 1, PAGES 243–268, FEBRUARY 1980

On Charting Global Ocean Tides

ERNST W. SCHWIDERSKI

U.S. Naval Surface Weapons Center, Dahlgren, Virginia 22448

This review article highlights the three-century development of our scientific understanding of ocean tides, culminating through myths, paradoxes, and controversies in a global tide model that now permits the prediction of the instantaneous total tide anywhere in the open oceans with an accuracy of better than 10 cm. All major aspects of tidal research, including empirical, mathematical, and empirical-mathematical methods, are considered. Particular attention is drawn to the most recently developed computerized techniques comprehending hydrodynamical dissipation and secondary tide-generating forces, finite-differencing schemes, geometric boundary and bathymetry modeling, and hydrodynamical interpolation of properly selected empirical tide data. Numerous computer experiments are mentioned that were carried out by various researchers in order to evaluate the magnitudes of the featured effects. Further possible improvements are mentioned, especially in nearshore areas, in the Arctic Sea, and near Antarctica, where empirical tide and bathymetry data are either rough or marginal.

CONTENTS

1. INTRODUCTION

As is reflected in art, music, and poetry, the restless sea with its surfing waves, swirling currents, and swelling tides has fascinated and frightened mankind from its beginning. All through the ages the mighty erosive and devastating power of ocean tides flooding shorelands and seaports troubled land developers, harbor engineers, and seamen. The observed periodic feature of tides very early led laymen tidalists to record, analyze, and predict high and low tides in coastal waters and to warn and prepare people to take appropriate precautions. Indeed, quite satisfactory tide tables were constructed by simple rules of thumb. Nowadays, in seashore areas, newspapers, radio and television broadcasts, and special bulletins publish daily, monthly, and longer-range tide predictions of interest to many users.

Up to recent years, practical interest in ocean tides was essentially confined to coastal waters. With the advancement of science and technology the need for extremely accurate tide predictions in all the world oceans has become an urgent problem. In fact, ocean tides represent fluctuating loads on the solid earth, which cause tilting of the crust and disturbances of the earth's stress and gravity fields. The precise knowledge of these loads permits researchers to determine important hydrodynamical parameters of the oceans and elastic parameters of the solid earth. Similarly, interactions between ocean tides and atmosphere, celestial bodies, and artificial satellites and missiles can be studied with high accuracy.

Recently, the National Aeronautics and Space Administration and the Department of Defense joined in the common mission to map the geoid at sea by satellite altimetry up to about 10-cm accuracy. However, the geoidal (rest) sea level is hydrodynamically disturbed by ocean tides and other currents due to pelagic density variations and atmospheric surface forces. This review deals only with 'ocean tides' and their corresponding 'tidal currents.' (Other 'general ocean currents or circulations' are mentioned when it is appropriate.) The tidal undulations of the sea surface topography reach amplitudes of more than 100 cm in open oceans and 200 cm in coastal waters. Thus ocean tides (and currents) need to be determined with an accuracy compatible with the desired geoid accuracy in order to provide effective corrections for altimeter measurements.

In brief, contemporary researchers of today in geophysics, geodesy, oceanography, meteorology, astronomy, and space technology all pose the same challenging problem: we need to know, for any given time and place of the world oceans, the tidal elevation of the sea surface over its geoidal level within 10-cm accuracy. It may be emphasized that this modern problem specifies an extraordinary concrete numerical accuracy for its solution. Therefore it is no longer purely an exclusive academic task which is solely concerned with the modeling of general qualitative features of ocean tides.

In the past three centuries, since *Newton* [1687], oceanographers, physicists, and mathematicians devoted an almost unbroken effort to develop realistic models of tides in the global or limited world oceans by (1) empirical methods, that is, by analyzing and extrapolating vast numbers of tidal measurements recorded at continental, island, and deep-sea stations around the world; (2) theoretical methods, that is, by tapping the wealth of analytical and numerical mathematical tools designed for the integration of physically defined hydrodynamical equations governing tidal currents; and (3) hydrodynamical interpolation methods, which combine the purely empirical and theoretical methods 1 and 2 into one consistent technique.

The continuous interest in the theory of ocean tides was enhanced by an international contest conducted by the Paris Academy of Sciences in 1738 [*Cantor*, 1901] with such leading mathematicians as Bernoulli, Euler, and Maclaurin competing for the honor of producing the best tidal model. Yet the understanding of the captivating phenomena of ocean tides progressed slowly through myths, paradoxes, and heated controversies. Still, in Grenoble an *International Working Group on Ocean Tides* [1975] sponsored by the International Association of the Physical Sciences of the Oceans, the Scientific Committee on Oceanic Research of the International Council of Scientific Unions, and Unesco dissolved itself after a 10-year unsuccessful attempt to map the global ocean tides. The constructed tidal charts predicted low tides for high tides and vice versa over large ocean areas and fell considerably short of the desired numerical accuracy [cf. *Cartwright,* 1977].

Nevertheless, the qualitative features of the tide models published by, for example, *Hansen* [1966], *Tiron et al.* [1967], *Pekeris and Accad* [1969], *Zahel* [1970, 1973, 1975], *Hendershott* [1972], *Gordeyev et al.* [1973], and *Estes* [1975] did suggest that significant progress in the understanding of the leading characteristic properties had indeed been made. Subsequently, *Estes* [1977], *Gordeyev et al.* [1977], *Zahel* [1977], *Parke and Hendershott* [1979], and *Accad and Pekeris* [1978] improved their earlier models by inclusion of secondary tide-generating forces. Supported by extensive computer experiments, *Schwiderski* [1976, 1978*a*, *b*, 1979*a*, *b*, *c*] improved the dissipative, geometrical, and numerical features of the Hansen-Zahel-Estes method and interpolated hydrodynamically over 2000 empirical tide data. The constructed semidiurnal principal moon (M_2) tide is believed to satisfy the desired accuracy anywhere in the open oceans. In fact, the *International Workshop on Interaction of Marine Geodesy With Inertial, Tidal, and High Frequency Aspects of Ocean Dynamics* [1978] concluded in its final report that techniques are now available to compute the total open ocean tides within the 10-cm accuracy limit.

The following treatise highlights the development of our scientific understanding of ocean tides, culuminating in a detailed qualitatively and quantitatively accurate global tidal model. The hydrodynamical-mathematical techniques used in ocean tidal research are, of course, closely related to those developed in the more general field of modeling laminar and turbulent motions of viscous fluids such as ocean currents and atmospheric circulations. The special field of ocean tides, which are distinguished by their simplifying periodic properties and by their widely observed boundary values, may be considered a proving ground for other more involved flow problems.

2. Elementary Approximations

a. Harmonic Decomposition of Equilibrium and Ocean Tides

Very early scientists linked the periodic feature of ocean tides to the apparent motions of the moon and sun. The first physical explanation of tides was given by *Newton* [1687], who introduced the still important 'equilibrium theory of tides' as a direct consequence of his newly discovered theory of gravitation. Owing to the inverse square law of mass attraction, the (true) moon (or sun) MT (see Figure 1*a*) is assumed to attract instantaneously the sea surface point A to A* and, somewhat less, point B to B*. At this stage it is commonly concluded that there should be one high tide per day, contrary to the vast majority of observations. However, since the center C of the solid earth is also attracted to about the midpoint C* between A* and B*, one finds, under conservation of mass, Newton's equilibrium tide shown in Figure 1*b*. It displays high tides on both sides of the earth as if they were generated by two opposing moons, a true moon MT and a virtual one MV. This simple theory resolved the first tidal paradox that there are two high tides daily, as is observed in most ocean areas. On the other hand, it supported an erroneous myth that high tide should occur directly under the moon.

Strictly, Newton's equilibrium tide provides the tidalist only with the exact astronomical tide-generating potential, to which the ocean responds hydrodynamically in a complex way. The equilibrium theory of tides was perfected in 1738 during the contest of the Paris Academy of Sciences mentioned in section 1. The theoretical and empirical investigations of ocean tides were greatly aided by Lord Kelvin [*Thomson,* 1868], who, following an earlier suggestion by Laplace, introduced the so-called 'method of harmonic analysis' into tidal research. The astronomical tide-generating potential or equilibrium tide is decomposed into an almost periodic (nonharmonic frequency spectrum) series of harmonic partial tides with each constituent featuring some property of the elliptic and inclined orbits of the moon and earth. *Darwin* [1883] extended this expansion up to 39 terms, and *Doodson* [1921] carried it on to 400 frequency modes by introducing his ingenious Doodson numbers. A novel nonharmonic and, subsequently, harmonic decomposition of the equilibrium tide was derived by *Cartwright and Tayler* [1971].

In directly applicable terms the equilibrium tide (or astronomical tide-generating potential) is decomposed into the series

$$\eta = \sum_{\nu=0} \eta_\nu(\lambda, \theta, t) \qquad (1)$$

with the following three major species of harmonic partial tides:

Semidiurnal equilibrium tides

$$\nu = 2 \qquad \eta = K \sin^2 \theta \cos (\sigma t + \chi + 2\lambda) \qquad (2)$$

Diurnal equilibrium tides

$$\nu = 1 \qquad \eta = K \sin 2\theta \cos (\sigma t + \chi + \lambda) \qquad (3)$$

Long-period equilibrium tides

$$\nu = 0 \qquad \eta = K(3 \sin^2 \theta - 2) \cos (\sigma t + \chi) \qquad (4)$$

In these equations the following notations are used:

- t universal standard time, seconds;
- λ east longitude;
- θ colatitude;
- η total or partial equilibrium tides, meters;
- $G\eta$ total or partial astronomical tide-generating potential;
- G gravity acceleration, equal to 9.81 m/s^2;
- K amplitude of partial equilibrium tide (Table 1), meters;
- σ frequency of partial equilibrium tide (Table 1), s^{-1};
- χ astronomical argument of partial equilibrium tide relative to Greenwich midnight (Table 1).

According to the decomposition of the total equilibrium tide (equations (1)–(4)), each harmonic component is completely determined by the three constants (K, σ, χ), listed in Table 1 for the 11 leading modes. As can be verified by other tabulations [e.g., *Dietrich,* 1963], all other tidal components fall below 4% of the dominating semidiurnal principal moon (M_2) tide and need not be considered in present global ocean tide models. At this point it may be mentioned that the magnitude of the partial equilibrium tide does not necessarily re-

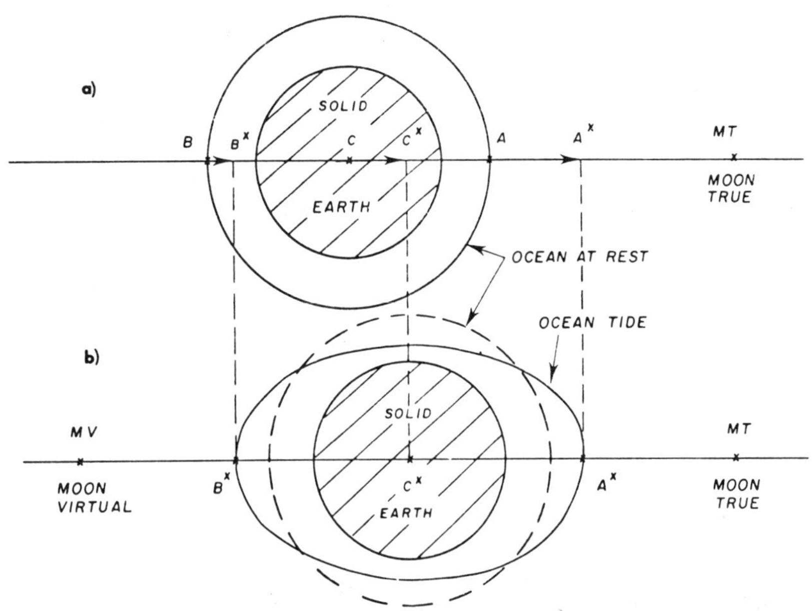

Fig. 1. Newton's equilibrium tide.

flect the magnitude of the corresponding ocean tide. Incidentally, the astronomical argument χ is actually (slightly) time dependent (see Table 1) and requires some updating at least annually.

The harmonic expansion of the tide-generating potential is evidently of fundamental significance for the mathematical analysis and prediction of ocean tides. If one assumes that ocean tides are governed by linear or almost linear equations of motion (equations (18) and (19) or (28) and (29)), then the total oceanic tide is also representable as a linear superposition of harmonic modes. Every harmonic component of the equilibrium tide (equations (2), (3), and (4))

$$\eta = K(\theta) \cos (\sigma t + \chi + \nu\lambda) \qquad \nu = 2, 1, 0 \tag{5}$$

generates through the ocean's response a similar oceanic partial tide

$$\zeta = \xi(\lambda, \theta) \cos [\sigma t + \chi - \delta(\lambda, \theta)] \tag{6}$$

of identical frequency σ differing only in the two harmonic constants $\xi(\lambda, \theta)$, the amplitude of the oceanic partial tide (in meters), which measures the half range between high and low waters, and $\delta(\lambda, \theta)$, the Greenwich phase of the oceanic partial tide, which measures the retardation time (in radians, degrees, or time units) of the oceanic tide ζ, say, relative to the forcing equilibrium tide η at Greenwich meridian $\lambda = 0$.

Owing to the assumed linearity of the ocean's response, every oceanic tidal component is hydrodynamically decoupled from all others and can be constructed without any knowledge of the others. Hence without loss of generality the following considerations may be confined to one particular tidal constituent unless otherwise specified. Under this convention it is convenient to neglect the astronomical argument ($\chi = 0$) in (2)–(6). This follows from the obvious shift of the origin of the universal standard time t by the individual substitution

$$t \rightarrow t - \chi/\sigma \tag{7}$$

TABLE 1. Constants of Major Tidal Modes

	Tidal Mode	K, m	σ, 10^{-4}/s	χ, deg
		Semidiurnal Species, $\nu = 2$		
M_2	principal lunar	0.242334	1.40519	$2h_0 - 2s_0$
S_2	principal solar	0.112841	1.45444	0
N_2	elliptical lunar	0.046398	1.37880	$2h_0 - 3s_0 + p_0$
K_2	declination luni-solar	0.030704	1.45842	$2h_0$
		Diurnal Species, $\nu = 1$		
K_1	declination luni-solar	0.141565	0.72921	$h_0 + 90$
O_1	principal lunar	0.100514	0.67598	$h_0 - 2s_0 - 90$
P_1	principal solar	0.046843	0.72523	$h_0 - 90$
Q_1	elliptical lunar	0.019256	0.64959	$h_0 - 3s_0 + p_0 - 90$
		Long-Period Species, $\nu = 0$		
Mf	fortnightly lunar	0.041742	0.053234	$2s_0$
Mm	monthly lunar	0.022026	0.026392	$s_0 - p_0$
Ssa	semiannual solar	0.019446	0.003982	$2h_0$

K is the amplitude of the partial tide, σ is the frequency of the partial tide, and χ is the astronomical argument of the partial tide. Here h_0, s_0, and p_0 are the mean longitudes of the sun and moon and the lunar perigee at Greenwich midnight: $h_0 = 279.69668 + 36{,}000.7689930485T + 3.03 \cdot 10^{-4}T^2$, $s_0 = 270.434358 + 481{,}267.883141 37T - 0.001133T^2 + 1.9 \cdot 10^{-6}T^3$, $p_0 = 334.329653 + 4069.0340329575T - 0.010325T^2 - 1.2 \cdot 10^{-5}\,T^3$, where $T = (27{,}392.500528 + 1.0000000356D)/36{,}525$, $D = d + 365(y - 1975) + \mathrm{Int}\,[(y - 1975)/4]$, d is the day number of the year ($d = 1$ for January 1), $y \geq 1975$ is the year number, and Int $[x]$ is the integral part of x.

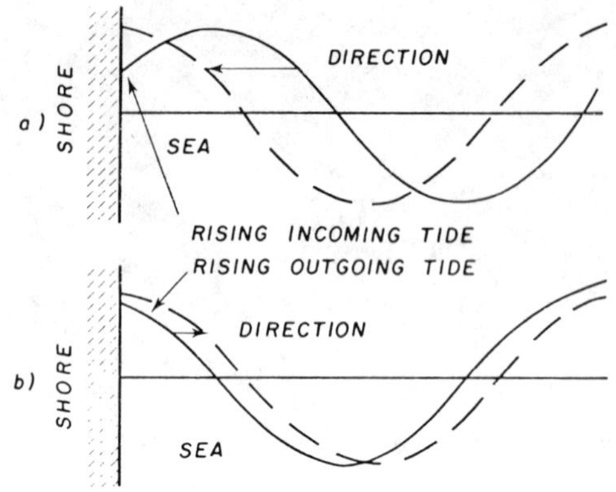

Fig. 2. Rising shore tides. (*a*) Incoming tidal crest. (*b*) Outgoing tidal crest.

Of course, when superpositions of partial tides are considered, the universal standard time and the corresponding astronomical arguments must be reintroduced.

In the new individual time count, all long-period equilibrium tides η ((4) and (5); $\chi = 0$) represent standing waves, which at $t = 0$ are at their highs everywhere in the middle latitude zone $|\phi| < 35.26°$ and at their lows in the outside polar zones with the separating nodal line at about $|\phi| = 35.26°$. All diurnal and semidiurnal equilibrium tides η ((2), (3), and (5); $\chi = 0$) represent irreversible strictly westward progressing waves that crest at $t = 0$ at Greenwich (half) meridian $\lambda = 0°$ or full meridian $\lambda = 0°$ and $\lambda = 180°$, respectively. The equator (not meridian) contour of any semidiurnal equilibrium tide is illustrated in Figure 1*b*.

Without any explicit knowledge of the numerical distribution of the amplitudes and phases (ξ, δ), one concludes from (6) ($\chi = 0$) that all forced oceanic partial tides ζ represent irreversibly progressing waves that crest at the specified time $t = t_0$ at the 'cotidal line': $\delta(\lambda, \theta) = \sigma t_0$. This simple observation clarifies a widely held myth concerning 'incoming and outgoing tides' at beaches. At shore points, ocean tides are either incoming or outgoing without reversals; that is, the tidal crests always move in from sea to shore (Figure 2*a*), or the tides swell to their crests first at the shore and then move out to sea (Figure 2*b*). Of course, the waters always come in from sea, but in a distinctly different fashion for rising incoming and outgoing tides (Figures 2*a* and 2*b*). The different curvatures of the rising incoming and rising outgoing tidal waves seem to indicate more stable tidal currents in the latter case. Hence rising incoming tides may display higher and rougher waters due to bottom disturbances and stresses of winds blowing from sea to shore.

Once the harmonic tidal constants (ξ, δ) are determined for a given geographical location (λ, θ) either by observation and harmonic analysis or by theoretical modeling, then the time-dependent ocean tide ζ can be predicted at any time t by (6). If (ξ_0, δ_0) and (ξ, δ) denote true and empirically or theoretically determined tidal amplitudes and phases, respectively, at any fixed location (λ, θ), then the time-dependent prediction error is

$$\tilde{\zeta} = \zeta_0 - \zeta = \tilde{\xi} \cos(\sigma t - \tilde{\delta}) \tag{8}$$

with

$$\zeta_0 = \xi_0 \cos(\sigma t - \delta_0) \qquad \zeta = \xi \cos(\sigma t - \delta) \tag{9}$$

$$\tilde{\xi}^2 = \xi_0^2 - 2\xi_0\xi \cos(\delta_0 - \delta) + \xi^2 \tag{10}$$

and

$$\tan \tilde{\delta} = \frac{\xi_0 \sin \delta_0 - \xi \sin \delta}{\xi_0 \cos \delta_0 - \xi \cos \delta} \tag{11}$$

Hence, the 'prediction error E' is defined by the error amplitude $\tilde{\xi}$ (maximum error) or more appropriately by the standard deviation

$$E = \mathrm{rms}(\tilde{\zeta}) = \tfrac{1}{2}2^{1/2}\tilde{\xi} \tag{12}$$

Some maximum errors are

$$\tilde{\zeta}_M = \tilde{\xi}_M = \xi_0 + \xi \quad \text{for} \quad \delta_0 - \delta = 180° \tag{13}$$

$$\tilde{\zeta}_M = \tilde{\xi}_M = \xi_0 - \xi \quad \text{for} \quad \delta_0 - \delta = 0° \tag{14}$$

$$\tilde{\zeta}_M = \tilde{\xi}_M = 2\xi \sin \tfrac{1}{2}(\delta_0 - \delta) \quad \text{for} \quad \xi = \xi_0 \tag{15}$$

and

$$\tilde{\zeta}_M = \tilde{\xi}_M = 0.52\xi \quad \text{for} \quad \xi = \xi_0 \quad \text{and} \quad \delta_0 - \delta = 30° \tag{16}$$

$$\tilde{\zeta}_M = \tilde{\xi}_M = \xi \quad \text{for} \quad \xi = \xi_0 \quad \text{and} \quad \delta_0 - \delta = 60° \tag{17}$$

From (16) one concludes the important fact that even if a tide model is exact in amplitude ξ, the maximum prediction error is larger than half the tidal amplitude itself for a phase error of 30° (about 1 hour for semidiurnal tides). For a phase error of 60° (equation (17)) the error exactly equals the tide and renders the tidal model completely useless for predictions. This rather severe result displays the extreme complexity needed to achieve the numerical accuracy specified in section

TABLE 2. Bermuda M_2 Tide Observations

Station	Geographic Coordinates °N	°W	ξ, cm	δ, deg	Reference
St. George's Island	32.55	64.70	36	359	*British Admiralty* [1977]
St. David's Island	32.37	64.65	34	355	*British Admiralty* [1977]
Great Sound	32.32	64.83	38	6	*British Admiralty* [1977]
St. George's Island	32.37	64.70	35	0	*National Ocean Survey* [1942]
St. George's Island	32.40	64.70	37	0	*Perkeris and Accad* [1969]
St. George's Island	32.37	64.70	36	359	*Zahel* [1970]
St. George's Island	32.40	64.70	36	358	*Zetler et al.* [1975]
Deep Sea (GOBI IV)	32.29	64.50	38	11	J. T. Kuo (personal communication, 1977)
Deep Sea (GOBI IV)	32.02	64.43	36	358	F. D. Malone (personal communication, 1978)

Here ξ is the tidal amplitude, and δ is the tidal phase relative to Greenwich.

1. Of course, in regions of sufficiently small amplitudes, any phase error is acceptable.

At this point it must be mentioned that in shallow coastal areas of rapidly varying ocean tides, nonlinear hydrodynamical effects may be significant and must be accounted for in one way or another (section 4*c*). There one must remember that any nonlinearity generates higher harmonic frequencies and causes involved interactions between different tidal modes and also other ocean currents. In this connection, one may compare the introductory remarks to the *British Admiralty* [1977] tide tables.

b. Empirical Methods

Over many years, massive tidal records have been taken with about 10,000 tide gauges placed around the world at continental and island stations. Some of these records contain tidal time sequences longer than half a century with more than 10 million hourly readings. In recent years, *Eyries* [1968], *Snodgrass* [1968], *Filloux* [1969], and others developed sensitive deep-sea tide (pressure) gauges that can be placed almost anywhere on the ocean floor to record tidal variations for up to 1 year. Such deep-sea instrument packages measure also tidal current velocities, water temperatures, and other oceanographic quantities.

Most of the voluminous data banks have been harmonically analyzed by classical least squares methods in order to extract the important time-independent amplitudes and phases of the leading oceanic partial tides. The results have been collected and are now available from, for instance, the *National Ocean Survey* [1942], the *International Hydrographic Bureau* [1966], and the *British Admiralty* [1977]. Unfortunately, the so-called 'harmonic' analysis is not equivalent to a rigorous Fourier analysis and hence is affected by subjective and controversial errors. In particular, the time sequences are noisy because of variable recording inaccuracies and other pelagic and atmospheric disturbances of the geoidal sea level. The nonharmonic spectrum of the total tide has no true fundamental period and consists of bands of closely clustered frequencies which are difficult to separate.

As is mentioned in the introduction to the *British Admiralty* [1977] tide tables, nonlinear interactions between tidal modes and other ocean currents may be strong and long lasting, especially in shallow coastal waters. Since rough tidal data are recorded uniformly in various corners of the world oceans (between Europe and the British Isles, Nova Scotia and Cape Code, and China and Korea; also northwest of Australia, on the Patagonian shelf, at the mouth of the Amazon river, etc.), one cannot simply dismiss such data as false recordings of 'certain' unreliable stations. These data clearly reflect the highly turbulent (eddying) nature of the oceanic tidal currents and must be properly accounted for in any realistic tidal model (section 4*c*). Indeed, significant distortions and retardations (some more than 60°) of ocean tides have also been recognized by researchers [e.g., *Harris,* 1904; *Bogdanov,* 1961*a*, *b*; *Defant,* 1961; *Gallagher et al.,* 1971; *Luther and Wunsch,* 1975; and *Schwiderski,* 1978*b*, 1979*b*] (see Tables 4*a* and 4*b*) near isolated islands and island chains or ocean ridges such as the Caribbean, Hawaiian, and Aleutian ridges. As was demonstrated by *Schwiderski* [1978*b*, 1979*b*, section 5*a*], the latter deformations have a profound effect on the global shape of open ocean tides.

Most of the mathematical shortcomings of the classical harmonic analysis can be considerably reduced by using sufficiently long tidal time sequences. However, with shorter observations, significant errors (especially in the phase constants) may be encountered (see Table 2). For example, *Irish and Snodgrass* [1972] mention the differences in the harmonic constants (e.g., 25° in the S_2 phase angle) which Radok arrived at by applying the harmonic analysis to two short (19 days in January and 11 days in May) tidal recordings taken by *Noye and Easton* [1967] at the same station on the shelf of southern Australia. Nevertheless, the great majority of the tide data listed, for example, in the most recent *British Admiralty* [1977] tables (see their introduction) probably permit a tide prediction with the 10-cm accuracy posed in section 1. Naturally, some discrepancies between different tabulations must be expected, but only surprisingly few were found by the author [*Schwiderski,* 1978*b*, 1979*b*] at island stations. In fact, to the author's pleasant surprise, almost all of the seemingly incoherent M_2 tidal data, recorded at numerous clustered and scattered islands of the world oceans, permitted an unexpectedly realistic hydrodynamical integration into one coherent tidal topography (section 5*a*). The high quality of the tidal observations and their analyses and recordings is also supported by more recent measurements reported, for instance, by *Luther and Wunsch* [1975] and K. Wyrtki (private communication, 1978) for the Pacific, which do not deviate significantly from available older listings.

As a partial project of the international working group mentioned in section 1, *Munk and Cartwright* [1966] introduced the so-called 'ocean response method' in order to analyze relatively short time records with sufficient accuracy. Though this method is basically nonharmonic, it too yields as its end product the most desired harmonic constants. The local time series of the ocean tide is considered as a convolution of the tide-generating potential and a transfer function (complex response weights) which is, a priori, not known and hence must be determined in an optimum sense [*Cartwright,* 1968, 1969; *Cartwright et al.,* 1969]). By using analyzed coastal reference stations, satisfactory results are being claimed for rather short recording times of 2 weeks and even 1 week. While such modern filter techniques can be very successful, they are highly subjective and sometimes controversial. In any case, they rely to a considerable degree on practical experience. In this respect the comparative studies conducted, for instance, by *Munk and Cartwright* [1966], *Nowroozi* [1972], *Larsen and Irish* [1975],) J. T. Kuo (private communication, 1977), and F. D. Malone (private communication, 1978) (see Table 3) are most instructive.

Analyzed deep-sea tidal measurements have been published by *Munk et al.* [1970], *Irish et al.* [1971], *Irish and Snodgrass* [1972], *Nowroozi* [1972], *Mofjeld* [1975], *Pearson* [1975*a*, *b*], *Zetler et al.* [1975], *Larsen and Irish* [1975], *Gill and Porter* [1979], and others. Additional measurements are being accumulated on a continued basis by D. E. Cartwright (private communication, 1978) around the British Isles (see also *Cartwright* [1971, 1977]), J. T. Kuo (private communication, 1977), F. D. Malone (private communication, 1978), S. K. Gill and D. L. Porter (private communication, 1978), off the U.S. Atlantic coast, and K. Wyrtki (private communication, 1978) at Pacific islands. Some of the deep-sea M_2 tide data known to the author are listed in Table 3 (section 5*a*) or illustrated in Figure 3. A complete collection in book form of all open ocean tide data (about 108 stations) known today has been

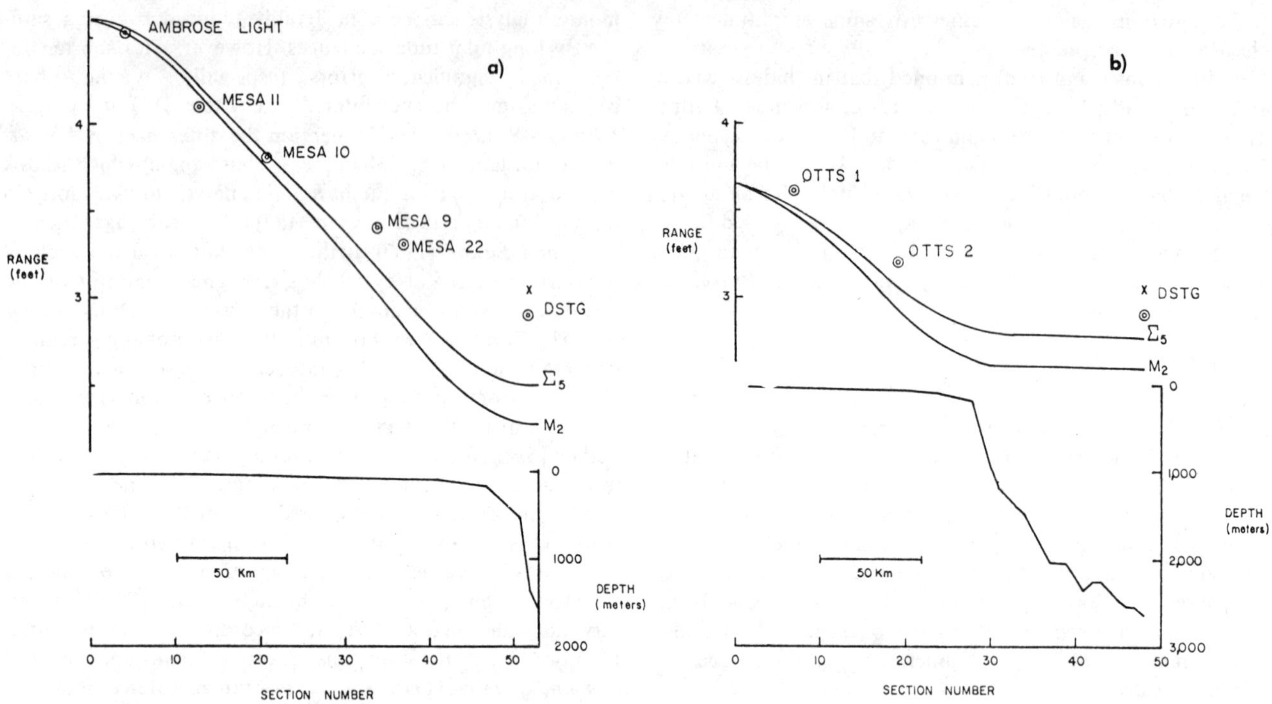

Fig. 3. *Gill and Porter's* [1979] profile measurements of decaying tidal ranges across a continental sneii. (*a*) Southeast of Sandy Hook, New Jersey. (*b*) Southeast of Bethany Beach, Delaware. The curve labeled M_2 is the Defant shelf model, and that labeled $\sum_5$ is the Gill-Porter shelf model. Crosses indicate the Schwiderski global model (average cell value (see Table 4*a*)).

prepared by *Cartwright et al.* [1979]. Finally, a large collection of coastal tide data around the islands of Japan compiled by *Miyazaki et al.* [1967] must be mentioned here.

By simple inspection of numerous observed ocean tide data and experienced intuition, 'cotidal' (equiphase) and some 'corange' (equiamplitude) maps have been constructed for marginal seas and partial or worldwide oceans by *Whewell* [1833, 1848], *Berghaus* [1845], *Harris* [1904], *Sterneck* [1920, 1921], *Dietrich* [1944*a*, *b*], *Villain* [1952], *Bogdanov* [1961*a*, *b*], *Luther and Wunsch* [1975], and others. All of these maps give a qualitative overview of the phenomena of ocean tides, and some of them have achieved a remarkable degree of reality, at least in certain ocean areas.

A more systematic empirical method to map large-scale ocean tides was proposed and demonstrated by *Kuo et al.* [1970*a*, *b*, 1977] and *Jachens and Kuo* [1973]. In their 'inversion method,' modern optimization techniques are utilized to fit low-degree polynomials simultaneously to observed ocean tide data and to measured oceanic tidal disturbances of the solid earth gravity field, which are correlated through well-known interaction equations. While this approach is definitely more reliable than purely intuitive attempts, it appears to be constrained by rather low-degree polynomials, which can hardly accommodate the various tidal undulations known from observations (section 5*a*). The method offers no physical insight into the tidal phenomena and yields no information on tidal currents, which is needed in certain applications. Aside from these shortcomings the method seems to yield promising results in limited ocean regions.

Open ocean measurements of tides are obviously of significant value to empirical and theoretical tide models, either as input or solely as calibration data. Future measurements should be encouraged, particularly in areas of high tidal amplitudes and even more in arctic and antarctic regions, where reliable measurements of tides (and depths) are sparse. Recently, *Gill and Porter* [1979] reported on the first profile measurements of tidal decay from shallow shore points to the edge of the continental shelf. They confirmed experimentally the long-believed drastic decay of tidal heights (Figure 3) downward on continental shelves. Precisely the same important phenomenon has been brought out independently by the author's hydrodynamical interpolation technique (see sections 4*c* and 5*a*, *Schwiderski* 1978*b*, 1979*b*], and Tables 4*a* and 4*b*). Clearly, more profile measurements of the same Gill-Porter type are needed (between Cape Cod and Nova Scotia, at the mouth of the Amazon river, on the Patagonian shelf, etc.) to improve and/or check the accuracy of tide models.

c. *Elementary Solutions of Laplace's Tidal Equations*

Although *Newton* [1687] clearly recognized the distinctly hydrodynamical nature of ocean tides, it was *Laplace* [1775] who formulated the first hydrodynamical equations of ocean tidal motions. The 'Laplace tidal equations' (LTE's) can be written in the form

$$U_t = \frac{GH}{R \sin \theta}(\eta - \zeta)_\lambda + 2\Omega V \cos \theta \qquad (18a)$$

$$V_t = \frac{GH}{R}(\zeta - \eta)_\theta - 2\Omega U \cos \theta \qquad (18b)$$

$$\zeta_t + \frac{1}{R \sin \theta}[U_\lambda - (V \sin \theta)_\theta] = 0 \qquad (19)$$

by using the following notations in addition to those introduced in section 2*a*:

R radius of the earth, equal to 0.637×10^7 m;

Ω earth angular velocity, equal to 0.72722×10^{-4} s^{-1};

H (static) ocean depth, equal to $H(\lambda, \theta)$;
(u, v) east north velocities;
$(U, V) \approx$ (depth integrated) east and north velocities, equal to $H \cdot (u, v)$.

These equations can be obtained from the more complete 'continuous ocean tidal equations' (COTE's) derived in full detail by *Schwiderski* [1978*b*, 1979*a*] by simply neglecting all eddy dissipation, bottom friction, and all terrestrial and oceanic tidal load effects in (28) and (29). The LTE's consider obviously an inviscid, incompressible, and single-layer ocean subject only to linearized inertial, potential, and Coriolis forces generated by the primary astronomical tide-producing potential $G\eta$ specified by (1)–(4). Of course, since the oceanic tidal motion is highly turbulent, the velocity (U, V) and the tidal height ζ must be considered as 'averaged' in some (usually tacitly bypassed) sense (see section 3*a* and *Schwiderski* [1978*b*, 1979*a*].

If one considers a single partial tide η specified by its frequency σ (equations (2)–(5)), one can eliminate the complex time dependence by the substitution

$$(\eta, \zeta, U, V) = (\bar{\eta}, \bar{\zeta}, \bar{U}, i\bar{V})e^{i\sigma t} \qquad (20)$$

and arrive at the complex amplitude equations

$$\sigma\bar{U} = \frac{iGH}{R \sin\theta}(\bar{\zeta} - \bar{\eta})_\lambda + 2\Omega\bar{V}\cos\theta \qquad (21a)$$

$$\sigma\bar{V} = \frac{GH}{R}(\bar{\eta} - \bar{\zeta})_\theta + 2\Omega\bar{U}\cos\theta \qquad (21b)$$

$$\sigma\bar{\zeta} = \frac{1}{R\sin\theta}[i\bar{U}_\lambda + (\bar{V}\sin\theta)_\theta] \qquad (22)$$

The reduced equations can obviously be combined into a single elliptic partial differential equation for the complex tidal amplitude $\bar{\zeta}$, which is everywhere regular except at the colatitudes θ determined by

$$\sin\theta = 0 \text{ (poles) or } \cos\theta = \sigma/2\Omega \qquad (23)$$

where regular solutions exist.

Numerous solutions of the LTE's were constructed by analytic and semianalytic methods for idealized ocean basins, estuaries, and channels. Realistic results were sought by many well-known researchers, such as *Laplace* [1775], *Airy* [1842], *Ferrel* [1874], *Thomson* [1879], *Hough* [1897], *Poincaré* [1910], *Lamb and Swain* [1915], *Proudman* [1915], *Proudman and Doodson* [1927], *Platzman* [1971, 1972*a*, *b*, 1975], and *Miles* [1974]. Most of this rich and sometimes controversial work has been collected and analyzed in many books and reviews [e.g., *Bartels,* 1957; *Defant* 1957; *Fomin,* 1964; *Hopfner,* 1931; *Lamb* 1932; *Neumann and Pierson,* 1966; *Stern,* 1975; *Stommel,* 1960].

The enormous difficulties, that were encountered with the LTE's, may be displayed by the following elementary solutions [*Schwiderski,* 1978*b*, 1979*a*; *Hendershott,* 1972]. As can be verified by simple substitution, the equations ((21) and (22)) for a stationary earth ($\Omega = 0$) of globally constant depth ($H = H_0$) yield the following exact solutions:

Semidiurnal tides ($\nu = 2$)

$$\eta = K\sin^2\theta c^{i(\sigma t + 2\lambda)} \rightarrow \zeta = -\hat{K}\eta \qquad (24)$$

Diurnal tides ($\nu = 1$)

$$\eta = K\sin 2\theta\, e^{i(\sigma t + \lambda)} \rightarrow \zeta = -\hat{K}\eta \qquad (25)$$

Long-period tides ($\nu = 0$)

$$\eta = K(3\sin^2\theta - 2)e^{i\sigma t} \rightarrow \zeta = +\hat{K}\eta \qquad (26)$$

where

$$\hat{K} = 6GH_0/(\sigma^2R^2 - 6GH_0) \qquad (27)$$

These simple integrals exhibit the following important properties of ocean tides:

Regularity and long periodicity of arctic tides. Near the north pole, arctic tides are regular and essentially of the long-period species. This significant feature holds true even for dissipative oceans on the rotating realistic earth [*Schwiderski,* 1978*b*, 1979*a*].

Resonance catastrophe of Laplacian hydrodynamic tides. On a stationary or rotating earth, Laplacian hydrodynamic tides are subject to resonance catastrophes at certain depths.

Second tidal paradox [Lamb, 1932]. On a stationary earth of modest ocean depth, all Laplacian hydrodynamic tides are inverted, i.e., lag 180° behind the corresponding Newtonian equilibrium (static) tides.

The dramatic 180° turnabout of Newton's equilibrium tide (section 2*a* and Figure 1*b*) by the Laplacian hydrodynamic tide with an inherent resonance menace is perhaps the most stunning result of the classical theory of ocean tides. The controversy between high and low tides under the moon (or sun) was, of course, quickly resolved by empirical studies and by considering more realistic ocean basins on a rotating earth. Since the forced ocean tides cannot reverse themselves (section 2*a*), they were found rotating around so-called amphidromic points (section 5*a*, Table 4, and Figure 8) due to the presence of the continents and the Coriolis force. Nevertheless, as can be verified by comparing the recent tidal maps published by experimental and theoretical tidalists (see, for example, the review collections of Hendershott [1973, 1977]), the fundamental controversy between high and low tides over certain significant areas (north central Pacific, South Atlantic, etc.) persisted as hot as ever till today.

Naturally, in realistic ocean basins, no eigensolution of the LTE's (18) can exactly match the forcing equilibrium tide, which precludes any total resonance catastrophe. Nevertheless, in the Laplacian theory of tides 'near resonance' is possible at certain ocean depths of the real ocean basin. Indeed, numerical tidalists [e.g., *Pekeris and Accad,* 1969; *Hendershott,* 1972, 1975, *Parke and Hendershott,* 1979; *Accad and Pekeris,* 1978] using the LTE's without any dissipation terms (sections 3*a*–3*c*) clearly felt the resonance threats as instabilities of their numerical procedures.

3. MODELING CONTINUOUS OCEAN TIDAL EQUATIONS

a. Continuous Ocean Tidal Equations

The difficulties with the Laplacian hydrodynamical theory of tides (section 2*c*) can only be attributed to the oversimplified LTE's (18) and (19). Indeed, it is hard to see how to overcome the resonance threats at certain ocean depths without assuming an artificial ocean basin, including density stratification, allowing vertical motions [e.g., *Lamb,* 1932; *Cartwright,* 1977], or reintroducing the neglected nonlinear inertial and viscous forces. Obviously, the last possibility yields the most natural resolution of the problem at hand. In fact, it is well known today [e.g., *Schwiderski,* 1972] that while nonlinear inertial forces cause instabilities and turbulence in viscous fluid motions, they also control them and prevent any kind of catastrophe by generating sufficient Reynolds stresses or eddy dissipation.

Even without the resonance problems of the LTE's the necessity for appropriate eddy dissipation forces follows directly from the enormous dimensions of the world oceans in which tidal currents of all 'resolvable' wavelengths are supercritically turbulent. For example, a tidal motion of 0.01-cm^2/s kinematic molecular viscosity, 100-km characteristic wave length (about 1° mesh size (section 4*a*)), and moderate characteristic velocity of 0.1 cm/s possesses a characteristic Reynolds number of 10^8, which [e.g., *Schlichting*, 1968] is well in the turbulent flow regime. Accordingly, any comprehensive study of ocean tidal currents must augment the LTE's (18) by two additional vector terms modeling eddy dissipation and bottom friction.

In his study of tides in the Red Sea, *Grace* [1930*a*] introduced the effects of the solid earth tide into ocean tidal research (see also *Munk et al.* [1970], *Hendershott and Munk* [1970], and *Cartwright* [1971]). Recent gravity measurements [e.g., *Kuo et al.*, 1970*a*, *b*] revealed a significant interaction between the oceanic and terrestrial tides. In spite of such earlier studies it was *Farrell* [1972*a*, *b*, 1973] who suggested that accurate global tide models must account for the mutual disturbances exerted by the terrestrial tide on the oceanic tide and vice versa. Consequently, the primary astronomical tide-generating potential of the LTE's (18) must be extended to include two major additional terms representing secondary effects due to the terrestrial and oceanic tidal loads.

With the important extensions of the LTE's (18) mentioned above, one arrives at the following most general continuous ocean tidal equations ever used in theoretical tide models:

$$U_t = \frac{GH}{r \sin\theta}(\eta - \hat{\eta} - \zeta + \mathfrak{z})_\lambda + 2\Omega V \cos\theta + A^\lambda + B^\lambda \quad (28a)$$

$$V_t = \frac{GH}{R}(\zeta - \mathfrak{z} - \eta + \hat{\eta})_\theta - 2\Omega U \cos\theta + A^\theta + B^\theta \quad (28b)$$

$$R \sin\theta\, \zeta_t\, U_\lambda - (V \sin\theta)_\theta = 0 \quad (29)$$

In these COTE's the notations of the considerably simpler LTE's (18) and (19) are retained with the following additions:

(A^λ, A^θ) lateral dissipation vector (to be modeled in section 3*b*);

(B^λ, B^θ) bottom friction vector (to be modeled in section 3*c*);

$\hat{\eta}$; $\mathfrak{z}$ terrestrial and oceanic tidal-load potentials, respectively (to be modeled in section 3*d*).

A formal complete derivation of the COTE's (28) and (29) from the Navier-Stokes equations of 'averaged' turbulent flow [e.g., *Schlichting* 1968; *Whitaker*, 1968] is given by *Schwiderski* [1978*b*, 1979*a*]. It suffices here to mention only the most important assumptions and simplifications.

1. The sea surface at rest is assumed to be perfectly spherical without any geoidal undulations or atmospheric and density-driven disturbances.

2. The ocean floor at rest is specified by some averaged (section 4*a*) bathymetry subject to perturbations due to the astronomical equilibrium tide and to the oceanic tidal load (section 3*d*).

3. The averaged turbulent velocity components and tidal height (or fluid pressure) are considered as some sort of 'generalized quantities,' which in their discrete form (section 4*b*) are physically defined as mean values ('divided mass fluxes') over some lateral or horizontal mesh cell area Δs (>0) and time interval Δt (>0). The area Δs and the time step Δt must be chosen in an optimum way to allow for a sufficient resolution of all significant wavelengths and tidal periods.

4. The 'stresslike' effects (Reynolds stresses) of the turbulent 'velocity fluctuations' (filtered-out velocity residuals) on the mean flow must be modeled as lateral eddy dissipation (section 3*b*) and bottom friction (section 3*c*) to achieve a macroscopically unique and physically realistic description of the microscopically undetermined turbulent motion under consideration.

5. The hydrostatic pressure assumption is invoked; that is, all quadratic inertial terms, centrifugal accelerations, vertical Coriolis and dissipation forces, and vertical motions are neglected.

6. A single-layer ocean is assumed; that is, because of the depth-independent driving body force, all velocity components are considered to be essentially depth independent.

Attention to the important concept of averaged velocities and tidal height (assumption 3) was drawn by the author [*Schwiderski*, 1978*b*, 1979*a*]. It was argued that the COTE's possess only a vague formal meaning, which helps the numerical analyst to set up mathematically well-defined finite-difference equations (section 4). In those physically approximate but sound discrete ocean tidal equations, all velocities and tidal elevations are rigorously defined as mass fluxes through mesh cell areas Δs during the time interval Δt divided by Δs, Δt, and the corresponding density. Obviously, in contrast to laminar flows the averaged turbulent flow variables depend strongly on the considered values Δs and Δt. The larger Δs and Δt are chosen, the more and larger fluctuating or eddying motions are filtered out from the remaining mean motions. Although the COTE's should always be recoverable from the finite-difference analog by a formal limit process, even in general laminar flows these limits must be understood in some generalized (distribution) sense [e.g., *Ladyzhenskaya*, 1969] with no pointwise meaning. Therefore the main effort of modeling oceanic tidal motions must focus primarily on the discrete ocean tidal equations (DOTE's) (60) and (61), which should represent the true macroscopical effects of the Reynolds stresses as closely as possible.

b. Lateral Eddy Dissipation

As was mentioned in section 3*a*, the COTE's (28) and (29) are derived from the Navier-Stokes equations of formally averaged turbulent flow. These equations contain the so-called Reynolds stress tensor [e.g., *Schlichting*, 1968] in which the velocity fluctuations (filtered-out residuals) manifest their influence on the remaining mean velocity. In spite of an enormous research effort of more than a century this stress tensor remains unknown. Obviously, it reflects the microscopically undetermined nature of turbulent motions, which has been pointed out, for example, by *Schwiderski* [1978*b*, 1979*a*].

Modern experiment [e.g., *Busse and Whitehead*, 1971] and mathematical [e.g., *Schwiderski*, 1972] research has clearly established that in the critical regime, laminar and turbulent motions are no longer uniquely determined by laminarlike differential equations and initial, boundary, and regularity conditions. The occurring motions appear to be governed by hysteresis and pure chance. Also, since the product of generalized (distribution) functions has no mathematical meaning [e.g., *Shilov*, 1968], there seems to exist no meaningful ordinary or generalized function that could possibly represent the Reynolds stress tensor which contains the pointwise undetermined fluctuating velocity residuals in quadratic form.

Nevertheless, the energy-dissipating stresslike effects of the filtered-out velocity fluctuations on the average flow are physically quite apparent and hence must be modeled in some macroscopical sense. The first model of such interior friction forces acting in a turbulent motion was introduced into fluid mechanics by *Boussinesq* [1877], who replaced the unknown Reynolds stress tensor by the known laminar stress tensor, changing only the molecular viscosity to the so-called 'eddy viscosity' with its magnitude at one's disposal. In oceanography this simple substitution was brought into practice by *Ekman* [1902] and was later effectively used by Munk in 1950 [see *Munk*, 1963]. In numerical modeling of general ocean currents and atmospheric circulations the Boussinesq idea has been picked up by many researchers, of whom only few may be named here because their important contributions with direct relevance to ocean tides: *Bryan* [1963], *Smagorinsky* [1963], *Friedrich* [1966, 1970], *Crowley* [1968, 1970], *Leith* [1968], *Cox* [1970], *O'Brien* [1971], and *Holland and Hirschman* [1972] as well as *Döös et al.* [1974] and *Reid et al.* [1975]. In numerical ocean tidal modeling, *Hansen* [1966] pioneered the use of lateral eddy dissipation with a constant eddy viscosity. He was soon followed by his co-workers *Brettschneider* [1967], *Trepka* [1967], and *Zahel* [1970, 1973, 1975, 1977] as well as by other independent workers such as *Marchuk et al.* [1969], *Marchuk et al.* [1973], and *Estes* [1975, 1977].

Ignoring the successes of Boussinesq's unique approach, some researchers still regard it as fictitious and controversial. In the absence of better solutions it seems idle to argue about the physical justification of the Boussinesq substitution. The fact remains that it represents the simplest constitutive equation including zero used 'artificially' by some objecting researchers. Above all, as with any other physical law, the Boussinesq eddy dissipation has successfully passed its crucial test in many practical applications in hydrodynamics, oceanography, and meteorology. The present oceanic tidal motion is no exception (see section 5*a*).

Apparently, the reluctant acceptance of the Boussinesq substitution orginates from the observation that the 'constant' eddy viscosity varies enormously in form and magnitude from investigator to investigator. While the kinematic molecular viscosity of water is of the order of 10^{-2} cm^2/s, oceanic eddy viscosities of the order of 10° to 10^{11} cm^2/s have been quoted [*Cox*, 1970; *Zahel*, 1970]. Nevertheless, the huge variation of the eddy viscosity is physically plausible (see discussion below), and some empirical evidence for the lower value has been discussed by *Munk* [1966]. In this connection it is of interest to know that the mean flow is quite insensitive to fairly large variations (say, 25%) of the eddy viscosity. This observation by *Munk and Palmén* [1951] was confirmed for oceanic tidal motions by the author's [*Schwiderski*, 1978*b*, 1979*a*, *b*] extensive computer experiments. It is probably related to the well-known fact that potential motions satisfy the complete Navier-Stokes equations of laminar flow with any constant viscosity.

As was first realized by *Prandtl* [1925], the eddy viscosity is a characteristic quantity of the turbulent motion under investigation and not just a fluid constant such as the molecular viscosity. In fact, a prescribed eddy viscosity actually specifies the macroscopical (averaged) features of the microscopically undefined turbulent flow. Consequently, the quality of the modeling of any turbulent motion is directly linked to the investigator's realistic choice of the eddy viscosity. The Boussinesq substitution permits the analyst to choose a suitable velocity-dependent or velocity-independent eddy viscosity either uniformly or separately for all three stress directions. If the velocity field were known a priori, then one could determine exact eddy viscosities a posteriori in many ways.

The physically sound meaning of Boussinesq's approximation of the Reynolds stresses was first illustrated by *Prandtl's* [1925] celebrated 'momentum austausch (exchange, mixing) theory' of turbulent flow. With the help of constant 'mixing-length parameters' [e.g., *Schlichting*, 1968] the eddy viscosity (momentum austausch coefficient) is found to be dependent on the averaged flow velocity; that is, the eddy dissipation appears as a quadratic phenomenon. In meteorology and oceanography, velocity-dependent eddy viscosities were derived on physical grounds by *Smagorinsky* [1963], *Leith* [1968], *Crowley* [1968, 1970], *O'Brien* [1971], and others.

The author's [*Schwiderski*, 1978*b*, 1979*a*] early computer experiments with a worldwide, constant eddy viscosity in an ocean of depth varying from 10 to 7000 m clearly indicated that the lateral eddy viscosity must depend linearly on the ocean depth. Eddy viscosities depending on $\frac{1}{2}$ and $\frac{3}{2}$ powers of the depth were also tried, but they failed to yield equally good results. A renewed look at the phenomenon of turbulent motion led to the introduction of a lateral eddy viscosity that depends linearly on a horizontal (cell or mesh size) and a vertical (cell size or depth) mixing length. In view of the planned grid discretization of the spherical ocean area a lateral eddy viscosity was defined by

$$A = (a/2)\, LH(\lambda, \theta)(1 + \mu \sin \theta) \tag{30}$$

where a is the reduced eddy coefficient (in s^{-1}), L is the chosen equatorial mesh length, and μ is the longitudinal grading parameter chosen to achieve more uniform spherical mesh areas (section 4*a*).

As is physically plausible, the novel eddy viscosity A depends on the mean lateral cross-section area

$$\Delta s = HL\,(1 + \mu \sin \theta)/2 \tag{31}$$

of the local flow cell of depth H. Since the mean tidal velocities in the discrete tidal equations (equation (60) and (61)) depend strongly on the mesh area Δs (see section 3*a*), the assumed dependence of A on Δs is physically well justified. Indeed, larger Δs values mean that more fluctuating or eddying motions are filtered out from the mean velocity, which results in more vigorous eddy dissipation. Also, in agreement with physical expectations one concludes from (30) that eddy dissipation is significant in deep ocean basins and negligible in shallow shelf regions. The remaining reduced eddy coefficient a depends probably in some way on the discrete time step Δt (section 3*a*). However, in any particular application, Δt is necessarily a constant, so that a remains to be determined, say, as a uniform constant over all oceans by trial and error to achieve best agreements with observations. In the author's tide model with a bathymetry varying from 10 to 7000 m the eddy viscosity varied as follows:

$$1.3 \cdot 10^3 < A < 1.3 \cdot 10^6 \text{ m}^2/\text{s} \tag{32}$$

which fits the customary range mentioned above very well. Thus the viscosity law (30) explains the huge magnitude differences in the constant eddy viscosity used by numerical analysts in different problems. The fact that the eddy viscosity must depend in some way on the mesh size (resolution) in discrete models of ocean currents was earlier noticed by *Cox* [1970], *Friedrich* [1970], *Holland and Hirschman* [1972], and *Zahel* [1975].

By applying assumptions 1–6 of section 3*a* to the Boussinesq substitution with the eddy viscosity specified by (30) one arrives [*Schwiderski*, 1978*a*, 1979*a*] at the lateral eddy dissipation terms of the COTE's (28) and (29):

$$A^\lambda = \frac{A}{R^2}\left[\frac{U_{\lambda\lambda} + \bar{H}_\lambda U_\lambda}{\sin^2\theta} + U_{\theta\theta} + (\cot\theta - \bar{H}_\theta + \tilde{H}_\theta)U_\theta - H^\lambda U\right] + \frac{A}{R^2 \sin\theta}[\bar{H}_\lambda \tilde{H}_\theta V - (2\cot\theta + \tilde{H}_\theta)V_\lambda] \qquad (33a)$$

and

$$A^\theta = \frac{A}{R^2}\left[\frac{V_{\lambda\lambda}}{\sin^2\theta} + V_{\theta\theta} + (\cot\theta - \bar{H}_\theta + 2\tilde{H}_\theta)V_\theta - H^\theta V\right] + \frac{A}{R^2 \sin\theta}[2\cot\theta U_\lambda - \bar{H}_\lambda U_\theta - (\cot\theta - \tilde{H}_\theta)\bar{H}_\lambda U] \qquad (33b)$$

where

$$H^\lambda = \frac{\bar{H}_{\lambda\lambda} + 1 + H_\lambda^2}{\sin^2\theta} + \bar{H}_{\theta\theta} - \bar{H}_\theta(\bar{H}_\theta - \tilde{H}_\theta) + (\bar{H}_\theta + \tilde{H}_\theta)\cot\theta \qquad (34a)$$

$$H^\theta = \frac{\bar{H}_{\lambda\lambda} + 1}{\sin^2\theta} + \bar{H}_{\theta\theta} + \bar{H}_\theta(\cot\theta - \bar{H}_\theta + 2\tilde{H}_\theta) \qquad (34b)$$

$$\bar{H}_\theta = H_\theta/H \qquad \bar{H}_{\theta\theta} = H_{\theta\theta}/H \qquad \theta \to \lambda \qquad (35)$$

$$\tilde{H}_\theta = \bar{H}_\theta + \mu\cos\theta/(1 + \mu\sin\theta) \qquad (36)$$

These equations include all lateral eddy dissipation terms used so far in ocean tidal models. In particular, for A = const, instead of (30) one obtains the considerably simpler terms used by *Hansen* [1966], *Zahel* [1970, 1973, 1975, 1977], *Marchuk et al.* [1973], *Estes* [1975, 1977], and others. Of course, for $A = 0$, all eddy dissipation is wiped out, as occurs in the COTE's used by most numerical tidalists, such as *Hansen* [1948, 1949], *Pekeris and Accad* [1969], *Hendershott* [1972, 1975], *Parke and Hendershott* [1979], and *Accad and Pekeris* [1978]. It may be mentioned that the author's [*Schwiderski*, 1978*b*, 1979*a*, *b*] extensive computer experiments clearly demonstrated best results for the eddy viscosity law specified by (30) instead of $A = 0$ or A = const >0.

c. Bottom Friction Laws

Through the integration of the simplified Navier-Stokes equations with the Boussinesq replacement of the Reynolds stress tensor (sections 3*a* and 3*b*) carried out over the depth of the single-layer ocean, one arrives (see *Schwiderski* [1978*b*, 1979*a*] for full details) at the bottom friction vector (B^λ, B^θ) entering in the COTE's (28). This stress vector depends directly on the vertical eddy viscosity and is, of course, not known. However, on the basis of hydrodynamical research on wall friction [e.g., *Schlichting*, 1968] the substitutions

$$B^\lambda = -Bu = -BU/H \qquad U = Hu \qquad (37a)$$

and

$$B^\theta = -Bv = -BV/H \qquad V = Hv \qquad (37b)$$

have been widely accepted in many successful applications. Similarily to the eddy viscosity (section 3*b*) the following velocity-independent and velocity-dependent 'bottom friction coefficients' B have been considered:

'Linear law' of bottom friction

$$B = \text{const in velocity } (u, v) \qquad (38)$$

'Quadratic law' of bottom friction

$$B = \gamma(u^2 + v^2)^{1/2} = \gamma(U^2 + V^2)^{1/2}/H \qquad (39)$$

where the values of B or γ, respectively, remain at one's disposal.

While the quadratic law is suitable for general (fast) motions, the linear law is appropriate only for so-called Stokes slow motions [e.g., *Schlichting*, 1968]. In fluid mechanics the physical significance of the quadratic law of wall friction was first pointed out by *Boussinesq* [1896]. Later, *Taylor* [1918] applied it advantageously in his study of tidal currents in the Irish Sea. Following Taylor the quadratic law of bottom friction was used in numerical modeling of ocean tides by *Hansen* [1966], *Zahel* [1970, 1973, 1975, 1977], *Marchuk et al.* [1973], *Estes* [1975, 1977], and others.

In order to retain the linearity of the tidal equations for their adopted time-independent numerical procedure, *Pekeris and Accad* [1969] used the linear law of bottom friction with a coefficient depending inversely on the ocean depth. These authors also called attention to the earlier work by *Grace* [1930*b*], who applied both the quadratic and the linear law of bottom friction to the problem of the tides in the Gulf of Suez. Grace experienced a slight preference for the linear law. Similarly to eddy dissipation (see discussion above) it appears that the linear law of bottom friction is more consistent with all other assumed linearizations of the equations of slow (Stokes-like) averaged tidal motion. Furthermore, the convergence characteristics of the time-stepping computations carried out and displayed by *Estes* [1975] with the quadratic law of bottom friction fail to exhibit any nonlinear symptoms.

On the basis of the earlier studies mentioned above, the author [*Schwiderski*, 1978*b*, 1979*a*] adopted the linear law of bottom friction as being preferable at least in open ocean areas. However, in contrast to the results of *Pekeris and Accad* [1969], the bottom friction coefficient was experimentally found to be more realistic without any dependence on the ocean depth. The depth-dependent coefficient chosen by *Pekeris and Accad* [1969] was evidently justified in their ocean basin, cut off artificially at the 1000-m depth level. Since turbulent flows display only a very thin boundary layer, a depth-independent friction coefficient is appropriate in an ocean of depth varying from 10 to 7000 m. As can be seen from the right-hand form of equations (37) with a constant B, bottom friction in the COTE's (28) decreases inversely with the first power of the depth H and becomes negligible in deep oceans. In the case of Pekeris and Accad the influence of bottom friction decreases inversely with the second power of the depth H, which is far too rapid in the 10- to-500-m depth range.

As was mentioned above, the bottom stress depends directly on the vertical eddy viscosity; hence a depth-independent bottom friction coefficient in equations (37) is consistent with the physical arguments presented in section 3*b* concerning eddy viscosities in discrete flow models. In complete analogy with those discussions the bottom friction coefficient B in equations

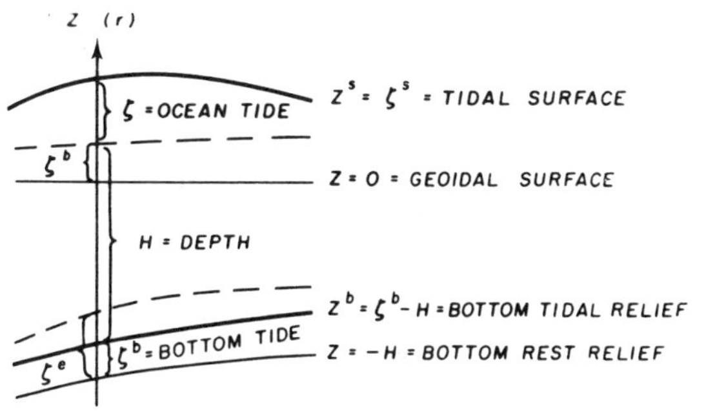

Fig. 4. Earth-ocean tidal interaction: ζ is the ocean tide. ζ^e is the earth tide, ζ^s is the surface tide. ζ^b is the bottom tide, $\zeta^{eo} = \zeta^e - \zeta^b$ is the earth dip response to the ocean tide, and H is the ocean depth.

(37) should depend linearly on the average bottom mesh cell area

$$\Delta s = L^2 \mu \sin \theta \tag{40}$$

i.e.,

$$B = \bar{b} L^2 \mu \sin \theta = b \mu \sin \theta \tag{41}$$

where L and μ are defined in section 3*b*. The remaining reduced friction coefficient b must be determined by trial-and-error computations for best results. The new friction coefficient B (equation (41)) was used by the author [*Schwiderski*, 1978*b*, 1979*a*, *b*] with only minor advantages over a simply constant B. At this point it may be mentioned that the author observed a considerably stronger dependence of the tidal model on the value chosen for b than on the corresponding reduced viscosity coefficient a of section 3*b*.

The author tested also bottom friction coefficients of the form $B \sim H^\epsilon$ with $\epsilon = \frac{1}{2}, -\frac{1}{2}, -1, -\frac{3}{2}$ but with poorer results. Other bottom friction laws were considered by *Johns* [1966] and *McGregor* [1972], as well as by *Kagan* [1971, 1972]. From the author's computer experiments it was concluded that in shallow ocean areas, certain nonlinear botton friction effects must be considered in addition to the linear law used (section 4*c*).

d. Secondary Tide-Generating Forces

All earlier numerical tidalists used only the primary astronomical equilibrium tide η (section 2*a*), mostly the M_2 component, for the driving force of global ocean tides. Recently, *Farrell* [1972*a*, *b*, 1973] called attention to significant interactions between the oceanic and terrestrial tides, which should be included in accurate ocean tide models. Farrell went on and derived the mathematical expressions of those secondary tide-modifying forces, assuming a dissipationless perfectly elastic earth subject to the astronomical body force and the oceanic tidal surface load.

In more detail the following five interaction effects are considered to be significant (see Figure 4): (1) ζ^e, $\zeta^{eo} = \zeta^e - \zeta^b$ are the earth tidal responses due to the astronomical equilibrium tide η and the oceanic tidal load ζ, respectively where ζ^b is the total ocean bottom tide; (2) η^e, η^{eo}, η^o are the perturbations of the equilibrium tide η in response to ζ^e, ζ^{eo}, and ζ, respectively.

According to their causes these perturbations enter in the COTE's (28) in the following grouped form:

$$\hat{\eta} = \zeta^e - \eta^e \tag{42}$$

(due to η) and

$$\hat{\zeta} = \eta^o + \zeta^{eo} - \eta^{eo} \tag{43}$$

(due to ζ). If η_n and ζ_n denote the nth-degree spherical harmonic components corresponding to η and ζ, then one has [after *Farrell*, 1972*a*]

$$\eta = \sum_\eta \eta_n \rightarrow \hat{\eta} = \sum_n (h_n - k_n)\eta_n \tag{44}$$

and

$$\zeta = \sum_n \zeta_n \rightarrow \hat{\zeta} = 3\alpha' \sum_\eta \frac{1 - h_n' + k_n'}{2n + 1} \zeta_n \tag{45}$$

where $\alpha' \approx 0.18$ denotes the ratio between the mean densities of the ocean water and the earth. The constant coefficients (h_n, k_n) and (h_n', k_n') are called Love numbers corresponding to (ζ^e, η^e) and (ζ^{eo}, η^{eo}).

Farrell derived some of the Love numbers on the basis of the Gutenberg-Bullen A earth model. Pekeris (in a presentation given in 1977 at the U.S. Office Naval Research) criticized Farrell's approach and went on to determine two new sets of Love numbers, one based on the Gutenberg and the other on the Dziewonski-Hales-Lapwood earth model [*Accad and Pekeris*, 1978]. However, all three sets display only minor differences which can be ignored in the present application.

Since all significant tidal species (section 2*a*) are purely spherical harmonics of degree $n = 2$, (44) reduces to

$$\hat{\eta} = 0.31\eta \tag{46}$$

where $h_2 - k_2 \approx 0.31$ in agreement with all three models of Farrell and Pekeris and Accad. This simple relationship reflects the drastically simplified properties of the earth model, which has been used to derive the representations for $\hat{\eta}$ and $\hat{\zeta}$ in (44) and (45) containing only real constant Love numbers. In particular, it has been assumed that lateral variations in the earth are negligible and that the terrestrial deformations (ζ^e, ζ^{eo}) and the gravity perturbations (η^e, η^{eo}, η^o) are all instantaneous responses to their causes η or ζ.

Equation (46) constitutes a significant 31% modification of the tide-generating forces in the COTE's (28), which definitely cannot be dismissed. Starting with *Hendershott* [1972], this correction has probably been included in all subsequent global tidal studies. All earlier linear or almost linear tide models [e.g., *Pekeris and Accad*, 1969; *Zahel*, 1970] with homogeneous boundary data (section 4*c*) can be corrected for this oversight by reducing all amplitudes with the uniform factor $\alpha = 1 - 0.31 = 0.69$. No adjustments of the phases are necessary.

The second correction $\hat{\zeta}$ (equation (45)) to the tide-generating forces in the COTE's (28) requires the knowledge of the oceanic tide ζ decomposed into its spherical harmonic components ζ_n. Assuming (for the moment) the a priori knowledge of ζ then one still faces a rather formidable numerical problem. The expansion of ζ must be carried out over the 'orthogonal range' of the spherical harmonics, that is, over the entire sphere with $\zeta = 0$ at land points. Thus, if a $1° \times 1°$ grid system (section 4*a*) is considered, 6.48×10^4 Fourier coefficients must be computed on 6.48×10^4 points to fit exactly all numerical ζ values. Since in practical computations only very few Fourier coefficients can be carried, the actual expansions of ζ and $\hat{\zeta}$ brush over all detailed features of the oceanic tide ζ (section 5*a*) and its response $\hat{\zeta}$. Moreover, no matter how many finite terms are being carried, the resulting ζ expansion is always contaminated with the well-known Gibbs phenomenon and yields only a generalized (δ type) derivative which is pointwise meaningless. These difficulties were experienced by *Estes* [1979], who expanded his tidal model into spherical harmonics and found poor agreement with the original numerical tide data.

Owing to the denominator $(2n + 1)$, the expansion for $\hat{\zeta}$ in (45) is absolutely convergent and yields ordinary derivatives as required by the COTE's (28). However, the expansions of those derivatives possess the same convergence and Gibbs phenomenon features as the oceanic tide ζ mentioned above. Using their own so-called zero-order tidal models of ζ, computed without the tidal loading effect $\hat{\zeta}$, *Hendershott* [1972, 1975], *Estes* [1977], *Zahel* [1977], and *Accad and Pekeris* [1978] evaluated by the equivalent (47) the response $\hat{\zeta}$ and its partial derivatives. Although rather crude approximations were used, the results confirmed the original observations by Hendershott: the secondary loading effect $\hat{\zeta}$ exceeds pointwise more than 50% of the primary equilibrium tide η, and its partial derivatives assume extremely large values compared to those of η. This seemingly stunning result lead to the prediction of huge ocean tidal loading effects on the oceanic tide. That puzzling experience was further enhanced when *Hendershott* [1972] included the estimated $\hat{\zeta}$ in his COTE's ((28) with $A = B = 0$) and constructed a new 'first-order' oceanic tide ζ', which differed drastically from the 'zero-order' tide ζ. A similarly unexplained result was recently reported by *Accad and Pekeris* [1978].

In an attempt to solve his COTE's with ocean loading effects $\hat{\zeta}$, *Hendershott* [1975] repeated his iteration step described above and computed successively $\zeta, \hat{\zeta}, \zeta', \hat{\zeta}', \cdots$ in an alternating procedure that failed to converge. In contrast, *Accad and Pekeris* [1978] followed Hendershott's iteration for a few steps and found no divergence symptons. These opposing results are mathematically most interesting because both models use the same COTE's (28) and (29) without any eddy dissipation ($A = 0$) and bottom friction ($B = 0$) and employ a similar time-independent numerical method (section 4).

The first successful iterative solution of the COTE's (28) and (29) including all terms A, B, $\hat{\eta}$, and $\hat{\zeta}$ was accomplished by *Estes* [1977] and *Zahel* [1977], who employed the well-known *Hansen* [1966] time-dependent numerical method (section 4). Both authors experienced no symptoms of divergence in their iteration. A similar result was reported by *Gordeyev et al.* [1977]. In order to overcome Hendershott's difficulties, *Parke and Hendershott* [1979] and *Parke* [1978] abandoned the search for an 'ordinary solution' of their COTE's and succeeded in finding a 'mean or weak' solution of these equations. Using earlier successive iteration solutions of ζ (converging or diverging) as base (called test) functions in a weighted linear superposition [*Hendershott*, 1977], they determined the weighting coefficients by fulfilling the COTE's or by matching empirical tide data in the mean. However, the quality of such modern mean solutions as ordinary solutions was not elaborated. Such an evaluation would be of great interest to numerical analysts, especially since only relatively few base functions have been used, which were permitted to diverge.

In order to understand the major mathematical difficulties involved in the above mentioned iterations the practical computation of the ocean loading effects $\hat{\zeta}$ from a known oceanic tide ζ (equations (45)) may be illustrated here in somewhat more detail. By introducing the well-known integral definition of the Fourier coefficients of ζ one finds the following integral representation for $\hat{\zeta}$ (equations (45)) [*Farrell*, 1972*a*]:

$$\hat{\zeta}(\lambda, \theta) = \int\int_\Sigma G(\lambda', \theta'; \lambda, \theta)\zeta(\lambda', \theta') \sin \theta' \, d\lambda' \, d\theta' \qquad (47a)$$

with the 'kernel (called Green's) function'

$$G(\lambda', \theta', \lambda, \theta) = \frac{3\alpha'}{4\pi} \sum_\eta (1 - h_n' + k_n')P_n(\cos \tilde{\theta}) \qquad (47b)$$

where Σ denotes the entire sphere, P_n the conventional Legendre ploynomial of degree n, and $\tilde{\theta}$ the angular spherical distance between the points (λ, θ) and (λ', θ').

In his presentation (see discussion above), Pekeris pointed out that the series of the kernel function (equation (47)) is only conditionally (not absolutely) convergent and yields no ordinary (only δ type) derivatives. Hence the evaluation of $\hat{\zeta}, \hat{\zeta}_\lambda$, and $\hat{\zeta}_\theta$ depends strongly on the number of terms carried and on the order of summation, integration, and differentiation. In particular, the partial derivatives of $\hat{\zeta}$ required in the COTE's (28) may be taken by (1) differentiating $\hat{\zeta}$ numerically, (2) differentiating $\hat{\zeta}$ analytically term by term, (3) differentiating G numerically, or (4) differentiating G analytically term by term. One concludes from the properties of the Fourier coefficients of ζ (see discussion above) that methods 1–3 are all mathematically pointwise acceptable. Method 1 can be expected to be fairly smooth without Gibbs phenomena. Methods 2 and 3 should produce rought values due to Gibbs phenomena. Method 4 is only correct in the generalized sense without any pointwise meaning.

With their important solutions, *Estes* [1977] and *Zahel* [1977] broke the incredible expectations attributed to the ocean loading term $\hat{\zeta}$ in comparison to the equilibrium tide η (see discussion above). Their and all subsequent tidal models displayed only moderate modifications (see discussion below) of the corresponding models with omitted ocean loading effects $\hat{\zeta}$. A closer look at the COTE's (28) reveals that $\hat{\zeta}$ could be compared to η or to ζ. Since $\hat{\zeta}$ is produced by ζ (equations (45)), only the latter case is evidently appropriate. In the light of this comparison, ocean loading effects emerge (as physically plausible) as truly secondary effects in agreement with all computed models. In fact, *Accad and Pekeris* [1978] compared their zero-order solution ζ with the corresponding $\hat{\zeta}$ and found (in agreement with all other models) that $\hat{\zeta}$ amounted only to some ('secondary') 10% of ζ. Moreover, ζ and $\hat{\zeta}$ displayed a very close agreement in their phases. These authors explained the moderate ocean loading effects by the observed close resemblance between the lateral features of ζ and $\hat{\zeta}$. In contrast to ζ or $\hat{\zeta}$ the equilibrium tide η covers essentially the entire ocean (Figure 1*b*) and should not be compared to $\hat{\zeta}$, which (although it is relatively large pointwise) is significant only within its amphidromic rotation systems similar to those shown in Figure 8 for the author's tide model. In other, more drastic, words, a high fountain of small diameter has little influence on the oceanic tide at large.

In order to avoid all mathematical difficulties involving the practical computation of the ocean loading term $\hat{\zeta}$, Pekeris in his presentation (see discussion above) suggested the simple relationship

$$\hat{\zeta} = \beta\zeta + \Delta \qquad (48)$$

on the basis of his observed resemblance between ζ and $\bar{\zeta}$. With $\Delta = 0$ this relationship obviously can be argued physically just as well as the analogous relation of (46). Treating Δ as being negligible, *Accad and Pekeris* [1978] determined the constant β in the mean by an iteration procedure that converged to

$$\beta = 0.085 \tag{49}$$

Following this ingenious approach, the author [*Schwiderski*, 1978*b*, 1979*a*, *b*] included (48) with $\Delta = 0$ in his COTE's (28), keeping β fixed at

$$\beta = 0.10 \tag{50}$$

as originally suggested by Pekeris. This modification eliminated all iteration requirements. Instead of (50), $\beta = 0.08$ and $\beta = 0.12$ were also tested with hardly any different results.

In agreement with all iterative solutions, *Accad and Pekeris* [1978] and *Schwiderski* [1978*b*, 1979*a*, *b*] found only moderate modification of the oceanic tide ζ by the ocean loading term $\bar{\zeta}$. Indeed, hardly any variations in the amplitudes were registered. Some (about 20°) differences in the phases were observed but not always toward improvement of the model. The result is well displayed by *Accad and Pekeris* [1978, Table 4], where some 1279 empirical tidal data are statistically compared with corresponding values computed with and without ocean loading effects. In particular, for the phases δ (= t in hours) they find the total averages

$$\langle \delta_o - \delta \rangle = 0.9 \qquad \langle \delta_o - \delta' \rangle = 0.2 \tag{51}$$

and

$$\langle |\delta_o - \delta| \rangle = 1.0 \qquad \langle |\delta_o - \delta'| \rangle = 0.9 \tag{52}$$

where δ_o stands for observed phases and (δ, δ') for computed values without and with ocean loading effects, respectively. The first equations only show that the phase errors seem to be more symmetrically distributed when ocean loading effects are included. The second equations show that there is a meager 10% absolute improvement. The extensive computer experiments of the author clearly demonstrated that there is some improvement with ocean loading effects, which should not (unnecessarily) be neglected. However, considerably more significant improvements (section 5*a*) were achieved through the inclusion of proper eddy dissipation and bottom friction terms in connection with a hydrodynamically defined ocean basin (sections 3*b*, 3*c*, and 4*a*).

e. Polar Singularities and Ice Caps

In spherical coordinates the COTE's (28) and (29) become singular at the north pole. Most authors avoid this difficulty by terminating the ocean basin at some convenient northern boundary. In order to check his numerical mehtod against an analytic solution in a circular arctic ocean basin, *Zahel* [1970] derived an analytic integral of the COTE's (28) and (29) with out eddy dissipation ($A = 0$). *Schwiderski* [1978*b*, 1979*a*] derived a unique second-order analytic solution of the complete COTE's (28) and (29) which included the special exact integrals discussed in section 2*c* (equations (24)–(26)). These solutions were matched with the numerical solution (section 4) which started at 4° colatitude southward. Thus all numerical difficulties with the north pole singularities were circumvented, and a more uniform grading of the numerically necessary grid system (section 4*a*) was achieved.

In all ocean tide models it is tacitly assumed that the polar ice covers have negligible effects on the oceanic tide and currents. This assumption is probably justified within the desired accuracy specified in section 1. It is certainly true at least near the north pole, where, according to section 2*c*, all semidiurnal and diurnal tides vanish with the second or first power of the distance from the pole. Furthermore, ice cap effects may be introduced indirectly into a tide model by using empirical tide data collected at stations of the Arctic Ocean and at Antarctica in the numerical procedure (section 4*c*).

4. MODELING DISCRETE OCEAN TIDAL EQUATIONS

a. The Gridded Ocean Basin and Bathymetry

With the advent of electronic computers, hydrodynamical numerical methods dominated the theoretical scene to map worldwide ocean tides. The modern, large-scale numerical investigations were pioneered by *Hansen* [1948, 1949, 1966], who realized that the complexity of natural ocean basins renders any analytic treatment of the COTE's (28) and (29) intractable. Moreover, as was pointed out by the author (see sections 3*a*–3*c*), the COTE's (28) and (29) themselves assume their true physical meaning only in some proper discrete form. Hansen was soon followed by a long line of researchers, such as *Gohin* [1961], *Accad and Pekeris* [1963, 1978], *Ueno* [1964*a*, *b*], *Bogdanov et al.* [1964], *Bogdanov and Magarik* [1967, 1969], *Brettschneider* [1967], *Tiron et al.* [1967], *Trepka* [1967], *Pekeris and Accad* [1969], *Marchuk et al.* [1969, 1973], *Zahel* [1970, 1973, 1975, 1977], *Hendershott* [1972, 1975, 1977], *Gordeyev et al.* [1973, 1977], *Estes* [1975, 1977], *Schwiderski*, [1976, 1978*b*, 1979*a*, *b*], and *Parke and Hendershott* [1979].

The definition of a finite-difference analog (DOTE's (60) and (61)) in place of the COTE's (28) and (29) requires a mesh discretization of the ocean area under consideration. While one would like to utilize a grid system as finely detailed as possible to achieve a high resolution of all significant tidal currents, one is, of course, quite limited by the memory capacity of the computer facility available and by the known bathymetric charts of the ocean basin. So far, all grid systems between 6° × 6° and 1° × 1° have been used by numerical tidalists with 'terrestrial' and 'oceanic' mesh cells distinguished by zero or nonzero depth data, respectively. Some authors [e.g., *Pekeris and Accad* 1969; *Parke and Hendershott*, 1979; *Accad and Pekeris*, 1978] terminated the ocean basin at some northern latitude and cast their COTE's into Mercator coordinates. Other analysts [e.g., *Zahel*, 1970, 1973, 1975, 1977; *Estes*, 1975, 1977; *Schwiderski*, 1978*b*, 1979*a*, *b*] retained the entire ocean basin with a grid system that is spherically graded toward the poles, in order to achieve a more uniform mesh area for higher numerical accuracy and stability (section 4*b*).

For example, the author used a 1° × 1° graded grid system in which each spherically rectangular mesh cell $S_{m,n}$ is bounded by the graded longitudes

$$\lambda_{m-\mu} = (m - \mu)^\circ \qquad \lambda_m = m^\circ \tag{53a}$$

and the colatitudes

$$\theta_{n-1} = (n - 1)^\circ \qquad \theta_n = n^\circ \tag{53b}$$

where

$$m = \mu, 2\mu, \cdots, 360 \qquad n = 1, 2, \cdots, 168 \tag{54}$$

The 'grading parameter' $\mu = \mu_n$ is defined by

$$\begin{array}{lll} \mu = 1 & n = 30\text{–}150 & \\ \mu = 2 & n = 15\text{–}29 & n = 151\text{–}168 \\ \mu = 4 & n = 8\text{–}14 & \\ \mu = 8 & n = 1\text{–}7 & \end{array} \tag{55}$$

The grading (equations (55)) was conveniently chosen so that

$$\mu \sin n^\circ \geq \tfrac{1}{2} \qquad n = 4, 8, 15, 30, 150, 165 \tag{56}$$

i.e., the southern mesh size of $S_{m,n}$ remains larger than half the equator mesh size

$$L = \pi R/180 \approx 111.1775 \text{ km} \tag{57}$$

This important condition is violated for $n = 1°, 2°, 3°; 166°, 167°, 168°$ without reducing the numerical stability of the author's finite difference analog of the COTE's (section 4*b*). All mesh cells south of colatitude $n = 168°$ are declared antarctic land cells.

The 'mathematically idealized' lateral boundaries of the oceans are generally defined to follow in an obviously zigzagging fashion the mesh lines of boundary cells. The latter are defined as oceanic cells with at least one neighboring terrestrial cell. Since the ocean boundaries shape the geometry of the tidal currents, the obviously somewhat subjective definition of boundary cells is most significant, especially in models of coarse mesh sizes. After numerous disappointing computations the author found it necessary to distinguish between boundary cells adjacent to continental and island land masses. In agreement with *Smith et al.* [1966], grid cells near continents were declared oceanic boundary cells if at least half of their mesh areas contained a water depth of more than 5 m; otherwise those cells were designated terrestrial cells. The same principle was applied to grid cells near islands large in area. However, grid cells crossed by elongated islands or island chains were declared terrestrial cells even if the island area was small in comparison to half the mesh area. Although a 1° × 1° grid system was used, the latter condition was found to be necessary in order to retain the hydrodynamically important separation properties of elongated islands and island chains. Remaining oceanic cells that were gridwise disconnected from the main oceans (e.g., the entire Mediterranean Sea) were redefined as terrestrial cells.

The artificial zigzagging boundary of the ocean basin caused some instability or resonance problems in the numerical procedures of some researchers [e.g., *Pekeris and Accad*, 1969; *Parke and Hendershott*, 1979; *Accad and Pekeris*, 1978] using nondissipative COTE's ((28) with $A = B = 0$). In order to overcome these difficulties, smoothed boundaries and depth data were artifically introduced. No such difficulties were reported by analysts using dissipative COTE's. The author's computer experiments with the full COTE's (28) and (29) displayed no discernible differences for rough or smoothed boundaries.

For the worldwide oceans, essentially 1° × 1° averaged depth data were compiled by *Dishon* [1964] and *Dishon and Heezen* [1968] and independently by *Smith et al.* [1966]. The latter collection has been somewhat revised by *Gates and Nelson* [1975], and both sets are available in tape form. In spite of specially collected depth data, all investigators prepared their own bathymetric charts corresponding to the grid system chosen. Those data were quite drastically curtailed by artificial limits on both the shallow and the deep ends of the depth range. Moreover, the stability requirements of the numerical schemes, especially in the dissipationless case, forced some numerical analysts [e.g., *Pekeris and Accad*, 1969; *Parke and Hendershott*, 1979; *Accad and Pekeris*, 1978] to apply drastic smoothing operations to the bathymetric data used.

Obviously, in such an ocean geometry, much of the absolutely necessary resolution of, say, narrow ocean ridges and continental shelves is lost. In fact, the corresponding computed tidal maps (see the collections by *Hendershott* [1973, 1977] and *Parke* [1978]) suffer most of all from the lack of proper resolution of bottom and boundary (see discussion above) irregularities. One may expect a 1° × 1° seamount or island to have little effect on the surrounding tidal wave, but a row of such obstacles over several mesh cells represents a hydrodynamical barrier separating the tidal currents on both of its sides.

In exploratory computations, *Schwiderski* [1978*b*, 1979*a*, *b*] used the dissipative COTE's (28) and (29) with the rough gridwise adjusted depth data of *Smith et al.* [1966] and detected no instabilities in the numerical method. However, although the preliminary tidal model [*Schwiderski*, 1976] satisfied some accuracy requirements in tidal applications [e.g., *Goad and Douglas*, 1977*a*, *b*], the model still failed to resolve the well-known (see sections 2*b* and 5*a*) large tidal deformations and retardations over narrow ocean ridges (e.g., Aleutian, Hawaiian, and Caribbean) and continental shelves. This experience led to a redefinition of the *Smith et al.* [1966] depth data in over 3000 grid cells on the basis of the following hydrodynamical principles in connection with the definition of oceanic and terrestrial cells given above [*Schwiderski*, 1978*a*].

1. In oceanic boundary cells near continents or large islands the averaged oceanic depth was assigned as the mean depth of the entire grid cell.
2. In oceanic cells crossed by shallow submerged or partly submerged ridges the average ridge depth was assigned as the mean depth of the whole grid cell.

It may be noted that the 'hydrodynamically defined' depth (1) replaces the 'statically defined' depth of *Smith et al.* [1966], which preserves the estimated water mass of the boundary cell and thus may introduce an artificial shelflike character to the cell. Exploratory computations with the new ocean bottom relief in connection with the novel depth-dependent eddy viscosity (30) and depth-independent bottom friction coefficient (41) dramatically reflected the improved resolution. The improvement of the tide model was even more spectacular when empirical tidal data were hydrodynamically interpolated (section 4*c* and 5*a*). Naturally, it must be kept in mind that the definition of depth values, in particular in boundary cells, is (even with the knowledge of numerous soundings) quite subjective and hence subject to some error. However, isolated inaccuracies of this sort were generally found ineffective in altering the overall quality of the tidal model.

b. Discrete Ocean Tidal Equations

On the gridded ocean basin (section 4*a*) the differential equations of oceanic tidal motions (COTE's (28) and (29)) can be converted into physically meaningful discrete finite difference equations for a computerized solution. Since the beginning of the computer age, numerous finite-differencing schemes with various characteristics have been developed and tested by numerical analysts working in applied mathematics, fluid mechanics, oceanography, and meteorology. The application of these techniques with or without possible variations to the present COTE's depends essentially on the numerical

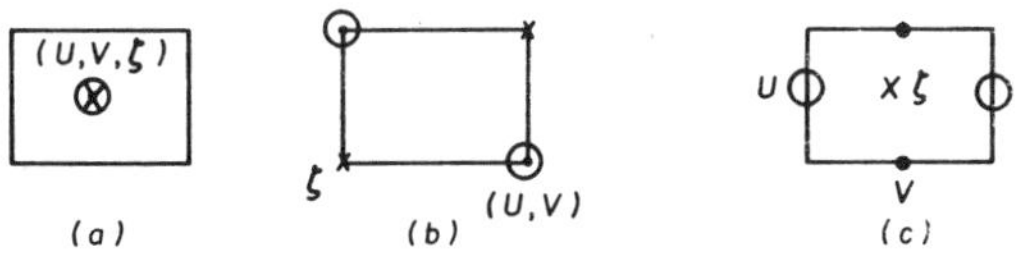

Fig. 5. Finite-difference schemes in space: (*a*) regular, (*b*) semi-staggered, and (*c*) (fully) staggered. Open circles are U points, solid circles are V points, and crosses are ζ points.

tidalist's empirically founded modeling skill and imagination, limited only by the available computer capacity.

Following the usual procedure, in the first step, all spacial derivatives in the COTE's (28), (29), (33), and (37) may be replaced by central finite differences corresponding to the grid system chosen. As was first pointed out by *Richardson* [1922], the differencing scheme may be chosen to be 'regular,' 'semi-staggered,' or (fully) 'staggered'; that is, the three dependent variables (U, V, ζ) may be computed and tabulated at collocated, semidislocated, or fully dislocated points, as shown in Figure 5.

If only a partial equilibrium tide η (section 2*a*) is considered as the driving force, the remaining equations may be integrated as a boundary value problem (BVP) or as an initial value problem (IVP). In the BVP approach one removes the complex time periodic factor by the substitution given in (20), provided the COTE's are strictly linear. (If nonlinear properties (e.g., square law of bottom friction (equation (39)) are modeled, infinite Fourier series could in principle be used, but these lead to an involved infinite system of equations.) In the general case (GBVP) the resulting three difference equations may be combined into two equations for the complex velocity amplitudes by eliminating the complex tidal amplitude, which is directly computable from the velocity field via the continuity equation (compare (21) and (22)). With sufficient velocity boundary conditions (section 4*c*) one winds up with an implicit complex system of linear equations that can be solved by modern computerized methods. If the higher-order eddy dissipation terms of (33) are included, the linear system to be solved is drastically simplified and has been successfully inverted by *Accad and Pekeris* [1963, 1978] and *Pekeris and Accad* [1969].

In the special boundary value problem (SBVP), where all eddy dissipation terms are neglected, the same three difference equations derived above can be combined into a single difference equation for the complex tidal amplitude (compare (21) and (22)). This alternative approach to the linear BVP is obviously advantageous whenever empirical tidal boundary data (section 4*c*) are introduced in place of the purely mathematical velocity boundary data required in the GBVP. The simple SBVP attracted the attention of several investigators, such as *Hansen* [1948, 1949], *Bagdanov and Magarik* [1967, 1969], *Hendershott* [1972, 1975], and *Parke and Hendershott* [1979], who also neglected bottom friction.

Since the basic tide model is time dependent, it may be integrated as an IVP starting from some arbitrary initial state and so offer itself as an experimental field for more involved oceanic or atmospheric circulation studies. In addition, this most natural approach leaves unrestricted room for full-scale realistic modeling of tidal currents, including linear or nonlinear eddy dissipation and bottom friction (sections 3*b* and 3*c*) as well as mathematical or empirical boundary data (section 4*c*). In fact, the resulting discrete ocean tidal equations (e.g., DOTE's (60) and (61)) permit a relatively easy examination (see discussion below) of the physically important decay and dispersion features of the oceanic motions modeled.

In the IVP approach, the first-order time derivatives of the COTE's (28), (29), (33), and (37), in which all spacial derivatives have been replaced by some finite-difference scheme (see discussion above), may be inverted by integration over one or two time steps Δt along the degenerated characteristic of the system i.e., parallel to the t axis. (It may be suggested here that an integration along the ruled lines of the characteristic cone of the system might be preferable, but this alternative has never been tried in modeling of ocean currents.) The remaining (not explicitly integrable) integrals are then replaced by some average (one, two, or three point) integration rule such as

$$\int_{t_j}^{t_{j+1}} U(t)\,dt = \Delta t[\kappa U^{j+1} + (1-\kappa)U^j] \tag{58}$$

where $\kappa(0 \le \kappa \le 1)$ denotes some 'difference parameter' and

$$U^j = U(t_j) \qquad t_j = (j-1)\Delta t \qquad j = 1, 2, \cdots \tag{59}$$

Depending on the chosen time integration formula, the obtained DOTE's must be solved either by an explicit or by an implicit time-stepping procedure that begins with some initial state. As was recognized by *Richardson* [1922], the time integration may utilize a regular, a step-over (or leapfrog), or an intermediary (or alternating) time-stepping scheme. In the regular scheme, all three (real) variables (U, V, ζ) are computed and tabulated at the same time points (Figure 6*a*). In Richardson's step-over (now often called leapfrog) method, both velocity components (U, V) are cotimed, but the tidal elevations ζ are computed and tabulated at intermediate time points (Figure 6*b*). In the intermediary (or alternating) procedure, all three variables (U, V, ζ) are computed and tabulated coincidently; however, both velocity components use computationally auxiliary intermediate time points in an alternating fashion (Figure 6*c*).

Although the step-over scheme is widely used in general ocean current and atmospheric circulation studies, the regular and intermediary schemes seem to have been preferred in tidal models. The intermediary method has been applied in a rather involved explicit-implicit form by *Marchuk et al.* [1969, 1973] and *Gordeyev et al.*) [1973] to tidal computations in border seas and in a global ocean model. In the now well-known 'hydrodynamical-numerical' method of *Hansen* [1966] the staggered scheme in space is combined with the regular scheme in time in an explicit and convenient fashion. This method was successfully applied by *Zahel* [1970, 1973, 1975, 1977] and *Estes* [1975, 1977] and in a modified version by the author [*Schwiderski*, 1978*b*, 1979*a*, *b*].

By applying the integration rule of (58), using $\kappa = 0$ and $\kappa =$ arbitrary in the momentum equations and $\kappa = 1$ in the continuity equation, one arrives at the following explicit, regular,

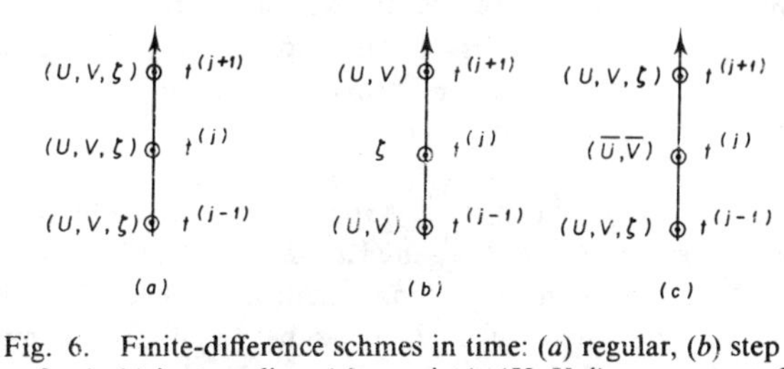

Fig. 6. Finite-difference schmes in time: (*a*) regular, (*b*) step over (leapfrog), (*c*) intermediary (alternating); (U, V, ζ) are computed and tabulated. ($\bar{U}$, $\bar{V}$) are computed but not tabulated.

and staggered DOTE's corresponding to the COTE's (28), (29), (33), and (37) and the grid system defined by (53).

$$\begin{aligned}
[1 + \kappa A_{m,n}{}^4]\, U_{m,n}{}^{j+1} &= A_{m,n}{}^1 \sin(\sigma\Delta t/2)(2j-1) \\
&+ A_{m,n}{}^2 \cos(\sigma\Delta t/2)(2j-1) \\
&+ A_{m,n}{}^3[\zeta_{m-\mu,n}{}^j - \zeta_{m,n}{}^j] + [1-(1-\kappa)A_{m,n}{}^4]U_{m,n}{}^j \\
&+ A_{m,n}{}^5 U_{m+\mu,n}{}^j + A_{m,n}{}^6 U_{m-\mu,n}{}^j + A_{m,n}{}^7 U_{m,n+1}{}^j + A_{m,n}{}^8 U_{m,n-1}{}^j \\
&+ A_{m,n}{}^9[V_{m,n}{}^j + V_{m,n-1}{}^j) - (V_{m-\mu,n}{}^j + V_{m-\mu,n-1}{}^j)] \\
&+ A_{m,n}{}^{10}[(V_{m,n}{}^j + V_{m,n-1}{}^j) + (V_{m-\mu,n}{}^j + V_{m-\mu,n-1}{}^j)] \qquad (60a)
\end{aligned}$$

$$\begin{aligned}
[1 + \kappa B_{m,n}{}^4] V_{m,n}{}^{j+1} &= B_{m,n}{}^1 \cos(\sigma\Delta t/2)(2j-1) \\
&+ B_{m,n}{}^2 \sin(\sigma\Delta t/2)(2j-1) \\
&+ B_{m,n}{}^3[\zeta_{m,n+1}{}^j - \zeta_{m,n}{}^j] + [1-(1-\kappa)B_{m,n}{}^4]V_{m,n}{}^j \\
&+ B_{m,n}{}^5[V_{m+\mu,n}{}^j + V_{m-\mu,n}{}^j] + B_{m,n}{}^6 V_{m,n+1}{}^j + B_{m,n}{}^7 V_{m,n-1}{}^j \\
&+ B_{m,n}{}^8[U_{m+\mu,n+1}{}^j - U_{m,n}{}^j] + B_{m,n}{}^9[U_{m+\mu,n}{}^j - U_{m,n+1}{}^j] \\
&+ B_{m,n}{}^{10}[U_{m+\mu,n+1}{}^j + U_{m,n}{}^j + U_{m+\mu,n}{}^j + U_{m,n+1}{}^j] \qquad (60b)
\end{aligned}$$

$$\begin{aligned}
\zeta_{m,n}{}^{j+1} = \zeta_{m,n}{}^j + [C_n{}^1(U_{m,n}{}^{j+1} - U_{m+\mu,n}{}^{j+1}) \\
+ C_n{}^2 V_{m,n}{}^{j+1} - C_n{}^3 V_{m,n-1}{}^{j+1}] \qquad (61)
\end{aligned}$$

Following standard principles [e.g., *Richtmyer and Morton*, 1967], *Zahel* [1970] and *Schwiderski* [1978*b*, 1979*a*] showed that the DOTE's (60) and (61) with 'locally constant' coefficients yield Fourier-type eigensolutions of the form

$$\begin{aligned}
U_{m,n}{}^j &= U_0 d^j \exp i[\gamma_1(2m - 2\mu)\Delta\lambda + \gamma_2(2n-1)\Delta\theta] \\
V_{m,n}{}^j &= V_0 d^j \exp i[\gamma_1(2m - \mu)\Delta\lambda + \gamma_2 2n\Delta\theta] \qquad (62) \\
\zeta_{m,n}{}^j &= \zeta_0 d^j \exp i[\gamma_1(2m - \mu)\Delta\lambda + \gamma_2(2n-1)\Delta\theta]
\end{aligned}$$

with an arbitrary wave vector (γ_1, γ_2) and some nonzero amplitude vector (U_0, V_0, ζ_0), provided d satisfies a cubic eigenvalue equation. Obviously, the DOTE's will be stable if the three eigenvalues (d_0, d_1, d_2) fulfill at all oceanic points the 'stability condition'

$$|d_k| \leq 1 \qquad k = 0, 1, 2 \qquad (63)$$

This requirement, which was found to be scrupulously binding, can be met by restricting the reduced eddy viscosity a (in (30)), the mesh length L, the time step Δt, and the differencing parameter κ to appropriate stability ranges. Within those ranges these parameters and the always stabilizing bottom friction coefficient b (in (41)) remain free to be determined by trial-and-error computations to achieve optimum results.

For a stationary earth ($\Omega = 0$) and zero bottom friction the author found d_0 to be real and positive. Hence the corresponding eigenwave (equations (62)) represents a standing but decaying wave with no phase shift. Under an additional condition imposed on the free parameters mentioned above, the other eigenvalues d_1 and d_2 became conjugate complex and defined a pair of eigenwaves progressing in opposite directions with the same decay and dispersion rates.

It is interesting to note that for a 1° × 1° grid system and sufficiently small time steps Δt (about 3 min) the amount of eddy dissipation was only limited from above and not at all from below. Hence the actual magnitude of the eddy viscosity (A in (32)) was solely determined by the quality of the computed tidal model and not by some artificial mathematical shortcoming. Furthermore, the rate of decay of all eigenwaves (equations (62)) was found to be directly dependent on the eddy viscosity $A = \bar{a}H$ (equation (30)). Thus for a fixed $\bar{a}$, waves in deep ocean basins decay faster than those in shallow regions if bottom friction is neglected. It was this physically realistic feature that led to the introduction of the novel depth-dependent eddy viscosity. For a constant A, waves would decay at the same rate in deep or shallow oceans even though no bottom friction were present. This latter situation is physically unacceptable in an ocean of strongly varying bathymetry and produced, indeed, unrealistic results.

At this point, attention of the more mathematically oriented reader should be drawn to the extensive and important investigations of *Marchuk et al.* [1969, 1973], *Gordeyev et al.* [1973, 1977], *Kagan* [1977], and *Marchuk and Kagan* [1977]. These authors used powerful modern methods of functional analysis to prove rigorously the stability and convergence properties of their unique finite-differencing schemes. A detailed analysis of these techniques exceeds, however, the scope of this review.

c. Lateral Empirical and Mathematical Boundary Data

In order to specify a unique integral of the ocean tidal equations in continuous or discrete form, proper lateral boundary data must be imposed. Attracted by the simplicity of the special boundary value problem (section 4*b*), which is reduced to a single equation for the complex tidal amplitude, *Hansen* [1948, 1949] supported his early tide models with empirical tidal boundary data known around the ocean basin considered. Although the SBVP has been applied by several investigators, it is restricted to lateral dissipationless linear models and leaves the fluid flow across the ocean boundaries without any control (compare hydrodynamical interpolation below). Such an unchecked violation of the 'stiff' condition of conversation of mass is obviously somewhat hazardous. For example, it played probably a major role in *Hendershott*'s [1975] divergent iteration attempt to include ocean-loading effects in his tide model (see section 3*d* and *Farrell* [1972*b*]). In fact, the convergent *Accad and Pekeris* [1978] tide model differs from the Hendershott model only in the imposed boundary conditions, where the former (equation (64)) places some indirect control on the allowed violation of conservation of mass.

Instead of empirical boundary values, *Accad and Pekeris* [1963] introduced the mathematical boundary condition of 'no flow' across the ocean shorelines, which is sufficient to specify a unique solution of their GBVP (section 4*b*) without lateral eddy dissipation. However, when higher-order eddy dissipation terms are included in the GBVP or in the complete IVP, two sets of boundary data are required to determine a unique integral. *Hansen* [1966], *Zahel* [1970, 1972, 1975, 1977], *Estes* [1975, 1977], and others used the complete mathematical boundary conditions of no flow across and 'free slip' along the ocean boundaries. Some authors [e.g., *Marchuk et al.*, 1969] replaced the free-slip condition by the 'no slip' condition, which is appropriately used to specify laminar flows. In turbulent flows with very thin boundary layers and large eddy viscosities the free-slip condition is generally preferred and is more consistent with all other simplifications of the tidal model (section 3*a*). The author's computer experiments with both the free-slip and the no-slip conditions exhibited only a slight preference for the first one.

Some numerical tidalists used a mixture of tidal and velocity boundary conditions. For example, *Tiron et al.* [1967] used empirical tidal elevations wherever they were available and the condition of no flow across the coastline everywhere else. *Accad and Pekeris* [1978] generalized an idea of *Proudman* [1952] and introduced a mixed mathematical tide-velocity relationship of the form

$$\zeta = r u_N \tag{64}$$

where u_N denotes the velocity normal to the boundary. The parameter r was derived for a simple sloping shelf wave model that neglects any contribution from the tangential motion along the shoreline and specifies no breadth of the horizontal shelf portion. The value of r was found to depend on the variable preshelf ocean depth H, on the horizontal shelf depth H', and on the width of the shelf slope S, where the last two quantities were arbitrarily fixed at $H' = 10$ m and $S = 0$, or 100 km. It may be noted that this boundary condition does not fulfill the exact continuity equation (equation (29) or (61)). Also, it appears only formally as a purely mathematical condition, because it depends directly on the assigned ocean boundary depth H, which is largely subjective (section 4*a*).

The author's [*Schwiderski*, 1978*b*, 1979*a*, *b*] experiments with the no-flow and free-slip boundary conditions demonstrated convincingly that the important localized distortions and retardations of tides at shore or bottom irregularities (see section 2*b*, Figure 3, and Tables 4*a* and 4*b*) could not be satisfactorily modeled on a 1° × 1° grid system without a proper direct inclusion of empirical tidal data. A unique 'hydrodynamical interpolation' technique was developed that incorporates into the tide model over 2000 empirical data selected (see discussion below) as being representative for oceanic boundary cells and oceanic cells with bottom anomalies such as small islands, seamounts, or coastal banks. Though this method is in principle applicable to all GBVP's and IVP's with bottom friction, it is technically most adapted to the author's modified version ($\kappa = 1$ in the DOTE's (60)) of Hansen's hydrodynamical-numerical difference scheme.

Assuming that for the oceanic grid cell $S_{m,n}$ the empirical tidal amplitude $\xi_{m,n}$ and phase $\delta_{m,n}$ are known, then the empirical value

$$\zeta_{m,n}^{j+1} = \xi_{m,n} \cos(\sigma \Delta t j - \delta_{m,n}) \tag{65}$$

replaces the corresponding mathematical value determined by the continuity equation (61), where the velocity entries are computed earlier by using the DOTE's (60) with $\kappa = 1$ and the no-flow and free-slip conditions for boundary cells. Of course, the substitution of (65) for the DOTE (61) constitutes a 'continuity gap'

$$\Delta\zeta_{m,n}^{j+1} = \zeta_{m,n}^{j+1} - \zeta_{m,n}^{j+1} \tag{66}$$

that signals a periodic violation of the sensitive condition of conservation of mass which must be checked (see discussion above).

The continuity gap was significantly narrowed by an indirect cellwise adjustment of the bottom friction coefficient b (equation (41)), which depends on the physical properties of the special cells considered, such as true cell size and bottom slope and roughness. The indirect adjustment of b was implemented by simply modifying the velocity entries in the DOTE (61) by using bounded multipliers that closed the continuity gap as much as possible while yielding optimum results. This technique is justified, since for $\kappa = 1$ the bottom friction coefficient b appears only on the left sides of the DOTE's (60) in the central coefficients $A_{m,n}^4$ and $B_{m,n}^4$, which additionally include only negligible depth-dependent eddy dissipation effects (equation (30)). By trial and error, best results were obtained when the uniform coefficient $b = 0.01$ m/s was allowed to vary to a maximum of $b = 0.04$ m/s. It must be mentioned that no physical justification was found to apply the same bottom friction adjustment technique to distant offshore oceanic cells with available deep-sea tide measurements (see Table 3).

In order to narrow further the continuity gaps in oceanic boundary cells the mathematically defined land-ocean boundary lines (section 4*a*) were implicitly replaced by more realistic shorelines. This redefinition was accomplished by allowing over the mathematical boundary segments limited periodic inflows or outflows (Figure 7) that reduced the continuity gaps to acceptable levels. In general, the continuity gaps (equation (66)) dropped significantly below half the corresponding empirical elevation change during the same time step Δt. The remaining uniformly moderate continuity gaps and velocity distortions could then be attributed to other physical inaccuracies of the tide model, such as faulty depth data (section 4*a*) erratic empirical tide data (see discussion below), and neglected nonlinear effects (sections 3*a* and 3*c*).

The novel hydrodynamical interpolation technique, which is described in full detail by *Schwiderski* [1978*b*, 1979*b*], has been tested and evaluated for its important smoothness characteristics, which are distinctly different from other direct (e.g., polynomial) interpolation methods. The adjustment of hydrodynamical parameters does not imply any smoothness of the interpolation, unless both the empirical data and the hdyrodynamical model are compatible with each other. As can be seen in Figure 3 and in Tables 2, 4*a*, and 4*b*, it is an important exclusive feature of the hydrodynamical interpolation that it preserves the physically realistic (i.e., observed) roughness of ocean tides near uneven shorelines, islands, and oceanic ridges.

Some objections against the use of certain unreliable empirical tide data have been raised by tidalists who experienced difficulties in selecting meaningful tidal boundary data from the available collections. As was pointed out in section 2*b*, the same criticism applies to some degree to all empirical tide data, none of which are exact measurements; all are mathematically treated by more or less subjective and controversial methods. Nevertheless, the bulk of the listed data were found to be sufficiently accurate for the corresponding stations and their immediate neighborhoods. The problem of choice (which resembles the selection of bathymetric boundary data (section 4*a*)) is certainly not overcome by excluding empirical tide data from the model but afterwards judging the modeled data against such unreliable values.

The crux of the matter is to select from the massive listings data that are representative for the corresponding oceanic cells. Although empirical tide data have been used for some time in direct or indirect applications to modeling of ocean tides, only *Zahel* [1970, 1973] and *Schwiderski* [1978*b*, and 1979*b*, *c*] published the complete sets of selected data. The author described also the detailed selection process involved. In principle, the collected data were subjected to an initial screening in order to eliminate obviously erratic and sheltered-station data. Out of the still remaining mass of sometimes rough and seemingly incoherent tidal constants, 'representative' data were determined by taking averages over the mesh area with weights biased toward more ocean exposures and more coherent relationships with values of adjacent cells. The application of such averaging principles is apparently

TABLE 3. Deep-Sea M_2 Tide Data for the Pacific, Atlantic, Mexican Gulf, and Caribbean Sea

	Geographic Coordinates		Observation		Model*		Error	
Station	°N	°W	ξ, cm	δ, deg	ξ, cm	σ, deg	$\Delta\xi$, cm	$\Delta\sigma$, deg
Pacific								
Middleton (1)	58.76	145.71	110	284				
Tofino (3)	48.97	127.29	99	239				
San Francisco	38.16	124.91	54	227				
Josie II	34.00	144.99	27	267	27	273	0	+6
Flicki	32.24	120.86	43	149				
Josie I	31.03	119.80	43	142				
Kathy	27.75	124.37	29	128	27	130	−2	+2
Filloux	24.78	129.02	19	107	18	105	−1	−2
Atlantic								
N.Y. Bight	39.32	64.36	44	350				
N.C. 1	32.69	75.62	48	356	46	358	−2	+2
Savannah B	31.95	80.68	88	15				
Scope	30.43	76.42	45	358	46	3	+2	+5
AOML 1	28.14	69.75	34	1	35	6	+1	+5
AOML 3	28.24	67.54	34	359	34	4	0	+5
MERT	27.99	69.67	34	360	34	6	0	+6
REIKO	27.97	69.67	35	1	34	6	−1	+5
EDIE-May	26.46	69.33	32	3	32	7	0	+4
EDIE-March	26.45	69.32	31	1	32	7	+1	+6
Mexican Gulf								
West Florida Shelf	26.71	84.25	7	97	7	92	0	−5
Deep Gulf	24.77	89.65	1.3	226	1.6	225	+0.3	−1
Caribbean								
Misteriosa Bank	18.88	83.81	8	84	9	89	+1	+5
Rosalind Bank	16.61	80.34	7	107	8	102	+1	−5
East Caribbean (6 month)	16.54	64.88	0.5	156	1.6	151	+1	−5
East Caribbean (1 month)	16.52	64.91	0.6	153	1.5	148	0.9	−5

Here ξ is the tidal amplitude and δ is the tidal phase relative to Greenwich
*Where there is no entry, ξ and δ are included.

consistent with the definition of mean turbulent tidal velocities and elevations given in section 3*a*, which, incidentally, is directly suggested by the discrete continuity equation (61). Indeed, the excess of fluid flux into a grid cell during a time interval Δt divided by the mesh area produces the increase $\Delta\zeta$ in the average tidal height ζ.

For example, the author determined from the various tidal constants listed in Table 2 for stations near Bermuda the values $\xi = 37$ cm and $\delta = 0°$ shown in Tables 4*a* and 4*b* for the 1° × 1° oceanic cell (BI) containing the small island of Bermuda. Of course, in areas of extremely rough tides the averaging process was much more critical. Nevertheless, such uncorrelated data must be considered as being indicative of intensely turbulent or eddying tidal motions produced by the open ocean currents on shallow shelfs, for example, in front of rivers or coastal corners (section 2*b*). Vice versa, the open ocean tides must be modeled to be flexible enough in order to accept rough empirical data with only moderate and localized backlash effects on themselves. As is shown in Figure 3 and in Tables 4*a* and 4*b*, the hydrodynamical interpolation technique described above meets this requirement to a satisfactory degree. In fact, perturbations were tested that were caused by deliberately created large 'errors' in 'empirical' amplitudes and phases of isolated cells. The computations remained stable, and only minor localized alterations were registered as if they were caused by a sudden birth of a volcanic island.

Naturally, in any selection the representative empirical tide data are strongly mesh size dependent and may become meaningless in very coarse grid systems. For instance, a 6° × 6° network covers the breadth of the Atlantic between Brazil and North Africa with only four cells, and two of them are supposed to carry empirical data. In such a grid system the resolution of the entire Atlantic width is about the same as the resolution of, say, the Davis Strait in a graded 1° × 1° network. Moreover, in contrast to coarser grid systems a 1° × 1° network allows for a clear differentiation between shallow and deep ocean waves. As the momentum equations (28) show, tidal slopes in deep and shallow oceanic cells are multiplied by the corresponding ocean depths of, say, $H = 5000$ m and $H = 50$ m. Hence small tidal variations in deep oceans appear to be more significant than, say, tenfold larger tidal slopes in shallow areas, which explains the observed (see discussion above) insensitivity of the open ocean tide with respect to locally large tidal boundary data.

5. DISCUSSION OF OCEAN TIDAL MODELS

a. Evaluation of Ocean Tidal Charts

Recent ocean tidal maps, which have been charted by various empirical and theoretical methods, are qualitatively compared and discussed in the review articles by *Hendershott* [1973, 1977] and *Parke* [1978]. Of course, every author discussed the quality of his own tide model by comparing it with earlier charts and with some empirical tide data either pointwise or statistically. Hence it suffices here to summarize the major characteristics of ocean tides and to describe briefly a new approach to graphical evaluation of the detailed numerical accuracy of a tidal chart.

In order to display the qualitative global features of open ocean tides the author's constructed M_2 tide is plotted in the usual map form in Figure 8. As was anticipated in section 2*c*, one finds the crests of the tidal waves rotating around amphidromic points of zero amplitudes. Although this general feature has been known for some time, the number and the detailed distribution of these amphidromic systems varied from model to model and caused considerable controversies. However, as was pointed out by *Parke* [1978], the recent models have reached a considerable degree of similarity, and many of the remaining differences emerged as being theoretically insignificant. Indeed, several seemingly major amphidromic systems that were displayed by some maps but failed to show up on other charts are now resolved by the author's model as double amphidromes or as low-amplitude tides.

For example, *Pekeris and Accad* [1969] discovered an amphidromic system in the north central South Atlantic (see Figure 8), which motivated *Cartwright*'s [1971] tidal measurements around Saint Helena that produced no definitive decision. The same amphidrome appeared later in slightly different positions in the model of *Zahel* [1970, 1977], *Gordeyev et al.* [1973], *Estes* [1975, 1977], *Schwiderski* [1976], and *Accad and Pekeris* [1978]. Earlier empirical and theoretical maps, as well as the *Hendershott* [1972] chart, failed to resolve this system completely. As was noticed by *Parke* [1978], the author's (Figure 8) new model resolves this controversy by a double amphidromic system, which settles in between the two opposing results and is qualitatively close to the low-amplitude tide found in the *Parke and Hendershott* [1979] chart. Similar situations can be identified, for example, in the Indian Ocean south of Australia, in the Pacific between the Marianas and Marshall Islands, and also near the Hawaiian Ridge. The last case played an important role in the development of the author's tide model, which has been fully discussed by *Schwiderski* [1978*b*, 1979*b*].

The display of ocean tides by maps of the form shown in Figure 8 is obviously inadequate to permit a thorough gridwise accuracy evaluation as specified in sections 1 and 2*b*. In order to eliminate these shortcomings, *Schwiderski* [1978*b*, 1979*b*] adopted a new method to display ocean tides graphically by computerized printing of all gridded amplitudes and phases in maplike tables. For example, Tables 4*a* and 4*b* exhibit vividly all the detailed features of the M_2 tide in the 30° × 50° area of the northwestern Atlantic Ocean, which the author's model has actually resolved. A brief glance at these tables reveals that the corresponding poriton of Figure 8 leaves all tidal features in this important area either unreproduced or completely blurred. It seems that for this reason, *Parke* [1978] suppressed most of these deficiencies in his reproduction of Figure 8.

In addition to the computed data the tidal Tables 4*a* and 4*b* depict also by special markings all empirical coastal and island data (subbars) and nearshore deep-sea data (subbrackets) that were included in the computations. Computed data near distant offshore deep-sea tide stations are marked by wavy underlines to be readily available for comparison with the corresponding empirical deep-seal data listed in Table 3. As can be seen in Table 3, the maximum amplitude and phase differences (possibly error estimates) between the empirical deep-sea data and the computed data are $\Delta\xi = 2$ cm and $\Delta\delta = 6° \approx 12$ min. At the Atlantic Scope station the tide prediction differences in the sense of (10) and (12) are $\xi_m = 4.1$ cm and $E = 2.9$ cm. These minor differences are probably within the empirical error due to short observation times and the use of the distant shore reference station Bermuda [*Zetler et al.,* 1975]. As is shown in Table 2, the various empirical tidal constants for shore and nearshore stations at Bermuda

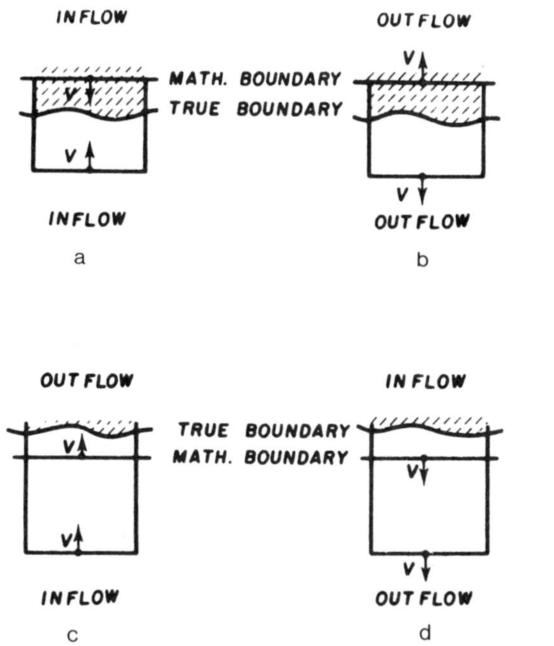

Fig. 7. Boundary cell inflow and outflow illustration. Figures 7*a* and 7*b* and Figures 7*c* and 7*d* are half periods apart. (The shaded regions are land areas.)

differ by 4 cm in amplitudes and 16° ≈ 32 min in phases. In this connection, attention may be drawn to the differences between the empirical deep-sea data determined by Kuo and his co-worker Malone without and with filter techniques for the nearby GOBI stations.

Since the phases specify the arrival times (in 1° ≈ 2 min (section 2*a*) of the tidal crests at the corresponding points after the moon's passage over Greenwich meridian, one can easily follow the motion of the tidal waves in Tabel 4*b* by starting, say, at the prominently visible 0° = 360° (or 100°) cotidal lines. Table 4*a* lists the corresponding water heights of the crests. Thus using both tables together, one gathers a vivid impression of the motion of tidal waves and their interactions with each other and with empirical data to which they must adjust themselves in a physically flexible way (see section 4*c*). In particular, one can see the tidal crests producing high and rough waters along uneven coastlines (e.g., between Cape Cod and Nova Scotia). An impressive reproduction of this recently empirically verified feature is depicted in Figure 3.

In the southern portion of Table 4*b* one can follow the tidal crest from the 0° cotidal line on its semidiurnal complete rotation around the amphidromic point (marked by a circled cross) in the Caribbean Sea southeast of the island of Puerto Rico (PRI). There along the Caribbean Ridge, one recognizes the remarkable resolution of the empirically long-known (section 2*b*) distortions and retardations which tidal waves suffer as they spill over narrow ocean ridges. As is described by *Schwiderski* [1978*b*, 1979*b*], the close reproduction of this important phenomenon along the Marianas, Nampo, Kuril, Aleutian, and Hawaiian ridges, which almost separate the tides in the Pacific from those in the Philippine, Okhotsk, and Bering seas, brought about a spectacular turnaround of the north central Pacific M_2 tide affecting the whole Pacific Ocean.

By extending the same detailed scrutiny outlined above to the entire constructed global M_2 tide, *Schwiderski* [1978*b*,

TABLE 4a. M_2 Tidal Amplitudes ξ (in Centimeters) of the Northwestern Atlantic Ocean

N \ M	280	281	282	283	284	285	286	287	288	289	290	291	292	293	294	295	296	297	298	299	300	301	302	303	304	305	306	307	308	309
32													UNGAVA BAY	400						88	92	89	88	86	85	83	82	80	78	77
33																				82	87	84	82	80	78	76	74	73	71	69
34						CANADA															75	79	77	74	72	70	68	67	65	63
35																						62	60	68	67	65	63	61	59	57
36																								57	52	58	57	55	54	52
37																										45	52	51	49	47
38																										40	48	48	46	44
39																										37	43	45	44	42
40																			GULF OF ST. LAWRENCE			51	52			35	40	41	41	39
41																			40	42	50	49	50				34	38	39	37
42																		35	25	27	31	48		NEW FOUNDLAND				34	37	37
43																	24	17	⊗ 6	20	30	43	48	54	57				35	39
44																			16	22		38	54	57	56	51	45	40	46	44
45																NOVA			SCOTIA		49	52	55	56	56	55	56	54	52	48
46														180	170			65	62	58	54	54	55	55	55	55	55	54	53	51
47										CAPE COD		135	155	132	130		80	66	61	59	57	S 52 I	54	54	54	54	53	52	51	50
48				UNITED STATES							110	91	84	81	81	100	66	62	60	58	57	55	55	54	53	53	52	51	50	48
49									LONG ISLAND			65	45	54	60	64	61	60	58	57	56	55	54	53	53	52	51	50	48	47
50								66	52	47	53	45	48	53	56	57	57	57	56	56	55	54	53	52	52	51	50	49	48	46
51								58	43	44	45	46	49	51	53	54	54	54	54	54	53	53	52	51	51	50	49	48	47	46
52							60	46	45	46	46	47	48	50	51	52	52	52	52	52	52	52	51	50	50	49	48	47	46	45
53						60	46	46	45	46	47	47	48	49	49	50	50	50	51	51	50	50	50	49	48	48	47	46	46	45
54				CAPE HATTERAS		56	46	46	45	46	46	47	47	47	48	48	48	49	49	49	49	49	48	48	47	47	46	45	45	44
55						55	45	45	45	46	46	46	46	46	46	47	47	47	47	47	47	47	47	46	46	45	45	44	44	43
56					56	45	45	45	45	45	45	45	45	45	45	45	45	45	45	45	45	45	45	45	44	44	43	43	43	42
57			72	66	48	46	45	44	44	44	44	44	44	44	43	43	43	43	43	43	43	43	43	43	43	42	42	42	41	41
58		80	59	52	49	46	43	43	43	43	42	42	42	42	42	40	B 37 I	40	41	41	41	41	42	41	41	41	40	40	40	40
59	88	64	56	51	48	44	42	42	42	41	41	41	40	40	40	39	39	39	39	39	39	40	40	40	39	39	39	39	38	38
60	70	60	54	49	45	42	41	41	40	40	39	39	39	38	38	37	37	37	37	37	37	38	38	38	37	37	37	37	37	37
61	58	54	50	46	43	41	40	39	39	38	38	37	37	36	36	35	35	35	35	35	35	36	36	36	36	35	35	35	35	35
62	51	49	46	43	41	40	39	38	37	37	36	35	34	34	33	33	33	33	33	33	33	34	34	34	33	33	33	33	33	33
63	FLORIDA	45	43	39	39	36	37	36	36	35	34	33	32	32	32	31	31	31	31	31	31	31	31	31	31	31	31	31	31	31
64		40	42	BAHAMAS	38	36	35	35	34	33	33	32	31	30	30	30	29	29	29	29	29	29	29	29	29	29	29	29	29	29
65		37	37	38		34	34	33	32	32	31	30	30	29	28	28	27	27	27	26	26	26	26	26	26	27	27	27	27	27
66	25	29	30			32	33	32	31	30	30	29	28	27	27	26	25	25	24	24	24	24	24	24	24	24	24	24	25	25
67	22	24	25	30	34		30	31	30	29	28	27	26	26	25	24	23	23	22	21	21	21	21	21	21	21	22	22	23	23
68				26	29	30		25	28	28	27	26	25	24	23	22	21	20	19	19	18	18	18	18	18	19	20	20	21	21
69	10	11	CUBA		28	27	29	30	26	26	25	24	23	22	21	19	18	17	16	16	15	15	15	16	16	17	17	18	19	20
70	10	11	11				25	24	24	23	22	23	21	19	19	17	16	15	14	13	12	12	13	13	14	15	16	17	18	19
71	10	10	12	17	15	14	13	15				21	19	17	15	15	14	12	11	10	10	10	11	12	13	14	15	17	18	19
72	10	10	12	JAM	6	9	12				HISPANIOLA			1	PRI	9	5	5	8	7	8	9	10	12	13	14	16	18	19	21
73	10	9	7	7	6	5	3	2	2	2	2	1	2	1	1	3	2	3	3	6	8	10	11	13	15	16	18	20	22	23
74	8	7	6	5	5	4	3	3	3	3	2	2	1	1	1	1 ⊗	2	3	4	7	11	12	14	16	17	19	21	23	24	26
75	7	7	6	5	5	4	4	4	3	3	3	2	2	2	2	2	2	3	5	7	15	16	17	19	21	22	24	26	27	29
76	7	7	6	6	5	5	5	5	4	4	3	3	3	3	3	3	4	5	7	6	17	19	21	23	24	26	28	29	31	33
77	7	7	7	6	6	6	6	5	5	4	4	4	3	3	4	4	5	6	8	8	21	B 24 B	26	28	29	30	31	33	35	37
78	8	7	7	7	7	7	6	6	6	5	4	5	4	4	4	5	6	8	10	10	25	31	32	33	33	34	35	37	39	40
79	8	8	7	7	7	7	7	7					5	5	5	6	7	8	13	25	28	36	38	38	38	38	39	41	43	44
80	8	6	8	8	8	8										6	7		CARIBBEAN		36	41	42	42	42	42	43	45	46	48
81	8	PAN	8	8	8						SOUTH AMERICA										40	45	47	46	46	45	46	48	50	52

TABLE 4*b*. M_2 Tidal Phases δ (in Degrees) of the Northwestern Atlantic Ocean

N\M	280	281	282	283	284	285	286	287	288	289	290	291	292	293	294	295	296	297	298	299	300	301	302	303	304	305	306	307	308	309
														UNGAVA	BAY															
32														70						317	309	306	304	302	300	297	295	292	289	286
33																				316	310	307	305	303	301	299	296	294	291	288
34																					313	310	308	305	303	301	299	296	293	290
35																						309	306	309	306	303	301	298	295	293
36						CANADA																		304	303	306	303	300	298	295
37																										302	305	303	301	298
38																										301	308	306	304	303
39																										300	311	309	306	307
40																						71	62			301	310	310	310	310
41																			131	91	79	64	57				303	311	312	314
42							UNITED STATES											190	110	69	49	38	NEW	FOUNDLAND				308	315	318
43																	242	281	331 ⊗	19	8	7	5	3	349				315	323
44																			296	351		359	3	359	355	341	332	324	338	338
45																		NOVA	SCOTIA		356	0	358	356	354	352	351	351	350	352
46														94	82			355	354	354	353	356	355	354	353	353	353	353	352	353
47								CAPE COD				102	98	71	62		10	355	353	354	355	352	354	354	354	354	354	354	354	353
48											101	95	76	62	40	31	357	356	355	355	355	355	355	355	355	355	355	355	355	355
49								LONG ISLAND				45	37	28	14	5	1	359	358	357	357	357	357	357	357	357	357	357	357	357
50								357	357	357	0	10	12	10	7	5	3	1	360	359	358	358	358	358	358	358	358	358	358	359
51								355	355	354	3	5	6	6	6	4	3	2	1	0	360	359	359	359	359	359	359	359	360	0
52							354	357	358	360	2	3	4	5	4	4	3	2	2	1	0	360	360	360	360	360	360	0	1	1
53						353	356	357	358	360	1	2	3	4	4	3	3	3	2	2	1	1	0	0	0	0	1	1	2	2
54			CAPE HATTERAS			351	357	358	359	0	1	2	3	3	4	4	3	3	2	2	2	1	1	1	1	1	1	2	2	3
55						352	357	358	359	0	1	2	3	4	4	4	4	3	3	2	2	1	1	1	1	1	1	2	2	3
56					350	355	357	359	360	1	2	3	3	4	4	5	5	4	3	3	2	2	1	1	1	1	1	2	2	2
57			358	354	354	357	359	360	1	2	3	3	4	4	5	6	6	5	4	3	2	2	1	1	1	1	1	1	2	2
58		6	358	357	358	358	359	1	2	3	3	4	4	5	5	4	0	4	3	2	2	1	1	1	1	1	1	1	1	1
59	15	1	1	2	2	1	1	2	3	3	4	5	5	5	4	3	3	3	3	2	2	1	1	1	1	0	0	0	0	0
60	22	7	6	5	3	2	3	4	4	5	5	6	6	5	5	4	4	3	3	2	2	1	1	0	0	360	359	359	359	359
61	19	10	8	6	5	5	5	5	6	6	6	7	7	7	5	4	4	3	3	2	1	1	0	360	359	359	358	358	357	357
62	16	13	12	9	8	7	7	7	7	7	7	6	5	5	4	4	4	3	2	2	1	0	360	359	358	357	356	356	355	355
63		15	16	10	10	10	9	9	9	9	8	6	3	4	4	4	4	3	2	1	1	360	359	358	356	355	354	353	352	352
64	FLORIDA	16	12		9	11	11	11	10	10	10	7	5	6	5	5	4	3	2	1	0	359	358	356	355	353	352	350	349	349
65		21	19	9	BAHAMAS	7	13	13	12	11	11	9	8	8	7	6	5	3	2	1	359	358	356	354	353	351	349	347	346	344
66	42	29	19			20	16	16	15	14	13	12	11	10	9	7	6	4	2	0	358	356	354	352	350	347	345	343	341	339
67	31	31	26	18	12		20	18	17	16	15	14	13	12	10	9	7	5	2	359	356	353	351	348	346	343	340	337	335	333
68				21	15	12		19	18	18	17	16	16	14	13	11	9	6	2	358	354	350	347	343	340	336	333	330	328	325
69	73	72	CUBA		17	17	15	14	18	19	20	19	19	18	16	14	11	6	1	356	350	345	340	335	331	327	324	321	318	316
70	76	74	72				18	18	19	20	21	22	22	22	21	17	13	8	1	352	344	336	329	323	319	315	312	309	306	304
71	78	75	67	55	50	45	40	34				22	19	28	18	22	18	15	3	347	332	321	313	308	303	300	298	295	293	291
72	84	81	87	JAM	80	45	50	HISPANIOLA						85	PRI	11	21	359	355	332	311	299	293	288	286	284	283	281	280	279
73	91	87	87	102	74	63	51	68	94	109	115	115	73	83	52	22 ⊗	8	318	300	298	283	276	272	271	270	269	269	269	269	268
74	103	102	102	102	96	92	91	102	111	119	125	129	133	135	142	95	153	257	246	250	260	257	256	257	257	258	259	259	260	259
75	116	116	116	115	112	111	111	113	119	125	131	136	144	153	165	183	212	220	218	230	246	244	245	247	248	250	251	252	252	253
76	131	129	127	126	122	121	120	121	125	129	134	141	149	158	168	180	193	202	206	219	239	235	238	240	242	244	245	246	247	247
77	142	137	134	132	129	126	125	126	128	131	137	143	151	160	168	177	187	197	204	217	227	224	235	236	238	239	240	241	242	243
78	147	142	138	135	132	129	127	127	130	138	161	161	155	161	166	175	183	194	202	217	223	234	233	234	234	236	237	238	239	240
79	152	146	140	136	133	130	128	124					159	157	155	173	179	200	210	218	218	231	230	231	232	233	234	236	237	237
80	155	148	141	137	134	137										168	180				220	228	228	229	230	231	232	233	234	235
81	166	PAN	141	139	138			SOUTH	AMERICA												229	226	225	227	229	230	231	231	233	233

M is longitude east (in degrees), *N* is the colatitude (in degrees), circled crosses are amphidromic points, subbars mark empirical input data, subbrackets mark input nearshore deep-sea measurements, and wavy underlines mark offshore deep-sea tide gauge stations with excluded measurements listed in Table 3.

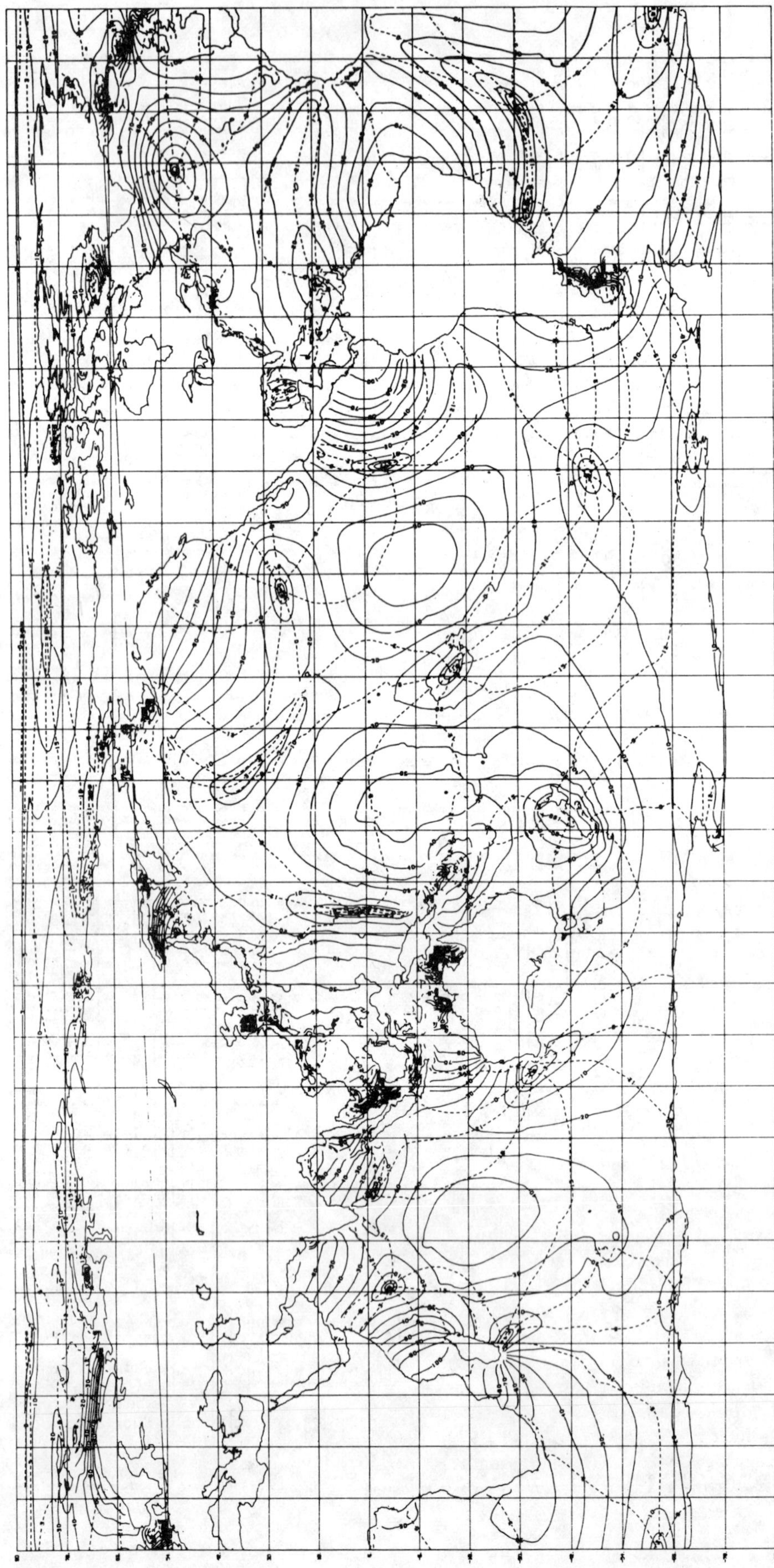

Fig. 8. Schwiderski global M_2 tide model (plot courtesy of R. H. Estes). Solid contours are amplitudes ξ in centimeters. Dashed contours are phases δ in $\frac{1}{2}$ lunar hours.

1979*b*, *c*] estimated that his model (schematically shown in Figure 8) permits a tide prediction with an error (equation (12)) of generally less than 5 cm anywhere in the open ocean Naturally, the achieved high accuracy may drop off somewhat near continental and island stations (e.g., in the Arctic Sea and near Antarctica) where only sporadic or erratic empirical tide and bathymetric data are know. Also, less accurate predictions must be anticipated in small border seas, bays, estuaries, and channels, where the $1° \times 1°$ network precludes a sufficient resolution. For example, Table 4*b* displays the well-known amphidromic point in the Gulf of Saint Lawrence [*Defant*, 1961], but the mesh size is much too coarse to permit a high-accuracy pointwise resolution.

b. Conclusions

Beginning with Newton, through three centuries the scientific charting of tides in the world oceans has captured the devotion of many scientists and mathematicians. Experimental and theoretical progress advance slowly throught paradoxes, myths, and controversies. In the preceding review sections this rich and fascinating work has been highlighted and analyzed up to the present state of the art. In agreement with the findings of the *International Workshop on Interaction of Marine Geodesy with Inertial, Tidal, and High Frequency Aspects of Ocean Dynamics* [1978] in Miami it is concluded that the total global ocean tide can now be computed with an accuracy which satisfies the 10-cm error bound set by modern tide users at least in open ocean regions.

The missing final step requires essentially only the actual computation of the 10 tidal components (S_2, N_2, K_2), (K_1, 0_1, P_1, Q_1), and (*Mf, Mm, Ssa*) which must be constructed with the same accuracy as the M_2 tide. Recent tidal maps of some of these constituents have been computed, for example, by *Zahel* [1973, 1977] for K_1; *Estes* [1977] for (S_2, N_2, K_2) and (K_1, O_1, P_1), *Parke and Hendershott* [1979] for S_2 and K_1, and *Accad and Pekeris* [1978] for S_2. The author [*Schwiderski*, 1980*a*, *b*, *c*,] has completed his own computations of the leading components S_2 and (K_1, O_1). The remaining minor partial tides are currently being processed. In agreement with Zahel and Estes the author detected no resonance or stability problems (section 2*c*) for any of the computed tides, which is obviously due to the modeled dissipation forces. It may be mentioned that the author noticed a distinctly tapered roughness in the empirical and computed data of the diurnal tides relative to the sometimes tumbling semidiurnal tides. This probably can be attributed to the doubled periods, which allow the diurnal tides to assume a calmer level.

While the successful charting of ocean tides was achieved by coordinating the various contributions and experiences made by numerous researchers in oceanography and related fields, the most dramatic improvement of the modeling techniques can be attributed to the direct inclusion of empirical tide data. As was observed by the author, in contrast to other significant improvements, which were often effective only in limited areas, the hydrogynamical interpolation of empirical tide data produced a correction uniformly over all oceans. The beneficial effects of empirical data are also apparent from the remarkable quality which *Parke and Hendershott* [1979] achieved with a rather coarse grid system. This important result is even more borne out by the limited-area maps produced by ordinary interpolation of empirical data using classical wave elements [*Munk et al.*, 1970], low-degree polynomials [*Kuo et al.* 1970*a*, *b*,], or simple rules of thumb [*Dietrich*, 1944*a*, *b*; *Bogdanov*, 1961*a*, *b*; *Luther and Wunsch*, 1975]. Ironically, the latter maps based on pure intuition and experience reflect a higher quality than some computed charts.

Though the charting of worldwide ocean tides has reached a first level of satisfactory numerical accuracy, the tidal phenomenon remains as alluring as ever. Future precise measurements of ocean tides and the search for their accurate harmonic analyses must be encouraged in order to check and, where it is necessary, improve the established models. In particular, improvements can be expected from new empirical data where present tide or bathymetric values are marginal, rough, or unreliable. In irregular ocean regions, for instance, in front of large rivers and coastal corners, the global tide models should be supplemented by limited-area models. Such models may include refined grid systems with more realistic dissipation forces. In fact, nearshore models of this kind are of immediate interest in many applications. New promising approaches to utilizing modern satellite technology for the experimental determination of open ocean tides have been reported by *Estes* [1979] and J. M. Diamante (private communication, 1978).

Acknowledgments. It is the author's most pleasant obligation to thank the Editor, J. W. Chamberlain, and the Associated Editor, F. Morrison, of *Reviews of Geophysics and Space Physics* for their invitation and patient encouragement to prepare this comprehensive review article. This project was supported by the Naval Surface Weapons Center (NAVSWC) Independent Research Fund and by a grant from the National Geodetic Survey/NOS/NOAA. Special acknowledgment is extended to Ralph A. Niemann, head of the Strategic Systems Department, R. J. Anderle, head of the Astronautics and Geodesy Division, and his associate, C. Oesterwinter, for their sustained and generous sponsorship of the author's entire work on ocean tidal modeling that included seemingly unending computer experiments. The involved computer programs were all prepared by L. Szeto in a competent and effective manner. Many critical and stimulating suggestions were gratefully received from the author's NAVSWC collegues C. J. Cohen and B. Zondek. The author wishes to thank R. H. Estes of Business and Technological Systems, Inc., for his courtesy in undertaking the difficult task of mapping (Figure 8) the author's tide model. Also, R. H. Estes communicated to the author copies of the prepublication papers by *Parke* [1978] and *Accad and Pekeris* [1978]. Other helpful papers, empirical data, and suggestions were received with gratitude from D. E. Cartwright, Institute of Oceanographic Sciences, Birkenhead, United Kingdom; S. K. Gill, NOS/NOAA; B. Kagan, Institute of Oceanography, Leningrad, USSR; J. T. Kuo and F. Malone, Lamont-Doherty Geological Observatory, Columbia University: H. O. Mofjeld, Pacific Marine Environmental Laboratory (PMEL/NOAA); W. H. Munk, Scripps Institution of Oceanography (SIO); M. E. Parke, Jet Propulsion Laboratory; C. A. Pearson, PMEL/NOAA; W. J. Pierson, Jr., Cuny Institute of Marine and Atmospheric Sciences, New York City College; D. L. Porter, NOS/NOAA; M. Schulkin, Applied Physics Laboratory, University of Washington; D. C. Simpson, NOS/NOAA; K. Wyrtki, University of Hawaii; B. D. Zetler, SIO; the International Hydrographic Bureau, Monaco; the Hydrographer of the Navy, United Kingdom; the Marine Environmental Data Service, Canada; and the Rand Corporation, Santa Monica, California.

REFERENCES

Accad, Y., and C. L. Pekeris, The K_2 tide in oceans bounded by meridians and parallels, *Proc. Roy. Soc. London, Ser. A, 278,* 110, 1963.

Accad, Y., and C. L. Pekeris, Solution of the tidal equations for M_2 and S_2 tides in the world oceans from a knowledge of the tidal potential alone, *Phil. Trans. Roy. Soc. London, Ser. A, 290,* 235, 1978.

Airy, G. B., Tides and waves, *Encyclo. Metrop. London, 5,* 241, 1842.

Bartels, J., Gezeitenkräfte, *Handb. Phys., XLVIII, Geophys. II,* 1957.

Berghaus, H., *Physikalischer Handatlas,* Gotha, East Germany, 1845.

Bogdanov, C. T., New charts of the cotidal lines of semidiurnal tide waves (M_2 and S_2) for the Pacific Ocean, *Sov. Oceanogr., Engl. Transl., 1,* 28, 1961*a*.

Bogdanov, C. T., New charts of the dirunal tide (K_1 and O_1) for the Pacific Ocean, *Sov. Oceanogr.*, Engl. Transl., I, 46, 1961*b*.

Bogdanov, C. T., and V. A. Magarik, A numerical solution of the problem of semiduirnal tidal wave distribution (M_2 and S_2) in the ocean, (in Russian), *Dokl. Akad. Nauk. SSSR, 172,* 1315, 1967.

Bogdanov, C. T., and V. A. Magarik, A numerical solution of the problem of tidal wave propagation in the world ocean (in Russian), *Izv. Akad. Nauk SSSR Fiz. Atmos. Okeana, 5,* 1309, 1969.

Bogdanov, C. T., K. V. Kim, and V. A. Magarik, A numerical solution of the hydrodynamical equations of tides on the BESM-2 computer for the world ocean area (in Russia), *Tr. Inst. Okeanol., Akad. Nauk. SSSR, 75,* 73, 1964.

Boussinesq, J., Théroie de l'écoulement tourbillant, *Mem. Presentes Acad. Sci., XXIII,* 46, 360, 1877.

Boussinesq, J., Expression du frottement extérieur dans l'écoulement tumultueux d'une fluide, *C. R. Acad. Sci., 122,* 1445, 1896.

Brettschneider, G., Anwendung des hydrodynamisch-numerischen Verfahrens zur Ermittlung der M_2-mittschwingungsgezeit der Nordsee, *Mitt. Inst. Meereskunde Univ. Hamburg, VII,* 1967.

British Admiralty, *Tide Tables,* vols. 1-3, Hydrographer of the Navy, London, 1977.

Bryan, K., A numerical investigation of a nonlinear model of a wind-driven ocean, *J. Atmos. Sci., 20,* 594, 1963.

Busse, F. H., and J. A. Whitehead, Instabilities of convection rolls in a high Prandtl number fluid, *J. Fluid Mech., 47,* 305, 1971.

Cantor, G., *Vorlesungen Über Geschichte der Mathematik,* vol. 3, W. de Gruyter, Leipzig, East Germany, 1901.

Cartwright, D. E., A unified analysis of tides and surges around north and east Britain, *Phil. Trans. Roy. Soc. London, Ser. A, 263, 1,* 1968.

Cartwright, D. E., Extraordinary tidal currents near St. Kilda, *Nature, 223,* 938, 1969.

Cartwright, D. E., Tides and waves in the vicinity of St. Helena, *Phil. Trans. Roy. Soc. London, Ser. A, 270,* 603, 1971.

Cartwright, D. E., Oceanic tides, *Rep. Progr. Phys., 40,* 665, 1977.

Cartwright, D. E., B. D. Zetler, and B. V. Hamon, *Pelagic Tidal Constants, Publ. Sci. 30,* International Association of Physical Sciences of the Oceans, 1979.

Cartwright, D. E., and R. J. Tayler, New computations of the tide-generating potential, *Geophys. J. Roy. Astron. Soc., 23,* 45, 1971.

Cartwright, D. E., W. H. Munk, and B. D. Zetler, Pelagic tidal measurements, *Eos Trans. AGU, 50,* 472, 1969.

Cox, M. D., A mathematical model of the Indian Ocean, *Deep Sea Res., 17,* 45, 1970.

Crowley, W. P., A global numerical model, 1, *J. Comput. Phys., 3,* 111, 1968.

Crowley, W. P., A numerical model for viscous, free-surface barotropic wind-driven ocean circulations, *J. Comput. Phys., 5,* 139, 1970.

Darwin, G. H., Report on the harmonic analysis of tidal observations, Brit. Ass. for the Advan. of Sci., London, 1883.

Defant, A., Flutwellen and Gezeiten des Wassers, *Handb. Phys., XLVIII, Geophys. II,* 846, 1957.

Defant, A., *Physical Oceanography,* vol. II, Pergamon, New York, 1961.

Dietrich, G., Die Gezeiten des Weltmeeres als geographische Erscheinung, *Z. Gesamte Erdk., 3,* 69, 1944*a*.

Dietrich, G., Die Schwingungssysteme der halb-und eintägigen Tiden in den Ozeanen, *Veroeff. Inst. Meereskunde Univ. Berlin, 41,* 1944*b*.

Dietrich, G., *General Oceanography,* Wiley-Interscience, New York, 1963.

Dishon, M., Determination of the average ocean depths from bathymetric data, *Int. Hydrogr. Rev., 41,* 77, 1964.

Dishon, M., and B. C. Heezen, Digital deep-sea sounding library, *Int. Hydrogr. Rev., 45,* 23, 1968.

Doodson, A. T., The harmonic development of the tide-generating potential, *Proc. Roy. Soc., London, Ser. A, 100,* 305, 1921.

Döös, B. R., et al., Modeling for the first Garp globel experiment, *Garp Publ. Ser. 14,* Int. Counc. of Sci. Unions, Pans, 1974.

Ekman, V. W., Om jordrotationens inverkan på vindströmmar i hafvet, *Nyt Mag. Naturvidensk., 20,* 67, 1902.

Estes, R., A computer software system for the generation of global numerical solutions of diurnal and semidiurnal ocean tides, *Rep. TR-75-27,* Bus. and Technol. Syst., Greenbelt, Md., 1975.

Estes, R., A computer software system for the generation of global ocean tides and crustal loading effects, Rep. TR-77-41, Bus. and Technol. Syst., Greenbelt, Md., 1977.

Estes, R., A simulation of global ocean tide recovery using altimeter data with systematic orbit error, *J. Mar. Geod.,* in press, 1979.

Eyries, M., Maregraphs de grandes profondeurs, *Cah. Oceanogr., 20,* 335, 1968.

Farrell, W. E., Deformation of the earth by surface loads, *Rev. Geophys. Space Phys., 10,* 261, 1972*a*.

Farrell, W. E., Global calculations of tidal loading, *Nature, 238,* 43, 1972*b*.

Farrell, W. E., Earth tides, ocean tides, and tidal loading, *Phil. Trans. Roy. Soc. London, Ser. A, 274,* 253, 1973.

Ferrel, W., Tidal researches, report, pp. 154, 239, 245, U.S. Coast and Geod. Surv., Washington, D. C., 1874.

Filloux, J. H., Bourbon tube deep-sea tide gages, in *Proceedings of the Symposium on Tsunamis and Tsunami Research, Honolulu,* Int. Union of Geod. and Geophys., Paris, 1969.

Fomin, L. M., *The Dynamic Method in Oceanography,* Elsevier, New York, 1964.

Friedrich, H. J., Numerische Berechnung der allgemeinen Zirkulation im Meere nach einem Differenzen Verfahren, Vornehmlich für den Atlantischen Ozean, *Mitt. Inst. Meereskunde Univ. Hamburg, III,* 1966.

Friedrich, H. J., Preliminary results from a numerical multilayer model for the circulation in the North Atlantic, *Deut. Hydrogr. Z., 23,* 145, 1970.

Gallagher, B. S., K. M. Shimada, F. L. Gonzales, Jr., and E. D. Stroup, Tides and currents in Fanning atoll lagoon, *Pac. Sci., 25,* 191, 1971.

Gates, L. R., and A. B. Nelson, A new (revised) tabulation of the Scripps topography on a 1° global grid, II, Ocean depths, *Rep. 1277,* Rand Corp., Santa Monica, Calif., 1975.

Gill, S. K., and D. L. Porter, Comparison of results from a simple Defant tidal model with observed offshore tides, *J. Mar. Geod.,* in press, 1979.

Goad, C. C., and B. C. Douglas, Determination of the M_2 ocean tide parameters from satellite orbit perturbations, *J. Geophys. Res., 82,* 88, 1977*a*.

Goad, C. C., and B. C. Douglas, Lunar longitude deceleration and tidal parameters estimated from satellite orbital perturbations, paper presented at the Eighth International Symposium on Earth Tides, Bonn, Federal Republic of Germany, Int. Ass. of Phys. Sci. of the Oceans, 1977*b*.

Gohin, F., Determination des denivellations et des courants de marée, *Proc. Conf. Coastal Eng. 7th, 2,* 485, 1961.

Gordeyev, R. G., B. A. Kagan, and V. Y. Rivkind, Numerical solution of the dynamic tidal equations in the global ocean (in Russian), *Dokl. Akad. Nauk SSR, 209,* 340, 1973.

Gordeyev, R. G., B. A. Kagan, and E. Polyakov, The effects of loading and self attraction on global ocean tides, the model and the results of a numerical experiment, *J. Phys. Oceanogr., 7,* 161, 1977.

Grace, S. F., The semidiurnal tidal motion of the Red Sea, *Mon. Notic. Roy. Astron. Soc., Geophys. Suppl., 2,* 274, 1930*a*.

Grace, S. F., The influence of the friction on the tidal motion of the Gulf of Suez, *Mon. Notic. Roy. Astron. Soc., Geophys. Suppl., 2,* 316, 1930*b*.

Hansen, W., Die Ermittlung der Gezeiten Beliebig Gestalteter Meeresgebiete mit Hilfe des Randwertverfahrens, *Deut. Hydrogr. Z., 1,* 157, 1948.

Hansen, W., Die halbtägigen Gezeiten im Nordatlantischen Ozean, *Deut. Hydrogr. Z., 2,* 44, 1949.

Hansen, W., Die Reproduktion der Bewegungsvorgänge im Meere mit Hilfe hydrodynamisch-numerischer Verfahren, *Mitt. Inst. Meereskunde Univ. Hamburg, V,* 1966.

Harris, R. A., Manual of tides, IV*b*, Report of the Superintendent, p. 313, U.S. Coast and Geod. Surv., Washington, D. C., 1904.

Hendershott, M. C., The effects of solid-earth deformation on global ocean tides, *Geophys. J. Roy. Astron. Soc., 29,* 380, 1972.

Hendershott, M. C., Ocean tides, *Eos Trans. AGU, 54,* 76, 1973.

Hendershott, M. C., Global numerical tide solutions with inclusion of ocean self-gravitation and of solid earth tidal loading, paper presented at XVI General Assembly, Int. Union of Geod. and Geophys., Grenoble, France, 1975.

Hendershott, M. C., Numerical models of ocean tides, in *The Sea,* vol. 6, p. 47, John Wiley, New York, 1977.

Hendershott, M. C., and W. H. Munk, Tides, *Annu. Rev. Fluid Mech., 2,* 205, 1970.

Holland, W. R., and A. D. Hirschman, A numerical calculation of the circulation in the North Atlantic Ocean, *J. Phys. Oceanogr., 2,* 336, 1972.

Hopfner, F., 1931. Die Gezeiten der Meere, in *Handbuch der Experimentalphysik, Geophysik,* vol. 2, Akademische, Leipzig, East Germany, 1931.

Hough, S., On the application of harmonic analysis to the dynamical theory of the tides, *Phil. Trans. Roy. Soc. London, Ser. A, 189,* 201, 1897.

International Hydrographic Bureau, Tides, harmonic constants, *Spec. Publ. 26,* Monaco, 1966.

International Workshop on Interaction of Marine Geodesy With Inertial, Tidal, and High Frequency Aspects of Ocean Dynamics, report presented at the International Symposium on Interaction of Marine Geodesy and Ocean Dynamics, Miami, 1978.

International Working Group on Ocean Tides, report presented at the XVI General Assembly, Int. Union of Geod. and Geophys., Grenoble, France, 1975.

Irish, J. D., and F. E. Snodgrass, Australian-Antarctic tides, in *Antarctic Oceanology, II, The Australian-New Zealand Sector, Antarct. Res. Ser.,* vol. 19, edited by D. E. Hayes, p. 101, AGU, Washington, D. C., 1972.

Irish, J. D., W. H. Munk, and F. E. Snodgrass, M_2 amphidrome in the northeast Pacific, *Geophys. Fluid. Dyn., 2,* 355, 1971.

Jachens, R. C., and J. T. Kuo, The O_1 tide in the North Atlantic Ocean as derived from land-based tidal gravity measurements, in *Proceedings of the Seventh Symposium on Earth Tides, Sopron, Hungary,* Akad. Kiado, Budapest, 1973.

Johns, B., Vertical structure of tidal flows in river estuaries, *Geophys. J. Roy. Astron. Soc., 12,* 103, 1966.

Kagan, B. A., Bottom friction in a one-dimensional tidal flow, *Izv. Acad. Sci. USSR Atmos. Oceanic Phys.,* Engl. Transl., *7,* no. 11, 1971.

Kagan, B. A., Resistance law of tidal flow, *Izv. Acad. Sci. USSR Atmos. Oceanic Phys.,* Engl. Transl., *8,* no. 5, 1972.

Kagan, B. A., Global interaction of ocean and terrestrial tides (in Russian), in *News and Problems of Science* Gidrometeoizdat, Leningrad, 1977.

Kuo, J. T., R. C. Jachens, G. White, and M. Ewing, Tidal gravity measurements along a transcontinental profile across the United States, *Comm. Observ. Roy. Belg.,* no. 9, 96, 1970*a*.

Kuo, J. T., R. C. Jachens, M. Ewing, and G. White, Transcontinental tidal gravity profile across the United States, *Science, 168,* 968, 1970*b*.

Kuo, J. T., R. C. Jachens, and S. S. Lee, The northeastern Pacific O_1 and the North Atlantic M_2 ocean tides as derived from inversion, paper presented at the Eighth International Symposium on Earth Tides, Bonn, Federal Republic of Germany, Int. Ass. of Phys. Sci. of the Oceans, 1977.

Ladyzhenskaya, O. A., *The Mathematical Theory of Viscous Incompressible Flow,* Gordon and Breach, New York, 1969.

Lamb, H., *Hydrodynamics,* Dover, New York, 1932.

Lamb, H., and L. Swain, A tidal problem, *Phil. Mag., 29,* 737, 1915.

Laplace, P. S., Recherches sur quelques points de système du monde, *Mem. Acad. Roy. Soc., 88,* 177, 1775.

Larsen, L. H., and J. D. Irish, Tides at Cobb Seamount, *J. Geophys. Res., 80,* 1691, 1975.

Leith, C. E., Two-dimensional eddy viscosity coefficients, in *Proceedings of the WMO/IUGG Symposium on Numerical Weather Prediction, Tokyo,* p. 140, World Meteorological Organization, Geneva, 1968.

Luther, D. S., and C. Wunsch, Tidal charts of the central Pacific Ocean, *J. Phys. Oceanogr., 5,* 227, 1975.

Marchuk, G. I., and B. A. Kagan, *Ocean Tides: Mathematical Models of Numerical Experiments* (in Russian), Gidrometeoizdat, Leningrad, 1977.

Marchuk, G. I., B. A. Kagan, and R. E. Tamsalu, A numerical method of calculating tidal motion in border seas, *Izv. Acad. Sci. USSR Atmos. Oceanic Phys.,* Engl. Transl., *5,* 694, 1969.

Marchuk, G. I., R. G. Gordeyev, V. Y. Rivkind, and B. A. Kagan, A numerical method for the solution of tidal dynamics equations and the results of its application, *J. Comput. Phys., 13,* 15, 1973.

McGregor, R. C., The influence of eddy viscosity on the vertical distribution of velocity in the tidal estuary, *Geophys. J. Roy. Astron. Soc., 29,* 103, 1972.

Miles, J., On Laplace's tidal equations, *J. Fluid Mech., 66,* 241, 1974.

Miyazaki, M., S. Kuronuma, and T. Inoue, Tidal constants along the coast of Japan, *Oceanogr. Mag., 19,* 13, 1967.

Mofjeld, H. O., Empirical model for tides in the western North Atlantic Ocean, *Rep. TR ERL 340-AOML19,* Nat. Oceanic and Atmos. Admin., Boulder, Colo., 1975.

Munk, W. H., On the wind-driven ocean circulation, in *Wind-Driven Ocean Circulation,* Blaisdell, New York, 1963.

Munk, W. H., Abyssal recipes, *Deep Sea Res., 13,* 707, 1966.

Munk, W. H., and D. E. Cartwright, Tidal spectroscopy and prediction, *Phil. Trans. Roy. Soc. London, Ser. A, 259,* 533, 1966.

Munk, W. H., and E. Palmén, Note on the dynamics of the antarctic circumpolar current, *Tellus, 3,* 53, 1951.

Munk, W. H., F. Snodgrass, and M. Wimbush, Tides offshore: Transition from California coastal to deep-sea waters, *Geophys. Fluid Dyn., 1,* 161, 1970.

National Ocean Survey, Tidal harmonic constants, report, U.S. Coast and Geod. Surv., Washington, D. C., 1942.

Neumann, G., and W. J. Pierson, Jr., *Principles of Physical Oceanography,* Prentice-Hall, Englewood Cliffs, N. J., 1966.

Newton, I., *Philosophiae Naturalis Principia Mathematica,* London, 1687. (Also in *Oceanography Concepts and History,* edited by M. B. Deacon, Dowden, Hutchinson, and Ross, Stroudsbourg, Pa., 1978.)

Nowroozi, A. A., Long-term measurements of pelagic tidal height off the coast of northern California, *J. Geophys. Res., 77,* 434, 1972.

Noye, J., and A. K. Easton, The computation of tidal constants, *Comput. Programmes 7,* Horace Lamb Centre, Flinders Univ. of S. Australia, 1967.

O'Brien, J. J., A two-dimensional model of the wind-driven North Pacific *Invest. Pesq., 35,* 331, 1971.

Parke, M. E., Open ocean tide modelling, paper presented at the 9th Geodesy, Solid Earth, and Ocean Physics Research Conference, Int. Union of Geod. and Geophys., Columbus, Ohio, Oct. 1978.

Parke, M. E., and M. C. Hendershott, M_2, S_2, and K_1 models of the global ocean tide on an elastic earth, *J. Mar. Geod.,* in press, 1979.

Pearson, C. A., Deep-sea tide and current observations in the Gulf of Alaska and northeast Pacific, *Tech. Memo. NOS 16,* Nat. Oceanic and Atmos. Admin., Rockville, Md., 1975*a*.

Pearson, C. A., Deep-sea tide observations off the southeastern United States, *Tech. Memo. NOS 17,* Nat. Oceanic and Atmos. Admin., Rockville, Md., 1975*b*.

Pekeris, C. L., and Y. Accad, Solution of Laplace's equation for the M_2 tide in the world oceans, *Phil. Trans. Roy. Soc. London, Ser. A, 265,* 413, 1969.

Platzman, G. W., *Ocean Tides and Related Waves, Lect. Appl. Math.,* vol. 14, American Mathematical Society, Providence, R. I., 1971.

Platzman, G. W., Two-dimensional free oscillations in natural basins, *J. Phys. Oceanogr., 2,* 117, 1972*a*.

Platzman, G. W., North Atlantic Ocean: Preliminary description of normal modes, *Science, 178,* 156, 1972*b*.

Platzman, G. W., Normal modes of the Atlantic and Indian oceans, *J. Phys. Oceanogr., 5,* 201, 1975.

Poincaré, H., *Leçons de Mécanique Céleste,* vol. 3, *Theorie des Marées,* Gauthier-Villars, Paris, 1910.

Prandtl, L., Über die ausgebildete Turbulenz, *Z. Angew. Math. Mech., 5,* 136, 1925.

Proudman, J., Free and forced longitudinal tidal motion in a lake, *Proc. London Math. Soc., 14,* 2, 1915.

Proudman, J., *Dynamical Oceanography,* Dover, New York, 1952.

Proudman, J., and A. T. Doodson, On the tides in an ocean bounded by two meridians on a nonrotating earth, *Mon. Notic. Roy. Astron. Soc., Geophys. Suppl., 1,* 468, 1927.

Reid, R. O., A. R. Robinson, and K. Bryan, *Numerical Models of Ocean Circulation,* National Academy of Sciences, Washington, D. C., 1975.

Richardson, L. F., *Weather Prediction by Numerical Methods,* Cambridge University Press, New York, 1922.

Richtmyer, R. D., and K. W. Morton, *Difference Methods for Initial-Value Problems,* Interscience, New York, 1967.

Schlichting, H., *Boundary-Layer Theory,* McGraw-Hill, New York, 1968.

Schwiderski, E. W., Bifurcation of convection in internally heated fluid layers, *Phys. Fluids, 15,* 1882, 1972.

Schwiderski, E. W., Preliminary M_2 tide, computer tape, Nav. Surface Weapons Center, Dahlgren, Va., 1976.

Schwiderski, E. W., Hydrodynamically defined ocean bathymetry,

Rep. NSWC/DL TR 3888, Nav. Surface Weapons Center, Dahlgren, Va., 1978*a*.

Schwiderski, E. W., Global ocean tides, I, A detailed hydrodynamical interpolation model, *Rep. NSWC/DL TR 3866,* Nav. Surface Weapons Center, Dahlgren, Va., 1978*b*.

Schwiderski, E. W., Ocean tides, I, Global ocean tidal equations, *J. Mar. Geod.,* in press, 1979*a*.

Schwiderski, E. W., Ocean tides, II, A hydrodynamical interpolation model, *J. Mar. Geod.,* in press, 1979*b*.

Schwiderski, E. W., Global ocean tides, II, The simidiurnal prinicpal lunar tide (M_2), report, Nav. Surface Weapons Center, Dahlgren, Va., in press, 1979*c*.

Schwiderski, E. W., Global ocean tides, III, The semidiurnal principal solar tide (S_2), report, Nav. Surface Weapons Center, Dahlgren, Va., in press, 1980*a*.

Schwiderski, E. W., Global ocean tides, IV, The diurnal luni-solar declination tide (K_1), report, Nav. Surface Weapons Center, Dahlgren, Va., in press, 1980*b*.

Schwiderski, E. W., Global ocean tides, V, The diurnal principal lunar tide (O_1), report, Nav. Surface Weapons Center, Dahlgren, Va., in press, 1980*c*.

Shilov, G. E., *Generalized Functions and Partial Differential Equations,* Gordon and Breach, New York, 1968.

Smagorinsky, J., General circulation experiments with the primitive equations, 1, The basic experiment, *Mon. Weather Rev., 91,* 99, 1963.

Smith, S. M., H. M. Menard, and G. Sharman, Worldwide ocean depths and continental elevations averaged for areas approximating one-degree squares of latitude and longitude, *Ref. 65-8,* Scripps Inst. of Oceanogr., La Jolla, Calif., 1966.

Snodgrass, F. E., Deep-sea instrument capsule, *Science, 162,* 78, 1968.

Stern, M. E., *Ocean Circulation Physics,* Academic, New York, 1975.

Sterneck, R. V., Die Gezeiten der Ozeane, 1, *Mitt. Sitzungsberg. Akad. Wiss. Wien, 129,* 131, 1920.

Sterneck, R. V., Die Gezeiten der Ozeane, 2, *Mitt. Sitzungsber. Akad. Wiss. Wien, 130,* 363, 1921.

Stommel, H., *The Gulf Stream,* University of California Press, Berkeley, 1960.

Taylor, G. I., Tidal friction in the Irish Sea, *Phil. Trans. Roy. Soc., Ser. A, 220,* 1, 1918.

Thomson, W. (Lord Kelvin), Report of committee for the purpose of harmonic analysis of tidal observations, Brit. Ass. for the Advan. of Sci., London, 1868.

Thomson, W. (Lord Kelvin), On gravitational oscillations of rotating water, *Proc. Roy. Soc. Edinburgh, 10,* 141, 1879.

Tiron, K. D., Y. N. Sergeev, and A. N. Michurin, Tidal charts for the Pacific, Atlantic, and Indian oceans, *Vestn. Leningrad. Univ.,* Engl. Transl., *24,* 122, 1967.

Trepka, L. V., Anwendung des hydrodynamisch-numerischen Verfahrens zur Ermittlung des Schelfeinflusses auf die Gezeiten in Modellkanälen und Modellozeanen, *Mitt. Inst. Meereskunde Univ. Hamburg, IX,* 1967.

Ueno, T., Theoretical studies on tidal waves traveling over the rotating globe, I, *Oceanogr. Mag., 15,* 99, 1964*a*.

Ueno, T., Theoretical studies on tidal waves traveling over the rotating globe, II, *Oceanogr. Mag., 16,* 47, 1964*b*.

Villain, C., Cartes des lignes cotidals dans les Océans, *Ann. Hydrog. Paris, 3,* 269, 1952.

Whewell, W., Essay towards a first approximation to a map of cotidal lines, *Phil. Trans. Roy. Soc. London, 1,* 147, 1833.

Whewell, W., On the tides of the Pacific and on the diurnal inequality, *Phil. Trans. Roy. Soc. London, 16,* 1, 1848.

Whitaker, S., *Introduction to Fluid Mechanics,* Prentice-Hall, Englewood Cliffs, N. J., 1968.

Zahel, W., Die Reproduktion gezeitenbedingter Bewegungsvorgänge im Weltozean Mittels des hydrodynamisch-numerischen Verfahrens, *Mitt. Inst. Meereskunde Univ. Hamburg, XVII,* 1970.

Zahel, W., The diurnal K_1 tide in the world ocean—A numerical investigation, *Pure Appl. Geophys., 109,* 1819, 1973.

Zahel, W., A global hydrodynamic-numerical 1°-model of the ocean tides: The oscillation system of the M_2-tide and its distribution of energy dissipation, paper presented at XVI General Assembly, Int. Union Geod. Geophys., Grenoble, France, 1975.

Zahel, W., The influence of solid earth deformations on semidiurnal and diurnal oceanic tides, in *Tidal Friction and Earth's Rotation,* edited by P. Brosche and J. Sündermann, p. 98, Springer, New York, 1977.

Zetler, B., W. Munk, H. Mofjeld, W. Brown, and F. Dormer, Mode tides, *J. Phys. Oceanogr., 5,* 430, 1975.

(Received February 22, 1979;
accepted July 19, 1979.)